Protection of Concrete

Other Books on Concrete Materials and Structures

Admixtures for Concrete - Improvement of Properties
Edited by E. Vazquez

Calcium Aluminate Cements
Edited by R. J. Mangabhai

Corrosion of steel in concrete
RILEM Report
Edited by P. Schiessl

Developments in Structural Engineering
Edited by B. H. V. Topping

Fracture Mechanics of Concrete Structures: From Theory to Applications
RILEM Report
Edited by L. Elfgren

The Maintenance of Brick and Stone Masonry Structures
Edited by A. M. Sowden

Reinforced Concrete Designer's Handbook
C. E. Reynolds and J. C. Steedmam

Rheology of Fresh Cement and Concrete
Edited by P. F. G. Banfill

Testing during Concrete Construction
Edited by H. W. Reinhardt

Publisher's Note
This book has been compiled from camera ready copy provided by the individual contributors. This method of production has allowed us to supply finished copies to the delegates at the Conference.

Protection of Concrete

Proceedings of the International Conference,
held at the University of Dundee, Scotland, UK,
on 11-13 September 1990

Edited by

Ravindra K. Dhir
Director, Concrete Technology Unit,
University of Dundee
and

Jeffrey W. Green
Industrial Consultant in Concrete Technology,
University of Dundee

E. & F.N. SPON
An Imprint of Chapman and Hall
LONDON . NEW YORK . TOKYO . MELBOURNE . MADRAS

UK   Chapman and Hall, 2-6 Boundary Road, London SE1 8HN

USA   Van Nostrand Reinhold, 115 5th Avenue, New York NY10003

JAPAN   Chapman and Hall Japan, Thomson Publishing Japan,
            Hirakawacho Nemoto Building, 7F, 1-7-11 Hirakawa-cho,
            Chiyoda-ku, Toyko 102

AUSTRALIA   Chapman and Hall Australia, Thomas Nelson Australia, 480 La
            Trobe Street, PO Box 4725, Melbourne 3000

INDIA   Chapman and Hall India, R. Seshadri, 32 Second Main Road, CIT
            East, Madras 600 035

First edition 1990

© 1990 E. & F.N. Spon

Printed in Great Britain at the
University Press, Cambridge

ISBN 0 419 15490 6
     0 442 31241 5   (USA)

British Library Cataloguing in Publication Data
Available

Library of Congress Cataloguing-in-Publication Data
Available

PREFACE

The Department of Civil Engineering at the University of Dundee has, over the last 15 years, become a major centre for research in the field of concrete technology. In 1988, the University formally established the Concrete Technology Unit (CTU), which then began to integrate its research activities with continuing education, postgraduate and vocational training, and advisory services for the benefit of the industry. The CTU is particularly active in promoting the efficient use of concrete, with emphasis on durability in order to minimise maintenance and repair cost. This three day Conference has been organised to promote this long-term commitment of the CTU to concrete construction. This book contains the papers accepted for presentation at the Conference.

The main purpose of the Conference was to provide an opportunity for the presentation of current thinking and possible future developments on means of protecting concrete and ensuring its satisfactory performance in the service environment. It also provided a forum for fruitful, international discussion of this vital area of concrete technology.

Eighty-two papers were presented from twenty-one countries, spread from China to the USA, and from the USSR to South Africa. The papers have been divided into six technical themes, covering separately such topics as coatings and linings, protection through design and construction, and the likely impact of new pan-European legislation.

The Opening Addresses were given by Christopher Chope, Minister for Construction at the Department of the Environment, and by Professor Michael Hamlin, Principal and Vice-Chancellor of the University of Dundee who welcomed the delegates. The Closing Address was given by Professor Peter Hewlett, Director of the British Board of Agrément.

A conference of this size and scope demands an enormous amount of work in planning and preparation. The organisers were fortunate in having an excellent Technical Committee to advise on the selection and review of papers, as well as the Conference programme. Ten Institutions supported the event indicating the appropriateness of the subject. The CTU has always enjoyed excellent working relations with industry and is a firm believer in the spirit of co-operation between academia and industry. The collective sponsorship of the Conference by nineteen companies, which was the first time that such wide ranging sponsorship had been provided, symbolised this partnership. It was also gratifying to have thirty-seven companies participate in the accompanying exhibition and thereby provide a suitable background atmosphere for the Conference.

The Organising Committee also appreciate the work undertaken by the Chairmen of the technical sessions, and the authors of the papers, particularly those who travelled to Dundee to present their papers personally. Amongst others who have provided invaluable help are the CTU staff (including research students) and the University Central Media, Residence, and Catering Services.

The editing of this voluminous Proceedings was most time consuming and the editors would like to express their gratitude to all those who have helped with this work. The book has been prepared from the manuscripts provided by the authors and whilst every care has been taken, the editors apologise for any errors or inaccuracies which may inadvertently have been overlooked.

Dundee
September 1990

Ravindra K Dhir
Jeffrey W Green

ORGANISING COMMITTEE

Ravindra K Dhir
(Chairman)

Jeffrey W Green
(Secretary)

Peter C Hewlett

M Roderick Jones

David S Swift

Concrete Technology Unit
Department of Civil Engineering
University of Dundee

TECHNICAL COMMITTEE

Mr Kevin J Coulman
Joint Managing Director
John Lelliot

Dr Ravindra K Dhir (Chairman)
Director, Concrete Technology Unit
University of Dundee

Dr John W Dougill
Director of Engineering
Institution of Structural Engineers

Mr Jeffrey W Green (Secretary)
Industrial Consultant in Concrete Technology
University of Dundee

Professor Peter C Hewlett
Director
British Board of Agrément

Mr Laurence H McCurrich
Technical Director
Fosroc Technology

Mr John Morley
Technical Development Manager
Shell International Chemicals

Mr Jack Rodin
Consultant
Building Design Partnership

Dr George Somerville
Director of Research and Technical Services
British Cement Association

Dr Graham P Tilly
Head of Structures Group
Transport and Road Research Laboratory

SUPPORTING BODIES

British Board of Agrément
Cement Admixtures Association
Concrete Society
Federation of Resin Formulators and Applicators
Institute of Concrete Technology
Institution of Civil Engineers
Institution of Structural Engineers
Royal Institute of British Architects
UK Certification Authority for Reinforcing Steels

SPONSORING ORGANISATIONS

Allied Bar Coaters

Ash Resources Ltd

Blue Circle Cement-Scotland

British Board of Agrément

BP Chemicals

Building Design Partnership

Cormix Construction Chemicals

Diespeker Concrete Company Ltd

ELE International

Elkem Materials Limited

Fosroc Construction Chemicals

Charles Gray (Builders) Ltd

Morrison Construction Group Ltd

Mott MacDonald Group

The Novus Group

Ready Mixed Concrete (UK) Ltd

Shell International Chemical Company Ltd

Sika Limited

STATS Scotland Ltd

EXHIBITORS

Allied Bar Coaters
Appleby Group Ltd
Ash Resources Ltd
Babtie, Shaw & Morton
Bierrum Structural Services Ltd
Blue Circle Cement-Scotland
Blyth & Blyth
BP Chemicals
British Board Of Agrément
Building Design Partnership
Castle Cement Ltd
Cormix Construction Chemicals
Dupont (UK) Ltd
ELE International
Elkem Materials Ltd
Exchem Mining & Construction Ltd
W A Fairhurst & Partners
FEB Limited
Fosroc Construction Chemicals
Charles Gray (Builders) Ltd
Huls (UK) Ltd
John Lelliot Ltd
Mackenzie Construction Ltd
G Maunsell & Partners
M B T UK Ltd
Morrison Construction Group Ltd
Morton International
Mott Macdonald Group
The Novus Group
Ready Mixed Concrete (UK) Ltd
R I W Ltd
Sika Ltd
Harry Stanger Ltd
STATS Scotland Ltd
Steinweg UK Ltd
Stirling Lloyd Ltd
UK Analytical Ltd

CONTENTS

THEME 1   **CONCRETE : THE CONSTRUCTION MATERIAL**
(Plenary Session)

Chairman   Mr L H McCurrich,
Fosroc Technology,
United Kingdom

THEME 2: **METHODS OF PROTECTING CONCRETE :
COATINGS AND LININGS**
(Parallel Sessions)

Chairmen   Mr D Kruger,
Rand Afrikaans University,
South Africa

Dr J Menzies,
Building Research Establishment,
United Kingdom

Dr T Oshiro,
University of the Ryukyus,
Japan

Mr V S Parameswaran,
Structural Engineering Research Centre,
India

THEME 3   PROTECTION OF STRUCTURAL CONCRETE
(Parallel Sessions)

Chairmen   Prof G E Blight,
University of Witwatersrand,
South Africa

Prof W H Chen,
National Central University,
China

Dr J Larralde,
Drexel University,
United States of America

Prof H R Sasse,
University of Technology
West Germany

THEME 4   **PROTECTION THROUGH DESIGN**
(Parallel Sessions)

Chairmen   Prof P C Kreijger,
Eindhoven University of Technology,
The Netherlands

Prof K Mahmood,
King Abdulaziz University,
Saudi Arabia

Dr W G Smoak,
United States Department of the Interior,
United States of America

Prof P Spinelli,
University of Florence,
Italy

THEME 5   **PROTECTION THROUGH CONSTRUCTION**
(Parallel Sessions)

Chairmen   Mr K A L Johnson
Fairclough Civil Engineering Ltd.,
United Kingdom

Dr M C Alonso,
Instituto Eduardo Torroja,
Spain

OPENING ADDRESS

Christopher Chope Esq. MP OBE
Minister for Construction

WELCOMING ADDRESS

Professor Michael J Hamlin F Eng FRSE
Principal and Vice-Chancellor,
University of Dundee

CONCRETE:
THE CONSTRUCTION MATERIAL

Chairman

Mr L H McCurrich

Technical Director,

Fosroc Technology,

United Kingdom

BENEFITS OF CONCRETE AS A CONSTRUCTION MATERIAL

C D Pomeroy

British Cement Association,

United Kingdom

ABSTRACT. It is not always appreciated that concrete is the most versatile of all buildings materials which is user-friendly and can be used in aesthetically pleasing ways. Many of the attributes of concrete are taken for granted but others may not be so obvious.

The diverse benefits are reviewed with particular attention to the less common ones and set against the conference theme, "The Protection of Concrete". The facility to tailor concrete products or structural designs economically both to give a long service life and to be attractive in appearance will be emphasised.

The simple name concrete does not indicate the wide variety of options available to the designer or architect. To benefit from the full family of concrete materials and products, it is necessary to have sound knowledge of what is available and what can sensibly be achieved. These questions will be addressed.

Keywords: Concrete types, Cover, Permeability, Durability, Repairs, Service life, Aggressive agents

Dr C Duncan Pomeroy is the Director of Standards and Information at the British Cement Association. Previously he was Director, Concrete Performance of the Cement and Concrete Association before it was amalgamated with the Cement Makers' Federation to form the BCA.

INTRODUCTION

Concrete is a most versatile building material and there can be
virtually no significant structure being built anywhere in the world
that is not using cement and concrete in one way or another. The use
of concrete is immediately apparent in reinforced concrete buildings,
in concrete bridges or in concrete dams. Other uses may be less
obvious. Steel-framed structures stand on sound concrete foundations
and are frequently protected against fire by concrete. Clay bricks
are bound together by cement-based mortars to build walls that may be
rendered. Even timber buildings usually stand on concrete footings
and have a concrete floor. So even buildings that are not considered
to be concrete rely heavily upon the versatility of concrete for their
very existence.

Despite this universal use, concrete often is blamed for the less
attractive aspects of modern architecture and it is worthwhile to
consider why this should be. In part, the versatility of concrete has
been its own worst enemy. The architect has been encouraged to
exploit the potential of concrete in the design of economical and
functional buildings at the expense of beauty. During a period in the
1960's, concrete structures proudly displayed the fact that they were
predominantly concrete and it was forgotten that large plain unpainted
areas of concrete have little aesthetic appeal, particularly on a damp
Autumn day when the greyness of the wet concrete shows up as black
sombre patches. No wonder the public has reacted so strongly against
these buildings. Moreover, it was assumed that concrete was
indestructible and regular repair and maintenance could be forgotten.
Sadly this is not true, and some structures now display deterioration,
often from the rusting or corrosion of the embedded reinforcing steel.

Within the theme of this conference, I shall be trying to dispel this
image and to show, properly used, the versatility that concrete should
give to structures that are economically viable, that will have a
long, low maintenance life and that will be attractive to the most
discerning of the public.

CONCRETE VERSATILITY

Before considering the diverse uses of concrete in detail, it is
useful to highlight some of the unique attributes of concrete in
comparison with other building materials. There are perhaps two
specific characteristics that make concrete so generally useful.
Firstly, concrete can be moulded into different shapes and sizes
either on site or in a precast concrete works at ambient temperatures
above freezing and, so long as the concrete remains wet, it will
harden and gain strength progressively. Its second dominant virtue is
the protection concrete can give to steel to inhibit rusting. In
fact, a high proportion of concrete failures can be attributed to the
corrosion of the reinforcing steel and this may be the consequence of
inadequate protection by a poor quality surface layer of concrete or
by the provision of inadequate cover to the steel. All other

beneficial attributes of concrete are secondary to these with the possible exception of cost, for concrete will normally not only provide the solution to a given structural problem but will do so cost effectively.

The full versatility of concrete is much greater in that by selection of the constituents of the concrete, the compressive strength may lie within the range of a few mega Pascales (MPa) to well over 100 MPa and with densities from a few hundred to over 3,000 kg/m^3. The lowest density materials are ideal for insulation, whilst the heaviest can provide shielding against nuclear radiation.

Although concrete is weak in tension and will normally be reinforced with steel, it is possible to use fibres of different kinds to raise the effective tensile strength by control of the growth of cracks and to change the brittle nature of concrete to one of ductility. Asbestos cement sheets and pipes provided the original exploitation of fibre concretes but as the health hazards associated with asbestos have become known, other more congenial fibres have been used, like glass, polymer or steel. Each type of fibre contributes particular characteristics thereby adding to the total versatility of concrete and providing opportunities for new uses of concrete products in construction.

Another way to expand the potential range of properties of concrete to satisfy the needs of the user is by the inclusion of polymer-based admixtures into the concrete mix. Water-based emulsions of rubber latex or of polymers or co-polymers like methyl methacrylate or styrene butadiene have been used. These will often increase the failure strain of the concrete but, of greater significance, will provide greater resistance to chemically aggressive environments and to acid attack.

The development of chemical admixtures to modify the properties of concretes to enable them to be used in even wider ways has been very rapid. In particular, the rheological properties of concrete can be significantly changed. Concretes can be made that flow freely with only the minimum of vibration to fill the formwork to a level surface. Moreover, such free flowing can be achieved with low water/cement ratios and with the minimum of segregation of the aggregate or of bleeding. Other chemical admixtures allow concretes to be pumped over long distances for placement at considerable heights in a structure or well away from the delivery point on site.

These few examples clearly demonstrate why concrete is so widely used and there are many more that could be cited. The importance of the appearance of concrete structures has already been emphasised and here, too, the concrete specialists have developed techniques that add aesthetic appeal to the engineering benefits offered by concrete. The manufacture of white cement, the use of stable coloured pigments and the exposure of aggregates at the surface of concrete are but a few examples.

Finally, from the public's point of view, it is possible to build in
concrete to an appropriate scale, using structural elements that do
not have to conform with a monotonous regularity, but which can be
adapted, economically, to add a human dimension to a structure. It is
against this background that modern concretes should be judged and
used. They can be tailored to a host of needs, functional, economic
and aesthetically acceptable.

HISTORICAL BACKGROUND

It behoves us to remember that concrete is not a new material. It is
known to have been used by the Egyptians some 2,000 years BC. The use
of lime-based mortars for the floors of huts has also been found in
excavations dated as long ago as 5600 BC. Probably the most
substantial existing concrete building from antiquity is the Pantheon
in Rome (27 BC). This has a 50 metre diameter unsupported roof that
is comprised of lightweight concrete segments. Volcanic pumice was
used as the aggregate and this was bound in a lime and pozzolana
matrix. The whole roof is designed to be in compression, so that
there is no need for any tensile reinforcement. The survival of this
roof for over 2000 years clearly shows that concrete can be very
durable. No doubt the absence of reinforcing steels that can corrode
has ensured this longevity.

Sadly the Roman skills were not passed down through the generations to
modern times. It was only during the eighteenth century AD that the
foundation of the modern cement industry can be said to have started.
In 1759 Smeaton built a new lighthouse on the Eddystone rock off
Plymouth on a concrete foundation. Although the lighthouse was
replaced in 1876, the foundations have survived. His cement consisted
of burnt blue lias, a limestone from South Wales and a pozzolana from
Italy.

This development was followed in the early 1800's with the invention
and patenting in 1824 of Portland cement by Joseph Aspdin. Since
then, Portland cements have been steadily improved, the chemical
constitution has evolved, combinations of Portland cement have been
made with other materials (fly ash, ground granulated blastfurnace
slag, silica fume and so on), and improvements have been made in the
establishment and control of the fineness.

These cements have been used to make the family of different concretes
used for diverse purposes, most of which are used in combination with
reinforcing or pre-stressing steels to carry the structural tensile
and shear loads. Thus since the early 1800's, there has been a steady
extension in the uses of Portland-cement based concretes, the growth
in its use being due not only to improvements in the material, but
also in construction practice.

TAILORING CONCRETES FOR SPECIAL NEEDS

The versatility of concrete has already been displayed. How is this to be used effectively? Firstly, it is essential to decide what is required from the concrete and whether there are any particular restrictions on its use. For example, construction in a confined space in a city centre may impose a restraint on the method of building that would be absent on a rural site. It could be essential to pump concrete from a single location to all parts of the building and to different levels, whereas multiple deliveries could be accepted elsewhere. The fact that pumpable concretes are readily available means construction with concrete becomes particularly attractive on a congested site. However, this is only one specific example of selecting the appropriate construction system and material for the job in hand.

As this conference is concerned with the protection of concrete, it is pertinent to question the requirements expected from the concrete. Firstly, what do we mean by protection? Is this the provision of a barrier that prevents contact between the aggressive environment and the concrete? Sometimes this is the only feasible way to protect concrete from attack by agents such as acids. Is it the selection of the constituents in a concrete mix to improve its resistance to a known reactant? The use of a sulphate resisting Portland cement in an aggressive sulphate environment, for example. Is it the use of concrete itself to provide protection? Increasing the cover to steel is likely to delay, if not preclude, the onset of corrosion, thereby extending the service life of the element or the structure.

It is essential to know the potential sources of deterioration of the concrete before a serious attempt is made to select a suitable concrete solution from the options available.

Table 1 lists some of the agents against which protection may be required. Some of these sources of aggression occur naturally in the environment, other relate more to the uses made from the concrete or to their introduction by man (e.g. the use of road salts to reduce the risk of ice formation on roads). If the expected environmental exposure is known, it is then encumbent to understand the mechanism of the aggression, its likely rate of progression and the required service life of the element or the structure.

In some instances, the slow erosion of the concrete surface may be acceptable, whereby the provision of a deeper concrete section is all that is required to give an extended and sufficient service life. This would be a solution to the gradual dissolution of a concrete surface by flowing neutral or acid water from a peaty catchment area. However, further advantage will be gained from the use of a dense, low permeability concrete made using a gravel or other inert aggregate.

TABLE 1   Concrete: Sources of aggression

I Natural agencies

Air-borne	Carbon dioxide
	Sulphur dioxide
	Sulphur trioxide
	Acid rain
	Sea spray
	Grit
Water-borne	Neutral waters
	Acid waters
	Sea water
	Mineral salts (including sulphates)
	Gravel
Environmental	Temperature cycles and extremes
	Drying and wetting cycles
	Freezing and thawing cycles
Internal	Alkali aggregate reaction
	Sulphate attack

II Foreign agencies

Road de-icing salts
Inorganic acids
Organic acids (including lactic and acetic)
Vegetable and mineral oils
Sugars
Silage
Sewage (including urine)
Beer
Mechanical wear

Generally, the less permeable the concrete the more slowly will it be eroded or distressed, except when cracking from alkali aggregate reaction is experienced. Here, a cement rich concrete mix can be dense, strong and of low porosity, but the higher cement content also means a higher alkali content which could, on some occasions, be sufficient to fuel a damaging reaction should there be a critical proportion of alkali-reactive aggregate present. However, the alkali levels of current Portland cements produced in the UK and the cement contents of most concretes used, combine to give concrete alkali levels well within the safe limits defined in BRE Digest 330[1].

There are, therefore, two distinct situations to address. Is the concrete to be used for a new structure or is it to be used in repair? If the former situation applies, one must ask whether the concrete can survive satisfactorily in the anticipated working environment. If the answer is no, there are several options:

(i)   Can the concrete mix and choice of cement be adapted to provide a more suitable concrete of sufficient durability for the job in hand?

(ii) Can the exposure conditions be changed, by diverting or diluting aggressive liquids flowing over the concrete, for example?

(iii) Should protective coatings be applied to the concrete or chemicals be incorporated in the concrete to make it more resistent to chemical attack?

(iv) Choose a non-concrete solution if concrete is clearly not the appropriate material.

Where aggressive environments will be experienced, it must be incumbent to allow for their effects carefully. For example, aggressive road salts should not be allowed to form stagnant pools or permeate unprotected areas of a concrete structure. Proper drainage must be provided and kept clear to ensure that the salts are safely directed away from the concrete. It will always prove best to remove the cause of deterioration wherever possible.

Another cause of deterioration of concrete results from the freezing and thawing of wet concrete. Although dense and virtually impermeable concrete can survive severe frost cycles, the most reliable protection is provided by the entrainment of air bubbles in the concrete to contain the expansions that take place when water freezes.

A third example that must be addressed is that of factory or warehouse floors. Concrete can be used to construct excellent, hardwearing and durable concrete floors, but much too frequently problems arise, not because the wrong material is being used, but because there is a great lack of understanding and application of the technology needed to guarantee the successful manufacture of an abrasion resistant floor. It is the construction methods and the length of the moist curing period that largely dominate the performance. This has recently been addressed by Chaplin[2], who showed that while the concrete mix and cements chosen are important, the construction practice and curing were much more significant. The choice of cement will interact with the construction practice, largely because a slow setting cement may preclude the completion of the laying of a good quality floor within a single working shift, thereby requiring expensive overtime or two shifts working to ensure the best quality of floor is constructed from the chosen concrete, as the correct finishing of the concrete surface is a vital necessity. Here it is worth noting that the settings of some cements are particularly temperature dependent, so that in winter, the delays between casting and finishing a floor are prolonged and the possibility of incorrect finishing is greatly enhanced. Other adaptations can be made to floor construction technology. Fibres can be added to provide additional cohesivity to the fresh mix, thereby allowing placement to be more easily controlled and also inhibiting the propagation of fresh shrinkage cracks that might be formed. Polymers can be added to make a floor less able to absorb liquids spilt, or paints or coatings can be applied for similar and aesthetic reasons.

Reverting to the list of aggressive agents in Table 1, it can be deduced that concretes can be designed and produced that will resist most of the naturally occurring agencies. Where high sulphate levels, neutral or acid waters or freezing and thawing are experienced, more care is needed both in the specification of the concrete mix and in its use. Properly tailored concretes will give good service life.

Where foreign agencies are present, the use of unprotected concrete will not always be sufficient and Lea[3] and the ACI[4] have tabulated sound advice for the concrete specifier and contractor. Some chemicals such as sugar or organic acids are particularly aggressive and concrete must be protected by a chemically inert and impermeable coating, usually an organic layer, though in some instances, stainless steel or other protection will be needed.

These examples of choosing not only the right concrete for a given use, but also ensuring the concreting contractors use sound construction practices, show that for new construction it is possible to tailor concretes to satisfy very demanding requirements but it is essential that the full demands to be imposed are known and allowance is made for them.

USE OF CONCRETES IN REPAIRS

The second very distinct use of concrete is in the repair of concrete. Most damage to structural concrete results from the corrosion or rusting of steel, the consequence of which is to initiate cracks in concrete and to cause surface spalling to occur. Other causes of damage are surface abrasion and wear of floors, and of structures subject to wind or water carried detritus and the chemical attack of concrete, either on the surface or from within the heart of the concrete itself (e.g. sulphate attack, high alumina cement conversion). As with new building the first questions, therefore, must always relate to the cause of distress, before any attempt is made to repair the damage.

One of the greatest contributions that concrete can make to a repair is the inherent alkalinity that provides the inhibition to rusting of the steel reinforcement. However, the superposition of a new concrete or mortar layer over carbonated, or more especially, chloride contaminated concrete could seal in a source of continuing corrosion. Once such carbonated or contaminated concrete has been removed though, the presence of Portland cement mortar with its reserve of alkalinity can be positively beneficial.

It is important that repair coatings are fully compatible with the existing structural substrates and this again becomes an area where the versatility of concrete can be used to select a suitable mortar or concrete that will act in an integral way with the original concrete. A significant difference in the elastic or thermal moduli between the repair and the original concrete can result in debonding between the layers. To avoid this the repair should have a larger strain capacity than the original concrete and the bond between the layers should be as strong as possible. Brown[5] showed that it was essential to have the substrate sufficiently damp before a mortar render is applied, but if this can be assured, excellent results are achieved that are at least as good as those resulting from chemical bonding agents.

Mechanical compatibility is only one requirement from a repair and similarly to new construction, it is necessary to choose the correct repair method in the light of the causes of deterioration and taking account of the future environmental and mechanical service loads.

If the damage has resulted from the loss of alkalinity in the concrete consequent upon atmospheric carbonation, it may be necessary to cut out all of the carbonated concrete down to and beyond the steel reinforcement before applying a surface layer of repair mortar or concrete. However, it has been shown that concrete can be re-passivated by the inward diffusion of alkalis from the repair coat though this may take an unacceptably long period and so be impractical. Moreover, the concrete would need to be sufficiently moist to permit counter diffusion to occur. Perhaps of greater importance would be the provision of a mechanically compatible layer that is virtually impermeable to the flow of oxygen, thereby removing one of the essential factors for corrosion to continue. To provide such a barrier involves good knowledge of concrete technology, good surface preparation and application of a mechanically compatible layer and adequate curing to ensure that the well compacted layer has hydrated sufficiently to become virtually impermeable. Again, it is stressed that the original concrete and the repair must act as a true composite. All too often a repair layer will separate and debond from the substrate as a consequence of the different responses of the different layers to moisture, thermal or frost actions. Fortunately the selection of appropriate mixes, aggregates and cements can minimise this risk.

If the cause of deterioration is the presence of chlorides, the remedial measures will be more demanding, since corrosion of the steel is likely to continue unless the chloride ions are driven or cut from the vicinity of the steel. Even if this is done and a good quality concrete surface coat is applied, all precautions should be taken to preclude the ingress of new chlorides. As with all repairs, our aim should be the provision of a low permeability layer, compatible with the substrate. Epoxy or polymer based mortars may prove useful, though if the underlying concrete is saturated, there could be debonding problems.

There is always the need for compromises when repairs are made. If concrete has cracked as a consequence of reaction between the alkalis in concrete and part of the aggregate, it would be unwise to provide a new source of alkalis from a cement rich repair. On the other hand, it is well known that steel does not rust in an alkaline environment. Thus it is vital to understand the causes of distress before choosing the best solution.

DISCUSSION

The principal message that follows from this superficial consideration of some of the problems of remedial work is that the best way to ensure good service life is to build correctly in the first instance. To do so requires full knowledge of the service conditions to be faced and the life expectancy from the structure. If it is known that severe conditions will be experienced, the concrete must be suitably tailored to the needs and the effects of the aggressive agents and must be designed out as far as is practicable. Good drainage must be provided and maintained if there is a continuous flow of an aggressive gas or liquid, stagnant accumulations must be avoided, but above all, the quality of the designated concrete must be adequate for its job, it must be correctly placed and be properly cured if the repair industry is not to be called prematurely into action.

The family of concretes available for the designer to choose from is vast and the achievable properties are well documented. These must be matched against prescribed needs and all service requirements must be considered at the design stage. It is insufficient to look primarily at structural loads and requirements in isolation, particularly where severe environmental requirements are to be expected or when a long service life is required.

CONCLUSIONS

Concrete is versatile, but to make full use of this versatility, it is essential to know what particular performance is expected from the structure in which it will be used. Some of the relevant factors have been discussed and particular issues have been highlighted.

Both natural and man-made sources of deterioration have been considered and some pointers have been given to particular solutions to particular problems. It is stressed that many of the effects of potential causes of deterioration will be minimised when good quality, dense and low permeability concrete is used. There are exceptions to this general rule, particularly when freezing and thawing are experienced. In general, it is better to invest in the provision of an adequately high quality concrete when the structure is built, than to rely on expensive maintenance and repair. Concrete usually has the capability for such a prospect.

REFERENCES

1.   Alkali aggregate reactions in concrete. Building Research Establishment, Digest 330, March 1988.

2.   CHAPLIN, R. G. The influence of GGBS and PFA addition and other factors on the abrasion resistance of industrial concrete floors. British Cement Association Report (in the press), 1990.

3.   LEA, F. M. The chemistry of cement and concrete. 3rd edition, Edward Arnold Ltd., 1970, Chapters 19 and 20.

4.   ACI Manual of Concrete Practice, Part 5, A guide to the use of waterproofing, damp-proofing, protective and decorative barrier systems for concrete". ACI 515 IR-79 (Revised 1985)

5.   BROWN, J. H. Renderings: modified adhesion to dense and to aerated concrete'. Cement and Concrete Association, Departmental Note DN 4037, October 1981.

PROTECTION AFFORDED BY CONCRETE

J E Krüger
B G Lunt

Division of Building Technology. CSIR,

South Africa

ABSTRACT.

The paper deals with the protection afforded by concrete. Firstly, the physical protection which concrete affords to man is considered. Secondly, it discusses how man's property, resources and environment can be protected by concrete. Thirdly, it points out that, if the concrete and the structures built with it are correctly designed, they can enjoy a significant degree of self-protection from physical and chemical distress.

Keywords: Protection, Concrete, Radiation, Impact, Containers, Fire, Environment, Physical and Chemical Distress.

Biographical notes

Dr Japie E Krüger joined the CSIR in 1949. After heading the Inorganic Materials Division of the National Building Research Institute from 1981 to 1987, he is now a project leader in the Building Materials Technology Programme of the Division of Building Technology of the CSIR. He has for many years been involved in research into inorganic building materials, notably blastfurnace slag and fly ash.

Dr Brian G Lunt, a registered professional engineer, is a project leader in the Engineering Structures Programme of the Division of Building Technology. He joined the CSIR in 1959 and has worked mainly on the structural behaviour of reinforced and prestressed concrete, with a special interest in prestressing steel, reinforcement detailing, shear strength of flat slabs, structural design against the effect of weapons, and novel building systems.

INTRODUCTION

The present estimated annual world production of concrete is four billion cubic metres. In one way or another, concrete structures that have been and are still being built serve to protect man, his property, a country's resources and the environment. When such protection comes to mind one is, quite naturally, inclined to think only of the protection afforded by concrete from hazards with catastrophic effects, such as nuclear radiation, explosions, earthquakes and toxic materials. The everyday protection which concrete structures give man and his property and resources is more often than not taken for granted - for example, structures for protection from the weather, structures for the safe disposal of industrial effluents and sewage, oil and fuel storage tanks, grain silos, water reservoirs, crash barriers on highways and even bridges for the safe transportation of goods across rivers and fiords.

Concrete is probably the most versatile material with which to build protective structures. It can be moulded into almost any shape and designed to the strength necessary to withstand predicted imposed dynamic or static stresses. For this purpose the engineer has at his disposal a vast fund of concrete technology that has been built up over the years. However, in many instances concrete itself needs protection to fulfil its protective role; this is the very theme of this conference. Fortunately, concrete can in most cases be designed or treated in such a way that it will be durable enough to fulfil its role without requiring extraordinary protective measures.

PROTECTION OF MAN

Protection from the Elements and Natural Forces

Man needs shelter, firstly for himself and secondly for his possessions, and over the ages he has acquired the knowledge to meet this need. Today he has sophisticated methods of design and construction to build durable shelters to exact specification. Concrete (both reinforced and unreinforced) lends itself admirably to the erection of structures to shelter man and his possessions and there is hardly any building (whether single-storey house, multi-storey office block or factory) that does not have some concrete in its structure.

In the design and erection of buildings the engineer not only has to deal with everyday situations, but he often also has to design the building to withstand gale forces or earthquakes. Fortunately, he has codes of practice and structural analysis at his disposal, enabling him to design concrete buildings[1] so that they will give adequate protection to their inhabitants against loss of property, life and limb. Much has also been published on the use of concrete for civil or military defence shelters, in particular those specifically designed to protect from nuclear radiation.

Radiation Shielding

The increased use of nuclear energy has given rise to an increase in artificially produced radiation and radioactive materials, and the design and construction of shields to protect man from the effects of radiation has never been more important.

Two kinds of nuclear emission, namely electromagnetic radiation (gamma radiation) and particle radiation (alpha, beta and neutron radiation) are of interest. Because of the low penetrating power of alpha and beta particles, such radiation is of little concern in the design of biological shields. Concrete is generally used for radiation shielding in structures built to attenuate gamma and neutron radiation, because of its versatility and cost effectiveness as, for example, in pressure and containment vessels for nuclear reactors, particle accelerators such as cyclotrons, and other biological shields. The importance of concrete as a radiation shield material is emphasised and dealt with comprehensively by Kaplan in his recent book, Concrete Radiation Shielding[2] and guidance on design can be found in British and American specifications and codes[3,4].

Apart from its favourable cost, concrete is considered to be an excellent material with which to build radiation shields because it can be cast, it is adaptable and good concrete requires little or no maintenance. Concrete is furthermore a commonly used material (if not the only one) that offers shielding from both gamma and neutron radiation.   Because it contains hydrogen (in the form of free water and hydration products) and other light nuclei, concrete is a good neutron shield. In addition, because of its relatively high density even normal concrete is a good material for attenuating gamma rays. When concrete is made with heavy aggregates, such as hematite, ilmenite or baryte, it is particularly effective for gamma-ray attenuation.

A disadvantage of concrete is its low thermal conductivity which retards the dissipation of heat generated in the concrete by the absorption of radiation, although concrete should be well able to sustain temperatures normally encountered[5]. In designing the shield, stresses due to temperature increases as a result of radiation absorption have, however, to be taken into account. Temperature increases may also result in the drying out of the concrete and thus reduce its ability to attenuate neutron radiation.

The above remarks also apply to shielding where concrete is used to build shelters against radiation resulting from nuclear detonations, containers to store radioactive waste and structures where radioactive isotopes are handled. In a recent article in Construction Weekly[6], the view was expressed that concrete and its constituent materials would in future find an attractive market in the construction of nuclear-waste treatment facilities and repositories.

Disposal and Containment of Harmful Substances

In our industrial, increasing urbanised world there is a constant
demand for the safe conveyance of harmful industrial substances to
facilities where they can be disposed of, or treated and safely
stored. The storage of harmful substances poses its own problems.

Concrete is probably the most popular material for these uses. The
disposal of sewage without concrete and fibre-cement pipes or a
concrete sewage purification works even though the concrete may have
to be lined with a chemically resistant lining is virtually
unthinkable. The same applies to hazardous industrial effluent,
whose transportation and safe containment usually depends on
facilities constructed from concrete. All this is aimed at
preventing pollution and protecting man from epidemic disease.

Figure 1:   A sewer pipeline and a section of the
             sewage purification works

Blast Protection

Structures that are specifically meant to afford protection from
blasts include missile silos, explosive stores, facilities where
explosives are handled and tested, factories where explosive
conditions can arise, and military and civil defence shelters.
Judging from publications on structures designed to afford
protection from blasts, concrete is a first choice - whether for an
underground structure or one within a normal building. Lunt[7],
for example, points out that reinforced concrete is eminently
suitable for civil defence shelters as continuous, monolithic

structures can readily be built with it. He also points out that reinforced concrete elements have a characteristic elastic-plastic response to loads that enable large plastic deformations to take place without seriously reducing structural integrity.

Other structures are those that must withstand catastrophic explosions like the one in Mexico City, a number of years ago, where, after a leakage in a PEMEX liquid gas plant several explosions and devastating fires occurred, almost destroying the plant and severely damaging the surrounding houses; according to official figures more than 500 people were killed and over 7 000 seriously injured[8].

In a paper delivered at the 1987 IABSE symposium, Bomhard[8] proposes the use of a passive safety system which rigorously restricts the damage caused by events such as the Mexico City explosion. He states that this can be achieved through structural safety, using reinforced and prestressed concrete components to form special containers. When operating pressures are high, the concrete container has the shape of a sphere. Figure 2 depicts the construction process in which a sphere consisting of several parts is expanded (and thus prestressed) and then closed to form a monolithic shell. Bomhard states that, where the contents of such structures may attack the concrete, the structures must be accessible for easy periodic checking. According to his calculations, concrete containers for this purpose are cheaper than steel containers of similar size.

The performance of concrete subject to impulse-loading can, if necessary, be further improved by reinforcement with synthetic fibres which serve as energy absorbers, although best results are obtained when both conventional steel reinforcement and random fibres are incorporated in the concrete[9].

Protection from Impact

There are numerous examples of concrete structures specifically designed to provide protection from impact. Any shortlist would include railings, balustrades, motorway crash-barriers, and casings around bridge columns to protect them from impact damage.

The Tjorn bridge disaster in Sweden in 1980, in which a ship collided with a bridge pier, causing its collapse, dramatically highlighted the vulnerability of such structures to impact and the value of wide protective bases - even though the remedy in that particular case was to build a suspension bridge with a long span, with its support towers on dry land, instead of having piers in the water.

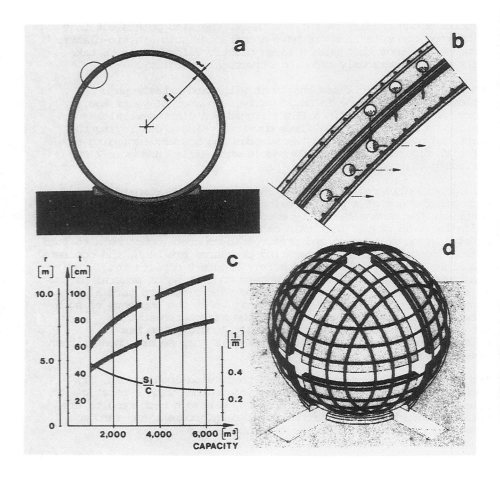

Figure 2 :   Spherical concrete pressure vessel[8].
(by permission of IABSE, ETH-Hönggerberg,
CH-8093 Zürich, Switzerland)
(a) Diametrical section, (b) Section through
wall, (c) Dimensional parameter relationship,
(d) General assembly.

An interesting use of concrete to protect a bridge from indirect
impact damage by ships is the wind barrier erected along the Caland
Canal near Rotterdam, Netherlands. It was found that ships with a
large windage could not be allowed to pass the Caland Bridge when
wind speeds exceeded a certain limit, without the risk of damaging
the bridge. The solution was to build a 25 m high wind barrier
consisting of a series of semi-cylindrical concrete shells, with 25
per cent permeability achieved by the spacing of the shells, along
the canal[10].

Figure 3:   Concrete crash barrier along a six-lane
             motorway

An advantage of using concrete for protection from impact is that
its impact resistance, like its blast resistance, can be
substantially improved by incorporating fibre reinforcement[9].

Protection from Fire and Thermal Radiation

There is no doubt that concrete will deteriorate when heated to
elevated temperatures and that it can be severely damaged by fire;
this damage can include loss in strength, partial collapse,
distortion, deflection, expansion, buckling of steel, spalling and
shattering[11]. The fire resistance of concrete is nevertheless
such that its failure will not be sudden and, in the case of a
concrete building, occupants should have safe exit.

Because of its low thermal conductivity, it is very likely that only
the surface of a concrete structure will be damaged in a fire, and
concrete is thus often used to provide a protective casing around
steel structures. Concrete made of lightweight aggregates is in
fact eminently suitable as a protective material with a high
resistance to fire[12].

THE PROTECTION OF MAN'S PROPERTY, RESOURCES AND ENVIRONMENT

Not only is concrete used to protect man, but it has also become one
of the most important materials for protecting his property and the

environment[5], as discussed below;   this includes indirect
protection.

Cryogenic Chambers

Cryogenic chambers used for the storage of liquid gases have
traditionally been constructed from special nickel-alloy steels.
However, since 1950, concrete - particularly prestressed concrete -
has been increasingly used for this purpose[11].   Apart from its
enhanced performance at very low temperatures, such as an increase
in both strength, and bond strength between steel and concrete,
Young's modulus improves and there is a reduced tendency to creep
with decrease in temperature[11].   Today there is a considerable
amount of information about the use of concrete which the engineer
could use for the economical and safe design of concrete cryogenic
chambers[13,14].

Oil and Liquid Fuel Storage Tanks

Concrete storage tanks for oil and liquid fuel are seen all over the
world and several articles on the topic have been published.   Many
of these publications deal with the influence of oil and its
derivatives on the performance of the concrete.   Although the
findings are somewhat at variance, it would appear that most of the
liquids do not have a significant, deleterious influence on good
quality, well-cured, dense concrete[15].   Most tanks (especially
those holding fuel of low viscosity) are, however, normally provided
with an impervious lining, mainly to make sure that no losses occur
under hydraulic pressure.

The following advantages that concrete tanks have over steel tanks
have been mentioned by researchers[16]:

(a)   Because concrete tanks have greater fire resistance than
        steel, they can be built nearer to one another and adjacent
        tanks are less likely to catch fire.

(b)   Concrete tanks can be built underground without much extra
        cost, so that parks can be created over them.   Several tanks
        with domed floors can be built on top of one another.

(c)   The sun will heat the contents of concrete tanks less than
        that of steel tanks, so that volatile materials will be less
        inclined to evaporate.   Only a moderate overpressure will
        therefore be required in concrete tanks to prevent evaporation
        loss.

Grain Silos

Reinforced concrete is an economically attractive material for

building large-capacity circular silos[17], like grain silos.   It
is, however, important that such silos be waterproof and adequately
airtight and that they are able to withstand static pressures as
well as the dynamic pressures imposed during filling and withdrawal
of their contents.   In addition, grain silos must also be vermin-
and insect-proof. Guidelines are available for constructing concrete
silos to fulfil their functional requirements[18].   It is
essential to use concrete of good quality.   If so, concrete silos
will be adequately airtight and water- and insect-proof.

Grain which is stored in concrete silos is protected to such a
degree against temperature fluctuations that there is no real risk
of excessive moisture loss[19].   It has also been mentioned that
the embryos of the peripheral layers of barley stored in a steel
silo can die, making the barley unsuitable for beer brewing.   This
is not the case with barley stored in concrete silos, where the
concrete protects it from overheating[20].   It has further been
claimed that, except in the case of small silos (15 000 m^3 and
less), concrete silos are the most economical to build[19].

FIGURE 4:   A row of grain silos on the Transvaal highveld,
             South Africa

Water Storage and Supply Facilities

Concrete is playing an indispensable role in the construction of
dams, other water reservoirs and water reticulations, for the
storage, protection and distribution of man's most vital resource.
One needs only to page through the proceedings of the various ICOLD
congresses and the numerous books and publications on concrete pipes

to realise the vital role that concrete is playing. One is also
struck by the research that has gone into concrete dam building and
the ingenuity of the engineering profession in its attempt to gain
optimum benefit from concrete in this field. For example, during
the 1970s it was realised that the building of conventional concrete
dams was becoming prohibitively expensive. As a result, the concept
of roller-compacted concrete (RCC)[21] was developed, again making
concrete a very competitive material. The advance of RCC was rapid,
and since 1980, some fifty large RCC dams have been built, again
placing concrete in the forefront as a dam building material.

Figure 5 :   The world's first roller-compacted concrete arch
            gravity dam being built by the South African
            Department of Water Affairs at Knellpoort, South
            Africa

Coastlines and Harbours

Concrete quays and harbours have been and are still being built all
over the world to handle and provide a safe moorage for the world's
ships.

Coastlines also have to be protected against erosion by wave action,
and here again concrete has fulfilled an important role. Several
interesting shapes of concrete elements, have been developed for
this purpose, including the 'tetrapod' and the very successful
'dolos', which was conceived by a South African harbour Engineer in
the early 1960s[22]. Today wave barriers consisting of three to
fifty ton dolos units[23] are seen along many a coastline,
preventing erosion of the shoreline.

Figure 6 : Concrete 'dolos' wave barrier along a coast

Another example of concrete as a protector of man and his property
from the sea, is the use of dykes and sea barrier/sluice gates along
the low-lying Netherlands' coast. Most of these structures are
examples of superb concrete engineering.

Retaining Walls

An important use of concrete is in retaining walls of the kind one
often sees in mountain passes (where cuttings have been made through
soft soil) to safeguard roads and traffic against landslides, or
where retaining walls protect lower lying property.

Concrete tunnel linings fulfil a similar role in preventing rock and
soil from caving in.

PROTECTION OF CONCRETE BY ITSELF

So far we have seen how man and his property enjoy the protection
afforded by concrete. There are however, many situations where
concrete itself needs long-term protection, in one way or another,
to remain functional. Fortunately, concrete can normally be made
durable by the judicious selection of the concreting materials and
the appropriate design of the concrete.

The following paragraphs illustrate some steps that could be taken
to build concrete structures that would afford themselves

protection without resorting to exotic types of treatment.

Water Ingress

Water penetration into concrete can be negligible if the concrete has a low permeability. Low permeability can be achieved by correct design, the use of workability aids, and proper compaction and curing.

Because concrete is relatively easy to cast, it lends itself to the construction of special watertight joints. Joints in in situ concrete generally make use of cast-in water-stops of various kinds, and special joints have been developed to deal with dimensional movement and water ingress between precast concrete elements[24].

The Effects of Temperature and Moisture Stresses

Concrete changes dimensionally when its temperature or moisture content changes and, if provision is not made to relieve the consequent stresses that may build up in a concrete structure, the structure may suffer cracking, distortion, spalling or related distress. Such distress can be overcome, however, if provision is made for dimensional movement in the structure by the incorporation of suitably spaced joints which will also limit moisture ingress, as mentioned earlier.

Freeze/Thaw Damage

Some exposed structures cannot always remain dry enough to avoid damage caused by the freezing of absorbed water. The freeze/thaw problem needs no introduction in many countries and it is well known that such exposed concrete should be cast with an adequate quantity of entrained air[25]. Entrained air bubbles normally remain empty of water, interrupt capillaries and can accommodate water forced out of capillaries during freezing, without the risk of unduly high pressures developing in them, thus allowing the concrete to protect itself.

Corrosion of Reinforcement

The corrosion of steel reinforcement causes serious problems in concrete. Not only does reinforcement corrosion cause cracking, spalling and the unsightly staining of concrete structures but, if the structural steel corrodes excessively it can endanger the integrity and safety of the structure. Much can, however, be done to let the concrete protect the steel from undue corrosion. In the first instance, any ingredient that might promote corrosion should be avoided. Secondly, a dense, impermeable, dimensionally stable concrete should be cast and, thirdly, the steel should be

protected by an adequately thick concrete cover. Such steps are
too often neglected in practice.

Chemical Attack

The various mechanisms of chemical attack from which concrete can
suffer, such as sulphate attack, alkali-silica reaction, soft water
attack[26] and biogenic sulphuric acid attack[27] are not always
well understood in the construction industry. However, the concrete
technologist and engineer can, in consultation with experts in these
fields, do much to design the concrete to protect itself, as
indicated in the following examples.

Where sulphate attack is a potential problem, a sulphate-resisting
cement can be used. Replacing part of the portland cement with fly
ash or ground granulated blastfurnace slag will also help greatly to
make concrete more resistant to such attack and less likely to
suffer deleterious expansion as a result.

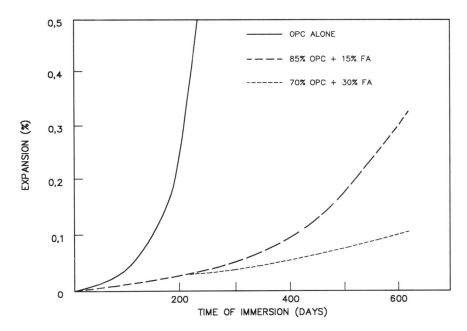

Figure 7 : The expansion of mortar prisms made with OPC and
blends of OPC and FA during immersion in a 5%
sodium sulphate solution

Concrete will not suffer deleterious expansion due to alkali-silica
reaction when an alkali-reactive aggregate has to be used, if a

low-alkali cement is used, or (when a high-alkali cement is used) if some 20 per cent, by mass, of the cement is replaced with a good quality fly ash, or some 30 per cent, by mass, of the cement is replaced with ground granulated blastfurnace slag[28].

Very little can be done to combat soft water attack on large existing concrete structures, except to treat the water or give the concrete a surface coating or surface impregnation, which measures are often impractical, expensive and at best a temporary solution. However, concrete products can be made significantly resistant to soft water attack if some 30 per cent, by mass, of the portland cement is replaced with fly ash and the products are then autoclaved; precast concrete elements such as pipes are suited to such treatment.

Biogenic sulphuric acid attack in sewers can be a costly problem, particularly in hot countries.   However, the lifespan of a concrete sewer pipe could at least be doubled if a calcareous aggregate is used instead of a siliceous aggregate, a measure that was first proposed by van Aardt[29] and is today used in many countries.

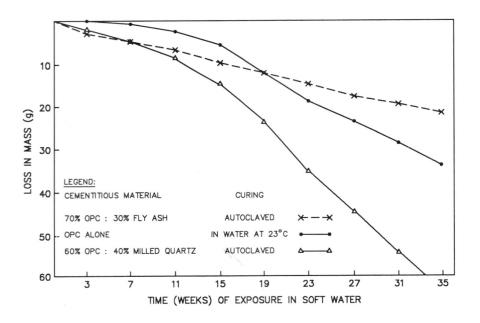

Figure 8 :   Mass loss curves for mortar prisms immersed in
              aggressive soft water (containing 500 ppm carbon
              dioxide). The prisms were brushed fortnightly to
              remove the attacked layer.

CONCLUSION

Concrete is the most important construction material used by modern
man. Vast volumes of it have been and are still being used to build
structures, most of which are erected to protect man, his property
and the environment.

Concrete is a versatile, adaptable material. Both the concrete and
the structures built with it can be designed to afford the desired
protection. Concrete itself, however, needs protection to function
most effectively. This point will be emphasised again and again
during the deliberations of this conference. In many instances this
protection can take the form of self-protection through judicious
selection of the concreting materials, and the appropriate design of
the concrete mix and the structure, without a need to resort to
exotic protective measures.

REFERENCES

1.   OKANO Y, TSUGAWA T, BESSHIO S, OKAWA J and ISOZAKI. Design
      and Construction of Tall Reinforced Concrete Buildings in a
      Seismic Country. Proceedings of IABSE Symposium on Concrete
      Structures for the Future, Paris - Versailles, 1987,
      pp 443-448.

2.   KAPLAN M F. Concrete Radiation Shielding (Nuclear Physics,
      Concrete Properties, Design and Construction). Longman
      Scientific and Technical, Harlow, UK, 1989, 457 pp.

3.   BRITISH STANDARDS INSTITUTION. BS 4975 : 1973. Specification
      for prestressed concrete pressure vessels for nuclear
      reactors, London 48 pp.

4.   AMERICAN SOCIETY OF MECHANICAL ENGINEERS. Code for Concrete
      Reactor Vessels and Containments. ASME Boiler and Pressure
      Vessel Code, Section III - Division 2, 1986, 401 pp.

5.   DHIR R K. Concrete Research : An International Perspective.
      Concrete Beton, No 54, 1989, pp 21-32.

6.   ANON. Treatment of nuclear waste will bring work.
      Construction Weekly, November 1989, pp 14 and 16.

7.   LUNT B G. Civil Defence Planning and the Structural
      Engineer. SAICE Convention, Pretoria, 1968, 12 pp.

8.   BOMHARD H. Concrete Structures as a Safe Engineering Response
      to Environmental Catastrophes. Proceedings of IABSE Symposium
      on Concrete Structures for the Future, Paris - Versailles,
      1987, pp 435-442.

9.   PORTLAND CEMENT INSTITUTE. Special techniques, types and
      applications. Fulton's Concrete Technology (Chapter 16). ED.
      B Addis, PCI, Midrand, South Africa, 1986, pp 772-898.

10.   SCHILPEROORD J and STRUIJS M. The Wind Barrier along the
      Caland Canal near Rotterdam. Proceedings of IABSE Symposium
      on Concrete Structures for the Future, Paris - Versailles,
      1987, pp 615-620.

11.   PORTLAND CEMENT INSTITUTE. Temperature effects and thermal
      properties of concrete. Fulton's Concrete Technology
      (Chapter 9). ED. B Addis, PCI, Midrand, South Africa, 1986,
      pp 515-575.

12.   BUILDING RESEARCH ESTABLISHMENT: Lightweight aggregate
      concretes - 3: Structural application. BRE Digest 111,
      Garston, England, November 1969, 7 pp.

13.   FEDERATION INTERNATIONALE DE LA PRECONTRAINTE (FIP).
      Preliminary recommendations for the design of prestressed
      concrete containment structures for the storage of
      refrigerated liquefied gases (RLG). FIP Commission on
      concrete pressure and storage vessels, Wexham Springs,
      England, 1982, 55 pp.

14.   TURNER F H. Concrete and Cryogenics. Viewpoint Publication,
      C & CA, Wexham Springs, England, 1979, 7 pp.

15.   LEA F M. The Resistance of Concrete to Various Organic and
      Inorganic Agents. The Chemistry of Cement and Concrete
      (Chapter 20). Edward Arnold Limited. London, 1970,
      pp 659-676.

16.   LEONHARDT F. Öl-und Treibstoffbehälter aus Beton. Beton- und
      Stahlbetonbau, Volume 36, No 2, February 1961, pp 25-32.

17.   MURTHY B J K and BHATNAGAR B S. Influence of construction
      imperfections on safety of circular silos. The Indian
      Concrete Journal, April 1989, pp 179-184.

18.   AMERICAN CONCRETE INSTITUTE COMMITTEE 313. Recommended
      Practice for Design and Construction of Concrete Bins, Silos,
      and Bunkers for Storing Granular Materials. ACI Manual of
      Concrete Practice, Part 4, 1988, pp 313-1 to 313-38.

19.   HENNY G E J. 'n Vergelyking tussen betonbuise en
      staalbuise. Proceedings of Grain Silo Symposium, Silverton,
      South Africa, Paper 9, 1976, 5 pp.

20.   PIETERSE T P. Private Communication with authors. Irrigation
      Engineering, Agricultural Development, South Africa, 1990.

21. AMERICAN CONCRETE INSTITUTE COMMITTEE 207. Roller Compacted Mass Concrete, ACI Materials Journal, September-October 1988, pp 400-445.

22. NATIONAL MECHANICAL ENGINEERING RESEARCH INSTITUTE. An investigation into the merits of the new 'Dolos' breakwater units. CSIR Report MEG 391, Pretoria, 1965, 40 pp.

23. ZWAMBORN J A. The dolos breakwater block. The Civil Engineer in South Africa, Volume 31, No 10, October 1989, pp 317-318.

24. LUNT B G. Joints in structures. Proceedings of Symposium on Movement in Concrete Structures, Paper 10. Concrete Society of Southern Africa. Johannesburg, Cape Town, Durban, CSA, Halfway House, South Africa, 1981, 17 pp.

25. NEVILLE A M. Durability of Concrete. Properties of Concrete (Chapter 7). Pitman Publishing Limited, London, England, 1973, pp 382-459.

26. OBERHOLSTER R E , VAN AARDT J H P and BRANDT M P. Durability of Cementitious Systems. Structures and Performance of Cements (Chapter 8). ED P Barnes, Applied Science Publishers, London, 1983, pp 365-413.

27. KRüGER J E and BOTHA J. Sewer Corrosion and its control : A literature survey of the status quo. Fibre-cement Association Pipeline Engineering Lectures, South Africa, October/November 1987, 13 pp.

28. OBERHOLSTER R E. Private communication with authors. Division of Building Technology, The CSIR, South Africa, 1990.

29. STUTTERHEIM N and VAN AARDT J H P. Corrosion of concrete sewers and some possible remedies. South African Industrial Chemist, Volume 7, No 10, 1953, 21 pp.

INHOMOGENEITY IN CONCRETE AND ITS EFFECT ON DEGRADATION : A REVIEW OF TECHNOLOGY

P C Kreijger

Eindhoven University of Technology,

The Netherlands

ABSTRACT. The role of inhomogeneities in mechanical, physical and chemical degradation processes of concrete is examined. In particular, the existence of various 'skin' layers within the concrete is identified. Factors such as sedimentation, microcracking, porosity, transition layers, thickness of cement cover and appearance are all examined with regard to inhomogeneities. The effects of atmospheric degradation processes are also reviewed with regard to the outer cement skin layer of the concrete. A coordinated interdisciplinary approach to further research is proposed.

Keywords: Acid Rain, Appearance, Atmospheric Attack, Bond Strength, Carbonation, Concrete Cover, Concrete Skin, Degradation Processes, Inhomogeneities, Micro and Macro Cracks, Mortar Skin, Porosity, Sedimentation, Transition Layers, Water Vapour.

Professor P C Kreijger was a full Professor at the Eindhoven University of Technology in Materials Science and Engineering. He retired in 1984 after devoting 35 years to research on all types of building materials with special attention to concrete technology.

Professor Kreijger has written more than 120 papers and given many invitation lectures at home and abroad. He has been a member, and Chairman of numerous national and international committees, several of them RILEM technical committees. He has been a RILEM delegate for 14 years, serving as President in 1976 and was subsequently awarded honorary membership in 1982. He received a certificate of appreciation for his contribution to the Science Program of NATO in 1980, and in 1985 the commemoration Medal of the Dutch Standardisation Institute. He still acts as an expert for the Dutch Council for Certification, the Dutch Committee for Reporting on Environmental Effects and the centre for Civil Engineering Research, Codes and Specifications (C.U.R.).

1. INTRODUCTION

Inhomogeneities in concrete are important with respect to durability
in that they can initiate and increase the rate of mechanical,
physical and chemical degradation processes. Such inhomogeneities
occur on both the macro and micro scale.

Macro heterogeneity results from the variations in bulk composition
due to batching deviations and changes in water content of the
aggregates, from sedimentation and segregation from wall effects,
and from transport of water into and out of the concrete.
Consequently the conditions, structure and properties of the outside
50mm of the concrete differs from those of the core. This outer
layer is designated the concrete skin. The outer most layer of the
concrete, and the layer in contact with rebars and coarse aggregate
particles is about 0.1-0.3mm thick and very rich in cement. This is
designated the cement skin. In addition to this, extending inward
for about 5mm from every boundary plane of the concrete is a layered
structure of mortar. This is designated the mortar skin.

The observed surface of concrete is in fact, part of the cement
skin, extending some 20 microns into the concrete and blemishes such
as differences in colour, blow holes, sand streaks and small cracks
are seen in many cases. Air, or water filled gaps can occur under
rebars or coarse aggregate particles if the method of compaction is
not suitable for the w/c ratio, or if sedimentation occurs. Even
where gaps do not appear, weakened layers 1mm thick may occur.
Generally there are always air bubbles irregularly distributed
throughout the concrete.

In Figure 1 some of the inhomogeneities are shown schematically.
The "skin-effect" leads to variations in cement-aggregate and water
content and in porosity over the surface layers, and therefore
influences durability. Consequently composition and properties of

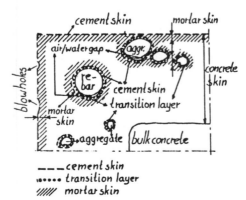

Figure 1: Schematic drawing of the skins in concrete.

the concrete skin are important in understanding processes such as
surface spalling, (micro) cracking, chemical attack and general
surface deterioration especially since the rebars are a constituent
part of the concrete skin.

During the hardening process, the hydration products in the bulk of
the concrete and the cement skin surrounding the rebars and coarse
aggregate are different. In the latter case a so called transition
or duplex layer develops, with a thickness which is measured in
microns and a higher w/c ratio at the interface. This allows ions
to diffuse more easily resulting in the formation of large crystals
and a higher final porosity in the interfacial layer.
Both the presence of micro cracks and the duplex layer increase
permeability to water and solutions (ie acid rain, chlorides) and
also permit gases such as CO_2 and SO_2 to diffuse more readily.

2.   INHOMOGENEITIES

2.1 Variations in Bulk Composition of Fresh Concrete

Using a coefficient of variability of 3.5% in cement and water
content, Figure 2 shows the variations in w/c and a/c ratios for two
compositions. It is apparent that although the coefficients of
variation in composition are quite reasonable, the resultant actual
variations are quite remarkable.

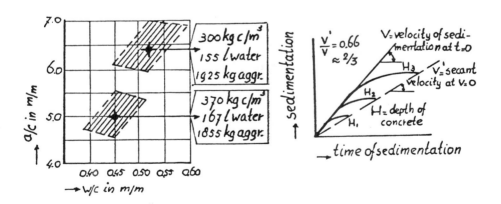

Figure 2: Variation of w/c and   Figure 3: Systematic patterns
a/c-ratio for two fresh concretes   of the sedimentation process.

34 Kreijger

2.2 Sedimentation, Macro Cracks and Adherence of Rebars

The sedimentation process presents certain systematic features as
shown in Figure 3 (Kreijger, 1977). For a concrete with a cement
content of about 300kgm^{-3} and w/c ratio≥ 0.47, the sedimentation
time in minutes is 0.265 H, where H is the height of the concrete in
mm. It follows therefore that sedimentation of concrete above
rebars and coarse aggregate particles will cease before that in the
concrete beside and below the rebars and coarse aggregate.

This differential sedimentation leads to visible cracks at the
upper side of the concrete (above the rebars) and gaps under the
rebars and coarse aggregate particles. It follows that the greater
the depth of concrete under such obstacles, the more serious the
problems become.

Since such gaps effectively lessen the bonding of the rebar to the
concrete, it seems wise to increase the anchorage for rebars which
are situated more than 100mm above the underside of the concrete
(most standards only do this for rebars 200mm above the underside).

2.3 Micro Cracks in Concrete

The deformability of concrete at rates comparable with those of
sedimentation and plastic shrinkage (up to 5mm hr^{-1}) varies with age
(Weirig, 1971) and a critical stage occurs between 3 and 24 hours
after placing. During this period deformations by thermal and
plastic shrinkage can occur as shown in Figure 4.

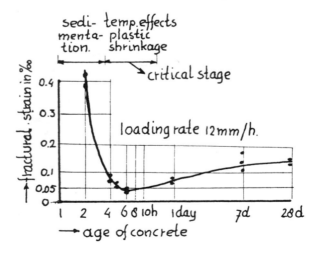

Figure 4: Deformability of concrete as function of age.

Figure 4 also demonstrates the probable causes for occasional macro
cracking, ie,
if the cover/concrete ratio is ≤0.2 and cracks are seen within 3
hours of placing, then sedimentation is the cause.
If cracks are seen after 3 hours, plastic and thermal shrinkage are
the cause (Kreijger 1977).

Micro cracks are also formed along the boundaries of the concrete
with rebars and aggregate (adhesion cracks). Some 10% of the
circumference of aggregate particles ≥1mm in size, is cracked after
about 24 hours. During loading mortar cracks are formed, which
connect the adhesion cracks. Such micro cracks are the cause of the
shape of the stress-strain curve illustrated in Figure 5.

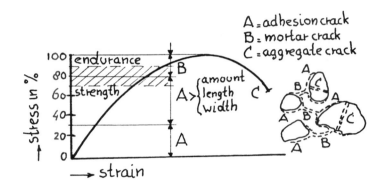

Figure 5:   Stress - strain diagram of concrete.

2.4 Porosity of Concrete

Apart from the macro pores in the concrete - some 1-3%after mixing -
the porosity of concrete is determined by the porosity of the cement
paste and that of the aggregates. Cement porosity depends on the
w/c ratio as can be seen in Figure 6 (Mehta et al., 1980) and is
closely related to the permeability (for pores ≥ 750A). The greater
the w/c-ratio the greater the pores and also the greater the
cumulative pore content. For w/c ≥ 0.05 there is a strong increase
in permeability.

On the other hand the porosity of the aggregates is smaller than
that of the cement paste. A combined picture of pore-size
distributions of cement paste, aggregates and (micro) concrete is
given in Figure 7 (Kayyali, 1985, 1986).

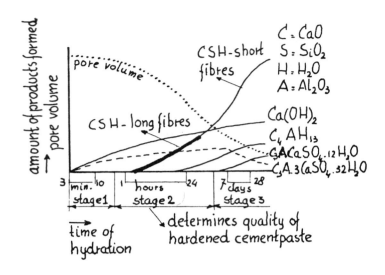

Figure 6:   Permeability and porosity of hardened
           cement paste as function of w/c ratio.

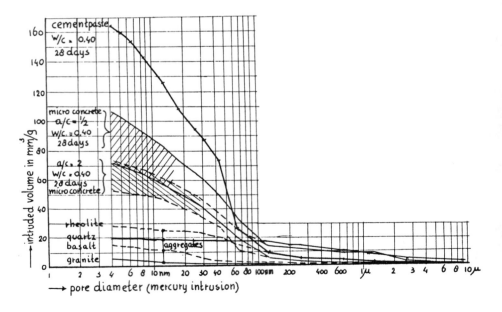

Figure 7: Pore size distributions of cement paste,
          concrete and aggregates.

2.5 The Skin of Concrete

From concrete prisms (325 kg/m^{3}PC, w/c=0.54. a/c=6.11, cured 7 days
in water and 21 in air at 20^{0}C, 65% R.H), 2mm thick slices were sawn
from sides and bottom and the composition and some properties
determined. The results are given in Figure 8 (Kreijger, 1984).

Special water absorption tests were carried out on the 2mm slices,
see Figure 9. It can be seen here that the coefficient of water
absorption strongly increases with the porosity becoming greater
than about 20%, a value which is likely to occur in the outer layers
of concrete (see Figure 8).

From stereological measurements on micro photographs of thin
sections (30-50µ) at various depths in concrete a better idea could
be formed about the ratio of aggregate diameters present in the
various layers, as shown in figure 10. At 0.5mm depth about 3 times
as many particles in the size range 0.033 – 0.070mm (average
0.048mm) are present than at 50mm depth in the concrete. At 0.5mm
and at 2.5mm about half as many particles in the size range 2-4mm
(average 2.8mm) are present with respect to a plane in the bulk of
the concrete.

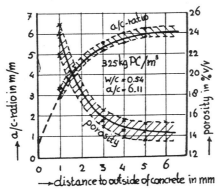

Figure 8: Porosity and a/c-ratio of mortar skin

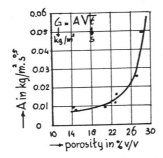

Figure 9: Coefficient of water adsorption versus porosity

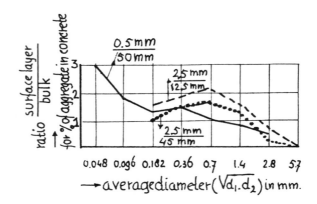

Figure 10: Ratio's of aggregate diameters in concrete layers.

By micro roughness measurements it is possible to get information about the amounts of air, cement paste and aggregates in a plane because in a ground or polished section, on a micro scale, the surface of cement paste is situated underneath the surface of the aggregates particles, while air bubbles show a steep slope in depth (Deelman, 1984).

With the help of this method it was possible to get more information about the composition of concrete over, especially the outer layer of concrete (Kreijger, 1984). Figure 11 gives the results for 3 concrete compositions with the average composition.

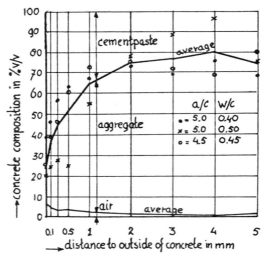

Figure 11: Composition of mortar skin.

One may conclude from this Figure that for concrete with
w/c = 0.04-0.05 and a/c = 4.5-5.0 the boundary plane is composed of:

    72% ± 10% (v/v) cement paste
    23% ± 10% (v/v) aggregates and
     6% ± 5% (v/v) air.

So at the boundary the a/c-ratio will be about 0.60 and in Figure 8
the extrapolated a/c-line is drawn to this value.

Since many properties of concrete are related to the %v/v of
aggregates in concrete, two such relationships have been used
(Kreijger, 1984) to calculate the properties as a function of the
distance from the boundary plane of concrete to the bulk. The
results are given in Figure 12. We must however keep in mind that
these data are true only for very well cured mortar and cement
skins.

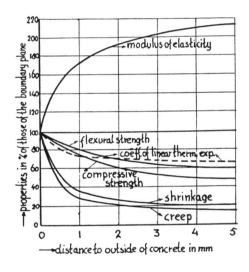

Figure 12: Estimation of properties of mortar skin.

So for the formation of the concrete skin the velocity of drying and
thermal shrinkage are important as well as the distance to which
these phenomena penetrate the concrete. Joisel (1968) calculated
the penetration depths, resulting in Figure 13.

It follows from Figure 13 that the depth for which the concrete
attains half of the maximum shrinkage is about x_s = 0.6 t (x_s in cm,
t in days). Consequently for the same time of exposure the depth
reached by thermal shrinkage is some 125 times greater than that
reached by drying shrinkage. Also for a certain concrete depth the
time to manifest drying shrinkage is about 1600 times greater than

that of thermal shrinkage. Therefore daily changes in (our) climate may influence about 30cm concrete depth for the thermal shrinkage and about 2mm for drying shrinkage. Seasonal changes might go as far as 6m in concrete for thermal shrinkage and about 40-60mm for drying shrinkage. The latter is the basis of the statement that the concrete skin is about 50mm thick, experimentally supported by Patel et al.,(1985). During the first days after placing the concrete however the danger of cracking is greatest because of the cooling down of the concrete, and the low tensile fractural strain (some 0.05-0.10%, see Figure 4).

2.6 Transition Layers Around Aggregates and Rebars.

Because the strength of the cement-aggregate bond is less than the strength of either paste or aggregate, this bond is the weakest link in the structure of concrete. Therefore it is not surprising that

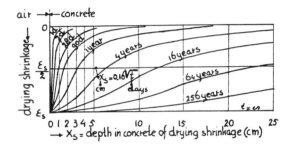

a: Pentration velocity of drying shrinkage in concrete.

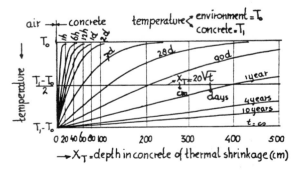

b: Pentration velocity of thermal shrinkage on concrete

Figure 13:   Penetration velocity of drying and thermal shrinkage in concrete.

literature on this topic was focused originally on the bond strength
between cement paste and aggregate, as the review by Alexander et
al. (1965) shows.   Later work gradually evolved the structure of the
transition layer, starting with the review of Struble et al. (1980)
followed by the work of Grandet et al., Idorn, Langton et al., Maso,
Massarazi et al., Nagataki et al., Sychev et al., Torgnon et al.,
and Torgnon et al. (Paris, 1980).   The most recent information
available to the author is given by Montiero et al. (1985) and by
Arliguie et al.(1985).   A reasonable idea now exists of the
structure of the transition layer between cement paste and
aggregates, which is summarized here.

The transition layer is characterized by a thin film of $Ca(OH)_2$ from
which CSH-particles appear to be growing.   In addition to CSH-
particles,larger crystals of $Ca(OH)_2$ appear to attach the
interfacial region to the bulk paste.   As the degree of hydration
increase more bulk paste becomes attached to the interfacial region,
making further characterization of the aggregate film contact
difficult, however it still appears as a two-layer or duplex zone as
Figure 14 shows, taken from Langton et al., (1980).

A nice illustration of the influence of the nature of aggregate is
given by micro hardness measurements, carried out by Lyubimosa et
al., (1962) and from which Figure 15 is taken.

As a consequence of the foregoing, the bond strength of aggregate-
paste varies by a factor of 2 for the various types of aggregate.
For extrusive rocks bond strength is directly proportional to the
silica content of the aggregates (Alexander et al., 1965).

The weak-layer effect and the adhesion cracks are the cause of the
increasing compressive/tensile strength ratio with increasing a/c-
ratio of the concrete composition.

Rebars normally are rusted and in concrete the rust reacts with
$Ca(HO)_2$ from the hydrating cement to calcium ferrite, which is the
protecting or passive layer of rebars against corrosion.   Inside the
cement paste again there is the interconnection between $Ca(HO)_2$ and
CSH-structure of the hardening paste giving rise to a transition
layer.   Little literature has been found by the author regarding
this layer since most investigations are carried out regarding the
corrosion of rebars amd the effect of lime and chlorides on this
process and the compounds formed (Murat et al., 1974; Gouda et al.,
1975), although here too an increased widening of the pores and a
more open structure are mentioned.

The bond strength is determined by this layer (Khalaf et al., 1979)
and failure was confined to a maximum depth of several microns from
the steel.   The mode of fracture is predominately one of adhesion
failure.   The bond strength increases with age for cement paste and
concrete.

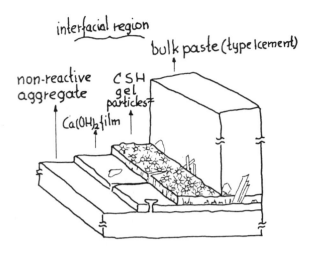

Figure 14:   Composition of transition layer around
             aggregates.

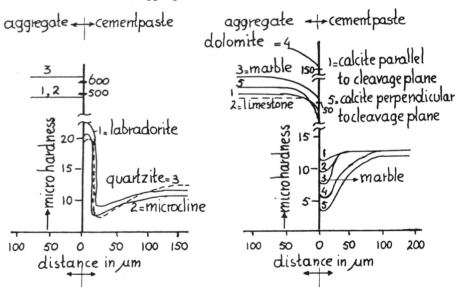

Figure 15:   Micro hardness at boundary of cement
             aggregate.

It is also important to realise that the transport of chloride ions, oxygen and water is quicker through the transition layer than through the denser cement paste.

The main conclusion may be (Tabor, 1981) that the adhesive bonds in mortar and concrete are provided by $Ca(OH)_2$ and CSH and probably the crucial "weak link" is the $Ca(OH)_2$ crystal.

2.7 Appearance of Concrete

The visual appearance of concrete is due solely to the outer layer some 20 microns thick. It is the composition of this layer that will determine the colour of the concrete. From Figure 12, it can be seen that the composition depends on the w/c ratio, and for a well cured concrete containing 300 kg m^{-3} cement this is given in Table 1.

Since the concrete colour is attributable to the colour of the unhydrated cement particles, it follows from Table 1, that for an increase in w/c ratio the colour becomes lighter. However an increase in cement content will lead to a darker colour owing to the increased amount of unhydrated cement. Therefore the cement is the most important factor in determining the final concrete colour.

Next in importance is the $Ca(OH)_2$ that is formed during hydration, which may change by reacting with CO_2 in air to $CaCO_3$ (with some bicarbonate and silica gel, see section 3.1).

Table 1   Composition of 20μ surface layer (in %m/m), determining the colour of concrete (300 kg P.C.m^{-1})

Component\ w/c ratio	0.30	0.40	0.50	0.60
unhydrated cement	10.3	0	0	0
hydrated cement	61.5	71.9	64.6	58.7
fine aggregate	16.7	14.6	13.2	11.9
calcium carbonate	11.5	13.5	12.1	11.0
capillary openings	0	0	10.1	18.4

2.8 Thickness of Concrete Cover.

As has already been stated, the reinforcement in concrete is an inhomogeneity and is situated in the concrete skin. The coefficient of linear thermal expansion at the outside of the concrete skin varies between 15-20 x 10^{-6} and decreases further inward according to Figure 15. Comparatively, the coefficient for steel is less (10-12 x 10^{-6}). An even greater difference exists between the thermal conductivity of concrete and steel (concrete 1.5-3W/m^{0}C; steel 50 W/m^0 C). The thickness of the concrete cover is also important because of the diffusion of gases and permeation of water (solutions).

To quantify the variability on actual concrete cover achieved, some 13000 measurements were taken on 13 types of structural members in 25 construction projects all over the Netherlands (de Lange et al., 1984). In summary, although the measured values were on average larger than the minimum cover specified, due to the variability some 10-40% fell short of the specified minimum cover values. There were no ascertainable differences between precast members and members cast in-situ.

The results of this research have lead to increase of the cover in the Dutch Concrete Standard and are laid down in NEN. 3880 (1984).

3. EFFECT OF INHOMOGENEITIES ON DEGRADATION PROCESSES.

Degradation processes may be mechanical, chemical, physical and involve environmental factors such as atmosphere, ground or natural waters, biological agents, stresses due to incompatibility of materials, and other effects arising from the use of special materials. Such processes are summarised in Table 2.

Table 2   Physical and chemical degradation process
of reinforced concrete

Material	Physical Processes	Chemical Processes
reinforcement		carbonation initiated corrosion; chloride initiated corrosion
concrete	diffusion of gases adhesion of particles, soiling; permeability to water/solutions; evaporation of moisture; thermal gradients; freezing action	carbonation, effect of dry deposits; reactions with $-Cl,-SO_4,-NO_3,FeS_2$, humic acids; alkali aggregate reactions

Since the effect of all these processes is felt firstly upon the
outer cement skin, with all its inhomogeneities and visual
consequences, some discussion of degradation processes mainly due to
the atmosphere is relevant.

3.1 Change of Concrete Appearance

There are a lot of blemishes that are likely to affect the
appearance of concrete surfaces, like differences in colour,
irregularities - such as sand streaks, small cracks, gravel pockets
etc. - and irregular distribution of blowholes of which many however
can be avoided for architectural concrete (Kreijger, 1966).
Requirements for the appearance of concrete surfaces are available
(RILEM Tentative Recommendations, 1982; CIB, -).

From section 2.7 it follows that the $Ca(OH)_2$ produced during cement
hydration is an important factor regarding the colour of concrete.
This compound is changed to $CaCO_3$ by reaction with CO_2 in the air,
both products being white. During hydration 100g portland cement
yields some 25g of $Ca(OH)_2$ and this can be transported by the
capillaries to the outside of the concrete. This means that if the
$Ca(OH)_2$ comes to the surface, the $CaCO_3$ content of the 20μ-layer
increases from 12% (Table 1, w/c=0.50) to 60% and thus a much
lighter colour may arise.
The conditions for $Ca(OH)_2$ transport are various and ill defined,
and are probably the cause of the differences in colour of a
concrete surface and of the fact that such a surface is difficult to
reproduce.

Table 3 gives an overall view of influences on the colour of
concrete,as described here. For the effects of numerous other cases
of colour differences and imperfections of the cement skin one is
referred to literature (ie. Kreijger, 1966; Trüb, 1973).

Table 3   Influences on the colour of concrete

Effect	Colour	
	light	dark
w/c-ratio	high	low
content of fine particles	high	low
content of unhydrated cement	low	high
content of hydrated cement	high	low
content of $Ca(OH)_2$ or $CaCO_3$	high	low
surface wetness	dry	wet
surface roughness	smooth	rough

46 Kreijger

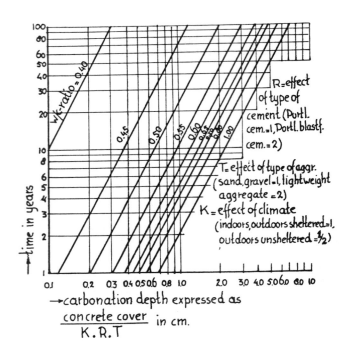

Figure 16: Estimate of carbonation time of the cover of concrete

3.2 Carbonation

The carbonation process has four stages (Suzuki et al., 1985)

(a)-Formation of CSH and Ca-modified silicagel
(b)-Formation of $CaCO_3$ from residual Ca^+-ions
(c)-Decomposition of CSH and release of silicate ions.
(d)-Change from modified silicagel to pure silicagel

Further action of CO_2 may give rise to bicarbonate which with the silicagel may be leached out by rain.

From Figure 16 the variation in carbonation time can be found also if the variation in w/c-ratio (assuming the data of table 1) is taken into account. The resulting carbonation times are a guideline for maintenance inspection. One may assume that if no repairs are done, cracking or release of cover will be reached some 3 years later and collapse of rebars due to corrosion some 12 years hence. Carbonation of the skin however also leads to a higher density and is a certain barrier against leaching.

3.3   Influence of Water Vapour.

Due to capillary condensation the moisture content of the concrete
skin in highly influenced by the R.H and temperature of the air.
Since the cement skin contains 70-80% cement paste, the porosity and
the moisture content is higher than that of the inner concrete. For
example at 90% R.H the skin may contain 13% moisture and at 75% R.H
about 11%, much higher than follows from the hygroscopic moisture
content as function of R.H.

Since nearly every night some condensation occurs, there is an
enhancement of the hygroscopic moisture content. This raised
moisture content directly promotes frost damage. The moisture
content in the cement skin also determines the rates of reaction
with gases like CO_2, SO_2 and Cl and also their diffusion.

The rate of carbonation strongly decreases at high and low R.H, with
a maximum rate at 40-75% R.H. This indicates that carbonation
occurs by dry deposition (see section 3.4).

The drying out of concrete, or the diminution of its moisture
content is one of the basic factors in shrinkage, crack formation
and creep phenomena. This problem was thoroughly investigated by
Pilhlajavaara (1963, 1965) et al. (1965), including the self-
dessication of cement paste and drying in combination with
carbonation.

During moist conditions there may exist a very thin and weak layer
of silicagel and calcium bicarbonate (see section 3.2) which can be
leached out by rain.

3.4   Effects of Gases and Acid Rain

Soiling is the result of four sources of contaminant, (Theissing,
1984) from the inner part of the concrete (efflorescence), from
surrounding building materials (and mostly cleaned just before
delivery), from pollutants (soot, reactive gases) and from microorganisms like bacteria, algae and fungi.

The external influence of the atmosphere and deposition can be
separated into dry deposition (deposition without intervention of
rain) of reactive gases (SO_2, NO_x, CO_2, H_2S, O_2, HCl, HNO_3) and
seasalts and the wet deposition of rain which can bring the dry
deposition into solution. It can attack the skin and provide the
wet conditions necessary for corrosion. The ratio between the
acidifying affect by wet and dry depositions may be estimated as
1:2.

The SO_2 can react with $Ca(OH)_2$ and O_2 from the air to form $CaSO_4$
which is able to adhere to soot and dust particles. Since H_2SO_4 is
a stronger acid than H_2CO_3, the SO_2 will expel the CO_2 from the
cement skin. So the outermost layer contains plaster, bicarbonate

and silicagel. The flux of SO_2 is small, about 30 g/m^2 per year, and if all the cement reached some 0.1mm of the top layer would be destroyed per year.

From acid rain data (SO_4, NO_3) it can be calculated that in one year about 0.05mm of the cement skin may be destroyed. So totally per year about half of the cement skin (0.1-0.3 mm) can be leached out by acid rain and dry deposits. This process however continues day by day and year by year.

Therefore after some five years most concrete looks sandstreaked, but also, because of frost action, the aggregate particles are gradually loosened by the chemical reactions. Imperfections of the surface (see sections 2.7 and 3.1) accelerate this weathering effect.

Chlorides react with lime and calcium aluminates of the hydrated cement and so form new hydrates which lead to expansion and cracking. The flux of chlorides at the shore is more than 100 times larger than inland. Transport in the direction of the reinforcement is possible only by diffusion in water. Transport will be facilitated by the transition layers (see section 2.6). If chlorides reach the rebars an accelerated corrosion process takes place; empirically one gains a critical percentage of 0.3-0.4% by weight of the cement before the corrosion starts. If the concrete is carbonated, the chloride diffusion greatly increases . For good quality concrete made with portland cement, diffusion coefficients have been measured of about 10^{-8}cm^2/s. This means that the critical Cl-concentration for concrete reaches a depth of 5mm after about 6 years. For lower quality concrete a depth of 10mm may be reached after only 2 years. The type of cement is important since portland blastfurnace cement has a coefficient of diffusion some 10 times lower than that of portland cement.

If the water system in the pores is discontinuous (at very low w/c-ratios) the chlorides will not penetrate and the salts stay at the surface. This may also be the case of the outer 3mm of the concrete has been made hydrophobic.

Degradation as described above is increased by temperature changes, on the one hand by cyclic loading, on the other by frost action (see section 2.5). The latter may also increase porosity, as indicated by Pigeon et al. (1986), who found that, especially in portland cement paste, Ca(OH)$_2$ can be gradually displaced leaving air voids, resulting in significantly increased paste porosity.

Since all agents act on the skin of concrete every imperfection there accelerates the degradation. Such imperfections are unequal distribution of cement and water, macro and micro cracks, bleeding effects, transition layers and gaps under coarse aggregates and rebars.

4.   CONCLUSION

Inhomogeneities in concrete can be classified according to where
they occur, ie in the cement skin (0.1-0.3mm from the surface), in
the mortar skin (up to 5mm thick) and in the concrete skin (some
50mm thick).

Each of them has their own imperfections and all have great
influence on the durability of concrete. It would be advisable to
carry out investigations on composition structure and properties of
the aforementioned skins with regard to their influence on
durability. Ideally this should be carried out in a multidiscipline
environment incorporating Universities, Institutions and Industry.

Such a materials science and materials engineering approach may take
concrete to the same level of detailed understanding as that of
metals, ceramics and polymers, which has developed greatly during
the last five to ten years.

5.   ACKNOWLEDGEMENT

The author would like to acknowledge the assistance given by
Dr D S Swift, in the preparation of this paper for presentation at
this conference.

REFERENCES

ALEXANDER, K M, WARDLAW J, GILBERT, B J.   Aggregate-cement bond,
cement paste strength and the strength of concrete. Int. Congress
on the Structure of Concrete, 1974, paper B2, 23p.

ARLIGUIE, G, GRANDET, J, OLIVIER, J P. Orientation de la
portlandite dans les mortiers et bétons de ciment Portland:
influence de la nature et de l'état de surface du support de
cristallisation. Materials and Structures. 1985, 106:263-267.

CIB-REPORT. Tolerances on blemishes of concrete, 8p. CIB,
Rotterdam.

DEELMAN, J C. Textural analysis by means of surface roughness
measurements. Materials and structures 101, 1984, pp 359-367.

GOUDA, V K, MIKHAIL, R S, SHATER, M A. Hardened portland
blastfurnace slag cement pastes, part 3 Corrosion of Steel
reinforcement versus pore structure of the paste matrix.   Cement and
Concrete Research. 1975, 5:99-102.

GRANDET J, OLIVER J P. Orientation of hydration products near
aggregate surfaces. Proc. 7th International Congress on the Chem.
of Cement. 1980 3.7, pp 63-68.

IDORN G M. Interface reactions between cement and aggregate in concrete and mortar. Proc. International Congress of the Chem. of Cement. 1980, 4, pp 129-156.

JOISEL A. Influence de retraits hydraulique et thermique sur la fissuration des bétons. Symp. The Shrinkage of Concrete, Madrid, 1968, Volume 2, 4b, 13 pp.

KAYYALI O A. Mercury intrusion porosimetry of concrete aggregates. Materials and structures, 1985, 106:259-262

KAYYALI O A. Porosity of concrete in relation to the nature of the paste-aggregate interface. Materials and Structures, 1986, to be published.

KHALAF M N A, PAGE C L. Steel-mortar interfaces: micro structural features and mode of failure. Cement and Concrete Research, 1979, 9:197-208.

KREIJGER P C. Techmologische invloeden op de scheurvorming in beton. (Inversitagtionof technological influences on the cracking of concrete). CUR Rapport 32, Betonvereniging, Zoetermeer, 1965, 143 pp.

KREIJGER P C. Schoon beton. (Architectural concrete). CUR Rapport 36, Betonvereniging, Zoetermeer, 1966, 124 pp.

KREIJGER P C. Scheurvorming van jong beton. (Cracking in young concrete). CUR Rapport 88, Betonvereniging, Zoetermeer, 1977, 48 pp.

KREIJGER P C. The Skin of Concrete, composition and properties. Materials and Structures, 1984, 100:275-283.

KREIJGER P C. Micro ruwheids aan betonoppervlakken. (Micro roughness measurements of concrete surfaces). Rapport BKO-MK-84-5, Eindhoven University of Technology, Eindhoven, 1984, pp.

LANGE de, W M J. Betondekking. (Concrete cover). CUR Rapport 113, Betonvereniging, Zoetermeer, 1984, 96 pp.

LANGTON C A, ROY D M. Morphology and microstructure of cement paste-rock interfacial regimes. Proc. 7th Int.   Congress of the Chem. of Cement, 1980, 3.7. pp 127-132.

LYUBIMOSA T Y, PISCES E R. Crystallation structure in the contract zone between aggregate and cement in concrete. Kolloidnyi Zhurnal 24, 1962, (5):491-498.

MASSARZI F, PEZZUOLI M. Cement paste/quartz bond in autoclaved concrete. Proc. 7th Int. Congress on the Chem. of Cement, 1980, 3.7, pp 16-21.

MASO J G. La liaisonentre les granulates et la pâte de ciment hydraté, principle report, 1980, , 1.7, pp 1-14.

MEHTA P K, MANMOHAN D. Pore size distribution and permeability on hardened cement paste. Proc. 7th Int. Congress on the Chem. of Cement, 1980, 3, pp 1-6.

MONTEIRO P J M, MASO J C, OLIVIER J P. The aggregate- mortar interface. Cement and Concrete Research, 1985, 15:953-958

MURAT M, CHARBONNIER M, HACHEMI AHNADI A, CUBAUD J C. Nature particuliére des oxydes de fer hydraté formés dans certaines conditions accelerée de l'acier dans le béton armé. Cement and Concrete Research, 1974, 4:945-952.

NAGATAKI S, TAKADA M. Effects of interface reactions between blastfurnace slag and cement paste on the physical properties of concrete. Proc. 7th Congress on the Chem. of Cement, 1980, 3.7, pp 90-94.

PATEL R G, PARROTT L J, MARTIN J A, KILLOCH D C. Gradients of microstructure and diffusion properties in cement paste caused by drying. Cement and Concrete Research, 1985, 15:343-356.

PIGEON M, REGOURD M. The effects of freeze-thaw cycles on the micro structure of hydration products. Durability of Building Materials, 1986, 4:1-9.

PILHLAJAVAARA S E. Notes on the drying of concrete. The State Institute for Technical Research, Helsinki, 1965, 110 pp.

PILHLAJAVAARA S E, RANTA MATTI A. A theoretical study on the effect of gravitation on drying with special reference to concrete. The state Institute for Technical Research, Helsinki, 1965, 61 pp.

PILLAJAVAARA S E. On the main features and methods of investigation of drying and related phenomena in concrete. Thesis, The state Institute for Technical Research, Helsinki, 1965, 142 pp.

R.I.L.E.M. Tentative Recommendation. Principle criteria for acceptance of precast concrete elements for building; surface appearance of precast concrete elements for building. Materials and Structures, 1982, 88:319-328.

SUZUKI K, NISHIKAWA T. Formation and carbonation of CSH in water. Cement and Concrete Research, 1985, 15:213-224.

SYCHEV M M. SVATOVSKAYA L B. Some problems on the chemistry of adhesion, cement hardening and the strength of cement stone. Proc. 7th Int. Congress on the Chem. of Cememt, 1980, 3.7 pp 81-84.

TABOR D. Principles of adhesion-bonding in cement and concrete. In Peter C Kreijger (Editor), Adhesion Problems in the Recycling of

Concrete, N.A.T.O. Conference Materials Science, 1981, Volume 4, Plenum Press, New York and London,pp 63-87.

THEISSING E M. Survey of 62-SCF Committee Soiling, Cleaning of Facades. Materials and Structures, 1984, 17:167-172.

TORGNON G P, URSELLA P, COPPETTI G. Bond strength in very high strength concrete. Proc 7th Int. Congress on the Chem. of Cement, 1980, 3.7. pp 75-80.

TORGNON G P, CANGIANO S. Interphase phenomenon and durability of concrete. Proc 7th Int. Congress on the Chem. of Cement, 1980, 3.7, pp 133-138.

TRÜB U. Die Betonoberflache, Bauverlag GMBH Wiesbaden, Berlin, 1973. 217 pp.

WEIRIG H J. Einige Beziehungen zwischen den Eigenschaften von grünen und jungen Beton und denen des Festbetons. Beton, 1971, 11:445-448, 12:487-490.

SOME ASPECTS CONCERNING CORROSION OF REINFORCEMENT

C Bob

Polytechical Institute of Timisoara,

Romania

ABSTRACT. The paper deals with the reinforcement corrosion in concrete elements under normal conditions. Two periods are to be considerated initial period and corrosion process period. The initial period occurs chiefly in two different ways : carbonation of the concrete and presence of chloride. An original formula was obtained for the depth of concrete carbonation, which was expressed as a function of : cement type, climatic condition, carbon-dioxide content, concrete compressive strenght and time. The corrosion rate of the reinforcement is presented as a function of : climatic conditions, crack state, class and covering of concrete, type of steel, air humidity. Some experimental data concerning the corrosion of steel in reinforced and prestressed concrete elements are given. A comparison between theoretical considerations and experimental data concerning the corrosion of reinforced and prestressed concrete elements is done : a satisfactory agreement was obtained.

Keywords: Concrete, Concrete Durability, Reinforced Concrete, Prestressed Concrete, Carbonation, Corrosion, Carbon Dioxide, Deterioration, Experimental Works, Permeability, Research.

Dr. Corneliu Bob is Assoc. Professor at the Faculty of Civil Engineering, The Polytechnical Institute, Timisoara, Romania. He has been most active in concrete behaviour, new special concrete types (fiber reinforced concrete, concrete containing polymers, pavement concrete, light-weight aggregate concrete), admixtures, assessment of concrete quality and durability of concrete. His current research work also includes methods of analysis and design of concrete structures He has published many papers and some books on various aspects of civil engineering. Professor Bob received his doctorate from Polytechnical Institute of Timisoara in 1971.

INTRODUCTION

A building has, in principle, a finite service life. After a number of years a building is judged to be obsolete and is then demolished or radically renovated. Deterioration of component parts of a building may also occur, leading to their repair or replacement. Remedial repairs to concrete structures have recently begun to represent an increasingly large proportion of the total expenditure on civil engineering works. Before repair of concrete structures is considered, it is imperative that the cause of the distress be firmly established. Without doing this, repairs may be completely ineffective and, in some cases, can be even counterproductive. The two most common causes for the need of repair are corrosion of reinforcement and deficient concrete as a result of deterioration damage from external or internal sources. This paper is devoted to analyse the corrosion of steel in the reinforced and prestressed concrete structures.

CORROSION OF REINFORCEMENT MECHANISM

In a reinforced or prestressed concrete element, a so-called passivation layer is formed on the surface of reinforcing bars; it is a thin but very dense layer of hydroxide - $Ca(OH)_2$ - which prevents further corrosion. This film may, however, be attacked by the surrounding concrete environment, so that electrical potential differences may develop along the bars: electrochemical corrosion will then take place. The progression of deterioration of an element with time is described by: initial period (time until deterioration starts); corrosion process period (time of deterioration).

Initial period

The initial period occurs chiefly in two different ways: carbonation of the concrete surrounding the reinforcement and presence of chloride.

The process of carbonation consists of several stages. First, gaseous CO_2 penetrates the surface of concrete, generally through the capillary pores. Second, the carbon dioxide reacts with the calcium hydroxide, as follows:

$$Ca(OH)_2 + CO_2 = CaCO_3 + H_2O \qquad (1)$$

As a result, the alkalinity of the concrete is lowered and the passivation can no longer be preserved, so that corrosion of the reinforcement is then possible. This neutralization process continues as further CO_2 enters the pore system. The permeability to air (CO_2) of the concrete is bound up with [5]: the element thickness; the type and content of the cement; the water-cement ratio; the particle size and grading of the aggregate; the curing of the concrete; the humidity; the content of the slag or/and ashes. The process of carbonation is dependent too, on the carbon dioxide in the air.

The depth of carbonation x was expressed as a function of time t by many authors [4][6][8][9]:

$$x = A(t)^{1/2} \qquad (2)$$

On the basis of the literature study, the author of this paper has suggested [1] a formula for the arverage value of the depth of carbonation:

$$x = \frac{150ckd}{f_c} (t)^{1/2} \qquad (3)$$

where c is the influence of cement, k is the influence of the climatic conditions, d is the influence of carbon dioxide content and f_c is the concrete compressive strength.
The following approximate values are stated [2][6][7]: c=1.0 for Romanian Portland cement class P40 and P45; 0.8 - P50 and P45; 1.2 for Portland cement with 15% ashes (slag); 1.4 for Portland cement with max. 30% ashes (slag); 2.0 for Portland cement with max. 50% ashes (slag);
k=1.0 under indoor conditions; 0.7 under protected outdoor conditions; 0.5 under average outdoor conditions; 0.3 for wet concrete;
d=1.0 for CO_2=0.03%; 2.0 for CO_2=0.1%.

In formula (3) the influences of the cement content, the water-cement ratio, the particle size and grading of the aggregate and the curing of the concrete are introduced by the concrete compressive strength. In most previous studies which have suggested a formula for the average value of the depth of carbonation, the constant A in equation (2) is expressed by the water-cement ratio, the cement content etc. The use of the concrete compressive strength in formula (3) was suggested because of:
(i) the concrete compressive strength is the major criterion when assessing the quality of a concrete class as well as the concrete permeability;
(ii) the concrete compressive strength is a conventional quantity and its values depends on a multitude of factors, among them, the quality and content of the cement, the water-cement ratio, the aggregates (grading, shape, surface quality, maximum sizes), the casting conditions etc.;
(iii) the factors mentioned influence the concrete permeability in reverse way than the concrete compressive strength so that it was put at the denominator of the formula (3).

The corelation between the depth of carbonation x and time t (carbonation period), in accordance with many studies, is shown in Figure 1, for Portland cement, water-cement ratio of 0.5, concrete compressive strength of 30 N/mm² and for CO_2=0.03%. It should be noted that curve in accordance with formula (3) yields as an average relationship. In Figure 2 is illustrated the correlation between the carbonation period t and the concrete compressive strength for Portland cement, depth x=20 mm and for CO_2=0.03%. It can be seen that

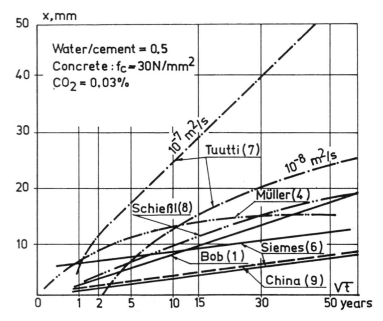

Figure 1 : The correlation between the depth of carbonation and time

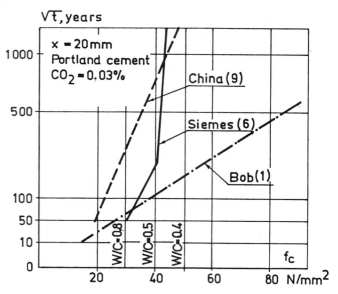

Figure 2 : The correlation between the carbonation period and the concrete compressive strength

TABLE 1   The quantitative effects of the reinforcement corrosion

Concrete characteristics			Corrosion effects	
Water Cement	Depth x mm	ϕ mm	λ_f mm	m_c mm
0.9	10	5	0.10 ÷ 0.15	0.25
		11	0.10 ÷ 0.20	0.19
0.9	30	5	0.05	0.48
		11	0.05 ÷ 0.10	0.23
0.5	10	5	0.05 ÷ 0.20	0.16
		11	0.05 ÷ 0.25	0.11
0.5	30	5	0.05	0.28
		11	0.05 ÷ 0.20	0.14

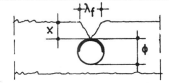 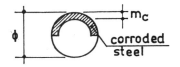

TABLE 2   Factors affecting corrosion rate

Factors		Influence on corrosion rate (CR) Qualitative	Quantitative v_c, mm/year
Envi- ronment	Indoor	No significantly corrosion process	–
	Outdoor	– Lightly CR in standard atmosphere	0.04
		– Averagely CR in industrial environmental conditions	0.10
	Intense	– Highly CR with salt solution	0.2 ÷ 0.3
		– Very highly CR with intense solution	0.6 ÷ 1.8
Cracks state		Width crack under 0.15 mm does not influence CR	–
Concrete covering		CR diminished with growth of concrete covering	$(v_c)_{x10}=$ $=1.10(v_c)_{x20}$
Concrete class		Inferior concrete class gives a rapid CR	$(v_c)_{c20}=$ $=1.3(v_c)_{c30}$
Type of steel		High strength steel gives a rapid CR	–
Air humidity		High humidity gives a rapid CR	$(v_c)_{U80}=$ $=(v_c)_{U70}$

Note: CR – corosion rate; x10 – concrete covering of 10 mm;
      C20 – concrete class of 20 N/mm2; U80 – air hunmidity of 80%.

formula (3) takes into account more correct the influence of the
concrete compressive strength and water-cement ratio on the
carbonation period.

Corrosion process period

After disruption of the passivation layer-whether by carbonation or by
chloride attack-corrosion of the reinforcing steel may occur, due to
the presence of water and oxygen. The corrosion products take up a
considerably larger volume than the original iron (approximately eight
times), which may result in fracturing of the concrete by expansive
pressure. Parallel cracks with reinforcing steel bars and spalling of
the concrete cover will take place. It was found out [7] that for a
parallel crack width of 0.1 mm the maximum limit of the reduction of
the steel cross-sectional areas is reached. The detrimental effects of
corrosion are given in Table 1 [7].
Some of many authors' results [2][3][4][9] concerning the corrosion
rate of the reinforcement are summarized in Table 2. In an indoor
environment there is generally not enough moisture in the concrete for
the corrosion process to occur at a significantly high rate. Concrete
in an outdoor environmen will, however, generally contain enough
moisture for this to occur and corrosion process period is relatively
shorter than initial period.

EXPERIMENTAL RESULTS

Some experimental data concerning the corrosion of the reinforcement
in concrete elements are given.

Prestressed concrete columns

It was found that some prestressed concrete columns, used for open-air
transmission line, were fractured after few years of used because of
reinforcement corrosion. To provide a good insight into the phenomena
concerning the durability of prestressed concrete columns, some in
situ testing were done. The test results concerning the bending
moments M_e as well as the theoretical bending moments M_t are given in
Table 3. The strength of the prestressed reinforcement (3∅3 mm) was of
810-1670 N/mm^2 and the concrete compressive strebgth was of 30 N/mm^2;
the results were deduced by tests on the samples extracted from
prestressed concrete columns. The data from Table 3 indicate a
diminished resistance of concrete columns number 2, 3 and 12 for which
$M_e/M_t=0.5$. By using the data from Table 1 and 2 as well as by appling
the formula (3) it was found the theoretical service life of a
concrete column as:

t=25+3=28 years for: c=1.2; k=0.5; d=1.0; f_c=30 N/mm^2;
                 x=15 mm; v_c=0.05 mm/year; m_c=0.16 mm
t=11+3=14 years for: c=1.2; k=0.5; d=1.0; f_c=30 N/mm^2;
                 x=10 mm; v_c=0.05 mm/year; m_c=0.16 mm.

TABLE 3 Prestressed concrete columns

No	Column age years	Experimental Me kNm	Theoretical Mt kNm	M_e/M_t M_e/M_e^5	Observations
1	14	42.8	47.4	0.90 / 0.77	
2	14	32.0	47.0	0.68 / 0.58	– Initial cracks – Corroded reinf.
3	14	23.1	42.0	0.55 / 0.42	– Two reinforcement with 60% corroded steel
4	9	48.9	47.4	1.03 / 0.88	– Initial cracks – Reinforcements with begining of corrosion
5	9	55.4	47.4	1.17 / 1.00	– Without damages
6	9	55.5	47.4	1.17 / 1.00	– Longitudinal cracks
7	9	73.9	47.5	1.55 / 1.33	– Without damages
8	9	42.1	47.4	0.89 / 0.76	– Longitudinal cracks – Superficial corrosion
9	9	53.6	48.5	1.10 / 0.97	– Cracks – Without begining of corrosion
10	9	53.0	48.0	1.10 / 0.97	–
11	9	39.8	47.4	0.84 / 0.72	– One reinforcement with complete corrosion
12	9	28.6	47.4	0.60 / 0.51	– One reinforcement with complete corrosion
13	9	45.9	47.4	0.97 / 0.82	– Cracks – Begining of corrosion

Note: M_e^5 is experimental bending moment for column 5

A satisfactory agreement between theoretical consideration and experimental results was obtained.

Reinforced and prestressed concrete beams

The stage of carbonation in reinforced and prestressed concrete beams
was analysed in laboratory by two methods: chemical method by
establishing the pH; phisico-chemical methods by infra-red analysis
(IR) and differential thermoanalysis (DTA). The concrete was extracted
with a mechanical device from different depths and was submitted to a
pulverizing process for analysis. Two prestressed concrete beams of 18
years were tested in different cross sections and zones. The depth of
carbonation was found; 5÷10 mm by chemical and physico-chemical
analysis and 8.4÷9.1 mm by using formula (3). The reinforced concrete
beams of an 80 years old bridge in Timisoara were also tested. The
depth of carbonation was found out to be over 30 mm by pH and DTA
analysis and 33.5 mm from formula (3). The corrosion of the
reinforcing steel was very large so that strength and rigidity of the
bridge beams were much diminished; this bridge was demolished and
rebuilt in 1987-1988.

CONCLUSIONS

The results obtained from the various sections in this investigation
assess some relevant aspects concerning the corrosion of the
reinforcing steel. Two periods of deterioration of a concrete element
are described: initial period and corrosion of reinforcement period.

The carbonation of the concrete surrounding the reinforcement is the
main factor which determins initial period. In this paper it was
suggested an original formula for the average value of the depth of
carbonation which depends on: cement type, climatic conditions, carbon
dioxide content, concrete compressive strength and time.

The corrosion process period has been analysed in accordance with many
experimental data. The corrosion rate is put function of: cilmatic
conditions, crack state, concrete covering, concrete class, type of
steel, air humidity.

Theoretical considerations were compared with experimental
investigation made in situ on the prestressed concrete columns and in
lboratory on reinforced and prestressed concrete beams. A satisfactry
agreement between experimental tests and theoretical proposal was
obtained.

REFERENCES

1   BOB C. The model of corrosion of the reinforcement in concrete (in
    Romanian). Simpozionul ICCPDC Timisoara, 1986,

2   BOB C. The check of quality, safety and durability of buildings
    (in Romanian). Editura Facla, Timisoara, 1989.

3   KASHINO N. A durability investigation of existing buildings.

RILEM, Testing in situ of concrete structures, Budapest, 1977.

4   MÜLLER K F. The possibility of evolving a theory for predicting
    the service life of reinforced concrete structures, RILEM, Long-
    term observation of concrete structures, Budapest, 1984.

5   NEVILE A N. Properties of Concrete (in Romanian), Editura Tehnica,
    Bucuresti, 1979.

6   SIEMENS A J., VROUWENVELDER A C W M and VAN DEN BEUKEL A.
    Durability of buildings: a reliability analysis, Heron, Delft
    University, Vol.30, No 3, 1985.

7   TUUTTI K. Service life of structures with regard to corrosion of
    embeded steel. Quality control of oncrete structures, Stockholm,
    1979.

8   CEB-Working Guide for Durable Concrete structures, Part 1,
    München, BRD and Durability of Concrete structures, Copenhagen,
    1983.

9   RESEARCH GROUP ON DURABILITY - People's Republic of China-
    Observation on reiforced concrete structures under long-term
    service and relevant experiments (see 3).

CORROSION OF CONCRETE BY SEQUESTRATING AGENTS OF DETERGENTS

J P Camps

Institut Universitaire de Technologie, de Rennes,

A Laplanche

Université de Rennes

K Al Rim

Institut National des Sciences Appliquées, Rennes,

France

ABSTRACT. Some additive agents of detergents are used to "lock up" polyvalent ions in their molecule and make these ions ineffective in water. The corrosion of concrete by two of these agents has been studied, these agents are N.T.A. and E.D.T.A., they are used to sequestrate the calcium ion Ca^{2+}. This ion is present in water but it may also be extracted from the concrete. Tests were done upon mortar samples immersed into solutions containing these sequestrating agents. They highlighted that concentrated solutions of E.D.T.A. are very aggressive and cause an important corrosion after a rather short period of time. N.T.A. solutions are less aggressive, but both solutions give an attack of the surfaces of samples and a significant fall in concrete strength.

Keywords. Corrosion, sewers, concrete, pipes, detergents, calcium, chelating agents, E.D.T.A., N.T.A..

Mr Jean Pierre Camps is Maître de Conférences and Head of Civil Engineering Department, Institut Universitaire de Technologie de Rennes, France. He works as a searcher at the laboratory of Civil Engineering, G.T.Ma., Institut National des Sciences Appliquées, Rennes, in the field of construction materials.

Dr Alain Laplanche is Maître de Conférences at Ecole Nationale Supérieure de Chimie, Université de Rennes. His current research work includes : treatment of drinking water, treatment of sewage, reactions and processes with ozone.

Mr Kamal Al Rim is a searcher at the laboratory of Civil Engineering, G.T.Ma., Institut National des Sciences Appliquées, Rennes. An engineer from Lattaquié University, Syria, last year he completed a Diplôme d'Etudes Approfondies about the topic dealt with in the present paper.

INTRODUCTION

The corrosion of concrete is a phenomenon which has been
widely studied because of its effects upon the durability
of civil engineering constructions.

Concrete pipes used in the sewer networks have a particular
behaviour to corrosion which is often an anaerobic process
where bacteria can produce S^{2-} ions in the waste waters.
When such a solution arrives in an aerated medium an
oxydation of S^{2-} into $SO_4{}^{2-}$ is possible giving sulfuric
acid, able to attack the concrete of the pipe.

This kind of damage is often observed at the end of a
discharge pipe when sewage which has been retained for a
while in the unaerated medium of the filled pipe of a
sewage pumping station flows again in an aerated medium.
Then an acidic corrosion of concrete may arise on the next
elements of the network [1], [2], [3].

If this is one of the main causes of corrosion of the
internal face of sewer pipes we may suppose that, because
of the great number of products flowing through the sewage
other aggressive agents can be encountered in waste.

Therefore we wondered about the possible effect of
detergents and mainly of one of their components, the
chelating agents, the function of which is to pick up and
trap calcium ions Ca^{2+} to make them ineffective in washing
waters.

This paper presents experimental results showing the
possibility of concrete corrosion by some of these calcium
sequestrating agents.

CHELATING AGENTS OF DETERGENTS.

The presence of solved calcium Ca^{2+} in washing waters
reduces the efficiency of detergents and may promote solid
dregs (scale) able to spoil the equipments for instance
warming parts of washing machines.
Some compounds, called sequestrating agents, are used in
order to make Ca^{2+} ineffective and unable to give solid
products.

Two molecules are particularly representative, they are :

- the Nitrilo Triacetic Acid (N.T.A.) the formula of which
is

```
              CH2-COOH
             /
HOOC-CH2-N
             \
              CH2-COOH
```

- the ethylene diamine tetraacetic acid (E.D.T.A.)

```
HOOC-CH2              CH2-COOH
        \            /
         N-CH2-CH2-N
        /            \
HOOC-CH2              CH2-COOH
```

In fact, it is their sodium salts which are used because of their better solubility. For instance the sodium salt of E.D.T.A. reacts with calcium to give the complex :

```
NaOOC-CH2              CH2-COONa
        \             /
         N-CH2-CH2-N
         |  \     / |
        CH2  \   /  CH2
         |    \ /   |
        O=C-O-Ca-O—C=O
```

The calcium seems to be locked up by a claw (chelos in Greek means claw) [4].

These agents could be the cause of damage in concrete by picking up calcium linked inside the concrete and solving it into the waste.

MATERIALS AND METHODS

The aim of this paper is to carry out a first study in order to measure the quantity of calcium extracted off concrete by sequestrating agents. The experiments are conducted in laboratory with synthetic solutions and mortar samples.

Solutions.

The studied chelating agents used in solution are the sodium salts of N.T.A.

$C_6 H_7 N Na_2 O_6$ Molar weight 235.11 g/mol

and of E.D.T.A.

$C_{10} H_{14} N_2 Na_2 O_8, 2 H_2O$ Molar weight 372.24 g/mol

for the first one concentrations of 0.2 g/l, 0.6 g/l,
10 g/l were studied in order to find a concentration giving
significant effects at the end of a rather short period of
time.

The second one was studied with a concentration of 10 g/l
and compared to the precedent at the same concentration.

Products are solved into tap water, this means that
initially there is a little amount of Ca^{2+} in the solution.

Lastly, three reference samples undergo all the treatment
of other samples but into drinking water without any
corrosive agent.

Samples.

Prismatic mortar samples were made whose composition is as
follows :

Portland cement (C.P.J. 45)	:	500 kg/m^3
Water	:	225 kg/m^3
Sand 0/5mm	:	1595 kg/m^3

Their size is 4 x 4 x 16 cm^3 according to the French
standards of cement testings [5].

After mixing and mouldering they are stocked, for 24 hours,
at a 20° Celsius temperature and 100% relative humidity ;
afterwards they are kept, during 7 days before testing, at
a 20° Celsius temperature and 50% relative humidity. Then
they are weighed and sound celerity through them is
measured (non destructive testing by dynamic testing
techniques).

Testing Procedure.

This procedure is a succession of 11 day cycles.
For 5 days, 3 mortar samples are immersed vertically in a
vessel containing 4 litres of the solution. Every day a
little volume of solution is sampled for further analysis
of calcium. After this first step, the mortar samples are
washed with water and slightly brushed to remove possible
corrosion products from the surface. Dried for 5 days at a
20° Celsius temperature and 50% relative humidity, the
samples are weighed. Then measurements are done with the
non destructive testing of samples by the dynamic testing
technique. At the end of this cycle, the samples are re-
immersed in a newly prepared solution and the procedure is
resumed.

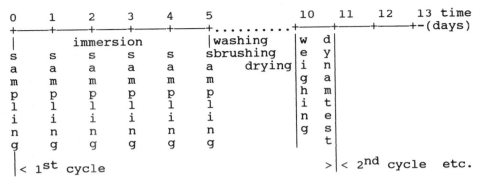

After 6 cycles in E.D.T.A. and 3 cycles in N.T.A. samples
were broken down to obtain the tensile strength (R_t)
measured by a bending test and the compressive strength
(R_c) measured by a compression test, according to the French
standards [6].

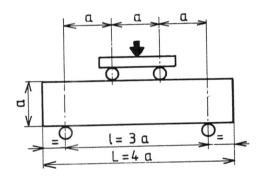

Figure 1
Bending Test Apparatus

Calling "a" the side of the cross section of corroded
samples, "l" the bearing distance of the bending apparatus,
"a_o" the length of sample crushed during the compression
test, "F_t" the failure load during bending test and "F_c"
the failure load during compression test, we have the
following formula :

$$R_t = \frac{F_t \cdot l}{a^3} \quad \text{and} \quad R_c = \frac{F_c}{a \cdot a_o}$$

with l = 12 cm and a_o = 4 cm

Analysis.

The analysis of calcium in solution is carried out by
atomic absorption spectroscopy with an air-acetylen flame.

The analysis wavelength is 422.7 nm and calibration samples
are 1, 2 and 4 ppm concentrated ones.

RESULTS AND DISCUSSION

0.2 and 0.6 mg/l solutions of N.T.A. allowed us to have a
first approach of reactions, showing that the dissolution
of calcium is more effective in these media than it is in
the tap water with the reference samples where the maximum
amount of calcium dissolved in water is about 5 mg/l and
this during the first cycle of the testing procedure. The
most spectacular effects have been observed for the samples
immersed into 10 g/l concentrated solutions. They are
presented below.

Dissolution Of Calcium.

Experimental points figurating the representation of the
concentration C of calcium in the solution v. time can be
fitted with an exponential curve like

$$C = C_{max} (1-e^{-Kt})$$

where
- C is the concentration of calcium in the solution at time
t (mg/l),

- C_{max} represents the maximum amount of calcium extracted
from the mortar that it is possible to sequestrate (mg/l) ;
It is the theorical concentration of calcium which may be
sequestrated by the chelating agent solution (1 mole/mole),
of course it depends on the concentration of N.T.A. or
E.D.T.A. in the solution and on the initial concentration
of Ca^{2+} ions in feed water (at time t=0),

- t is the time when sampling was carried out (days).

- K is a parameter proportional to the amount of calcium
"C_A" which chelating agent may access to ($K = k.C_A$) thus
the expression is similar to the Bohart and Adam's or
Thomas' [7] equation of the rate absorption modelising
(example : actived carbon).

With this assumption the rate of the reaction is given by

$$\frac{dC}{dt} = K.(C_{max}-C)$$

Figures 2 and 3 represent the curves C(t) in two cases
figure 2 for the worst fitting obtained and figure 3 for
the best one.

For each experience, the value of the constant K is
obtained by computing the slope of

$$Ln\left[\frac{C_{max}-C}{C_{max}}\right] = K.t$$

by means of a least square method.

It is interesting to have these computed values because
they allow to evaluate an important parameter which is the
initial rate of sequestration of calcium given by :

$$\frac{dC}{dt}_{t=0} = K.C_{max}$$

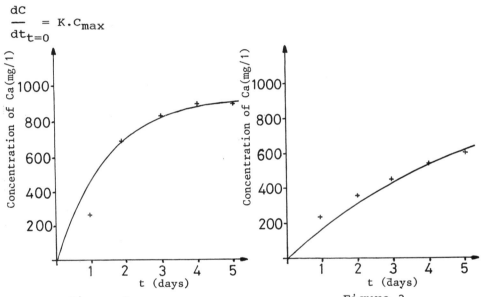

Figure 2
Curve obtained in E.D.T.A.
10 g/l solution at the
first cycle.

Figure 3
Curve obtained in E.D.T.A.
10 g/l solution at the
sixth cycle.

The results of experiments are reported in Table 1 for
E.D.T.A. and in Table 2 for N.T.A. both being 10 g/l
concentrated solutions.

It is seen , during the different cycles, that the term C_A
is decreasing because easily accessible calcium is already
sequestrated. Moreover, if we compare the chelating agents,
the action of E.D.T.A. is faster than the action of N.T.A.
is.

Cycle number	C_{max} mg/l	K day^{-1}	$\dfrac{dC}{dt}_{t=0}$ mg/l.day
1	940	.714	671
2	992	.286	284
3	1016	.231	235
4	1019	.226	230
5	1014	.189	192
6	1027	.193	198

Table 1

Cycle number	C_{max} mg/l	K day^{-1}	$\dfrac{dC}{dt}_{t=0}$ mg/l.day
1	1616	.240	388
2	1614	.122	197
3	1570	.050	79

Table 2

Age (days)	Reference NTA		N.T.A. 10 g/l		Reference EDTA		E.D.T.A. 10 g/l	
	m (g)	c (m/s)	m (g)	c (m/s)	m (g)	c (m/s)	m (g)	c (m/s)
8	553.0	3712	551.5	3687	541.2	3419	536.0	3376
18	556.1	3756	550.2	3704	548.2	3620	538.5	3516
28	556.4	3765	547.6	3721	550.4	3756	533.4	3653
38	556.6	3774	547.3	3729	550.8	3865	524.9	3756
48					550.7	3874	516.7	3774
58					550.6	3846	505.2	3747
68					550.4	3810	494.7	3712

Table 3

(m : mass of the mortar sample ; c : celerity of sound)

Mechanical Properties

Sound celerity through the samples was measured using a non destructive dynamic testing equipment. These measures give indications upon the compared state of cracks or other defaults in the material. It must be noticed that this measurement is not very accurate due to the small size of the samples.

At the end of each testing procedure, tensile (R_t) and compressive (R_c) strength determination gave the results reported in table 4 and 5, where the reference sample has the same age as the corroded sample.

If we consider the mass variation of samples, it is observed that there is a diminution of this parameter either in N.T.A. or in E.D.T.A. solution. Respectively the relative variations of mass have a maximum value of 0.8% (N.T.A.) and 8.1% (E.D.T.A.). This weight loss is due to particles of mortar which are extracted by corrosion and that are found precipitated at the bottom of the vessel. Assuming that the relative loss of mass is uniform on the whole sample, the relative volume variation having the same value, it is possible to calculate the new length "a" of the side of the cross section of samples, assuming that the loss δL of the total length L is equal to the diminution δa of the side of the cross section.

Calling v the volume, we have $v = a^2.L$ and the relative variation

$$\frac{\delta v}{v} = \frac{\delta m}{m} = 2\frac{\delta a}{a_0} + \frac{\delta L}{L_0} = \delta a\left[\frac{2}{a_0} + \frac{1}{L_0}\right] = 0.5625\ \delta a$$

with $a_0 = 4$ cm, $L_0 = 16$ cm.

it is possible to calculate the new value $a = a_0 - \delta a$ in order to compute R_t and R_c of corroded samples

The variation R_t and R_c in both cases is quite significant but it can be seen,once again, that corrosion is more important with E.D.T.A. than it is with N.T.A.. The reduction in resistances computed on the actual cross section of the samples shows that agressive effects are not localized on the surface of the mortar and that the attack weackens the material internally .

The photography of figure 4 shows the appearance of samples after exposure to different solutions. The reference sample has the smooth surface it had when removed from the mould, it is possible to see the grains of sand on the surface of the N.T.A. sample and, as far as the E.D.T.A. sample is concerned, its rough surface can be seen due to the pronounced attack of the solution.

		Strength of		Variation NTA/Ref (%)
		Reference sample	N.T.A. 10g/l sample	
R_t	(N/mm^2)	8.97	8.24	8.1
Rc	(N/mm^2)	41.2	36.7	10.9

Table 4

		Strength of		Variation EDTA/Ref (%)
		Reference sample	E.D.T.A.10g/l sample	
R_t	(N/mm^2)	10.00	7.5	25.0
R_c	(N/mm^2)	38.1	29.8	21.8

Table 5

E.D.T.A.   N.T.A.   Reference
Sample   Sample   Sample

Figure 4

CONCLUSION

This is a first approach of a possibility of concrete corrosion in sewage. The experiments led on show significantly that the studied products, N.T.A. and E.D.T.A., may be considered as corrosive agents even if the concentration studied in this work looks unusually high. It makes possible to see significant damage in a rather short period of testing. The effect of such products in sewer pipes would surely be much slower but it would be interesting to study them combined with corrosive products usually found in sewers such as hydrogen sulfide and sulfuric acid.

REFERENCES

[1] COLIN F., MUNK-KOEFED N., "Formation de l'H$_2$S dans les réseaux d'assainissement. Conséquences et remèdes", Agence de bassin Rhône-Méditerranée-Corse, Ministère de l'Environnement, Paris, 1987, p 1-132

[2] THISTLEWAYTE D.K.B., "Control of sulphide in sewerage systems", Ann Arbor Science Publishers Inc., 1972, p 1-173.

[3] POMEROY R.D., "Process design manuel for sulphide control in sanitary sewerage systems", Rapport E.P.A. 625/1-74-005, 1974, p 1-123.

[4] DAVIDSON A., MILWIDSKY B.M., "Synthetic Detergents", 5th Edition, Leonard Hill, London, 1972, p 72-79.

[5] Norme NF P 15-401, AFNOR, Paris.

[6] Normes NF P 15-451 et NF P 18-407, AFNOR, Paris.

[7] THOMAS H.C., "Heterogeneous ion exchange in a flowing system", Journal of American Chemical Society, 1944, 66, p 1664-1666.

THE EFFECTIVENESS OF A CARBONATED OUTER LAYER TO CONCRETE IN THE PREVENTION OF SULPHATE ATTACK

G J Osborne

Building Research Establishment,

United Kingdom

ABSTRACT.
A series of concretes containing Portland cements and blends of the same Portland cements with blastfurnace slags were stored in magnesium and sodium sulphate solutions following different early curing regimes. 100mm concrete cubes demoulded at one day then stored in air at 20°C and 65% RH for 27 days had carbonated to some extent and were shown to have achieved a significant protection against the ingress of sulphate ions and were highly resistant to sulphate attack after 5 years storage.

The same concretes moist-cured for one day and then stored in water for 27 days, so as to attain their prescribed 28 day target strengths prior to immersion in sulphate solutions, were shown at 5 years to have deteriorated due to sulphate attack, in some cases extensively.

The practical implications of these findings are considered in terms of the important role that a controlled amount of carbonation has on concrete elements and structures prior to placing in highly sulphated environments.

KEYWORDS Protection, Concrete, Sulphate Attack, Carbonated Layer, Carbon Dioxide, Curing, Sulphate Resistance Test, Strength, Portland Cement, Blastfurnace Slag, Durability.

Mr Geoffrey J Osborne is a Principal Scientific Officer and Head of Concrete Durability Section in the Inorganic Materials Division at the Building Research Establishment, Garston, England. His main research interests include, properties of blastfurnace and gasifier slags, their use as cement replacement materials, and the long term durability of concretes.

INTRODUCTION

The factors responsible for sulphate attack on concrete below ground level are discussed in BRE Digest 250 'Concrete in sulphate-bearing soils and groundwaters'[1]. The digest recommends what can be done, by suitable selection of cement type and concrete quality, to resist attack by naturally-occurring sulphates. Aqueous solutions of sulphates attack the hardened cement in concrete, the chemical reactions occurring depending upon the nature of the sulphate present and the type of cement. The rate of attack depends greatly on the permeability of the concrete and its quality. Concrete of low permeability is essential to resist sulphate attack and so the concrete must be fully compacted and the mix designed to have sufficient workability yet have as low water/cement ratio as possible. There are other factors which influence the sulphate attack and these include: the amount and nature of the sulphate present, the water table level and its seasonal variation, the flow of ground water and soil porosity, the form of construction and significantly, the way in which the concrete is cured.

For classification purposes sites have been divided into five categories of increasing severity, based on the sulphate contents of the soil and ground water. The recommended type of cement and the corresponding minimum cement content for each of these classes are given in Table 1 of the present Digest 250[1]. The divisions between these classes are somewhat arbitarily drawn and the recommendations are judgements based on previous data and knowledge. More recent small-scale accelerated test data[2] and the results of long term concrete durability studies carried out on 100mm cubes at BRE are currently being assessed and will be taken into account now that the revision of BRE Digest 250 is being undertaken. The classification of sites will also take note of water movement and acidity[3] and their individual effects on various forms of cast-in-situ and precast concrete constructions. The revision of Digest 250 will combine the site classification with any new recommendations as to the type of cement which should be used and on compaction, cement content, water/cement ratio and the curing regime of the concrete.

The onset of sulphate attack occurs at the surface of concrete and Biczok[4] discussed in detail the usefulness of certain important types of surface protection for preventing attack by corrosive mixtures. He concluded that a significant step is taken towards avoiding corrosion by preventing water and any associated ions from penetrating into the interior parts of the concrete. This is the consideration underlying all methods of surface treatment but also applies in part to barriers which may form just below the surface of the concrete. Biczok noted that after the formwork is stripped from fresh concrete and the latter is exposed for a few days to the atmosphere, the carbon dioxide present in the air combines with the calcium hydroxide of concrete to form calcium carbonate. Continued aeration results in a carbonate layer of increased thickness, which has been shown to afford protection against sulphate attack. Kind[4] showed as long ago as 1955 that the sulphate resistance of Portland cement mortar specimens stored in 5% sodium sulphate solution was improved by pre-curing for two days in a moist atmosphere followed by six days in carbon dioxide prior to immersion in the sulphate solution, compared with the poor performance of those mortars pre-cured for 8 days in the moist atmosphere. These findings have been confirmed in recent studies at BRE where concretes cured in air at 20°C and 65% relative humidity for 27 days [5] have exhibited excellent sulphate resistance when stored in sulphate solutions for up to 2 years.

Portland and blastfurnace slag cement concretes subjected to different early curing regimes, prior to their immersion in high strength sulphate solutions (1.5% SO_3), were chosen for this comprehensive study as there was little or no up to date information on the subject. The paper identifies the important role that the pre-curing of concrete has, in particular that due to the formation of a carbonated outer layer, on the long term durability of these concretes after 5 years of storage in aggressive sulphate environments.

EXPERIMENTAL

In the present study a series of concretes containing Portland cements and blends of the same with granulated and pelletized blastfurnace slags was stored as 100mm cubes in magnesium and sodium sulphate solutions following different early curing regimes. The sulphate resistance of the 100mm cubes was assessed by a range of techniques which included visual assessment, measurement of the sulphate attack rating, mineralogical examination, compressive strength development and determination of ingress of sulphate into the concrete.

The important factors which were found to influence the sulphate resistance properties of these concretes, namely the early curing regime, the tricalcium aluminate content of the Portland cement and the alumina content of the ground glassy blastfurnace slag have been elucidated and their relative importance is discussed.

MATERIALS

Three Portland cements were selected on the basis of their tricalcium aluminate (C_3A) contents as this compound is known to be the predominant cause of sulphate attack on ordinary Portland cements (OPC). OPC 850 (medium C_3A = 8.8%), OPC 814 (high C_3A = 14.1%) and SRPC 853 (low C_3A = 0.6%) were used, the sulphate resisting Portland cement of proven properties acting as a control. It is recognised that the level of 14.1% C_3A for OPC 814, calculated from the Bogue equation, is very high for modern day Portland cements, but when these studies began over five years ago there were examples of high C_3A OPC's such as this being produced in the UK.

Several blastfurnace slags, granulated and pelletized were also
chosen; the main criteria being their availability and their alumina
(Al_2O_3) contents eg granulated slag M286(Al_2O_3 = 14.7%) and pelletized
slag M302 (Al_2O_3 = 7.5%). Various authors [6-9] have shown that ground
glassy blastfurnace slags with high Al_2O_3 contents could adversely
affect the sulphate resistance of blended cements, particularly at the
lower levels of replacement. The use of slags with lower Al_2O_3 levels
(less than 12%) was shown to be particularly beneficial for improving
the poor sulphate resistance of high C_3A OPC.[2] The chemical
analyses and main physical data of the cements and slags have been
given in more detail in a previous paper[2].

Concrete Mixes

Concrete mixes were prepared to an appropriate mix design which
satisfied (a) the minimum requirements of BS 8110; Part 1, 1985[10]
for the 'Structural use of Concrete', when placed in conditions of
severe exposure and (b) the requirements for concrete exposed to
sulphate attack in Classes 3-5 conditions of sulphate in BRE Digest
250[1].

The concrete mix design is given in Table 1 with the mean values of
the wet concrete properties, including cement content, wet density and
slump. A more detailed description of these properties can be found
in earlier work[11].

Table 1 Concrete Mix Design and Wet Concrete Properties

CONCRETE MIX PROPORTIONS					FRESH CONCRETE PROPERTIES (MEAN VALUES)		
THAMES VALLEY AGGREGATES			"CEMENT"	"WATER"	"CEMENT" CONTENT	WET DENSITY	SLUMP
20-10mm (67%)	10-5 mm (33%)	<5mm	(SLAG+PC)	TOTAL W/C (free W/C)	(kg/m3)	(kg/m3)	(mm)
2.91	1.75	1.0		0.5 (0.45)	380	2340	75

The specimens were made in BRE's concrete laboratory at 20°C using
dried Thames Valley coarse aggregates and sand with the appropriate
Portland cements and ground glassy slag additions. A series of
concretes was cast containing the three Portland cements alone and
blends of the medium and high C_3A OPCs with 60, 70 and 80% of the
low and high Al_2O_3 content blastfurnace slags (by weight). Slags with
moderate levels of alumina have been used with the high C_3A OPC in a
later series of concretes.

Curing Regimes

All concretes were vibrated into 100 mm moulds and stored for 24 hours below a cover of damp hessian and polythene sheet to maintain an initial curing condition close to 100% relative humidity. The concrete cubes were then demoulded, randomly numbered and five batches of 9 cubes set aside from each concrete mix for sulphate resistance and water control purposes. The batches of cubes were subjected to three curing regimes, (Table 2), prior to storage in tanks of either magnesium or sodium sulphate solutions for assessment at different ages of test.

TABLE 2 CURING AND STORAGE REGIMES

No of cubes	Curing regime after demoulding at 24 hours	Sulphate storage solution (or water)
9	Water at 20°C for 27 days (W.27d)	Water (CONTROL)
9	Water at 20°C for 27 days (W.27d)	Sodium sulphate (1.5%SO3) (SOLUTION I)
9	Water at 20°C for 27 days (W.27d)	Magnesium Sulphate (1.5%SO3) (SOLUTION E)
9	Air at 20°C/65%RH for 27 days (A.28d)	SOLUTION E
9	No further curing (A.1d)	SOLUTION E

SULPHATE TANK TEST

The sulphate resistance of the 100 mm concrete cubes was measured according to the method established at BRE following procedures described by Steele and Harrison[12]. The cubes were stored in tanks of sodium and magnesium sulphate solutions containing the equivalent of 1.5% SO3 by weight (designated I and E), and in water as control. The solutions were changed and renewed at 3 monthly intervals and maintained at a constant temperature of approximately 20°C. The performances of batches of three cubes at each age were assessed following immersion for periods of 1, 2 and 5 years. The high strength sulphate solutions represented Class 5 conditions of sulphate as classified by BRE Digest 250 and provided a severe form of test. The choice was deliberate as the test was intended as a relatively short-term performance test (2-3 years) for appraising and comparing the sulphate resistance of concretes containing different cements.

Measurement of Sulphate Attack

The sulphate resistance was measured by comparing the compressive strength of the cubes after storage in each sulphate solution with the compressive strength of the water-stored control specimens at the same age and by recording the cubes visual appearance photographically and in terms of an "attack" rating. The extent of deterioration at each corner of the struck face and the opposite face was measured diagonally in mm for each of the three concrete cubes and the sulphate attack rating per cube face calculated as:- the sum of the loss in mm of 8 corners of each of 3 cubes divided by 6. The results of these measurements are given in Table 3.

Measurement of Compressive Strength

The compressive strength was determined in accordance with BS 1881, Part 116 (BSI 1983) for both the sulphate-stored and water-stored (control) specimens following their visual assessment. The relationship between the load applied before failure for the cubes stored in sulphate solution with the compressive strength of the equivalent water-stored control specimens at the same age (% strength retained) provided an assessment of the sulphate attack which had occurred. The compressive strength measurements made should not really be regarded as the true compressive strengths but rather as the "load applied before failure" because, in cases where severe attack and spalling occurred, the area of the cube surface presented to the steel platens of the compression testing machine was substantially reduced though no attempt was made to quantify this reduction in surface area. The strength data are given in Table 3.

Ingress of Sulphate

A selection of concrete specimens was used to determine the ingress of sulphate ions following storage in both the sodium and magnesium sulphate solutions for 5 years. One 100mm cube from each batch of 3 cubes was set aside for this purpose, whilst the remaining 2 cubes were assessed for wear and compressive strength. The concretes chosen contained those cements where the high C_3A, OPC814, (C_3A=14.1%) had been blended with the high alumina slag (M286; Al_2O_3=14.7%) and the low alumina slag (M302; Al_2O_3=7.5%). The 100mm cubes were drilled to 5 depths using a 10mm masonry bit in a percussion drill and the dry powder samples collected for sulphate and lime analyses. Each cube was drilled to obtain representative samples at the depth ranges; 1-6mm, 6-11mm, 11-16mm, 16-21mm, 21-26mm, discarding the surface top one mm. The dust samples were analysed for calcium oxide (CaO) and total sulphate (SO_3) in accordance with the procedures given in BS 1881: Part 124, 1988. However, the sulphate analyses were modified to take into account the presence of sulphur compounds from the blastfurnace slag components and expressed as the percentage by weight of each cement composition; it was assumed that the aggregates were non-calcareous. The results are given in Table 4.

RESULTS

Fresh Concrete Properties

All the concretes had similar workabilities with compacting factors in
the range 0.92–0.96, slumps of 50–100mm and there were no significant
differences in the fresh concrete properties other than the marginal
water reductions produced by the use of the ground blastfurnace slag.
The mean wet density was 2340kg/m^3 and the nominal Portland cement or
'OPC + slag' cement content was 380kg/m^3.

Physical Appearance

Following 5 years storage in the sodium and magnesium sulphate
solutions, the four sets of 3 cubes from each concrete mix were
assessed visually for any physical effects of sulphate attack. One
cube from each set was photographed prior to their measurement for
corner wear and some of the photographs are illustrated in Figure 1.
These clearly showed the benefits of 28 days of air curing prior to
storage in the magnesium sulphate solution, as most of these cubes
have sharp edges, smooth surfaces and had little or no signs of
sulphate attack. The high C_3A OPC concretes were severely damaged,
with the exception of the specimens which had been pre–cured in air,
but the degree of deterioration due to sulphate attack was reduced as
the level of slag as cement replacement increased from 60 to 80%.

Sulphate Attack Rating

The results of the measurements of corner wear and reduction in
specimen size due to the sulphate attack are given in Table 3 and
correspond with the photographic evidence in Figure 1. The benefits
of pre–curing in air, as against in water, prior to storage in
sulphate solution and to the use of higher levels of slag as
replacement, particularly for the high C_3A OPC are demonstrated. A
rating of less than 50mm per face at 5 years is deemed to signify good
sulphate resistance and nearly all specimens pre–cured in air or
containing 70 and 80% of slag, fell into this category. The SRPC
concretes (water–cured) were also highly resistant to sulphate attack.

Compressive Strength Data

The data in Table 3 are presented as the 'compressive strengths' of
specimens recovered from the different sulphate solutions after
storage for 5 years. These values are also expressed as percentages
of the compressive strength of the equivalent water–cured specimens.
This percentage figure has been termed 'the percentage strength
retained'. These results are illustrated in Figure 2 and the
relationship established with the corresponding 'sulphate attack
rating'. A good correlation was obtained between the % strength

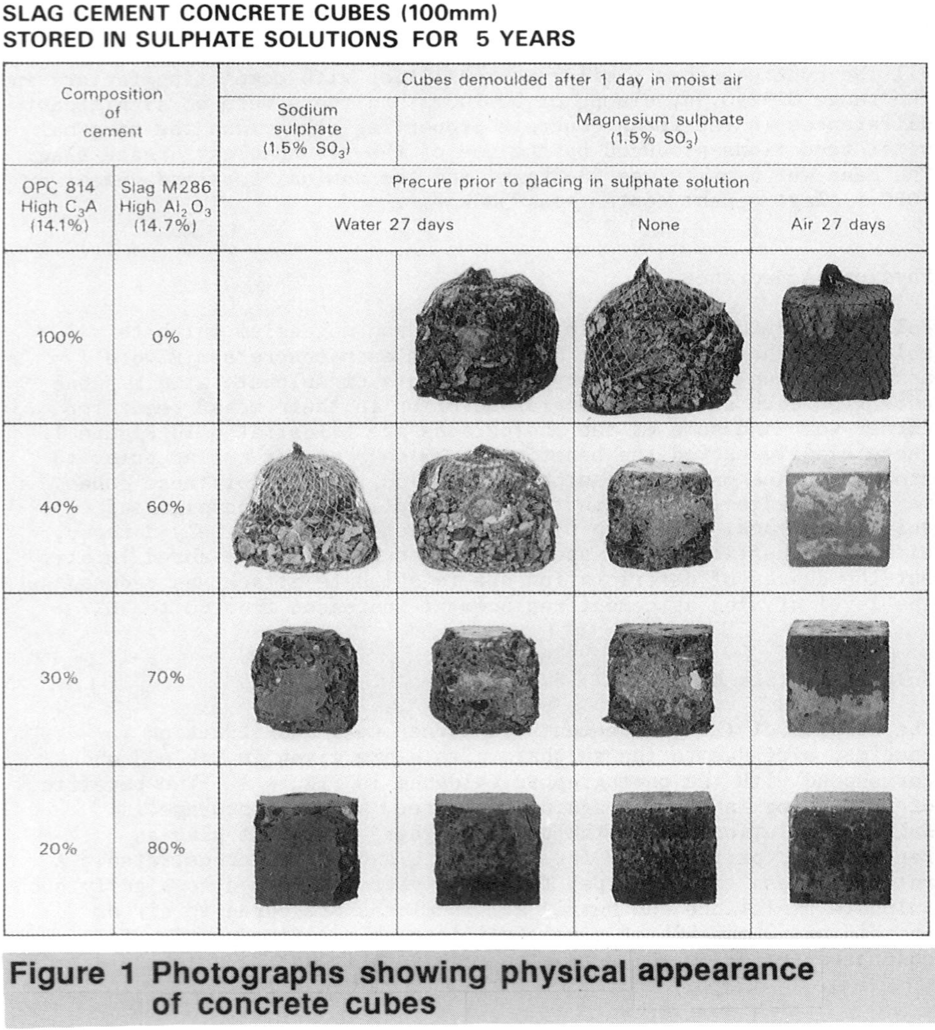

**SLAG CEMENT CONCRETE CUBES (100mm)
STORED IN SULPHATE SOLUTIONS FOR 5 YEARS**

Composition of cement		Cubes demoulded after 1 day in moist air			
		Sodium sulphate (1.5% SO$_3$)	Magnesium sulphate (1.5% SO$_3$)		
OPC 814 High C$_3$A (14.1%)	Slag M286 High Al$_2$O$_3$ (14.7%)	Precure prior to placing in sulphate solution			
			Water 27 days	None	Air 27 days
100%	0%				
40%	60%				
30%	70%				
20%	80%				

**Figure 1 Photographs showing physical appearance
of concrete cubes**

TABLE 3.   COMPRESSIVE STRENGTH DATA AND ATTACK RATING OF PORTLAND AND SLAG CEMENT CONCRETE CUBES (100 MM) FOLLOWING STORAGE IN SULPHATE SOLUTIONS FOR 5 YEARS

CEMENT COMPOSITION			W	STRENGTH (MN/m^2)				(%) STRENGTH RETAINED				ATTACK RATING (%)			
R No	OPC	SLAG		I (W.27)	E (W.27)	E (A.1)	E (A.28)	I (W.27)	E (W.27)	E (A.1)	E (A.28)	I (W.27)	E (W.27)	E (A.1)	E (A.28)
1	850 M286 (100)	(O)	75.5	51.0	28.0	31.0	49.5	68	37	41	66	35	74	16	32
4	(40)	(60)	61.0	23.0	10.0	29.5	42.0	38	16	48	69	119	138	66	22
3	(30)	(70)	61.5	48.5	23.5	46.5	42.0	79	38	76	68	37	81	14.5	25
2	(20)	(80)	48.0	44.5	28.0	25.0	28.5	93	58	52	59	12	37	46	34
5	814 (100)	(O)	69.0	—	—	6.0	47.5	—	—	8	69	—	216	229	38
8	(40)	(60)	65.0	4.5	4.0	15.0	46.5	7	6	23	72	208	204	119	24
7	(30)	(70)	60.5	6.5	8.0	23.0	42.0	11	13	38	69	171	141	93	15
6	(20)	(80)	50.5	21.5	32.5	35.5	39.5	43	65	70	78	102	35	26	9
9	M302 (40)	(60)	61.0	47.0	31.5	33.0	47.0	77	52	54	77	14.5	49	30	10
10	(30)	(70)	55.0	48.0	38.0	35.5	43.0	87	69	65	78	12	22	31	8
11	(20)	(80)	47.5	42.0	29.5	25.5	38.0	88	62	54	80	5	28	24	8
GS11	SRPC 853 (100)	(O)	71.0	55.5	57.5	N.D.		78	81	N.D.		19	5.5	N.D.	

N.D. = NOT DETERMINED

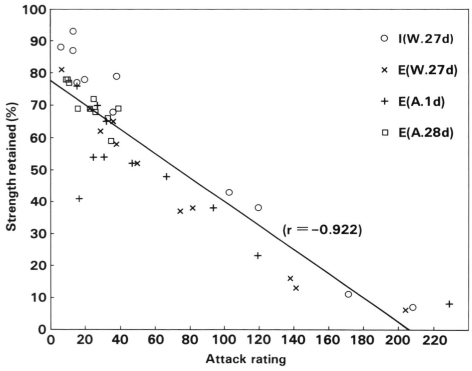

Figure 2 Relationship between (%) strength retained and sulphate
attack rating for Portland and blastfurnace slag cement
concretes (100mm cubes) after storage in sodium (I)
and magnesium (E) sulphate solutions for 5 years

retained and the attack rating with a correlation coefficient of
r = - 0.922. The results showed that the least sulphate attack and the
highest retention of strength occurred in either those specimens which
had been pre-cured in air at 20°C and 65% RH for 27 days, or in those
which contained higher levels of slag or had been made with sulphate
resisting Portland cement. These findings will be discussed in more
detail later.

TABLE 4.   INGRESS OF SULPHATE INTO CONCRETE CUBES (100mm) STORED IN SULPHATE SOLNS FOR 5 YEARS

				SULPHATE CONTENT (SO3 AS % CEMENT) AT DIFFERENT DEPTHS			
					SULPHATE SOLUTIONS (PRE-CURE)		
SR NUMBER	CEMENT OPC	COMPN. SLAG	DEPTH (mm)	SOLN I (w.27d)	SOLN E (w.27d)	SOLN E (a.1d)	SOLN E (a.28d)
1	850 100%		1-6 6.-11 11-16 16-21 21-26	9.83 4.81 4.16 2.86 2.76	N.D.	N.D.	N.D.
5	814 100%		1-6 6-11 11-16 16-21 21-26	N.D.	N.D.	N.D.	10.78 4.0 2.41 6.34 3.13
8	814 40%	M286 60%	1-6 6-11 11-16 16-21 21-26	N.D.	N.D.	N.D.	4.88 3.77 2.31 3.46 3.03
7	814 30%	M286 70%	1-6 6-11 11-16 16-21 21-26	N.D.	3.96 7.35 5.74 - -	N.D.	6.62 4.69 2.89 3.3 2.11
6	814 20%	M286 80%	1-6 6-11 11-16 16-21 21-26	3.35 1.56 1.54 2.06 1.83	6.66 3.65 2.13 - -	3.95 2.8 2.64 2.35 1.72	6.99 2.87 1.12 1.56 0.78
9	814 40%	M302 60%	1-6 6-11 11-16 16-21 21-26	3.83 4.12 2.48 - -	8.63 10.47 4.6 4.4 5.59	5.09 6.11 4.69 - -	6.17 3.33 0.55 1.07 -
10	814 30%	M302 70%	1-6 6-11 11-16 16-21 21-26	3.98 2.00 1.19 0.98 1.54	4.69 4.36 6.38 6.38 5.69	5.74 4.7 4.66 3.43 5.27	4.47 2.89 1.84 1.73 1.35
11	814 20%	M302 80%	1-6 6-11 11-16 16-21 21-26	2.91 1.71 1.38 0.96 1.39	5.58 5.1 3.96 3.52 1.28	4.56 5.07 3.11 5.12 -	5.45 3.44 1.78 1.35 1.99

N.D. or- Not Determined (Cube either badly deteriorated or broke up on drilling)

Ingress of Sulphate Ions

The results of the sulphate ion ingress at different depth ranges for the concrete specimens containing blends of the high C_3A OPC and the high and low alumina slags after 5 years storage in sulphate solutions are given in Table 4. The relationship between the % sulphate ingress and the depth of sample for the range of high C_3A OPC/high Al_2O_3 slag blended cements (air-cured for 28 days) clearly shows that the least ingress of sulphate occurred with the specimens containing 80% slag whereas the highest levels of sulphate penetrated the plain OPC concretes. The air-cured specimens had consistently been shown to have the least amount of sulphate penetrating the concrete (see Figure 3) with 'A.28 days' the least, 'A.1 day' the next and the water-cured specimens providing the highest sulphate ingress.

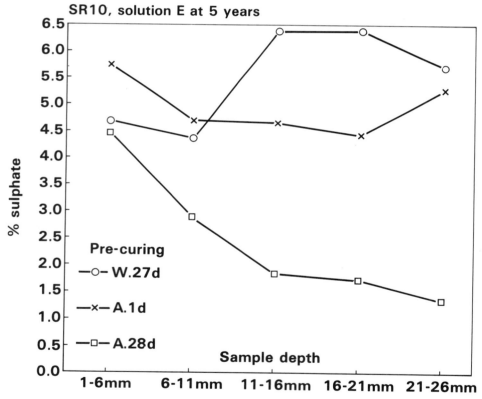

Figure 3 Ingress of sulphate (%SO$_3$) at different sample depths for sulphate stored concrete cubes

DISCUSSION

Effect of Curing Regime on Sulphate Resistance

The most significant finding of this study was the beneficial effect that derived from pre-curing the concrete specimens, (Portland and slag cements), in air at 20°C and 65% RH prior to storage in sulphate solution. Even a 'token' storage in air for a short time, probably for only several hours, following the demoulding of the concrete cubes at 24 hours, was sufficient air-curing to provide an improved sulphate resistance compared to the water-cured specimens. The performance of those specimens placed directly into water after demoulding, particularly for the high C_3A OPC concrete, was considerably reduced. This can be attributed to the fact that no carbonated layer had been allowed to form even though the optimum early strength development and concrete design strength had been achieved by water-curing.

Evidence for pore blocking was demonstrated by the way gypsum crystals formed as needles on the surface of some concretes due to the hydration products reacting with sulphate ions which were unable to permeate the fine pore structure of the outer layer of concrete. The effectiveness of 27 days of air curing was clearly demonstrated by the excellent physical appearance, strength retention and low sulphate attack ratings observed with Portland and blastfurnace slag cement concretes alike. There is a need for petrographic studies to be carried out to determine the precise nature of the hydration products and for practical evidence of any reductions in permeability to be sought.

Previous work by Osborne [11] showed that blastfurnace and gasifier slag cement concretes, especially those with high levels of slag as cement replacement, had carbonated to a depth of several millimetres following air storage for 28 days. It is believed that the pore blocking effect of this layer is the means by which the sulphate attack was effectively prevented. In earlier studies Biczok [4] contended that without the development of a carbonate layer at its surface, the durability of concrete exposed to aggressive surroundings would be appreciably diminished. Hannawayya [13] stated that the products of the carbonation- hydration process contained calcium carbonate, rather than calcium silicate hydrate (CSH) which normally forms as a product of hydration and the CO_2 retards the ability of the calcium aluminate and ferrite components of cement to react with sulphur trioxide (SO_3) to form ettringite. Neville [14] observed that the resistance of concrete to chemical attack was increased by allowing it to dry out before exposure. He concluded that precast concrete was therefore generally less vulnerable to attack than concrete cast in situ. In recent unpublished work by Harrison [15] from the results of a 15 year old study of concretes buried in London's highly sulphated clayey soils at Northwick Park, he concluded that access by some of the concrete specimens to air had probably made a greater contribution to sulphate resistance than any other single factor. This observation has been confirmed in the present studies

where air-storage for 27 days has dramatically enchanced the sulphate
resistance of the high C_3A cement concretes. Precast concretes
probably receive a curing regime somewhere between that achieved via
demoulding at 1 day and air-curing for up to 28 days, whereas the
worst situation for concrete could be where it is immediately
subjected to running water containing different concentrations of
sulphate (ie: as when placed in water storage straight from
demoulding).

Effect of Type of Sulphate Solution

Sodium sulphate solution was generally less detrimental to the
concrete specimens than the equivalent strength magnesium sulphate
solution.   Only the concretes containing the high C_3A OPC and 70% or
less of high alumina slag had poor sulphate resistance in the sodium
sulphate solution.   Magnesium sulphate was significantly more
aggressive than sodium sulphate for concretes pre-cured in water prior
to storage, even for the combinations of the high C_3A OPC and low
alumina slag which were rated as sulphate resistant.   Although the
SRPC concretes in this study were highly sulphate resistant in both
sodium and magnesium sulphate solutions (80% strength retained and an
attack rating of <20 at 5 years), Frearson[16] has shown, using sodium
sulphate for small scale methods of test, that their effectiveness can
vary as can that of certain slag cement blends.

Effect of Tricalcium Aluminate of Portland Cement

The trialcium aluminate (C_3A) content of the Portland cements was one
of the predominant chemical factors governing their sulphate
resistance.   The low C_3A (0.6%) sulphate-resisting Portland cement had
very good sulphate resistance, whereas the medium C_3A (8.8%) OPC and
the high C_3A (14.1%) OPC had moderate and very poor resistance
respectively.   A good correspondence was obtained between the results
from the small scale accelerated tests carried out previously[2] with
these sulphate tank test data.

Effect of Alumina Content of Blastfurnace Slag

The alumina content of the blastfurnace slag component in slag cements
has a significant effect on their sulphate resistance, [2,6-9] and the
present studies confirm these findings.   The results show the benefit
of using pelletized blastfurnace slag with the low alumina content
(Al_2O_3 = 7.5%).   The poor sulphate resistance of the high C_3A OPC (C_3A
= 14.1%) was improved by blending with this slag to give a very good
rating for all pre-cure conditions, but particularly for those
specimens air-cured prior to storage in sulphate solutions.   However,
the high alumina granulated slag (Al_2O_3 = 14.7%) was far less
effective and 70 and 80% levels of replacement of the medium C_3A OPC
were required to achieve good sulphate resistance.

CONCLUSIONS

The sulphate resistance of Portland and blastfurnace slag cement concretes was shown to be dependent upon the following criteria:

(i)   Irrespective of cement type, the early curing of concrete during the first few weeks of manufacture was the most significant factor.

(ii)   Concretes cured in air at 20°C and 65% RH prior to immersion in high strength sulphate solution, producing a carbonated layer at the surface, were extremely resistant to sulphate attack.

(iii)   Concrete specimens demoulded after 1 day and placed in sulphate solution within the next few hours exhibited better resistance to attack than those specimens water-cured for up to 28 days.

(iv)   Sulphate resisting Portland cement concretes were highly resistant to attack as were combinations of the high C_3A ordinary Portland cement with the low alumina slag.

(v)   Sulphate attack was greatest when both the C_3A content of the Portland cement and the alumina level of the blastfurnace slag were high.

(vi)   The replacement level for OPC by slag was also important with 70 and 80% proving most beneficial, particularly for high C_3A OPCs.

(vii)   Magnesium sulphate was a more aggressive agent than an equivalent strength sodium sulphate solution.

The BRE tank test using water-cured concrete specimens is a severe form of test which has provided a useful assessment of the sulphate resistance of Portland and slag cement concretes. The procedure merits consideration as a performance test for concrete. The present studies when completed, with the results from tests on concretes containing slags with moderate levels of alumina and the high C_3A OPC, will provide an important input to the revision of BRE Digest 250.

It is important that the curing of concrete should be relevant to its application in construction. The depth of carbonation within the concrete surface, caused by the natural carbon dioxide in the atmosphere (0.03%), may attain only a few millimetres. The thickness of the layer could be increased by a brief period of artificial carbonation involving vacuum treatment and the subsequent introduction of carbon dioxide under pressure[4]. However, the alkalinity of the concrete would also be reduced and could result in an increased chance of corrosion of the reinforcing steel. Artificial or accelerated carbonation of reinforced concrete structures if contemplated, should therefore be controlled or monitored with great care or even avoided in favour of the natural process of carbonation.

ACKNOWLEDGEMENTS

The work described has been carried out as part of the research programme of the Building Research Establishment of the Department of the Environment, and the paper is published by permission of the Director.

REFERENCES

1   BUILDING RESEARCH ESTABLISHMENT. Concrete in sulphate-bearing soils and ground-waters. BRE Garston, 1986, BRE Digest 250.
2   G J OSBORNE. Determination of the sulphate resistance of blastfurnace slag cements using small-scale accelerated methods of tests. Advances in Cement Research, 1989, 2, No 5, Jan 21-27.
3   W H HARRISON. Durability of concrete in acidic soils and waters. Concrete, Vol 21, No 2, February 1987.
4   BICZOK I. Concrete corrosion and concrete protection. Publishing House of the Hungarian Academy of Sciences, Budapest, 1967, Chapter V. Protective Measures, 2, 428-431.
5   OSBORNE G.J. The durability of concretes made with gasifier slag cement. Durability of Building Materials, 4, (1986), 151-177.
6   DUCIC V and MILETIC S. Sulphate corrosion resistance of blended cement mortars. Durability Building Materials, (1987), 4, 343-356.
7   EMERY J.J. Sulphate resistance of Standard's slag cement. Standard Slag Company, Fruitland, Ontario, 1983, H3816-ATS, Private communication.
8   VAN AARDT J.H.P. and VISSEN S. The behaviour of mixtures of milled granulated blastfurnace slag and Portland cement in sulphate solutions. National Building Research Institute, Pretoria, 1967, Bulletin 47, CSIR Research Report 254.
9   TANEJA C.A. Effect of alumina content and fineness of blastfurnace slag on the sulphate resistance of Portland blastfurnace cements. Zement-Kalk-Gips, 1975, 28, No 2, 76-79.
10 BRITISH STANDARDS INSTITUTION. BS 8110, 1985. The structural use of concrete. B.S.I. London.
11 OSBORNE G.J. Carbonation of blastfurnace slag cement concretes. Durability of Building Materials, 4, (1986), 81-96.
12 STEELE B.R. and HARRISON, W.H. Immersion tests on the sulphate resistance of concrete. Proceedings of a RILEM International Symposium on Durability of Concrete, Prague 1969. Prague, Academia, 1969, Preliminary report, part II, C163-C186.
13 HANNAWAYYA F. Microstructural study of accelerated vacuum curing of cement mortar with carbon dioxide. Part 1, World Cement, (Nov. 1984): 326-334: Part II World Cement, (Dec 1984): 378-384.
14 NEVILLE A.M. Properties of Concrete. Pitman Publishing, London,

Second Edition, 1973, pp 398-399.

15   HARRISON W.H. Long-term tests on sulphate resistance of concretes at Northwick Park: Third Report 1989. To be published as a BRE report.

16   FREARSON J.P.H. Sulphate resistance of combinations of Portland cement and ground granulated blastfurnace slag. Proceedings of Second International Conference on the use of ash, silica fume, slag and natural pozzolans in concrete, Madrid, Spain, 1986, American Concrete Institute, SP91-74, Volume 2, 1495-1524.

CARBONIC ACID WATER ATTACK OF PORTLAND CEMENT BASED MATRICES

Y Ballim

M G Alexander

University of the Witwatersrand,

South Africa

ABSTRACT. Concretes based on portland cement (OPC) and blends of OPC with slag and fly ash were subjected to carbonic acid water attack. Measurements and observations included mass loss, depth of attack, leachate water composition and changes in surface water absorption properties. The paper aims to highlight some of the problems related to the characterisation of concrete deterioration. It is shown that, depending on the method of characterisation used, different conclusions can be drawn about the relative performance of various concretes exposed to carbonic acid water attack. This is relevant to other durability studies using similar test methods with different aggressive environments.

Keywords: Durability, Carbonic Acid, Characterisation, Fly Ash, Blastfurnace Slag, Water Absorption, Mass Loss, Atomic Absorption Analysis.

Mr. Yunus Ballim is the Portland Cement Institute Research Fellow in the Department of Civil Engineering, University of the Witwatersrand, Johannesburg, South Africa. His research is directed mainly at the durability of cover concrete exposed to various aggressive agents.

Prof. Mark G Alexander is an associate professor in the Department of Civil Engineering, University of the Witwatersrand, Johannesburg, South Africa. His research interests include durability of concrete, fracture mechanics and aggregates for concrete.

INTRODUCTION

This project was undertaken as the first part of a larger study into the effects of aggressive carbonic acid water on the durability of concrete. The aims of this part of the study were as follows:

(a) to assess the suitability of the test methods used, and
(b) to determine suitable means of characterising the deterioration of concrete exposed to carbonic acid water.

The report discusses the characterisation aspect of the project, critically assessing the techniques used for characterising deterioration.

Because of their low dissolved ion content, pure or soft waters will leach calcium hydroxide from the hardened cement paste phase of concrete. If this process is allowed to continue, the loss of lime will cause the cement hydrates to become unstable and decompose. Also, when calcium hydroxide is being lost around the reinforcing steel, the pH of the pore water will drop to a level where the passivation of the steel is destroyed, exposing the reinforcement to possible corrosion. The aggressiveness of pure water is considerably increased if it also contains dissolved carbon dioxide (CO_2), forming carbonic acid water. This CO_2 assists in the conversion of calcium hydroxide to the more soluble calcium bicarbonate, thereby increasing the rate at which calcium hydroxide is removed from the concrete [1].

In order to compare the relative performance of different concretes exposed to carbonic acid water, it is necessary to obtain a measure of deterioration. This measure must be applicable to all concretes, regardless of binder type. Such a measure or characterisation parameter must also be able to assist in the prediction of the potential deterioration of a given concrete when exposed to carbonic acid water. In addition to improving our understanding of the deterioration process, this will provide designers of structures in these environments with the means to ensure that the expected service life of such a structure is not compromised.

Some researchers [2],[3],[4] have used engineering properties of concrete, such as strength or elastic modulus, as a means of characterising deterioration. It was felt that this would not be adequate for this project since these are measures of the bulk condition of the material while the deterioration process affects only the surface layer. It was therefore decided to investigate the following possible characterisation techniques:

(a) sample mass loss,
(b) atomic absorption analysis of the leachate water,
(c) depth of deterioration, and
(d) surface water absorption.

EXPERIMENTAL DETAILS

Concrete Materials and Mixes

Concrete mixes were prepared using ordinary portland cement (OPC), a 50/50 blend of OPC and fly ash (FA) and a 30/70 blend of OPC and ground granulated blastfurnace slag (GGBS). In terms of chemical composition, the materials used represented typical South African binder materials. A local crushed dolerite 19 mm stone and crusher sand (fineness modulus = 3,4) were used as aggregate throughout. Table 1 shows the mix proportions used. All the mixes were designed to have a slump of 25 ± 10 mm and the same 28 day strength of 30 MPa.

TABLE 1: Concrete Mix Proportions (kg/m^3)

	MIX NUMBER:		
MATERIAL	BS1	FA2	PC3
OPC	104 (30)*	170 (50)	312 (100)
GGBS	242 (70)	–	–
FA	–	170 (50)	–
Crusher Sand	845	820	895
19 mm Stone	1200	1250	1200
Water (W)	190	180	190
Binder/W Ratio	1,82	1,89	1,64

* Values in brackets show the % blend proportions of the various binder types.

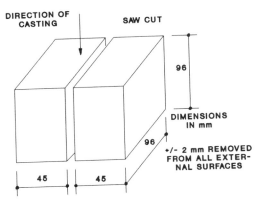

Figure 1: 100 mm Cube Cut and Faced.

Sample Preparation

Six 100 mm cubes had ± 2 mm removed from all external surfaces on a high speed facing machine. This was necessary to remove the distorting influence of different amounts of mould oil and laitance on the surfaces. The cubes were then halved as shown in Figure 1. This resulted in 12 'slab' samples which were divided into 3 groups of 4 slabs each to be treated as follows:

> Group 1: exposed to carbonic acid water; periodically brushed.
> Group 2: exposed to carbonic acid water; unbrushed.
> Group 3: control group exposed to lime saturated water.

Care was taken to ensure that no group contained two halves of the same cube.

Aggressive Water Environment

Carbonic acid water was manufactured by bubbling CO_2 through distilled water until the pH of the water was reduced from approximately 5,9 to 4,1 ± 0,05. Because the solubility of CO_2 in water is dependent on the water temperature [5], distilled water was brought into the test room at least 24 hours before acidifying to ensure that the water temperature was at equilibrium with the test room temperature, which was held constant at 23 ± 1 °C.

The samples to be exposed to carbonic acid water were placed on edge in 20 ℓ domestic PVC buckets, one sample group per bucket. One end of the underside of each sample was supported on a PVC strip to allow free movement of water around all surfaces. Five litres of carbonic acid water were added to each bucket giving a concrete surface area to aggressive water ratio of approximately 28 mm^2/ml. Before the buckets were sealed with air-tight lids, the air above the water surface was purged with CO_2. This was necessary to minimise the tendency for CO_2 to leave the water to establish equilibrium with the air in the bucket. All the samples in buckets, as well as the control samples in lime saturated water were stored in the test room mentioned above.

TEST METHODS

Mass Loss

Samples were exposed to carbonic acid water 110 days after casting. On the day of first exposure, all samples were weighed in a saturated, surface dry condition to an accuracy of 0,1 g to obtain the original sample mass. The samples were weighed in this state throughout the project.

Initially at weekly intervals and later at two-week intervals, the buckets were opened and, using a medium stiffness nylon brush, the Group 1 samples were hand brushed until no further solid material could be seen to be removed. These samples were then surface dried and weighed. The Group 2 samples were weighed unbrushed with great care being taken in the handling and surface drying so as to avoid removal of the deteriorated material. The water in the buckets was discarded and replaced with fresh carbonic acid water after each weighing operation. At the end of the test period, the unbrushed samples were brushed and weighed to determine the extent of surface deterioration for this exposure condition.

Atomic Absorption Analysis

Each time the buckets were opened for mass loss determination, a 25 ml sample of the discard water was drawn off for atomic absorption analysis of the calcium (Ca), silicon (Si) and magnesium (Mg) content. In the case of the brushed samples, this water sample was drawn before the samples were brushed so that the water did not contain any material that was physically removed from the concrete surface.

Water Absorption Tests

At the end of the aggressive water test programme, the Group 2 samples, which were then brushed, were used to determine the surface water absorption characteristics of the concretes. Six consecutive slices, each 3 mm thick, were cut parallel to a 45 x 96 mm face of one sample from each mix. The outside edges of each slice were removed so that the resulting slice measured approximately 50 x 20 mm. The blade used to cut these slices was 2 mm wide.

The slices were oven dried at 105 °C for 24 hours, after which they were weighed to an accuracy of 0,001 g. They were then immersed in lime saturated water such that the water surface was approximately 1 mm above the top of the slice. Each slice was weighed at 1 minute, 10 minute and 24 hours after immersion to determine the water absorption. The cement content of each slice was then determined using an HCl digestion method based on BS 1881: Part 124: 1988.

RESULTS AND DISCUSSION

Mass Loss

Figure 2 shows the results of the measured mass loss for the brushed and unbrushed samples. Note that the results have been normalised by the original sample mass. It is clear from this figure that the OPC mix shows the greatest mass loss for most of the test period. On the other hand, the GGBS mix (BS1) shows the lowest mass loss for most of the test period despite having the lowest compressive strength at

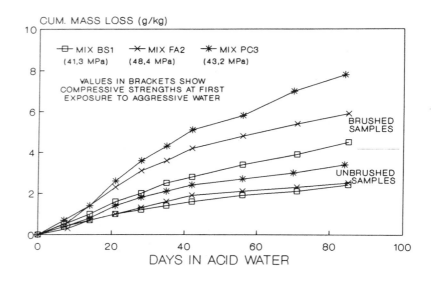

Figure 2: Cumulative Mass Loss for Brushed and Unbrushed Samples

the time of first exposure to the aggressive water. For the brushed samples at 84 days, the OPC and FA samples respectively showed 75 % and 32 % more mass loss than the GGBS samples. In the case of the unbrushed samples, there is very little difference between the mass loss of the GGBS and FA mixes while the OPC shows 44 % more mass loss than the GGBS samples. On the basis of mass loss, it appears that the GGBS and FA improve the resistance of concrete to carbonic acid water attack.

However, caution is required when interpreting these results with the intention of comparing the relative resistance of different mixes. The results shown in Figure 2, do not account for the different amounts of aggregate removed when the samples are brushed. Hence a sample which presents more fine sand at the surface for removal will show a greater mass loss even though the binder matrix is not necessarily more readily deteriorated by the aggressive agent. Also, the method used for exposing the samples to aggressive water may introduce a distortion. It can reasonably be assumed that the concretes used in this project would release calcium hydroxide at different rates when exposed to the leaching action of carbonic acid water. This being the case, the aggressive water in the buckets would be neutralised at different rates. It could therefore be argued that, since the aggressive water was replenished at fixed

Table 2: Comparison of the Mass Loss of Samples Regularly Brushed with   that of the Samples Brushed at the End of the Test Period.

| | TOTAL MASS LOSS (g/kg) FOR SAMPLES BRUSHED: | | |
MIX No.	AT REGULAR INTERVALS (a)	AT END OF UN- BRUSHED PERIOD (b)	RATIO OF (a)/(b)
BS1	4,5	3,3	0,73
FA2	5,9	4,7	0,80
PC3	7,8	6,9	0,88

intervals, the different concretes had not been exposed to aggressive water for the same time period. This could have an influence on the differences in the results obtained in this project.

Table 2 shows the mass loss results obtained when the unbrushed samples were brushed at the end of the test period. For comparison, the total mass loss of the continuously brushed samples is also shown.

If it is taken that the difference between these two results is an indication of the degree of protection afforded by the deteriorated skin layer, then the calculated ratio shown in Table 2 indicates that the deteriorated skin layer of the GGBS mix has shown the greatest degree of protection. It is interesting that the mass loss upon brushing the unbrushed samples shows a similar trend in terms of the performance of the different concretes when compared with the mass loss results in Figure 2.

Atomic Absorption Analysis

Figure 3 shows the cumulative results obtained from the atomic absorption analysis. Note that, for clarity, the Si and Mg results have been magnified 5 and 10 times respectively. There appears to be some correlation between the amount of Ca leached from the concretes and the mass loss results shown in Figure 2. However, the increased loss of Ca shown by the OPC mix can be explained by the higher amount of Ca which OPC concrete presents for leaching which would not necessarily result in a greater degree of deterioration. Figure 3 therefore suggests that the concretes may have suffered a similar degree of deterioration over the 84 days of exposure to carbonic acid water.

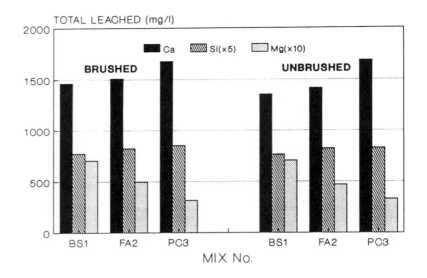

Figure 3: Cumulative Atomic Absorption Analysis Results

Water Absorption

Of the six slices tested, very little difference was noted between
the water absorption results of the third to the sixth slices.
Figure 4 therefore shows the results only for the first three
slices, representing a concrete depth of 11,5 mm. The results are
presented as mg of water absorbed per g of cement in the slice.

The absorption at different times can be related to different pore
sizes or structures. The 1 minute absorption results indicate the
volume of large, easily accessible pores while the 10 minute results
relate to smaller and less accessible pores. The 24 hour results can
be taken as a measure of the total pore volume of the sample. On
this basis, Figure 4 shows that almost all the pore volume of the
OPC mix is made up of large, accessible pores. Hence this mix shows
a high 1 minute absorption but not correspondingly high 10 minute
and 24 hour absorptions. On the other hand, the GGBS and FA mixes
show a more continuous grading of pore size. The FA mix shows the
highest 10 minute and 24 hour absorptions. The 10 minute and 24 hour
absorptions of the GGBS mix is very similar to those of the OPC mix.
Because the water absorption tests showed uniform results for the
4th, 5th and 6th slices, it was assumed that the concrete at these
depths had been unaffected by the exposure to carbonic acid. The
average of the results obtained for these three slices was therefore
taken as the water absorption of the undeteriorated concrete. To

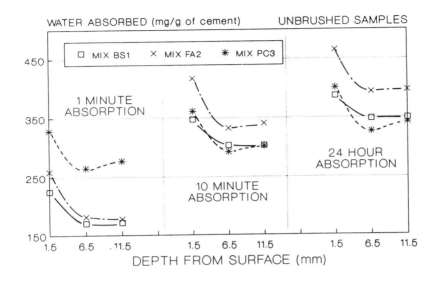

Figure 4: Water Absorption Results vs Depth into Concrete

obtain a measure of the degree of deterioration, the results shown
in Figure 4 are plotted in Figure 5 relative to the corresponding
average value for the unaffected concrete. Here it can be seen that,
on the basis of the 1 and 10 minute absorption results, the OPC
concrete has not suffered as much relative damage as the other two
concretes. In terms of changes in total pore volume, the 24 hour
absorption results show that the concretes have shown similar
relative resistance.

It is interesting to note that some form of densification has
occurred in the OPC concrete at 6,5 mm from the surface as evidenced
by the decreased water absorption at this level. This feature is
similar to that encountered by Valenta and Modry [6] who investigated
the variation in calcium hydroxide concentration with depth from
the surface of concrete dam structures exposed to soft water with
dissolved CO_2. The reason for this feature is unclear at present and
will require further investigation when a more complete test program
is undertaken. Also, Figures 4 and 5 indicate that the depth of
deterioration for all the concretes is approximately 5 mm. From a
characterisation point of view, the depth of deterioration therefore
indicates that the concretes have presented similar resistance to
the penetration of carbonic acid water attack.

The water absorption results have confirmed that the effect of FA
and GGBS in concrete is to refine the pore size in the hardened

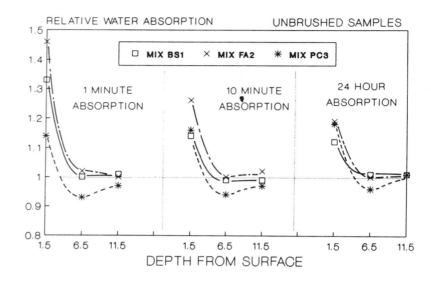

Figure 5: Relative Water Absorption vs. Depth from Surface

binder paste when compared to a plain OPC concrete. It is however, unclear what the implications of this pore refinement are for the durability of concrete exposed to the leaching action of carbonic acid water. While a good correlation exists between the one minute absorption and the mass loss results, the trend is reversed when considering the 10 minute and 24 hour absorption results. Given the problems associated with the use of mass loss to characterise deterioration, the correlation obtained with the one minute absorption results is not an indication that either method is to be accepted as a characterisation technique. Further testing using a larger number of samples and concretes of different strengths is required before any positive statements can be made regarding the use of these techniques for characterising the deterioration of concrete.

CONCLUSIONS

The following conclusions are drawn on the basis of the results and discussions presented above:

For the range of concrete strengths used in this project, there does not appear to be any relationship between concrete strength and resistance to carbonic acid attack. However, there is a need to test a wider range of concrete strengths in order to confirm this.

When investigating the relative resistance of different concretes to deterioration, very different conclusions can be drawn depending on the method used to characterise the deterioration.

By itself, mass loss is not a sufficient means of characterising the deterioration of concrete when the intention is to compare the relative performance of concretes made with different binder types. There may, however, be structural applications where the mass loss of the concrete is an important parameter to be measured.

It appears that water absorption is a more appropriate means of characterising deterioration since it relates directly to the porosity and pore structure of the concrete. It is a parameter which is therefore more closely related to the movement of water and dissolved ions between the concrete and the external aggressive environment.

ACKNOWLEDGEMENTS

The authors wish to thank Slagment (Pty)Ltd. for the permission to publish the results presented in this report.

REFERENCES

1   BASSON J J. Deterioration of Concrete in Aggressive Waters. Concrete Durability Bureau, Portland Cement Institute, Midrand, South Africa. 1989.

2   BERTACCHI P. Deterioration of Concrete Caused by Acid Water Containing Carbon Dioxide. RILEM Int. Symposium, Durability of Concrete. Prague, 1969. Final Report, Part II. pp C159 - C168.

3   EFES Y. Influence of Blastfurnace Slag on the Durability of Cement Mortar by Carbonic Acid Attack - Problems Connected with Tests on Corroded Specimens. Durability of Building Materials and Components. ASTM STP 691. Sereda and Litvan (Eds). ASTM. 1980. pp 364 - 376.

4   HALSTEAD P E. An Investigation of the Erosive Effect on Concrete of Soft Water of Low pH Value. Mag. of Concrete Research. Vol. 6, No. 17. Sept. 1954. pp 93 - 98.

5   MOSKVIN V. (Ed). Concrete and Reinforced Concrete Deterioration and Protection. Chap. 7. Mir Publishers, Moscow. 1983.

6   VALENTA O and MODRY S. A Study of the Deterioration of the
        Surface Layer of Concrete Structures. RILEM Int. Symposium,
        Durability of Concrete. Prague, 1969. Final Report, Part I.
        pp. A55 - A64.

METHODS OF PROTECTING CONCRETE COATINGS AND LININGS

Session A

Chairmen

Dr J Menzies

Director, Geotechnics and Structures

Building Research Establishment,

United Kingdom

Mr V S Parameswaran

Deputy Director, Concrete Composites Laboratory,

Structural Engineering Research Centre

India

METHODS OF PROTECTING CONCRETE - COATINGS AND LININGS

P C Hewlett

British Board of Agrément

United Kingdom

ABSTRACT. Coatings and linings are defined and reviewed as protectants for concrete.

The protectant function is primarily one of a physical barrier rather than chemical conditioner, although the latter are mentioned as are coatings for embedded steel.

Defining the function should assist selection, however, achieving performance depends on characterising the surface of concrete and applying the materials in such a way as to obtain intimate contact and adhesion. Bonding criteria are identified and recommendations given.

Relevant coatings' properties are stated and methods of measurement reviewed.

Areas of doubt are identified as are research needs.

Since much information is spread throughout the literature a comprehensive reference list is given.

Keywords: Sealers, Impregnants, Coatings, Waterproofers, Damp proofers, Linings, Renders, Adhesion, Corrosion, Reinforcement

Professor Peter C Hewlett is Director of the British Board of Agrément, an independent certification authority dealing with proven performance of building and construction materials and systems.

He is also Visiting Industrial Professor in the Department of Civil Engineering at the University of Dundee.

Current research interests cover admixtures, surface and quality characterisation of concrete, durability criteria and transport mechanisms at and within the surface of concrete.

INTRODUCTION

Concrete based upon Portland cement is ubiquitous and as a
consequence subject to the widest range of weather and service
conditions, including abuse. As such it has and is providing a
unique contribution to the built environment. Unique because no
other engineering material can be placed and formed at ambient
workaday temperatures to produce in situ massive structural
articles. As a result of its versatility and physical similarity
to rock and masonry, concrete is often assumed to be permanent and
inert. However that assumption is untrue since unlike masonry and
rock, concrete has not evolved from natural geochemical processes
and as such is a geologically immature material. It is therefore
to be expected that a detailed understanding of degradation
mechanisms has not, as yet, developed. Even the oldest studied
concretes, the so-called geo-polymeric concretes ([1], [2]) said to
comprise the Pyramids and great Sphynx of Egypt, have not told us
how to make a long lasting concrete.

Therefore we are left with protecting the outer layers of concrete
by interposing within it and/or on top of it, materials that
prevent and repel the ingress of damaging agents.

The present concern about concrete's degradability, durability,
life span, maintenance and repair have focussed attention on the
processes of concrete decay and the means of slowing them down and
perhaps designing and specifying for their elimination.

Until quite recently the function of concrete has been maintained
by way of its bulk properties such as section, strength, mass and
density. However it is now accepted that most degradation
processes occur from the "outside in" requiring ingress of
moisture, active gases and deleterious ions.

The state of knowledge of these invasive mechanisms is very
partial, there being more concern with providing some form of
practical protection or palliative. As a result a "barrier
technology" is evolving.

Such barriers may also impart additional benefits such as improved
chemical and wear resistance and appearance.

Barriers in this context include coatings, impregnants, sealers,
repellants, overlays or linings.

Much advice already exists ([3]).

In his book entitled "The Science of Surface Coatings" Chatfield
([4]) nearly 30 years ago outlined the requirements for successful
coating of concrete. Account was taken of such factors as surface
moisture, alkali resistance, vapour transportation, antipathetic
forces (cohesion vs. adhesion) and film formation. A

science-based approach was adopted that contrasts markedly with the empiricism of today. It is timely to reconsider whether enough is known about concrete surface condition and chemistry as well as the dynamics of wetting, film formation and stability.

On a more practical level the requirements for coating for concrete have been suggested by Treadaway et al ([6]) and Browne ([7])

However, a recent ACI committee report ([8]) dealing with functional and decorative barrier systems for concrete highlights the need for tests that can establish the surface quality of concrete before a coating is applied. A simple issue such as surface cleanliness - important for long term adhesion - cannot be quantified.

CIRIA Report 130 ([9]) attempts a coatings classification and a schematic representation is given in Fig. 1 below.

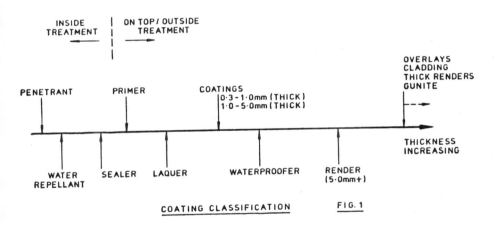

The CIRIA report gives and excellent review (table 2 of reference [9]) of 6 main performance properties together with 48 sub-properties used for conformity testing and certification and attempts to place limits on moisture vapour resistance and performance.

Firstly then the surface nature of concrete since it is concrete we are trying to protect.

SURFACE NATURE OF CONCRETE

This topic has been considered previously by the author ([10]).

The application of coatings to concrete are very dependent on surface condition as well as imparting their own conditioning.

Most surface studies to date have been more concerned with the durability of concrete rather than aspects of appearance. ([9], [12], [13])

P.C. Kreijger ([11]) has shown very graphically that the conventional properties of porosity, bulk density, cement content, aggregate: cement ratio are notably different at the surface compared to the bulk.

Kreijger initially used the term "skin" to describe this zone and suggested it was 30-40mm deep comprising,

cement skin   0.1mm thick
mortar skin   5mm thick
concrete skin 30mm thick

However, more recently this zone within the surface has been reduced to 5mm, and whilst such a value is arbitrary the smaller figure is more readily recognised.

Concrete and mortar surfaces are non-ideal and broadly may be described as

1. comprising fresh concrete that has hardened but is still immature (>one month but <six months old)
2. aged concrete
3. fractured concrete with exposed aggregate that itself may be fractured
4. prepared concrete

Such variability raises the problem of creating a "standard" concrete surface for laboratory assessment of coatings and linings.

Surfaces 1 - 4 above may also be porous, rough and differ in chemical make up and hardness. By any standards the surface of concrete is most complex, but for convenience we may consider it from two standpoints, namely microscopic and macroscopic.

Microscopic

Kreijger's work ([11]) shows that the outermost layer of concrete
has the lowest quality. Porosity almost doubles and the average
pore size (calculated) is approximately 15 microns, whereas for
the bulk it is 10^{-4} microns, a factor difference of 150,000.
Cement content is higher by 50% and the air content is up by 50%
with the aggregate:cement ratio varying from 6.1 to 3.2.

The dominant phase of a concrete surface may be hydrated cement or
aggregate depending upon whether the surface results from being
cast or fractured. If cast, then hydrated Portland cement will
dominate. Itself characterised by calcium silicate hydrate, some
Portlandite with relatively minor amounts of ettringite and/or
mono sulphate.

Such a structure is porous (approximately 30%) and recent mercury
porosimmetry measurements ([14]) indicate a range of values from 3 -
2 nm (.003 - .002 microns) for opc pastes, 3 - 40 nm (.003 - .04
microns) for opc/pfa mixes for "continuous" pores. Gel pores by
comparison are approximately 2nm (.002 microns). Suffice to note
that porosity and pore size governs accessibility and this is
variable and limiting.

Porosity is graded by scale, for instance, the capilliary pores
are substantially bigger than the fine gel pores and may be
accessible whilst voids resulting from entrained and entrapped air
are by comparison very large and contribute to the micro roughness
of the surface (x500,000 gel pore size).

Concrete is likely to have an ultra micro roughness also, resulting
from hydration products that consist of sheets, platelets and
needles. Such micro roughness is of the order of tens of nm
whilst the macro roughness is several orders larger (micro metres)
(Figure 2).

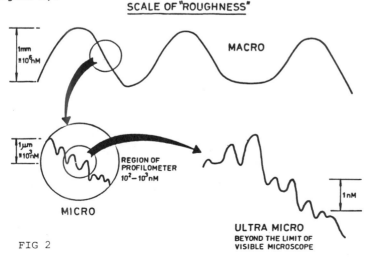

SCALE OF "ROUGHNESS"

MACRO

1mm $\equiv 10^6$ nM

REGION OF
PROFILOMETER
$10^2 - 10^3$ nM

1μm $\equiv 10^3$ nM

MICRO

1nM

ULTRA MICRO
BEYOND THE LIMIT OF
VISIBLE MICROSCOPE

FIG 2

For coatings and linings to adhere and react with the substrate
intimate contact is necessary.

Even if coatings and linings are within themselves adequate for
the task of protection without good adhesion, they will fail.

Characterising the surface condition

The ultimate aim is to characterise the outermost concrete in
order to make a judgement on its durability with or without a
surface treatement.

In selecting the means of protecting concrete one may also have to
make a judgement about its surface condition.

The proliferation of test methods that depend upon measuring liquid
uptake or passage into and through a concrete specimen raise
questions about what is really being measured. For instance, if
concrete is dry and water makes contact, it is pulled into the
specimen by capilliary attraction. This driving force is
significant, being equivalent to approximately 15m head plus
standing head for pores of 1 micron or less. The _rate_ at which
water is taken into the specimen is the sorbtivity, the saturated
value is the absorption. This property should be distinguished
from permeability that measures the ease of flow _through_ a
speciment that is saturated to begin with.

Sorbtivity follows a Fickian form with depth of penetration
proportional to the squareroot of time.

$$\text{Depth} = S \; x\sqrt{T}$$

The constant of proportionality (S) being the sorbtivity.
Permeation on the other hand obeys D'Arcy's law.

A very elegant yet simple test method has been developed by S
Kelham of Blue Circle Industries PLC [15] that allows the absolute
permeability to be measured as well as the sorbtivity by measuring
the progressive weight increase of a small inundated sample.

The merits of measuring the intrinsic permeability [13] rather
than air permeability by means of an index [16] have been
discussed at length. Characterising the outer most layers of
concrete by means of absorbtivity or permeance in some form has
been or is an active area of interest and research [12, 13, 17,
18, 19, 20]

Apart from practical problems of repeatability and interpretation there is the issue of relating the measured values to a "quality" condition of the concrete. References 20 and 21 attempt to address this problem.

For instance, Figg ([19]) has correlated the so-called air index values with a propensity for the concrete to carbonate and compiled a tentative table for "concrete protective quality" covering the ranges obtained. For instance, index values <30 - >1000 corresponding to poor (porous mortar) through to excellent quality (polymer modified)

Figg extends the comparison to absolute air and water permeability and reproduced here is his table 6 for comparative values.

Table 1. Deducing concrete quality from intrinsic permeability

Concrete protective quality	Intrinsic permeability $K(m^2) \times 10^{-14}$	
	air	water
Poor	>170	>170
Moderate	50-70	25-60
Fair	20-50	12-25
Good	5-20	2-12
Excellent	<5	<2

These are very useful indicators when deciding whether or not to coat concrete. The values obtained are dependent upon knowing the moisture content of the concrete. For in situ measurements this may prove very difficult.

The establishment of accepted measured ranges is of high priority.

Roberts ([21]) has described the technique of measuring carbonation depths and in a recent publication ([22]) BRE describes how to establish alkalinity profiles and link these to composition, curing strength and density. It is possible to deduce "K" factors from carbonation rates and relate these values to inferred quality.

COATINGS AND SURFACE TREATMENTS

Much surface treatment is concerned with physically changing the surface by reducing porosity or increasing its strength using a chemical (polymer) to achieve this change. However there are

occasions where chemical reconditioning is also required. Such a
case is the reinstatement of alkalinity in carbonated concrete.
For instance, sodium silicate solution is highly alkaline

$$Na_2SiO_3 \underset{\xrightarrow{2H_2O}}{\rightleftharpoons} H_2SiO_3 + 2NaOH \qquad \text{(strongly alkaline)}$$

The alkalinity will be reduced somewhat in contact with carbonated
concrete.

$$H_2SiO_3 + CaCO_3 \rightleftharpoons CaSiO \downarrow + H_2CO_3 \qquad \text{(weakly acid)}$$

$$2NaOH + CaCO_3 \rightleftharpoons Na_2CO_3 + Ca(OH)_2 \qquad \text{(alkaline)}$$

Whilst this approach appears attractive and cheap, sodium silicate
will also react with carbon dioxide/carbonic acid to give sodium
carbonate that may appear as a white powdery deposit. Such crystal
growth chemicals have been shown to vary considerably in
performance. [23]

The most extensive and confusing range of surface treatements and
in particular water repellents, are those based upon the
substituted silicon atom. They take many forms [24]

As the akyl group increases in size from methyl (CH_3-) to bytyl
(C_4H_9-) or octyl ($C_8H_{17}-$) the alkali resistance increases. Water
repellency increases with molecular size but penetrability
decreases. A compromise is required with isooctyl trialkoxy
silane and oligomerous isoocytyl alkoxy siloxane and isobutyl
methoxy silane being favoured [25]. However the latter is very
volatile and surface temperatures may affect the depth of
penetration. [26]

After application of whatever variety repellency is imparted by
the resulting silicone (polysiloxane) resin.

It is worth considering for a moment how such compounds attach
themselves to the surface of siliceous materials such as concrete.
Rodder [27] suggests that condensation reaction occurs between
surface silanol groups and those within the repellent and the
repellent within itself.

$$R - \overset{\displaystyle OR'}{\underset{\displaystyle OR'}{Si}} - OR' + H_2O \rightarrow R - \overset{\displaystyle OH}{\underset{\displaystyle OH}{Si}} - OH$$

Alkyl alkoxy
  silane

```
        OH                          |              O         O
        |                           |              |         |
 R  -  Si  -:OH      HO ---{   Si -      R  -  Si  - O - Si -
        |   '.........'             |              |         |
        OH                          |              O         O
                                    O   --->       |         |
                                    |
                                    |
        OH                          |              |
        |          ,----- .         |              O
 R  -  Si  -|OH      HO ---{   Si -      R  -  Si  - O - Si -
        |                           |              |         |
        OH                          |              O         O
                                                   |         |
```

Treatment   Siliceous
chemical   surface

The surface bonds are covalent and strong (J=88.2kcal/mol) which is about 50x the dipole interaction.

The silanes are small and penetrating being similar in size to the water molecule (.5-1nm) and form the basis of several proprietry treatments ([28]). It has been suggested ([25]) that as far as penetrability is concerned,

   akylalkoxy silanes > silicones > acrylics > epoxy

How effective are repellent treatments?

It will depend upon uniformity of treatment and the pore size distribution within the surface of the concrete. Calculation using the Washburn equation ([29]) and assuming a natural contact angle of 120° implies that for pores in the range 0.6-.6 microns such treatment should be able to resist 40-140 metres water pressure. In practice the levels obtained are substantially lower than this due likely to non uniform distribution of the treatment throughout the treated zone of concrete.

Water repellents are readily applied, colourless, allow reuse of surfaces within a few minutes of application. However the conditions of application should be carefully controlled and selection of treatment take regard of the chemical nature of the substrate.

Surface treatments for concrete and masonry

Concrete in many ways resembles masonry, it being a consolidated but porous mass containing dissolved salts. They differ in their chemistry and also in their maturity. All materials respond to the environment in which they are put and so even stable masonry can be troubled when a change in its natural environment occurs. This is manifestly true for historic structures that are decaying. Attempts to arrest such decay have lessons for concrete and are worth commenting upon.

BRE Digest 125 ([30]) outlines the functional requirements, namely watershedding capability coupled with moisture transmission. The success of the treatment depends upon the internal moisture state of the masonry. Moisture movement resulting in dissolved salts concentrating at the surface where they may be washed away, can be impeded by water repellent treatment such that salts concentrate beneath and back from the surface resulting in spalling. (ref Fig 3).

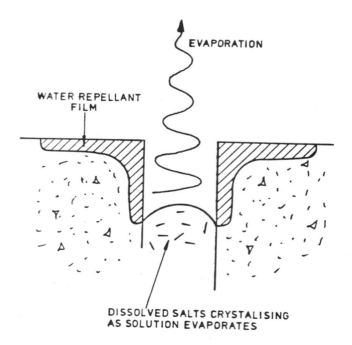

EVAPORATION

WATER REPELLANT FILM

DISSOLVED SALTS CRYSTALISING AS SOLUTION EVAPORATES

INTERNAL CRYSTALISATION RESULTING FROM
WATER REPELLANCY WITHIN PORE   FIG. 3

The preconditioning and timing of application are important and conditions of application can be quite product specific.

Early water repellents were solutions of emulsions of waxes, oils, resins, fats and metallic soaps such as aluminium stearates.   A limited by effective life of such treatments is about 10 years and reapplication can be problematic if the treatment history is not known.

A very comprehensive survey of brush applied masonry preservatives reported by Clarke and Ashurst ([31]) covers 7 types of treatment from lime water, silicones, siliconates, silicate esters and various combinations showed over a 6 year period none of the treatments were satisfactory.   However the trialkoxyalkyl silane designated Brethane ([32]) appears to have been more successful. This is a three component composition mixed just before use that penetrates several centimetres and serves two functions.

1. To bond friable and decayed stone together
2. To prevent ingress of aggressive chemicals and pollutants

The original moisture/salt condition of the masonry can compromise the treatment.

Some treatments are very substrate specific.   For instance, a silicone/petroleum spirit mixture may be suitable for brick, stone, concrete, cinder blocks and stucco and yet not be suitable for limestone and gypsum.   Whereas an aqueous solution of sodium methyl siliconate may be used to treat limestone ([33]).   The correct selection and application conditions are paramount.

There is still much trial and error in the field of masonry treatment with some favouring complete encapsulation others preferring a vapour permeable treatment.   Deep penetration is however desirable.   It would seem according to Fidler ([34]) that no treatment is a panacea and the prospect of retreatment has to be considered.   These lessons should be regarded when considering concrete surface coatings and linings.

Intercoat compatibility and/or treatment removal can be very difficult.   It raises the matter of maintenance records and the reduction of long term expectation resulting from such treatment. Our knowledge of degradability mechanisms is minimal for all the materials used.

Metal surfaces (external and embedded)

External

Since much concrete has metal buried within it to act as load
carrying reinforcement or is used adjacent to metal in composite
construction, in considering how to protect concrete it is prudent
to include protective systems for metals and in particular steel.

The protection of steel has evolved from replacement of iron by
wrought iron and replacement of steel by coated steel. Composite
mixed coating systems are used as a result of disappointing
performance of some traditional coatings in extreme off-shore
conditions [35].

It is noteworthy that some manufacturers are claiming service
lifes of 35 years with a maintenance cycle of approximately 20
years under extreme exposure conditions [36].

A detailed review covering the painting of steel work has been
published by CIRIA in 1982 [37] and includes a comprehensive
statement on corrosive mechanisms, substrate preparation,
specification and selection of materials together with practical
problems of application. The report identifies those external
agencies that affect coating durability, e.g. chemical, mechanical
and biological.

What expectations can reasonably be demanded?

  Short term - less than 5 years
  Medium term - 5-10 years
  Long term - 10-20 years
  Very long term - greater than 20 years

The longer the life expectancy the more viable the seemingly
expensive system becomes.

The report covers metals other than steel.

The arguments about galvanising vs. multi coat paint
systems remain unresolved [38]. Work reported by Andrade et al
showed that the presence of chloride radically changed the
corrosion behaviour of galvanised steel [39]. Galvanising was
considered better than nothing but not good enough.

An interesting range of inorganic (potassium and sodium silicate) based paints are available together with organic silicates based on solvented ethyl and isopropyl silicates. Solvent evaporation produces the alcohol and that in turn evaporates leaving polysilisic acid and zinc in a manner similar to zinc rich primers. The conditions of application have to be closely defined since inorganic silicates require dry conditions but the organic silicates require some moisture.

Much exposed steel has to cope with acid industrial environments where in the presence of oxygen and moisture the sulphuric acid is constantly regenerated as a result of the corrosion it induced to begin with.

$$4H_2SO_4 + 4Fe + O_2 \rightarrow 4FeSO_4 + 4H_2O$$

$$4FeSO_4 + O_2 + 6H_2O \rightarrow 2Fe_2O_3 . H_2O + 4H_2SO_4$$

The coating functions as an isolating barrier for the steel and its gas and solution permeability are controlling factors. The Paint Research Association Report TR/2/87 ([40]) looks in some detail at sulphur dioxide transport processes through coatings and how that depends upon the chemical structure of the coatings particularly in the presence of negatively charged groups.

Embedded

Much of the concern about the maintenance of concrete stems from the consequences of reinforcement corrosion. It being estimated that 75% of some 200 DTp road bridges have sufficient chloride in them to cause corrosion ([41]) therefore any technique that reduces the likelihood of such corrosion has to be of interest to designers, specifiers and engineers at large. Practical use of coatings to prevent and/or delay rebar corrosion has occurred since the mid 1960s in the USA but similar use in the UK is quite recent with varying levels of acceptance and doubt ([41], [42], [43]). In the USA usage is estimated at 200,000 tons of coated rebar and increasing at 12% per year ([43]).

Some of the earlier attempts at rebar coating had the merit of being low cost and amenable to site application. Compositions such as Ferrotek 900 showed considerable early promise ([44]) being based on polyvinyl butyrate, special pigments and other specified polymers together with a 20% phosphating acid, applied at 20-30 micron thicknesses.

At the present time ASTM A775-81 provides the basis of conformance for rebar coatings since no British Standard exists other than in draft form. The latter being arguably more demanding on coating thickness (95% should be within 150-250 micron range with an optimum of 200 microns with 100% within the 130-300 micron range). Below 200 microns there is the chance of pinholing and above 200 microns thickness reduces the ability to accommodate bending.

The coating technique uses powder epoxy resins electrostatically sprayed* on to pre heated and textured rebar (texturing measuring approximately 70 microns).

Some 15 years ago Clifton, Beeghly and Mathley published a comprehensive and painstaking review of product selection, properties and performance testing that to this day is still relevant and unsurpassed. ([45])

This report covers 47 commercially available coatings, 36 of which were epoxy resin based (21 liquid and 15 powder). Whilst the review concerns itself about chloride penetration/diffusion it takes no regard of carbonation. Attempts are made to establish acceptance criteria. For instance, chloride transmissions are expressed as grams/day/exposed area/film thickness. A "good result" equals less and 2×10^{-6} and a "poor" one equals 3×10^{-4} Such tests can be used in a "discriminatory way" but highlights the need for agreed and consistent methods of tests allowing data comparisons to be made.

*Tribostatic or friction charging affects the powder only and not the air resulting in greater uniformity of coating.

TESTING AND ASSESSMENT OF COATINGS

The requirements for the painting of buildings is well documented ([46]) containing recommendations for wood, iron and steel, non ferrous metals as well as plaster, external rendering and to a far lesser extent, concrete. Indeed concrete appears to have little need for protection other than in acidic atmospheres. Views have changed in the last 10 years.

Testing may be by individual test or by more comprehensive assessment such as Agrément ([47]).

Browne and Robinson ([48]) have suggested that assessment should be aimed at indicating an "effective life" of a building component resulting from treatment by an anti carbonation coating. Having established the thickness to be applied for a given performance that in turn relates to time for embedded steel corrosion to occur then various coating systems can be compared on a cost effective basis.   Using a cementitious coating as the comparison the requirements shown in Table 2 were established by those authors.

Table 2. Coating effectiveness versus cost

Coating type	Minimum effective thickness (microns)	Cost/m^2 (£)
Acrylic (solvent based)	125	1.66
Acrylic (water based)	125	2.15
Cementitious coating	4300	1.0 *
Epoxy paint	125	1.49
Styrene acrylate	800	1.8
Urethane sealer	125	2.0

* Actual cost = 0.75, all others have been adjusted

If value = performance then a high performance, high cost coating
           cost
may be the better choice. So how does one define performance in
relation to concrete's ability to protect reinforcement? Brown et
al suggest:

Anti-carbonation coatings

1. Carbon dioxide diffusion resistance
2. Oxygen diffusion resistance
3. Water permeability
4. Water vapour transmission

Reinforcement coatings

1. Resistance to chloride ion migration
2. Resistance to oxygen diffusion
3. Moisture transmission
4. Rust creep behaviour
5. Alkali resistance
6. Moisture uptake

Beckett ([49]) and Ho and Lewis ([50]) also considered coating
performance in relation to concrete quality.

We have to distinguish between tests that give engineering data
but are of an arbitrary nature and those that quantify the
processes of change that occur when concrete is exposed.

Performance of coatings

In order to formulate or select a coating that will give adequate
barrier performance to ingress of carbon dioxide, chlorides or
sulphates, it is necessary to be able to measure and specify
appropriate properties. Studies to determine such limits vary
from the simulation type of test to those based on more absolute
measures of diffusion coefficients and vapour permeabilities.

The programme of Rajagopalan and Chandrasekaran is an early
example of the empirical approach ([51]) where 43 compositions
comprising 13 materials groups were coated onto concrete cubes
containing reinforcement and subject to wet/dry cycling and
accelerated (heating) tests.

Despite the crudity of the test it allowed inadequate coatings to
be identified. However, adequacy could not be established alone
by such initial screening. Of the organic coatings, chlorinated
rubber performed well and of the inorganic materials, lime:cement
mortar.

An equally empirical approach has been used more recently by the
BRE to judge effectiveness of surface coatings as carbonation
barriers ([58]). Such coatings are not considered suitable when
there is a combination of carbonation and chloride ingress.

14 generic coatings (29 in all) were tested. The most effective
materials were epoxides, polyurethanes and least effective were
cementitious coatings and clear silanes. Pigmentation was
beneficial (found by other workers) and a mean coating thickness
of 200 microns was required. Film defects such as pinholing were
readily apparent.

This type of test regime can indicate comparative trends but
little more.

The alternative is to measure more absolute properties such as
carbon dioxide, oxygen and chloride diffusion coefficients as well
as vapour permeability and relate these properties to a "concrete
factor" usable by engineers. Factors that are currently used are,

1. Equivalent concrete cover
2. Equivalent air layer thickness

For instance Robinson ([53]) assessed 71 coating systems by means of carbon dioxide diffusion resistance ( )* - the coatings being applied to a porous ceramic. Water vapour transmission was determined using a Payne cup method.

$$* = \frac{D\ air}{D\ coating}$$

where D = diffusion coefficient for transported gas or vapour

Assuming Ficks law applies then diffusion coefficients for gases and vapours can be deduced from the coating/substrate combined response. The purpose of characterising coatings in this way is to determine performance. Having measured   it in turn can be related to an equivalent air layer thickness denoted as R for CO2 where R =   x S (coating thickness in metres), or alternatively SD (for water vapour) where SD =   x S (coating thickness in metres). Even so these values have to be related to optimum values and these have been stipulated by Klopfer ([54]).

where

R = > 50 metres for an anti carbonation coating
SD <= to 4 metres for a vapour transmitting barrier

Some representative values are given in Table 3 taken from Concrete Journal ([55]).

Table 3 - Typical carbon dioxide and water vapour diffusion resistances ([55])

Coating system	Equivalent air layer Thicknesses (m)	
	R CO_2	S D
Coloured methacrylate	308	1.72
Ethylene copolymer	1100	0.15
Hypalon	1625	21.2
Chlorinated rubber	4477	15
Water dispersed epoxide	101	61.5
Epoxy (solvent dispersed)	179	12.6
Concrete ($35N/mm^2$)	3.58	0.3
Concrete ($35N/mm^2$) silicone impregnated	2.67	0.3
Thin layer cementitious render	2.40	0.2
Standard exterior emulsion paint	2.07	0.46

Alternatively since equivalent air layers are difficult to visualise in engineering terms one can quote performance limits for the diffusion coefficients at a given film thickness. For instance 200 microns D_{CO2} =< $5 \times 10^{-7} cm^2/secs$

D water = >$1.3 \times 10^{-5} cm^2/secs$

As an alternative to air layer thickness it is more meaningful to relate an equivalent thickness to concrete itself (Sc). However we need to establish acceptable equivalent concrete layer thicknesses as was done for R and S_D.

Some typical values are,

1. Water based acrylic (good performance as $CO2$ barrier and acceptable water vapour transmission

   $D = 1.03 \times 10^{-7} cm^2/sec$
   $R = 436m$
   $Sc = 109 cm$
   $D_{H2O} = 7.53 \times 10^{-5} cm^2/sec$
   $S_D = 0.54m$

By way of comparison,

2. Polymer modified cementitious coating that did not perform too well and had the following characteristics

   $D_{CO2} = 1.29 \times 10^{-5} cm^2/sec$
   $R = 9.3m$
   $Sc = 2.32cm$
   $D_{H2O} = 3.35 \times 10^{-4} cm^2/sec$
   $S_D = .36m$

3. Solvented epoxy resin based penetrating sealer

   $D_{CO2} = 1 \times 10^{-7} cm^2/sec$
   $R = 149m$
   $Sc = 37.3 cm$

It is not necessarily beneficial to have a μ value any greater than that required to give an acceptable R value for a given thickness. Other coating properties such as durability may reduce as μ increases. The actual value of μ depends upon resin binder, type of filler and whether it is fibrous, lamella or spherical ([56]).

CRITERIA FOR ADHESION

The perspective and relationship between macroscopic vs. microscopic is a question of scale. It is a most pertinent factor that has to be considered when looking at practical bonding. In the field of bonded attachments, the usual ideal rules that govern adhesion between one substrate and another apply only in part to materials such as concrete. For instance the Young-Dupre equation to predict wetting of the substrate by an adhesive relies upon defining interfacial tension between the two materials.

$$\gamma_S = \gamma_L \text{ COSINE } \theta + \gamma_{LS}$$

$\gamma_L \gamma_S$ = surface tension of liquid and solid respectively
γ_{LS} = interfacial tension
θ = angle of contact between liquid and solid

For surface tensions we may substitute the numerically equivalent surface energy (F) and this is more appropriate for solid substrates. If the contact angle θ = 0

$$F_S = F_L + F_{LS}$$

This inbalance is assisted causing natural spreading of the liquid on the surfaces if F_L is low and F_{LS} = 0 or even negative.
The resulting adhesive bond strength is enhanced the higher the work of adhesion.

$$W_A = \gamma_A + \gamma_B - \gamma_{AB} \qquad \text{(Dupre)}$$

WA is high (good bond) if γ_{AB} is low or negative
(interfacial penetration)

With porous substrates the liquid adhesive may be absorbed into
the substrate and the concept of a clearly defined interface is
no longer valid. The interfacial tension becomes negative. Such
a condition both assists wetting and the work of adhesion and
should be a condition associated with the surface of concrete.

We have to decide whether as a result of the concrete surface
being so non-ideal consequential adhesion is more dependent upon
macroscopic effects rather than microscopic effects. Between
mechanical interlocking mechanisms rather than short range
interatomic and intermolecular bonding ([57]).

To obtain good thermodynamic adhesion one must permit the
molecules of the substrate and the adhesive (coating) to approach
one another within normal bonding distances - a few or less nm.
In practice actual bond strengths are always considerably lower
than theoretical since surfaces will be irregular with crevices
and other defects causing stress concentrations and points of
weakness. To obtain good adhesion a penetrating primer may use
these defects to advantage that might overwise cause early
detachment of the bonded resin. However, they have to be
penetrated in order to be made use of.

Obtaining molecular contact between adhesive and the CSH phase may
have to accommodate the detailed chemical form that is thought to
exist near to the surface of concrete. For instance Tabor ([58])
suggests when concrete is placed against an interface such as
formwork or a mould, the presence of a solid surface restricts the
packing of the cement particles causing a density reduction nearer
to the surface (contrary to Kreijger's work). Segregation also
occurs resulting in a greater concentration of Portlandite at the
interface. This layer is intimately bound with CSH to form a
"duplex" film (Fig. 3). The Portlandite is in the form of lamella
that act as a point of weakness resulting in cleavage so that
actual strengths are around $10N/mm^2$ whereas a potential cohesive
strength of $10^3N/mm^2$ should be attainable. Similar structuring is
referred to in the later papers of Crumbie, Scrivener and Pratt
([59]) as well as Keer and Shabaan ([60]).

To take advantage of the ultra micro porosity, a coating must be
able to penetrate it and make intimate contact or wet it.

In relation to size and penetrability
isobutyl silane>siloxane>linseed oil>silicone>acrylic>epoxy

However such values can only be used comparatively.

"DUPLEX" LAYER STRUCTURE

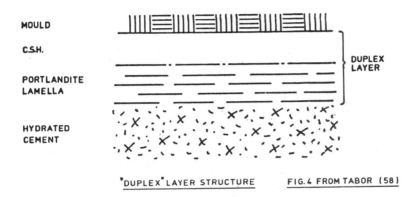

MOULD

C.S.H.

PORTLANDITE
LAMELLA

HYDRATED
CEMENT

DUPLEX
LAYER

"DUPLEX" LAYER STRUCTURE   FIG. 4 FROM TABOR (58)

We should either be concerned about using the macroroughness to effect adhesion or preconditioning the surface in some way as to create a substrate that is reproducible and amenable to good adhesion. In some ways the application of a penetrating primer/sealer to the concrete surface before the application of the coating, creates a "standard" substrate that removes the effects of ultramicro and microroughness and even nullifies the effect of surface composition and porosity (Fig. 4).

The concrete surface is not so much as it is as how we make it what we want. Priming with a water dispersible epoxide may be a better solution so long as an average droplet size does not exceed 100nm (.1 microns) so allowing penetration of the micro porosity.

SURFACE WITH MACRO, MICRO AND
ULTRA MICRO "ROUGHNESS"

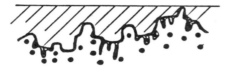

SOME MICRO AND ULTRA MICRO "ROUGHNESS"
AFTER ABSORPTION OF PENETRATING SEALER

PENETRATING PRIMER PLACED OVER SEALER
BEFORE APPLICATION OF THE ADHESIVE

FIG.5

GENERAL SPECIFICATION (SUGGESTED) FOR CONCRETE SURFACE TREATMENT

Assuming that the surface has been cleaned and all loose material
removed, the following is recommended for coating porous and rough
concrete surfaces.

1. Apply a penetrating sealer/primer

The resin should be of low molecular size/weight.   Effective
diameter less than .01 microns.

The resin/solvent mix should be of low viscosity, less than 1-5
cps

The resin/solvent mix to retain its fluidity for long enough,
greater than 60 mins and less than 180 mins to allow penetration
to pores of diameter less than or equal to 0.1 microns to a depth
of a few millimetres at pressures not greater than the standing
head of liquid sealer.   Solvents should have a medium vapour
pressure 100-300 mm of mercury.

The resin/solvent mix should have a zero contact angle and low
interfacial tension against CSH gel.   It should be chemically
constituted to have an affinity for hydroxyl groups so as to form
hydrogen bonds with the CSH gel.   Even some form of formal
chemical bonding (condensation) could have a most significant
effect.   This is the basis of the silane adhesion promotors.

$$H\delta + \qquad\qquad\qquad \delta+$$
$$- \underset{\delta-}{\overset{|}{O}} - - - - - - - - - - - - H - O^{\delta-}$$

or

$$- OH + HO - \rightarrow - O - H_2O$$

As an alternative to long range chemical interaction the inclusion
of surfactants that will assist displacement of surface moisture
and associate with the negative charges (electro static) at the
cement interface will aid adhesion.

2. Apply primer coat (bonding)

A low viscosity resin (η less than 500 cps) should be
applied and comprise the same generic resin as the sealer/primer
i.e. epoxide or acrylic. The hardeners for these materials should
also be compatible since this will minimise interfacial tension
and allow "interdiffusion" across the sealer/bond interface.

The adhesive is now applied to the bond coat before the bond coat
has fully cured.

In this way surface micro topography is minimised as well as
chemical variability and macroroughness. The surface of the
substrate is preconditioned so that the interface is monolithic
with the adhesive.

RESEARCH ITEMS AND RECOMMENDATIONS

CIRIA Report 130 ([9]) identifies some ten topics worthy of further
work. It is pertinent and a little disappointing that the
suggestions are mainly concerned with data gathering, compilation
and presentation. Attention must also be given to research of a
more basic nature such as that dealing with transport mechanisms
at the concrete/air interface, degradation processes, preferential
displacement as well as bonding methods. The "standard" test
method approach must give some ground to the more speculative,
scientific method aimed at comprehending as well as solving.

Coatings and linings do fail resulting in blistering, pinholing,
cracking and detachment. Gunther and Hilsdorf ([61]) have
calculated and measured osmotic pressures as high as 45 bar and
capilliary pressures of 2 bar resulting from the coating acting as
a semi-permeable membrane causing migration of water across the
coating into the concrete, as well as migrating from the pore
fluid within the concrete into the coating itself. This migration
results from the coating containing pockets of unreacted polymer
and/or saponification products. The stronger, higher modulus
coatings cause higher osmotic pressures. Flexible coatings
relieve this stress with reductions by two orders of magnitude.

Our knowledge of such long term transport mechanisms is poor.

Permeation of surface capilliaries driven by surface tension can cause stresses in liquid coatings. In the capilliary range 1nm-10^5 (.001 micron - 100 microns) the pressure ahead of the miniscus of a penetrating liquid can reach 1 bar (145N/mm^2) assuming the compressed air does not burst through the coatings. This back pressure can cause ultimate bond stress.

Dissolved salts within concrete can be disruptive due to expansive crystallisation but they may also be microscopic and retain dampness. Controlling dissolved salt movement is desirable.

Surface preparation is very important if coatings and linings are to adhere and high pressure water jetting shows considerable promise [62]. However how does one judge the level of surface quality achieved? There is no concrete equivalent to the Swedish standard S1S05 5900 for steel preparation prior to coating.

Ambient condition can be important during application. A high ambient temperature can cause water vapour penetration into a porous mass and should the temperature fluctuate then inter caplliary condensation may occur making bonding difficult if the applied coating is not water compatible.

Exactly what happens when concrete containing chloride also carbonates is poorly understood. Roberts [21] implies that bound chlorides can be released that may result in corrosion.

Transport mechanisms through coatings and their dependence upon coating chemical structure are very speculative when one disregards the physical effects of pigments and fillers.

Most concrete surfaces that are to be coated are already cracked or crack after the coating has been applied. The interfacial stress patterns that result together with latent and actual crack bridge capability need study. In particular the effects of dynamic load cycling.

The relevance of endosmosis, particularly at the steel/coating boundary could have serious implications for coated rebars. The rate of water transport across membranes under these conditions is allegedly increased.

Coatings may also be subject to environmental restraints and water based systems might be preferred to solvented ones. Our knowledge of coherent film formation from emulsions and dispersions and the intercoat effects of retained moisture are not known.

In making such choices we have to be sure that removing one problem does not create another.   For instance, some water dispersible resins contain reactive diluents such as glycol diglycidylether that have potential to be carconogenic.[63]

The use of carbon dioxide as a spray medium in place of solvents [64] is an interesting prospect.   It is suggested that 1kg of solvent is released for 2 litres of coating (paint) applied and in the USA alone 1.6 billion litres of coatings were applied in 1988.

Has the coating of concrete reached a point where the term "surface engineering techniques" [65] would describe better the combination of properties from the bulk of concrete to imposed surface response?

REFERENCES

1. DAVIDOVITS, J. "Ancient and Modern Concretes : What is the real difference". Concrete International December 1987 pp 23-35

2. REGOURD M., KERISEL J, DELETIE P, and HAGUENAUER B.   "Micro structure of mortars from 3 Egyptian Pyramids".   Cement and Concrete Research, Vol 18, 1988 pp 81-90

3. Current practice sheet No 41, Part 1, Concrete, December 1978 pp 31-32
    Current practice sheet No 41, Part 2, Concrete, January 1979 pp 31-33
    Current practice sheet No 41, Part 3, Concrete, February 1979 pp 33

4. CHATFIELD H W. "The Science of Surface Coatings" 1962 Ernest Benn pp 594

6. TREADAWAY K W J, ROTHWELL G W and DAVIES H. "The history and review of research into surface coatings for concrete" ICE Seminar "Improvement in the durability of reinforced concrete by additives and coatings" 26 February 1987

7. BROWNE R D and ROBINSON H L. "Surface coatings for reinforced concrete" ICE Seminar "Improvement in the durability of reinforced concrete by additives and coatings" 26 February 1987

8. "A guide to the use of waterproofing, dampproofing protective and decorative barrier systems for concrete" ACI Committee 515 1985 pp 44

9. "Protection of reinforced concrete by surface treatments" CIRIA Report TN130 1987 pp 72

10. HEWLETT P C. "Perceptions of concrete - the surface" The Third Sir Frederick Lea Memorial Lecture, 16th Annual Convention of the ICT, Coventry 18-20 April 1988

11. KREIJGER P C. "The skin of concrete, composition and properties" Materiaux et Constructions, Vol 17 No 100 July/August 1984 pp 275-283

12. DHIR R K, HEWLETT P C and CHAN Y N.   "Near surface characteristics of concrete - assessment and development of in-situ test methods" Magazine of Concrete Research, Vol 39, No 141 December 1987 pp 183-195

13. DHIR R K, HEWLETT P C and CHAN Y N. "Near surface characteristics of concrete: intrinsic permeability" Magazine of Concrete Research, Vol 41, No 147 June 1989 pp 87-97

14. HUGHES D C "Pore structure and permeability of hardened cement paste" Magazine of Concrete Research Vol 37, No 133, December 1985 pp 227-233

15. KELHAM S. "A water absorbtion test for concrete" Magazine of Concrete Research, Vol 40, No 143 June 1988 pp 106-110

16. CATHER R, FIGG J W and O'BRIEN T P. "Improvements to the Figg method for determining the air permeability of concrete" Magazine of Concrete Research Vol 36, No 129, December 1984, pp 241-245

17. BERISSI R, BONNET G and GRIMALDI G "Measurement of the porosity (communicating voids) of hydraulic concrete" Bull Liaison Lab Pont Chausse N142 March - April 1986 pp 59-67

18. DHIR R K, HEWLETT P C and CHAN Y N "Near surface characteristics of concrete: prediction of carbonation resistance" Magazine of Concrete Research Vol 41, No 148, Sept 1989 pp 137-143

19. FIGG J W "Concrete surface permeability: measurement and meaning" Chemistry and Industry, 6 November 1989 pp 714 - 719

20. "Permeability testing of site concrete - a review of methods and experience" Concrete Society Report CSTR 31, 1987 pp 95

21. ROBERTS M H " Carbonation of concrete made with dense natural aggregates" BRE Information Paper IP6/81, April 1981 pp 4

22. CURRIE R J "Carbonation depths in structural - quality contrete: an assessment of evidence from investigations of structures and from other sources", BRE publication 1986 pp 19

23. KEER J G and GARDINER G M "Crystal growth materials as surface treatments for concrete" TRRL report October 1985 pp 67

24. ROTH M "Siliconates - silicone resins - silanes - siloxanes" origination not known but thought to emanate from Wacker Chemie

25. DUMBLETON B "Chemicals keeps concrete pores dry" New Civil Engineer 26 November 1987 pp 18-19

26. SHAW J D N "Water repellent systems" Construction Repair, April 1989 pp 8-9

27. RODDER K M "Protecting buildings by impregnation with water repellant silanes" Chemische Rundschau, 22/77 pp 3-7

28. Dynamit Nobel Chemicals Aktiengesellschaft, TH Goldscmidt Ltd

29. Private communication PC Robery (Taywood Engineering) and P C Hewlett 10 December 1987

30. BRE Digest 125 "Colourless treatments for masonry" January 1971 pp 1-3

31. CLARKE B L and ASHURST J "Stone preservation experiment" BRE Publication 1972 pp 78

32. "Brethane Stone Preservative - an appraisal" BRE News No 53 Spring 1981 p5

33. ICI Data sheets for R221 and R222, May 1975

34. FIDLER J "No panacea to arrest the decay of stone" Building Construction May 1981 pp 8-9

35. BERKOVITCH I "Progress in protective coatings" Civil Engineering October 1981 pp 39-41

36. Desmodur/Desmophen (urethane) based coating systems, Bayer UK Ltd, Newbury, Berks RG13 1JA or MTM Services Ltd, Specialist Coatings Division, 24 Sedling Road, District 6, Wear Industrial Estate, Washington, Tyne & Wear, ME38 9BZ

37. HAIGH I P "Painting steel work" CIRIA Report No 93, 1982 pp 124

38. STONEMAN A "Protective coatings for steel: initial costs in perspective" Civil Engineering, Sept 1986 pp 35-37

39. GONZALEZ J A, VASQUEZ A J, JAUREGUI G and ANDRADE C "Effect of four coating structures as corrosion kinetic of galvanised reinforcement in concrete" Materieux et Constructions Vol 17, No 102 1984, pp 409-414

40. SHERWOOD A F "The protection of steelwork against atmospheric pollution" Paint Research Association Report TR/2/87 March 1987 pp 24

41. "Chloride attack may hit chunnel linings" New Civil Engineer, 22 November 1989 p 5

42. "Epoxy coatings earns share of European market" Concrete Journal November 1988, pp 14-15

43. RIDOUT G "Eroding confidence" Building Magazine 22nd July 1989 pp 62-63

44. CORK H A "Coating treatment for reinforcing steel" Concrete Journal January 1977, pp 31-33, also refer Metprep Industrial Products Ltd, Denbigh Road, Bletchley, Milton Keynes MK1 1PB

45. CLIFTON J R, BEEGHLY H F and MATHELY R J "Non-metallic coatings for concrete reinforcing bars" Building Science Series 65, National Bureau of Standards August 1975 pp 34

46. BS 6150: 1982 (formerly CP 231) code of practice for painting of buildings"

47. "Paints and stains performance assessment" RIBA Journal June 1989 pp 79-97

48. BROWNE R D and ROBINSON H L "Surface coatings for reifnorced concrete" ICE Serminar 26th February 1987 "Improvement in the durability of reinforced concrete by additives and coatings"

49. BECKETT D " Specifying coatings for reinforced concrete" as above in 51

50. HO DWS and LEWIS R K "The water sorbtivity of concretes: the influence of constituents under continuous curing" Durability of Building Materials, Vol 4 1987 pp 241-242

51. RAJAGOPALAN K S and CHANDRASEKARAN S, Indian Concrete Journal Sept 1970 pp 411-416

52. DAVIES H and ROTHWELL G W "The effectiveness of surface coatings in reducing carbonation of reinforced concrete" BRE Information Paper 7/89 May 1989

53. ROBINSON H L "Evaluation coatings as carbonation barriers" proceedings of the 2nd International Colloquim, Materials Science & Restoration, Technische Akademie Esslingen Sept 2-4 1986

54. KLOPFER H "The carbonation of external concrete and how to combat it" Bautenschutz and Bausanierung 1 no 3 1978, pp 86-97

55. OVENS A "Repair and protection of reinforced concrete in high rise buildings" Part II, Concrete, June 1984, pp 44-45 (refer also Part I, May issue)

56. "The protection and restoration of concrete" Harlow Chemical Co Ltd Publication author K Sellars?

57. SASSE H R and FIEBRICH M "Bonding of polymer materials to concrete" Materiaux et Construction Vol 16 No 94 1983 pp 293-301

58. TABOR D "Principles of adhesion - bonding in cement and concrete" Conference proceedings "Adhesion problems in the recycling of concrete" Plenum Press 1981, pp 63-87 Editor P C Kreijger

59. CRUMBIE A K, SCRIVENER K L and PRATT P L "Microstructural gradients in the interfacial zone of concrete" Institute of Metals seminar "The micro structure of cement and concrete" University of Oxford, 20-21 September 1988

60. KEER J G and SHABAAN M "Observations on the microstructural and hydration characteristics of polymewr modified cement mortars" - as in 66

61. GUNTER N and HILSDORF H K "Stresses due to chemical actions in polymer coatings on a concrete substrate" ISAP 1986 "Adhesion between polymers and concrete" Sept 16-19, Aix-en-Provence, pp 8-21 Chapman and Hall

62. DONALDSON M "Good surface preparation - the key to effective concrete repairs" Concrete Journal Oct 1987, pp 24-27

63. Ciba-Geigy Plastics data sheet No R 101B relating to Araldite PY340-1 June 1986

64. BEARD J "Carbon dioxide paints a clearer picture" New Scientist 23 Sept 1989 p 35

SURFACE TREATMENTS FOR THE PROTECTION OF CONCRETE

M Leeming

Ove Arup and Partners,

United Kingdom

ABSTRACT. There are a large number of surface treatments on the market that claim a variety of properties. Their behaviour in protection against deterioration of concrete due to corrosion of the reinforcement is not well established and there are few standard test methods to assess their likely performance. Generic classifications do not help in selection of materials. More research is required to provide for certification of materials, 'benchmarks' against performance in practice and for acceptance tests after application. Performance in service generally indicates the benefits of treating concrete surfaces but few well documented case histories are available. We need to know how good a treatment needs to be and how long it will afford that protection.

The paper describes the state of the art review and definition of research needs carried out by Arup Research and Development on behalf of the Construction Industry Research and Information Association (CIRIA).

Keywords: Concrete, Surface coatings, Non-destructive testing, Carbonation, Chlorides, Deterioration.

Mr Mike Leeming is a Chartered Civil Engineer, has worked for Local Authorities and Consultants on the design and construction of motorway and other major prestressed concrete bridges in the UK and abroad. He then joined CIRIA as Technical Manager of the Concrete in the Oceans research programme directing research into marine concrete for North Sea Oil structures. In 1984 he moved to Arup Research & Development to manage and carry out research and investigatory projects relating to the diagnosis and cure of deteriorated concrete, surface coatings on concrete, repairs and non-destructive testing.

INTRODUCTION

Cave paintings are evidence that man has wanted to decorate his
environment since earliest times. With modern organic materials,
surface coatings can be applied to concrete to achieve more than
just decorative effect. Protection can be provided against the
ingress of injurious substances which cause reinforced concrete to
deteriorate. There are a large number of exterior paints now
available which are aggressively marketed in terms as glowing as the
hues available but giving little factual information on which to
base a choice. The Construction Industry Research and Information
Association has produced guidance on this Subject[1,2]. This paper
does not try to provide a `Which' for these materials but seeks to
give the basis on which to make an informed choice.

DETERIORATING INFLUENCES

All chemical/physical mechanisms of deterioration of concrete
involve water in either transporting injurious substances,
supporting the chemical reactions or providing expansive forces[3].
Such deterioration mechanisms are sulphate attack, freeze/thaw
damage, alkali silica reaction, salt crystallization and salt
scaling.

Corrosion of the reinforcement is also highly influenced by moisture
(see Figure 1) limiting gaseous diffusion such as oxygen and carbon
dioxides, transporting chlorides to the surface of the steel or
providing the electrolyte for the corrosion cell. Surface
treatments have an important role in limiting water penetration into
concrete.

While a coating on the surface of concrete can inhibit the ingress
of injurious substances into concrete, it cannot generally help with
problems within the concrete itself such as alkali silica reaction
and calcium chloride contaminated concrete. However, the coating
may influence the moisture content of the concrete, reducing the
range of wetting and drying, but no research has yet been published
which determines whether the moisture content is stabilised or
whether it is increased or decreased.

However, there is no substitute for properly designed and
constructed concrete where the water/cement ratio and cement content
suits the exposure, where the cover is correctly specified and
maintained during placing of the concrete and proper curing is
carried out. Surface coatings are very thin organic films on an
inorganic substrate and as such have relatively short lives in
engineering terms. Modern specialised coatings can be expected to
last longer than normal household paints, for at least ten or
possible fifteen years. The deterioration of protective properties
with time is not well researched. The most likely cause of
deterioration is mechanical breakdown due to embrittlement and loss
of adhesion.

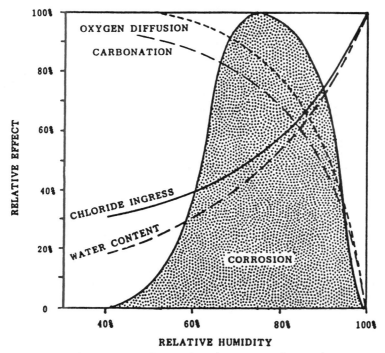

Figure 1: Effect of Moisture on Corrosion

A wide range of surface coatings are available and can be chosen to provide progressively increasing protection. All coatings should at least provide resistance to water in its liquid state, while more impermeable coatings also resist the ingress of carbon dioxide and the most impermeable coatings additionally provide a barrier to water vapour. It is generally assumed that chloride ions can only be transported into concrete, or diffuse through concrete, in liquid phase water. A hydrophobic coating such as a silane will therefore generally hinder the ingress of chlorides but may still allow the concrete to carbonate. It has been argued that this type of coating, by keeping the concrete in a less saturated condition, will allow it to carbonate faster. However, corrosion is less likely in the drier concrete. Water vapour permeance (breathability) is generally considered to be a desirable quality in a coating as it allows moisture in the concrete to escape. Complete impermeability to water vapour can lead to failure of the coating due to loss of adhesion if water vapour pressures can build up behind the coating.

WHY SURFACE TREATMENTS FAIL

The causes of failure of surface coatings are many. Loss of
adhesion can be due to poor surface preparation. A coating will
soon fail if it is applied to a dusty friable surface or one which
is contaminated with oil or dirt. Capillary or osmotic forces can
be set up behind the coating to cause blistering. With breathable
coatings, evaporating water can leave salts behind at the interface
with the concrete causing loss of adhesion. Freezing of water
trapped behind the coating can cause similar problems. The surface
of the concrete receives the full force of temperature variations
and this is particularly severe for a surface coating which probably
has a different temperature coefficient to concrete. Concrete
cracks under load, due to shrinkage and thermal movements, and
therefore elasticity is a desirable property allowing the coating to
bridge the cracks.

PROTECTIVE ACTION

Surface coatings range from materials that are intended to penetrate
the pores of the concrete, through sealers to coatings of increasing
thickness until they are more properly termed renderings. A wide
range of organic polymers are used as sealers and coatings while the
most widely used penetrant materials tend to be siliceous which line
the pores of concrete forming silicone resins providing protection
through their water repellent properties. There is another class of
penetrant materials whose action is to block the pores of the
concrete such as epoxy resin impregnation or crystal growth
materials. The former require expensive provisions to get adequate
penetration and the latter have yet to be proved to be effective.
Penetrant materials provided adequate penetration has been achieved,
are less vulnerable to weathering as they lie below the surface of
the concrete.

Sealers, coatings and renderings protect by providing a barrier film
on the surface of the concrete of various thicknesses. Like all
paints they have up to four main constituents, the base film former
or binder, the inert fillers and pigments, the liquid solvent or
dispersant and finally additives to provide various properties.
Coatings dry or cure in many ways, the two major methods being loss
of solvent-dispersant and chemical reaction as a result of a
catalyst or hardener. The large number of possible formulations
means that it is not possible to classify materials into generic
groups as, for instance, one epoxy can behave very differently from
another. The type of solvent or dispersant, and whether the coating
is pigmented or not seems to be much more important in the
performance of these materials than the type of polymer[4]. A re-
analysis of this work has shown that most solvented unpigmented
materials meet the criteria for breathable membranes but not for

carbon dioxide inhibitors, see Figure 2. The pigmented aqueous
dispersions (emulsions), however, met both criteria. Pigmented
solvented materials had a wide range of properties from almost
impermeable to those that provided little protection.

Most deteriorating influences on concrete occur more severely under
alternate wetting and drying[3]. On wetting, capillary absorption
has a dominant effect before other forms of "Permeability" such as
diffusion take effect. On drying, vapour diffusion is the dominant
factor evaporating from a reducing reservoir of moisture so that it
is not steady state.

CRITERIA FOR SELECTION

Few standard tests exist that adequately test the properties of
surface coatings on concrete. In Germany[6] a lot of work has been
done on measuring the carbon dioxide diffusion rate through coatings
and the results have been expressed as an equivalent thickness of
air; anything greater than 50m is considered to be a carbon dioxide
retardant. This method of measurement of protective properties is
not considered to be very scientific as air is a very changeable
medium. The author prefers to use flux, grams per square metre per
day, as it is much more meaningful to the average engineer. It
requires standard conditions of temperature and humidity and is a
property of a specified thickness of the material. It is important
that properties of surface coatings be quoted for a complete system
and not given as fundamental properties of unit thickness of the
material. A surface treatment system comprises various layers of
levelling coats, primers, undercoats and topcoats, all with
different properties and all contributing to the protective
properties. Coats can only be applied in certain thicknesses
depending on the type of paint. On the rough surface of concrete
even coatings of the specified thickness are hard to achieve and
measuring and monitoring during application is not reliable. The
equivalent resistance of greater than 50m of air is a flux of less
than 5 grams per square metre per day. There are a number of other
units used to present the protective properties of surface coatings
which are correct from a scientific point of view but have little
meaning to the average engineer. These obscure the problem and are
best not used. See appendix to CIRIA Technical Report 130[1].

Tests for protective properties of coatings are often carried out on
a paper substrate with unnatural relative humidities and therefore
the results do not relate directly to their performance on concrete.
Comparison of various research results on two particular materials,
see Table 1, show a wide variation and indicate that method of test
and substrate are important. Only when standardised tests have been
agreed can reliable choices be made between different materials.

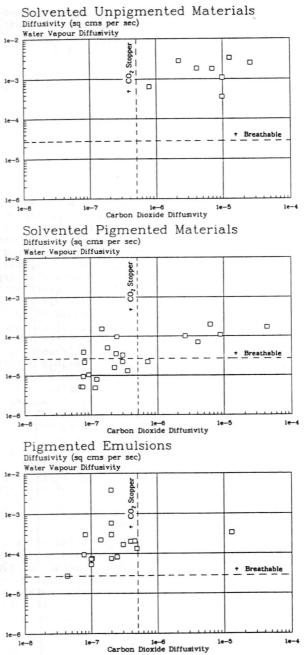

Figure 2: Relative Performance of Types of Coatings

Table 1: Comparison of Research Results

	Substrate	Coating A Thickness Microns	Coating A Resistance MNs/g	Coating B Thickness Microns	Coating B Resistance MNs/g
To Carbon Dioxide					
PRA	Free film & paper	180	10,700	210	13,500
BRE	Mortar Block	210	730	210	600
Klopfer	Free film & paper	140	1,194	160	4,218
Taywood	Ceramic	400	2,824	300	1,644
To Water Vapour					
PRA	Free film & paper	180	50.0	210	5.7
Klopfer	Free film & paper	140	9.4	160	2.4
Taywood	Ceramic	400	25.2	300	5.4

Where possible these tests should be carried out on a concrete substrate to reflect as closely as possible performance in service. The criteria for effective protective properties are now widely accepted as given in Table 2.

Table 2: Protection Criteria for Surface Coatings on Concrete

	Equivalent Air thickness	Resistance[16] MNs/g	Flux g/m²/day
Carbon dioxide ingress	> 50m	>190	< 5
Water vapour transmission	< 2m	< 11	> 20

The first of these criteria has been calculated by Englfried[6] to give a sufficiently low rate of carbonation to give an adequate life for a normal structure with typical cover depths. It may not be adequate where the carbonation zone has nearly reached the level of the reinforcement. Vapour transmission rates through various building materials for walls and the like have been used as a criteria for `breathability' of concrete. These criteria are only guide values and have not been set as a result of intensive research.

Table 3: Typical values for the ISA Test

Concrete	ISAT values at time t (mL/m²/s)				Absorption Rate
Absorption	10mins	30mins	1hr	2hr	g/m²/√hr
low	< 0.25	< 0.17	< 0.10	< 0.07	750
average	0.50-0.25	0.35-0.17	0.20-0.10	0.15-0.07	1500-750
high	> 0.50	> 0.35	> 0.20	> 0.15	1500

The rate of liquid water transmission through a coating has not up till now been given a criterion. However, most reasonably effective coatings should have a rate of water absorption of less than 35 grams per square metre per root hour which compares with a concrete with low absorption, see Table 3, of 750 grams per square meter per root hour.

Total immersion of treated and untreated cube specimens in salt solutions[7,8] has shown that untreated concrete can become nearly saturated within half a day while surface treatments can lengthen this period to 21 days or longer. As the rate of capillary absorption reduces exponentially with time the greater amount of water take up occurs within the first few hours during the period of a rain storm.

METHODS OF TESTING

The simplest method of measuring capillary flow or absorption is to take a sample of concrete, seal the sides, place it face down on capillary matting or in a shallow depth of water and to weigh it at predetermined intervals. This is the method adopted by DIN 52617[9]. The results are expressed in a graph as water gain plotted against the square root of time which usually gives a straight line, the slope of which is the water absorption coefficient. This method was used to compare the relative performance of surface coatings on concrete in inhibiting the ingress of water containing deicing salts for the Transport and Road Research Laboratory. In this instance a 15% salt solution was used and all faces of the specimen except the test surface were sealed. The test ran for 21 days and measurable gains of weight were recorded for all the coated specimens. The uncoated plain concrete control specimens, however, rapidly gained weight to near saturation within a few hours. This method is only suitable for use in a laboratory but has the advantage that the specimens can be preconditioned relative to a known humidity before the test. Other similar tests[10] measure the depth of penetration of water against time.

While the above test methods measure the amount of water that has passed through the test surface, the Initial Surface Absorption Test (ISAT) (BS 1881: Part 5: 1970)[11] monitors the volume of water entering the test face. A narrow capillary tube is used to measure the volume of water absorbed by a test area of 5000mm² at a head of 200mm. This test, however, gives results in units of ml/m²/s at fixed intervals of 10 mins, 30 mins, 1 hour and 2 hours. These four figures are not very meaningful except in a relative sense for comparison with other results on concrete. However, on the assumption that water absorption is directly related to root time these results can be expressed as a water absorption coefficient in grams per square metre per root hour by calculating the slope of the line at each of the fixed intervals. The result can then be expressed as one figure by taking the average of the four individual results.

Typical values for concrete taken from the Concrete Society Technical Report No 31[12] are given in Table 3.

The tests on the much less absorbent surface coatings have been carried out by a modification of the test specified in BS1881: Part 5: 1970[11] which is effectively a Cumulative Initial Surface Absorption Test. The standard set up is used until the point when the valve to the reservoir is closed after 10 minutes. From that point on the valve to the reservoir remains closed and the total movement of the meniscus in the capillary tube is recorded from the start point to the various recording times. As above slopes of the lines of the relationship of absorption to root time can be calculated for the periods 10-30 mins, 30-60 mins and 60-120 mins. Again, an average of the three results provides one figure for comparison with the standard method.

Tests have been carried out using the ISAT apparatus on the same specimens that were used for the capillary test and reasonable correlation of the results were obtained. Results so far have shown some variability but the averaged value is well within the accuracy of the method. However, the capillary test is basically a uniaxial flow test while the ISAT method includes radial components. As a result the simple root time relationship may be less valid. The ISAT method can be used both in the laboratory and on site although it is not widely used as the equipment is cumbersome, can take up to 2 hours to carry out a test and achieving a good seal with the concrete surface can be difficult. However, the principle of the test is sound and some of the above problems can be overcome. More work is required in improving the test to make it more convenient for site use.

The initial surface absorption test has changed little since Levitt[13] carried out the initial work on it in 1969. Small holes in concrete for chloride determinations, measurement of carbonation depth and for the Figg test are accepted as non destructive. These holes can be utilised with an expanding bolt to clamp the ISAT equipment to the face of the concrete. The equipment itself can be made more compact and manageable for site use. The method used for electronic measurement which has been developed for use in the Figg water permeability equipment is equally applicable to the ISATest. The main problem lies in achieving a quick and reliable seal around the test surface. The original method was to use plasticine mixed with vaseline. This is messy and not always reliable. Experiments have been made with silicone bath sealant but this takes an hour or so to set and again is not 100% reliable. Various proprietary sealing strips such as Arboseal, 'E' shaped rubber draught excluder strip and foamed rubber have been used with some success.

There are other versions of the initial absorption test some of which apply various pressure heads to the water which are reviewed in the Concrete Society Technical Report No 31[12]. As capillary suction forces of up to 20 metre head of water have been measured in concrete[14], the application of small heads to the water on the surface are not significant. A simple test that has simple equipment and is easy to do has an advantage.

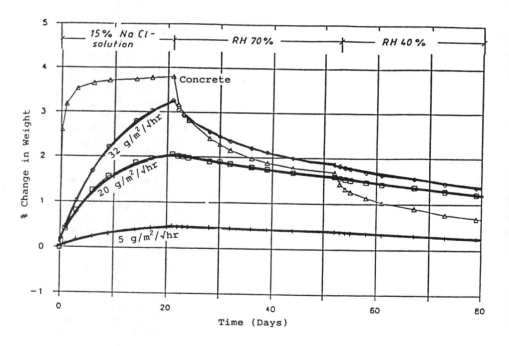

Figure 3: Performance of Typical Coatings under Immersion

Tests have been done in the USA[7] and Finland[8] where 100mm concrete cubes have been coated and submerged in water or salt solutions for up to 21 days followed by a period of drying. See Figure 3 for results of typical coatings[8] which show the much better performance of coatings (less than 35g/m²/√hr) compared with the reference concrete which had an absorption rate of 370 g/m²/√hr. This Figure also shows the slower rate of drying compared with the rate of absorption. However, these tests have also shown some surface treatment materials to have no advantage over bare concrete but it must be more difficult to get a defect free coating on six sides of a cube than on a single face of a slab. Furthermore, complete submergence is not typical of real exposure of an atmospheric surface and capillary absorption in all directions into a cube is a much more complex mechanism than that into a single surface.

Breathability is an equally difficult criterion to measure. This parameter is of importance in understanding the mechanisms of water transport into and out of concrete in the environment but there are few tests to measure it. Vapour diffusion is a mechanism obeying Fick's first law. A water vapour diffusion test is described in the Concrete Society Technical Report No 31[12], section 4.6, the Dry Cup test. Another variant of the same test is the Wet Cup test which may be more relevant to actual conditions in that the relative humidity difference is from 100% RH to say 50% RH as opposed to the Dry Cup test where the difference is from say 50% RH to 0% RH. The test can only be carried out in the laboratory on specially prepared specimens with the paint film laid down on an artificial substrate. Again the results suffer from the inability of artificial substrates to represent concrete.

Determining the rate at which coated concrete dries out after the capillary absorption test is a better method more akin to what happens in practice. In the tests described earlier on surface coatings, the specimens were taken off the capillary matting and placed face up in a chamber at 50% RH. Weight loss over the next 21 days was recorded. This loss was found to be up to 5 times slower than weight gain for some specimens. However, the moisture content of the specimens at the start of the test differs from coating to coating depending on the amount of absorption in the first phase of the test. Results are not therefore truly comparable. It is clear from this and other research[6] that the rate of absorption is much higher than the rate of drying. While little research has been done to determine to what extent breathability is desirable.

Schonbin and Hilsdorf[15] describes a similar test to the Figg Air Permeability Test which is applied to the surface of the concrete. However, problems of getting an effective seal to concrete surface are more acute than Figg and ISATests. Preliminary trials of this method have been carried out which indicate that it may be a useful tool for both laboratory and on site to compare the air 'permeability' of treated and untreated concrete surfaces.

Table 4: Protective Performance of Coatings in Service

Applied to New Concrete On	Age at Test	Carbonation Depth	
		Plain Concrete	Treated Concrete
Flats	15 yrs	8 to 15mm	< 1mm
Underground Station	8½ yrs	11 to 15mm	2.3 to 2.7mm

Comparing the relative performance of adjacent treated and untreated concrete after a number of years in-situ in a real situation is a relatively simple method of determining the protective merits of surface treatments. However, there are very few case histories of such research. Those that have been carried out do clearly demonstrate the advantage of surface treatments[1]. See table 4.

APPLICATION

The success of a protective treatment is very dependent on conditions during application. Good adhesion can only be achieved by careful preparation of the surface by grit or water blasting. The moisture content of the concrete is an important factor, most materials require the concrete to be dry while some require it to be moist. Coatings need protection from wind blown dust and many need protection from the rain for several days after application. Few coatings can be applied successfully in temperatures below 5°C, the upper limit being 35°C. Relative humidity should not exceed 85%. Coatings should not be applied under conditions of heavy condensation, the temperature of the concrete should be 3°C above the dew point. The thickness of a coating can not be easily measured and therefore it is necessary to monitor coverage rates to determine whether the required thickness has been achieved. Careful choice of the colours of successive coats and the use of trial panels for comparative purposes is the best way of achieving an even coating of the specified thickness.

Coatings can be expected to last at least 10 years and there are examples of coating that are up to 30 years old which still appear to be performing satisfactorily. Recent un-published research has, however, shown that the performance of some coatings can deteriorate after a period under artificial weathering. There is insufficient research and experience on the long term performance of surface coatings and therefore the application of a surface coating must be considered as an obligation to future maintenance.

CONCLUSIONS

There are no codes or standards at present other than that produced by The British Board of Agrément – MOAT.33: 1986[16] that can provide any guidance in the selection of suitable surface coatings for concrete. The claims made by suppliers for their products need to be sifted carefully and quoted test results need to be closely examined for their relevance to performance on concrete.

The lack of satisfactory methods for monitoring application parameters suggest the use of reliable applicators who are experienced in applying the chosen system. Good surface preparation is essential before application to avoid early deterioration of the paint film

While surface treatment materials are expensive in themselves they are very thin so a little material goes a long way. However, getting access to the surface of the concrete, with scaffolding for instance, is a large proportion of the cost of any surface treatment. Surface preparation and the labour of application are required by all surface treatments, hence small differences in material costs represent a much smaller percentage of the overall cost. It is cost effective in the long run to concentrate on good workmanship and materials that have a good track record.

The reasons why the concrete is deteriorating needs to be carefully examined and an appropriate surface treatment chosen to deal with the problem. A surface treatment can protect only from external deteriorating influences on concrete. If the concrete is already heavily contaminated there is little a surface coating can do to help.

Surface treatments can be both decorative and protective and research has shown that most coatings of adequate thickness are a considerable improvement on bare concrete.

REFERENCES

1   Protection of reinforced concrete by surface treatments. CIRIA Technical Note 130. 1987.

2   LEEMING M B. CIRIA review and the future for surface treatments. Conference on Permeability of Concrete and its Control. Concrete Society, London, December 1985.

3   LEEMING M B. Keeping water out of the concrete – the key to durability. International Conference on Bridge Management. Guildford, March 1990.

4   ROBINSON H L. Evaluation of coatings as carbonation barriers. 2nd International Colloquium on Materials Science and Restoration. Esslingen, September 1986.

5   LEEMING M B. Concrete in the Oceans Programme - Co-ordinating
     Report on the Whole Programme. Offshore Technology Report
     OTH 87 248 HMSO. 72, 1989.

6   ENGELFRIED R. Preventative protection by low-permeability
     coatings. Conference on Permeability of Concrete and its
     Control. Concrete Society, London, December 1985.

7   PFEIFER D W and SCALI M J. Concrete sealers for protection of
     bridge structures. National Co-operative Highway Research
     Program. Washington DC, Report Number 244, December 1981.

8   A comparative study on coating agents. Technical Research
     Centre of Finland. Report Number 258/0/87/BET, 1988.

9   DIN 52617 (1984). Determination of the water-absorption
     coefficient of building materials. Deutsches Institute fur
     Normung e.v. 1984.

10   BAMFORTH P B, POCOCK D C and ROBERY P C. The sorptivity of
     concrete. Our world in Concrete and Structures Conference.
     Singapore, 27-28/8/85, 1985.

11   BS 1881: Part 5: (1970). Initial Surface Absorption Test
     (ISAT). Methods of Testing Concrete: Part 5: Methods of
     testing hardened concrete for other than strength. Clause 6:
     Test for determining the initial surface absorption of
     concrete. BSI. (due to be revised and issued as BS 1881:
     Part 208 in due course).

12   Concrete Society Technical Report No 31. Permeability Testing
     of Site Concrete: A Review of Methods and Experience, ISBN
     0946691 21.5, 95, 1988.

13   LEVITT M. Non-destructive testing of concrete by the initial
     surface absorption method. Proc. Symp. on Non-destructive
     Testing of Concrete and Timber. London 11-12 June 1969. Inst
     Civ Eng 23-28.

14   GUNTER M and HILSDORF H K. Stresses due to physical and
     chemical actions in polymer coatings on a concrete substrate.
     ISAP 86. Conference on Adhesion between polymers and
     concrete. RILEM Aix-en Provence. ED H R Sasse. Chapman and
     Hall, 1986.

15   SCHONBIN K and HILSDORF H. Evaluation of the effectiveness of
     curing of concrete structures. The Katharine and Bryant
     Mather International Conference of Concrete Durability.
     American Concrete Institute ACI SP100-14. 1.207, 1987

16   The assessment of masonry coatings. MOAT 33:1986. British
     Board of Agrément.

SURFACE COATINGS TO PRESERVE CONCRETE DURABILITY

R N Swamy

University of Sheffield,

United Kingdom

S Tanikawa

Toagosei Chemical Industry Co. Ltd.,

Japan

ABSTRACT. Surface coatings have a significant role to play in protecting and preserving new and existing structures, and particularly those that are damaged and detriorating by controlling the ingress of aggressive agents into concrete. In evaluating the performance characteristics of such coatings, it is shown that certain basic engineering requirements such as crack bridging ability, elasticity, strain capacity, adhesion strength and fatigue resistance are also essential for the successful protection of concrete. A highly elastic acrylic rubber type coating (Aron Wall) with an overall thickness of about 1000μm is reported. The basic chemistry of the coating is discussed, and it is shown that the coating satisfies many of the engineering requirements which make surface coatings effective. Test results are also presented on the coating's weathering resistance and diffusion resistance to water, air, chloride ions and water vapour. All the data presented indicate excellent performance characteristics of AW coating in resisting the intrusion into concrete of a wide range of aggressive agents.

Keywords: Concrete, Durability, Surface Coatings, Crack Bridging Ability, Fatigue Resistance, Weathering Resistance, Diffusion Resistance, Chloride Ions, Performance Characteristics.

R. Narayan Swamy is a member of the Material and Structural Integrity Unit at the University of Sheffield. His main research interests are in concrete materials and concrete structures. He is the recipient of the George Stephenson Gold Medal from the Institution of Civil Engineers and the Henry Adams Diploma from the Institution of Structural Engineers. He is the Editor of Cement and Concrete Composites and also Editor of the book series on Concrete Technology and Design.

Shin Tanikawa is a research engineer at the research laboratories of Toogosei Chemical Industry Co. Ltd., Nagoya, Japan. He is a member of the American and Japan Concrete Institutes.

INTRODUCTION

Concrete can be the most durable construction material, and provide a safe and protective alkaline environment to steel reinforcement embedded in it, provided care and control are exercised in the selection of its constituent materials, and in fabricating, placing, compacting and curing the material. There is extensive evidence that concrete structures with excellent durability and long service life can be designed and constructed, in almost any part of the world.

Material and structural degradation have, nevertheless, become increasingly common, and it is now recognised that concrete can deteriorate as a material and lead to corrosion of steel embedded in it for a variety of reasons. When material and structural distress occur due to deficiencies in the constituent materials, or due to internal or external sources of deterioration, then the ambient environment has a profoundly significant influence in enhancing and accelerating the deterioration process, particularly when the structure is exposed to corrosive or even just unfriendly environments.

The three major factors influencing the service and design life of concrete structures are cracking, cover and the quality of concrete. These three parameters have an interactive and inter-dependent, almost synergistic, effect in controlling the intrusion into concrete of external aggressive agencies such as air, water and chloride ions. Carbonation and chloride ions are now recognised to be the two major agents affecting the performance and durability characteristics of concrete structures, and cover, the most critical parameter in determining the electrochemical stability of steel in concrete (1-3).

However, the degradation process of concrete is not the result of one single factor or process. Many factors such as CO_2 induced carbonation, chloride ions and freezing/thawing generate not only their own individual deterioration processes, but they also exercise an overall synergistic effect. On top of all these, is the further synergistic effect of atmospheric pollutants, the effects of which are not readily known or appreciated. Atmospheric gases of sulphur and nitrogen can reduce pH values to as low as 2.5, and create expansive compounds on reacting with cement hydration products, and cause disintegration of concrete; 1 ppm of SO_2, for example, can accelerate the progress of carbonation front due to CO_2 alone (4). The cumulative synergistic effect of atmospheric pollutants and other deteriorating agencies is probably one of the main reasons for the unexpected and premature deterioration of concrete structures observed often at a much earlier age than that predicted by service life models.

Environment is thus probably the one single aggressive factor that can create an alarming degree of deterioration in a short time, and critically decide the stability and service life of concrete structures. Experience shows that environmental conditions can unpredictably accelerate the process of damage of structures subject to deterioration processes such as those arising from alkali aggregate reactions, freezing and thawing, carbonation, and chloride intrusion. The loss of strength of high alumina cement concrete is one typical example of the insidious and destructive effect of ambient environment on the performance characteristics of concrete structures arising from deficiencies of one of the constituent materials of the material (5).

SURFACE COATINGS

Of all the methods of protecting and preserving the existing stock of structures into serviceable and usable elements, the use of surface coatings has the unique advantage that it can be applied to protect existing structures, new construction or as part of a programme of refurbishment of damaged and deteriorating structures. Surface or barrier coatings have also a long history of application to concrete to protect it and the embedded steel from various aggressive agents and external environments. A wide range of vapour barriers, vapour permeable coatings, and surface penetrating sealers have also been evaluated in published literature and available in the market (6-11).

Basic Requirements

One of the main difficulties with barrier coatings is that, with the wide range of such coatings available on the market, it becomes extremely difficult to choose the right type of coating since coatings of similar generic types are known to have considerably different diffusion characteristics (12). The nature and severity of exposure, i.e. environmental conditions, as pointed out earlier, is the major factor determining the performance characteristics of barrier coatings. Vapour barrier coatings, for example, do not always exclude passage of oxygen; on the other hand, coatings based on silane/siloxane are known to have poor resistance to carbon dioxide penetration, although they may have good resistance to water absorption and chloride penetration (11, 12). Similarly, a coating with variable water diffusion resistance may be required for different exposure conditions. Guides to the use of surface coatings should thus be used with considerable engineering judgement and knowledge of the short and long term performance characteristics of the coating, and the exposure regimes of the structure to which the coating is applied (8, 9).

Selection Criteria

The basic technical requirements of barrier coatings can be summarised as follows:

Chemical resistance
  - Resistance to salt (salt spray, dry/wet salt spray cycle,
    immersion in sea water)
  - Acid resistance
Diffusion resistance
  - Water, air (oxygen, carbon dioxide), chloride ions, water
    vapour
  - Air diffusion at high pressure
Weathering resistance
  - Outdoor exposure (ultra violet light, ozone, variable
    temp/humidity)
  - Heat resistance
  - Resistance to water permeability
Resistance to expansive forces
  - Freezing and thawing
  - Alkali silica reaction

There is considerable published information on the performance
characteristics of a wide range of surface coatings detailing their
behaviour under conditions described above (7, 10–13).

However, there are two other major sets of considerations, which
have not always been readily recognised in literature. Almost all
concrete structures are in a cracked state – the cracks arising
from inherent hydration characteristics of the cement such as plastic
shrinkage, drying shrinkage, heat of hydration and thermal/moisture
movements, as well as stress-induced cracks. Cracking may also arise
from deteriorating processes such as alkali aggregate reactions
and freezing/thawing. If surface coatings are to be effective in
existing as well as new construction, then several engineering
requirements need also to be satisfied. These requirements are:

Crack bridging ability
Adequate elastic/fatigue resistance (to resist crack
  opening/closing)
Strain capacity – stress-strain behaviour
Adhesion strength (film continuity)
Abrasion resistance
Thermal compatibility

The coatings should also have the following economic requirements:

Ease of application
Long service life
Pleasant appearance

It will be readily seen that barrier coatings have to satisfy a
number of stringent performance specifications, and although some
generic types will fulfil most of the demands, the big variations
within each type makes critical evaluations very essential (14).

This paper reports the development of a highly elastic acrylic rubber type coating. The Aron Wall (AW) coating is made by chemically reacting acrylic ester, the main raw material, with other components to form an acrylic rubber emulsion in which a crosslinking reagent is further added to enhance performance characteristics so that the dry film coats have excellent resistance to aggressive agents, high durability, elasticity and flexibility even at low temperatures . The coating has been systematically evaluated over a number of years for its performance characteristics against a variety of aggressive environments. This paper presents data on the basic chemistry of the coating and its engineering properties, weathering resistance and diffusion characteristics. Resistance to carbonation and intrusion of chloride ions, acid resistance, resistance to expansive forces and other relevant properties of the coating will be reported elsewhere.

CHEMISTRY OF ARON WALL

The acrylic rubber coating consists of a primer, base coat and top coat with an overall thickness of about 1000μm. The acrylic rubber contains 54% of acrylic polymer by weight, the rest being made up of inorganic filler and pigment. The main component of the base coat is 2- Ethyl hexyl acrylate with the formula

$$\left[\begin{array}{c} \overset{\displaystyle H}{\underset{\displaystyle |}{C}} - \overset{\displaystyle H}{\underset{\displaystyle C}{\underset{\displaystyle |}{}}} - \\ \underset{\displaystyle \underset{O}{\|}}{C} - O - CH_2 - CH_2 - CH_2 - CH_2 - CH_2 - CH_3 \\ \hspace{4cm} C_2H_5 \end{array} \right]_n$$

Fig.1 shows the schematic structure of the base coating. The uniqueness of the base coating is that it mainly uses a soft monomer which has a low glass transition temperature (-65°C) which gives it remarkable elastic properties at low temperatures, and enables this elasticity to be sustained for a long time.

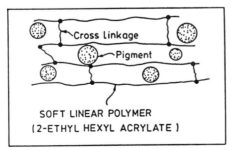

Figure 1. Schematic structure of base coat.

ENGINEERING PROPERTIES

It is important that surface coatings intended to protect concrete structures have adequate engineering properties in terms of strain capacity, stress capacity, elasticity, ability to bridge cracks, ability to accommodate movements and the opening and closing of cracks, and adequate adhesion strength.

Stress-strain Behaviour

Figure 2 shows the load-elongation and stress-strain properties of the base coating. These results show that the coating has large extensibility and very high strain capacity before rupture occurs. The coating is also able to resist tensile stresses in excess of 20 MPa.

In many practical situations, concrete structures may have to withstand a wide range of external temperatures, and the coating should be able to preserve both its strength and elasticity throughout this temperature range. The effect of temperature variations on the tensile strength and elongation properties of the base coat is shown in Figure 3. It is readily seen that the coating can accommodate large temperature changes, -50°C to +70°C, with adequate strength and elongation properties.

Crack-bridging Ability

The crack bridging ability of the coating was determined by testing coatings in tension, mounted on a slate board 150x250mm in size and 5mm thick. The size of the coating was 100x200mm. The thickness

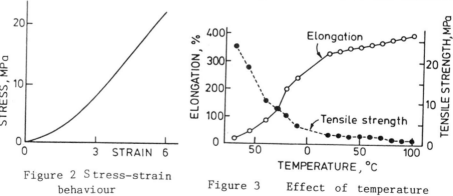

Figure 2 Stress-strain
behaviour

Figure 3   Effect of temperature

of the base coating was varied from 0.2mm (0.34 kg/m²) to 2.5mm (4.3 kg/m²), whilst the primer and top coating thicknesses were kept constant at 0.03mm (0.3 kg/m²) and 0.1mm (0.3 kg/m²) respectively. In general terms, applying the base coat at 1.7 kg/m² corresponds to dried membrane thickness of 0.9mm so that the total thickness at this level is about 1030μm. The results of these tests are shown in Figure 4 which shows the enormous crack bridging ability of the coating.

Fatigue Resistance

In practice, the cracks in a concrete structure also open and close with both applied stress and duration of the applied load, as happens for example, in a highway bridge as well as with movements or internal forces occurring within the concrete. It is therefore important to evaluate the ability of the coating to withstand these stress/crack width variations with time. Cyclic tension tests were conducted on an AW coating thickness of 1.0mm (primer + base + top coating). Both the widths of the crack and the temperature of the test were varied. The details of the test sequence are given in Table 1 – at each range of crack width, the specimens were subjected to 667 cycles each at 20°C, 5°C and –10°C, so that for each range of crack width, the test specimens were subjected to a total of 2001 cycles. At the end of the first 2001 cycles, the range of crack width was increased from 1.0 to 2.00mm, and the 2001 cycles repeated at 20°C, 5°C and –10°C. The last 2001 cycles were carried out at a crack width of 2.0 to 5.0mm. The coating was carefully examined at the end of each 667 cycle, and finally, at the end of 6003 cycles.

The results of this series of cyclic tests are shown in Table 1. The results show that in the range of 0.5 to 1.00mm and 1.0 to 2.0mm crack widths, cracks or pinholes occurred only in the top coating. At very large crack widths of 2.0 to 5.0mm, part of the base coating cracked, but in no case did a complete break down of the coating occur. Figure 4 and Table 1 confirm that Aron Wall coating has the ability to bridge a wide range of crack widths and protect the structure under both static and dynamic conditions.

Adhesion Strength

The adhesion strength of AW coating to concrete was determined under various exposure conditions. These tests are described below.

External exposure tests

These tests were carried out on 200x200x300mm reinforced concrete specimens which were coated on all sides with a coating thickness of 1.0mm consisting of the primer + base + top coating. The concrete mix for these tests consisted of 1·0:2·38:2·91:0·58 (cement: sand: coarse aggregate: w/c ratio) and a water-reducing agent. The concrete mix also contained 0%, 0.6% and 1.0% salt by weight of concrete. Two types of external exposure tests were carried out – half immersion in sea water and open air outdoor exposure. The tests were carried out up to 8 years.

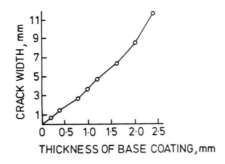

Figure 4   Crack bridging ability of base coat.

Table 1 Fatigue resistance to crack opening and closing

Width of Cracks	Temperature during test			Number of cycles
	20°C	5°C	−10°C	
0.5 1.0mm	* ┤┼·—·┤	* ┤┼—┤	** ┤┼—┤	2001
1.0 2.0mm	** ┤┼·—·┤	** ┤┼·—┤	** ┤┼·—┼	2001
2.0 5.0mm	+ ┤┼·—⊣	+ ┤┼·—⊣	+ ┤┼·—⊣	2001

*   No change
**   Cracks or pinholes in top coating
\+   Part of base coat cracking

Table 2   Adhesion strength of Aron Wall (MPa)

	Exposure years	Salt content in concrete		
		0.0%	0.6%	1.0%
Immersion in sea water	1.5	1.6	1.5	1.5
	5.0	1.6	1.5	1.5
Outdoor exposure	5.0	1.6	–	1.7

The results of the tests up to 5 years are shown in Table 2. It is readily seen that neither continuous immersion in sea water nor continuous outdoor exposure had any effect on the adhesion properties of AW coating on concrete. The adhesion strength remained virtually unaffected up to 5 years exposure.

Effect of water content on adhesion strength

These tests were carried out to assess the effect of water in concrete on the adhesion strength of Aron wall coating. The concrete mix used for these tests was a 1:2·60:2·98:0·55 (cement: fine aggregate: coarse aggregate: w/c ratio) and contained a water-reducing agent. The concrete was cast in PVC boxes (thus effectively scaling the four sides and underside against moisture) and cured at various temperature and humidity conditions as shown in Table 3 for 1, 3 or 7 days. At the end of the curing period, the top face was coated with AW coating (primer + base + top coating) and left in the curing regime. The adhesion strength test was carried out at the age of 21 days.

The water content in the concrete at the time of placing the concrete in the boxes was 8.99% by weight. The water content was again measured at the time of applying the coating. The results are shown in Table 3.

The results show that adhesion strength is reduced as the water content in the concrete increases. In general terms, the adhesion strength appears to vary approximately linearly (Figure 5), and for water contents in the range of 6.0% to 5.5%, the adhesion strength is generally in the range of 1.0 to 1.5 MPa. Below a water content of about 5.5%, these and other data not shown here indicate that adhesion strength is unaffected by water content and remains constant at about 1.5 MPa. Comparing the data in Table 2 and 3, it is clear that some precautions need to be taken when applying AW coating on freshly cast concrete surfaces; on the other hand, it is unlikely that when applied to mature concrete, there will be any reduction in adhesion strength, even when exposed to extreme environments.

Table 3 also shows the modes of failure of the coating. At early ages and high water contents in concrete, the coating failed at the interface between the concrete surface and the primer coating. With maturing, the coating showed a duel breakdown-failure at the concrete-primer interface and cracking of the base coat. This progressively turned into failure of the base coating as the concrete aged, and adhesion strength reached values of about 1.5 MPa.

158   Swamy, Tanikawa

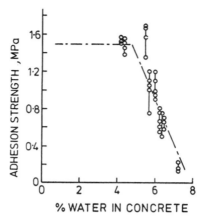

Figure 5   Effect of water content on adhesion strength.

Table 3   Effect of water in concrete on adhesion strength

Curing Condition	Term	Curing Period (days)		
		1 day	3 days	7 days
	Water (%)	7.26	6.54	5.68
5°C 50% RH	Adhesion strength (MPa)	0.13 (C) 0.22 (C) 0.15 (C) 0.12 (C) 0.15 (C)	0.57 (C) 0.64 (C) 0.69 (C) 0.64 (C) 0.69 (C)	0.74 (A&C) 0.98 (A&C) 1.08 (A&C) 1.18 (A&C) 1.03 (A&C)
	Mean	0.15 (C)	0.65 (C)	1.00 (A&C)
	Water (%)	6.37	6.00	4.42
20°C 60% RH	Adhesion strength (MPa)	0.49 (C) 0.54 (C) 0.74 (C) 0.59 (C) 0.55 (C)	0.93 (A&C) 0.88 (A&C) 0.96 (A&C) 1.08 (A&C) 1.18 (A&C)	1.35 (A) 1.50 (A) 1.43 (A) 1.47 (A) 1.50 (A)
	Mean	0.58 (C)	1.01 (A&C)	1.45 (A)
	Water (%)	6.25	5.50	4.20
30°C 20% RH	Adhesion strength (MPa)	0.54 (C) 0.59 (C) 0.79 (C) 0.74 (C) 0.64 (C)	1.32 (A&C) 1.53 (A) 1.63 (A) 1.65 (A) 1.50 (A)	1.50 (A) 1.50 (A) 1.53 (A) 1.47 (A) 1.50 (A)
	Mean	0.66 (C)	1.53 (A)	1.50 (A)

C – Failure at concrete – primer interface
A – Cracking of base coat

WEATHERING RESISTANCE

When surface coatings are applied to concrete surfaces which are exposed to external environments, it is important to ensure that the coating film has adequate stability and long life under the exposure conditions of the concrete structure. To evaluate the weathering resistance and performance characteristics of AW coating, free films of the base coating were subjected to two kinds of external exposure, and these tests are described below.

Environmental Exposure

In these tests, the base coat membrane was left exposed in the open facing south at an angle of 45° for a period of 10 years. At this orientation, the coatings were exposed to the most severe solar radiation, the annual mean radiation (in Langley) over a five year period being the following:

Ultraviolet	6,155.8
Visible ray	62,965.5
Infra-red ray	62,989.6
Total	132,110.9

The test specimens were dumb-bell shaped and 1mm thick and were tested in tension at regular intervals.

The exposure site was also chosen to have a high variation of temperature, humidity and rainfall. The mean monthly temperature and relative humidity over a period of 12 months, varied from about 2.5°C to 27°C and 60-80% RH respectively, with the mean monthly rainfall over the same period varying from about 50 to 200mm. These would be seen as severe exposure conditions. The results of the tension tests after various periods of exposure to the above conditions are shown in Figure 6. The tensile strength increased with exposure time, whereas the elongation was reduced from over 300% to about 170%. Even with the loss in elongation, the results confirm that the base coating was able to maintain both adequate strength and elasticity for a long period of time.

Heat resistance

In addition to the environmental exposure tests, similar test specimens were also exposed continuously to temperatures of 80°C and 100°C. The results of these tests are shown in Figure 7. At both 80°C and 100°C the strength continued to increase; the elongation, on the other hand, showed very little deterioration at 80°C, although at 100°C, the elongation had reduced to about half the original value, about 50% after continued exposure for over 100 days. 100°C is an extremely severe exposure condition, unlikely to be met in practice; temperatures of 50°C to 70°C, on the other

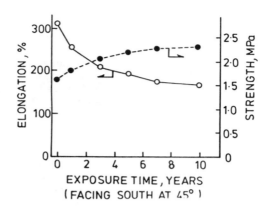

Figure 6   Weathering resistance of base coat.

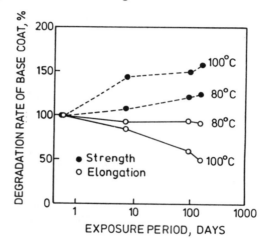

Figure 7   Heat resistance of base coat.

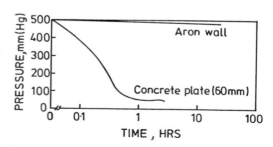

Figure 8   Air diffusion resistance of AW coating under high pressure.

hand, are likely to occur on the external surfaces of buildings in the summer. The results of Figures 6 and 7 show that the AW coating can resist severe exposure conditions for a long period of time, and remain stable and serviceable.

DIFFUSION RRESISTANCE

The diffusion resistance of a barrier coating to water, air, chloride ions and water vapour is an important property directly related to its performance characteristics in practice. However, it is not sufficient to test the coating in its natural state: when applied to concrete surfaces, the coating will be subjected to tensile forces and stretched over crack openings due to various movements and expansive forces occurring within the concrete. It is therefore important to carry out the tests in both these stress states. In the tests reported here, the AW coating (1.0mm thick, primer + base + top coating) was tested (a) in initial unstretched condition, and (b) stretched to 100% elongation.

The test methods used to establish the diffusion characteristics are shown in Table 4. The test results obtained for AW coating were compared with those of concrete for water, air and chloride ion diffusion, and with that of mortar for water vapour diffusion. The tests on concrete were conducted on concrete boards, 30mm thick, at the age of two months, after curing 1 month in water and 1 month in air. The mix proportions for the concrete were 1:2·38:2·91:0·58 (cement: sand: coarse aggregate: w/c ratio). The mortar board for the water vapour test was 10mm thick and consisted of 1:2 (cement: sand) mix.

The results of the tests for both AW coating and concrete/mortar boards are shown in Table 4. As would be expected, the diffusion resistance of the coating in the stretched condition was less than that in the unstretched condition. This is an important factor to be appreciated when evaluating the performance characteristics of surface coatings. The results show that even in the stretched condition, the AW coating has a very high diffusion resistance against water, air, chloride ions and water vapour.

The permeability rates for the AW coating and concrete/mortar boards are also compared in Table 4. The ratio of the permeability rates of concrete to AW coating are respectively 720, 3300–5000, 1100–4300 and 40 for water, air, chloride ion and water vapour diffusion. These values are of course obtained from laboratory tests under controlled environmental conditions and cannot be translated directly to field performance. What is significant is that the acrylic rubber membrane has very high diffusion resistance to those elements that cause considerable damage to concrete. In a later paper, the authors hope to show how with this type of cross linked soft acrylic polymer, very high resistance to the ingress of aggressive agents into concrete can be obtained in practice in reinforced concrete elements.

Table 4 Diffusion resistance of Aron Wall coating

	Water repellence	*Air diffusion	**Chloride ion	Water vapour
Test method	water	air vacuum pump	3%NaClaq. soln. distilled water	water vapour Anhydride $CaCl_2$
Unit	ml	cc/m/hr/mmHg	cm²/sec.	g/m²/day
Aron Wall a.	0.0ml	2.8×10^{-4}	1.4×10^{-11}	12.5g/m²/d
Aron Wall b.	0.0ml–0.1ml	4.3×10^{-4}	5.3×10^{-11}	–
Concrete	72ml	1.4	6.0×10^{-8}	50.4g/m²d, mortar
Ratio	720	3300 – 5000	1100 – 4300	40

Aron Wall a. Initial state. b. Stretched at 100% elongation.

Table 5 Effect of base coating thickness and type of top coating on diffusion resistance

			Diffusion of chloride ion (mg/cm²/day)	Diffusion of water vapour (g/m²/day)
Free film of base coating only	Thickness of base coating	0.5mm	1.89×10^{-2}	135.0
		1.0mm	1.97×10^{-4}	60.0
		1.5mm	1.46×10^{-4}	38.5
Multilayer membrane base coating + top coating	Types of top coating	Acrylic resin	2.7×10^{-4}	15.5
		Acrylic urethane	1.17×10^{-4}	12.5
		Acrylic silicone	1.35×10^{-4}	32.3

One of the advantages with this type of coating is that apart from the variable nature of the thickness of the base coat, the type of top coating can also be varied to suit a particular situation. Tests similar to those shown in Table 4 have been carried out to evaluate the effects of the thickness of the base coat and the type of top coat on the diffusion resistance to chloride ion and water vapour. In these tests, three base coat thicknesses, 0.5mm, 1.0mm and 1.5mm and three kinds of top coating, with a constant base coating of 1mm, were used as shown in Table 5.

The results of these tests are also summarised in Table 5. The diffusion resistance should obviously increase with the coating thickness and this is generally validated for both chloride ions and water vapour. However, with multilayer coatings, the type of top coating has also a definite influence on the permeability to water vapour. These results thus show that the AW coating can be designed to suit a given set of environmental expousre conditions as well as the type of internal deterioration processes in concrete which need to be contained. The coating has thus some very unique properties in its ability to be modified to contain a wide range of concrete deterioration problems.

Diffusion under high pressure

The diffusion resistance of surface coatings under high external pressure is also an important property, particularly where structures are in an exposed area. The air diffusion resistance of AW coating under high pressure was determined in a special rig where the coating was subjected to an external pressure of 1 atm ; this was compared to that of a concrete plate 60mm thick. The results are shown in Figure 8. The data show very low diffusion of air through the AW coating from outside, compared with very much higher diffusion of air through the concrete plate. The results again confirm the very superior performance characteristics of AW coating in resisting the penetration into concrete of a wide range of aggressive elements.

CONCLUSIONS

The application of effectively impervious surface coatings to concrete surfaces to prevent the ingress of potentially harmful agents such as water, air and chloride ions is a very attractive solution to protect new and existing concrete structures. This paper reports the development of a highly elastic acrylic rubber type of coating (Aron Wall) with an overall thickness of about 1000μm. The chemistry of Aron Wall (AW) is briefly discussed. It is shown that surface coatings should satisfy certain basic engineering requirements such as crack bridging ability, adequate strain capacity and elasticity, fatigue resistance and adhesion strength. Data are presented to show that AW coating fully satisfies many of these requirements. Data are also presented on weathering resistance and diffusion resistance to water, air, chloride ions and water vapour.

The paper shows that AW coating has excellent performance characteristics in resisting the penetration into concrete of wide range
of aggressive agents.

REFERENCES

1.   WALLBANK, E.J.   The Performance of Concrete Bridges, Department
     of Transport, Her Majesty's Stationery Office, London, April
     1989, pp.96.

2.   BROWN, J.H.   Factors affecting steel corrosion in concrete
     bridge substructures, IABSE Symposium, Lisbon 1989, Vol.2,
     1989, pp.543-548.

3.   SWAMY, R.N.   Durability of rebars in concrete, To be published.

4.   CHANDRA, S.   Influences of pollution on mortar and concrete,
     Document D6:1990, Swedish Council for Building Research, 1990,
     pp.83.

5.   SWAMY, R.N.   Behaviour of high alumina cement concrete under
     sustained loading, Proc. Instn. Civ. Engrs., Vol.57, Dec. 1974,
     pp.651-671.

6.   MITCHELL, W.G.   Effect of a waterproof coating on concrete
     durability, Journal, American Concrete Institute, Vol.29, No.1,
     July 1957, pp.51-57.

7.   SNYDER, J.M.   Protective coatings to prevent deterioration
     of concrete by deicing chemicals, National Cooperative Highway
     Research Programme, Report 16, Transp. Res. Board, 1965.

8.   ACI COMMITTEE 515   Guide for the protection of concrete against
     chemical attack by means of coatings and other corrosion
     resistant materials, Journal, American Concrete Institute,
     Vol.63, No.12, Dec. 1966, pp.1305-1391.

9.   ACI COMMITTEE 515   A guide to the use of waterproofing, damp-
     proofing protective and decorative barrier systems for concrete ,
     Concrete International: Design and Construction, Vol.1, No.11,
     Nov. 1979, pp.41-81.

10. PFEIFER, D.W. and SCALI, M.J.   Concrete sealers for protection
     of bridge structures, National Cooperative Highway Research
     Programme, Report 244, Transp. Res. Board, 1981.

11. McCURRICH, L.H., WHITAKER, G. and HUMPAGE, M.L.   Reduction
     in rates of carbonation and chloride ingress by surface
     impregnation, Proc. Second Int. Conf. on Structural Faults
     and Repair, 1985, pp.133-139.

12. HANKINS, P.J.   The use of surface coatings to minimise carbona-

tion in the Middle East, Proc. First Int. Conf. on Deterioration and Repair of Reinforced Concrete in the Arabian Gulf,
Vol.1, 1985, pp.273-285.

13. HANSEN, J.H. and VILLASDEN, C. Design of surface coatings
for protection of reinforced concrete structures, Proc. 3rd
Int. Conf. on Deterioration and Repair of Reinforced Concrete
in the Arabian Gulf, Vol.1, 1989, pp.647-664.

14. DAVIES, H. and ROTHWELL, G.W. The effectiveness of surface
coatings in reducing carbonation of reinforced concrete, BRE
Information Paper IP 7/89, 1989.

PERFORMANCE OF HIGH BUILD COATING MATERIALS FOR CONCRETE STRUCTURES FOR PREVENTING CORROSION DAMAGE

M Yasui

M Fukushima

Nippon Paint Co. Ltd.,

Japan

ABSTRACT:
Coating on concrete surface is promising countermeasure for preventing corrosion of reinforcing steel in concrete. The quality and performance of coating systems have been evaluated with laboratory tests and with field tests.
We selected new types of high build materials, soft type epoxy, polymer rich cement and others as coatings for concrete structure. We have recommendation on the concrete coating for preventing salt damage of structure, and some structures have been applied with these coating system. We will present important test results and the follow up inspection results.

KEYWORDS: Concrete structure, Salt damage, Coating, Anti-corrosive, Softness

MR. M. YASUI is Chief at the Department of Engineering, Nippon Paint Co., Ltd., Tokyo. He is research into concrete surface coating for preventing corrosion of reinforcing steel in concrete.
MR. M. FUKUSHIMA is Senior Manager of Engineerings, Nippon Paint Co., Ltd., Tokyo. He has undertakes studies anti-corrosive coating for concrete or steel structures.

OUTLINE:
The concrete structure damaged by corrosion, etc. shows numerous cracks of concrete due to the corrosion of reinforcing rods. It there is one method to shut off the permeation of moisture, oxygen and salt as corrosive factors from the outside, by covering it with concrete-surface coating. Covering/repairing materials are requested for such performances as (1) higher sealing property against the permeation of moisture, oxygen and salty, (2) flexibility to follow the crack of concrete.
In this report, description is made on the present state of measures against the salt damage with the surface coating of concrete in Japan, introductions of each performances and coating specification of flex-ible thick-film paints having both of highly sealing performance and following performance to the crack, and the application examples for the concrete structure, as new covering materials.

PRESENT STATE OF MEASURES AGAINST SALT DAMAGE OF CONCRETE STRUCTURE IN JAPAN

(1) State of concrete structure by salt damage
   The salt damage basically means the phenomenon to show peeling or crack on the concrete by swelling and corrosion of reinforcing rods after the permeation of salt with sea splash or sea breazes, (salt damage also occurs by freezing preventive agents).
   Corrosion occurs numerously on the seaside of Japan Sea and Okinawa liable to be influenced by the sea splash, etc. According to the investigations of Ministry of Construction in 1983 (920 brings having a total length of more than 15m constructed within 500m apart from the seashore by 1980).
   The repairing objective quantity for concrete bridges is 25% of total bridges showing the salt damage or its sign, 5% of bridges showing numerous peeling defects of concrete with crack and 5% of already repaired bridges.

(2) Present state of measures against salt damage
   Salt damage is prevented with such measures as surface coating of concrete, increase in covering thickness with concrete, the use of reinforcing rods coated with powdered epoxy resin in case of new concrete structure.
   Among them, corrosion-preventive effects were tested on the con-crete-surface coating with exposure-test performed at a marine facility, Tokyo Bay, belonging to the Ministry of Construction.
   The results are shown in Fig. 1 and Fig. 2.
   As a result in the case coating the paint of epoxy-resin, it was possible to reduce the corrosion area as compared with the case taking the concrete thickness over 12cm. Also, even in the result measuring the salt-containing amount, coated film induced compre-hesively less amount of contained salt as compared with that of coated film, thus, the coated film showed the excellent ion-shielding capacity.

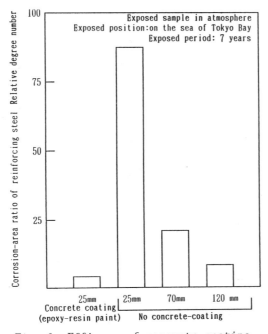

Fig. 1   Efficacy of concrete coating

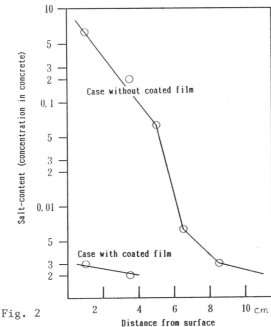

Fig. 2

As a result of the test using free film for the permeability of
water, oxygen, salt, as corrosive factors of iron in addition to
the various coating materials, excellent performances were observed with coated film of epoxy-resin paint (excluding the emul-
sion-type) and glass-flake containing epoxy-resin or vinyl ester
resin paint. Fig. 3 shows the measurement results of permeability
of steam.

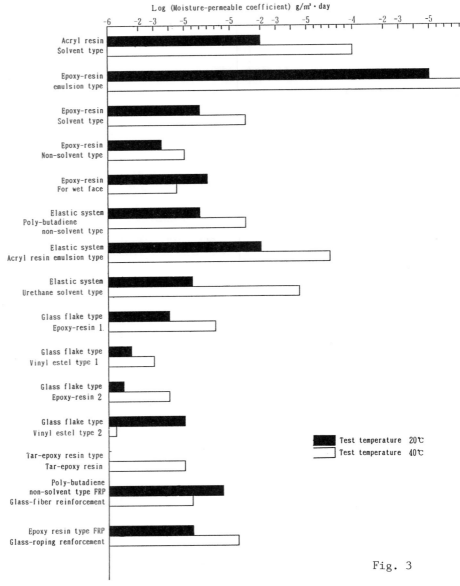

Fig. 3

(3) Paint against salt-corrosion
Now, quality-specifications for paints to prevent the deterioration of concrete are shown in "Guidelines against corrosion by salt on road-bridge" (1984) for the newly established bridges as measures against salt-corrosion as issued by Japan Road Association, Judicial Corp. In this guideline, there are 3 types of paint-specifications, and all types adopt the paint of epoxy-resin as shielding film against salt, while poly-urethane resin paint is used by its excellent weatherability as finishing coat. Table 1 shows the coating system of A-type used numerously in the actual result.

Table 1 A-type Coating System

Process		Using materials	Coating conditions			Coating interval
			Target-film thickness	Standard using amount	Coating method	
Pretreatment	Primer	Epoxy-resin primer	–	0.10	Airless spray (Brush roller)	Interval of each process shall be within 10 days over 1 days as standards.
	Putty	Epoxy-resin putty	–	0.30	Spatula	
Medium coating		Epoxy-resin paint medium coating	60	0.32 (0.26)	Airless spray (Brush roller)	
Finish coating		Epoxy-resin paint finish coating	30	0.15 (0.12)	Airless spray (Brush roller)	

NEW COATING MATERIALS THE SURFACE OF CONCRETE:

(1) Flexible paint of high-functional thick film type
As performances of concrete-surface coating materials, it is requested to prevent or inhibit the permeation of oxygen, moisture and salty ingredients as deteriorating factors from the outside, and to have full weatherability or adhesiveness with concrete and reinforcing or repairing materials, as well as crack-following property. For fulfilling these requests, improvement has been made for epoxy-resin paint used for new bridges with countermeasures against corrosion by salt while maintaining the higher (excellent) shielding performance, and development has been made for high-performance thick film type paints such as rubber-type resin paints using the resin of poly-butadiene and highly elastic and flexible epoxy-resin paints provided with higher flexibility. Fig. 4 shows the measured results of elongation rate of coated film and the result of test to following the crack in between these paints and the conventional paints.

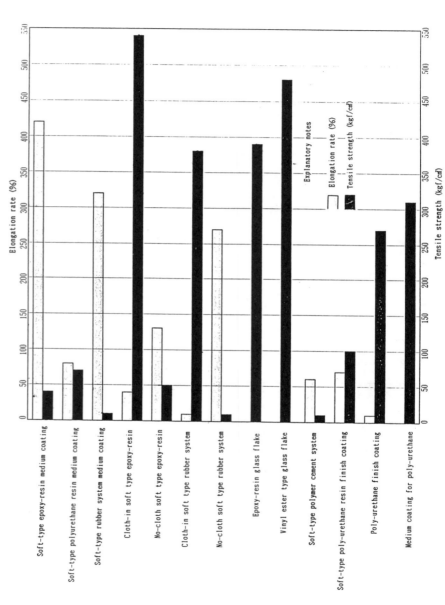

Fig. 4   Result of test on crack following property

(2) Highly weatherable finish coating
Finish coating paints have been developed as with silicon-resin paints, fluorine-resin paints by giving the weatherability in order to extend the repainting cycle, and those provided with flexibility. These are the paints using organic resins, and in addition, there is mortar-lining technique. In the past, polymer-cement type coating materials were inferior to the organic resin type paints in performances such as water permeability, salt shielding and softness.
However, some polymers having the softness with improvement of these performances are also developed.

(3) Concrete-falling defect preventing paints
There are the covering paints to prevent the peeling defect at the positions liable to show the concrete falling trouble by peeling/falling defect of concrete. These materials are reinforced by synthetic fibers or glass-cloth to be bearable for the weight at defective part of concrete, and it is used by impregnating and solidifying the material in the use of flexible epoxy-resin adhesive paints. If features the combined effects of reinforcement with cloth and proper crack-following property by the flexible coating film.

(4) Quality standard on the repairing paints for the salt damage
The quality of these new materials for coating the surface of concrete should meet the quality shown in Table 2.

Table 2 Quality of Concrete Surface Covering Materials

QUALITY OF CONCRETE SURFACE COVERING MATERIALS

Coating Item (Test method)		Class 1 Case of strict control for corrosive environment, difficulty for re-repair.	Class 2 Other than the Class 1
Weatherability (JIS K 5400 6.17)		After testing for accelerated weatherability test for 300 hours, there should be no chalking, crack and peeling on the coating.	
Salt shielding		Cl ion permeable amount should be less than 1×10^{-3} (mg/cm^2/day)	Cl ion permeable amount should be less than 1×10^{-2} (mg/cm^2/day)
Anti-alkalization (JIS K 5400 7.4)		After soaking in saturated solution of calcium hydroxide for 30 days, there should be no abnormality in the coated film.	
Adhesivenss with concrete (JIS A 6910)		There should be no damage on interface.	
Crack following (Bending load test of sample)	For PC	There should be no occurrence of defect on the coating with crack width up to 0.1mm of concrete.	
	For RC	There should be no occurrence of defect on the coating with crack width up to 0.2mm of ooncrete.	

(5) Coating system for high built soft type paint
    Fig. 5 shows the general operation procedures for the coating of
    concrete surface in the use of these high built soft type paints.

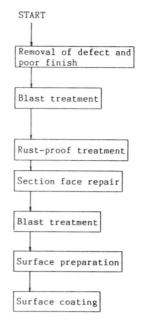

START

Removal of defect and poor finish	Remove the concrete containing the salt or crack caused by salt damage.
Blast treatment	To improve the adhesiveness between repairing part and exisiting concrete, remove the salt with the blast treatment the rust on the exposed steel material.
Rust-proof treatment	Make rus-proof treatment for steel material by using epoxy primer after blast treatment.
Section face repair	Recover the concrete by prepacked concrete technique, construction joint technique and mortar technique.
Blast treatment	Remove the salt by blast treatment on whole surface of concrete to improve the adhesiveness with coated film.
Surface preparation	To coat the coating paint uniformly, prepare smooth surface of concrete by using a primer or a putty.
Surface coating	Make the coating layer by the use of concrete surface coating paint.

Next, Table 3 shows one example of coating methods to make the
concrete surface coating.

Table 3   Coating Methods of High Build Soft Type Paint,
         Soft Type Epoxy Resin Coating   (500μm)

Process		Material	Coating conditions				Coating interval
			Target film thickness (μm)	Standard using amount (kg/m²)	Coating method		
Pretreatment	Primer	Epoxy resin primer	–	0.10	Brush, roller		Standard interval of each process should be over 1 day within 7 days.
	Putty	Epoxy resin putty	–	0.50	Lancet		
Medium coating	1st layer	Soft type epoxy resin paint medium coating	160·	0.35	Brush, roller		
	2nd layer	Soft type epoxy resin paint medium coating	160	0.35	Brush, roller		
	3rd layer	Soft type epoxy resin paint medium coating	160	0.35	Brush, roller		
Finish coating		Soft type polyurethane resin	30	0.12	Brush, roller		

These coating new materials for concrete surface are already used
to repair the concrete bridges at salt-damage area in Okinawa Pref.

CONCLUSION:

We found that the concrete surface coating was effective as counter-measures against the salt damage of concrete structure, because coat-ings can shield the corrosive factors such as moisture, salt and oxygen from the outside, causing the corrosion of reinforcing rods in concrete.

Coating method for the countermeasures against the salt damage of con-crete is stipulated by Ministry of Construction, etc. The paint used for this coating method has an excellent shielding performance and a crack follow-up performance.

Thus, it is already coated on concrete bridge, etc., and its efficacy has been confirmed.

As countermeasures against the salt damage, updated coating materials have been developed. In addition to the higher shielding performance, theses new coatings have a higher softness for a higher crack follow-up performance, and it is called a high built soft type paint. Also, fiber reinforced paints for the prevention of the exfolication of con-crete and high weatherability finish coating paints have been devel-oped. Even for repairing the salt-damaged part, coating is playing a vital role, and its efficacy is confirmed by coating on concrete bridges, etc.

REFERENCES:

1) Concrete coating manual:
   Civil Engineering Research Institute, Ministry of Construction.
   Feb., 1984.

2) Guideline (Draft) for measures against corrosion by salt of bridges
   with explanations:
   Japan Road Association, Judicial Corp.
   Feb., 1984.

3) Development of technology and elevation of durability of concrete:
   Ministry of Construction.
   May, 1989.

DEVELOPMENT AND OPTIMIZATION OF IMPREGNATION MATERIALS

H R Sasse

D Honsinger

Institut fur Bauforschung,

West Germany

ABSTRACT. Laboratory tests have been used to study the effectiveness of impregnation materials applied to some types of porous substrata. New approaches of accelerated testing were carried out in cooperation between the chemical industry and interdisciplinary collaborating scientific institutions. Preferably agents on polymer base have been developed and are subject to further optimization. As a base for the research activities reported, a catalogue of requirements for protective agents has been elaborated. The assessment of the effectiveness of new impregnation materials is carried out by comparison of physical test results and accelerated aging test results on treated and untreated samples. A number of important properties, which are to be optimized, has been quantitatively and qualitatively evaluated. Related to the new concept of natural stone protection that has been developed at the Aachen University first experimental results concerning concretes are exemplary presented here.

Keywords: Impregnation, Concrete, Natural Stone, Polymer Microlayer, Protection, Strengthening

Prof. Dr.-Ing. H. R. Sasse is Director at the Building Research Institute, The University of Technology, Aachen, Germany. He has been most active in polymer concrete materials, adhesion of polymers to mineral substrata, ageing of polymers, surface properties of building materials and repair of concrete structures.

Dipl.-Ing. D. Honsinger is Research Engineer at the Building Research Institute, The University of Technology, Aachen, Germany, where he is actively engaged into various aspects of impregnation of porous materials.

INTRODUCTION

Air pollution and other natural deteriorating agents show a strong synergism which has produced increasing collapses of many natural stone monuments and sculptures. Sandstone is susceptible to considerable weathering processes and the conservation of this type of rock, which is characteristic for very many old buildings and monuments in Germany and other countries, causes serious problems. Also concrete constructions show early weathering effects of surfaces, particularly due to the action of climatic agents and following premature rusting of reinforcement.

The main reasons are too thin concrete cover and imperfection to conform to the rules of good workmanship in making and using the concrete.

A preventive surface treatment like impregnation may be useful to reduce carbonation progress and to protect reinforcing bars near to the surface of the structural element.

Additional purpose of the impregnation of exposed concrete structures are to reduce the susceptibility of moisture and the corrosion of the concrete itself and to increase the strength. Development and optimation of impregnation materials are basing on a new fundamental theoretical model and new types of polymeric materials. The combination of both has given very promising laboratory results in the protection and the structural consolidation of porous natural stones like sandstone.

TECHNOLOGICAL REQUIREMENTS

As a base for the research activities a "catalogue of requirements" for protective agents has been elaborated in interdisciplinary cooperation /1/. The technical demands can be summarized into four main criteria:

- aesthetics of the surface
- effectiveness and durability of protective and strengthening means
- building site usability
- compatibility with ecological demands.

These main criteria are specified in detail by several subcriteria. The subcriteria are discussed in /1/.

CONCEPT FOR PROTECTION AND CONSOLIDATION

The so called "supporting corset" model is understood as a protecting and strengthening polymer microlayer, coating

the internal pore surfaces of the substratum. The microlayer is supposed to be

- water repellent (hydrophobic) and water resistant
- impermeable against water
- restraining the water vapour diffusion
- resistant against standard chemical and biological agents
- rubber elastic through a wide range of temperature (-30°C
  to 80°C).

Decayed components of the internal micro-structure are
expected to be evenly coated by the microlayer and relinked
to each other by polymeric bridges. As a result of the
planned mechanical properties the polymer microlayer dis-
plays the function of a "supporting corset". There are no
significant changes in water vapour diffusion within the
bulk volume of the substratum, because large capillary
pore-channels remain open. In simplified form this concept
is explained for example at sandstone in fig. 1.

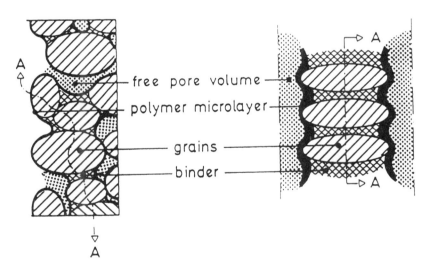

Fig. 1: The "supporting corset", protective polymer
microlayer coating the pore walls

This concept differs from the well known hydrophobic treatment, because of the distinctive formation of a protective
and impermeable microlayer, coating the inner surfaces of
the substratum. In addition, disadvantages of increasing
the modulus drastically (as by conventional polymer
impregnation) and/or forming a secondary pore network due
to adhesive or cohesive failures (as by silica ester
impregnation) do not appear.

SELECTED TEST RESULTS

Aims

On the basis of the technological requirements, effectiveness tests for existing products have been carried out.
In an iterative process, specifications were elaborated as
a basis for developments of new polymer treatment systems
which are able to protect mineral substrata against harmful
atmosphericals and to strengthen the microstructure. New
types of polymeric materials do not only mean absolutely
new materials but also developments of standard materials
used for other purposes than impregnation.

Tested products, substrata and application-methods

About 150 cold curing polymeric products for stone treatment were studied during preliminary test series. The products were based on solvent containing

- polyurethanes, both hardening from ambient humidity
  and from special chemical components (PUR)
- epoxy resins, 2-component systems (EP)
- modified acrylic resins (AY)
- silicon organic compounds, silanes, polysiloxanes (SIL)
- modified silica ester (SE)
- fluoro ethylene (FE)
- unsaturated polyesters (UP).

In addition, some preliminary tests were carried out using
hydrous dispersions on the basis of polyurethane, acrylic
and polyester resins. All products have very low viscosities and can be used successfully for deep impregnation of
concretes with medium and with high porosity. Suitable
treatment materials must have:

- low viscosity coefficient less than about 10 mPa*s
- moderate surface tension higher than about 20 mN/m and
  lower than 30 mN/m
- moderate interrelation between polar and unpolar groups
  inside the resin molecules (in relation to the polarity
  of the substratum)
- low average molecular weight
- low reactivity, very slow increase of viscosity
- low solvent content.

Substrata were different types of sandstones and concretes.
Three different quarry sandstones and two different grades
of concrete were selected for their typical features, such
as mineral content and pore structure. The treatments were
carried out on

- Ebenheider Sandstone   (EH)
- Obernkirchener Sandstone   (OK)
- Sander Schilfsandstone   (SS)
- concrete grades B15 and B35   (B15, B35).

The samples had the dimension of 5cm x 5cm x 10cm. Concrete samples were cured for 28 days according to German standards, i.e. 7 days moist, thereafter storage at 20°C and 65% R.H. up to 28 days. Sandblasting was used for the concrete front and back sides (5cm x 5cm) preparation. In order to simulate the situation of a large element, treated and weathered from the surface, an 1.0 mm EP sealing was applied around the sample lateral faces. The front and back sides remained free. Before application the samples were exposed to different climates (8°C/60% R.H., 23°C/50% R.H., 28°C/95% R.H.) until reaching constant humidity.

The laboratory tests covered measurements on

- liquid impregnation components and free polymer layer properties (e.g. viscosity, surface tension, non volatile contents, polarity, reactivity, hardness)

- porous material properties (e.g. water adsorption and desorption, pore structure)

- impregnated porous material properties (e.g. penetration depth of the liquid components, pore structure, modulus of elasticity, permeability to water vapour and their alteration under extreme service-live conditions)

TEST PROCEDURES AND SELECTED TEST RESULTS

Penetration depth test

The penetration tests were carried out by different methods /2/. Dependent on type of substratum, weathering grade, and physical properties of the liquid agents, penetration depths from some mm up to 70 mm have been achieved. The correlation between penetration depth and viscosity or surface tension respectively are illustrated in /3/. In this case the different climates have no significant influence on the penetration depths of the protective liquids. The influence of polarity has not been clarified yet.

Fig. 2 shows penetration depths in relation to contact time of the liquid agents with the treated surfaces. The different liquid agents point out different sucking velocities over time. Systems with a very high initial sucking velocity will retain the rapid gradient during the whole application time only, if the reactivity of the resins is very low.

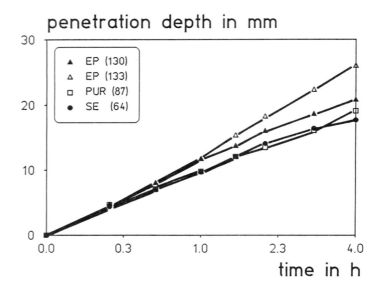

Fig. 2: Penetration depth in relation to contact
time at SS

If the rate of polymerization is too rapid, gel forming
will start before a useful degree of penetration has been
attained. The agents will not only penetrate deep enough,
but will instead tend to seal the substratum at its surface, creating in case a serious source of future decay.

Suction of water and drying behaviour

Suction of water is understood as a result of capillarity
or adsorptive forces when wetting surfaces without external
pressure. The polymer treated surface was immersed into
water and suction was vertically upwards. The time related
increase of weight was measured until reaching constance of
weight and was reported in percent of the maximum capillary
water content of the untreated specimens.

The drying behaviour is important because liquid water
sometimes accumulates behind the treated zone due to condensation or moisture transport from hidden sources. Drying
behaviour was determined after suction of water for 28 days
by measuring the weight during the drying process under
23°C/50% R.H.

In the case of sandstones, the time related water absorption and -desorption processes, the samples treated with

modified silica ester (SE) show a significantly hydrophobic character throughout the whole time of exposition. Samples treated with polyurethane (PUR) show a moderate rise of water sorption after 7 days, meaning that after that time of exposition the hydrophobic, water repellent effect is lost. As a principle it is advantageous to have as short as possible desorption (drying) times compared to adsorption times. This property was not fairly met by the selected polymer systems /5/.

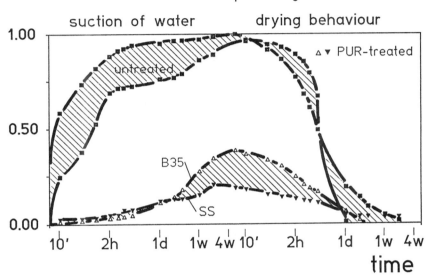

Fig. 3: Suction of water and drying behaviour: untreated B35, SS and PUR-treated B35, SS

The graphs in fig. 3 show suction of water on medium strength concrete B 35 with a similar effect. The difference to sandstone lies in the fact that the hydrophobing effect shows a moderate rise of water sorption not until after 1 day. This is an important point in protecting reinforcement from corrosion, because the concrete remains dry for a long period and therefore the steel will not corrode due to normal raining periods.

Contact angle and laboratory weathering

Testing the contact angle a drop of water of a defined volume is deposited on a horizontally adjusted sample surface. Using a goniometer microscope the contact angle can

directly be determined. Above 90° degrees, a surface can be
defined as hydrophobic. Laboratory weathering tests allow a
comparative evaluation of the durability of the different
products. As a result of exposure the polymer microlayers
at the macro surface are partly destroyed, leading to an
activation of the surface energy, and thus to smaller contact angles. These surfaces can be wetted, but with durable
products no suction of water into the inner capillary
system is possible /2/.

Water vapour diffusion

The polymer treated substratum is expected not to act as a
barrier towards the flow of water vapour. This is especially important for moisture exchange processes between
the building structures and the surrounding atmosphere.
Water vapour transmission rate (WVT) was tested according
to DIN 52615 /4/. The treated samples had a cross section
of 50 mm x 50 mm and a thickness of 10 mm. The test was
carried out according to the so called "wet cup method".

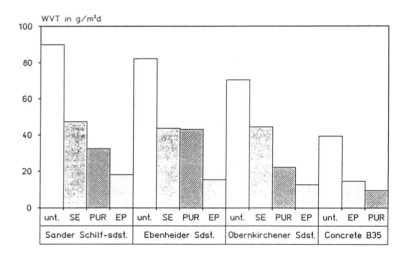

Fig. 4: Water vapour transmission rate on treated and
        untreated sandstones and concrete B 35

The applied products reduce WVT for a factor of 3 to 4,
although the microlayers on the inner pore walls only
affect porosity in the desired way, fig. 4. The results can
be attributed to the hydrophobic effects of the products,
inducing a reduction of the adhesive moisture. Due to that
fact, transportation of water molecules within the pore

system is reduced. In this case, surface diffusion and capillarity only moderately attribute to the diffusion of water vapour. The results of WVT tests seem tolerable but the product's hydrophobic features should nevertheless be optimized in future.

Water vapour adsorption

The isotherm of sorption characterizes the amount of moisture being adsorbed on the specific inner surfaces in equilibrium with the relative humidity of the surrounding air. Hydrophobing treatments effect a reduction of water adsorbing features of the substratum. This should be mainly reached by sealing pores which do not contribute essentially to the moisture transport mechanisms of the substratum. Compared to the untreated samples the adsorption of water on the polyurethane and epoxy treated specimens is principally lower for all moisture grades /5/. The larger amount of adsorbed moisture in untreated stones can be explained by capillary condensation within pores smaller than 0.1 micron in diameter. The significantly reduced water content of the PUR and EP impregnated samples can be explained as a function of a smaller inner surface resulting from the polymer sealing of intercrystalline gaps. Results from SEM investigations confirm these statements. In addition, hydrophobic effects will play a role.

Biaxial flexural strength and elastic modulus

Strengthening behaviour of the integral compound between polymer and substratum is largely governed by the mechanical properties of the polymer itself. Various chemical structures, particularly those of polyurethanes and epoxies have proved to be very suitable for flexible connections.

The strength of decayed border areas has to be increased, if necessary, so that occurring internal stresses can safely be accepted. A remarkable increase of the elastic modulus is not desired in order to prevent dangerous thermic and hygroscopic induced internal stresses. Strength and elastic modulus measurements, before and after treatment, in small slices from the surface to the inner part of the substratum are therefore central to the assessment of any consolidant. Such measurements until now have been impossible or at least gave a great scattering in results. A technique attributed to /6/ was tried instead. This test procedure allows to determine the relative strength and the relative elastic modulus in depth-profiles. From the original samples (5cm x 5cm x 10cm) cylinders of 4cm diameter were drilled and these were cut into slices of 4mm thickness. The slices were tested in a special device in biaxial

bending. The relative elastic modulus in depth-profiles of
untreated and treated sandstone are plotted in fig.5.

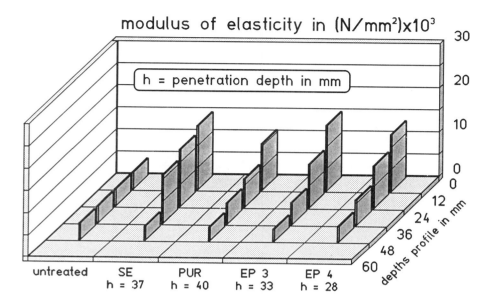

Fig. 5: Relative elastic modulus in depth-profiles before
        and after polymer treatment at sandstone

Impregnation causing protection in addition to structural
consolidation of quarry stone samples with entropy-elastic
polymers, increases only the elastic modulus of the treated
stones moderately. Samples treated with PUR show a gradually decreasing profile due to a successful consolidation
procedure.

Investigation on the pore structure

The description of the pore structure is a very important
requirement for the assessment of impregnation treatments
/7/. In investigating the quality of the formation of the
polymer corset at the treated inner surfaces and cracked
sections, on which the main interest is focussed, the aid
of the scanning electron microscope (SEM) proved to be
particularly suitable /8/. The recognized thicknesses of
the polymer microlayer range between some microns (μm) and
nanometers (nm), depending upon local geometric situations
and type of substratum. Using specially developed sample
preparation methods, the topography of the inner surface
features can be studied down to nanometer structures /9/.

The visual shape of treated sandstone samples is characterized by homogeneous continuous films bonding together the quartz crystals. The polymer does not seal the capillary pores, but only the micropores. It tends to leave a fine supporting bandage with strengthening grain-to-grain contacts. fig. 6 shows that the adhesive capability between the substratum and the polymer microlayer is very good and moreover the ability of the liquid product to penetrate and completely fill micropores.

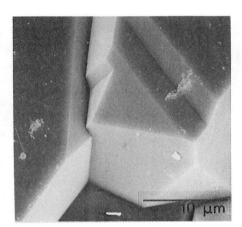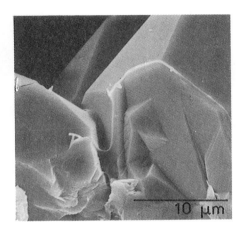

Fig. 6: Quartz crystals inside a sandstone before (left) and after treatment (right)

Fig. 7: Vermicular stacks, the characteristic structure for kaolinite, before (left) and after (right) treatment

Fig. 7 clearly shows that the polymer is able to form a
nearly continuous film covering also the crystals of clay
minerals. The thickness of the polymer films mainly ranges
below 1 micron. The polymer microfilms not show fracturing
or blistering. Thus, the micro-mechanical stresses due to
short term periods of alternating moistening and drying of
swellable clay minerals, attributing substantially to the
physical decay of the stones, are reduced to a great extend.

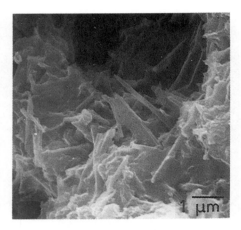

Fig. 8: Concrete B35, before (left) and after (right)
        treatment

As regards fig. 8, the resin forms a continuous microlayer
and bridging the micropores of the concrete substratum and
covering the cement crystals.

SEM studies are supported by mercury-porosimetry to measure
even slight changes within the distribution of pore radii
as a result of the treatment. Differences in pore volume
are expressed as percentage over the distribution of pore
radii and show a clear mark, which can be related to the
effectiveness of the treatment. In the case of the sandstones, pore radii smaller than 5µm are clearly reduced,
while there are no significant changes within the larger
pore radii ranks. The large pore channels are still available for a gaseous transport of humidity through the treated volume of the stone or concrete /5/.

EVALUATION OF EFFECTIVENESS

The overall effectiveness of a treatment can only be
characterized by multiple effectiveness features. In order
to carry out a computer supported, quantitative evaluation

of the complex effectiveness, criteria are differentiated
into ranks. At present status only 7 effectiveness criteria
of about 20 are taken into account for optimizing the
treatment systems. The quantitative intensities of these
features allow the determination of an effectiveness coefficient. For the most promising products further developments are discussed with the producers of the polymers /2/.

CONCLUSIONS

The research is aimed at establishing the most adequate
consolidant and protective methods for different porous
materials. Investigations on the pore structure indicate
that some epoxies and polyurethanes show good penetration
into the substrata and form nearly continuous microlayers
covering the mineral pore walls. They do not seal capillary
pores, but only the micropores of the stone. They tend to
form a fine supporting bandage and do not drastically
effect the elastic modulus or the colour, moreover they
give strength and protection including increased water
repellent property. On the basis of these results a selection of the most effective products was made. Some test
products seem to be worth for further development. In
connection with the chemical industry these products will
be developed and tested in a more sophisticated way. Of
these, studying the polymer treated material under natural
weathering conditions and in complex short term laboratory
simulation tests, will be the most important work in future
investigations.

REFERENCES

/1/ SASSE, H. R.: Polymere als Schutzstoffe für Natur-
stein. Bautenschutz und Bausanierung Sonderausgabe,
(1) 1987, S.65-88. 1. Statusseminar des BMFT, Mainz,
17./18. Dez. 1986

/2/ HONSINGER, D.; SASSE, H.R.: Neue Wege zum Schutz und
zur Substanzerhaltung von Sandsteinoberflächen unter
Verwendung von Polymeren. Bautenschutz und Bausanierung, Heft 6, 1988

/3/ HONSINGER, D.; FIEBRICH, M.: Polymers for consolidation and preservation of natural stone. Proceedings.- 2nd International Conference on Engineering
Materials. Bologna, 19.-23.06.1988

/4/ DIN 52615 11.87. Wärmeschutztechnische Prüfungen.
Bestimmung der Wasserdampfdurchlässigkeit von Bau- und
Dämmstoffen

/5/ HONSINGER, D.; SASSE, H.R.: Alteration of micro-
    structure and moisture characteristics of stone
    materials due to impregnation. In: Proc. Vth Inter-
    national Conference on Durability of Building
    Materials and Components. Brighton, U.K. 7.- 9.11.1990

/6/ WITTMANN, F.H.; PRIM, R.: Mesures de l'effet
    consolidant d'un produit de traitement. In: Materiaux
    et construction (RILEM), 16 (1983), Nr.94. S 235-242

/7/ ROSSI-DORIA, P.: Pore Structural Analysis in the
    Field   of Conservation. State of the Art and Future
    Developments, Proceedings. RILEM/CNR International
    Symposium. Principles and Applications of Pore
    Structural Characterization, Milan 1985

/8/ HONSINGER, D.: Die Bedeutung von Strukturuntersuchun-
    gen im Elektronenmikroskop für die Qualitätssicherung
    von imprägnierten Sandsteinen. In: Tagungsband,
    14.Vortragsveranstaltung des Arbeitskreises "Raster-
    elektronenmikroskopie in der Materialprüfung",
    25.-27.04.1990. Berlin 1990

/9/ BURCHARD, W.G.; CLOOTH, G.; SASSE, H.R.; HONSINGER, D.:
    Rasterelektronenmikroskopische Strukturuntersuchungen
    an polymergetränkten Naturwerksteinen als Hilfsmittel
    zur Wirksamkeitsbeurteilung von Schutzmaßnahmen. In:
    Beiträge zur elektronenmikroskopischen Direktabbildung
    von Oberflächen 20 (1987). 20. Kolloquium des Arbeits-
    kreises für elektronenmikroskopische Direktabbildung
    und Analysen von Oberflächen (EDO)

PROLONGING THE LIFE OF REINFORCED CONCRETE STRUCTURES BY SURFACE TREATMENT

LP McGill
M Humpage

Fosroc Construction Chemicals,

United Kingdom

ABSTRACT This paper provides an indepth evaluation of the various types of surface treatments that can be utilised in the protection of concrete structures. Penetrants are low viscosity liquids that penetrate into the concrete, line the pores and provide an effective barrier to chloride salts. The majority of commercially available penetrants are based on either silanes or siloxanes. Silanes are claimed to be more penetrative than siloxanes but from the work undertaken at Fosroc no discernable difference in penetration could be found between the systems except when the concrete samples tested were completely dry. The lower volatility of siloxanes means that under' site conditions more active material is retained in the surface of the concrete. Acrylic/methacrylic pigmented coatings provide a very effective physical barrier to carbon dioxide and other acidic gases. In addition these have excellent UV resistance, colour stability and a minimum life expectancy of 15 years. The most effective method of ensuring long term protection to a reinforced concrete structure is to use a combination system which consists of a siloxane/acrylic blend primer and a pigmented acrylic topcoat. This system acts synergistically to produce the equivalent of an extra 500mm of concrete cover over the reinforcement and, in addition, enhances the appearance of the structure.

Keywords: Concrete, Silane, Siloxane, Carbonation, Coating, Sealer Penetrant, Acrylic, Methacrylic, Flexible.

Paul McGill is a Marketing Manager working in the Concrete Repair and Protection business sector at Fosroc Construction Chemicals. He has spent over six years in the construction chemical industry. He is currently responsible for the promotion of the company's coating products worldwide.

Mick Humpage is the Resin Laboratory Manager at Fosroc Technology. He has 14 years experience in coatings and construction chemicals development and technical service. He is currently responsible for Fosroc's resin based products, principally coatings systems.

INTRODUCTION

Concrete reinforced with steel provides man with an ideal building material for the modern environment. It is a low cost, hard, durable product that is easy to manufacture at the point of construction and has excellent compressive and flexural strength. However it is this very simplicity in application, coupled with poor quality control, that has led to the enormous problems of deterioration now being experienced world-wide. Theoretically, the combination of steel and concrete is complementary, with the steel providing the strength and the alkaline nature of the concrete providing passivation of the reinforcement and preventing oxidation[rusting]. In practice, minimal attention paid to the amount of cover over the reinforcement, as well as the inadequate design and control over the quality of the concrete mix, has meant that the boom in the use of concrete over the last thirty years has created enormous problems which are only just coming to light. The present cost of repairing corrosion-damaged concrete is estimated to be approximately £6 billion world-wide. In the UK, nearly 40% of the total annual construction bill of £35 billion [1] is spent on maintenance and refurbishment, of which a significant part is spent on repairing deteriorated concrete.

The main agents responsible for the corrosion of reinforced concrete are chlorides, carbon dioxide and other acidic gases. These materials, in the presence of oxygen and moisture, break down the passivating film on the reinforcing steel produced by the alkalies in the concrete and therefore lead to corrosion. Chlorides may be present in the concrete from contamination of raw materials or possibly from accelerating additives. Alternatively, chlorides in solution can also leach into the concrete via the capilliary pores on the surface. Thus potential problem areas are structures in coastal regions and bridges that are regularly exposed to de-icing salts.

Carbon dioxide and other acidic gases are present in the general environment and are paticulary concentrated in polluted or built-up areas. These gases diffuse into the concrete, reducing the inherent alkalinity and thus lead to corrosion. The rate of corrosion is, in part, dependent on the porosity of the concrete, with low permeability mixes giving up to ten times the protection of poorer quality concrete. Even with a well designed concrete mix, poor site practice can lead to low cover, poor compaction, inadequate curing, high permeability and cracks.

Once corrosion of the reinforcement occurs, rust is formed which occupies many times the size of the original volume of steel. Once these expansive forces exceed the tensile strength of the concrete cover, cracks will occur which will allow further ingress of the corrosive agents and acceleration of the problem. The visible outward signs of this underlying distress are cracking, rust stains, and spalling of the concrete. The final result, if left unchecked, is loss of structural integrity and possible collapse.

Although there is no universal answer to the mechanisms of corrosion, there is a growing awareness of its causes. This has led to much greater attention being paid to detail during new construction. Specifications have become much clearer with the importance of admixtures and cement replacements being much more widely recognized and used.

This paper evaluates the various types of surface treatments that are available to protect reinforced concrete from environmental attack. The most effective results are obtained when the product is generally one component of a good quality concrete repair system that can be used to reinstate even heavily deteriorated structures to their original design life. It should not be overlooked, however, that such treatments can also be used in preventative maintenance programmes to retard or halt the ingression of aggressive chemicals into structures before the corrosion process begins. The systems reviewed are based on a wide range of polymer types and provide significantly different levels of protection. They have been classified under five: headings penetrants, sealers, coatings, cement finishes and combination systems.

PENETRANTS

These materials are generally low viscosity liquids that penetrate into the concrete and line the pores. They are colourless and therefore have little effect on the appearance of the structure. The systems are hydrophobic and thus repel water and water-containing chloride salts. They do not block the pores of the concrete and so allow the passage of potentially harmful water vapour and other gases which may otherwise remain trapped within the structure. The majority of commerically available penetrants are based on either silane, siloxane or silicone resins.

Silanes and Siloxanes

Silanes are more correctly described as alkylalkoxysilanes. The most widely used "monomeric silane" for the protection of concrete is Iso-Butyl Trimethoxysilane. Silanes become reactive in the presence of moisture with the speed of reaction being governed by the surrounding pH. Thus in normal alkaline concrete the silane will react with the pore lining quite rapidly, but if the substrate is neutral eg., brick, stone etc, or there is no moisture, no reaction can take place. Since much of the concrete requiring the treatment has lost its alkalinity due to the action of acidic gases, the reaction time is frequently slow. While the reaction takes place the volatile silane will continually evaporate and to maximise the chances of success, it is therefore necessary to use very high concentrations of silane (up to 100%).

Siloxanes   are   more   correctly   described   as   oligomeric
alkylalkoxysiloxanes.   They have virtually all the advantages of
silanes with respect to reactivity and water repellency but, in
addition, have low vapour pressure. Consequently they have been used
for a number of years in North America for the protection of concrete
highways and structures [1,3].   They may, however, under very dry
substrate conditions exhibit slightly less penetration than silanes.

In tests performed at Fosroc Technology air cured concrete blocks
150*100*50 mm were acid etched and placed under water for 16 hours.
They were then stored at 20°C and 65% RH and application of
penetrants were made after certain time intervals. After curing for
a further three days, penetration depths were measured by applying
ink to the broken surfaces of the blocks. From the results in it can
be seen that the monomeric silane was only slightly more penetrative
when the blocks were completly dry. In all other situations there was
no discernable difference between the penetration of the oligomeric
siloxane and the monomeric silane.

Table 1   Comparison of Penetration Depths of Monomeric Silanes
and Oligomeric Siloxane

| | | Penetration Depths (mm) | |
Drying Time	Moisture Content(%)	Monomeric Silane	Oligomerous Siloxane
8 hours	3.9	1-2	1-2
1 day	3.7	1-2	1-2
5 days	3.2	2-3	2-3
9 days	2.4	3	3
28 days	1.5	5	3

Notes.   A control block was dryed at 120 °C to determine the
moisture   content.

Concrete Used

	Kg/m3
OPC	300
20-25 mm Aggregate	1263
Sand	625
Water/Cement	0.5

Further tests were undertaken to determine the amount of reactive material that could be expected to be retained on a cementitious substrate under simulated site conditions. In these tests circular mortar discs 48*5mm were prepared and cured in sealed bags. When cured they were air-dried to constant weight at the required test temperature. The penetrants were then applied and the discs were weighed at intervals during drying at 20 °C, 35 °C and 20 °C in a wind tunnel (air movement 6m/s). From Table 2 it can be seen that only 14% of monomeric silane was retained in the pores of the substrate at 20 °C. If the temperature was raised to 35 °C or if the sample was placed in a wind tunnel then the amount retained was reduced to less than 7%. In contrast 68% of the oligomerous siloxane was retained in the pores at 20 °C and a similar percentage was retained at 35 °C and in the wind tunnel. Thus it can be concluded that under site conditions siloxanes are more likely to be retained in a cementitious substrate and therefore give a more controlled degree of protection. With the inherent variability of concrete and application conditions there must be some uncertainty as to how much protection monomeric silanes can give.

Table 2 Comparison of Volatility of Monomeric Silane and Oligomerous Siloxane Under Simulated Site Conditions

Test Conditions	20 °C no wind	35 °C no wind	20 °C wind speed 6m/sec
Retention of applied active material (%)			
Monomeric Silane	14±5	6±5	7±5
Oligomeric Siloxane	68±5	68±5	68±5

Mortar Used   3:1 Sand: Cement and 0.15 water: powder ratios

Silicone Resins

These materials are much higher in molecular weight than silanes or siloxanes and thus are not penetrative; ie. penetration of more than 0.1mm is considered unlikely under site conditions. They are not normally reactive materials and dry by solvent evaporation to leave a surface film of resin. The substrate has to be air-dry before application of the product and the systems can be susceptible to dirt pick up and weathering. Their use has declined in recent years.

SEALERS AND COATINGS

Sealers are intermediate between penetrants and coatings and work by blocking the pores of the substrate. They are more viscous than penetrants and generally form a film on the surface of the concrete. This group includes silicates, silica fluorides and linseed oils, and their effectiveness as a barrier against environmental attack has been shown to be limited.

Coatings are more viscous than sealers and provide protection to the structure by forming a relatively thick film on the surface (up to 1mm). These materials generally consist of binder, pigments, fillers and a carrier. The binder is the polymer which can be of various types. It imparts the majority of the properties of the system including anti-carbonation resistance and, in some cases, flexibility. It is generally acknowledged that a degree of flexibility is needed in coating systems so that the anti-carbonation properties are not reduced in areas of the structure which are subjected to movement, ie. if cracks appear due to, for example, thermal movement then a flexible film is needed to bridge the gap to maintain an integral barrier. The pigments provide colour while the fillers provide thickness and surface texture. The carrier is either water or solvent and is incorporated into the system to reduce the viscosity and aid application. Solvents generally have a much greater effect at reducing the viscosity and improving early age washout resistance as well as improving low temperature application and drying properties. However, concern is increasing about the long-term health risks associated with using solvents with the inevitable increase in interest in water-based systems. A large number of polymer types have been used in coatings including chlorinated rubbers, urethanes, epoxies and acrylics.

Chlorinated Rubber

Chlorinated rubber coatings have been available for many years and have good moisture, alkali and anti-carbonation resistance. They also show a degree of flexibility but have poor UV resistance and yellow easily. Because these coatings are externally plasticised they also have poor dirt shedding properties.

Polyurethanes

Polyurethane coatings can be either one component, ie. moisture curing, or two component systems. They can be formulated to produce flexible anti-carbonation coatings with excellent UV and abrasion resistance. Generally the two component systems have improved chemical and physical properties in comparison to the one pack. The main disadvantage of polyurethanes is their limited 'breathability' and thus may be more susceptible to Freeze/Thaw damage. The systems are also sensitive to moisture with damp substrates being particularly problematic. Their excellent solvent resistance means that urethanes are particularly useful for anti-graffiti systems in exterior locations.

Epoxies

Epoxy coatings can be either in emulsion form, solution form or
solvent free. They are almost always two component and based on
liquid resins and hardeners. The solution epoxy systems are generally
of lower viscosity than the solvent free and can therefore penetrate
more effectively into the substrate.   Water borne systems are
available but these are not as penetrative. As a class of paints,
the epoxies have excellent chemical and abrasion resistance as well
as good anti-carbonation properties.   The major disadvantages of
these systems is a tendency to chalk when subjected to UV light in
external situations. Recent improvements in formulation technology
have reduced the chalking significantly but still the problem
remains. Epoxies have limited breathability and are not generally as
flexible as polyurethanes.

Acrylics

Acrylic and methacrylic resins have been successfully used for a
number of years as the base polymer of a wide variety of masonry
coatings. They are light in colour and thus it is possible to produce
non-pigmented systems. However 'clear acrylics' generally darken the
substrate and because of their low film build do not give as good
anti-carbonation resistance as conventional pigmented systems.
Pigmented products based on methyl methacrylate resins diluted with
a range of different solvents have excellent anti-carbonation
resistance, ie. many times greater then the generally recognised
industry standard of 50m of air (Klopfer Method) which is equivalent
to 125 mm of 30N/mm2 concrete. They are also durable, have excellent
dirt shedding and UV properties and have a life expectancy well   in
excess of 15 years.

The limitation of this standard type of system is the lack of cured
film flexibility. Thus in areas where the concrete is susceptible to
cracking or movement, a quite common situation, the coating film can
itself crack and thus allow direct access of external corrosive
elements. The traditional method of overcoming the lack of
flexibility is by adding an external plasticier. The flexibility is
increased but at the expense of the dirt shedding properties.
This is because the flexibiliser is not incorporated into the
backbone of the polymer and the surface of the coating continues to
exude the plasticiser even after the system has cured. This attracts
dirt.

Flexible acrylics which incorporate the flexibility in the base
polymer (ie. are internally plasticised) have all the properties of
conventional acrylics in terms of anti-carbonation resistance,
breathability and UV stability, as well as the ability to span
incipient cracks from 0 - 3mm. By the incorporation of a catalyst,
the systems can UV cure on the surface and thus minimise long-term
dirt pick up.

CEMENT BASED FINISHES

There are a variety of cementitious finishes available ranging from simple washes through to high-build trowel or sponge-applied materials. These systems will generally provide a cosmetic and, to some extent, protective finish. Perhaps the most versatile of these are the polymer modified materials which give a comparatively low-cost finish with reasonable durability combined with the ability to bond to damp surfaces. Problems do occur however when very low-cost polmers such as polyvinyl acetate, which is not hydrolysis resistant are used.With these materials failure can be embarrassingly rapid. An interesting recent development is the use of highly flexible acrylics which, with the cementitious component, give a product that has a small degree of flexibility and enables the finish to cope with very fine movement in any hair-line cracks that may exist in the underlying concrete.

Some cementitious finishes containing "crystalline growth materials" which are purported to migrate into the pores of the substrate and crystallise, thus blocking the pores, have been proposed as chloride and carbon dioxide barrier treatments. However, there is doubt about their performance.

COMBINATION SYSTEMS

Combination systems have been formulated to provide the maximum protection possible to concrete structures. These products were initially developed in North America and have now been used successfully for a number of years [4,5]. They combine excellent chloride resistance with outstanding anti-carbonation properties. One of the most sucessful systems consists of a silane/siloxane acrylic blend primer with a pigmented acrylic topcoat. In this system the low viscosity, low volatility silane/siloxane-containing primer penetrates the substrate and gives the system its resistance to water soluble chloride salts. In addition, the small percentage of acrylic resin contained in the system acts as a masonry stabiliser and conditions porous or substrates. The topcoat is a pigmented acrylic which film-forms over the surface of the concrete and gives the system is resistance to carbonation. The combination of the two has been shown to exhibit a synergistic effect etc while still providing breathability, thus allowing potentially harmful water vapour to escape from the structure. Two coats of the topcoat are required to minimise the likelihood of pin-holing which would significantly reduce the overall properties of the system.

To provide optimum performance, and to ensure than the aesthetics of the structure are always enhanced, it is necessary to have a range of combination systems in various colours to provide as wide a choice as possible. Flexible systems are needed for areas that are subject to movement and textured coatings are required to mask imperfections on rough substrates. It should be noted that textured systems when used alone generally fail to provide a sufficient barrier to carbonation due to the limited film-build between the high profile peaks. Thus, if anti-carbonation resistance is required, textured coatings must be used in conjunction with a single application of a smooth topcoat.

Table 3 shows a comparison of the performance of untreated $30N/m^2$ concrete with a combination coating system consisting of a siloxane penetrant and a flexible acrylic topcoat. It can be seen that 1 mm of coating gives the same chloride resistance as over 1 m of concrete and the same anticarbonation resistance as over 300 mm of concrete. In addition the combination coating system is unaffected by freeze/thaw cycling whereas under the same test regime the concrete was destroyed.

Thus it can be concluded that the combination of a penetrating primer and topcoat gives significantly improved resistance to chloride and carbonation attack. In addition these systems are expected to have a minimum working life of 15 years.

Table 3   Comparison of Performance of Untreated Concrete with a
Combination Coating System Consisting of a Siloxane
Penetrant and a Flexible Acrylic Topcoat

	Combination Coating System	Ratio of Performance CombinationCoating:Concrete
Coverage m2/litre primer	2.5	--
Topcoat	1.25	--
Carbon Dioxide Diffusion Resistance (Klopfer Method) Air Equivalent Thickness (m)	>50	>300:1
Chloride Ion Resistance (Reduction in Penetration in comparison to mortar control %)	>91	>1000:1
Crack Bridging Ability (Modified ASTM C836-84)	10 mm	--

	Combination Coating System	Concrete
Freeze Thaw Resistance (ASTM C672 50 Cycles)	Unaffected	Destroyed

REFERENCES

1   McLain   L.Constructing A Cure For An Epidemic - Financial
     Times 18.1.90

2   Bradbury A, and Chojnacki R, "A Laboratory Evaluation of
     concrete surface sealants" Ministry of Transport and
     Communications, Toronto, Ontario, Canada 1985.

3   Rizzo E and Bratchie S "The use of penetrating sealers for the
     protection of concrete highways and structures. Journal of
     protective coatings and linings January 1989.

4   Pfeifer and Scali " MJ National Co-operative Highway Research
     Program Report 244 "Transportation Research Board, Washington
     DC, December 1981.

5   McCurrich LH and Jeffs P The use of penetrating sealers for
     the protection of concrete highways and structures, conference
     on concrete in transportation, Vancouver, Canada September
     1986.

SURFACE COATINGS - SPECIFICATION CRITERIA

P C Harwood

Sika Limited,

United Kingdom

ABSTRACT. The progress made in investigation and repair techniques of reinforced concrete structures has rapidly increased during the last decade and will continue to do so as the construction industry learns from past and present mistakes. The owner or party responsible for the maintenance of a reinforced concrete structure in 1989 now has ample and informed sources from where he may seek advice concerning the degree of existing deterioration and the likely rate of future failure. Surface coatings, whilst playing a vital role in protecting against original failure of the new structure or the on-going deterioration of the repaired structure, have had insufficient field . study to take into account the severe exposure in many parts of UK and overseas. This paper discusses the criteria needed before specification or use of any surface coating by considering several examples of different structures with different specific needs.

Keywords: Carbonation, Deterioration, Phenolphthalein, Depassivation, Diffusion, Impregnation, Methylmethacrylate, Ethylene Copolymer, Siloxane, Cavitation.

Mr Peter Harwood is International and Special Projects Manager at Sika Limited, Welwyn Garden City, Hertfordshire, where he has concentrated on repair and coating protection of reinforced concrete structures for the last ten years in both UK and a number of overseas countries.

INTRODUCTION

The degree of erosion or weathering of any given structure, be it an occupied building or civil structure, will depend on a combination of factors, including concrete mix design, curing, detailing and exposure.

Should the structure be painted with a protective coating of any sort during its service life, it is absolutely essential that careful and informed consideration is given as to the precise reason for such a move.

Life itself is a compromise and no less so the selection of a repair and protection package. Should the structure have been built using poor quality or contaminated materials, with bad workmanship and detailing, then the application of any surface coating must necessarily be considered to be a short-term compromise.

Work carried out by Smith and Evans (Ref 1) indicates the variation in deterioration of similar structures where detailing, concrete quality and exposure affect the rate of deterioration.

Figure I

Figure I shows a poorly detailed car park structure in a severely exposed marine environment where both the presence of air bourne chlorides and the high relative humidity have, together with very high levels of concrete carbonation, caused severe deterioration.

Coatings applied within two years of construction have assisted partially, but have not been of sufficient cohesive strength to accept both thermal and dynamic movement of the structure.

It is therefore essential that whoever is responsible for the use or specification of the surface coating on the repaired or new structure must not expect inherent problems to vanish, simply because they are now covered up; indeed the reverse may be the case! If, for example, the concrete contains chloride ions and is continued to be "fed" with water from behind the coating (from leaking sanitation, storm water, etc.), corrosion activity could well increase rather than decrease from application of any coating.

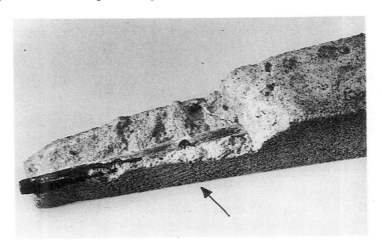

Figure II

Figure II shows a (35 n/mm2 w/c ratio 0.35) concrete prism cast with deliberately variable cover over the reinforcing bar which was cleaned prior to casting. After three year's exposure in a mild rural environment of very low relative humidity, the concrete prism was cut open and sprayed with phenolphthalein indicator to determine the variation in carbonation. The small, apparently insignificant crack just visible on the lower middle of the block has clearly contributed towards depassivation around the reinforcing steel locally.

At the end of the block where the concrete has carbonated from both sides, the corrosion is greater as indicated from Smith and Evans (Ref 3) findings in the accelerated deterioration of slender, more architectural structures.

Although corrosion activity, as evident on the steel reinforcing bar in the centre of the experimental prism, is slight and is unlikely to contribute to expansion stresses to cover concrete, the significance of this minor defect in the depassivation around the reinforcing steel cannot be ignored. This deterioration will of course proceed far more rapidly in poor quality concrete in a chloride laden environment.

Work by both Smith and Evans (Ref 1) and Fattuhi (Ref 2) confirm carbonation of concrete in Arabian Gulf States, for example, is considerably higher than in Northern Europe and, therefore, care and thought in the specification of any surface protection must be greater.

Surface coatings must, therefore, form part of a considered approach to the protection of the given structure and only after several preceeding steps have been taken.

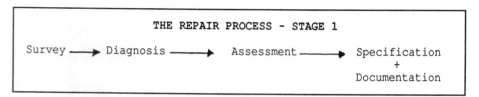

Figure III

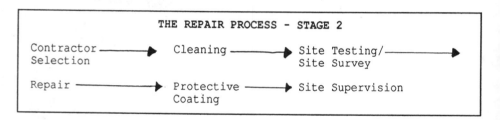

Figure IV

Figures III and IV suggest this approach must also include provision for inspection and maintenance. Nothing manmade lasts forever and there are many structures the author has inspected where deterioration has continued as a result of damage in the surface coating which simple, routine, programmed maintenance would have arrested.

SURFACE PREPARATION

Preparation of any substrate prior to the use of a surface coating is fundamental. The failure of the surface coating in a number of cases can be wholly attributable to the non-existent surface preparation.

All concrete surfaces, whether new or weathered, will have weak cement laitance, oil, grease or surface contamination deposits and if these are not removed, the coating will fail sooner or later.

All concrete substrates, new or weathered, contain surface flaws, pinholes or areas of grout loss and these must be treated (dependent on size) prior to the use of any surface coating.

A surface blowhole or pocket is always larger inside than it appears on the surface and even if the surface coating does apparently bridge or fill the blowhole, the air contained in the blowhole will expand with temperature and very often form a break in the coating.

Depending on location, this may only represent a visual defect, but in most cases, the whole purpose of the coating is a protection against deterioration and a defect will usually reduce the effectiveness of the protection.

Should the surface coating be applied onto a rendered substrate, the bond or effectiveNEss of the render is just as important as the concrete substrate.

Water blasting with or without nozzle introduced fine abrasive or controlled abrasive blast cleaning may be used to clean and prepare the concrete to expose surface flaws.

Removal of existing defective surface coatings will require the use of controlled abrasive blast cleaning ensuring the abrasive is clean chloride free propriatory material.

COATING SELECTION

The choice and selection of the surface coating is most important, depending on the specific requirement of the structure.

Fattuhi (Ref 2) found that one coating in his evaluation programme, an ethylene copolymer, reduced carbonation rates by almost thirty five fold.

Both Robinson (Ref 3) and Hankins (Ref 4) make it clear that there are considerable differences between the diffusion characteristics of similar generic types of coating and it is essential therefore that the specifier has a clear idea of precisely what he is trying to achieve by the application of the coating, and in specifying such a coating gives as much detail as possible.

Important work by Taywood Engineering Ltd has indicated the significance of weathering on the performance of the coating.

Weathering from ultra violet light, changes in humidity and temperature will deteriorate the binder in the surface coating and reduce the resistance to thermal stresses and reduce progressively the dry film thickness of the coating.

An indication of the resistance of the coating to water ingress or carbonation early in its exposure life may therefore be quite misleading to the client, who clearly must be interested in the continued protection of his building in year five or year ten.

CRITERIA FOR CONCRETE COATING SELECTION

- Water Ingress
- Carbon Dioxide Diffusion Resistance
- Chloride Ingress
- Water Vapour Diffusion Resistance
- Ultra Violet Light Resistance
- Elastic/Crack Bridging Abilities
- Chemical Resistance
- Abrasion Resistance
- Ease of Application
- Life Required to First Maintenance
- Ease of Overcoating
- Aesthetic Appearance

Figure V

Figure V gives a basic checklist for the specifier to select a coating according to the specific properties he requires.

It almost certainly will be the case that in any given structure there will be several different properties required from a coating which may not be met from one material only.

Great care is also required by the specifier to take into account a variety of factors not least of which is the suitability for the location in question. Materials, for example, which work well and may be applied easily in European conditions may be quite unsuitable for locations elsewhere in the world.

COATING EXAMPLES

Three Year Old Two-Storey Housing Accommodation

These units comprise pre-cast panel construction with insitu concrete frame and long cast-insitu beams which run along ground, first and second floors. Location is a UK coastal town.

The long-term requirements of the surface coating are summarized as follows:

- Resistance to moisture and wind bourne chloride ions.
- Elastic crack-bridging coating on beam locations.
- Resistance to carbonation (R Value).
- High water vapour transmission (SD value).
- Ease of application (factory application to pre-cast units).
- Resistance to UV light breakdown of the binder (chalking).
- Ease of overcoating when required with a variety of locally
  available materials.

Over all of these structures, it is essential to provide a maximum
protection against both carbonation and chloride ions. The latter may
be air bourne and with the relatively high coastal humidity,
particularly at night, small defects in the concrete will provide the
starting point of loss of resistivity and passivation with consequent
corrosion risk to embedded reinforcement.

An impregnation based on a solvented siloxane to form a positive
barrier to water/chloride ions followed by at least 200 microns dry
film thickness of a coloured methylmethacrylate coating.

The siloxane will also provide a benefit in lining fine surface
blowholes and plastic shrinkage cracks in the surface with the
methylmethacrylate coating giving excellent protection against
carbonation and weathering.

Klopfer (Ref 5) indicates minimum and maximum "R" and "SD" values for
carbonation and water vapour transmission, and in practise, most high
quality proprietary methylmethacrylates exceed these values by many
times.

Great care should be taken to ensure that weathered results for the
coating selected for specification are requested and inspected. Work
carried out by Taywood Engineering Ltd clearly indicates that coatings
of identical generic form deteriorate differently in accelerated
weathering conditions. "R" and "SD" values, therefore, may markedly
differ from time zero results.

Robinson (Ref 3) suggests that water based coatings and solvented
based coatings of identical generic form offer similar "R" and"SD"
values. Current development progress by coatings manufacturers will
increase the availability of water based equivalents of existing
successful solvented systems with improved ecological acceptance.

The horizontal beams may benefit from a crack-bridging protection to
abridge fine cracks opening at night with a drop in temperature. The
rise in humidity will increase the risk of condensation on the
underside of the beam and the resultant risk of chloride induced
corrosion.

A correctly formulated ethylene copolymer will bridge cracks pro rata
with its dry film thickness and provide good "R" and "SD" values
respectively.

Should the beam be used for foot traffic by window cleaning services, then further attention must be given to abrasion protection.

A New Motorway Link Bridge in a Coastal Environment

A link tunnel is part of this development which has increasing heavy industrial expansion.

The long term durability of concrete surfaces here will be substantially dependent on the waterproofing durability of the bridge deck. Failure of the waterproofing of the deck or the joints of the bridge will inevitably result in chloride ions and possibly corrosion producing chemical spillage from adjacent industrial output leaking through the deck. An elastic membrane waterproofing system is highly recommended as a priority here over surface coatings.

Summary of the Long-Term Requirements of the Surface Coating

- Resistance to moisture and air bourne chlorides.
- Resistance to carbonation.
- Crack-bridging properties (areas subject to dynamic movement).
- Abrasion resistance (lower columns adjacent to traffic movement).
- High UV resistance.

An impregnation using a solvented siloxane will serve as a proven protection against water and air bourne soluble or precipitated corrosion promoting materials. This will leave the appearance unchanged.

A hydrophobic impregnation, however, such as a siloxane will not protect against carbonation. Should carbonation resistance be required, use either a solvented or water based methylmethacrylate or for crack-bridging protection on areas subject to dynamic or thermal crack movement, an ethylene copolymer coating may be used.

Both these systems have high durability in UV light.

Lower areas of the bridge structure subject to abrasion from traffic movement may be protected with a sovlent free epoxy coating having high resistance to chloride ingress and mechanical damage. A high degree of light reflection from the tunnel is desirable and proof of this will be required of the coating manufacturer.

Concrete Spillway of a Dam

As water speeds of up to 30 metres per second could be flowing over the surface, it is absolutely essential that a completely smooth, pore/defect free surface is achieved to avoid the potential errosion and cavitation effect from the water flow.

Most conventional reaction cured epoxy or polyurethane surface coatings will be unsuitable in the long term for protection of such a spillway, and it is therefore fundamental to ensure the proven service life of any surface coating protection.

The original concrete specification, placing and curing is also critical, in order to achieve concrete with as few voids as possible.

After a minimum curing period of 28 days, abrasive mechanical preparation of the concrete substrate will be necessary by either water or dry abrasive blast cleaning in order to achieve a surface free of cement laitance and with all surface imperfections exposed.

If surface blowholes/imperfections are not exposed at this stage, be completely assured exposure and consequent cavitation will result after early to medium usage of the structure.

After preparation, it is necessary to completely level the substrate using a scraping/levelling application of a proprietary reaction cured, slightly thixotropic epoxy binder of fine aggregate combination. The application of such a material is important in order to completely fill and correct the surface by an unbroken scraping/levelling coat over 100% of the surface area.

If this is not carried out absolutely, fine air voids, virtually invisible to the naked eye, will become rapidly visible as air bubbles/voids after the application of the finishing surface coating.

The final surface coating must be of a thixotropic proprietary reaction cured epoxy or equivalent in two or more applications by either brush, roller or airless spray to achieve a final dry film thickness in excess of 1000 microns (1.0 mm).

Thixotropy is important in order to avoid the flow of uncured material down the inevitable steep fall of the spillway and the final dry film thickness must be representative of identical material in identical exposure conditions.

It will be virtually impossible and disproportionately costly to take the spillway/s out of commission for remedial works if either specification or application are found wanting.

Summary of surface coating requirements:

- Proven resistance to abrasion from 30 m/sec water flow
- Ease of conventional application under steep fall/vertical locations
- NWC certification (potable water)
- Overcoatable/repairable with the minimum of surface preparation
- Proven safety of application in confined spaces.

CONCLUSIONS

The random selection of surface coatings chosen for initial cost or appearance considerations alone are very likely to result in early failure.

Care is needed to consider precisely the individual requirements of the given structure, taking into account the local conditions of exposure.

All parties involved in the specification or use of the surface coating protection of a given structure must consider the foregoing, therefore, to avoid quite preventable failures.

REFERENCES

1.   D G E Smith and A R Evans of Scott Wilson, Kirkpatrick & Ptrs. "Purple Concrete in a Middle East Town". Concrete Magazine, February 1986.

2.   N I Fattuhi, University of Kuwait. "Concrete Carbonation and the Influence of Surface Coatings". Structural Faults and Repair, London 1987.

3.   H L Robinson, Taywood Engineering Ltd. Evaluation of Coatings as Carbonation Barriers. 2nd International Colloquium on Materials Science - Esslingen, West Germany, September 1986.

4.   P J Hankins. The Use of Surface Coatings to Minimize Carbonation in the Middle East. 1st International Conference Deterioration and Repair of Reinforced Concrete in the Arabian Gulf - Bahrain 1985.

5.   H Klopfer "The Carbonation of External Concrete and How to Combat it". Bautenschutz and Bausanierunig No.3 Pages 86-97, 1978.

METHODS OF PROTECTION CONCRETE COATINGS AND LININGS

Session B

Chairmen

Mr D Kruger

Department of Civil Engineering,

Rand Afrikaans University,

South Africa

Professor T Oshiro

Department of Architectural Engineering,

University of the Ryukyus,

Japan

DURABILITY OF SURFACE TREATED CONCRETE

P A M Basheer

F R Montgomery

A E Long

M Batayneh

Queen's University, Belfast

United Kingdom

ABSTRACT. The permeabilities, to both air and water, of the surfaces of laboratory prepared specimens of concrete which had been coated with a range of sealants were measured at various ages using the 'Clam'[1,2], a site portable, non destructive permeability testing equipment developed at The Queen's University of Belfast. The sealants used were mainly silane or silane based in combination with other materials. The prepared samples were then subjected to a regime of freeze/thaw testing to further assess the effectiveness of the coatings. It was found that the performance of the various materials was determined by the quality of the base concrete, that silane is an effective means of reducing low pressure water permeability if applied without an alcohol, and that to reduce air permeability, silane must be augmented with other material such as acrylic. Freeze/thaw performance was found to be closely related to surface water permeability and not to air permeability. From these results it is seen that the likely effectiveness of a sealant may be measured on site after application.

Keywords: Concrete, Permeability, Sealants, Silane, Acrylic, Permeability Test, Freeze/Thaw.

Mr P A M Basheer is a postgraduate student at the Department of Civil Engineering, The Queen's University of Belfast, UK. He is on leave of absence from the Regional Engineering College, Kerala, South India, where he is a lecturer in civil engineering.

Dr F R Montgomery is a lecturer at the Department of Civil Engineering, The Queen's University of Belfast, UK.

Prof A E Long is Director of the School of the Built Environment and Dean of the Faculty of Engineering at The Queen's University of Belfast, UK.

Mr Malek Batayneh is a Research Assistant at Oxford Polytechnic, UK. Until recently he was a postgraduate student at the Department of Civil Engineering, The Queen's University of Belfast, UK.

INTRODUCTION

It is now accepted that concrete deteriorates with time, not so much as a result of any lack of strength, but usually as a result of an inadequacy in one of its other properties. It may be that the surfaces exposed to the environment were inadequately cured and, though the bulk of the material may be adequate, the surface is not up to the task of surviving the ravages of time and pollution. On the other hand it may be that the curing was adequate but that the mix was specified on strength only, often a recipe for trouble. The end result is the same, a sad looking structure, old and deteriorating before its time.

It is our belief, and commonly held, that the deterioration of concrete structures due to environmental factors is determined almost completely by the ability of the surface to keep out the harmful agents in the environment. To survive, a surface must be able to resist the ingress of such things as carbon dioxide, chloride ions dissolved in water and even water itself in excess quantities. While many of these things are not of themselves harmful to the concrete, they cause problems of corrosion if they can penetrate to any reinforcing steel present under the surface. It is possible, indeed it is relatively easy, to make concrete with all the survival properties needed but it costs a bit more than has been acceptable hitherto. It costs very little more in terms of money but it does cost more in terms of care during manufacture, casting and curing. The result is that a lot of concrete has been manufactured with a less than adequate built in defence against the environment. This usually manifests itself as a high permeability at the surface.

Sealants and water repellants have been used for some time to try to give to a concrete surface the properties of impermeability and resistance to ingress of gasses and water borne salts that it needs. Some have been successful in many ways. Many have not. However, there are a number that have found favour in recent years in Europe and in North America. Perhaps the best known among these is silane. It is the purpose of the work reported here to compare the performance of silane with that of a few other newer materials, some of which incorporate silane either as part of one material or as one coat of a multi-coat system. This is based on the measured permeabilities, to both air and water, of the surfaces of laboratory prepared specimens of concrete which had been coated with a range of sealants.The prepared samples were then subjected to a regime of freeze/thaw testing to further assess the effectiveness of the coatings and to compare their actual performance with the expected performance as may be predicted from the results of the permeability tests.

SAMPLE   PREPARATION

Two different concrete mixes were used throughout with all tests being performed in parallel on both mixes. The mixes differed only in the quantity of water used in each. The proportions used were 1 of cement to 2.18 of fine aggregate to 3.96 of coarse aggregate by mass of carefully prepared, surface dry material. The cement was Ordinary Portland. The fine aggregate was washed river sand in standard zone 2. The coarse aggregate was crushed natural rock graded with 20mm and 10mm particles. Water was added to produce the two different water/cement ratios of 0.55 and 0.7.

The samples for test were cast as 150mm cubes, two cubes for each test. Uncoated control samples and eight different coatings were used and each sample type was subjected to permeability tests and to freeze/thaw tests so, for each mix, 36 cubes were cast. The cubes were compacted in two layers on a vibrating table for a standard 12 seconds for each layer. The cubes were demoulded at one day and were cured under wet hessian for a further 5 days.

They were then uncovered and were heated at $105^{o}C$ for just sufficiently long to ensure a dry surface. After cooling they were coated as required and left uncovered in the laboratory during the remainder of the testing.

In parallel with the preparation of these samples, 9 small 215mm square by 75mm thick samples were prepared from each of the two mixes. Eight were coated and one used as a control. These samples were used as a check on the effect of ultra violet light on the performance of the various coatings.

COATINGS   USED

The various coatings used were supplied by Fosroc Technologies Ltd to whom we are indebted. They are listed in Figure 1 together with a layout plan of the experimental programme. The silane was NITOCOTE SN511 and the silane-siloxane was NITOCOTE SN502. The silane-acylic compound was a development product as yet unmarketed. The 100% silane is as supplied by the manufacturer and was applied without a carrier. The 40% silane was applied as 40% dilute with 60% ethyl alcohol. The coatings were applied as per the manufacturers instructions.

TEST   PROCEDURE

Four different tests were performed on each mix/coating combination. The air permeability and sorptivity tests were performed at sample ages of 28 days and 56 days. The freeze/thaw tests were completed at 300 cycles. The UV light tests were terminated at 8 weeks.

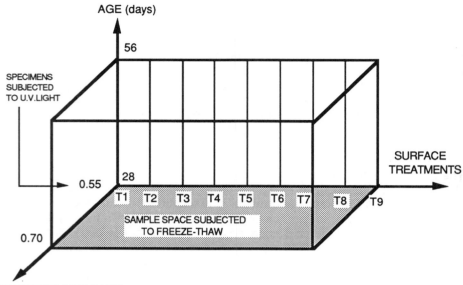

NOTE:
T1 = CONTROL
T2 = 100% SILANE 1 COAT
T3 = 100% SILANE 2 COAT
T4 = 40% SILANE 1 COAT
T5 = 40% SILANE 2 COAT
T6 = SILANE-SILOXANE 1 COAT
T7 = SILANE-SILOXANE 2 COAT
T8 = DEKGUARD
T9 = SILOXANE-ACRYLIC COMPOUND

VARIABLES NOT MARKED IN FIGURE
FREEZE-THAW CYCLES
ULTRA-VOILET RADIATION
VARIATES MEASURED
AIR PERMEABILITY
WATER SORPTIVITY
WEIGHT LOSS

FIG. 1 LAYOUT OF EXPERIMENTAL PROGRAMME

Air Permeability

Using the 'Clam' machine[1], an air pressure of 0.1 bar was applied to the surface of the sample within a ring bonded to the surface. The rate of decay of this pressure was continuously recorded for a period in excess of 20 minutes. It is known[2] that this produces a relationship of the form - logarithm of pressure decay is proportional to the time. The proportionality constant is referred to here as air permeability index and may be directly related to the intrinsic permeability of the material. Two readings were taken on each cube, one on each of two opposite moulded faces. The results reported later are, therefore, the mean of four for each value.

Sorptivity

Again using the 'Clam' machine[1], a water pressure of 0.01 bar was applied to the surface of the sample. Maintaining this pressure constant, the rate of flow of water into the surface was continuously recorded for a period in excess of 20 minutes. It is known[2] that this produces a relationship of the form - volume flow is proportional to the square root of time in the time interval 10 minutes to 20 minutes of test. It is effectively measuring the very low pressure absorption or sorptivity of the surface, the property controlled by the capillary attraction in the surface. Two readings were taken on each cube, one on each of two opposite moulded faces. The results reported later are, therefore, the mean of four for each value.

Freeze/Thaw

These tests were performed in a computer controlled environment chamber which was programmed to follow the cycle detailed in Figure 2 for a total of 300 times. The freeze/thaw durability was obtained by brushing off the loose, broken surfaces of the samples and weighing the portion remaining intact. This weight was compared with weights recorded before the test.

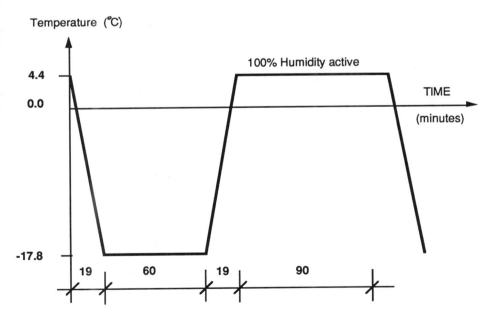

FIG. 2 DETAILS OF FREEZE-THAW CYCLE

Ultra Violet Light Exposure

This test could not claim to be very scientific but was included to give some preliminary results only. The samples were continuously subjected to UV light under a standard UV lamp for a period of 8 weeks. Air permeability and water sorptivity were measured on each sample at the beginning and end of this period.

RESULTS

The results for the air permeability and for the water sorptivity are summarised in Tables 1 and 2. Figures 3 to 6 show these results expressed for clarity as bar diagrams. Figure 7 is an extra result obtained at the end of the permeability tests by breaking off pieces of the surfaces of the samples to record the depth of penetration of the coatings. This was quite easy to do by simply wetting the broken surface. The coated margin was then seen as a slightly different colour. Figure 8 shows, as a bar diagram, the weight loss from each sample type after the exposure to the 300 cycles of freeze/thaw.

TABLE 1 SUMMARY TABLE FOR AIR PERMEABILITY INDEX

W/C RATIO	0.55		0.70	
AGE (days)	28	56	28	56
SAMPLE NO	AP INDEX (bar/min)			
T1-CONTROL	0.181	0.744	0.565	0.868
T2-100%S 1C	0.214	0.436	0.355	0.680
T3-100%S 2C	0.166	0.371	0.316	0.586
T4-40%S 1C	0.326	0.542	0.628	0.824
T5-40%S 2C	0.275	0.491	0.512	0.780
T6-SS 1C	0.532	0.583	0.287	0.726
T7-SS 2C	0.458	0.453	0.304	0.649
T8-DG (2C)	0.094	0.070	0.132	0.456
T9-SAC 1C	0.019	0.194	0.253	0.206

TABLE 2 SUMMARY TABLE FOR SORPTIVITY

W/C RATIO	0.55		0.70	
AGE (days)		56	28	56
SAMPLE NO	SORPTIVITY (m^3/√min x 10^-7)			
T1-CONTROL	0.985	0.898	3.000	1.850
T2-100%S 1C	0.112	0.268	0.253	0.191
T3-100%S 2C	0.119	0.073	0.079	0.164
T4-40%S 1C	0.178	0.066	0.389	0.804
T5-40%S 2C	0.096	0.047	0.440	0.360
T6-SS 1C	0.692	0.148	0.066	0.597
T7-SS 2C	0.109	0.082	0.676	0.103
T8-DG (2C)	0.026	0.017	0.044	0.095
T9-SAC 1C	0.079	0.027	0.072	0.032

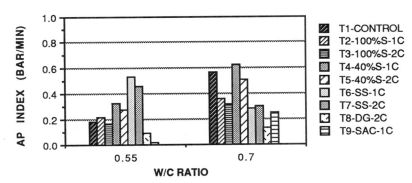

FIG 3. COMPARISON OF AIR PERMEABILITIES AT 28 DAYS

DISCUSSION   AND   CONCLUSIONS

Air Permeability

In Figures 3 and 4 it can be seen that air permeability is less for the low water/cement samples, as expected. What was not expected is that it would be more on the mature samples. It is thought that this is explained by a drying out process in the samples as they sat for the two months in the dry laboratory. It is known from previous gas permeability tests with CO_2 that a gas does not easily displace water in the pores of concrete at the low pressures used so, if a surface layer is drier to a greater depth, the intruding gas will be able to disperse to a greater depth with less resistance. Silane alone does not improve the air permeability of concrete very much. This is clearly seen and understood since silane does not block pores to molecules of the size found in air. However, the two materials containing acrylic do show an improvement in air permeability. They seal the surface much better to small molecules. This is of importance if the protection required is from carbon dioxide and carbonation.

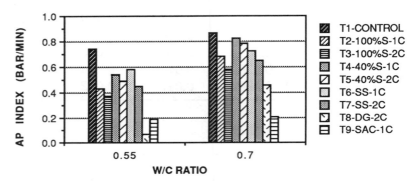

FIG 4. COMPARISON OF AIR PERMEABILITIES AT 56 DAYS

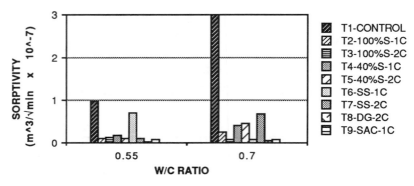

FIG 5. COMPARISON OF SORPTIVITIES AT 28 DAYS

Water Sorptivity

Figures 5 and 6 show a reduction in sorptivity with increasing age and with decreasing water/cement for the control specimens, as expected. There is some remaining ambiguity in the results but it can be seen that pure silane is better than dilute silane, particularly for the higher water/cement. For good concrete, it seems that there is not much advantage in applying the second coat of pure silane. The acrylic containing treatments are seen to perform well here too but we cannot be sure if this is because of the acrylic or not. It is certain, however, that the silane or siloxane part will contribute a lot to the water resistance in these cases. The observations from Figures 5 and 6 are of most relevance when the protection required is against salt ingress in solution in water.

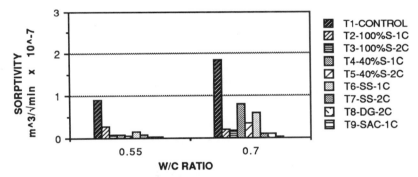

FIG 6. COMPARISON OF SORPTIVITIES AT 56 DAYS

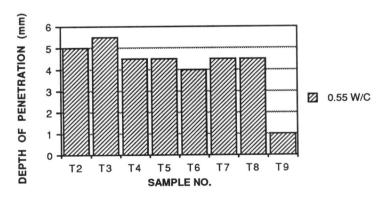

FIG 7. DEPTH OF PENETRATION OF COATINGS

Depth Of Penetration of Coatings

The depth of penetration achieved with all the materials is seen in Figure 7 to be remarkably similar, except for the combination material containing siloxane and acrylic. It is thought that the problem lies not with the siloxane but with the acrylic. It has a much larger molecule size and cannot penetrate far and prevents the siloxane from penetrating too. On the other hand the Dekguard is a two coat system with the first coat of silane which can penetrate before the second coat of acrylic is applied on top. This may have relevance for the longevity of a system when its penetrating ability could ensure a longer working life.

Freeze/Thaw   Resistance

Two thing stand out from Figure 8, the natural ability of a good concrete to withstand freezing and the improvement which can be gained by a poor concrete if the correct treatment is applied. Comparing Figure 8 and Figure 6, one can see immediately that the ability of a sealant to assist a poor concrete to resist freezing may be related directly to the reduction in sorptivity achieved and it is quite a significant help.

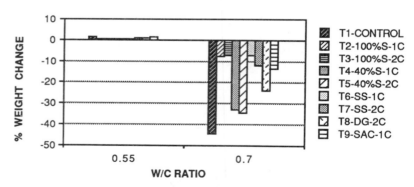

FIG 8. FREEZE-THAW DURABILITY AFTER 300 CYCLES

Ultra Violet Light Exposure

No significant differences were recorded in air permeability or water sorptivity between those samples which had been exposed to UV light and those which had not.

Test Techniques and Equipment

The test techniques and equipment used in this work to measure air permeability and water sorptivity have been designed and developed to be used on site and, indeed, have been extensively so used. The equipment used for these tests contains its own automatic control and recording devices and the data obtained were taken directly from the recording facility via a computer link in the equipment and used to produce the results without any manual calculation. This would seem to provide a means to assess the need of any concrete for a protective system and the means to check its effectiveness once applied. It may also be possible to continue to monitor that effectiveness with the passage of time.

REFERENCES

1. MONTGOMERY, F R, ADAMS, A. Early Experience With a New Concrete Permeability Apparatus, Proceedings, International Conference, Structural Faults and Repair '85, London, 30 April - 2 May 1985, Engineering Technic Press, Edinburgh, pp 359 - 363, ISBN 0 947644 05 9.

2. BASHEER, P A M. Development and Application of Universal Clam for Determining Durability of Concrete, PhD Thesis, The Queen's   University of Belfast, 1990.

CAN CONCRETE STRUCTURES BE WATERPROOFED TO INHIBIT ASR?

G E Blight

University of the Witwatersrand,

South Africa

ABSTRACT. The hypothesis is advanced that if concrete can be dried out so that the relative humidity in its pores is less than about 97%, expansion resulting from alkali silica reaction (ASR) will be inhibited or eliminated. The paper then goes on to explore the possibility of maintaining reinforced concrete structures that are exposed to the weather, in the necessary state of desiccation. The observations raise severe doubts as to the practicality of successfully applying the hypothesis, especially to structures that have already suffered damage from ASR.

Keywords: Alkali silica reaction, moisture, humidity, waterproofing, drying.

Geoffrey E Blight is Professor of Construction Materials and Head of the Department of Civil Engineering at the University of the Witwatersrand, Johannesburg, South Africa. His research interests include the behaviour of ASR - affected structures, loading on bins and silos, the disposal of mining and industrial solid waste and the properties and behaviour of tropical and residual soils.

INTRODUCTION

Forty years ago Vivian[1] showed that the amount of expansion that
occurs in cement mortars as a result of alkali silica reaction (ASR)
depends on the amount of removable water in the mortar. Vivian's
original results are reproduced in Figure 1. If the removable water
was less than 4% by dry mass, no expansion due to ASR occurred.
Once the removable water exceeded 4% the expansion became directly
proportional to the excess of removable water over 4%. Vivian
called this excess the available water. Removable water was defined
as the water lost after prolonged storage over calcium chloride (a
relative humidity of 32%). The available water is that part of the
total water that is loosely held in capillaries in the mortar. Note
from Figure 1 that the available water may be contained within the
concrete ab initio (sealed specimens) or be allowed to penetrate the
mortar from outside (unsealed specimens) after some drying has
occurred. Assuming that Vivian's results on mortar are applicable
to concrete, it would appear that if the available water in concrete
can be kept to zero, expansion by ASR will be minimized.

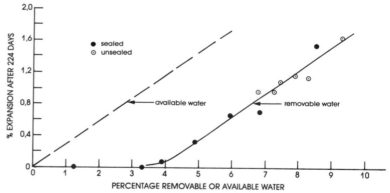

Figure 1: Vivian's observed relationship between removable water
           and expansion

Other evidence[2,3,4,5] has shown that if the relative humidity in
the atmosphere surrounding a concrete structure can be maintained at
below 95%, ASR will be inhibited. However, the relative humidity of
the surroundings is unlikely to be the same as the relative humidity
in the pores of the concrete where the ASR takes place. Water in
the pores of concrete maintained at a relative humidity of less that
100% is in a state of tension or suction. The pore water tension is
balanced by compressive stress in the solid components of the
concrete. The net effect is that any expansive stresses are
counterbalanced, partly or completely, by the pore water tension or
suction.

The relative humidity (RH) in a porous material can be related to the pore water suction p~ by the Kelvin equation[6] :

$$p^\sim = 311000 \log_{10}(RH) \text{ in kPa}$$

where RH is expressed as a fraction of unity, eg 0.95.

At a RH of 0.95, the suction is nearly 7000 kPa. Even if the suction is not completely effective in compressing the concrete there must, at RH = 0.95, be an isotropic compressive stress in the concrete of several thousands of kPa.

Hence one obvious way of inhibiting ASR is to dry the interior of the concrete out until the suction exceeds the swelling pressure developed by the ASR, and then maintain it in that condition by means of a waterproofing layer or coating. Based on observed values of the swelling pressure exerted by ASR, it appears that if the suction within the concrete can be maintained at above 5000 kPa (RH below 0.97) expansion caused by ASR should be eliminated or much reduced.

This paper describes two field experiments to probe the practicality of maintaining full-scale structures, exposed to the weather, in the required state of desiccation.

It should be emphasized that the climate of Johannesburg, where the tests were performed, is particularly favourable for this type of approach. The highest short period atmospheric relative humidity is 96% while the average maximum is only 75%. During the dry winter months, the average maximum relative humidity is below 65% and the average minimum relative humidity is less that 40%.

FIELD EXPERIMENTS INVOLVING ARTIFICIAL WETTING FOLLOWED BY DRYING

A group of columns supporting an overhead motorway was selected for the first field trial. The column shafts are completely protected from impinging rain by the overhead structures they support. This situation was chosen deliberately so that the moisture condition of the column surfaces could be controlled artificially. The columns were cleared down with high pressure water jets. Four columns, selected at random were each waterproofed with a different waterproofer, under the supervision of the suppliers of the preparations. A fifth column was left untreated as a control. The waterproofers consisted of :

Coating 1:   A Portland cement - based resin - modified slurry.

Coating 2:   An aqueous emulsion of synthetic resins.

Penetrant 1:   A silicone in a hydrocarbon solution.

Penetrant 2:   An alkyl alkoxy silane solution.

All preparations were applied by brush.

The effect of wetting and drying was assessed by inserting thermocouple psychrometer probes to a depth of 100mm into holes drilled into the columns. A psychrometer probe[7] consists of a thermocouple sealed into a cavity in the concrete. The air in the cavity comes to thermal and moisture equilibrium with the atmosphere in the concrete. A current is passed thorough the thermocouple which cools by the Peltier effect until the temperature falls below the dew point of the moisture in the surrounding air. Moisture then condenses on the thermocouple junction. The cooling current is interrupted and the condensed moisture starts to evaporate at the dew point temperature. The thermocouple is used to measure this temperature and hence the relative humidity in the cavity and the related pore water suction ($p^{\sim}$) in the concrete can be determined via the Kelvin equation. The measuring range of the psychrometer is from RH = 1.00 to about RH = 0.96. Below RH = 0.96 it is not possible to condense water out of the atmosphere by cooling the air. The range covers the RH range of interest of 0.97 to 1.00.

Figure 2:   Test column that was waterproofed and subjected to
            artificial wetting. Note cracks from ASR in shaft and
            mushroom slab.

Figure 2 shows one of the columns that were the subject of the experiment and Figure 3 shows a psychrometer installed in one of the columns. Figure 4 shows the results of two sets of measurements on the untreated control column and the column treated with Penetrant 1 (silicone). Not all of the columns are in the same physical condition. Some are completely sound, but others show signs of ASR attack, in the form of cracks in the columns, or in the mushroom slabs they support, or both. As it happened, the control column is completely sound whereas the silicone-treated column has a badly cracked mushroom slab, although the column itself has only minor cracks in it.

Figure 3:   Psychrometer probe installed in side of column, and protected from water by clear plastic cylinder

The left hand section of Figure 4 Shows results of preliminary measurements on the two columns to establish the range of $p^{\sim}$ or RH to be expected. $p^{\sim}$ in the control column was very high (more than 5000 kPa) and the measurements eventually went out of range. $p^{\sim}$ in the silicone-treated column, however, was unexpectedly low (2000 kPa) and after a spell of heavy rain, reduced even further. A few days later streaks of white ASR gel appeared through cracks in the column and ran down the surface. This showed that water was entering the column via the structure it supports and causing further ASR. The silicone treatment was quite ineffective as the water was not entering through the surface of the column. In a second series of tests conducted during very hot dry weather, initial $p^{\sim}$ readings were taken and the columns were then watered continually by hose over a period of six hours on three consecutive days to represent as severe a condition of wetting as is ever likely to occur on an outside structure.

The wetting caused $p^{\sim}$ in the control column to fall from beyond the range of the psychrometer to only 350 kPa (Figure 4). After 3 days of drying, the $p^{\sim}$ was again out of range.

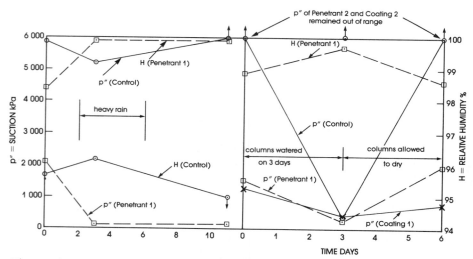

Figure 4:   In-situ psychrometer measurements in two columns
supporting overhead structure

p~ in the silicone treated column started at 1500 kPa, fell to 250
kPa after 3 days watering and then rose again to 1900 kPa after 3
days of drying.   This indicated that water was entering the concrete
through pre-existing cracks, even though the surface had been
treated.   There was no rain during the test and the water could not
have been entering through the superstructure.   However, atmospheric
drying removed the water relatively quickly.

These tests led to the following conclusions:

1. If there is no internal source of water from, for example a
   blocked internal drainage pipe, etc, sound concrete in a dry
   climate will quickly dry out after being surface wetted.   It
   probably does not need to be waterproofed.

2. If there is an internal source of water it is pointless to
   waterproof the exterior of the concrete.

3. Water will probably enter the surface of cracked concrete even if
   the surface has been treated with a water repellent.

The column treated with Coating 1 is also cracked by ASR.   p~ in the
column started at 1200 kPa, but fell to 300 kPa after the period of
watering.   Three days later p~ has increased but was still at a
relatively low 800 kPa.   These results confirm that water can enter
a concrete member through surface cracks, even if the surface has
been treated.   The rate at which the structure dries again will
depend on the type of treatment and the climate.

The column treated with Coating 2 is in sound condition. That treated by Penetrant 2 is cracked, but the psychrometer probe was inserted in an area of sound concrete. p^- recorded in these two columns started out of measuring range and remained out of range even after the watering. This shows that waterproofing treatments can assist in preventing concrete from absorbing water. However, a previous investigation by the author[8] has shown that many preparations sold as "waterproofers" for concrete are not really effective as such.

FIELD OBSERVATIONS OF A STRUCTURE EXPOSED TO THE WEATHER

For the second stage of this study, variations of suction with time were observed within portion of a reinforced concrete portal frame that had deteriorated severely as a result of ASR. Full details of this structure have been published elsewhere[9]. The variation of suction in the damaged concrete was observed for six months, after which the larger surface cracks were filled using quick-setting Portland cement-based mortar and the finer cracks were painted over with a slurry of the same material. Figure 5 shows the structure as it appeared after application of the surface coating.

Figure 5:   Knee of portal frame that is subject of measurements shown in Figure 6. Dark areas show where large cracks were filled.

Psychrometers were installed from faces of the portal oriented north, west and south, at depths of 150mm and 400mm from each face. As the thickness of concrete in the north-south direction is 1250mm, the deeper psychrometers were close to the axis of the structure.

Figure 6 summarizes the observed variations of suction. In this
diagram 150S, 400S etc represent the depth of the psychrometer
(150mm or 400mm) and the orientation (N,S,W) of the face from which
it is installed. The first two months of observation were dry and
suctions recorded at 150 mm depth rose to about 2000 kPa. At 400mm,
however, the suction did not rise above 1000 kPa. Rain at the end
of October caused the suction to plummet to the region of 500 kPa.

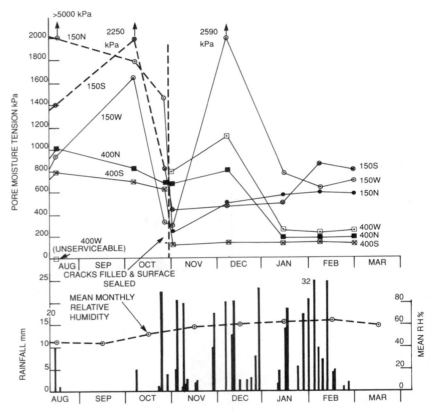

Figure 6:   Variations of suction observed in the knee of the portal
frame shown in Figure 5

At the end of October the crack repairs and surface sealing were
carried out. At this time the psychrometers were overhauled and
400W which had been unserviceable, was replaced. After an initial
period of rather wide fluctuation, the suctions have now settled
down into two groups. At 150mm depth the suction is 400 to 800 kPa,
while at 400mm depth it is much lower, at about 200 kPa. There
seems to be little doubt that sealing the cracks when suctions were
low has entrapped moisture within the concrete. Over the four
months since sealing, this moisture has distributed itself and the
moisture regime within the concrete is now relatively stable.

Figure 6 shows that the suction is essentially unaffected by incident rainfall, even if this is heavy and of frequent occurrence.

However, it will be noted that during the dry months, the suction did not reach the level of 5000 kPa considered necessary to inhibit ASR. Now that the surface has been sealed, the suction has stabilized at well below the level necessary for inhibition. Hence, it does appear that even if the sealing had been carried out before the first rain of the season, when the suction was at its maximum, the resultant suction in the sealed structure would still have been well below 5000 kPa.

Hence it may be concluded that even in the relatively dry climate of Johannesburg, it is unlikely that a structure that has been damaged by ASR and which is exposed to the weather, will dry out sufficiently in dry weather to reach a suction sufficient to inhibit ASR. Sealing the surface of an exposed structure damaged by ASR will result in stabilization and evening out of the moisture within the structure, but a suction level sufficient to inhibit ASR is unlikely to be reached.

GENERAL CONCLUSIONS

The hypothesis has been advanced that a reinforced concrete structure subject to ASR attack can be protected by waterproofing it, thus maintaining the concrete in a dry enough state to inhibit the ASR.

However, the measurements described in this paper show that there are serious practical difficulties in attaining this ideal condition. The only really practical way of protecting an ASR – damaged structure appears to be to shield it from incident rainfall by means of a specially constructed canopy or enclosure. The provision of such external protection would allow the concrete to dry out progressively over a period of several years, until a level of moisture sufficiently low to inhibit ASR is reached.

REFERENCES

1   VIVIAN H E. The effect on mortar expansion of amount of available water in mortar. Studies in Cement-Aggregate Reaction XI, CSIRO, Australia, Bulletin 256, 1950.

2   GUDMUNDSSON G and ASGEIRSSON H. Parameters affecting alkali expansion in Icelandic concretes. Proc. 6th International Conference Alkalis in Concrete, Copenhagen, 1983, pp 217-222.

3   NILSSON L O. Moisture effects on the alkali-silica reaction. Proc. 6th International Conference Alkalis in Concrete, Copenhagen, 1983, pp 201-208.

4   LUDWIG U. Effects of environmental conditions on alkali-aggregate reaction and preventative measures. Proc. 8th International Conference Alkali-Aggregate Reaction, Kyoto, Japan, 1989, pp 503-596.

5   HAWKING M R (Chairman). Minimizing the risk of alkali-silica reaction. Report of Working Party, Cement and Concrete Association, UK, 1983.

6   BLIGHT G E. Strength characteristics of desiccated clays. Journal Soil Mech and Found Div, ASCE, 92, SM6, 1966, pp 19-37.

7   SAVAGE M J and CASS A. Measurement of water potential using in-situ thermocouple hygrometers. Advances in Agronomy, 37, 1984, pp 73-126.

8   BLIGHT G E. Experiments on waterproofing concrete to inhibit AAR. Proc. 8th International Conference Alkali-Aggregate Reaction, Kyoto, Japan, 1989, pp 733-740.

9   BLIGHT G E, ALEXANDER M G, RALPH T K and LEWIS B N. Effect of alkali-aggregate reaction on the performance of a reinforced concrete structure over a six-year period. Magazine of Concrete Research, 41 (147), 1989, pp 69-77.

EVALUATION OF THE RESISTANCE OF CONCRETE COATINGS AGAINST CARBONATION AND WATER PENETRATION

A M Garcia
C Alonso
C Andrade

Eduardo Torroja Institute of Science and Construction,

Spain

ABSTRACT. Concrete carbonation is one of the processes which may develop rebar corrosion, and therefore the service life of the structure decreases. Carbonation is in general a slow process, so that to study its consequences is necesary to apply accelerated tests in order to obtain significant results in a period of time rasonably short.
Concrete coatings are increasingly being used to delay the risk of corrosion, but there is not a standarized method to evaluate their efficiency.
In the present paper a new testing methodology is described. A semiaccelerated carbonation test is employed and the results are compared with those obtained from the accelerated one.
The first one allows a partial classification of the coating to CO_2 resistance but the accelerated method is more selective and takes shorter time. Water permeability of the coatings is also studied.

Keywords: Concrete coatings, carbonation, water permeability rebar corrosion.

A.M. García is a Research Student at the E. Torroja Institute of the High Research Council of Spain. She is doing studies on the subject "Methodology to evaluate repair materials used in corrosion damaged structures".

C. Alonso is a Researcher at the E. Torroja Institute. She has been working in corrosion of rebars in concrete since 1982. She is author of about 20 papers.

C. Andrade is a Research Professor at the E. Torroja Institute. She has been working in corrosion of reinforcements since 1969. She is author of about 70 papers.

INTRODUCTION

Concrete carbonation is one of the processes which may produce rebars corrosion. Although carbonation is a slow process, there is an increase number of structures corroding due to this fact.

Some laboratory experiments (1) have demonstrated that when a structure corrodes because of the carbonation of the concrete cover, the corrosion rate of the rebars can develop high and dangerous levels of deterioration and, in fact, seriously affect the service life of a structure in a relatively short period of time.

The corrosion attack of the rebars has been delayed using different methods of protection, one of them is the use of impermeable concrete coatings such as paintings, membranes etc, (2)(3)(4).

Concrete paintings have been used for years with a decorative function but since seventy's they started to be employed as protectors against rebars corrosion and chemical attack of concrete. One of the first applications with this idea was in offshore structures to delay chlorides diffusion. Actually they have become widely employed either in new than in repaired structures.

This increasing application of concrete coatings has favoured the appearance in the market of a great number of them, which sometimes makes difficult an appropriate selection. Specially because there is not a standard methodology to test these products.

Different testing methodologies are being employed (5)(6)(7) which makes difficult a suitable comparison between the results obtained.

In present paper a testing methodology is described. The method allows to know and classify the carbonation resistance of different concrete coatings. It is discussed also the advantages and suitability in the use of an accelerated testing methodology.

EXPERIMENTAl

Materials

Mortar specimens of 2 x 5.5 x 8 cm and 5 x 5 x 5 cm size were made with Ordinary Portland Cement, c/s = 1/3 and w/c = 0.5. The curing period was 28 days at 100% R.H. and 20 ± 2ºC and 15 days at 50% R.H.

Later, each specimen was covered with two coats of paint. The type of coatings were:

1 - Blank (free of coating)
2 - Colloidal Silica Based (Col. S.B.)
3 - Epoxi Resin Based (ERB)
4 - Cement Based (C.B.)

5 - Acrylic Resin Based (ARB)
6 - Ethylen polymer Resin Based (Et PRB)
7 - Methyl Metacrylate Resin Based (MMRB)
The prismatic specimens also have embedded two steel bars and a graphite rod for corrosion control. The cover was approximately 0.5 cm.

Procedure

A set of specimens (2 x 5.5 x 8) was submitted to a partially accelerated carbonation process in a chamber at 60% RH. Nitrogen with 1% CO_2 was flowed through during 43 days, going on with Nitrogen + 5% CO_2 during 163 days, to finish with 100% CO_2 till full carbonation of the specimens. The other set of specimens were accelerated carbonated bubbling 100% CO_2 for 8 days from the beginning of the test.

The carbonation progress was monitored measuring the weight of the specimens.

The changes in Ecorr, Icorr and concrete resistance were also measured. The potential measurements were made using a calomel saturated electrode. The Icorr evolution was determined by means of the polarization resistance technique, wich methodology has been descrived elsewhere (8). The changes in the ohmic drop of the concrete were measured using the possitive - back of the potentiostat.

The set of cubic specimens (5 x 5 x 5) were submitted to an accelerated carbonation bubbling 100% CO_2 for eight days checking the weight increase during the process.

RESULTS

Accelerated Carbonation

The variation in weight of the cubic specimens submitted to accelerated carbonation (100% CO_2) are given in figure 1.

The behaviours of the specimens with paints 4 (CB) and 5 (ARB) are practically the same and slightly better than the specimen without coating. In these cases, the specimens have a sudden increase in weight, exponential trend, followed by a much softer one.

On the other hand, paints 3 (ERB) and 6 (Et.PRB) have exhibit the highest resistance to CO_2 penetration since the increase in weight is smaller and no exponential, being practically none during the first hours of the carbonation.

An intermediate behaviour is observed with the specimen coated with paint 7 (MMRB).

After four days of carbonation, the increase in weight of the specimens are less remarkable and then, this period of time would be enough to state a first classification of the protective capacity against CO_2 penetration of these paints:

$$3 > 6 > 7 > 5 \simeq 4 > 1$$

More protective $\longrightarrow$ less protective

This classification was also confirmed with the phenolphthalein.

Regarding the electrochemical measurements carried out on the reinforcements of prismatical specimens, figures 2, 3 and 4, it can be observed that after 4 days of carbonation, the specimens with paints 3 (ERB) and 6 (Et PRB) had lower Rohm than the rest of the specimens tested. This fact could be an indication that they store larger amounts of humidity favouring the corrosion process. The Icorr is also high and the potential more negative than in the rest of the specimens, since the latter ones lost easier the water generated in the carbonation process, which restrict the corrosion progress.

Semiaccelerated carbonation

In figure 5 it is observed a decrease in weight of the specimens during the carbonation with 1% CO_2. A drying of the specimens is produced due to the low humid atmosphere they are. The carbonation process is too slow and the rate of water generated is smaller than the evaporated.

With 5% CO_2 the specimens coating with ERB, nº 3, and Et PRB, nº 6, still loss weight, whereas the rest increase. The carbonation process is more favored now and the formation of water is faster than its evaporation. When pehnolphthalein was used in a set of specimens after 5% CO_2 carbonation, it was observed that specimens 3 and 6 were partially carbonated and the rest totally carbonated.

With 100% CO_2, a sudden increase in weight is detected in specimens nº 3 and 6 whereas the rest were not altered.

The classification for carbonation resistance stated from these results is: 3 (ERB) and 6 (Et PRB) coatings are more protective than the rest tested.

Similar information give the Rohmic measurements took out from results of figure 6.

In figures 7 and 8 the Icorr and Ecorr are represented. They indicate negligible corrosion rate during carbonation with 1% CO_2. When the specimens are completely carbonated and the humidity inside then is low, the corrosion is also low, but when CO_2 reaches the rebar and the water content is high enough the depassivated rebar corrodes.

The semiaccelerated carbonation also gives information about water permeability. The slowness of the test allows the equilibrium between the water formed in the carbonation process and the evaporation. The coating 3 (ERB), results the more resistant.

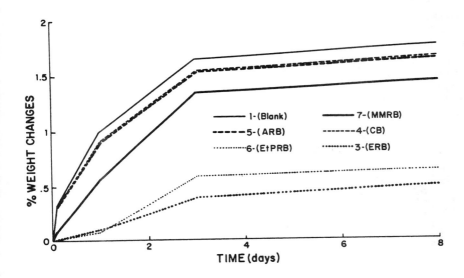

Figure 1. – Weight changes of the specimens during accelerated carbonated.

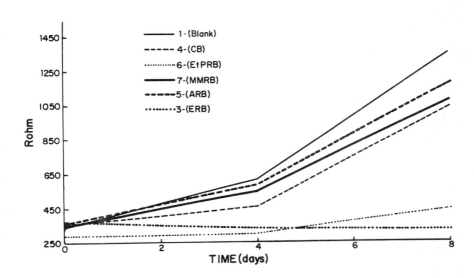

Figure 2. – Rohm changes of coated mortar specimens during accelerated carbonation.

238   Garcia, Alonso, Andrade

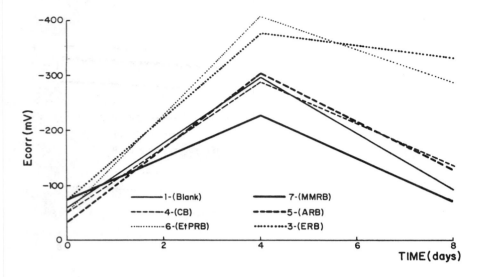

Figure 3. - Ecorr evolution measured on rebars during accelerated carbonation.

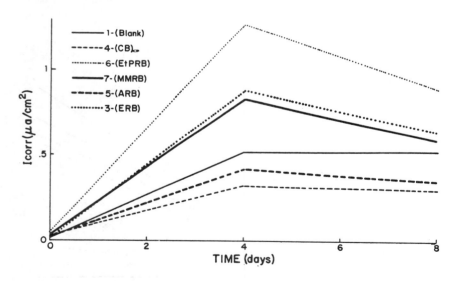

Figure 4. - Icorr evolution measured on rebars during accelerated carbonation.

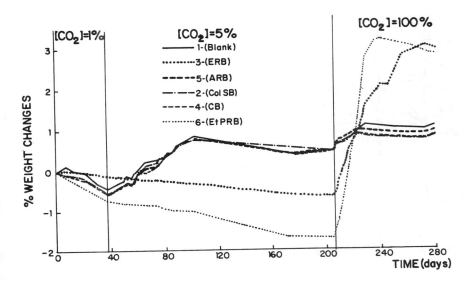

Figure 5. — Weight changes of the specimens during semiaccelerated carbonation.

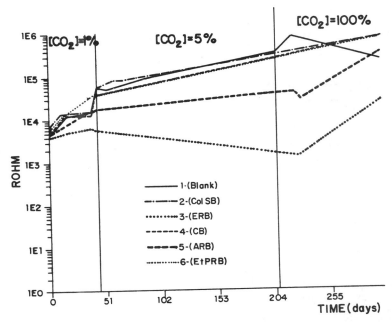

Figure 6. — Rohm of coated mortar specimens during semiaccelerated carbonation.

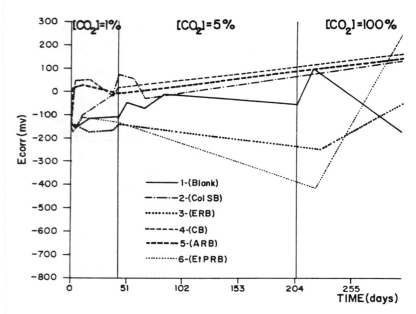

Figure 7. – Ecorr evolution measured on rebars during semiaccelerated carbonation.

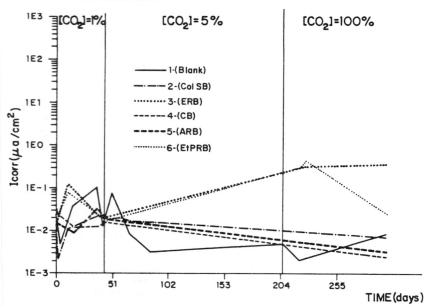

Figure 8. – Icorr evolution measured on rebars during semiaccelerated carbonation.

DISCUSSION

Suitability of the Methodologies employed

The control of the changes of weight in the specimens during accelerate carbonation gives a reliable and fast information wich permits to obtain a practically complete classification of the permeability to CO_2 in all the coatings tested.

When the test is semiaccelerated, a partial classification is only obtained. The semiaccelerated test is a slower process in which the specimens may reach an inner equilibrium of humidity with the outside environment and therefore the variations of weight are as a result between the water evaporation and the penetration effect of CO_2. However the test help to know simultaneously the permeability of the coatings to water vapour. In the case of accelerated carbonation that is possible after carbonation, with later and specific test aimed to study the permeability to water.

On the other hand, the employment of electrochemical techniques allow to carry out a parallel study to nalyse the state of rebars embedded in the specimens: the way the rebars react against the presence of aggresive agents and the influence of the humidity content.

Both methods have as main advantage their non-destructive property that enables to carry out a close study on the same specimen.

The study of the progress of carbonation by means of the fracture of the specimens and their study with phenolphthalein has as disadvantage its destructive character, along with the fact that it only provides an aproximate idea on the degree of carbonation This indicator changes colour a pH around 8, and then although only a small amount of CO_2 reaches the reinforcement, this could be locally depassivated and, however be unrecongnizable with the phenolphthalein.

Classification of the coatings tested

The different types of coatings tested in this work make up a wide range of those most usually employed. Among them, those having as base epoxi resin and ethylen polymer resin seem to be the most resistant to CO_2 penetration whereas that with methyl metacrylate resin has an intermediate position. Those with colloidal silica, cement based and acrylic resin present the lowest resistance.

This classification does not imply that all the coatings with similar base than the ones tested here, will behave alike, since the impermeable capacity of coatings greatly depends upon their pore size, their stability against CO_2 and their application procedure.

From these results the time that carbonation in atmosphere conditions takes to reach a given depth, according to the type of coating, could also be calculated.

Ats it has been mentioned, the tests carried out also enable to know the permeability of these coatings to water vapour. In this context the epoxi resin is the most impermeable. To know this variable is very important, since when the aggressive agent has penetrated into the reinforcement if the coating is low permeable, it accumulates high humidity inside for longer period of time developing high corrosion rates and therefore reducing its protective character.

It is more convenient the employment of a coating less permeable to water which favours a faster drying of the structure.

CONCLUSIONS

1.- The Accelerated Carbonation in mortar specimens is a valid and quick laboratory test by which it can be checked the protective capacity of coatings. In addition, it can be obtained a relative classification in a short time.

2.- The Semiaccelerated Carbonation enables to state the differences of the permeability to CO_2 but not completly. However, it permits to obtain simultaneously the permeability to water vapour.

3.- The determination of the variation of weight in the specimens during the carbonation process is an easy-handle technique providing reliable information.

4.- The parallel measurement of electrochemical parametres, such as Ecorr, Icorr and Rohm in the study of coatings for concrete, allows to know the harmful or beneficial effect on the reinforcements.

ACKNOWLEDGMENTS

Authors are grateful to Spanish CICYT for the financial support of this research.

REFERENCES

1.- C. Alonso and C. Andrade - Effect of cement type and cement proportion in the corrosion rate of rebars embedded in carbonated mortar. Materiales de Construcción. Vol. 37 nº 205 (1987) 5-15.

2.- F. Dutruel and J.P. Le Wair. Nouvelles Techniques de protection du Beton manufacture par impregnation. Beton, nº 364, sept. (1978) 91-103.

3.- D.W. Pfeifer and W.F. Perenchio. Cost-effective protection of rebars against chlorides, sealess or overlays? - concrete construction. May (1984) 503 - 507.

4.- B. Lindberg - Protection of concrete against aggressive atmospheric deterioration by use of surface treatment (painting). 4th Int. confer. on Durability of Building Materials and Components. Singapore (1987) 309-316

5.- H. Weber - Methods for calculating the progress of carbonation and the associated life expectancy for reinforced concrete components. Betonwerk + Fertigteil + Technik, 8 (1983) 508-514.

6.- R.D. Browne and P.C. Robery - Practical experience in the testing of surface coatings for reinforced concrete. 4th Int. Conf. on Durability of Building Materials and Components. Singapore (1987) 325-333.

7.- A. Baba, O. Senbu. A predictive procedure for carbonation depth of concrete with various types of surface layers. 4th Int. Conf. on Durability of Building Materials and Componentes. Singapore (1987) 679-685.

8.- C. Andrade and J.A. González - Quantitative measurements of corrosion rate of reinforcing steels embedded in concrete using polarization resistance measurements. Werkstoff und Korrs, 29 (1978) 515-519.

CRACK-BRIDGING BEHAVIOUR OF PROTECTIVE COATINGS SUBJECTED TO NATURAL WEATHERING

B Schwamborn

M Fiebrich

Institut fur Bauforschung

West Germany

ABSTRACT. The crack-bridging properties of protective polymer coatings on certain concrete structures are a basic supposition for the durability of protection measures. The technological requirements and testing concept for protective coatings with crack-bridging properties - as described in german regulations - are presented. First test results with commercial polymer coatings subjected to artificial weathering and crack-bridging test are discussed.

Keywords: concrete, coating, crack-bridging behaviour, artificial ageing procedure, test techniques

Dipl.-Ing. B. Schwamborn is member of the working group - polymers I and bearing technology - at the Institute for building research of the Technical University in Aachen, Germany, where he has undertaken studies into various aspects of coating for concrete surfaces.

Dr.-Ing. M. Fiebrich is leader of the working group - polymers I and bearing technology - at the Institute for building research of the Technical University in Aachen, Germany.

INTRODUCTION

Crack-bridging properties are basic supposition for the durability of protective coatings on concrete surfaces subjected to natural weathering conditions. Most of the commercial coatings based on polymeric dispersions or solutions show a limited crack-bridging especially at low temperatures (-15° to -20° C). For special applications e.g. coatings on surfaces of exposed concrete structures (bridges, off-shore, facades etc.) a new generation of dispersion coatings with high crack-bridging capacity at low temperatures was developped. In order to assess the serviceability of new systems basic experimental investigations are neccessary.

TESTING TECHNOLOGY OF CRACK-BRIDGING PROPERTIES

According to german regulations protective polymer coatings have to meet certain requirements as far as their resistance against

- UV-radiation
- temperature variation
- freeze and thawing cycles

are concerned. The crack-bridging properties as well as the adhesion strength of the coating to the concrete substrate are determined after they have been subjected to defined artificial weathering conditions.

The crack-bridging behaviour of coated mortar specimen are evaluated after they have been subjected to UV-radiation combined with temperature and rain variation. The artifical weathering is carried out according to DIN 53 384 /1/ or DIN 53 231 /2/. The procedure according to DIN 53 231 is realized by filtered xenon radiation and cyclic rain.

The used testing concept had been elaborated nearly 10 years ago. The apparatus (provides space for about 45 prismatic specimen) has been designed with four xenon radiation units in horizontal position. The principle view is given in figure 1. Detailed description of the weathering design is given in /3/.

Before placing the specimen into this apparatus the side areas of each mortar prism were sealed by an epoxy resin and the back side was impregnated with a hydrophobic agent to protect the mortar against penetrating water. In order to prevent corrosion of the prestressing bars a coating was applied onto the steel.

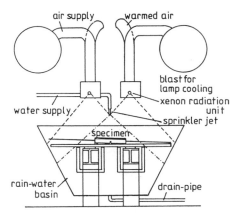

Figure 1: Principle view of the ageing testing device

Figure 2: Weathering equipement with specimen
for crack-bridging test

The following 2 hours-cycle was repeated automatically:

- water spraying: 18 minutes
- dry period: 102 minutes
- relative humidity during the dry periods: 60 - 80% r.h.

The overall duration of the weathering procedure is 2525 hours, about 105 days. Figure 2 shows the weathering equipment with 45 specimen prepared for crack-bridging tests.

The above mentioned test procedures and requirements have been elaborated by Federal Ministry for Traffic (BMV) and the German Society for Reinforced Concrete (DAfStb).

At the european level the technological requirements on the service properties of polymer coatings are discussed in two different CEN (Comité Européan de Normalisation) TC (Technical Commitees):

- CEN TC 104 deals preferably with concrete and its admixtures as building material,

- CEN TC 139 represents the community of the producers of paintings predominantly used for facades.

Task group 1 (TG 1) in working group 8 of TC 104 is involved with protective polymer coating systems for concrete structures. Presently in both technical committees a final decision about the testing procedures and technological requirements for coatings has not yet been fixed. However, there are some ideas predominantly proposed by the french experts, which are summarized in NF T 30-049 (4.85) "Peintures et vernis, Revêtement à usage extérieur, Essai de vieillissement artificiel."

EXPERIMENTAL INVESTIGATIONS

Aims

Experimental investigations aim to evaluate the crack-bridging properties of protective coatings. In order to prove the usability of polymer based dispersive or reactive systems also for exposed structures it was neccessary to define an artificial weathering procedure prior to the crack-bridging test. The results of preliminary crack-bridging tests without any previous artificial weathering cycle will be subject to be reported below. It is intended to discuss crack-bridging test results including the artificial weathering at the oral presentation of this paper.

Substrate

Mortar prisms (40 x 40 x 160 mm^3) were used as substrate specimen, which consisted of :

- portland cement : PZ 55 (DIN 1164)
- aggregate : quartz sand, grain size 0 - 2 mm
- water cement ratio: w/c = 0.50.

Figure 3 shows the shape of a prism prepared for a crack-bridging test. Different to the specimen used for cement test according to DIN 1164, in the middle axis of each prism a prestressing steel bar (1420/1570 N/mm^2 with a diameter of 8 mm) was put into the formwork before casting the fresh mortar. The length of the bars amounts about 800 mm. The middle 90 mm of each steel were covered by a polymeric tube for preventing the transfer of stresses from the prestressing bar to the hardened mortar in the middle part of the specimen.

The test specimen had been stored in water up to an age of 7 days, after that in normal climate conditions of 23° C and 50 % relative humidity. One surface of each prism was prepared by light sandblasting 21 days after casting.

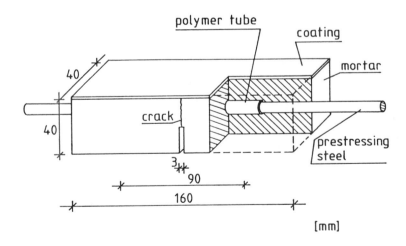

Figure 3: Prism test specimen for crack-bridging test

Prior to the application of the coating materials on the sandblasted surface a saw cut was inserted in the middle of the opposite side. The depth of the cut amounts about 16 mm, so that the bars were not hurted by the saw blade.

Coating systems

The tested systems were mainly polymeric dispersions or solutions based on acrylic binders (AY). The test programme was completed by some 2-component coating systems, polymethylmethacrylate (PMMA) or polyurethane modified epoxy resins (EP-PUR).

The mentioned coating systems shall be applied on free weathered concrete surfaces in the areas of sprayed, with deicing salt contaminated water and ensure at least low crack-bridging behaviour.

Crack-bridging test

The tests were executed according to the technical regulations for surface protecting systems of the german ministry of traffic /4/. In these regulations the large number of coating systems is divided up into 4 crack-bridging classes with different requirements relating to the crack width to be bridged. Additional the index T or T+L indicates whether the crack movement is induced only by temperature variations or by temperature variations and load. In Table 1 these crack-bridging classes are listed.

Table 1: Crack-bridging classes /4/

crack-bridging class		test conditions							
sign –	name –	temp. °C	w_{tr} mm	f_{tr} Hz	c_{tr} –	w_{si} mm	f_{si} Hz	c_{si} –	w_{st} mm
I_T	low	-20	0.1-0.15	0.03	10^3	--	-	-	0.15
II_T	medium	-20	0.1-0.30	0.03	10^3	--	-	-	0.30
II_{T+L}	medium	-20	0.1-0.30	0.03	10^3	±0.05	5	10^6	0.30
III_T	high	-20	0.2-0.40	0.03	10^3	--	-	-	0.40
III_{T+L}	high	-20	0.2-0.40	0.03	10^3	±0.05	5	10^6	0.40
IV_T	very high	-20	0.2-0.40	0.03	10^3	--	-	-	1.00
IV_{T+L}	very high	-20	0.2-0.40	0.03	10^3	±0.05	5	10^6	1.00

w = crack width   indizes: T   = temperature
f = frequency   T+L = temperature and load
c = cycles   tr = trapezium cycle
                                  si = sine cycle
                                  st = static

The test procedure consists of a dynamic part with a varying crack width and a static part with a fixed crack in in a servo hydraulic tensile testing machine (INSTRON 8034) using a temperature box with -20° C. Before placing the specimen into the testing apparatus a first crack in the mortar was induced by a slight bending at room temperature. At one side of the specimen an extensiometer was attached in order to measure the crack width continuously with an accuracy of at least 0,001 mm. The testing machine was controlled by the signals of this strain measuring element.

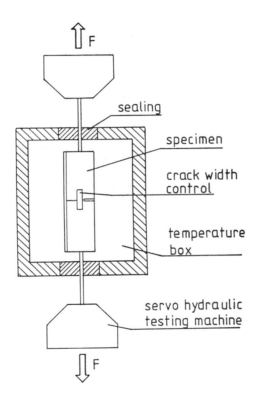

Figure 4: Principle of the crack-bridging test equipment

As mentioned before one distinguishes protective coating systems for temperature induced crack movements (Index T) and additionally load induced crack movements (Index T+L). The crack width cycle in the tests for the classes I_T, II_T, III_T and IV_T as function of time is shown in Figure 5. The period for a test consisting of 1.000 cycles is 9.25 h.

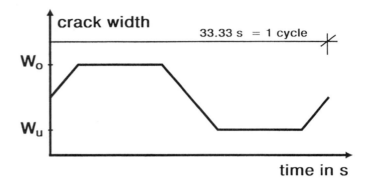

Figure 5: Crack width cycle as function of the time for the crack-bridging classes with index T

In the tests for the crack-bridging classes with the index T+L the function plotted in figure 5 was superposed by two periods with higher frequency (5 Hz). These additional dynamic movements were conducted during the periods without crack width change, the plateaus in the upper function. The total duration of the crack-bridging test was constant. The corresponding crack width function for the classes with index T+L is illustrated in figure 6.

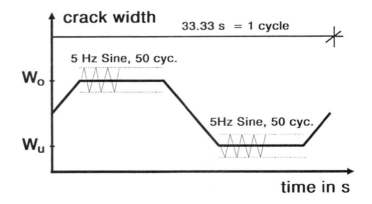

Figure 6:   Crack width cycle as function of the time for the crack-bridging classes with index T+L

During the test period the coated surface was evaluated visually for cracks and delamination every hour. The cycling procedure was interrupted if a crack occured. If not, after 1.000 cycles the temperature was raised up to 23°C and a crack width w_{st} was opened for the static part. This width was fixed by glueing a steel plate onto both sides of the mortar prism. After hardening of the glueing material the specimen was taken out of the machine and stored for 7 days in a temperated box with 70°C. After this ageing the specimen was evaluated again for cracks and delamination. The requirements for a successful crack-bridging test are, that no delamination or crack occurs on the coating after the dynamic and static loading have been finalized.

RESULTS AND EVALUATION

Up to the current state 12 commercial crack-bridging coating systems for protection of concrete surfaces were tested in our laboratory. In figure 7 the results of the dynamic crack-bridging tests are illustrated, all tests with index T. The coating thickness varies between:

- 0.3 mm and 1.2 mm with dispersive systems (AY) and
- 1.5 mm and 5.5 mm with reactive systems (PMMA or EP-PUR).

The static tests were carried out successfully with these 12 systems.

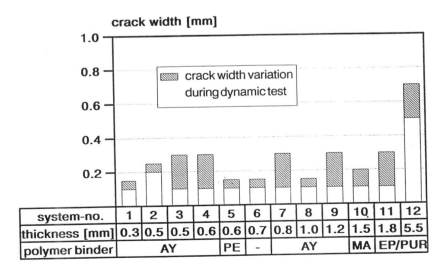

system-no.	1	2	3	4	5	6	7	8	9	10	11	12
thickness [mm]	0.3	0.5	0.5	0.6	0.6	0.7	0.8	1.0	1.2	1.5	1.8	5.5
polymer binder	AY				PE	-	AY			MA	EP/PUR	

Figure 7: Results of the dynamic tests

For dispersive coating systems with low crack-bridging capacity a thickness of at least 0.3 mm is requested in /4/, the surface protecting class D (OS-D). The system has to pass successfully at least a test according to crack-bridging class I_T for being included in the list of tested materials recorded by the german "Bundesanstalt für Straßenwesen". Coating systems with higher crack-bridging capacity (OS-E) require a thickness of at least 1.0 mm and a crack-bridging class II_{T+L}.

CONCLUSIONS

With respect to the results of the crack-bridging tests the following conclusions can be drawn:

a) The highest crack-widths are bridged by polymer systems with reactive binders (EP, PMMA, PUR).

b) Viewing the results of system 11 and 12 (figure 7) the already known correlation (e.g. from /5/,/6/) between the coating thickness and bridged crack width is confirmed. However the question, if there is a linear dependency between thickness and bridged crack width is not yet answered. This will be subject to further investigations.

c) Even polymer coatings with dispersions as binders reaches bridged crack width worth while mentioning (0.1 to 0.3 mm at test temperatures of -20°C).

The most interesting question which will be subject to future investigations is the influence of the ageing on the crack-bridging properties of protective polymer coatings.

REFERENCES

/1/ DIN 53 384 (1989) Testing of plastics; artificial weathering or exposure in laboratory apparatus; exposure to UV-radiation.

/2/ DIN 53 231 (1987) Paints, varnishes and similar coating materials; artificial weathering and irradiation of coatings in apparatus using filtered xenon-arc radiation.

/3/ KWASNY, R. Alterungsprüfung mechanisch beanspruchter Kunststoffproben. Materialprüfung. 26 (1984), No.6, pp. 179 - 184.

/4/ BUNDESMINISTER FÜR VERKEHR ; Abteilung Straßenbau. Technische Lieferbedingungen / Technische Prüfvorschriften für Schutzsysteme von Betonoberflächen im Brücken- und Ingenieurbau. Stand 02.1990.

/5/ TEEPE W. Mechanische Beanspruchung der Kunststoffe im Massivbau. Kunststoffe. Band 52, Heft 11. 1962, pp. 658-666.

/6/ RIECHE G. Rißüberbrückende Kunststoffbeschichtungen für mineralische Baustoffe. Farbe + Lack. 85. 10/1979, pp. 824-831.

FUSION BONDED EPOXY (FBE) POWDER COATING FOR THE PROTECTION OF REINFORCING STEEL

K McLeod

Courtaulds Coatings,

N Woodman

Allied Bar Coaters,

United Kingdom

ABSTRACT. The application of Fusion Bonded Epoxy (FBE) powder coatings to reinforcing bars (rebar) has a successful track record of some sixteen years in North America. The technology developed to meet the challenge presented by demands for an increased maintenance-free lifetime for bridge decks. Premature corrosion of rebar in reinforced concrete bridge structures was prevalent in areas where severe winters necessitate the liberal use of de-icing salts.

The technology has now spread outside of North America and custom coating plants are established in Europe, Middle-East and Far East.

The following review will consider: Key Requirements for FBE Powder Coatings, Plant Design, Product Specification, World Trends, Examples of Projects, and Economics.

Keywords: FBE, Powder Coating, Rebar, Concrete, Corrosion, Chlorides, Application, Specification, World Trends.

Dr K McLeod is Market Project Manager of Courtaulds Coatings, Gateshead. He joined Courtaulds Coatings in 1984 as a Research Chemist and in 1986 joined the Marine Division. In 1988 he moved to Powder Coatings Division in a marketing role. He obtained his D.Phil (Physical Sciences) from Oxford University.

Mr N Woodman has been General Manager of Allied Bar Coaters, Cardiff since 1988. He joined Allied Steel and Wire in 1982 after graduating from Sheffield University. In 1987 he became ASW's Materials Development Manager.

INTRODUCTION

The staining, cracking and spalling of concrete associated with the corrosion of reinforcing steel, may now be the most prevalent form of metallic corrosion visible in daily life.

The presence of chlorides in reinforced concrete can alter the local electrochemical environment. This results in the reinforcing steel becoming non-passive. The source of chloride can be external (e.g. marine salt spray, de-icing salts). Alternatively, chlorides can originate internally (e.g. chlorides in admix, saline aggregates). The process is commonly accelerated by a reduction in the pH of the system by the ingress of carbon dioxide and other atmospheric pollutants. For guidance, last year some 6mt of de-icing salt was used on highways in Europe and 10mt in the USA. Indeed, 100,000T are used on British motorways every winter [1].

The premature corrosion and deterioration induced in reinforced concrete bridges in the UK by the use of salt as a de-icing agent and anti-icer has recently received much publicity. This follows the latest report by the Midlands Links working party [2], a report by G Maunsell and Partners on behalf of the DTp [3] and frequent press editorial.

The Midlanks Links are a classic case. The eleven viaducts, stretching over thirteen miles, were built in 1972. In 1974, the first signs of decay attributable to the action of de-icing salts, were clearly visible. To date, £45m have been spent on repairs compared to a construction cost of £28mn. Repair costs exceeding £120mn are estimated for the next fifteen years [4]. The severe chloride contamination of the A19 Tees-viaduct, which necessitated large-scale repairs, was highlighted in a highly critical, National Audit office investigation of UK road repairs [5] and a simple calculation for the Thelwall viaduct quantifies the magnitude of the problem [1]. The twenty-five year-old crossing already shows serious corrosion even though salt application was at a reduced level of 15T/yr over the 1.3km elevated stretch. The replacement cost is estimated at £10mn or £27,000 for every tonne of salt used on the structure to date.

A similar situation was recognised in the USA in the early 1970's. This prompted an extensive research programme entitled "Non-Metallic Coatings for Concrete Reinforcing Bars" performed by the NBS on behalf of the

Federal Highways Authority (FHWA). In the final report [6], the FHWA concluded that only FBE powder coatings could be recommended as suitable systems for the protection of rebar in concrete. FHWA Notice N5080-48 (1976) set out the first prequalification standards for organic coatings for steel reinforcement. The technology now has a track record of over fifteen years in North America with over forty of the State Highway Authorities in the USA routinely specifying FBE coated rebar.

The success of this policy is evident in a Pennsylvania study [7]. Twenty-two bridge decks were examined, eleven with uncoated rebar and eleven with FBE coated rebar. A detailed study of four of these decks revealed extensive deterioration of those containing bare steel, but no deterioration in those containing FBE coated bar. Of the other bridges, about 40 percent of those with bare steel were in the initial stages of reinforcement corrosion, but none of the decks with FBE coated rebar showed any evidence of steel corrosion. Bridges built in the USA before the use of FBE coated rebar still present problems. The FHWA concluded in 1989 that of 576,025 bridges over 6M in length, a total of 239,568 were structurally deficient or functionally obsolete [8].

At the present time, FBE coated rebar offers a simple and effective solution when compared to alternative approaches to protection.

Improvements in concrete quality and guaranteed achievement of specified concrete cover, may provide a solution, but the construction industry, from senior consultants to site foremen, does not believe current site practice allows this as a viable option.

Waterproofing by use of liquid paint coatings applied direct to concrete or more commonly by the use of waterproof membranes, can be effective. However, the Maunsell report [3] clearly states that waterproof membranes are not providing suitable protection at joints in bridge decks.

Cathodic protection systems, whether impressed current or sacrificial, can be highly effective in protecting reinforcing steel and have the obvious merit that they can be retro-fitted to high risk structures. Questions remain over selection of protection potentials and provision of uniform protection.

FBE coated rebar is a high quality, factory finished product with a proven North American track record. The construction industries in Europe, Middle East and Far East now have access to a proven effective solution to chloride induced rebar corrosion.

CHEMISTRY

FBE coatings have outstanding anti-corrosive and chemical resistance properties having for many years been well established as the optimum protection for steel pipeline in the oil and gas industries worldwide, including sub-sea applications.

By careful screening of epoxy resins, a thorough understanding of curing agent chemistry and selection of pigments and inorganic fillers of known morphology and particle size distribution coupled with advanced formulation technology, allow the production of FBE coatings suitable for application to rebar.

The coatings must be compatible with high speed application, have optimum levels of flexibility and adhesion to allow post forming (cutting and bending of straight bar) and display high levels of impact and wear resistance in order to withstand on-site handling.

APPLICATION

Rebar powder coatings are specifically designed for application to rebar on high speed coating plants. After powder coating, the straight bar can be fabricated by cutting and bending as for normal bar without cracking or delamination of the coating. Custom built straight rebar coatings plants operate at line speeds of the order of 15 metres/minute. Straight bar, typically in 12 metre lengths, is conveyed through grit blasting and then through induction heaters which raise the temperature of the bar to over 235°C. Powder is spray-applied to the hot bar through a multiple gun array. Typically, only 5 seconds will elapse between powder deposition and the bar touching a first wetted pick-up roller. In this short time, the powder must melt and flow to provide a uniform 200 micron cover to the reinforcing bar and also develop a gel strength sufficient to prevent removal of the coating by the rollers. Precise formulation and control during powder manufacture ensures precise control over the melt viscosity of the system during this phase. A subsequent time of only 20-25 seconds is typically the time available for cure using residual heat in the bar, prior to water quenching.

FILM PROPERTIES

Anti-Corrosion Performance

A factory-applied FBE coating with a film thickness of 200 ± 50 microns will provide optimum anti-corrosive performance. Indeed, the ultimate protection possible with an organic coating. Rebar coatings must pass a most severe accelerated corrosion test. When working to the British Specification BS 7295 (10), the FBE coating must meet cathodic disbondment requirements of BS 3900 (11), and when working to ASTM A775/A775M-88 compliance to an Applied Voltage Resistance Test is required, (ASTM G8-72). Two sections of coated bar are placed in a 7% NaCl solution and an applied voltage of 2V is passed between them. No hydrogen evolution at the cathode or corrosion at the anode, must be visible around a drilled holiday placed in the coating 24 hours before the end of the test.

Chemical Resistance

The FBE coating must withstand the alkaline chemical environment presented by the surrounding concrete. Coated bars are routinely tested (to BS 7295 Part II and to ASTM G20-77) by immersion in distilled water, 3M Calcium Chloride, 3M Sodium Hydroxide and saturated Calcium Hydroxide for 45 days at 24°C. No disbonding, blistering or holidays must be evident. In addition, no creep must be visible around intentional holidays.

Flexibility/Adhesion

Coated lengths of bar must be cut and bent to the contractor's specification before delivery to the site. Bars are bent to a 4D ratio (D = r (radius of mandrel)/d (diameter of bar)) in the USA and to 3D in the UK to meet the standard UK requirements for bending reinforcement BS 4466 (12). Prequalification test requirements in West Germany however may be as severe as 2D. The coating shall resist cracking, ductile tearing and adhesion loss upon bending. Otherwise, anti-corrosive performance will be impaired. Of the many generic types of organic coating

available,   only a correctly formulated FBE powder will
satisfy this demand.

Mechanical Properties

Factory-finished FBE coated rebar is tough and durable
and will withstand reasonable damage.   Bar coaters
should provide a complete cutting and bending service
for their coated product, as well as advising the client
on the handling and storage of materials. Good site
practices are the use of padded slings and plastic-
coated tie-wire.   A comparison of the   mechanical
properties of an FBE rebar coating with a common wet
paint anti-corrosive, a coal-tar epoxy (CTE), appears
below.   This clearly shows the superior impact and
abrasion resistance of an FBE coating :

	FBE	**CTE**
Impact Resistance (ASTM GH14-72)	> 18 J	2J
Taber abrasion resistance loss (ASTM D1044, CS10 wheel, 1 kg load, 1000 cycles	Max 25 mg loss	Max 120mg
Knoop hardness (ASTM D1474-68, 10 gm)	> 16	12-14

BOND STRENGTH TO CONCRETE

Using a high yield deformed rebar with a consistently
uniform 200 micron coating of FBE, then the conventional
requirements for bond strength in concrete are readily
achievable. When tested in accordance with British
Standard 4449 [13], this states that the rebar shall
exhibit a free end slip not exceeding 0.2mm when loaded
to its characteristic strength.

Whilst FBE coated rebar exhibits greater free end slip
than uncoated bars, the measure slip is consistently
less than the level allowed by BS 4449 [13].   Typical
results are as follows :-

Bar Diameter	Free End Slip (mm)	
	Coated	Uncoated
20	0.06	0.005
32	0.08	0.02

Experimental work comparing a range of FBE coated rebar with uncoated rebar [14] found that a 10% reduction in bond strength was typical as a result of coating.

WORLD SPECIFICATIONS

The American ASTM specification is the most commonly accepted performance standard worldwide for the applied coating on high yield deformed rebar, although Japanese industry operates within the dictates of JCSE EP30 [15]. However, in Europe, specifications and standards which are more stringent than ASTM and more acceptable to European consulting engineers have been introduced. British Standard BS 7295 [10] can be compared with the ASTM Standard as follows :-

	ASTM	BS 7295
Surface Preparation		
Surface Cleanliness	Sa 2.5	Sa 2.5 [16]
Surface Texture	None Specified	70 micron
Coating		
Coating Thickness	90% 130-300 microns	95% 150-250microns 100% 130-300 microns
Bend Test	120°C bend around a 4D mandrel at temperatures up to 30°C	180°C bend around a 3D mandrel at temperatures not greater than 15°C
Quality Assurance	Not Specified	Consistent with BS 5750 part 2
Holidays	6 per metre	5 per metre

The British Standard BS 7295 places absolute limits on coating thickness, the 130 micron minimum ensuring a minimum level of corrosion protection, and the 300 micron upper limit ensuring that coating retain optimum flexibility.

The specification of a required surface texture ensures that the FBE coating bonds to the steel substrate and, consistency with BS 5750 Part II/(ISO 9000) [17], ensures that the FBE coating application is controlled in accordance with European Quality Assurance criteria.

In the Netherlands, TNO/CUR have proposed a set of guidelines for the use of coated rebar [18] and in Germany, a "Quality Club" is being formed under the auspices of the Building Research Institute, Aachen.

WORLD TRENDS

The use of FBE technology for the protection of steel reinforcement is well established in North America where twenty nine straight bar coating plants [19] produce 300,000T of coated bar per annum. consumption is still growing annually at approximately 10%.

The spread of coated bar technology is now rapidly spreading worldwide. Three custom-built plants are now operating in Europe (Allied Bar Coaters - Cardiff, Norsk Jernverk-Norway and Socotherm-Italy) with plans for another in Finland [20]. Two plants service the Arabian Gulf market (Protech-Abu Dhabi and Khansaheb-Dubai) with plans for further developments in Saudi Arabia.

The Japanese market, like Europe, is still in its infancy with only one dedicated coating line (Ajikawa Iron and Steel-Osaka) [21]. However, coated rebar has been used in other developing areas of the Far East. The Pan-Hu Island Bridge and Tai-Chung Harbour projects in Taiwan, consumed limited quantities of Japanese coated rebar.

Whereas, in North America, the application has primarily been the protection of highway bridge decks from aggressive de-icing salts, in Europe and elsewhere, FBE coated rebar is finding a wide range of applications. These include sea defences, chemical plant, cooling towers and railway bridges.

One of the most innovative uses of FBE powder coating is the protection of welded reinforcement cages for the concrete tunnels linings of the Danish Storebelt link [14,22]. This comes at a time when reinforcement corrosion problems in other sub-sea tunnels (Dartford (UK), Dubai Creek (UAE) are emerging [23] and has drawn criticism of the protection specified by Eurotunnel for the cross-channel link.

APPLICATIONS

The emerging UK market serves to show a diversity of applications. FBE coated rebar was first used in the UK in the early 1980's with a number of small trial applications. The largest use of FBE coated rebar in the mid-1980's was the 'Cardiff Peripheral Distribution Link Road', where over two hundred tonnes of coated rebar was used in the parapets of a 1.5km dual carriageway viaduct. Since the commissioning of Europe's first purpose built rebar coating plant in the UK in 1987, the range of construction projects using FBE coated rebar has grown steadily, both in number and diversity. Examples of applications using FBE coated rebar manufactured in the UK in accordance with British Standards under a CARES approved quality management scheme, are as follows :-

Road Rail Bridges

FBE coated rebar has been substituted for conventional reinforcement in the vulnerable areas of a number of railway bridges. In those parts of the bridge likely to be contaminated by salt carried in the spray from passing traffic, namely the soffit and parapets, FBE coated rebar was used. The use of FBE coated rebar in this instance adds only 1.5% to the cost of a typical bridge structure.

Dowel Bars

Many of the transverse joints in concrete carriageways often need repair because chlorides from de-icing salts have penetrated the joint and resulted in corrosion of the steel dowel bars. FBE coating of dowel bars provides practical corrosion protection without the need for any major redesign of the jointing system.

Marine Environments

FBE coated rebar has been used in a number of coastal projects ranging from sea walls and cliff stabilisation projects to outfalls for treated effluent.

Chemical Plant

Chemical manufacturing processes often necessitate the handling and storage of potentially corrosive materials. FBE coated rebar can provide a very simple and cost effective solution. On the construction of a chemical storage silo, FBE coated rebar added less than 3% to the civil engineering costs. However, the durability of the basic reinforced concrete structure was fundamental to support the relatively expensive process engineering equipment which was subsequently installed. Thus, FBE coated rebar added less than 1% to the overall project cost.

Water Retaining Structures

FBE coated rebar recommended in BS 8007 [24] had been used for both water storage and treatment tanks, conventional designs and concrete covers to be maintained without compromising the overall durability of the structure.

Environments Where Stray Electrical Currents May Be Induced

A concrete protection structure built around a cathodically protected oil pipeline, used FBE coating to ensure electrical insulation of the steel reinforcement. The excellent insulating properties of epoxy coating effectively prevent stray eddy currents from being set up within a mat of steel reinforcement.

The majority of structures to have used FBE coated rebar have been in specialist niches of the construction industry, a marked contrast to North America where use has been centered around the decks of road bridges.

CONCLUSION

Since its first commercial applications in 1974, FBE coated rebar has developed rapidly to establish itself as an economic and practical solution to the corrosion of steel in concrete.

Recent developments in coating technology and powder formulation, combined with stringent product standards, mean that today, FBE coating can be specified to protect reinforcement in a wide range of concrete structures.

REFERENCES

[1]   New Civil Engineer - 12 October (1989)

[2]   Working Party Report - Midlands Links.
       DTp, W S Atkins & Partners, G Maunsell & Partners
       (1988)

[3]   The Performance of Concrete Bridges
       DTp, HMSO (1989)

[4]   New Civil Engineer - 28 February (1989)

[5]   New Civil Engineer - 14 December (1989)

[6]   Cliffton, Beeghly, Mathy. Final Report
       "Non-Metallic Coatings for Concrete Reinforcing
       Bars" FHWA-RD-74-18 (1974)

[7]   R E Weyers, P D Cady. "Deterioration of Concrete
       Bridge Decks from Corrosion of Reinforcing
       Steel". Concrete International - January (1987)

[8]   Construction Repair - September (1989)

[9]   Standard Specification for Epoxy Coated
       Reinforcing Steel Bars. ASTM A775/A775M-88

[10]   British Standard BS 7295. Fusion Bonded Epoxy
       Coated Steel Bars for Reinforcement of Concrete

[11]   BS 3900 Methods for Test of Paints

[12]   BS 4466 Bending dimensions and Scheduling of
       Reinforcement for Concrete

[13]   BS Specification for "Carbon Steel Bars for the
       Reinforcement of Concrete"

[14]   Epoxy Coated Reinforcement Cages in Precast
       Concrete Segmental Tunnel Linings - Preliminary
       Testing and Specification. Ecob C R, King E,
       Rostam S, Vincentsen L. FBE Coated Reinforcement
       - Sheffield, 17 May (1989)

[15]   Standard Specifications and Test Methods for
       Coatings of Epoxy Coated Reinforcing Steel (JCSE)

[16]   SIS 05-5900. Swedish Standards Institute.
       Pictorial Surface Preparation Standards for
       Painting Steel Surfaces

[17]   BS 5750 Quality Systems

[18]   CUR Rapport 89-2 (1989)

[19]   CSRI Membership Listings (1988)

[20]   Epoxy Coated Concrete Reinforcement.
       Nordisk Betong (5) (1989)

[21]   Fujisawa H, Polymers Paint Colour Journal.
       Vol. 179 - 18 October (1989)

[22]   ENR - 14 December (1989)

[23]   New Civil Engineer - 23 November (1989)

[24]   BS 8007. British Standard Code of Practice for
       "Design of Concrete Structures for Retaining
       Aqueous Liquids"

STUDIES OF THE PERFORMANCE OF FUSION BONDED EPOXY COATED REINFORCEMENT DURING THE CONSTRUCTION PROCESS

H Davies

Building Research Establishment,

United Kingdom

ABSTRACT. Problems of deterioration of steel reinforced concrete have created considerable interest in alternative reinforcement for use in concrete structures where there is a risk of chloride initiated corrosion. One material with potentially greater durability than carbon steel is fusion-bonded epoxy-coated reinforcement.

BRE has carried out a site trial to study the effect of normal construction site practices on coated reinforcement in order to assess the implications for long term reinforcement durability.

Results are outlined and discussed in relation to the new British Standard. Implications for design, specification and construction of concrete reinforced with epoxy coated steel are discussed. The findings of this trial indicate that further studies are required to evaluate fully the effect of coating damage on long term durability.

A British Standard for fusion-bonded epoxy-coated reinforcement has recently been published and the material is being specified increasingly for use in conditions where reinforcement is exposed to chlorides. Integrity of the coating is essential for effective corrosion protection and coated bars must be handled with greater care than uncoated bars.

Keywords: Concreting Practice, Durability, Epoxy-Coated Reinforcement, Site Practice.

Mr Hywel Davies is a Senior Scientific Officer in the Organic Materials Division at the Building Research Establishment, Garston, Watford.

INTRODUCTION

Corrosion of reinforcing steel in concrete is now a widespread phenomenon in the UK and elsewhere. The causes of reinforcement corrosion are now more widely known and have been comprehensively detailed elsewhere [1-3]. It has been established that steel embedded in chloride-free uncarbonated concrete forms a passive surface layer within the first few days after casting and remains passive indefinitely [1]. Loss of durability only occurs when the protection to the steel normally afforded by the concrete is lost.

Durable reinforced concrete must therefore be designed to resist carbonation and to exclude chlorides from any source. Reinforcing steel should be embedded in concrete specified in accordance with BS 8110 [4]. In particular the mix design and minimum cover prescribed must be observed. In many cases this will provide sufficient corrosion protection to the steel, provided that the concrete is correctly placed, compacted and cured and that the specified cover is obtained in practice.

There is significant evidence [5] that some of these conditions have not been met during construction, and that a proportion of the current problem of premature concrete deterioration is due either to inadequate design or to incorrect site practice.

There are circumstances in which it is very difficult to achieve the specified design life without recourse to additional corrosion protection measures. Most common is the problem of designing reinforced concrete for service conditions which include likely exposure to high concentrations of chloride. Marine structures, structures where large quantities of de-icing salts may be used, structures partially exposed to saline groundwaters and structures in chemical works all pose problems of designing the concrete to provide adequate durability.

Designers may consider modifications to the concrete mix design such as addition of pulverised ful ash, ground granulated blastfurnace slag or silica fume. Coatings and surface treatments to limit chloride ingress into the concrete and the use of corrosion protected reinforcement may also be considered. Corrosion protected reinforcement may take the form of galvanised steel, stainless steel or coated steel.

Fusion-bonded epoxy-coated steel (FBECR) was first proposed in the USA as an answer to the problems of reinforcement corrosion in the US roads network. A comprehensive study [6] of non-metallic reinforcment coatings was undertaken by the then US National Bureau of Standards. FBECR was shown to be the most promising material for the protection of steel reinforcement in concrete.

The use of FBECR grew rapidly in the US; by 1977 at least 17 State Highway Departments were specifying epoxy-coated steel as standard for the top mat of bridge decks. Ontario, Canada, introduced the material

in 1978 and published its own standard [7]. An ASTM standard
specification for the material was published in 1981 [8].
The material was not available in the UK until 1981, when BRE
initiated its programme of testing FBECR using bars imported from the
USA. UK production of FBECR did not begin until 1988. In contrast to
the developments in the USA the material is not currently permitted in
UK road structures, although it is being specified for a range of
other projects in which significant exposure to chloride is
anticipated.

BRE has maintained a programme of research into the long term
performance of various types of reinforcing steel in concrete for over
twenty years. Results of this work to date have been published
elsewhere [9-11]. Initial trials included carbon steel, various
stainless steels and galvanised steel. When FBECR became available in
the UK an additional exposure trial was begun to evaluate the
performance of this material. The material was imported from the USA
and manufactured to the ASTM Standard Specification.

Findings from this trial [12] as well as other information
highlighting drawbacks in the ASTM standard were available at the time
of drafting of the new British Standard [13,14]. The thickness
requirements, coating process specification, the QA procedure and the
extent of allowable damage to bars when they leave the plant are all
more stringent in BS 7295 than in the ASTM document.

A significant concern of potential specifiers of FBECR is the
possibility that material which arrives on site in a form which
satisfies the BS may be damaged during periods of storage on site,
during site handling or during the fabrication of the reinforcement
cage and placing of concrete. Particular concern was expressed about
impact and abrasion damage to the coating during pouring and placing
of the concrete.

To assess whether there was serious cause for concern about effects of
the construction process on FBECR, BRE devised a site trial. This
aimed to determine the effect on the reinforcement of various stages
in the process, and to investigate the extent of damage likely to
occur between delivery of the reinforcement and placement of concrete.

RESULTS OF SITE TRIALS OF EPOXY COATED REINFORCEMENT

The site trial was established in order to assess:
(i)   the extent of damage to FBECR arising from transport to the
        construction site and storage on site.
(ii)   the extent of damage caused during the fabrication of
        reinforcing cages, mats, etc.
(iii)   the extent of damage sustained by the coating during the
        pouring and compaction of concrete.

Site Trial Procedure

A site was selected where the client intended to use FBECR for a substantial part of the works, and agreed to the conduct of a trial of the material. In consultation with the designer and the contractor a large vertical panel was designed to be incorporated into a doorway in the works. The 2000 x 800 x 200mm panel (figure 1) was designed to provide access for inspection after fabrication, to allow a realistic height of pour and to allow rapid removal of the formwork and concrete after placing. One edge was formed by existing concrete and the other by a stop end. The panel was designed to be poured in one lift.

Figure 1. The reinforcement cage prior to placement of concrete.

The design was fabricated using bars from the bending schedule for the main works. This was done to ensure that the bars tested were drawn at random from the general supply of bars for the whole works, and had to be found and removed from the store of coated bars by the steel fixers in the normal way. This approach ensured that the bars tested were from the manufacturer's normal production run and not prepared specifically for the trial.

The reinforcement cage was designed to provide a high density of reinforcement within the panel, illustrated in figure 2. This increased the quantity of FBECR exposed, thereby increasing the likelihood of damage caused by impact and abrasion by aggregate particles during pouring and compaction and from the use of poker vibrators.

Figure 2. The density of reinforcement in the trial panel.

The cage was designed to incorporate straight bars and bends of 90° and 180°. Two bar diameters were used, 12mm and 16mm. The smaller diameter bar is the more difficult bar to coat and is therefore more susceptible to manufacturing defects.

The bars used were manufactured, cut and bent in the UK in accordance with BS 7295 and the manufacturer's CARES approval. The bars were confirmed as compliant with BS 7295 Part I by the manufacturer and were accepted as such by the contractor and resident engineer. Exact sizes, dimensions and bending details are given in Table 1. Spacers, chairs and tie wire were also in accordance with the requirements of BS 7295.

TABLE 1. Bending schedule of bars used. (Dimensions in mm)

Shape* Code	Bar Length	Bar diameter	Dimension* A	B	Number of Bars
20	1890	12	Straight	–	10
35	925	16	715	–	20
35	1125	12	920	–	8
39	1150	12	550	110	12

* As specified in BS 4466

Results of Site Trial

The FBECR was delivered to site in March and April 1989. The bars remained in site storage until July, when the trial cage was fabricated and concrete poured. The coating suffered a certain amount of damage in transit from the plant to the site and during the period of storage. When inspecting the reinforcement prior to pouring the concrete this damage was readily identified because of rust staining of exposed steel. Further damage occured during fabrication, including removal of the bars from the site store, and was evident by the clean and bright nature of the steel exposed by the damage. After fabrication and prior to placement of concrete, the cage was thoroughly inspected for damage of either type, and the bars were marked where this had occured.

Damage to FBECR is not always clearly visible to the naked eye. Holidays and discontinuities in the coating may be identified using an electronic continuity tester. This was used particularly to test bends in a number of bars because previous studies have shown that bends are particularly vulnerable to such damage. The holiday test was carried out after the cage was fabricated and so attention was focussed on the bends to bars of shape code 35 as these were most accessible. In total 10 of the bent bars were tested, and holidays detected at 40% of bends (8 bends out of 20). Testing straight sections of bar revealed no holidays. None of the bars were found to have sufficient holidays to fail the requirements of BS7295.

Cut ends have been shown to be a weakness in the earlier exposure site study[12]. Cut ends of bars were carefully examined. Several had deteriorated during the period of site storage, figure 3. When the pre-placement inspection had been completed the formwork was secured and the concrete poured. This was done using a standard skip of 1m^3 capacity. The concrete was poured from a height of 250mm – 500mm above the formwork. Compaction was achieved by a compressed air driven poker vibrator. When the concrete had been fully compacted in the usual way the formwork was removed from one side of the slab and the concrete washed out with a water jet. This method was employed because it is the least damaging method of removal and was chosen in preference to removal from hardened concrete. The reinforcing cage was then removed, washed to remove as much of the concrete as possible and removed from the site to the laboratory for further examination.

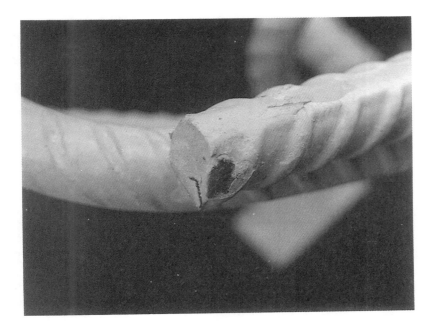

Figure 3. Damage to the epoxy coating at a bar end, showing staining
which had developed in storage.

The cage was dismantled carefully in the laboratory and the bars
marked to indicate orientation and location within the cage. They
were examined visually for damage which occured during concrete
placement and tested using a holiday detector.

Visual Examination of the Coated Steel

Each bar was examined thoroughly after dismantling the reinforcing
cage and all damage to the coating was marked. Each defect and the
process during which it occured was then counted and recorded.
The results are summarised in table 2. All points separated by more
than 5mm were counted as separate defects.

Figure 4 shows an example of damage to the coating. Vibration was
carried out by an operative standing on a platform placed in front of
the cage. Damage on the front of the cage and at lower depths of
cover was worse than on the back and near the centre of the cage.
Numerous scorch marks were observed. These are thought to be due to
the poker vibrator abrading the surface of the coating. Later
investigation showed that they were not complete breaks through the
film to the steel. Such defects do not constitute a defect as defined
in BS 7295 and so were not counted in the count of defects in Table 2.

TABLE 2. Summary of Coating Damage.

Shape Code (No.)	Bar Size (mm)	Defects occuring during						Total Defects	
		Transport		Fabrication		Concreting			
		No.	%*	No.	%*	No.	%*	No.	%*
20	12	5	3%	14	9%	143	88%	162	100%
35	16	21	11%	16	8%	152	81%	189	100%
35	12	2	4%	5	10%	42	86%	49	100%
39	12	10	13%	6	8%	62	79%	78	100%

* % of total defects per bar

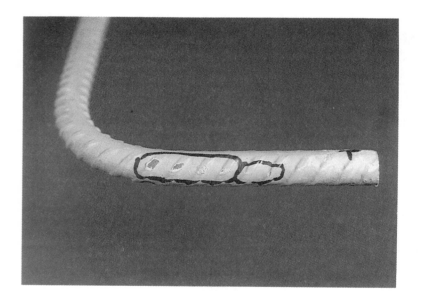

Figure 4. Damage to the ribs of a bent bar caused during placement of the concrete.

Where two bars crossed at right angles it was noted that there was some damage to both bars at the point of intersection. This is thought to arise from the pressure between the two bars from the tie-wire.

An examination of table 2 reveals that the average damage to bars of any shape and size was 7.8 defects per metre. The 16mm bars sustained an average of 9.9 defects/m; the 12mm bars sustained 6.8 defects/m. This level of damage is clearly considerably in excess of that which is permitted for bars leaving the factory.[1]

The results do not indicate a significant difference in the performance of 12mm and 16mm bars. The 16mm bars were placed on the outermost faces of the cage and suffered the most damage due to impact and abrasion from the poker vibrator. The greater damage to these bars is apparently due to their position rather than their diameter.

Electrical Examination of the Coated Steel

A number of specimens (including those tested after fabrication and before concreting) were tested using a holiday detector of the type specified in BS 7295. Defects which had been counted under the previous count were not included again. It was found that, on average, there were three times as many visual defects as holidays. Over 80% of the bars examined electrically had more holidays per bar than allowed by the standard in a bar leaving the factory, excluding all the visible defects.

The average level of all defects rises to 10.4 defects/m if holidays are included. For 16mm bars the level rises to 13.2 defects/m and for 12mm bars to 9.0 defects/m.

DISCUSSION

The results clearly indicate that a certain amount of damage can occur during site handling and steel fixing. In more than 90% of cases the extent of this damage was below the acceptable limit set in the BS 7295 and in most cases would have been repairable prior to concreting. The principal exception was the damage occuring where bars were tied together, the significance of which is the subject of further investigation.

The results also indicate that a much higher level of damage occurs during the process of pouring and compacting concrete on and around a reinforcing cage made of FBECR. This process is clearly detrimental to the integrity of the bar coating. Moreover it occurs at a stage which prevents any repair work or even inspection of the damage.

1.BS 7395 states that the clauses prescribing permitted levels of visible damage and holidays apply to material leaving the factory and are not criteria for rejection on site. Part I, Appendix D, "Guidelines for Site Practice" suggest a much higher allowable level of damage on site of 1% of total surface area of the bar. However, it should be noted that this is a guide and not a prescribed limit.

The occurence of significant numbers of defects is significant because published data on the performance of epoxy-coated reinforcement indicates that where there are defects in the coating then under certain conditions corrosion may occur, starting from defects in the film. These data were obtained using bars which conformed to to ASTM 775-81, but not to BS 7295. The applicability of these data to bars which conform to the BS is not clear, but it raises concern that the performance of FBECR may be adversly affected by the damage occuring during concreting.

In concrete which contains a level of chloride ion higher than 1% by weight of cement, chloride attack may occur at the points where the coating has been damaged. Since pitting attack on uncoated steel is random the exact location of attack cannot be predicted. The area affected and the depth of pitting increase with the chloride ion concentration. The likelihood of corrosion occuring at defects in FBECR therefore increases with the chloride ion content of the concrete.

The implications of these findings for long term durability are clear. The performance of FBECR as a corrosion protected reinforcement may be adversly affected by concreting. It is hard to assess or predict the likely extent of this effect.

FBECR has been in use in North America for about ten years. Reports which have reached the UK on its performance have tended to emphasise problems and indeed particular structures and projects. These may not be representative of the wider North American experience. The fact that a significant number of State authorities have continued to use the material during the last decade must be taken to indicate that the material has not given great cause for concern. The ban on epoxy-coating by Florida must be set against its continued use by many more States.

A recent survey of a number of multi-storey car parks in Canada in which FBECR had been used in the top reinforcing mat indicated that corrosion of the coated reinforcement was not a problem and that it was performing satisfactorily as corrosion protection after nine years.

The normal design life of buildings and structures is in the range 60 to 120 years. In this context FBECR is still a new material. In the absence of sufficient experience of its longer term performance and durability there is clearly a need for caution in its use. It must be recognised by designers and specifiers that material produced in the UK to BS 7295 is not identical to the material which has been used widely in North America. Care must be exercised in applying North American experience with epoxy coated reinforcement to the UK.

CONCLUSIONS

Further studies are required to confirm the results reported here and for the full significance to be assessed. But the results of this field trial suggest that:

> damage to fusion-bonded epoxy-coated reinforcement in transport, site handling and fabrication falls within the limits of allowable damage set in BS 7295;

> the extent of damage sustained during concreting is considerably in excess of the permissible damage;

> this damage includes both visible defects and holidays;

> care must be taken in design of reinforcing cages to allow access for poker vibrators;

> further work is required to evaluate the effect of increased damage on long-term durability of FBECR.

ACKNOWLEDGEMENTS

This paper forms part of the research programme for the Department of the Environment, carried out at the Building Research Establishment, and is published by permission of the Director.

BRE acknowledges the assistance of PSA Wales and Allied Bar Coaters in the conduct of this research.

REFERENCES

1   BUILDING RESEARCH ESTABLISHMENT. The durability of steel in concrete: Part 1. Mechanism of protection and corrosion. Digest 264. BRE, Garston, 1982.

2   TREADAWAY K W J AND PAGE C L. Corrosion of Steel in Concrete. Nature 297, (5862), pp 109 – 115, 1982.

3   PULLAR-STRECKER P. Corrosion damaged concrete: assessment and repair. CIRIA 1987, Butterworths.

4   BRITISH STANDARDS INSTITUTION. Structural use of concrete Part 1. Code of practice for design and construction. BS 8110:Part 1:1985. London, BSI.

5   MAROSSZEKY M, SADE D W, CHEW M Y L AND GAMBLE J. Site investigation – Quality of reinforcement placement on buildings and bridges. Proc. 4th Int. Conf. on Durability of Building Materials and Components, Pergamon, Singapore, 1987.

6   CLIFTON J R, BEAGHLEY H F AND MATHEY R G. 'Non-metallic coatings for concrete reinforcing bars. Coating materials', US National Bureau of Standards, Technical Note 768, Washington, 1973.

7   ONTARIO MINISTRY OF TRANSPORTATION COMMUNICATION, 'Provincial highways; guidelines for inspection. Patching and acceptance of epoxy-coated reinforcing bars at the job site', EM-69, 1983.

8   ASTM Standard Specification for Epoxy-coated reinforcing bars. A-775-81, Philadelphia, 1981.

9   TREADAWAY K W J, COX R N AND BROWN B L. Durability of corrosion resisting steels in concrete. Proc. Institution of Civil Engineers, Part I, Volume 86, pp305-331, 1989.

10   FLINT G N AND COX R N. Resistance of Stainless Steel partly embedded in concrete to corrosion by seawater. Magazine of Concrete Research, Volume 40, Number 142, pp13-26, 1988.

11   TREADAWAY K W J, COX R N AND BROWN B L. Durability of Galvanized steel in concrete. ASTM Special Technical Publication 713, Philadelphia, 1980.

12   TREADAWAY K W J AND DAVIES H. Performance of Fusion Bonded Epoxy Coated Steel Reinforcement. The Structural Engineer, Volume 67, Number 6, pp99-108, 1989.

13   BRITISH STANDARDS INSTITUTION. Fusion Bonded Epoxy Coated Steel Bars for the Reinforcement of Concrete. Part I: Coated Bars. BS 7295:1990:PartI.

14   BRITISH STANDARDS INSTITUTION. Fusion Bonded Epoxy Coated Steel Bars for the Reinforcement of Concrete. Part II: Coatings. BS 7295:1990:PartII.

PROTECTION OF REINFORCED CONCRETE ELEMENTS AGAINST CORROSION BY POLYMER COATINGS

M M Kamal

Building Research Center,

Cairo

A E Salama

Helwan University,

Egypt

ABSTRACT. This research was carried out to investigate the potentiality of using polymers as protective coatings for reinforced concrete elements exposed to ammonium nitrate salts. Tests were carried out on hinged concrete beams bond test specimens and mild steel bars. The main variables taken into consideration were the concentration of the salt in the surrounding medium, the concrete cover and the method of protection. Bond behavior under loading and up till failure of hinged reinforced concrete beam specimens, the concrete compressive strength, the degree of corrosion and the tensile strength of steel bars exposed to ammonium nitrate salts are presented and discussed. Ammonium nitrates is of destructive effect on concrete and steel. Polymer coatings showed to be of great efficiency in protecting reinforced concrete structures against deterioration of concrete and corrosion of reinforced concrete.

Keywords: Concrete, Steel, Bond Test, Ammonium Nitrate, Concentration, Concrete Cover, Protective Coatings, Epoxy, Corrosion.

Dr. Mounir M. Kamal is an associate Professor at The Building Research Center, Cairo, Egypt. He has been actively engaged in research into various aspects of concrete behavior and has published extensively. His current research work includes: Polymer and fibre concrete, Lightweight concrete, repair and strengthening of R.C. Structures.

Prof. Amr E. Salama is a professor of Strength of Materials at the Department of Civil Engineering, Faculty of Engineering, Helwan University , Cairo, Egypt. He has been most active in concrete technology for number of years by means of research publishing. His current research work includes assessment of concrete quality and durability of concrete.

INTRODUCTION

During the last few decades the problem of the corrosion of reinforced concrete structures which exposed to sever conditions greatly increased. Among the most severe exposed structures are: marine structures, piles, chemical factories, railway underpasses, swimming pools and constructions in agricultural land. Chemical fertilizers have been widely used in Egypt for more than twenty five years. Ammonium nitrates is one of the most commonly used chemicals which showed destructive effects on the concrete structures in the agricultural areas. Extensive research has been carried out on the concrete corrosion and concrete protection[1-6].

This research was carried out to determine the potentiality of using epoxy resin as a concrete coating to resist the destructive effect of ammonium nitrates salt on the deterioration and strength of concrete, the corrosion of the steel reinforcement and the bond between concrete and steel.

EXPERIMENTAL WORK

Seventy two hinged concrete beam specimens with a simple roller hinge were casted and tested to determine the bond behavior and bond strength between concrete and steel after exposure to ammonium nitrate solutions with different concentrations. These beams represented two groups of tested specimens (A and B). Group (A) was exposed to aggressive medium without surface protection, while group (B) was coated with epoxy resin before exposure to the aggressive medium. The dimensions and the method of testing of the hinged beams used for bond test are shown in figure (1). The bond tests were carried out on twenty four sets of beams. Table (1) shows the scheme of bond tests. The main variables were the method of the concrete protection, the depth of the concrete cover and the concentration of the ammonium nitrates solution.

The concrete mix used was made from Ordinary Portland Cement, sand and gravel obtained from El-Yahmoum quarries near Cairo. The nominal maximum size of coarse aggregate was 20 mm. The mix properties by weight were 1 C: 1.765 S :3.505 G and the water/cement ratio was 0.5 by weight. The polymer used as a protective coating is locally available in the Egyptian market as a solvent-free epoxy resin. The epoxy used was of two components consisted of a resin with white colour and a hardener with a green colour. These components were mixed with 100 parts by weight resin to 60 parts by weight hardener.

Surface treated plywood forms were specially made for the bond test specimens. The form was divided into two prisms by means of polystyrene block fixed to the bottom, through which 13 mm diameter steel bar passed and a plate supported on the sides of the form rested on the polystyrene block. The form dimensions were checked and the sides were coated with oil before concreting. The steel bars were placed at different heights from the bottom of the moulds.

Immediately after mixing, concrete was placed in the moulds of the various test specimens in layers and compacted mechanically on a vibrating table. In no case was segregation or laitance allowed to occur. The test specimens were left in the forms for 24 hours after casting after which the sides of the form were stripped, and the test specimens were removed and kept in a curing room. The upper plate was removed. The bond test specimens as well as concrete control specimens were cured by sprinkling water for 28 days. At the age of 27 days the test specimens of Group (B) were left to dry in the laboratory atmosphere, the concrete surface was cleaned from dust and then coated with solvent free epoxy resin. Two coats of the solvent free epoxy resin system were applied. The second coat was applied 10 hours after the first to achieve good adhesion between them. The specimens were left to cure in the laboratory atmosphere for 24 hours before exposure to ammonium nitrate salts.

At the age of 28 days the bond test specimens as well as the steel test specimens were immersed in tanks of ammonium nitrates solutions with different concentrations. The solution covered 100 mm height of the bottom side of the tested specimens. The test specimens were left in the solution for twelve months.

After twelve months of exposure to the ammonium nitrates solution the test specimens were taken from the tanks and left for a few hours in the laboratory atmosphere until their surfaces were dry. Degree of corrosion activity of the steel bars embedded in the concrete beam after exposure to ammonium nitrates solutions were recorded using a copper-copper sulphate half cell potentialmeter before carrying out the bond test as shown in figure (2). Two steel plates with a pin between them were carefully inserted in the top groove of the test beam to form a hinge. Two dial gauges were fixed at the free ends of the test bar and rested on the vertical sides of the beam to measure the relative movement of the bar.

The bond test was carried out using 10 tons testing machine. Loading was applied at constant rate and the load was recorded at first movement of the dial gauge. The dial gauges were removed when the ultimate load was reached and registered by the testing machine.

Losses in area of the steel bars immersed in the ammonium nitrates solution as well as those embedded in concrete were recorded and compared with the area of the steel bars left in the laboratory atmosphere. Tension test was carried out to determine the effect of exposure to ammonium nitrates solutions with different concentrations on the tensile strength of the steel bars.

Schmidt hammer readings were taken on different heights of the concrete beam bond test specimens after holding slightly with compression test machine to determine the effect of exposure to ammonium nitrate solution on the concrete compressive strength. Crushing using compression testing machine was also carried out on one part of the concrete beam bond specimens to determine their compressive strength.

ANALYSIS AND DISCUSSION OF TEST RESULTS

Corrosion of Steel Bars

Steel test specimens immersed in ammonium nitrates solutions indicated that corrosion and consequently loss of area increased with the increase of the salt concentration. Bars with 13 mm diameter indicated losses of 15%, 22% and 30% in the area when immersed for twelve months in ammonium nitrate solutions with 5%, 10% and 15% concentration, respectively. Brown desposits of the salt and the rust covered the steel bars which were removed by wet cloth before measuring the diameters. These deposits were also existed on the parts of the steel rods immersed from the concrete beam bond test specimens.

Tension test carried out on the steel bars emersed in the ammonium nitrates solutions showed that the loss of their tensile strength increased with the increase in the solutions concentration. Bars with 13 mm diameter indicated loss in the tensile strength of 22%, 30% and 55% when immersed for twelve months in ammonium nitrates solutions with 5%, 10% and 15% concentrations, successively. This destructive effect is much higher than the percentage of loss of area.

Tests carried on steel bars embedded in concrete beam specimens indicated that a considerable amount of corrosion of about 20% of the area occurred for bars without coating covered with concrete of 10 mm thickness and exposed to the highest degree of ammonium nitrates concentration of 15% of water by weight. Less corrosion was recorded for other specimens. For coated beams no signs of steel corrosion were recorded.

Tension tests carried out on the steel bars taken from the concrete beam bond specimens exposed to the highest solution concentration without any surface protection indicated about 30% reduction in the steel tensile strength.

Deterioration of Concrete

Visual inspection of the concrete beam bond test specimens immersed in the ammonium nitrates solution without surface coating showed discolouring of the concrete to the brown colour and salt stains in the area close to the steel bar embedded in concrete. The degree of discolouring and the salt stains increased with age and with those specimens immersed in the solutions with higher concentrations as shown in figure (3). When removing the concrete beams from the solutions after exposure for twelve months thick salt disposits were existed in the part of the beam in direct contact with the solution. This salt deposits were also existed but with less degree in the near by area over the level of the solution. Deterioration of the concrete was quite clear after removal of the salt stains and small friction showed quite abrasion of the concrete surface. Hair cracks were also observed in the layer of concrete immersed in the solution. The number of cracks increased in the specimens immersed in ammonium nitrate solution of higher concentration.

Concrete Specimens coated with epoxy resin showed no signs of brown discolouring, deterioration or cracks after exposure for twelve months to ammonium nitrates solution with concentrations up to 15%. Salt deposits were existed on the surface of the concrete specimens which were easily removed by washing with water.

Schemidt hammer readings taken on the sides of the concrete beam specimens without surface coating at different heights after removal of the weak surface layer subjected to the solution indicated that the concrete strength greatly decreased by exposure to the ammonium nitrate solution. Higher decrease in concrete strength was found with those specimens exposed to solutions of higher concentrations. Figure (4) shows the variation of strength along the concrete beams height measured by the schimdt hammer.

Schemidt hammar readings taken on the sides of the concrete beam specimens coated with epoxy resin and exposed for twelve months to ammonium solutions of different concentrations indicated negligible differences in the concrete strength along the different heights of the specimens. Generally speaking a slight increase in the compressive strength of the concrete specimens coated with epoxy was recorded in comparison with those specimens cured in air without exposure to ammonium nitrate salts.

Compression test carried out using 300 ton testing machine on half beam lower concrete compressive strength was observed with concrete specimens without epoxy coating which were exposed to ammonium nitrate solutions of higher concentrations. Those beams with epoxy coating indicated that exposure to ammonium nitrate solution with concentration up to 15% had no destructive effect on concrete compressive strength as shown in table (2).

Bond between concrete and steel bars:

Figure (5) shows the the results of the bond tests carried out on concrete beam bond specimens exposed to different concentrations of ammonium nitrate solutions for twelve months in comparison with those specimens cured in air in laboratory atmosphere. Generally speaking, those beams without epoxy coating which were exposed to the solution showed lower bond stresses at the first slip of the free end of the bar and lower ultimate bond strength than those cured in air. Lower values of bond stresses were recorded for beams with smaller concrete cover. Higher decrease in bond stresses were observed with those beams exposed to the ammonium nitrate solutions with higher concentrations. The reduction in the bond strength within this period of time is mainly referred to the deterioration and cracking of the concrete cover. Exposure to the ammonium nitrate solution with high concentration might led to more destructive corrosion of the steel bars and consequently higher reduction of bond strength with concrete.

Concete beam bond specimens coated with epoxy resin and exposed to ammonium nitrate solutions with different concentrations for twelve months showed negligible difference in bond stress at different stages of loading and up till failure in comparison with those cured in air in the laboratory atmosphere.

CONCLUSIONS

Exposure of reinforced concrete structures in agricultural areas or chemical factories and stores to the ammonium nitrate solutions causes deterioration of the concrete and corrosion of steel. The reduction in the concrete compressive strength and the bond between steel and concrete is higher for higher solution concentrations and lower concrete cover.

Increasing the concrete cover increases the protection of steel reinforcement against corrosion and consequently decreases the loss in bond between steel and concrete.

Solvent free epoxy resin coating showed high protection for concrete structures exposed to ammonium nitrate solutions.

REFERENCES

1   Lewis, D. A, "Some Aspects of the Corrosion of Steel in Concrete", Proceedings of the 1st international Congress, Metallic Corrosion Butterworths, London, 1962.

2   Biczk, I. "Concrete Corrosion Concrete Protection", Akaole Kiado, Budapest, Eigth edition, 1972.

3   The Siria guide to concrete construction in the gulf region, Special publication 31, Construction Industry Research and Information Association, London 1984.

4. Kamal, M. M. Tawfik, S. Y. and Noseeir, M. H., "Polyester Mortar", Jornal of applied Polymer Science, Vol. 33, 1987.

5. Kamal, M. M. and El-Refai, F. E., "Durability of Steel Fiber Reinforced Concrete", 4th Int. Conf. on Durability of Buil. Mat. and Components, Singabore, Bulletin No. 12, Nov. 1987.

6. Kamal, M. M., El-Ibiari Sh. N. and Abd El-Rahman A. M., "Corrosion of Reinforced Concrete Structures" 3rd Int. Conf. on Deterioration and Repair of R. C. in the Arabian Gulf, Vol. 1, Oct. 1989.

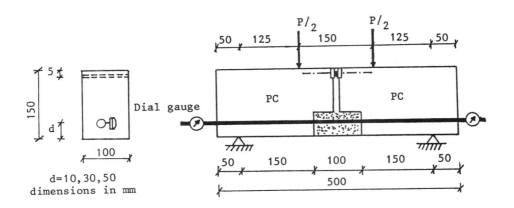

Figure 1 Hinged beam used for bond test

Figure 2 Corrosion measuring device

a – Reference Beam

b – Concentration of the solution 5%

specimens exposed to
ammonium nitrate solution
with different
concentrations

c – Concentration of the solution 10%

d – Concentration of the solution 15%

Figure 3: Discolouring and salt stains on a concrete beam
bond test specimens (without coating)

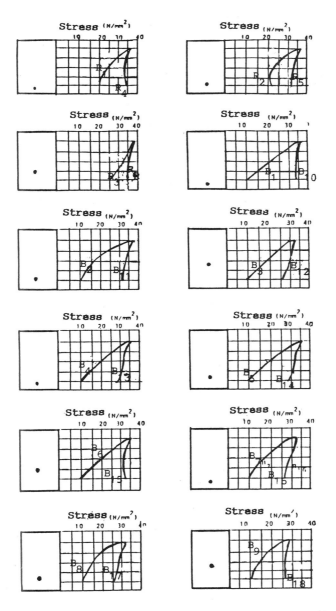

Figure 4 Compressive strength distribution measured by Schmidt hammer at different heights of the concrete specimens exposed to ammonium nitrates solution with differnt concentrations

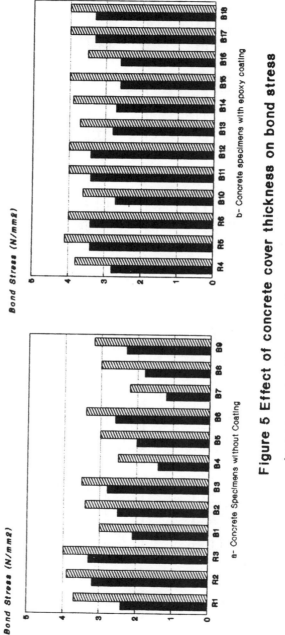

Figure 5 Effect of concrete cover thickness on bond stress between steel bars and concrete exposed to ammonium nitrates solution

TABLE 1 Scheme of Bond Tests

Group	Surface Protection	Sub-Group No.	Set No.	Concrete cover d´(mm)	Concentration of the solution (by wt of water)	concrete mix proportions (by weight)	Dia.of steel bar (mm)	Age of testing	Tested property
A	Without Coating	A1	R1	10	----	1 Cement 1.765 Sand 3.505 Gravel 0.5 W/C		Twelve Months	- Visual inspection - Steel corrosion - concrete strength - bond stress at first slip of the bar at free end - Ultimate bond stress
			R2	30					
			R3	50					
		A2	B1	10	5%				
			B2	30					
			B3	50					
		A3	B5	30	10%				
			B6	50					
			B7	10					
		A4	B8	30	15%				
			B9	50					
			R4	10					
B	With Epoxy Coating	B1	R5	30	----				
			R6	50					
			B10	10					
		B2	B11	30	5%				
			B12	50					
			B13	10					
		B3	B14	30	10%				
			B15	50					
			B16	10					
		B4	B17	30	15%				
			B18	50					

TABLE 2 Bond Test Results

Group	Surface Protection	Sub-Group No.	Set No.	Concrete cover d´(mm)	Concentration of the solution (by wt of water)	Bond stress at first slip N/mm²	Ultimate Bond stress N/mm²	concrete compressive strength N/mm²
A	Without Coating	A1	R1	10	----	2.4	3.7	32
			R2	30		3.2	3.9	34
			R3	50		3.3	4.0	34
		A2	B1	10	5%	2.1	3.0	23
			B2	30		2.5	3.4	27
			B3	50		2.8	3.5	28
		A3	B4	10	10%	1.4	2.5	20
			B5	30		2.0	3.0	22
			B6	50		2.6	3.4	23
		A4	B7	10	15%	1.2	2.2	18
			B8	30		1.8	3.0	20
			B9	50		2.3	3.2	22
B	With Epoxy Coating	B1	R4	10	----	2.8	3.8	35
			R5	30		3.4	4.0	36
			R6	50		3.4	4.0	36
		B2	B10	10	5%	2.7	3.6	33
			B11	30		3.4	4.0	33
			B12	50		3.4	4.0	34
		A3	B13	10	10%	2.8	3.7	34
			B14	30		2.7	3.9	35
			B15	50		2.6	4.0	34
		A4	B16	10	15%	2.6	3.5	33
			B17	30		3.3	4.0	32
			B18	50		3.3	4.0	33

TECHNICAL PROPERTIES AND DURABILITY OF CEMENT MORTARS WITH ADDITION OF ASPHALT EMULSION

A M Grabiec

W Lecki

Technical University, Poznan,

Poland

ABSTRACT. Cement mortars based on quartz sand and four types of cement with admixture of an asphalt emulsion in amount of 1, 3, 5, 7, 10 and 15% at the rate of mixing water were tested. The influence of emulsion on setting time, water absorbability, compressive strength and corrosion resistance was described. Two solutions: 5% acetic acid and 5000 mg/dm3 sodium sulphate were selected for immersion of mortar spicemens. The prolongation of setting time, decrease in water absorbability, compressive strength and increase of corrosion resistance of mortars with emulsion were found.

Keywords:Cement Mortars, Corrosion Resistance, Setting
         Time, Water Absorbability, Compressive
         Strength, Asphalt Emulsion.

Dr Anna M. Grabiec is a Graduate of the Chemistry Department, the Technical University, Poznan, Poland. She works as a Lecturer at the Land Reclamation Department, the Academy of Agriculture, Poznan, Poland. Her scientific work deals with technology of concrete. Her recent studies concern the influence of plasticizing admixtures for concrete on its properties.

Dr Wlodzimierz Lecki is a Graduate of the Civil Engineering Department, the Technical University, Poznan, Poland. He works as a Lecturer at the Land Reclamation Department, the Academy of Agriculture, Poznan, Poland. His research interests are in the field of corrosion protection of concrete structures, particularly agricultural.

INTRODUCTION

The purpose of the work reported here was to describe the influence of asphalt emulsion as an admixture for cement mortars on some technical properties of these mortars, particularly on corrosive action of acid and sulphate media.

Tests were carried out on the specimens of cement mortars based on the four different cements. The following two solutions were selected for immersion of mortar specimens: 5% acetic acid and 5000 mg/dm3 sodium sulphate.

Asphalt anionic emulsion used as an admixture for cement mortars is a diphase system of water and high-quality asphalts. Water is a continuous phase and asphalt is a dispersed phase in this emulsion. The asphalt content in emulsion is 50% by mass. The size of asphalt particles does not exceed 10 micrometers. Addition of emulsifiers and stabilizers secures the durability of the emulsion system.

The tests of basic properties and further tests of corrosion resistance were made on rectangular prisms measuring 4x4x16 cm and prepared of cement mortar with composition of 1 m3 as follows: cement - 585 kg, sand - 1755 kg, water - 292.5 kg.

Properties of cements used for tests are presented in Table 1. Quartz sand of 0.08 - 1.6 mm size, with 33% content of fraction above 0.5 mm was used for tests. Asphalt emulsion was proportioned by volume and replaced 1, 3, 5, 7, 10 and 15% of water.

Test results of mortars with emulsion were related to the results of control mortars without emulsion.

TEST OF BASIC PROPERTIES

Tests of basic properties were carried out in accordance with Polish standards (3). They included: setting time of cements, compressive strength and water absorbability of cement mortars.

Tests of setting time were made on cement pastes, using the Vicat apparatus. It was found that admixture of asphalt emulsion prolonged the initial and final setting time of cement. Prolongation of setting time was proportional to the amount of emulsion. Variations in setting time are shown in Figure 1.

TABLE 1 Properties of cements used for tests

TYPE OF CEMENT	SPECIFIC AREA (cm2/g)	COMPRESSIVE STRENGTH AFTER 28 DAYS (N/mm2)	CHEMICAL COMPOSITION (%)			
			CaO	SiO2	Al2O3	Fe2O3
A BLAST-FURNACE CEMENT 25 "WARTA"	2631	25.9	57.40	28.16	5.32	2.71
B PORTLAND CEMENT 35 WITH ADDITIVES "GORAZDZE"	3086	35.1	51.90	27.40	7.41	3.78
C PORTLAND CEMENT 35 "CHELM"	3152	38.6	65.70	21.16	5.05	1.85
D PORTLAND CEMENT 45 "CHELM"	3648	46.3	63.72	21.32	5.16	2.07

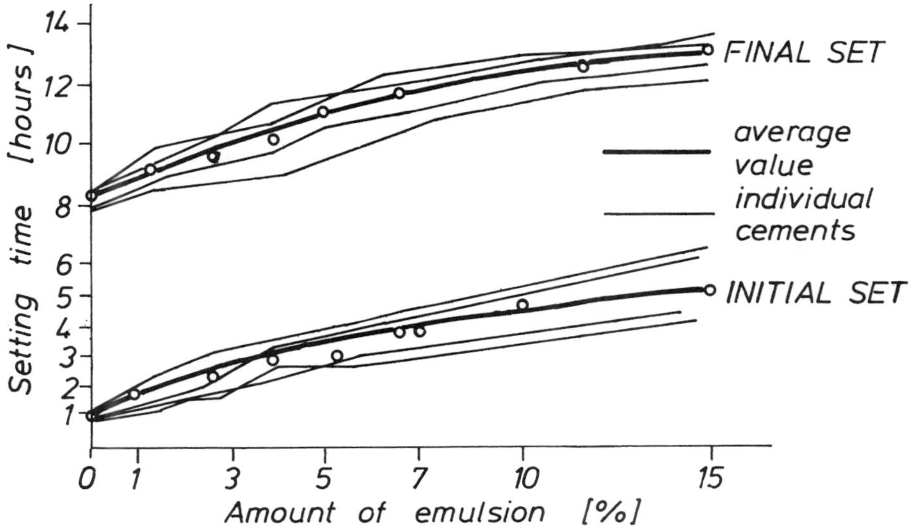

Figure 1: Influence of amount of emulsion on setting time

Differences in changes of setting time between
individual cements are slight. Prolongation of initial
setting time was about 40 minutes at 1% amount of
emulsion. Prolongation of final setting time was 27
minutes. Initial setting time was prolonged 183 minutes
and final setting time was prolonged 243 minutes at 15%
amount of emulsion.

Tests of water absorbability were made on specimens
cured for 28 days. Results are shown in Table 2.

TABLE 2 Water absorbability in % of mass

TYPE OF CEMENT	ABSORBABILITY IN % OF MASS AMOUNT OF EMULSION (%)						
	0	1	3	5	7	10	15
A BLAST-FURNACE CEMENT 25 "WARTA"	7.6	7.4	7.3	6.8	5.9	5.1	4.9
B PORTLAND CEMENT 35 WITH ADDITIVES "GORAZDZE"	6.8	6.5	6.3	6.0	5.8	5.3	5.1
C PORTLAND CEMENT 35 "CHELM"	6.4	6.2	6.0	5.7	5.5	5.2	4.9
D PORTLAND CEMENT 45 "CHELM"	6.3	5.9	5.9	5.3	5.1	4.9	4.7

Absorbability of specimens with 1% emulsion decreased
from 2.64 to 6.35% and from 23.44 to 35.53% at 15%
amount of emulsion.

Tests of compressive strength were carried out on halves
of rectangular prisms measuring 4x4x16 cm, after 28 and
90 days of hardening.   The influence of the admixture of
emulsion on the compressive strength is shown in Figures
2 and 3.   Emulsion at the amount of 1 and 3% decreased
the compressive strength slightly.   A decrease in
compressive strength was greater for higher amount of
emulsion. For instance it was 50% at 15% amount of
emulsion.

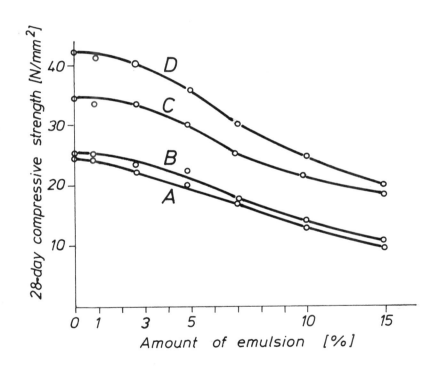

Figure 2: 28-day compressive strength
         of mortars with emulsion

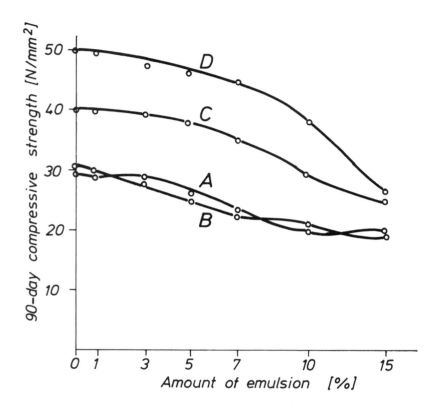

Figure 3: 90-day compressive strength
of mortars with emulsion

After 28 days of hardening the decrease in compressive
strength for specimens with 1% amount of emulsion was
from 2.31 to 3.74% and for specimens with 15% amount of
emulsion was from 56.19 to 67.33%. Decreases in
compressive strength after 90 days were smaller - from
2.22 to 2.87% at 1% amount of emulsion and from 34.60 to
51.56% at 15% amount of emulsion.

Test of the corrosion resistance of mortars were carried
out by 1, 3 and 6 month exposure of specimens to two
media: water solution of 5% acetic acid (pH about 4) and
water solution of 5000 mg/dm3 sodium sulphate. Control
specimens were kept in water. All specimens were cured
before staying in water for 28 days. Specimens were
rectangular prisms measuring 4x4x16 cm. Compressive
strength was measured at ages of 1, 3 and 6 months of
staying in corroding media. The percent decrease of

compressive strength was a measure of the corrosion
resistance:

$$\% \ Rs = \frac{Rss - Rsk}{Rss} \ 100,$$

where:
% Rs is a percent decrease in compressive strength,
Rss is a compressive strength of control specimens,
Rsk is a compressive strength of specimens   exposed
     to aggresive medium.

Results of tests are presented in Table 3 and Table 4.

TABLE 3 Decrease in compressive strength of
        specimens exposed to 5% acetic acid

TYPE OF CEMENT	PERIOD OF TESTS (months)	%DECREASE OF COMRESSIVE STRENGTH						
		SPECIMENS WITH EMULSION						
		0%	1%	3%	5%	7%	10%	15%
A BLAST FURNACE CEMENT 25 "WARTA"	1	27.4	27.0	26.2	24.8	20.6	19.3	15.2
	3	44.1	30.6	31.2	29.8	28.7	25.4	20.8
	6	64.1	40.3	40.1	36.8	34.2	29.8	25.4
B PORTLAND CEMENT 35 WITH ADDITIVES "GORAZDZE"	1	31.2	30.6	31.2	28.4	26.2	22.1	20.9
	3	47.6	40.0	38.6	36.2	33.2	34.1	32.7
	6	72.3	50.8	53.2	48.1	43.2	40.8	40.7
C PORTLAND CEMENT 35 "CHELM"	1	27.6	26.4	25.8	23.2	20.8	18.3	16.7
	3	41.2	39.4	35.3	34.7	33.2	29.1	27.6
	6	66.2	43.2	37.1	33.2	34.6	30.5	29.0
D PORTLAND CEMENT 45 "CHELM"	1	30.8	29.1	29.2	27.3	26.5	24.1	20.7
	3	50.3	38.6	36.2	34.1	33.3	30.0	24.7
	6	68.3	50.6	49.2	48.8	44.3	41.1	36.5

TABLE 4 Decrease in compressive strength of specimens
exposed to 5000 mg/dm3 sodium sulphate

TYPE OF CEMENT	PERIOD OF TESTS (months)	%DECREASE OF COMPRESSIVE STRENGTH						
		SPECIMENS WITH EMULSION						
		0%	1%	3%	5%	7%	10%	15%
A BLAST-FURNACE CEMENT 25 "WARTA"	1	-19.0	-16.2	-17.3	-10.6	-8.9	-9.3	-7.2
	3	17.6	8.4	6.3	4.4	3.8	2.7	1.6
	6	30.2	21.3	14.4	12.3	16.7	12.1	14.4
B PORTLAND CEMENT 35 WITH ADDITIVES "GORAZDZE"	1	-20.3	-19.3	-15.8	-16.3	-12.6	-13.0	-12.1
	3	20.4	12.3	7.4	6.3	5.9	4.0	4.0
	6	36.2	24.2	20.6	19.3	17.1	12.0	9.4
C PORTLAND CEMENT 35 "CHELM"	1	-24.2	-20.1	-18.7	-17.3	-14.2	-16.1	-8.8
	3	22.1	13.5	12.6	10.2	8.8	7.6	6.8
	6	34.2	24.6	19.7	18.6	14.2	12.3	11.7
D PORTLAND CEMENT 45 "CHELM"	1	-26.1	-20.8	-18.3	-17.0	-16.3	-12.2	-13.0
	3	24.1	12.9	13.2	6.8	9.4	6.8	5.4
	6	37.2	24.9	19.9	18.1	14.6	12.7	9.8

"-" means, that strength of control specimens was smaller

DISCUSSION OF TEST RESULTS

The tests showed a great influence of an asphalt
emulsion on properties of cement pastes and mortars.
The admixtures causes:
- prolongation of initial and final setting times
  of cements
- decrease in water absorbability of cement mortars
- decrease in compressive strength of cement mortars.

Changes of technical properties were dependent on the amount of asphalt emulsion. The higher amount of emulsion was, the larger prolongation of setting time, the lower absorbability and compressive strength were. Similar results were obtained by Czerkinskij and Slipczenko (1). Essential differences in the influence of asphalt emulsion on technical properties of cement mortars according to the type of cement were not found. The differences of standard deviations of the obtained results in individual series were not smaller in comparison to standard deviations of differences between individual series.

The main purpose of our tests was to describe a corrosion resistance of mortars with admixture of an asphalt emulsion. Obtained results showed that emulsion increased this resistance to a destruction caused by aggressive media: acid and sulphate. After 6 months of exposure to acid solution decrease in compressive strength of specimens with 15% amount of emulsion was from 39.6 to 56.3% of strength decrease of specimens without emulsion. Analogical specimens exposed to the solution of sulphates showed 25.9 - 40.7% of compressive strength decrease of specimens without emulsion.

The cause of these changes is presence of asphalt particles in the structure of mortar. Asphalt emulsion as a constituent of liquid mortar, increases the thickness of the wetting layer of cement and aggregate particles and changes in this way consistence and workability. Simultaneusly asphalt particles covering cement grains block the access of water molecules and make difficult the hydration of cement. The decrease of setting rate and compressive strength is the effect of this. On the other hand asphalt particles tighten the structure of mortars decreasing in this way absorbability and increasing corrosion resistance. The decrease in water absorbability was found by Czerkinskij and Slipczenko (1) and Wieczorek with co-authors.(2)

CONCLUSIONS

The addition of asphalt emulsion for cement mortars causes the prolongation of setting time, decrease in water absorbability, decrease in compressive strength and increase of corrosion resistance. The increase of corrosion resistance is conditioned upon about 50% decrease in compressive strength.

REFERENCES

1   CZERKINSKIJ J, SLIPCZENKO G. Latex-cement sand
    aggregate concrete with improved properties.   Beton
    i Zelezobeton, Number 5, 1973, pp. 13-14.

2   WIECZOREK G, Gust J, Bobocinska A. Conbit - mortar
    or concrete of increased corrosion resistance.
    Prospects of the use in building engineering.
    Part I. Przeglad Budowlany, Number 4, 1988, pp.
    176-178 and 184.

3   POLISH STANDARD. PN-88/B-04300. Cement. Test
    methods for physical properties.

SURFACE TREATMENT OF HYDRATED CEMENT PASTE USING "XYPEX"

B S Terro
D C Hughes

University of Bradford,

United Kingdom

ABSTRACT. The use of XYPEX, a pore blocking active surface treatment which may be applied to existing structures, is described. Poor quality cement paste samples were water cured for 4 weeks and subsequently conditioned for a further 18 weeks at 88% RH. Samples were treated with Xypex and conditioned at approximately 100% RH for periods up to 26 weeks prior to assessment. This was performed using propanol counter diffusion, chloride diffusion and water permeability. Pore structure was characterised using propanol adsorption. Results show that Xypex reacts rapidly in the substrate, yielding improvements to the engineering properties. However, this assessment is test dependent. Comparison with published data suggests that the application of Xypex is equivalent to a reduction in w/c from 0.8 to approximately 0.5.

Keywords: Cement Paste, Surface Treatment, Pore Blocker, Xypex, Diffusion, Permeability, Pore Structure, Mineralogy.

Mr Bassam S Terro currently works for Bullen and Partners. This study was undertaken whilst he was a research student at the Department of Civil Engineering, The University of Bradford, West Yorkshire, UK.

Dr David C Hughes is a lecturer at the Department of Civil Engineering, The University of Bradford, West Yorkshire, UK. His current interests include the engineering properties of grouts containing PFA and GGBS and more interactive methods for teaching Civil Engineering undergraduates.

INTRODUCTION

Infrastructure renewal is a major element of modern civil engineering. Of particular concern is the large number of highway structures suffering chloride induced degradation. Suitable existing structures may be maintained by the application of surface treatments which modify the ingress of deleterious agents. So-called pore blockers form a category of such treatments.

The paper describes the application of an organic mortar treatment, XYPEX, to cement paste samples. Its efficacy has been assessed by measuring propanol counter diffusion, chloride diffusion and "water" permeability. Pore structure was monitored by a propanol adsorption technique. In order to isolate the effect of the active ingredient within Xypex, the control programme included both untreated samples and those treated with a special Xypex preparation in which the active ingredient was omitted.

SAMPLE PREPARATION

In order to mimic possible "realcrete", the substrate should possess high permeability and be unsaturated. The former was achieved by a combination of a low C_3S OPC and a high w/c ratio; the latter by conditioning the samples at 88% RH.

Weardale OPC (47.5% C_3S, 23.9% C_2S) was used to produce pastes at a w/c of 0.65 for moulding into 34 mm diameter cylinders. These were sealed and rotated overnight. Subsequent curing was in water which had been previously conditioned by the immersion of similar hcp samples. At an age of 28 days the samples were removed and sawn in half to yield surfaces for subsequent treatment. They were placed in sealed polyethylene boxes together with saturated $BaCl_2$, for 88% RH, and soda lime to minimise carbonation. After a period of 18 weeks the rate of weight loss had substantially reduced and the samples were treated. The samples for permeability determination were prepared in conical moulds as previously described (1).

Xypex is supplied for application as a mortar. As such it contains Portland Cement, very fine silica sand, lime and the active ingredient. The prepared hcp faces were moistened, rather than "saturated" as the manufacturer suggests. Xypex mortar was prepared according to the manufacturer's instructions and applied as a 2 mm thick coat. The control samples were left either untreated or similarly treated with the special mortar. All samples were restored in the sealed boxes but with the $BaCl_2$ replaced by a substantial quantity of water in order to yield optimum conditions for the hydration of the active ingredient. The samples were stored over the water with the treated face vertically aligned to prevent ponding on the treated surface. They were occasionally sprayed with water during the first 3 days.

Samples were taken for test at ages up to 26 weeks following treatment. For the diffusion and microstructural assessments 3 mm discs were cut at 2 depths within each sample, namely 0-3 mm and 6-9 mm from the original treated surface after the Xypex coat had been removed. The permeablity measurement was made on the complete 15 mm thick sample and hence is a bulk property.

An additional set of control samples were cured continuously in the conditioned water to provide a datum for the potential development of the substrate given ideal curing conditions. These samples were evaluated at the same time as those previously described.

EXPERIMENTAL PROCEDURE

Propanol Counter Diffusion

Propanol counter diffusion was measured at the time of surface treatment and after a further 6, 13 and 26 weeks had elapsed. The alcohol used in this study was propan-2-ol, Analar grade. At each test age 4 No. 3 mm discs were prepared, from both depths to be assessed, and placed in a small quantity of conditioned water. This was stored in a vacuum overnight to ensure full disc saturation. After establishing SSD weights and volumes, each set of discs was immersed in approximately 400 ml of propanol. The subsequent weight loss was measured until equilibrium was achieved. The propanol was replaced when 80% of the anticipated exchange had occured.

Propanol exchange in hcp may be considered a non-reacting counter diffusion process (2) although it remains a matter for debate (3,4). As such, it may be described by a diffusion coefficient (5), D_P :-

$$D_p = \left(\frac{\pi}{16} \right) \times R^2 \quad (m^2/s)$$

where

$$R = \frac{(Wt_2 - Wt_1)}{Wt_c} \times \frac{l}{(t_2^{0.5} - t_1^{0.5})}$$

Wt_1, Wt_2 are the weight losses per unit volume at times t_1, t_2 respectively
W_c is the weight loss per unit volume at complete exchange
l is the average sample thickness (m)

Chloride Diffusion

Chloride diffusion was measured at the same ages as that for propanol. The cells are shown in Figure 1. Discs were prepared as previously described and attached to the 50 ml cells by means of Expandite Expocrete UA, taking care not to contaminate the diffusing surface. This process was performed with the assembly immersed in a saturated $Ca(OH)_2$ solution. They were further sealed with a heat shrink moisture proof tape. The cells were then immersed in 5 l of 1M NaCl, saturated $Ca(OH)_2$ solution and removed at intervals of 1, 2, 3 and 4 weeks. The chloride concentration of each cell was measured using a $AgNO_3$ titration technique (6).

The chloride diffusion coefficient, D_{Cl}, may be calculated using Fick's law:-

$$D_{Cl} = \frac{C_2 \times V \times l}{C_1 \times A \times (t_2 - t_1)} \quad (m^2/s)$$

where

C_1, C_2 are the chloride concentrations in the cell at times t_1, t_2 respectively
V is the cell volume (m^3)
A is the area of the diffusing surface (m^2)

Permeability

Permeablity was measured at the time of surface treatment and after further periods of 11, 17 and 26 weeks had elapsed; the average of 4 samples for each determination. The outflow was measured for samples subject to an inflow pressure of 6.9 N/mm^2. The permeating solution, prepared to simulate the pore solution (7), is shown in Table 1. Precise details of the technique may be found elsewhere (1).

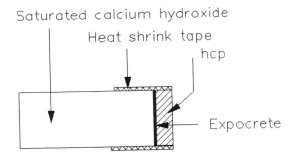

Figure 1: Chloride diffusion cell

Table 1: Composition of permeating solution
(Concentration expressed as oxide weight g/kg solution)

CaO	Na$_2$O	K$_2$O
1.9	43.6	146.7

The outflow was measured over a period of 8 days, with the Darcy coefficient, k, being calculated from the average flow rate of days 6, 7 and 8:-

$$k = \frac{Q}{A \times (dH/dx)} \quad (m/s)$$

where

Q is the outflow (m^3/s)
A is the mean cross section area (m^2)
dH/dx is the hydraulic gradient between inlet and outlet

Pore Structure

Pore structure was measured using those samples for which propanol counter diffusion had been established. Thus, after weighing the propanol saturated SSD discs, they were placed in n-Pentane until full exchange had occurred. Samples were transferred to a desiccator and held at room temperature overnight. They were then placed in an oven at 40°C for a further 3 weeks. The atmospheres in both the desiccator and oven were circulated through silica gel and soda lime.

The pore structure was determined using the adsorption technique detailed by Parrott (8). In this study 3 standard porous glasses were utilised with controlled pore diameters of 7.9, 36.4 and 148.9 nm respectively. Thus, it is possible to classify porosity in terms of total "water" and propanol porosities and cumulative porosities in pores of diameter less than or equal to 7.9, 36.4 and 148.9 nm. A typical adsorption profile from the 7.9 nm glass is shown in Figure 2. Precise determination of the point of saturation in the glass is not possible due to the incremental nature of the adsorption cycle. This is a problem common to all the glasses used. Hence, it was decided to make the assessment at propanol adsorption in the glasses of 0.2, 0.15 and 0.07 g/g in ascending order of pore size. The intercept of the comparator with the adsorption profile is projected onto the sample axis to determine a porosity.

RESULTS

The following shorthand has been adopted to describe the sample treatment and disc location:-

CW - cured in conditioned water only
X   - Xypex treatment
XF - "Xypex Free" treatment (mortar lacking the active ingredient)
U   - untreated
3   - disc representative of a layer 0-3 mm from the treated surface
6   - disc representative of a layer 6-9 mm from the treated surface

Propanol Counter diffusion

The data in Table 2 shows that the application of Xypex results in a rapid reduction in D_P, of approximately 50%, in both layers examined. Indeed, it would appear that Xypex has yielded a lower diffusion coefficient than was achieved by those samples which were continuously cured in ideal conditions (CW). However, the benefit cannot be wholly ascribed to the active ingredient within the Xypex mortar. Whilst the X series consistently show the lowest values of D_P, reductions were also registered by the XF and U series although achieved at a slower rate. Hence, two mechanisms may be identified, ie. a rapid Xypex induced reaction and a

slower hydration of OPC. It must be recognised that
additional curing of 26 weeks at 100% RH is unrealistic and
whilst further OPC hydration is of academic interest, it is
unlikely to be of practical significance. Indeed, of more
interest is the relative performance of D_P at the earliest
test age of 6 weeks between the X and XF, U series.

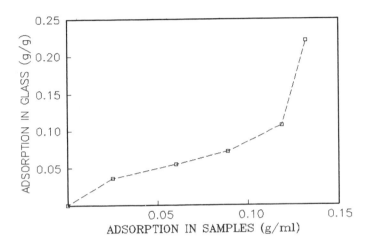

Figure 2: Typical adsorption profile for hcp using 7.9 nm
glass standard

Chloride diffusion

The application of Xypex results in a rapid reduction in
D_{cl} within 6 weeks (Table 3) to a value below that of the
CW control samples. The reduction is more pronounced in the
surface layer than at depth. However, in contrast to D_P,
the XF,U series do not apparently show a long term
reduction in D_{cl}. Indeed following treatment the CW,XF and
U series show quite similar values of D_{cl}.

Permeability

The benefit of Xypex is not apparent. Whilst there is a
reduction in k, the values displayed by X, XF and U
series are similar, given the known variability of
permeability data. In contrast to the diffusion data no
permeability sample indicates an improvement upon the
ideal curing conditions.

Table 2: Propanol counter diffusion data

TEST AGE FOLLOWING TREATMENT	DIFFUSION COEFFICIENT $(D_P \times 10^{-12} \ m^2/s)$						
Disc	CW	X3	XF3	U3	X6	XF6	U6
0 weeks	6.2	----	10.8	----	----	11.0	----
6 weeks	6.7	5.1	9.1	9.6	6.1	9.0	9.5
13 weeks	5.9	5.1	7.6	8.0	5.3	8.4	9.0
26 weeks	6.0	4.9	6.3	6.4	5.3	6.0	7.9

Table 3: Chloride diffusion data

TEST AGE FOLLOWING TREATMENT	DIFFUSION COEFFICIENT $(D_{Cl} \times 10^{-12} \ m^2/s)$						
Disc	CW	X3	XF3	U3	X6	XF6	U6
0 weeks	13.9	----	16.7	----	----	15.1	----
6 weeks	12.0	9.3	12.0	14.1	11.0	12.8	14.3
13 weeks	12.3	8.7	11.7	12.0	10.0	12.4	14.0
26 weeks	14.5	9.4	14.2	14.7	12.0	14.9	16.5

Pore Structure

Table 5 shows the pore structure data and the following notation has been adopted. Total water porosity - P_w;

total propanol porosity - P_P; propanol porosities in
pores **greater than or equal to** the specified diameter -
$P_{7.9}$, $P_{36.4}$ and $P_{148.9}$.

It is apparent that a complete analysis of the pore
structure data is outside the scope of this paper. However,
by the majority of comparisons, the X series yields the
finer pore structure at a given age. This is particularly
so for test ages of 6 and 13 weeks; these being the ages
which showed the greater difference in D_P between X and
XF,U samples. Whilst Xypex reduced D_P to a value below that
of the CW samples there is no such consistent correlation
with the pore structure.

Table 4: Permeablity data

TEST AGE FOLLOWING TREATMENT	PERMEABILITY COEFFICIENT ($k \times 10^{-13}$ m/s)			
Disc	CW	X	XF	U
0 weeks	0.9	----	17.0	----
10 weeks	1.2	8.1	9.5	8.0
16 weeks	1.8	5.7	5.9	2.2
26 weeks	1.4	3.0	9.1	3.4

DISCUSSION

This study has shown that, for the conditions examined,
Xypex is capable of modifying the pore structure of hcp.
Reductions in both propanol counter diffusion and chloride
diffusion are attributable to the active ingredient in the
product. In contrast, permeability measurements are not
able to positively identify any benefit from the treatment.
There may be a number of reasons for this apparent
discrepancy.

Each test measures something different. Propanol counter
diffusion is assumed not to react significantly with the
pore structure. Also, it utilises the complete pore
structure whether it is continuous or not. In contrast,
chlorides are known to react with the aluminates in
particular. Further, whilst chlorides can diffuse into both
continuous and discontinuous pores connected to the high
concentration inflow surface, the outflow test only
measures those free chloride ions which diffuse through
completely continuous pores without reacting with the hcp.
The latter pores are also those utilised in the
permeability test. However, the test regime is different
with the driving agent being a high pressure of 6.9 N/mm².
It is not currently known whether Xypex generated minerals
are capable of withstanding such pressures. Should this
prove to be so, an appropriate test regime needs to be
developed. Hence, careful test selection is important if a
particular application is to be assessed.

Comparing D_P and D_{c1} of the various series gives a measure
of the efficacy of Xypex. In order to obtain a preliminary
engineering assessment of the significance of these figures
it would be useful to relate them to the more familiar w/c.
Feldman (4) has reported values of D_P for w/c in the range
0.3 - 0.8 (Table 6). He claimed that the values for w/c of
0.3 and 0.4 were unrealistically high due to difficulties
of manufacturing such pastes. With this proviso, it could
be suggested that the application of Xypex reduces the
apparent "effective" w/c from 0.8 to 0.5. Although this is
obviously a crude comparison, materials and curing
conditions being different, it may be useful in allowing
the engineer to make a comparison between the potential of
a repaired concrete and an alternatively designed original
mix.

Table 6: Values of propanol counter
diffusion (4)

D_P		w/c			
($\times 10^{-12}$ m²/s)	0.3	0.4	0.5	0.6	0.8
	7.3	8.6	4.2	8.7	12.2

Table 5: Pore structure measured by propanol exchange and adsorption

TEST AGE FOLLOWING TREATMENT	MEASURE OF POROSITY	POROSITY (%)						
Disc		CW	X3	XF3	U3	X6	XF6	U6
0 weeks	P_w	44.9	----	44.4	----	----	44.4	----
	P_p	43.3	----	43.8	----	----	43.4	----
	$P_{7.9}$	26.4	----	25.1	----	----	24.9	----
	$P_{36.4}$	12.5	----	12.9	----	----	13.1	----
	$P_{148.9}$	8.2	----	9.8	----	----	9.5	----
6 weeks	P_w	46.2	41.2	43.6	46.0	42.5	44.1	45.9
	P_p	45.0	39.7	42.5	45.5	41.0	42.8	45.0
	$P_{7.9}$	28.5	23.5	26.6	29.6	25.8	26.1	28.9
	$P_{36.4}$	15.2	13.7	15.5	18.6	14.1	15.7	18.7
	$P_{148.9}$	6.8	9.0	10.1	13.3	9.3	10.3	12.3
13 weeks	P_w	45.7	41.4	43.8	44.4	42.7	44.3	44.8
	P_p	44.8	40.5	42.7	43.3	41.8	43.5	43.6
	$P_{7.9}$	27.0	24.3	27.6	28.8	26.4	26.0	27.8
	$P_{36.4}$	14.7	12.0	13.4	15.5	13.2	14.2	15.1
	$P_{148.9}$	7.6	7.8	9.5	9.8	8.8	9.8	9.9
26 weeks	P_w	44.7	40.9	42.9	43.7	42.8	43.4	44.0
	P_p	42.9	39.0	41.1	42.7	41.4	42.3	42.3
	$P_{7.9}$	22.1	25.2	25.9	24.5	25.7	25.5	24.1
	$P_{36.4}$	12.3	13.5	13.6	13.7	13.7	13.8	14.3
	$P_{148.9}$	8.2	7.4	7.7	8.6	7.9	8.7	8.6

Figure 3 shows that D_p of the X, XF and U samples following treatment is broadly related to the volume of coarse pores ($P_{36.4}$). However, this relationship is dependent upon the curing regime, which is highlighted by comparing $P_{36.4}$ values for CW and the substrate at treatment (Table 5). Despite exhibiting similar porosities the diffusivity of both samples is very different (Table 2). Simple measures of pore structure are unable to account for the complexities of the structure of hcp such as pore shape, conductance or tortuosity (9). Such features are important consequences of the "water", 88% RH and 100% RH regime adopted in this work. Hence, care must be taken when considering the implications of microstructural changes.

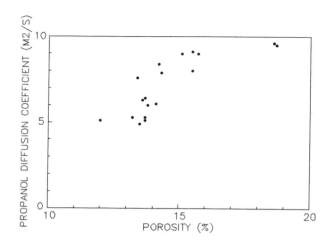

Figure 3: Propanol diffusion as a function of volume of
pores greater than 36.4 nm

CONCLUSIONS

Assessment of the efficacy of Xypex is test dependent.
Measurements of chloride diffusion and propanol counter
diffusion indicate substantial benefits whilst permeability
testing is inconclusive. Appropriate test selection is
important if a particular application is to be assessed.

Improvements in propanol counter diffusion are related to
modifications to the pore structure. These arise from
reactions between the active ingredient in Xypex and hcp
together with the consequences of the ingress of water to a
partially dehydrated substrate.

REFERENCES

(1) MARSH, B.K. Relationships between engineering
properties and microstructural characteristics of hardened
cement paste containing PFA as a partial cement
replacement. PhD thesis, Hatfield Polytechnic,1984.

(2) PARROTT, L.J. Novel methods of processing cement gel to
examine and control microstructure and properties. Phil.
Trans. R. Soc., London - A310, 1983, 155-166.

(3)TAYLOR, H.F.W. and TURNER, A.B. Reactions of tricalcium
silicate paste with organic liquids. Cem. Concr. Res., vol
17, 1987, 613-623.

(4) FELDMAN, R.F. Diffusion measurements in cement paste by water replacement using propan-2-ol. Cem. Concr. Res., vol 17, 1987, 602-612.

(5) CRANK, J. The mathematics of diffusion. Clarendon Press, Oxford. 1975, 236-241.

(6) VOGEL, A.I. Vogel's textbook of quantitative inorganic analysis, including elementary instrumental analysis. Longman, London. 1978, 31-33.

(7) LONGUET, P., BURGLEN, L. And ZELWER, A. La phase liquide du ciment hydrate. Revue des Materiaux et Construction, 1973, 35-41.

(8) PARROTT, L.J. Thermogravimetric and sorption studies of methanol exchange in an alite paste. Cem. Concr. Res., vol 13, 1983, 18-22.

(9) HUGHES, D.C. Pore structure and permeability of hardened cement paste. Mag. Concr. Res., vol 37, 1985, 227-233.

SURFACE COATINGS FOR IMPROVING PERFORMANCE OF CONCRETE UNDER MARINE ENVIRONMENTS

T Fukute

H Hamada

Port and Harbour Research Institute,

Japan

ABSTRACT. The Naminoue bridge is a prestressed concrete type bridge. It is a bayside bridge constructed in 1983 in Okinawa prefecture, Japan. Okinawa is southernmost island in Japan, and is in the sub-tropical zone. Because of this, the environmental conditions affecting the bridge are very severe, especially salt which attacks the prestressed concrete members. In order to prevent deterioration of the members from salt attacks, prestressed members have to be coated with a special coating material within a few years. But, in Japan, neither a simplified classification standard for environments for concrete structures nor suitable standards for selecting coating materials have been established. A series of tests whose object is to evaluate the effects of several coating systems on the prevention of salt attack is carried out. Tests are all carried out using concrete specimens – first, an exposure test under the real environment ; second, an exposure test in a simulated splash zone ; last, an accelerating laboratory test under drying and saturating cycles. In this paper, test results are presented and suitable materials for the prevention of the salt attack are discussed. Also, a simplified evaluation diagram for coating systems is suggested.

Keywords: accelerating laboratory test, chloride, coating, concrete durability, diffusion, exposure test, marine environment, salt attack

Tsutomu Fukute is a chief of the Materials Laboratory, Port and Harbour Research Institute, Ministry of Transport, Japan. He received his doctorate in engineering from the University of Nagoya in 1985. Currently much of his research work has been on the durability of concrete and the repair of airport pavement, etc.

Hidenori Hamada is a reserch engineer at the Materials Laboratory, Port and Harbour Research Institute, Ministry of Transport, Japan. He received his masters degree in engineering from the University of Kyusyu in 1986. His recent research has been on the durability of concrete including the sulfate problem, salt problem and aggregate reactions, etc.

INTRODUCTION

Concrete is a widely-used construction material. The need to protect
concrete structures from deterioration is surveyed along with various
protective approaches that may be taken to protect them. This paper
focuses on the use of protective coatings as the means of protecting
concrete structures. In certain environments, concrete is susceptible to
chemical attack or to physical and mechanical damage. The intrusion of
salt and water into concrete will cause premature corrosion of reinforcing
steel and eventual cracking and spallation of the concrete. Coating
concrete provides protection against reinforcement corrosion, chemical
attack, water and salt intrusion. Concrete is also painted for decorative
purposes to enhance appearance along with other functions.

The Naminoue bridge, the most important port-side bridge in Okinawa
prefecture, is located at 26° N, 128° E and the total length is about 500m.
A panorama of this bridge is shown in Fig.1 From investigations it was
found that chloride ions supplied from seawater had accumulated in the PC
(prestressed concrete) members. It was, therefore, considered that within
a few years some PC members should be coated with a special coating. In
order to make a suitable choice of coating system for this bridge and to
establish an evaluation system for coating systems, a series of test
programmes were started at the actual bridge site in Okinawa Pref. and the
Port and Harbour Research Institute (PHRI) in Kanagawa Pref. Fig.2 shows
the locations of Okinawa Pref. (the Naminoue bridge) and Kanagawa Pref.
(the PHRI).

There is a dearth of information as to the effects of protective coatings
on the prevention of salt attack under severe marine environments such as
in the splash zone. Also in Japan there are no standards for evaluating
coating systems. The aim of the research reported here is to gather
quantitative data on the performances of several coating systems under the

Fig.1 The Naminoue Bridge

Fig.2 The location of
      Okinawa and Kanagawa

marine environment. The tests are all carried out using reinforced
concrete specimens whose all surfaces are coated. Three kinds of tests are
performed, (1) an exposure test in the real environment, (2) an exposure
test in a simulated splash zone and (3) accelerating laboratory tests under
drying and saturating cycles. Six kinds of testing coating systems are
selected based on local interests. During and after the exposure period,
several kinds of test data are gathered. Half-cell potentials are recorded
by the single electrode method. The change in appearance, chloride content
in the concrete and corrosion of the reinforcements, etc, are investigated.
This paper discusses the relative performances of the protective-coated
reinforced concrete specimens. Further, the effectiveness of color
measurement and electro-chemical measurements are discussed. Also, a
simplified evaluation diagram for coating systems is suggested.

COATING MATERIALS

Six kinds of testing systems are selected making use as much as possible of
local interests. Table 1 presents these systems and their specifications.

Table 1   Coating systems

System	Process	Material	number of paintings	Standard Coverage (kg/m^2)
A	coat 1	Silane	2	0.3
	coat 2	Acryl modified cement mortar	1	2.0
	coat 3	Chlorinated poly-olefine resin	1	0.2
	coat 4	Silicon emulsion	1	1.0
	coat 5	Silicon resin	1	0.15
B	coat 1	Lithium silicate	2	0.4
	coat 2	Corrosion inhibitor (nitrites)	2	0.5
	coat 3	SBR modified cement mortar	1	22.5
C	coat 1	Epoxy resin	1	0.1
	coat 2	Epoxy resin	1	0.6
	coat 3	Elastic epoxy resin	1	0.35
	coat 4	Elastic poly-uretane resin	1	0.12
D	coat 1	Polymer modified cement	1	1.5
	coat 2	Epoxy resin	2	0.3
	coat 3	Acryl rubber	2	2.0
	coat 4	Acryl urethane resin	2	0.5
E	coat 1	Rubber latex resin modified cement paste	1	3.75
	coat 2	Denatured epoxy resin	1	0.12
	coat 3	Shloroprene rybber	3	0.75
	coat 4	Chlorosulfonie polyethylene	2	0.5
F	coat 1	Epoxy resin	1	0.1
	coat 2	Epoxy resin	1	0.3
	coat 3	Vinylester resin containing glassflake	2	1.1
	coat 4	Acryl Urethan resin	1	0.12

TEST PROCEDURE

Specimen

The outline of the specimen is shown in Fig.3. The specimen is a beam type
of 800mm in length, 150mm in height and 150mm in width. In the specimen,
two steel bars (ϕ 13mm, round bar) are embedded at a cover depth of 20mm.
The two steel bars are welded with steel bar (ϕ 6mm), therefore they are
electrically connected. Also, an electrical lead is connected to these two
bars for the electro-chemical measurements. For concrete mixing materials,

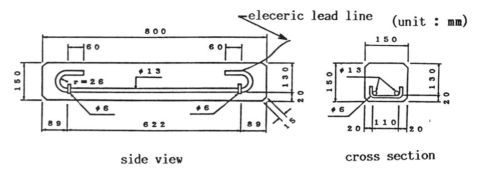

side view   cross section

Fig.3 The outline of the specimen

Table 2 Summary of the specimen

Test Series	Specimen No.	W/C	coating system	note
Exposure test at real site (Series 1)	RE-A-1 RE-B-1 RE-C-1 RE-D-1 RE-E-1 RE-F-1 RE-O-1	37 37 37 37 37 37 37	A B C D E F non-coating	all specimes are non-cracked
Exposure test at simulated splash zone (Series 2)	SE-A-1,2 SE-B-1,2 SE-C-1,2 SE-D-1,2 SE-E-1,2 SE-F-1,2 SE-O-1,2	37 37 37 37 37 37 37	A B C D E F non-coating	"1" is non-cracked "2" is cracked
Acceleration test of cycles of drying and saturating (Series 3)	A-A-1,2 A-B-1,2 A-C-1,2 A-D-1,2 A-E-1,2 A-F-1,2 A-037-1,2 A-050-1,2	37 37 37 37 37 37 37 37	A B C D E F non-coating non-coating	"1" is non-cracked "2" is cracked

(in total 37 specimens)

ordinary portland cement, ordinay aggregate (to JIS, JIS=Japan Industrial Standard, non-reactive) and tap water are used. Water cement ratio, maximum aggregate size, slump value and compresive strength at 28 days are 37%, 20mm, 8cm and 400 kgf/cm^2 respectively. The specification of this concrete is the same as the concrete used in the Naminoue bridge. All surfaces of the specimen are coated with the coating systems presented in Table 1. As presented in Table 2, thirty seven specimens are fabricated. Seven specimens are used in the real site exposure test (series 1), fourteen specimens are used in the splash zone simulating exposure test (series 2) and sixteen specimens are used in the acceleration test (series 3). In some specimens, cracks are induced by bending after coating. The maximum crack width is 0.2mm under load.

Fig.4 Exposure condition
(Real site)

Fig.5 Exposure condition
(Simulated splash zone)

Exposure Test under Real Environment (Series 1)

Fig.4 shows the exposure condition of the specimens. As shown in this figure, specimens are attached under the bridge deck, about 6m from sea level. Specimens are occasionaly splashed by seawater and always exposed to the sea breeze. They are not exposed to severe sunshine.

Exposure Test in Simulated Splash Zone (Series 2)

Fig.5 shows the exposure conditions for the simulation system. As shown in this figure, this system splashes the specimens with seawater. Splashing cycle is 3 hours per 1 splashing and 2 splashings per day. In the daytime, the upper surfaces of the specimens are exposed to strong sunshine (direct rays).

Acceleration Test under Drying and Saturating Cycles (Series 3)

Fig.6 shows the drying and saturating cycles of this test. As shown in this figure, 1 cycle is 7 days which is composed of 3 days' hot-seawater saturation (60℃) and 4 days' air drying (20℃). In total, 20 cycles are carried out in this test series.

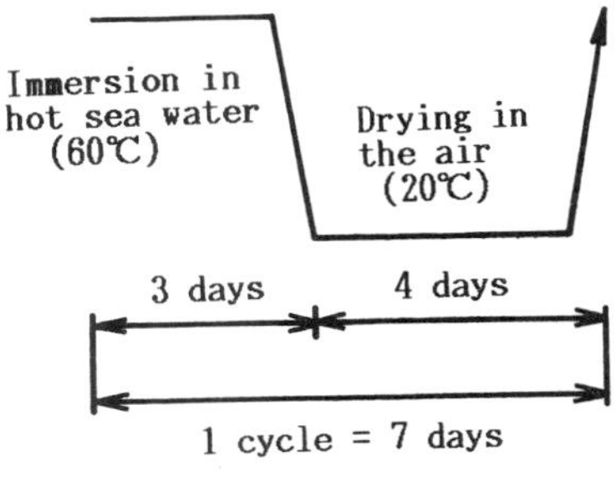

Immersion in hot sea water (60℃)

Drying in the air (20℃)

3 days   4 days

1 cycle = 7 days

Fig.6 Accelerating cycle

Test Items and Test Methods

Table 3 presents a summary of test items.

Specimens are washed with tap water before the visual inspection. In the visual inspection, photographs are taken and surface detreioration are sketched. After this, colour measurement is carried out with colorimeter to collect tristimulus values.[1]

Fig.7 shows the measuring method of electric resistance. The resistance of coating is obtained by measuring the resistance between embedded steel bars

Table 3 Test items

Test Series		Test Items
Exposure test at real site (Series 1)	1 2 3 4	Visual inspection Half-cell potential of embedded steel Adhesive strength of coating layer Chloride content in concrete
Exposure test at simulated splash zone (Series 2)	1 2 3 4 5 6 7 8	Visual inspection Colour measurement of coating surface (psychophysical color specification) Electrical resistance of coating layer Half-cell potential of embedded steel Potentiodynamic anodic polarization curve Potentiostatic step-wave current density Chloride content in concrete Corroded area of embedded steel
Acceleration test of cycles of drying and saturating (Series 3)	1 2 3 4 5	Visual inspection Half-cell potential of embedded steel (periodically during the accelerating test) Potentiodynamic anodic polarization curve Chloride content in concrete Corroded area of embedded steel

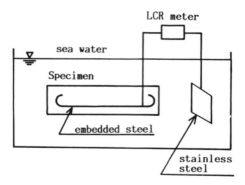

Fig.7 Measuring method of electric resistance

and counter electrode (stainless steel). Resistance is measured by using an alternating current whose frequency is 1kHz or 0.1kHz.

Fig.8 shows the method of electro-chemical measurements of embedded steel. Ag-AgCl electrode is used as reference electrode, as counter electrode stainless steel is used and seawater is used as electrolite. The measurement is carried out under the condition that the concrete is perfectly wetted.

Fig.9 shows an example of potentiodynamic anodic polarization curve. As shown in this figure, the potential is swept between +1000mV and -1000mV (v.s Ag-AgCl electrode) at the speed of 40mV/min. During sweeping, potential and current are recorded continuously. From this polarization curve, "Grade of passivity" is judged in conformity to a standard shown in

Fig.10. [2]   The grade of passivity is ranked into 1,2,3,4 and 5. It does not have unit (abstract number). The Grade 5 means best passivity and Grade 1 means worst one.

Fig.11 shows an example of the result of step-wave current measurement. In this measurement, the potential of the embedded steel bar is kept at -1000mV (v. sAg-AgCl electrode), and current is recorded continuously. Fig.11 shows an   example of the result of this measurement. After about 10 hours from the start, the current shows constant value. This value depends on the oxygen supply to the embedded steel bar.

Adhesive strength is measured in conformity to JIS-A-6910-1988.

After the tests mentioned above, specimens are crushed and embedded steel bars are taken out. Rust condition is observed and sketched. From this sketch, corroded area is calculated.

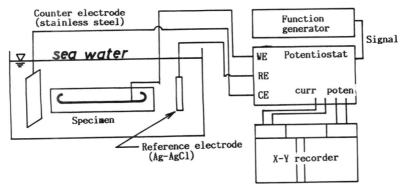

Fig.8 Measuring method of electro-chemical measurement

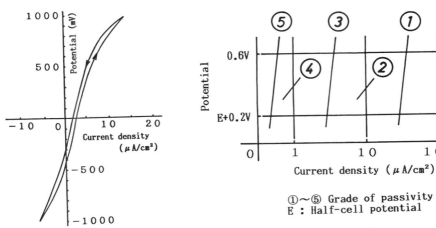

①～⑤ Grade of passivity
E : Half-cell potential

Fig.9 An example of
     polarization curve

Fig.10 Judgement of
     "Grade of passivity"

324 Fukute, Hamada

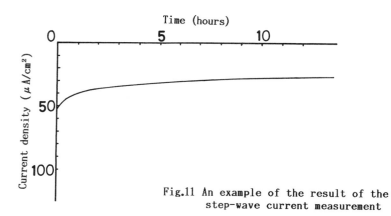

Fig.11 An example of the result of the
step-wave current measurement

Sampling of concrete for chloride content measurement is carried out at
several cover depth by using sampling machine. Water-soluble chloride
content is measured in conformity to the JCI (Japan Concrete Institute)
method.

TEST RESULTS

Exposure Test under Real Environment

In all specimen surfaces, some stains are found, however no surface
deterioration is found.

Test results of the the half-cell potential of embedded steel bars after 1
year' exposure are presented in Table 4. They are all relatively noble,
and they do not belong in the corrosion zone stipulated by the the ASTM
standard. In this standard, the corrosion zone is below -228mV v.s Ag-AgCl
electrode.

The results of adhesive strength of coating layers are presented in Table
4. Coating system "A" shows a slightly low value. In this "A" system, the
upper coat is torn from the middle coat, however the bottom coat is not
torn from the concrete surface. From this, it can be said that the
adhesive str ength of all surfaces is sufficient after one year's exposure.

Table 4 Test results (Series 1)

(after 1year exposure)

Specimen No.	Half-cell potential (mV)	Adhesive strength (kgf/cm²) initial	after 1year exposure			Chloride content (%) Cover depth (mm) 5	15	25	35
RE-A-1	+ 12	13.1	6.2	5.0	4.3				
RE-B-1	+ 1	21.9	20.6	21.8	31.2 31.2	0.001	0.001	0.001	0.001
RE-C-1	+ 40	18.1	15.5	34.3	30.0 21.8				
RE-D-1	+ 4	7.7	7.1	7.5	6.5				
RE-E-1	- 43	10.6	22.5	25.0	32.5				
RE-F-1	- 94	28.7	18.4	40.6	40.6 18.7				
RE-0-1	- 4					0.091	0.004	0.002	0.002

(v.s Ag-AgCl electrode)

(NaCl v.s Concrete in weight)

The test results of chloride content of the concrete are presented in Table 4. The chloride content of the coated specimens was smaller than that of the non-coated specimens.

Exposure Test in Simulated Splash Zone (Series 2)

In the case of specimens SE-B-1 and SE-B-2, after only two months' exposure large cracks developed on the coating surfaces and immediate spalling took place on all the coatings. In other specimens, no surface deterioration was found, however the color of the coating surfaces changed to yellow from white.

The test results of colour measurement of coating surface are presented in Table 5. The "upper surface", "side surface" and "bottom surface" refer to the exposed surface as reference. Each value in this table is an   average of twelve data measurements.

The electrical resistance of coatings are presented in Table 6. The resistance of the coated specimens was higher than for the non-coated ones.

Table 5   The results of colour measurement (Series 2)

Specimen No.	upper surface			side surface			bottom surface		
	Y	x	y	Y	x	y	Y	x	y
SE-A-1	31.25	0.36	0.37	37.76	0.32	0.34	40.50	0.31	0.32
SE-B-1	22.77	0.32	0.33	31.94	0.34	0.35	30.52	0.32	0.32
SE-C-1	28.80	0.40	0.39	34.60	0.34	0.35	34.94	0.35	0.35
SE-D-1	31.62	0.39	0.38	37.20	0.35	0.36	39.95	0.34	0.34
SE-E-1	46.62	0.31	0.33	39.89	0.34	0.36	42.99	0.33	0.34
SE-F-1	34.09	0.40	0.39	39.42	0.36	0.37	48.88	0.32	0.33
SE-0-1	20.97	0.33	0.35	20.13	0.33	0.36	18.58	0.35	0.36

$x = X/(X+Y+Z)$, $y = Y/(X+Y+Z)$, X,Y,Z = psychological color specification

Table 6 Test results (Series 2)

Specimen No.	Half-cell potential (mV)	Resistance (kΩ)		Grade of passivity	Current density (μA/cm^2)	Corroded area (%)
		1 kHz	0.1 kHz			
SE-A-1	-184	0.31	0.41	4	2.8	0.00
-2	- 50	0.70	1.23	5	0.7	0.00
SE-B-1	-466	0.04	0.04	2	10.8	0.07
-2	-494	0.04	0.04	2	10.8	2.43
SE-C-1	-547	0.47	0.76	4	0.2	0.00
-2	-450	0.51	1.63	4	1.2	0.00
SE-D-1	-258	0.58	1.55	5	3.4	0.21
-2	-265	0.56	1.41	4	0.1	0.00
SE-E-1	-463	0.07	0.10	3	0.4	0.00
-2	-466	0.10	0.14	4	2.7	0.00
SE-F-1	-493	0.23	0.22	5	0.1	0.00
-2	-330	1.22	7.15	5	0.1	0.00
SE-0-1	-487	0.04	0.05	3	4.3	1.91
-2	-411	0.04	0.04	2	10.8	3.55

(v. s Ag-AgCl electrode)

The test results of the half-cell potential of embedded steel after one year's exposure are presented in Table 6. They are all relatively base and belong to the corrosion zone as stipulated by the ASTM standards (below -228mV v.s Ag-AgCl electrode).

The values of the Grade of passivity drawn from the polarization curves are presented in Table 6. The values for the coated specimens except "B" are higher than those for the non-coated specimens. From this, it can be said that the conditions of passivity for the coated specimens are better than those for non-coated specimens. As shown in Table 6, the values of the Grade of passivity for specimens SE-B-1 and SE-B-2 are low because of the spalling of the coatings.

The test results of potentiostatic step-wave current measurement are presented in Table 6. The current density of the non-coated and coated with "B system" specimens is about 10 μ A/cm^2, however that of the coated specimens except "B" is about 0.1 $\sim$3.5 μ A/cm^2. From this it is said that the oxygen supply to the embedded steel is inhibited by the good coating systems.

Corroded area of embedded steel bars is presented in Table 6. Embedded steel bars in the non-coated specimens were slightly rusted. However those in the coated specimens except "B" were not rusted.

Chloride content in concrete is presented in Table 7. The chloride content of the non-coated specimens is larger than for the coated specimens. As shown in this table, the chloride content around the rusted steel was larger than in other parts. On average, the chloride content for this series (series 2) was larger than for series 1.

Table 7 Chloride content (Series 2)

Specimen No.	Cover depth (mm)			around the rusted steel
	30	50	70	
SE-A-1	0.003	0.005	0.005	
-2	0.005	0.003	0.003	
SE-B-1	0.009	0.003	0.005	
-2	0.009	0.003	0.003	0.476
SE-C-1	0.005	0.005	0.005	
-2	0.006	0.003	0.005	
SE-D-1	0.005	0.005	0.005	
-2	0.005	0.005	0.005	
SE-E-1	0.005	0.005	0.005	
-2	0.005	0.005	0.005	
SE-F-1	0.005	0.003	0.005	
-2	0.005	0.005	0.005	0.009
SE-0-1	0.013	0.005	0.006	0.641
-2	0.017	0.006	0.005	0.440

(NaCl v.s Concrete in weight)

Acceleration Test under Drying and Saturating Cycles (Series 3)

In the case of specimens SE-B-1 and SE-B-2, after only 10 cycles, large cracks developed on the coating surfaces and immediate spalling of all surfaces took place. In other specimens no surface deterioration was found. However the color of the coating surfaces changed to yellow from white.

Test results of the half-cell potential of embedded steel bars after 20 cycles are presented in Table 8. Generally speaking, the potential of the non-rusted steel bar is almost constant, otherwise the rusted steel bar shows a tendency to drop. However, the values are dispersed widely.

Table 8   Test results (Series 3)

Specimen No.	Half-cell potential (mV)	Grade of passivity	Corroded area (%)	Chloride content cracked part	Chloride content non-cracked part
A-A-1	-125	5	0.0		0.005
-2	-119	4	0.1	0.015	0.008
A-B-1	-563	3	0.0		0.020
-2	-576	2	0.0	0.123	0.016
A-C-1	-294	4	0.0		0.005
-2	-417	4	0.1	0.037	0.008
A-D-1	-89	5	0.0		0.005
-2	-244	5	0.0	0.008	0.006
A-E-1	-236	5	0.0		0.005
-2	-265	5	0.3	0.061	0.006
A-F-1	-514	5	0.0		0.005
-2	-330	4	0.5	0.113	0.019
A-037-1	-42	4	0.0		0.053
-2	-245	2	1.9	0.152	0.131
A-050-1	-301	2	1.0		0.126
-2	-304	2	4.1	0.330	0.276

The values of the Grade of passivity drawn from the polarization curves are
presented in Table 8. These values for the coated specimens are higher
than for the non-coated specimens.
The corroded area of embedded steel bar is presented in Table 8. Corrosion
was not found in the coated specimens. However in the non-coated
specimens, the steel bars were slightly corroded.
Chloride content in concrete is presented in Table 8. The chloride content
of the coated specimens was smaller than for the non-coated specimens. The
chloride content around cracks is larger than at other parts. On average,
the chloride content for this series (series 3) is almost the same as for
series 2.

DISCUSSION

Effect of Coatings on The Prevention of Salt Intrusion into Concrete

The performances of the specimens coated with system "B" are very bad. In
the test series 2 and 3, large cracks developed on the coating surfaces and
immediate spalling of all surfaces took place within a short service
period. This premature deterioration may be due to the cycles of wetting
and drying. Such deterioration is not found in test series 1, because in
series 1 wetting/drying cycles are not as severe as in series 2 and 3. The
performances of the specimens coated with other systems are good. In these
specimens, chloride intrusion into the concrete is almost completely
prevented, and the embedded steel bars are not corroded.

Effectiveness of Colour Measurement of Coating Surface

Figs. 12 and 13 show examples of chromaticity diagram of the coating
surfaces. The colour of the upper surface of the specimens coated with
system "A" (shown in Fig.12) changed to yellow from white. The color of
the non-coated specimens (shown in Fig.13) did not change. The trend in
the color change from white to yellow is clearly indicated in this figure.
From this, it is clear that the chromaticity diagram is effective to
quantify color changes of the coating surface.

Effectiveness of Electro-chemical Measurements of Embedded Steel

Figs.14 ~16 show the relation between corrosion area and results of the electro-chemical measurements. The relationship between the half-cell potential and the corrosion area is not clear. However the relationships between the Grade of passivity and the corrosion area, and between the current density and the corrosion area are relatively clear. From these results, the authors suggest a diagram shown in Fig.17 to determine the corrosion of the embedded steel non-destructively. In this diagram whose indexes are Grade of passivity and current density, a corrosion zone can be given as shown in Fig.17.

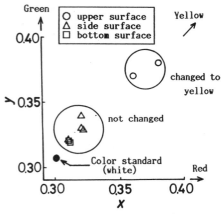

Fig.12 Chromaticity diagram
(Coating system A)

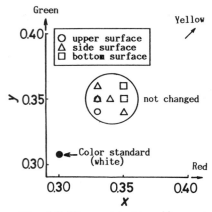

Fig.13 Chromaticity diagram
(non-coating)

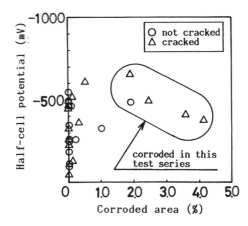

Fig.14 Relationship between half-cell
potential and corroded area

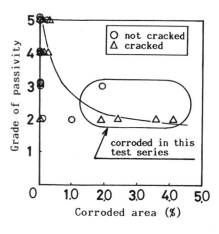

Fig.15 Relationship between Grade of
passivity and corroded area

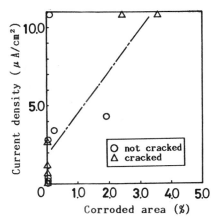

Fig.16 Relationship between current density and corroded area

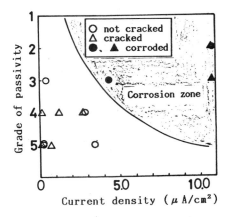

Fig.17 Judgement of the corrosion of embedded steel

Some Factors Affecting The Corrosion of Embedded Steel

Fig.18 shows the relationship between the resistance of the coating and the corrosion area. Corrosion is not found in the high-resistance zone. Fig.19 shows the relationship between the chloride content and the corrosion area. In this figure a good relationship is found. From this, it is clear that the resistance of the coating and the chloride content of the concrete are important factors affecting the corrosion of embedded steel.

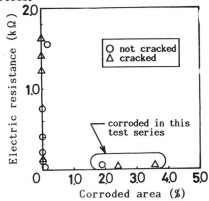

Fig.18 Relationship between electric resistance and corroded area

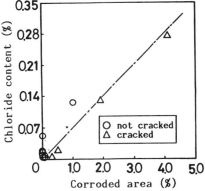

Fig.19 Relationship between chloride content and corroded area

Evaluation Method of Coating Systems

In order to evaluate the coating systems, an index of weather resistance and an index of salt resistance are necessary. An index of weather resistance can be determined by colour change, adhesive strength and surface deterioration, and so on. In cold regions, needless to say, the freeze resistance is also important. An index of salt resistance can be determined by the ability to prevent chloride and oxygen intruding into the concrete. In order to evaluate this ability, electro-chemical measurements used in this study or the diffusion cell test are effective. As an evaluation diagram, the authors suggest the example shown in Fig.20. As shown in this figure, the diagram is divided into four zones and coating systems are classified into three ranks. In this paper, it has not been possible to suggest numerical indexes for weather resistance and salt resistance.

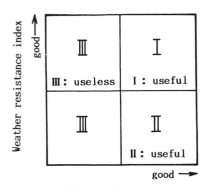

Fig.20 Evaluation diagram of coating systems

CONCLUSIONS

Concrete surface coating is effective in protecting concrete structures against salt attack.

Electro-chemical measurements of embedded steel are effective in detrmining corrosion conditions non-destructively, and also in evaluating the coating systems indirectly.

Colour measurements are effective in quantifying color changes in the coating surface.

The chloride content of the concrete and the electric resistance of the coating are important factors affecting the corrosion of the embedded steel.

The acceleration test method (cycles of drying and saturating) used in this study is an effective test method.

Coating systems should be selected with the environmental conditions of the structure clearly in mind.

In order to make an adequate evaluation of the coating systems, an index of weather resistance and an index of salt resistance are necessary.

REFERENCES

1   MITSUO IKEDA, Basic of Chromatics, Asakura-shoten, Tokyo, 1988, pp92 ～ 109 (in Japanese)

2   NOBUAKI OTSUKI, Resarch on the Influence of Chloride on Corrosion of the Embedded Steel Bars in Concrete, Report of The Port and Harbor Research Institute, Vol.24, No.3, 1985, pp183   ～283 (in Japaneses)

PROTECTION OF STRUCTURAL CONCRETE

Session A

Chairmen

Professor G E Blight

Department of Civil Engineering,

University of Witwatersrand,

South Africa

Professor W H Chen

Department of Civil Engineering,

National Central University,

China

PROTECTION OF STRUCTURAL CONCRETE

R K Dhir
M R Jones
J W Green
University of Dundee,

United Kingdom

ABSTRACT. An overview is provided of the current practice of concrete construction and how concrete can be protected from premature deterioration. The characteristics of environments to which concrete are exposed are examined together with their effects on concrete materials and reinforcement. The influence of method specifications for concrete are considered in terms of their approach to durability. It is suggested that a move towards performance specifications will help provide better protection for concrete. It is argued that tests for this purpose should concentrate on the cover concrete. The concept of design life and whole life costs are introduced. Options for providing durable concrete are discussed. The role of education and training is raised.

Keywords: Overview, Concrete Durability, Method Specification, Performance Specification, Environmental Characteristics, Permeation Tests, Cover Concrete, Education and Training.

Dr Ravindra K Dhir is Director of the Concrete Technology Unit of the University of Dundee. The main areas of research pursued include the various aspects of concrete durability, curing effects, use of cement replacement materials, chemical admixtures and lightweight concrete. He has published and travelled widely. A past chairman of the Concrete Society, Scotland, he is a member of several technical committees in the UK and abroad.

Dr M Roderick Jones is Research Fellow in Concrete Technology, University of Dundee. His main research interests are concerned with cement replacement materials, particularly pulverized fuel ash, and corrosion of reinforcement.

Mr Jeffrey W Green is Industrial Consultant in Concrete Technology, University of Dundee. Prior to his present position he was employed in a major UK consultancy practice.

INTRODUCTION

Concrete is a major construction material. It is cheap, widely available and relatively easy to use. It can be formed on site into a range of shapes, sizes and finishes and this flexibility has helped in the development of concrete construction.

Concrete is inherently durable, requiring little maintenance. Indeed, its basic building block components, ie Si, Al and Fe oxides, are the same as those of durable natural rocks. Consequently, there are few natural environmental effects that can rapidly degrade a good quality concrete. However, the characteristics of good quality concrete are not easily defined, with the added difficulty in that during construction a non-durable concrete may visually appear to be similar to a durable concrete.

The problem of concrete durability has long been apparent, indeed even in 1837 Vicat was concerned that poor workmanship may cause premature deterioration (1). In recent years concrete durability or lack of it has attracted a great deal of concern. Concrete durability may be viewed at different but interdependent levels.

- Whole structure

- Material macrostructure

- Material microstructure

- Reinforcement/concrete interface.

The main causes of durability problems in concrete structures are highlighted in Figure 1 (2). Design and construction faults rather than the materials themselves are shown to be responsible for a high proportion of the failures. Cementitious materials are, however, chemically complex and even today, after over 150 years of use, details of the hydration of Portland cement are not fully understood. The use of chemical admixtures, cement additives and cement replacement materials, such as pulverized-fuel ash (pfa), ground granulated blast furnace slag (ggbs) and microsilica (ms) have brought both technical and economic benefits but also introduced further complications. Frequently there are confusing, even contradictory, specifications, which do not differentiate sufficiently clearly the requirements for their efficient use.

Figure 2 demonstrates the faults typically found in concrete structures (3). Of most importance is corrosion of reinforcement and not the deterioration of concrete itself. In this case the cover concrete is expected to provide protection to the steel both as a physical and a chemical barrier to aggressive agents. However, engineers now realise that the cover concrete can be very much poorer than the heart concrete, Figure 3 (4). Since the site dependent effects of workmanship and curing cannot be predicted, specifications which require a particular minimum strength, minimum

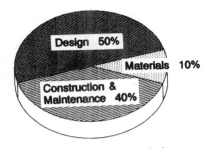

Fig 1 Sources of durability problems (ref 2).

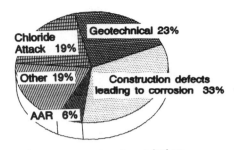

Fig 2 Typical concrete durability problems (ref 3).

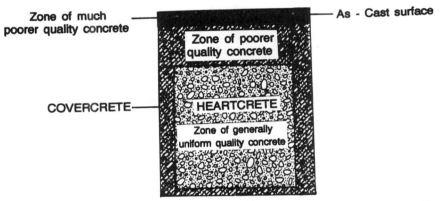

Fig 3 Variations of concrete quality across a section (ref 4).

cement content and maximum water/cement ratio will on their own be inadequate for assuring durability.

This paper provides an overview of the current practice of concrete construction and how concrete can be protected from premature deterioration. The agents adversely affecting the durability of concrete and the role of specifications are discussed. The concept of performance specifications and life cycle costs are introduced.

CHARACTERISING THE EFFECTS OF EXPOSURE CONDITIONS ON CONCRETE

Concrete serves in all environments, each of which has its own particular set of effects on concrete and/or reinforcement. Deterioration is inevitable but the engineer is concerned with the rate at which concrete degrades. This can be defined as,

$$\text{Rate of concrete deterioration} = f\left[\begin{array}{ll} \text{Type and severity,} & \text{Concrete resistance} \\ \text{of exposure} & \text{to deterioration} \end{array}\right]$$

This suggests that the estimation of potential durability requires a knowledge of the concrete characteristics and of the exposure environment. The factors affecting concrete durability are shown in Figure 4 . These factors are diverse and extensive and more often than not several will occur concurrently. Overall, these factors may be divided into four broad categories:

1. Internal material failure, eg ASR, HAC conversion.

2. Fluid and/or ion ingress from the environment which may attack concrete and/or reinforcement, eg chloride ingress, carbonation, sulphate attack.

3. Direct wear and impact, eg abrasion, cavitation.

4. Mechanical disruption due to excessive strains, eg thermal load, flexural cracking, wetting and drying, frost action.

Category 1 tends to arise unexpectedly, often as an unforseen consequence of changes in the production of concrete materials or due to insufficient knowledge of material limitations, as in the case of high alumina cement (HAC). However, broadly speaking, solutions to such problems are relatively easy to apply, even if some details remain disputed, eg reducing the risk of ASR (5).

However, as readily available material sources become scarce, the use of alternative, recycled and lower quality materials will become necessary. These materials may produce concrete with different performance characteristics to traditional materials and it may not be appropriate to treat them in the same way. Thus, the engineer will appreciate that from durability considerations more care should be exercised in the use of such materials.

Category 2 is by far the major source of agents which produce durability problems. The ingress of moisture containing dissolved ions, particularly chlorides, or carbon dioxide can lead to reinforcement corrosion. The problem is that there has been traditionally no simple way of assessing the resistance of concrete to the ingress of these agents.

Category 3 is associated with wheeled traffic on highway pavements and warehouse floors. The rate of wear is a function of the characteristics of both the mortar and coarse aggregate. There are well established tests for assessing the potential wear resistance of aggregates, such as the polished stone value (BS 812), which may be used to assess the potential wear resistance of the concrete (6).

Category 4 factors can be sub-divided into two additional groups, ie those directly affecting the macrostructure of concrete, eg freeze/thaw attack, and those which affect the concrete as part of a structure, eg thermal, flexural and moisture related strains. The former can be overcome by air-entrainment whilst the latter can be avoided by careful design and detailing.

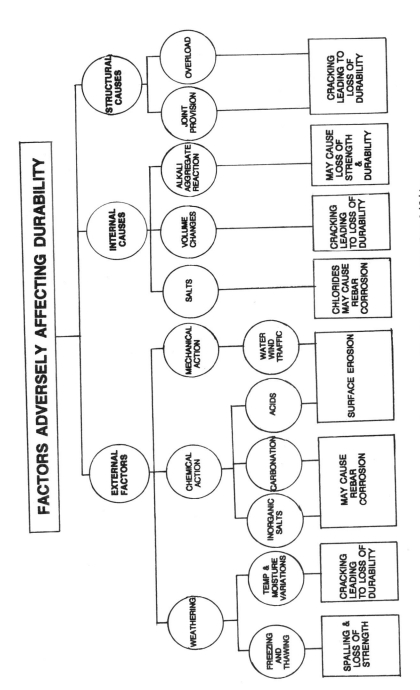

Fig 4 Factors adversely affecting concrete durability.

A particular durability problem currently facing the industry is reinforcement corrosion due to chloride attack (7). There has been a significant increase in the application of de-icing salts to highways. This, together with poor construction and inadequate maintenance, is estimated to cost approx £800m in repairs to chloride-induced damage to concrete bridge structures in England and Wales (8). Concrete structures located in soils containing saline groundwater have been noted to deteriorate rapidly (9).

In hot climates carbonation-induced reinforcement corrosion has been noted within 5 years of the completion of structures (10).

Mather (11) discussed the problem of the interaction of different changes in the characteristics of exposure environments which may not be apparent to the specifier. He illustrated this point by describing the damage occurring from a major alkali-silica reaction that took place inside a building. The problem was traced to a change in the cleaning of the building by wet rather than dry mopping. This example also highlights possible latent problems facing the engineer when a structure is required to undergo a change of use, particularly where this entails a change in exposure, either internally or externally.

INFLUENCE OF SPECIFICATIONS

The main code of practice for the structural use of concrete in the UK is BS8110 which is typical of method specification (12). In terms of durability BS8110 has five categories of exposure for structural concrete,

- Mild   protected against weather or
              aggressive conditions

- Moderate   sheltered or not subject to   Non-chloride
              severe conditions   environment

- Severe   severe rain or condensation,
              freezing or cyclic wetting

- Very severe de-icing salt contact, severe
              freezing or sea water spray   Chloride
                                                environments
- Extreme   sea water carrying solids,
              abrasion or acid water contact

In an attempt to address the question of durability for each exposure condition, BS8110 states requirements of concrete in terms of minimum concrete grade, cement content, cover and maximum water/cement ratio as shown in Figure 5. In addition, it recommends a series of minimum curing periods for different concrete types depending on the ambient conditions. Three ambient exposure conditions are covered, depending on the atmospheric humidity, as:

- Poor   (<50%RH, not protected from sun and wind)
- Average   (intermediate between poor and good)
- Good   (>80%RH, protected from sun and wind).

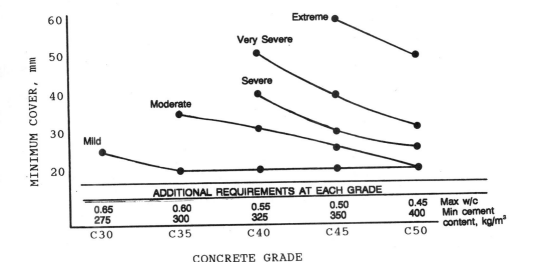

CONCRETE GRADE

Fig 5 Durability requirements of BS 8110.

The recommended curing periods vary up to 10 days, and may be provided by,

i)   maintaining formwork in place, or

ii) covering the surface with impermeable sheeting, or

iii) applying a curing membrane, or

iv) applying a damp absorbent material, or

v)   continuous or frequent application of water.

There are a number of limitations to this approach. Firstly, the class of exposure does not define the aggressive agents in any detail. This is particularly so for the chloride environments where the rate at which chlorides ingress into concrete is a function of their surface concentration. Secondly, if no active curing is provided as is allowable under certain conditions there is an increased risk of deterioration due to carbonation. Thirdly, although curing methods are described, no guidance is given as to their relative efficiency. Moreover, there is no way of assessing how effective the specific curing method has been.

Compliance Testing

Compliance testing for durability is mainly concerned with the characteristics of the constituent materials, eg limits on deleterious components, and generally have worked well. However, the real problem is testing the concrete itself and as yet none of the standard tests in use give any measure of potential durability performance. For example, whilst the air test for fresh concrete provides a measure of the total air content, it does not deal with the critical void parameters, such as bubble size and spacing which control the resistance of concrete to freeze/thaw attack. More frequently, specifications have requirements which cannot be tested such as 'shall be fully compacted'.

Alternative Forms of Specifications

In order to overcome the limitations of method specifications, European documents currently being drafted are, where possible, shifting towards performance specifications. This approach has merit as by testing directly in-situ concrete an estimation of potential performance can be established more realistically. This also allows the contractor the scope of providing the required concrete in a manner most suited to him and using the materials and methods he can best adopt. However, as with cube strength, this type of testing is retrospective and places the responsibility onto the contractor. A further limitation is that such specifications may not suit contractors with less experience of concrete construction.

Specifying Concrete in Performance Terms

The most significant concrete parameters defining the resistance of concrete to deterioration are the permeation characteristics of the surface and near-surface concrete. Not only does the cover concrete interface and react directly with the environment but it is the concrete which has the poorest quality and most likely to suffer.

Permeation can be divided into three distinct but connected transportation phenomena for moisture vapour, dissolved ions, gases and aqueous solutions:

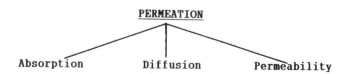

Absorption describes the process by which concrete takes in a liquid, normally water or aqueous solution, by capillary attraction. The rate at which water enters is termed absorptivity (or sorptivity) and can be described by,

$$\text{Absorptivity} = f \begin{bmatrix} \text{capillary size, capillary interconnection,} \\ \text{moisture gradient.} \end{bmatrix}$$

The moisture may contain dissolved salts, such as chlorides or sulphates and dissolved gases, such as oxygen, carbon dioxide and sulphur dioxide. The transportation of ions is therefore often a combination of absorption and diffusion. The concrete will tend to absorb moisture containing ions perhaps to a depth of 20 to 25mm in typical situations and thereafter, the ions penetrate deeper into the cover by diffusion.

Absorption of water often leads to wetting and drying strains and can form, for example, part of the action that may lead to ASR or freeze/thaw attack. Absorption may be controlled by reducing the size and interconnection of capillary pores by using a low water/cement ratio and providing good curing.

Diffusion is the process by which a vapour, gas or ion can pass through concrete under the action of a concentration gradient. Diffusivity defines the rate of movement of the agent and can be described by,

$$\text{Diffusivity} = f \begin{bmatrix} \text{concentration gradient, amount of reaction} \\ \text{of agent with hydrate structure, capillary} \\ \text{size and interconnection.} \end{bmatrix}$$

Diffusion is the mechanism by which carbonation occurs and also characterises the ingress of chlorides and other ions. It is, therefore, closely linked to reinforcement corrosion problems. Diffusion can be controlled in a similar manner to absorption but also important in this case is the 'binding capacity' of the hydrates, particularly for ions such as chlorides. Cement replacement materials such as pfa and ggbs are very effective in providing binding sites for chlorides and thus reducing diffusivity of concrete.

Permeability is defined as the flow property of concrete which quantitatively characterises the ease by which a fluid will pass through it, under the action of a pressure differential. This contrasts with absorption and diffusion which are caused by a concentration differential. Permeability can be described by,

$$\text{Permeability} = f \begin{bmatrix} \text{pressure gradient, capillary size and} \\ \text{interconnection.} \end{bmatrix}$$

Thus, strictly speaking, permeability is a characteristic which is applicable in specific situations only, such as for dams and tunnel linings where significant heads can develop. Permeability can be controlled in a similar manner to absorption.

The transportation processes are influenced by a large number of material and environmental parameters, for example, as:

Primary parameter ──────▶   Secondary parameter ------▶

This demonstrates that there is an extremely complex interaction between the material properties, environmental conditions during casting, curing and workmanship and the resulting cover permeation characteristics. It is clearly not possible to determine the permeation characteristics from the normally specified concrete parameters. To overcome this problem a number of tests have been developed to measure directly the permeation characteristics.

- Absorption
  - Initial Surface Absorption Test (BS 1881: Part 5, ref 13)
  - Covercrete Absorption Test (CAT, ref 14)
  - Water Absorption Test (BS 1881: Part 122, ref 13)

- Diffusion
  - Accelerated carbonation (eg ref 15)
  - Accelerated chloride diffusion (eg ref 16)

- Permeability
  - Water permeability (eg ref 17)
  - Gas permeability (eg ref 18)

Using these methods to test in-situ concrete, it is possible to account implicitly for all the materials and environmental effects on the cover zone and it has been possible to use the data to predict the durability of concrete, for example carbonation using Figg Air Index and abrasion using ISAT, as illustrated in Figures 6 and 7 (19, 20). Further work underway at Dundee University suggests that it should also be possible to use such tests to predict other durability properties, such as chloride ingress and frost attack.

The development of such tests can allow the engineer to apply quality control to durability for in-situ concrete in a similar manner to that applied for strength. Such tests can provide a scope for the development of performance specifications for concrete. However, a great deal of further development work will be necessary before this approach could be ready for adoption on site.

There has to be a fundamental change in the client's understanding of design life. In the UK engineers are now becoming aware of the implications of a finite design life. Figure 6 illustrates the typical design life requirements for different structures (21).   The provision of the concrete to satisfy these design lives will be different. Consideration of this concept leads to a notional design and operation model which indicates the optimum route to the provision of the most economical structure as shown in Figure 7.

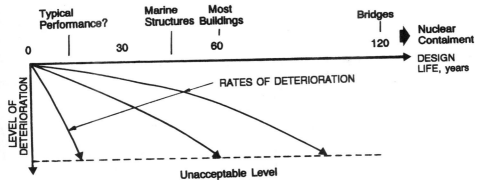

Fig 6 Design life of different structures (modified from ref 7).

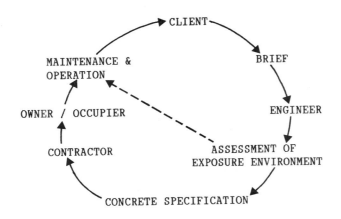

Fig 7 Design and operation model for durable concrete.

If any part of the 'circle' is not fulfilled then either the
required design life may not be achieved or the concrete may not be
economically exploited. For this circle to be complete there is a
need for the client to define the brief adequately and for the
provision of simple and reliable quality control for durability.

OPTIONS FOR PROTECTING CONCRETE

Concrete can be produced in a large variety of forms, with strengths
up to 150N/mm^2 and densities from 500 to 4000 kg/m^3. However, where
durability is concerned it is suggested that in essence there are 5
main options for protecting concrete,

- Ordinarycrete

- Supercrete

- Coatedcrete

- Coatedrebarcrete

- Efficientcrete

Ordinarycrete is produced to current specifications but is likely
to require repairs and maintenance at intervals to restore
serviceability, as shown in Figure 8. Ordinarycrete is all too
often a result of inadequate site practices and is therefore
unlikely to comply with the design life requirements of a structure
without additional expenditure.

Supercrete is produced to current specifications, satisfying
durability requirements but is otherwise greatly in excess of
structural requirements. This can be produced by specifying a high
cement content, say >400 kg/m^3 and low water/cement ratio, say <0.4.
This, depending on curing, is likely to give in-situ strengths in
excess of 50N/mm^2. Problems may, however, arise due to heat of
hydration associated with the high cement content.

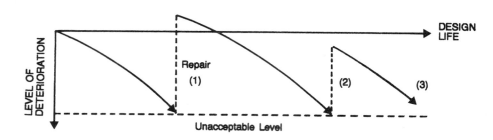

Fig 8 Typical service life of Ordinarycrete (modified from ref 21).

Coatedcrete is designed to satisfy structural requirements and is then coated to provide adequate durability. Over the past five years the use of silane and siloxane penetrating coating systems have been seen as one possible answer to the corrosion of reinforcement problem. However, there are questions as to longevity of the coating system, its cost and to the performance of recoated concrete. It is also difficult to tell whether the whole structure has been coated, to what depth and the consequent resistance to further attack.

Coatedrebarcrete is designed to satisfy structural requirements and is used where protection of reinforcement from corrosion only is required. In contrast to the above, the reinforcement itself is coated, eg by zinc or fusion bonded epoxy resin. The close control that can be exercised during manufacture of the coated reinforcement gives rise to more confidence of the in-situ performance of concrete. However, the reinforcement is expensive and needs care during handling and fixing. Any damage may not be detected and could give rise to intense local corrosion attack. There are also unanswered questions regarding bond strength.

Efficientcrete is produced to the state of the art using the full potential of the constituent materials by improved construction, workmanship and curing practices, as for example demonstrated in Figures 6 and 7. Such a concrete requires no special consideration other than fully exploiting the potential of already existing materials. It would seem likely that efficientcrete would utilise chemical admixtures and cement replacement materials, together with the use of a high efficiency curing membrane to ensure proper curing. Such a material would not only be cheap in terms of whole life costs in relation to the other options but also only requires improvements and not wholescale changes to current practice.

ROLE OF EDUCATION AND TRAINING

Recent problems with concrete structures, together with the research and development work that has been taking place, has finally brought home the fact that concrete technology is a complex subject. Equally important has been that education and training in this area has been sadly lacking and must be improved. It is suggested that this can be achieved by,

   i)   Improving undergraduate syllabuses and providing
         specially tailored postgraduate courses.

   ii)   Making provision for continuing professional develop-
         ment of practising engineers.

   iii) Improving the training of site personnel.

Undergraduate and Postgraduate Courses

This should provide an in-depth rather than superficial coverage of concrete and its role in construction. Clearly, this will require a shift in emphasis within the undergraduate curriculum. At present, teaching of concrete technology tends to suffer as being the poor relation to structural engineering and this tended to be the main stumbling block in providing an adequate coverage of concrete technology at undergraduate level.

Specially focused postgraduate courses provide an additional method to broadening and increasing the depth of training in concrete technology and such a course can be offered on a full-time, part-time and distance learning basis in order to meet the needs of employers and engineers alike.

Continuing Professional Development

It has been estimated that 10 years ago a degree course remained current for about 10 years, and that with rapid development in technology this figure in the 1990's will reduce to about 6 years.

The Institution of Civil Engineers in the UK recognises that there is a real need for continuing professional development and encourages this through its Chilver scheme which is a requirement to becoming a chartered engineer. Many organisations are also increasing their own training provision. However, on average, UK engineers still receive less in-service training than their European counterparts.

Site Personnel

Unlike in Europe and the USA, site personnel in the UK generally receive no formal training in concrete technology. To uplift concrete practice it requires operatives to become craft trained. Although this may, on a direct basis, cost contractors more in terms of wages, an overall cost benefit should be accrued in the long-term due to a reduction in remedial work.

CONCLUDING REMARKS

Recent experience with concrete construction highlights the fact that there is a problem with durability and it is recognised that this needs to be urgently addressed. This improvement must take into account both the complexity of material usage and the influences of often inadequate design and construction practices.

It can be argued that national codes of practice make adequate provision for durability through a method specification dealing with material quality requirements, concrete cement content, water/cement ratio, strength and cover to reinforcement, placing and compaction and curing. However, the final quality of concrete in-situ can vary

greatly due to site practices. Another difficulty is that even
where specifications are met in certain instances the required
durability may not be achieved.

It is suggested that a move towards performance specifications will
help to provide a better protection for concrete. A number of
suitable tests have been established for assessing the potential
durability of concrete and can be developed for use on site. These
tests concentrate on cover concrete, which is required to protect
reinforcement and interfaces directly with the environment and is
most vulnerable to the effects of site practices and curing.
This will require a great deal of national effort and investment in
order for performance specifications to be successfully implemented.

There is a need for clients to change their understanding of the
design-life of structures and to recognise that the required
durability of concrete cannot be infinite. The provision of
concrete to suit, for example, office structures is very different
to that required for a bridge.

There are a number of options to providing protection to concrete
ranging from supercrete to coatedcrete. Each has its individual
merits and may be advantageous in particular exposure circumstances.
Overall, further work is necessary to identify which option is most
suitable for the general situation and for each of the many
exposures in which concrete structures serve.

There is an inadequacy in the provision of education and training in
the field of concrete technology. This needs to be improved in
order to achieve better value for money in concrete construction.

REFERENCES

1 VICAT L J. A practical and scientific treatise on calcareous
   mortars and cements, artificial and natural. Translation by
   J J Smith. American Concrete Institute SP-52. 1976. pp 334.

2 PROPERTY SERVICES AGENCY. Quality Assurance. HMSO 1987, pp 10.

3 NATIONAL AUDIT OFFICE. Quality control of road and bridge
   construction. HMSO. December 1989.

4 DEWAR J D. Concrete durability: Specifying more simply and surely
   by strength. Concrete, Volume 15. No, 2, Feb 1982, pp 19-21.

5 CONCRETE SOCIETY. Reducing the risk of alkali-silica reduction.

6 BRITISH STANDARDS INSTITUTION. BS 812 Testing aggregates. Part 3:
   Methods for determination of mechanical properties. 1975, pp 28.

7 BROWNE R D. Durability of reinforced concrete structures.
   Proceedings, Pacific Concrete Conference. Vol 3 1988, pp 847-886.

8 WALLBANK E J. The performance of concrete in bridges. Report by G Maunsell and Partners for Depart. of Transport. HMSO, 1988.

9 ROSE H, ANDERSON G M. Some practical aspects of concrete investigations in the South East Gulf region. Proceedings, 1st International Conference on Deterioration and Repair of Reinforced Concrete in the Arabian Gulf. Volume 1, Oct 1985, pp 529-542.

10 CIRIA. Guide to concrete construction on the Gulf Region. CIRIA Special Publication 31, 1984, pp95.

11 MATHER, B. How to make concrete durable. Construction Specifier. Volume 42, November 1. January 1989, pp 48-51.

12 BRITISH STANDARDS INSTITUTION. BS 8110 Structural use of concrete. Part 1: Code of proctice for design and construction. 1985, pp 124.

13 BRITISH STANDARDS INSTITUTION. BS 8110 Testing concrete. Part 5: Methods of testing hardened concrete for other than strength. 1970, pp36. Part 122: Method for determination of water absorption. 1983. pp 4.

14 DHIR R K, JONES M R, MUNDAY J G L. A practical approach to studying carbonation of concrete. Concrete. Oct 1985, pp 32-34.

15 DHIR R K, HEWLETT P C, CHAN Y N. Near-surface characteristics of concrete: Assessment and development of in-situ methods. Magazine of Concrete Research. Volume 39, No 141, Dec 1987. pp 183-195.

16 DHIR R K, JONES M R, AHMED H E H, SENEVIRATNE A M G. Rapid estimation of coefficient at chloride diffusion. To be published in Magazine of Concrete Research. 1990.

17 DHIR R K, MUNDAY J G L, HO NY, THAM K W. Pfa in structural precast concrete: Measurement of permeability. Concrete. Volume 20, Number 12, December 1986, pp 4-8.

18 DHIR R K, HEWLETT P C, CHAN Y N. Near surface characteristics of concrete: Intrinsic permeability. Magazine of Concrete Research. Volume 41, Number 147, June 1989, pp 87-97.

19 DHIR R K, HEWLETT P C, CHAN YN. Near-surface characteristics of concrete: Prediction of carbonation resistance. Magazine of Concrete Research. Volume 41, No 148, Sept 1989, pp 137-143.

20 DHIR R K, HEWLETT P C, CHAN Y N. Near-surface characteristics of concrete: Abrasion resistance. To be pub Materials & Structures.

21 BROWNE R D. Degradation of concrete structures and the extent of the problem. Proceedings, Cathodic Protection of Concrete Structures. Insititution of Structural Engineers. London, 1988.

THE PROTECTION AND IMPROVEMENT OF CONCRETE

O V Smirnov

Tujmen Civil Engineering Institute,

A F Yudina

Leningrad Civil Engineering Institute,

USSR

ABSTRACT. The authors describe the results of investigations of the effect of electro treatment on processes such as armature corrosion protection, release of shuttering from concrete, electro-osmotic dehydration and activation of the concrete mix. The process of treating the mixing water by an external electrical field is discussed in detail. The use of electro-treated water in the concrete mix causes an improvement in physico-mechanical and technological concrete mix properties accompanied by an improvement in durability. They propose a theoretical mechanism for the observed changes in concrete mixes using electro heated water. This process leads to an improvement in concrete micro-dispersion, a decrease in porosity and a decrease in variability.

Keywords: Activation, Electro-treatment, Concrete, Mixture, Mixing Water, Structure, Components, Porosity, Hydration products.

Dr O V Smirnov is the Head of Department of the Tujmen Institute of Civil Engineering. He is concerned with the investigation of dispersive systems. He is a senior scientific worker of the USSR North Problem Investigation Institute and is Chairman of the Tujmen Ecological Committee. He has published some reviews and monographs.

Mrs A F Yudina is an assistant professor of the Building Production Technology Department of the Leningrad Institute of Civil Engineering. She is concerned with research into the electro treatment of concrete, mix preparation technology involving the evaluation of mixing devices, and complex automation. She has published her work extensively.

INTRODUCTION

Activation of the concrete mix can lead to improved concrete properties and durability which must be viewed in perspective with the different process stages of concrete mix preparation and its use in construction. This is particularly relevant in the case of concrete production with regard to the development of all-year round monolithic house building, reconstruction of civil and public buildings, and underground structure erection. For foundations the same conditions apply for both new building and renovation work.

The volume of monolithic concrete usage is projected to increase by $50,000m^3$ in Leningrad and by as much as 20% of total building in Siberia. Electro-treatment of the concrete mix components reduces setting times and increases strength, resulting in a faster technological cycle. At last concrete work can be carried out faster, without impairing quality.

The activated material contains a thermodynamically and structurally unstable disposition of crystal lattice elements, which differ by increased free enthalpy valves as a result of active centres with accompanying electric microfields. The mechanism by which these centres occur is essentially dislocational. However, the acceleration of normal diffusion processes is also possible, which accelerate chemical reaction by the rapid transfer of reagent from the surface to the interior of the crystal.

In the early stages of hydration the electro static interaction forces greatly influence the structure forming process, the magnitude of which is determined by the electro-kinetic potential of the concrete in particles. Zeta-potential can be measured using M A Pohl's effect: the mechanical generation of electrical charge, eg under the influence of acoustic frequencies.

The negative zeta-potential variation curve, which represents a positive charge during the period from initial mixing to the end of coagulation structure formation, may be approximated into sections. From the moment of mixing until stabilisation some 5-10 minutes later, the charge decreases sharply until a value of 15-35 mV is reached. During the stabilisation period there is a slow decrease in Zeta potential, which can be associated with the appearance of new formations, which in turn influence surface potential variation. The zeta-potential then decreases very rapidly up to the moment of setting. During this period, the diffusion fluxes are decreasing and interlocking structures are formed as a result of interaction between concrete cement particles and the new, positively charged formations. This sequence allows the correct time for repeated mechanical activation to be chosen.

Review of Current Thinking

Electrical treatment methods, are an innovative practice in building construction.

There are many applications, such as armature corrosion protection, shuttering adhesion inhibition, concrete mix electro-osmotic dehydration, cement activation, aggregate and mixing water activation and others.

For electrical concrete protection methods a superficial electrical field is generally used. The methods employed depend on the phenomena taking place in the interelectrode space, but may be classified by taking into account the electrical treatment and the peculiarities of the superficial field (ie frequency, non uniformity etc). The following methods are examples: electrodialysis, electrolysis, electro-chemical coagulation, electro-flotation electrophoresis, electro-coagulation, electric discharge (infinitesimal power), high voltage pulse discharge and composite electrical effect. These methods are arranged in order of increasing electric field intensity (from $E=0.5Vm^{-2}$ to 10^4Vm^{-2}).

Investigations concerning the electrical current used for armature corrosion protection by the deposition of a coating on the surface showed that this resulted in improved properties of reinforced concrete structures, which is attributable to the improved bond between the armature and the concrete. It was shown that when a constant electrical current was passed through fresh layed concrete, thin and dense concrete fractions and cement hydrate products were formed on the negative electrode surface. These thin dense deposits confer a high degree of corrosion protection to the metal.

Concrete protection from surface damage during striking of formwork, and cleaning of metal formwork by use of a constant electric current, allows an improvement in the quality and design of monolithic concrete structures. Protection from surface damage during striking is achieved by passing through the concrete, in its setting and hardening phase, an electrical field of $1-2Vm^{-2}$ intensity. This method is used for non reinforced or poorly reinforced thin walled concrete constructions. Compound coagulation of the concrete mix is achieved with out electrolyte additions. The benefits of this method are two fold: it lowers the specific consumption of materials and it lowers the amount of surface defect repair.

The avoidance of surface damage by the shutter requires adequate cleaning. This is achieved by two methods. Firstly when the shuttering is full of concrete, and secondly when the shuttering is empty, in order to prevent repeated surface damage. The concrete mix can be with or without electrolyte additions. In the first case, the shutter surface is positive ie anodic, and in the second case, where cohesion is being prevented, the shutter surface is negative ie cathodic. For shuttering cleaning during the concreting

process the current density in the cleaning surface, work stress and
treatment time depends on the density and thickness of the concrete
stone soiling layer, density of structural interactions with
cleaning surfaces and on the electrolyte type and concentration.

Investigations of increased durability due to electro-osmotic
dehydration of the concrete mix have shown that the concrete mix can
be dehydrated by a constant current to achieve an optimum water-
cement ratio. Consequently the general porosity may be reduced
leading to increased concrete frost resistance, and overall
durability.

Electrochemical activation of concrete, including the influence of
variable polarity at constant current on the concrete mix, permits
faster hydrate structure formation and a simultaneous increase in
hardened concrete durability.

Experimental investigations of hardening concrete to test the
influence of electro-treatment on durability were carried out in
semi-site conditions. The hardening test treatment was carried out
by high voltage and heterogenous electric fields. The results of
the experiment have shown that treatment by high voltage electric
fields promote the strongest hydration of cement materials, as
indicated by the effect on the hardened concrete durability index.
During treatment by inhomogeneous electric fields degradation of the
solidification process can take place owing to active water
transport into the specimen, resulting in decreasing strength
indexes. Experimental investigation of the high-voltage electric
field treatment of the products of ceramic dehydration has also been
undertaken.

Many methods of activating concrete raw material components are
known. One of these, is by variation of the pH of the concrete mix
liquid phase, during the mixing period. Such liquid phase
activation has a remarkable effect, and was achieved using a
standard electrodialyser, adapted for producing water with pH≤3 and
pH≥11. The use of membrane electrolysers allowed us to obtain an
alkaline medium with a pH≥12.

The combination of physical and chemical factors in defined
conditions produced an effect which was not the actual sum of the
effect of each individual influence. Activation of the concrete mix
components, ie aggregate, cement and water, in defined conditions
raised the degree of cement hydration and consequently the concrete
mix strength. Research in to the complex activation of concrete mix
materials are often undertaken. The effect regarding strength
increase can be obtained by adjusting the impurity content in the
mixing water, and also by the influence of an electric field and
ultrasound upon them. The processes involved are increased
crystallisation centres, accelerated dissolution of the calcium
silicates in the concrete mix, changed alkali medium, decreased
water surface tension, decreased concrete mix water demand, and

finally acceleration of the chemical reactions with the cement minerals.

THE EXPERIMENT

The first reports of water dispersion electro-treatment can be traced back to the end of the last century[1]. The authors have investigated the treatment of the concrete mixing water by a uniform electric field. A stable positive influence of physico-mechanical and technological properties was found. The effect of concrete mix ingredients, with water treatment conditions and parameters were also determined. The water was treated in an electrical treatment cell with an electric field strength from $10\text{-}35Vm^{-2}$ for durations of 1-15 minutes. The electrical treatment unit was a hermetic cell with aluminium electrodes connected to a direct current source.

The improvement of concrete qualities by the activation of the concrete mix by mechanical means may be investigated by using dry and wet finish grinding mills (from ball mills to disintegrater mills), vibro and ultrasonic processing, thermal, chemical and radioactive means. Recently electrical fields, with impulses of more than 500V amplitude (F=10-100 Ghz) and modulation of such impulses have been used to activate the concrete mix components and the concrete mix itself. The electric processing of all the components essentially changes the structure forming process both at early and late ages. The average porosity of such structural elements determined by nuclear magnetic resonance and porometry methods show a decrease from 17.5% to 16.8%.

The X-ray structure, differential thermal analysis and the 35% cement consistency test, indirectly confirm the increase of hydrorotation and dispersity. During the first three days the hydration kinetics can be changed by activation of the dry cement. The tensile strength is increased by 100% and the compressive strength by up to 30% accompanied by earlier setting. In terms of durability the 1.4-1.5 times decrease in structural inhomogeneities, combined with an increase in strength of up to 40% is of considerable interest.

The processes which lead to uniformity are limited by the unbalanced and uncontrolled dissipative formation of hydrate structure and ultimately by the hardened concrete.

It is possible to carry out an electrochemical concrete modification which will bring it nearer to a polymer cement concrete composition from the point of view of its structure.

While using polymer-cement concretes dispersed polymer particles can form a reliable protective precipitate around the reinforcement and also on the external surfaces of concrete structures because of electrical change predominance. In experiments carried out by the authors using pulsed high-voltage fields such a protective layer was formed during seconds.

Specimens of cement paste in the form of 3x3x3 cm cubes, with a
water-cement ratio of 0.32, were prepared. The specimens were aged
under natural hardening conditions and tested for strength at
various ages. The strength values were compared with those of
specimens prepared similarly with ordinary water. The test results
showed an increase of strength in individual cases of up to 33%; the
changes of strength at different ages decreased with time, ie.,
mixing water treated in a static electric field of soluble
electrodes accelerates hardening processes during the early stages.

The results of tests on concrete mixes showed that mixes made with
treated water have higher plasticity. For example a mix with a
water-cement ratio of 0,5. made with ordinary water, gave a slump of
6-7 cm. while the slump of a mix made with treated water was 12-14
cm.

The influence of water treated in the electric field of soluble
electrodes on cement hydration was investigated by X-ray phase
analysis with the DRON-0:5 diffractometer. Cement specimens of
different ages were studied.

Examination of the diffractograms showed that the intensities of the
diffraction lines of calcium hydroxide (4.93, 3.11, 2.63 A) on the
diffractograms of specimens mixed with treated water differ from the
intensities of the control intensities. The same applies to the
intensities of the alite diffraction lines (3.03, 2.61, 1.77 A).
Specimens mixed with untreated water were used as controls.

It can be inferred from the results of X-ray phase analysis that
hydration processes are intensified in specimens made with water
treated in a static electric field.

Microscopic analysis of the specimens (magnification x 50-250)
showed that the structure of specimens made with treated water is
denser and more homogeneous. The particles of new phases are
considerably finer than in the controls. Specimens made with
treated water have a more uniformly dispersed and finer structure,
which is consequently stronger than that of the control specimens.

The usage of nuclear magnetic resonance for building technology
quality control with the application of concrete mix component
activation allows us to maintain objective differences in the
building materials structure versus the modes of concrete component
electro-treatment.

THE THEORETICAL PREREQUISITES

The principle hardening process in cement mixed with water treated
in a static electric field are linked with electrode processes
(during electrical treatment of water) and electro-kinetic effects.
Colloid-chemical aspects play an important part in the formation of
concrete mixes.

Water treated in an external electric field with soluble electrodes differs from ordinary water by a higher content of H^+ and OH^- ions, and also by a higher concentration of multivalent aluminium hydroxide ions in the range of 40-60 mg/litre[4,5]. This water is a saturated ionic solution containing polymerizing aluminium hydroxide chains having a permanent dipole moment. Since transformation of a "gel" structure into a crystalline structure is possible not only as a result of "recrystallization" as is commonly believed, but also as a result of an increase of interaction forces in newly formed crystalline hydrates, it is likely that mixing water treated in a static electric field acts directly on particle interaction of the crystal size.

An increase in the concentration of multilent aluminium hydroxide ions leads to compression of the diffusion part of the electric double layer and the thickness of the ion atmosphere decreases. At the same time, compression of the ion layer results in an increase in the depth of the secondary minimum on the potential curves of particle interaction, leading to an increase of long-range aggregation changes and degree of dispersion of the newly formed products.

When cement is mixed with water, its components dissolve to form solutions supersaturated with respect to newly formed hydrates, which separate out in the form of crystalline hydrates. These crystals then intertwine and partially coalesce to form strong hardening structures[6].

During its treatment in the electric field the water becomes saturated with aluminium hydroxide, which is highly active, finely dispersed, and is also a weak acid.

When cement is mixed with such water, an active reaction occurs between calcium hydroxide (a strong base) and aluminium hydroxide (a weak acid). This results in formation of calcium aluminate hydrates which, in our opinion are distributed uniformly in the system to form an open network structure reinforced by chain aggregates. These aggregates serve as available crystallization sites. The newly formed products, which precipitate from the supersaturated solutions during interaction of the cement with water, crystallize more easily and rapidly on these sites. This process leads to accelerated hardening. Through its hydroxyl groups, aluminium hydroxide is absorbed more easily on the cement grains and the hydration products, and this in turn , makes the cement paste plastic when the cement is mixed with water treated in a static electric field of soluble electrodes.

CONCLUSIONS

Electro-treatment of concrete mix components leads to the formation of an activated dissipative structure which increases the concrete strength by up to 33%.

The mix component electro-treatment results in the formation of polarised polymeric chains from the generated particles of electrode material hydroxide. This leads to armouring of the concrete with a structure of the PCS type (periodical collision structures) which may be considered to be similar to that found in polymer cement concrete.

The formation of monolithic hidispersion concrete structures with decreased porosity demands the application of X-ray structural analysis and nuclear magnetic resonance for high quality concrete control. Such a system would allow the expectation of increased concrete stability with regard to mechanical and temperature factors and particularly resistance to corrosion and degradation caused by atmospheric and gaseous mediums containing CO_2, and other vapours and aggressive media.

REFERENCES

1   WEBSTER, W I R. British Patent. 27.01.1887, Number 1.333.

2   EVDOKIMOV, V A, SMIRNOV, O V, and YUDINA, A F. Technology of Erection of Buildings and Constructions, Sb. Tr. Leningrad, Inzh. Stroit, Inst., 1984, pp 5-13.

3   YUDINA, A F. Organisation, Design and Management in the Building Industry. Sb. Tr. Leningrad, Inzh. Stroit. Inst., 1983, pp 115-119.

4   SMIRNOV, O V. Electrical Treatment of Systems with Liquid Dispersion Media. Author's summary of doctoral thesis, Leningrad, 1975.

5   SVETLISKII, A S. Clarification of Water in an Electric Field with the Aid of Soluble Electrodes. Author's summary of candidate's thesis, Leningrad, 1980.

6   SYCHEV, M M. Hardening of Cements (in Russian), Stroizdat, Moscow, 1974, pp 10-52, 72-79.

RECENT DEVELOPMENT OF PROTECTION TECHNIQUES FOR CONCRETE STRUCTURES IN JAPAN

K Katawaki

Public Works Reasearch Institute,

Japan

ABSTRACT. New technologies related to corrosion protection for maritime concrete structure have been developed in the civil engineering field in Japan. New technologies include not only the innovation of protective and concrete materials, but also the development of evaluation techniques for durability of materials.

Recently we constructed a new large scale test site in the ocean, and we have performed exposure test for evaluating concrete quality and protective materials. With these test results, we are preparing to make a guideline on the protection design of large scale maritime concrete structures.

Keywords: Corrosion, Protection, Concrete, Maritime Structure, Epoxy, Coating, Polymer Impregnated, Cathodic Protection, Carbon Fibre Composite.

Dr Kiyoshi Katawaki is the director of Chemistry Division, Public Works Research Institute, Ministry of Construction, where he has been most active in concrete technology for 18 years by performing researches, committee membership, and managing research activities. His current research work includes: protection of concrete structures, maintenance and diagnosis of structures, hi-tech materials, and computer aided designing.

INTRODUCTION

Japan is a long seaside country with an oceanic climate, stretching from sub-arctic through the sub-tropical zones. By typhoons from summer to autumn and seasonal winds in winter, the districts facing seacoast are affected by splash of sea water of salt particles transported with sea wind. Sea salt particles sometimes fall on the areas about 10km from the coast.

To prevent deterioration, the reinforcement should be covered with good quality and thick cement paste, and the generation of crack must be controlled perfectly, or directive protective methods. The protection technology for concrete structure is required, so, it has been carried out that a developmental study of highly advanced corrosion preventive technology intended for oceanic structures, especially the concrete structures constructed in splash zones, as well as long-term durability evaluation.

DETERIORATION PROCESS

Deterioration process involves many phenomena: chloride penetration, chloride accumulation, start of steel corrosion, crack initiation and crack propagation. The deterioration process due to corrosion is proposed that chain of deterioration process can be explained in Figure 1.

Although crack propagation seems to be the most influential to the structure failure, the chloride penetration is the first stage of the tragic chain reaction.

Corrosion of Concrete Beam

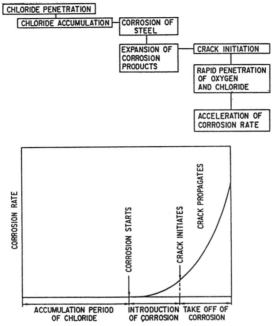

Figure 1: Deterioration process of Concrete Structure

Concrete structure built in an environment subject to the influence of sea water or sea wind induce cracks and deterioration of concrete. To prevent such corrosion, it is necessary to take care in the design and the mixing and placing of the fresh concrete and to take precaution crack generation. At some cases, corrosion of the reinforcement may be avoidable without additional direct protective measures.

Many kinds of protection methods have been developed recently and they show specific features and application area. They are classified according to the protective function.

o epoxy coated reinforcing steel
o concrete surface coating
o concrete surface sealing agent
o polymer impregnated concrete
o cathodic protection
o silica fume concrete
o carbon fibre composite cable for prestressed concrete

EPOXY COATED REINFORCING STEEL

This test is intended to test the corrosion preventing performance of epoxy-coated steel bars and how to apply practically. Epoxy-coated steel bars can be expected to prevent corrosion because even if surrounded by the chloride ions. Three blue colour-type epoxy-coated steel bars were selected to be tested, which are standardized under the tentative guideline. Galvanized steel bars will also be tested for comparison.

Development of corrosion Preventative Materials for PC Tendon replace the conventional steel sheaths having no corrosion resistance. Paying attention to the quality of polyethlylene, attempts will be made to test the corrosion preventing effect and applicability of polyethylene sheaths. In the present test, polyethylene sheaths having a nominal diameter (inside diameter) of 30mm, outside diameter of 42mm, thickness in the range from 0.8mm to 1.2mm and spiral pitch of 9.5mm are used as the post-tension members. For the erection bars, epoxy-coated steel bars are used.

CONCRETE SURFACE COATING

To prevent or inhibit the permeation of chloride ions into concrete by applying a coating on the concrete surface, the test is carried out to research the effect of inhibiting permeation of chloride ions by the coating systems with respect to various types of coating materials.

The coatings are classified into three groups as described below. Group 1 is for prestressed concrete members, and the coating membrane is not required, so high crack followability (1% or more). Group 2 is for reinforced concrete members, and the membranes required crack followability (4% or more). Group 3 is for long term corrosion preventive coatings intended for application on the concrete structure in which particularly severe corrosion is anticipated. For these three groups, it has been decided to carry out the test with respect to 15 types of coating.

The test for verifying the protective effect of the coating materials was conducted for four years in coastal site affected directly by sea water splash, and confirmed that these coating materials could show high performance to intercept chloride and waterproof effect. The weathering performance of most of them will be maintained for a long time to come.

Concrete beam coated with epoxy
(precast beam fabricated in factory)

Concrete beam coated with urethane
(painted in field site)

CONCRETE SURFACE SEALING AGENT

The purpose of application of the sealing agents is same as water proof membranes described in Practical application of water proof membranes. Because of their characteristics, the sealing agents intended for application on the prestressed concrete members of which crack followability is not required, and it has been decided to carry out the test on 7 types of sealers.

POLYMER IMPREGNATED CONCRETE

Polymer impregnation involves the formation of a water intercepting layer by impregnating into concrete with synthetic monomer as MMA, to fill pores in concrete surface layer.

Samples of concrete with impregnated at different depths were prepared and the protective effect of them was measured. An instrument to accelerating corrosion was set up for simulating the corrosion condition for splash in coastal areas. With this test result, 3 cm impregnated layer thick is sufficient for controlling corrosion of reinforcing steel.

The tests for choosing the suitable drying and impregnation conditions was followed for thin actual member (forms), prestressed concrete and reinforced concrete. The result showed that it is possible to determine the impregnation depth in advance from a figure showing the relationship between a monomer impregnation rate and a specific surface area in concrete.

This study confirmed that surface polymer impregnation could be as a practical and more economical, also could determine the treatment conditions for forming impregnated layer with a good water interception performance.

CATHODIC PROTECTION

We have conducted field tests using large structural and has collected basic data. The applicability and the effects of cathodic protection were tested on a bridge which has been in use for 50 years.

Since a number of cracks detected on the concrete surface were due to the corrosion of the reinforcement, the surface concrete was partially chipped, and the surface of the reinforcing bars were polished with sand. Then, the chipped sections were repaired with non shrinkage mortar.

A cathodic protection method called the external power supply method, which employs a combination of linear electrode (directly connected to power supply) and covered electrode (for better distribution, of the current), was applied. For the cathodic protection two systems were used: one is a combination of conductive resin linear electrode and conductive mortar (system A), and the other is a combination of platinum plated titanium linear electrode and conductive paint (system B). In order to identify the protective effects, lead reference electrodes and reinforcement probes were embedded into the concrete for monitoring the reinforcement potential and the current, respectively.

From the test results, it was concluded that the job is relatively easy, and the values measured in the cathodic protection tests coincide with the predicted values. It was also determined that the maintenance cost of protection would not be too great, and consequently a guideline for cathodic protection design was proposed by PWRI.

Laboratory test for cathodic protection
(measuring protective potential)

Concrete beam applied cp
(under construction)

SILICA FUME CONCRETE

To prevent the penetration of chloride ions into concrete members by
making the concrete structures fine by adding as internally sealing
mixture. Silica fume is permissible mixture and was adopted to be
test.

CARBON FIBRE COMPOSITE CABLE FOR PC

In alternative of steel cable for tensioning concrete member, we
developed carbon fibre composite cable and apply new concrete bridge
in Ishikawa October 1988.

Concrete bridge firstly applied with CFRP

LONG TERM EXPOSURE TEST SITE AND PROCEDURE

The exposure test has just started and the period of study is
scheduled for ten years at present.

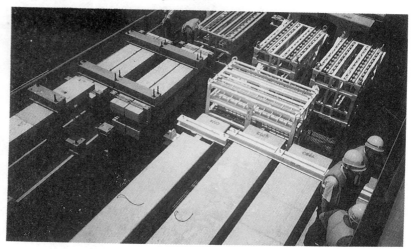

Concrete test beams exposed at Suruga Test Station

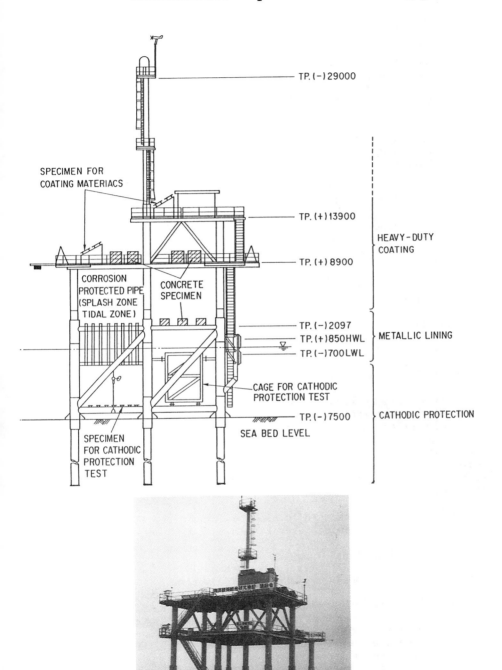

TP. (−) 29000

SPECIMEN FOR
COATING MATERIACS

TP. (+) 13900

HEAVY-DUTY
COATING

TP. (+) 8900

CORROSION
PROTECTED PIPE
(SPLASH ZONE
TIDAL ZONE)

CONCRETE
SPECIMEN

TP. (−) 2097
TP. (+) 850 HWL
TP. (−) 700 LWL

METALLIC LINING

CAGE FOR CATHODIC
PROTECTION TEST

CATHODIC PROTECTION

SPECIMEN
FOR CATHODIC
PROTECTION
TEST

TP. (−) 7500
SEA BED LEVEL

Suruga Test Station - Long Term Exposure Test in Ocean

The test pieces were set up in the Ministry of Construction's Ohigawa Research Station installed off River Ohi in Suruga Bay and exposure tests have already started.

This study mainly consists especially of long term exposure tests in splash and atmospheric zone of actual marine atmosphere.

A total of 154 test beams were prepared in April 1984. In addition, a total of 264 cylinders (150 x 300) also were prepared to provide an auxiliary role.

TENTATIVE GUIDELINES TO PREVENT CORROSION DAMAGE

A tentative guideline of countermeasure of road bridges in sea side area to prevent corrosion was published from the Japanese Road Association. The brief contents of the guidelines are shown as follows;

Japan is divided into three areas according to the severity of the environment and classification of countermeasure is ruled which three countermeasure classifications are set to the area and the distance from the coast.

Three methods are adopted in the guidelines, which are the increasing cover depth of reinforcement, the use of epoxy coated reinforcement, and the coating of concrete.

When the minimum cover depth according to the counter measure classification can not be adopted, the use of epoxy coated reinforcement or the coating on concrete will be chosen.

When epoxy coated reinforcement is used, allowable bond stress for design is 80% of normal one, and it is required to have enough workability and durability in-situ, enough bond strength to concrete and alkali resistance.

When the coating on concrete is used, coating material will be chosen among the following three kinds:

A Epoxy and polyurethane
B Soft type epoxy or soft type polyurethane
C High build type epoxy or glass-flake polyester

A is applied to the prestressed concrete structure, B is applied to reinforced concrete and C is coated for important concrete structure or one which repair will be difficult. The use of concrete surface coating for concrete is featured by the easy acquisition of coating materials and ease of workability. (However, the use of soft coating materials is recommended for RC structures, and relatively hard coating materials for PC structures.) The life of coats of this type is thought to be at least fifteen years, which is not as short as was first expected.

Cathodic protection will be an essential means of protection system, however, at that time the tentative guideline did not include this method since cathodic protection requires maintenance and care after installation, in addition to its application to new structures, it can also be used for the protection of existing structures.

CONCLUSION

This paper describes briefly the state of the development of new protection technology for concrete structure in marine atmosphere, which has been developed recently.

While the establishment of protective measure from corrosion is demanded urgently, it depends on the progress of material technology and engineering. We plan to accelerate the development of protection engineering related to concrete structure and will prove helpful to the improvement of durability of marine structure.

This topic is fast becoming a major issue in Japan, and will increase in importance along with the further construction of marine structures.

REFERENCES

1   KATAWAKI K. The survey of deterioration of concrete bridges damaged with chloride. The Technical Report of PWRI, (1986).

2   KATAWAKI K. Research on epoxy coated reinforcing steel materials test of epoxy for reinforcing bars. The Technical Report of PWRI, (1986).

3   KATAWAKI K and KOBAYASHI S. Research on epoxy coated reinforcing steel performance of reinforced concrete in marine environment. The Technical Report of PWRI, (1986).

4   KATAWAKI K, FUKUSHIMA M, TSUJI T and HASHIMOTO S. Research on concrete surface coating materials performance. The Technical Report of PWRI, (1986).

5   KATAWAKI K and NISHIZAKI I. Research on polymer impregnated concrete treatment technique of PIC. The Technical Report of PWRI, (1986).

6   KATAWAKI K. Research on polymer impregnated concrete corrosion resistant of PIC. Technical Report of PWRI, (1986).

7   KATAWAKI K, SAKAMOTO H, TOMURA J and SHIMIZU K. Research on cathodic protection for concrete structure a guideline of cathodic protection design. Technical report of PWRI, (1986).

8   KATAWAKI K and SAKAMOTO H. Research on cathodic protection
       for concrete structure a guideline of cathodic protection
       design. Technical Report of PWRI, (1986).

9   Large scale exposure test for improving protection technology
       of maritime structure, The Science and Technology Agency,
       (1988).

A FIELD TRIAL OF WATERPROOFING SYSTEMS FOR CONCRETE BRIDGE DECKS

A R Price

Transport and Road Research Laboratory,

United Kingdom

ABSTRACT. A wide range of waterproofing systems considered suitable by manufacturers for concrete bridge decks have been laid under experimentally controlled conditions on a concrete slab, then overlaid with hot asphaltic surfacings. These surfacings were removed and the effects of hot asphalt, bond, durability and waterproofing integrity of the membranes monitored over a three year period.

Many of the systems, including some of those with a Roads and Bridges certificate for use, were found to be unsuitable or ineffective. Severe damage was often caused during the laying of the base course asphalt due to penetration of the hot aggregate into the membrane. The use of a sand carpet to protect membranes overcame this problem, but the interface bond between membrane and sand carpet was weaker than that for base course road surfacing materials. Where membranes were punctured or leaked, water transmission was much greater where there was poor bond between the concrete and the membrane. Resinous primers gave a better bond to the concrete than their bituminous counterparts.

Keywords: Waterproofing Systems, Experimentally Controlled Conditions, Concrete slab, Integrity, Ineffective, Penetration, Aggregate, Bond, Resinous, Bituminous.

Mr A R Price is a Senior Scientific Officer in the Bridges Division of the Transport and Road Research Laboratory. Currently he is responsible for TRRL research on bridge deck waterproofing. He has worked on several major research projects which have included bridge expansion joints, bridge bearings and falsework and he has written a number of research reports on these topics.

INTRODUCTION

Waterproofing systems are applied over bridge decks to prevent road de-icing salts in solution penetrating the concrete substrate and corroding the embedded steel reinforcement. The systems normally used consist of either sheet materials or a liquid coating bonded to the deck surface to form a waterproof membrane.

In 1965 the waterproofing of bridge decks became mandatory for motorway and trunk road bridges and has been applied to most other bridges in the intervening years. Since 1971, waterproofing materials and systems have had to meet the requirements of Departmental Standard BE 27 (1) with the Specification for Road and Bridge Works (2) and later the Specification for Highway Works (3) providing contract requirements. Since 1975 these materials have also been subject to a series of tests in order to obtain a certificate for use. Despite meeting these requirements there is evidence that many waterproofing systems are not performing effectively in service.

To investigate the problems associated with waterproofing systems a field trial was undertaken at the Transport and Road Research Laboratory. A wide range of proprietary waterproofing systems were laid in individual bays on a concrete slab by manufacturers or their chosen representatives. The membranes were then overlaid with two parallel strips of hot base course and sand carpet asphalt repectively. The sand carpet was to enable an assessment of the benefits of protecting systems with this material.

The trial provided data on which to base recommendations for changing the current approval certification procedure and the Department of Transport technical memoranda. It also enabled an assessment of the suitability of the systems used for waterproofing bridge decks.

The membranes tested covered the full range of materials available at the time and included all those with a Roads and Bridges Certificate. In addition a number of membranes which were under development and others which had only been used on multistorey car park decks were included.

CLASSIFICATION OF WATERPROOFING SYSTEMS

Table 1 shows the principal membrane groups tested.

The membranes used were classified as follows:

Sheet Systems

These included the various preformed factory manufactured rolls and boards which are bonded to the bridge deck to form a continuous membrane.

TABLE 1

Principal membrane groups

		Number of systems Laid
SHEET MEMBRANES		
Bituminised Fabrics		9
Polymeric		12
Elastomeric		2
Bitumen laminated boards		5
Bitumen protective sheets		4
LIQUID MEMBRANES		
Bituminous	- Solutions / compositions	2
	- Keg mastics	4
Resinous	- Urethane resin based	7
	- Epoxy resin based	5
	- Acrylic resin based	2

Sheets were categorised into bituminised fabrics, polymer and elastomer based systems and bituminised laminated boards; the latter were also used to protect membranes. Mineral dressed protective sheets were also included in this category (figure 1).

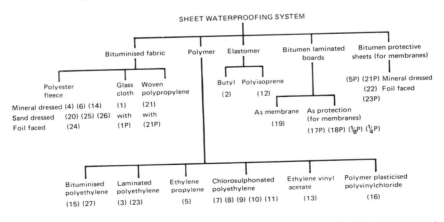

Fig.1 Sheet waterproofing systems used in the trial

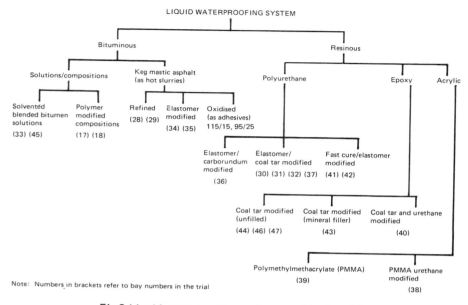

Note: Numbers in brackets refer to bay numbers in the trial

Fig.2 Liquid waterproofing systems used in the trial

Liquid Systems

These consisted of one or two component moisture or chemically curing solutions which were applied to the concrete surface.

Liquid membranes were categorised into bituminous and resinous systems. Bituminous systems were subdivided into bituminous solutions or compositions and keg mastics, the latter required heat to convert them to a liquid. Resinous membranes were subdivided into urethane, epoxy and acrylic resin based systems (figure 2).

EXPERIMENTAL

A concrete slab 24 x 6 x 0.15m with a surface finish representative of many bridge decks was used for the trial. The surface was cleaned and divided into bays separated by wooden slats. Each bay was 6 x 0.5m (figure 3).

Siliconised release paper was laid across parts of the bay to produce areas where the membrane would not be bonded to the concrete so that it could be removed later for examination and testing. Manufacturers or their chosen reprsentatives then laid their waterproofing system or systems onto their allotted bay. Mineral dressed protection sheets were laid in 1m wide strips over many of the membranes (figure 4).

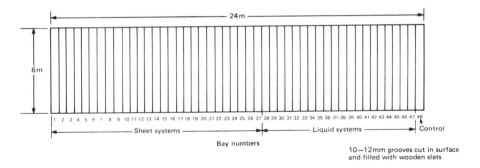

Fig.3 Layout of concrete slab for waterproofing membranes

Four strips of bituminised release paper, 750mm wide were laid over the bays containing the waterproofing systems. This was to prevent the surfacing bonding to the membrane in places so that the effects of hot asphalt could be observed.

Two 3m wide strips of asphaltic surfacing were laid on the waterproofing membranes, one consisted of a base course asphalt and the other was a sand carpet asphalt. The surfacings were compacted to a thickness of 50mm.

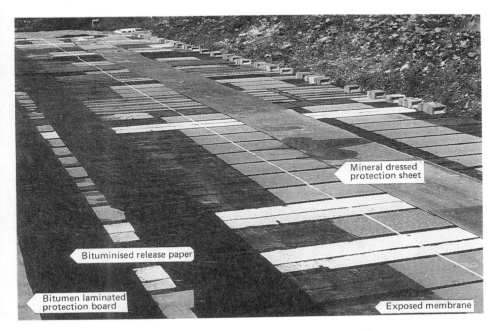

Fig.4 Completed installation of waterproofing systems

After the asphalt had been laid and cooled an asphalt/concrete cutting machine cut the surfacing into sections which were then lifted to observe the condition of the membrane.

PERFORMANCE OF MEMBRANES DURING LAYING AND PRIOR TO ASPHALTING

Pinholes

During laying, pinholes appeared in several of the liquid systems as air or moisture rose up through the curing membrane. Most pinholes occurred on one coat pitch urethanes and epoxies, mastic asphalt and bitumen-in-solvent systems with up to 80 pin holes within a test area of 100 x 100mm, many of which went through to the concrete substrate. The occurrence of pinholes appeared to be related to the moisture content of the concrete and weather conditions during laying. Systems which used two coats or had a finished thickness greater than 5mm (excluding mastics), the fast curing acrylics and modified polyurethanes, were generally free of major pinholing.

Blisters

Blisters were observed on a pitch urethane system, a bitumen-in-solvent system and on an elastomeric mastic system.

Cure

Some liquid membranes took several days to cure fully, particularly the bitumen-in - solvent systems and one of the pitch epoxy systems. Several pitch epoxy systems showed signs of embrittlement after they had cured.

Debond

Debonding of sheet systems was generally confined to the edge of the concrete slab. It was prevalent with oxidised bitumen adhesives and some self adhesive sheets. Latex adhesives debonded from both surface and the edge of the slab.

Most liquid systems bonded well. The exception was a bitumen-in-solvent system which debonded over most of its surface area.

Damage

The membranes were left exposed for about four weeks before the surfacing was laid. In this period some membranes were damaged by pedestrian traffic, particularly liquid systems which were less that 1mm thick. Membranes protected by mineral dressed sheets generally remained undamaged during this period.

CONDITION OF MEMBRANES ON REMOVAL OF SURFACING

Membranes overlaid with base course asphalt

The hot aggregate in the asphalt in combination with the application and compaction of the mix by paver and roller during laying tended to rupture, deform or severely reduce the thickness of some membranes, particularly where their softening point temperature had been exceeded. Bitumen and polymer based sheet membranes were highly susceptible to this form of damage.

Mineral dressed sheets were less likely to be punctured but in most cases the dressing did not prevent penetration. Most liquid coatings less than 2.0mm thick were damaged. Epoxy coatings were embrittled by the high temperatures of the asphalt mix. Some fast cure polyurethane systems and the acrylic systems were generally undamaged except for a few pock marks.

Bituminised mineral dressed protection sheets and bitumen laminated boards were softened in a similar manner to bitumen membranes. The result was that the board was ruptured or severely reduced in thickness by the hot aggregate. Protected membranes with a low softening point were also damaged.

Membranes protected by sand carpet

Some bituminous and polymer based sheets and all the mastic asphalt membranes showed signs of melting under sand carpet but no further damage was observed. Pitch epoxies embrittled but most membranes were undamaged.

BOND ON REMOVAL OF A SURFACING

Bond of membranes to the concrete substrate

Most bituminous primers were used in conjunction with an oxidised bitumen adhesive. In most cases this gave a very poor bond to the concrete with failure at the concrete/primer interface. Latex primers had negligible bond. Resinous primer/membrane combinations generally gave very good bond to the concrete.

Bond of membranes to asphaltic surfacings

In most cases base course asphalt gave a better bond than sand carpet. This was due in part to the composition of the mix and to the aggregate particles embedding into the membrane.

The sand carpet had a weak, sand rich layer at the interface and bond failure normally occurred at this layer. Most bituminous and polymer sheet systems which softened under the hot asphalt had a good fused bond. Resinous systems had a generally weaker bond which was improved by using a tack coat.

TEMPERATURES ON MEMBRANES

Asphalt temperatures on membranes

Asphalt delivery temperatures were approximately 180°C and rolling temperatures were approximately 160°C for base course and 120°C for sand carpet. These temperatures were measured with a probe. Temperatures were also recorded at the interface between the membrane and the asphalt using thermocouples bonded to the surface of the membrane. Membranes received a thermal shock and reached a peak temperature 3-4 minutes after the asphalt had been laid (figure 5). For membranes less than 1.5mm thick, peak temperatures occurred immediately the asphalt was applied.

Temperatures on membranes often exceeded their softening points for up to 10-15 minutes. This was a major cause of aggregate penetration, particularly for bitumen and polymer based materials.

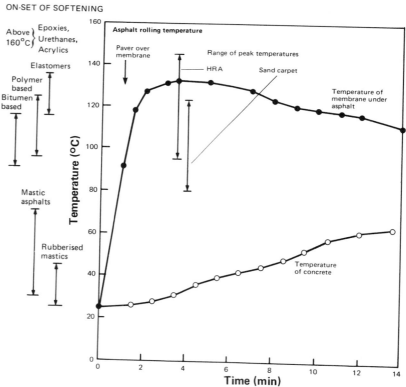

Fig.5 Typical asphalt application temperatures on waterproofing membranes

Ambient temperatures on membranes

During the trial maximum and minimum temperatures on exposed membranes and on those covered by various thicknesses of asphalt were recorded. The highest temperatures on exposed membranes were 2½ to 3 times the shade temperature. It was found that even under asphalt it was possible to exceed the softening point of some membranes on dry sunny days.

During cold periods the temperatures on membranes can fall below zero so membranes with a high water absorption were at risk of damage from freezing water. Some polymer sheets increased in stiffness and pitch epoxies embrittled. Most resinous systems were unaffected by ambient temperatures. Table 2 provides a method of estimating the temperature on the surface of a membrane from a knowledge of the shade temperature.

TABLE 2

Estimate of temperture on membrane from ambient shade temperature

Ambient shade °C	Exposed membrane °C	Under 50mm asphalt °C	Under 100mm asphalt °C	Under 150mm asphalt °C	Weather conditions
30	70-75	50-55	40-45	35-40	Clear Skies
25	55-60	40-45	30-35	25-30	"
20	40-45	25-30	20-25	20	"
15	30-35	15-20	15	15	"
10	20	10	10	10	"
5	8	5	5	5	"
-5	-5	-5	-5	-2	Dull and
-10	-10	-8	-6	-4	overcast
-15	-15	-11	-8	-6	"
-20	-18	-14	-10	-10	"

These temperatures are for the consistent weather conditions shown

WATERPROOFING INTEGRITY

The ability of membranes to resist the transmission of water was investigated by applying a standard head of water. This was done by bonding 50mm diameter tubes to the surface of the membrane (figure 6) and filling them with 500ml of water.

Fig.6 Waterproofing integrity tests on membranes

The results showed that where a membrane was fully bonded, undamaged and had low water absorption characteristics there was little or no transmission over a period of 80 days.

Fully bonded membranes with pinholes or which had been damaged by hot aggregate had a continuous transmission of water with time. Where membranes were debonded and punctured water rapidly flowed beneath them and spread out across the surface of the concrete. Some membranes were partially self sealing or were sealed by deposits in the water or by aggregate particles. This reduced the volume of water passing through the membrane (figure 7).

LONG TERM DURABILITY

During the three years over which the trials were conducted the condition of most systems under the asphalt covering did not deteriorate. The exceptions were the very thin liquid applied systems which flaked from the concrete where they had been damaged by aggregate, and a urethane modified acrylic system which cracked over an existing crack in the concrete test slab.

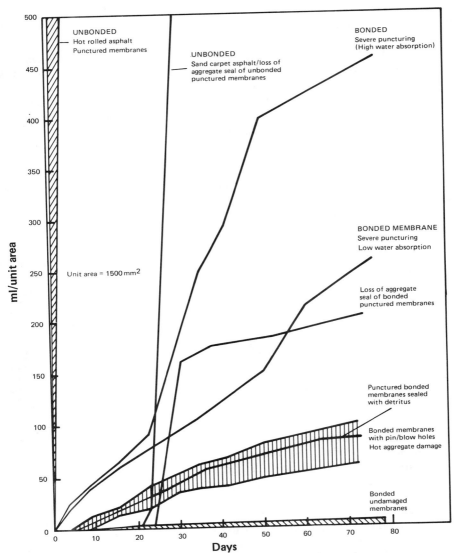

Fig.7 Water transmission/absorption through membranes and surfacings

Where membranes were exposed to weathering and ultra violet radiation, some highly rubberised coal tar modified systems and an unfilled polyurethane system degraded with a significant reduction in tear strength. The former also delaminated from the concrete.

LONG TERM BOND

Bitumen Primers/adhesives

Most bituminised fabric and polymeric sheet systems using bituminous primers in conjunction with oxidised bitumen adhesives had a negligible bond to the concrete after 3 years. The time taken for the bond to fail appeared to be related to the maximum temperature attained in the boiler for the oxidised bitumen. Where the temperature exceeded 270°C debond occurred within six months. It also appeared that compatibility with the sheet affected bond and this requires further research.

Latex primers/adhesives

Two systems used latex primer/adhesives and these had a very poor bond which rapidly deteriorated to no bond at all.

Resinous Primers

Resinous primers gave a very good bond to the concrete which remained unchanged for the duration of the trial.

CONCLUSIONS AND RECOMMENDATIONS

The trial has shown the shortcomings in performance of some of the waterproofing systems considered suitable by manufacturers for concrete bridge decks. It is likely that these systems may not be performing effectively in service.

It was found that a number of key factors influenced the performance of the waterproofing membranes. These are listed below..

> Method of application and compaction of hot asphalt .
> Size, shape and temperature of hot aggregate in the mix.
> Weather and temperature during laying.
> Moisture content of the concrete.
> Condition of the concrete substrate.
> Preparation, workmanship and site procedure.
> Degree of bond between concrete and asphaltic interfaces.
> Type of bonding adhesive and primer used.
> Durability of the waterproofing system.
> Thermoplastic properties of the membranes.
> Thickness of the membrane.
> Use of protection materials

Often it was a complex combination of these factors which influenced the effectiveness of a waterproofing system.

The hot aggregate in the asphaltic surfacing can rupture or severely damage many different types of waterproofing system, providing a pathway for water to reach the concrete deck. The unprotected bituminous and polymer sheet systems less than 2.5mm thick and liquid coatings less than 2.0mm thick, were particularly vunerable.

Damage was related to the temperature of the mix and the softening point of the membrane. Limiting the rolling temperature would considerably reduce the risk of the softening point of the membrane being exceeded.

The sand carpet asphalt protection considerably reduced damage to the membrane. However the sand carpet had a very poor bond with most non-bituminous materials with failure at the sand rich layer which forms at the interface. An alternative is needed which has better durability and bond but still provides the necessary protection.

Bituminous mineral dressed protection sheets and bitumen laminated boards gave good protection to membranes after laying and prior to asphalting but were unsatisfactory in protecting against hot aggregate. Improvements are needed to raise their softening point.

All membranes are at risk from cold loose aggregate compacted into their surface during normal site activities. Membranes should be capable of resisting this form of damage without the benefit of protection as experience has shown that this is often not applied immediately the membrane is laid.
Resinous primers and membranes had a better long term bond to the concrete substrate than their bituminous counterparts, but most resinous membranes had an inferior bond to the asphaltic surfacing. There is a need to develop compatible tack coats.

Unmodified bitumen solvent primers, latex and oxidised bitumen adhesives gave a poor bond which deteriorated to negligible bond with time. These require improving or replacing.

The current approval certification tests need revision with more emphasis on the conditions prevailing on site. Consideration should also be given to the approval of installers of waterproofing systems and long term performance of the materials used.

REFERENCES

1   DEPARTMENT OF TRANSPORT Waterproofing and Surfacing of bridge decks. Technical Memorandum (Bridges) No BE 27 1970.

2   DEPARTMENT OF THE ENVIRONMENT Specification for road and bridge works. HMSO London 1976

3   DEPARTMENT OF TRANSPORT Specification for highway works Part 5 HMSO 1986.

ACKNOWLEDGEMENTS

The work described in this paper forms part of the programme of the Transport and Road Research Laboratory and the paper is published by permission of the Director.

Crown Copyright. The views expressed in this Paper are not necessarily those of the Department of Transport. Extracts from the text may be reproduced, except for commercial purposes, provided the source is acknowledged.

APPLICATION OF POLYMERS TO WATERPROOF MEMBRANES FOR PROTECTION OF CONCRETE BRIDGE DECKS AND ITS EVALUATION

A Miyamoto

Kobe University,

T Maeda

Shimizu Corp.,

N Wakahara

Konishi Co. Ltd.,

H Fujioka

Sunkit Corp.,

Japan

ABSTRACT. The waterproof membranes for protection of the concrete decks are exposed to more severe conditions than the linings used for repair of deteriorated concrete structures in both the internal environments, which is the high-temperature at placing of hot asphalt concrete, and the external environments, which are the combination of distributed shear and peeling stresses and repeated crack opening into interaction surface between waterproofing layer and asphalt pavement under heavy moving traffic load in service. This paper discusses not only the determination of properties such as thermal compatibility at high-temperature, crack bridging capability, etc., required for application of polymers to waterproof membranes but also the material evaluation of some newly developed polymers for waterproof membranes.

Keywords: Polymer, Waterproof Membrane, Concrete Deck, Material Evaluation, Full-scale Test, Finite Element Analysis

Ayaho MIYAMOTO, born 1949, is an Assoc. Prof. at the Department of Civil Eng., Kobe Univ., Kobe, Japan. He received his Doctor of Eng. Degree from Kyoto Univ. in 1985. His recent research activities are in the areas of structural safety evaluation of concrete structures and investigation of the selection criteria of new materials.

Toshiya MAEDA, born 1964, is a Research Engineer at Civil Eng. Div., Shimizu Corp., Tokyo, Japan. He received his Master of Eng. Degree from Kobe Univ. in 1990. He has been involved in research projects concerning the waterproof membranes for protection of concrete structures.

Naoki WAKAHARA, born 1947, is a Senior Chemist at Konishi Co., LTD., Osaka, Japan. He graduated in Industrial Chemistry at Doshisha Univ. in 1970. His research interests include new organic materials development and repair materials evaluation for engineering field.

Hidehiro FUJIOKA, born 1946, is General Manager, Sunkit Corp., Kobe, Japan. He graduated in Civil Eng. at Kobe Univ. in 1975, and is now interested in development of material design and concrete protection technique by new materials.

INTRODUCTION

As the deterioration mechanisms of reinforced concrete slabs in service have been made clear, there is an increased interest in the usage of waterproof membranes for protection of concrete decks during both new construction of slabs and repair of pavements in Japan.[1] The waterproof membranes for protection of the concrete decks are exposed to more severe conditions than the linings used for repair of deteriorated concrete structures in both the internal environments, which are a severe traveling load of the spreader and the high-temperature at placing of hot asphalt concrete, and the external environments, which are the combination of distributed shear and peeling stresses and repeated crack opening into interaction surface between waterproofing layer and asphalt pavement under heavy moving traffic load in service.

This investigation is concerned with not only the determination of properties required for application of polymers to waterproof membranes but also the material evaluation of some newly developed polymers for waterproof membranes. In determinating properties required for application of polymers to waterproof membranes, several experimental tests on full-scale models and the analytical studies on two-dimensional elasticity solutions by use of the finite element method were carried out. On the other hand, the performance of six waterproof membrane systems of polymers which are classified as the Hot-melt type and the Wet type are assessed by requirements such as thermal compatibility at high-temperature, bond and shear strength between waterproofing layer and asphalt pavement, permeability of waterproofing layer, crack bridging capacity of waterproofing layer and skid resistance.

OUTLINE OF WATERPROOF MEMBRANE SYSTEM AND ASPHALT PAVING TEST

Six types of waterproof membrane systems were placed on real-size concrete decks according to the specified construction procedures. The names, specification of material properties and cross-sectional profile of the waterproof membrane systems are shown in Table 1. The membranes can be classified as Dry type, Hot-melt type and Wet type. The Dry type of membranes (B and C) were placed two days before placing of asphalt concrete and curing was also carried out. On the other hand, the Hot-melt type and Wet type of membranes (A, D, E and F) were placed immediately before placing of asphalt concrete.

After placing of the waterproof membrane systems, real-size machineries were used to place hot asphalt concrete in order to check the effects of high temperatures, dump truck and asphalt spreader and finisher on the amount of damage to the membranes and also workability during construction and finally also to measure the thermal hysteresis of the membranes. Two layers of asphalt concrete, which consist of a coarse aggregate layer (4cm in depth) and a fine aggregate layer (3cm in depth), were placed. The sequence for the layering procedure was the same as a normal in-situ construction procedure, with the dump truck, asphalt spreader and finisher moving consequently from membrane A to F, and finally

Table 1 List of waterproof membrane systems

Symbol of waterproof membrane system	A	B	C	D	E	F
Type of waterproof membrane material	Hot melt	Epoxy resin emulsion (Post cure type)	Epoxy resin I (Room temperature cure type)	Epoxy resin II (Post cure type)	Epoxy resin III (Post cure type)	Epoxy resin IV (Post cure type)
Appearance of waterproof membrane at hot asphalt concrete placing	Dry	Dry	Dry	Wet	Wet	Wet
Curing condition	Softening point 100 degrees	140 degrees 1 hrs.	20 degrees 7 days	20 degrees 7 days	20 degrees 7 days	20 degrees 7 days
Tensile strength (kgf/cm2)	35	62	105	146	30	16
Tensile elongation (%)	830	350	47	95	155	183
Cross section of waterproof membrane	Waterproof membrane / Primer	Waterproof membrane / Primer	Waterproof membrane / Tack coat / Silica sand / Primer	Waterproof membrane / Intermediate coat / Primer	Waterproof membrane / Primer	Waterproof membrane / Primer

Table 2 Bond (adhesion) strength test results (kgf/cm2)

Symbol of waterproof membrane system	Specimen of asphalt spreader passed	Specimen of asphalt spreader non-passed
A	8.0	6.1
C	3.8 (0.4)	3.3 (3.0)
E	6.6	5.1
F	4.4	2.6

compaction by a roller compacting machine was carried out. The thermal hysteresis of membranes after placing of hot asphalt concrete were measured using electronic thermal sensors which were arranged on each membrane system.

Results and Discussions

The results for experiments carried out on the effects of placing of hot asphalt concrete can be summarized as:
① If the construction period for the waterproof membranes are taken into consideration, the Hot-melt type (A) and Wet type (D, E and F) membranes would be suitable for application where the process have to be completed just before the placing of hot asphalt concrete. However, when considering the construction process, workability and also the complexity of the handling processes involved, the Wet type (D, E and F) membranes would then be the most appropriate.
② The Hot-melt type (A) and Dry type (B and C) membranes are more suitable from the point of view of workability and reliability during placing of asphalt concrete because no weaknesses in the waterproof membrane due to tires and caterpillar tracks of the layering machineries during placing of hot asphalt concrete will be contained.
③ During the tests, the surrounding temperature was approximately 5°C while the peak temperature for asphalt concrete in the truck was about 170°C. From measurements taken from the thermal hysteresis of the waterproof membranes, a peak temperature of about 78°C and a maximum thermal energy of about 9700°C·min. was obtained. Therefore, it can be assumed that the peak temperature transferred to the waterproof membranes is 100°C while the duration can be set as 100min. This value is smaller than the peak temperature of 180°C that is specified by the Japanese Ministry of Construction, etc.

EVALUATION OF DEFECT RESISTANCE IN WATERPROOF MEMBRANES

Three kinds of tests for investigation of defect resistance during placing of asphalt concrete are considered here, that is bond(adhesion) strength test of membranes, isolation (peel off) test between concrete and asphalt and permeability test of the membranes. The resistance capability of the membranes are compared to evaluate the performance during construction.

Tests for the bond(adhesion) strength were carried out by cutting off blocks with dimensions of 30×30×20cm from the portions where the tires are known to pass over and also from portions not affected by the tires. Four cores were then selected and extracted from each piece of test block. Each core is made up of asphalt concrete, a layer of waterproof membrane and also about 1cm in depth of concrete from the surface of concrete decks. The cores were then subjected to tension tests according to the procedures specified by the Building Research Institute of Japanese Ministry of Construction.

For the isolation (peel off) tests, the asphalt concrete layer from test blocks of 30×30×20cm were separated by using gas burners and chemicals. The assessment was then carried out by visual inspection during isolation, which was simulated by chipping off little portions by chisels. Permeability tests were carried out by enclosing 30×30×20cm blocks with see-through acryl boxes. The gap between the acryl boxes and test blocks were sealed with waterproofing sealants from the portions below the asphalt concrete layer. Assessment was then carried out by checking for permeability of colored ink from the asphalt concrete surface to the concrete surface.

Results and Discussions

The evaluation results of the bond(adhesion) strength, isolation and permeability tests are given in Tables 2, 3 and 4, respectively. The Quantification theory I was used to assess the degree of influence attributed by the items of isolation tests on bond(adhesive) strength. The items consist of the inspections carried out during the isolation tests while the bond strength was selected as an external criterion. The results are as shown in Table 5. A similar evaluation was also carried out for the permeability tests by using the Quantification theory II. The result for the evaluation is given in Table 6. The results from both of the tests can be summarized as:
① It can be considered that the amount of discoloration in the waterproof membrane during placing of asphalt concrete is related to the bond strength. The bond strength shows a tendency to be smaller as the amount of discoloration increases.
② The bond strength increases as the amount of saturation into asphalt concrete layer increases. This can be attributed to the resin seeping and also filling up gaps between the asphalt concrete layer.
③ As the amount of resin saturation into the asphalt concrete layer increases, the coefficient of permeability increases too, causing the waterproofing effect to decrease.
④ The coefficient of permeability increases as the amount of discolored area during placing of asphalt concrete increases, that is the interception of water decreases.

The results of the three types of tests and also the assessment by Quantification theory are compiled together to give an integrated evaluation of the waterproof membrane systems. An evaluation based on the Scoring Method is employed here, with the assigned points for the Scoring Method being as shown in Table 7. And also, the results of the evaluation are given in Table 8. According to the Scoring Method, the waterproof membrane system with a higher average score would be more appropriate. It should be noted that complete isolation for the type A resin was not possible, and thus the scoring values for items such as surface condition, isolated area and discolored area are predicted values from visual inspection. The total result for this evaluation would then be: C > A > E > F > B > D. The crack bridging capacity, which will be dealt with later, was then evaluated based on the concrete prism bending tests for the top 3 (C, A and E) types of waterproof membrane systems.

Table 3 Isolation test results

Items — Waterproof membrane system	Bond strength	Isolation time 1	2	Passed or non-passed 1	2	Surface condition 1	2	3	Isolated area 1	2	3	Starve joint 1	2	3	Discolored area 1	2	3	4	Saturation into asphalt 1	2	3
B−B1	0.0	○		○				○	○			○						○	○		
B−B2	0.0	○			○		○			○		○						○	○		
B−B3	0.0	○		○			○				○	○					○		○		
B−B4	0.0	○			○		○				○	○						○		○	
C−B1	3.3		○	○				○	○			○			○				○		
C−B2	3.8		○		○			○	○			○			○				○		
C−B3	3.3		○	○				○	○			○				○			○		
C−B4	3.8		○		○			○		○		○				○				○	
D−B1	0.0	○		○		○			○			○					○		○		
D−B2	0.0	○			○	○			○			○					○		○		
D−B3	0.0	○		○		○				○		○						○	○		
D−B4	0.0	○			○	○			○			○						○	○		
E−B1	5.1		○	○				○	○				○		○					○	
E−B2	6.6		○		○			○	○				○	○						○	
E−B3	5.1		○	○				○	○				○			○					○
E−B4	6.6		○		○			○	○				○		○						○
F−B1	2.6		○	○				○	○				○	○					○		
F−B2	4.4		○		○			○	○				○		○					○	
F−B3	2.6		○	○				○	○				○	○					○		
F−B4	4.4		○		○			○	○				○		○						○

Remarks:

- kgf/cm²
- Isolation time: 1 Natural isolation, 2 Forced isolation
- Passed or non-passed: 1 Non-passed, 2 Passed
- Surface condition: 1 Smooth, 2 Normal, 3 Rough
- Isolated area: 1 Small, 2 Moderate, 3 Large
- Starve joint: 1 None, 2 Small, 3 Large
- Discolored area: 1 None, 2 Small, 3 Moderate, 4 Large
- Saturation into asphalt: 1 Small, 2 Moderate, 3 Large

Table 4 Permeability test results

Symbol of waterproof membrane system	Colored area	Symbol of waterproof membrane system	Colored area	Symbol of waterproof membrane system	Colored area
A−1	Small	B−1	None	C−1	None
A−2	Small	B−2	None	C−2	Small
A−3	Small	B−3	Large	C−3	Moderate
A−4	Small	B−4	Moderate	C−4	Small
D−1	Small	E−1	Small	F−1	Small
D−2	Moderate	E−2	Moderate	F−2	Moderate
D−3	Moderate	E−3	Large	F−3	Small
D−4	Small	E−4	Moderate	F−4	Small

Table 5 Results of discriminant analysis for categorical data (Quantification theory I)

Items	Categories	Number of data	Category Score	Distribution of score	Score range
Isolation time	1. Natural isolation	8	1.39		0.75
	2. Forced isolation	12	4.12		2.73
Passed or non-passed	1. Non-passed	10	0.0		0.21
	2. Passed	10	0.28		0.28
Starve joint	1. None	12	0.0		0.66
	2. Small	4	0.74		2.11
	3. Large	4	2.11		
Discolored area	1. None	4	0.0		
	2. Small	7	-0.78		0.59
	3. Moderate	6	-1.68		1.68
	4. Large	3	-1.43		
Saturation into asphalt	1. Small	8	0.0		0.30
	2. Moderate	9	-0.42		0.61
	3. Large	3	0.19		

Table 6 Results of discriminant analysis for categorical data (Quantification theory II)

Items	Categories	Number of data	Category Score	Distribution of score	Score range
Isolation time	1. Natural isolation	8	0.0		0.08
	2. Forced isolation	12	0.42		0.04
Passed or non-passed	1. Non-passed	10	0.0		0.43
	2. Passed	10	-0.16		0.16
Starve joint	1. None	12	0.0		0.64
	2. Small	4	-0.36		0.45
	3. Large	4	-0.45		
Discolored area	1. None	4	0.0		
	2. Small	7	0.09		0.59
	3. Moderate	6	0.26		0.49
	4. Large	3	0.49		
Saturation into asphalt	1. Small	8	0.0		0.82
	2. Moderate	9	0.23		0.98
	3. Large	3	0.98		

Table 7 Assigned points for Scoring Method

Evaluation items	Evaluation methods	Maximum point	Minimum point
Isolation time	When core sample picked up, asphalt core separated from concrete :0 point. Others :10 points.	10	0
Isolated area	No isolated area :10 points. 3 ranks evaluation according to the isolated area.	10	0
Surface condition	Maximum grade of roughness :10 points. 3 ranks evaluation according to the grade of roughness.	10	0
Starve joint	No starve joint :10 points. 3 ranks evaluation according to the starve joint area.	10	0
Discolored area	No discolored area :10 points. For asphalt melt into waterproof membrane, 4 ranks evaluation according to the asphalt area. On the other hand, for waterproof membrane itself discolored, 4 ranks evaluation according to the discolored area.	10	0
Saturation into asphalt	No saturation :10 points. 3 ranks evaluation according to the degree of saturation into asphalt.	10	0
Bond strength	Maximum bond strength (8.0 kg/cm²) :10 points.	10	0
Interception of water	No coloration :10 points. 4 ranks evaluation according to the coloration area.	10	0

Table 8 Results of Scoring Method

Items Waterproof membrane system	Bond strength	Isolation time	Passed or non-passed	Surface condition	Isolated area	Discolored area	Saturation into asphalt	Interception of water	Total score	Average score
A-1	8	10	10	5	10	6	5	6	60	
A-2	10	10	10	5	10	6	5	6	62	61
A-3	8	10	10	5	10	6	5	6	60	
A-4	10	10	10	5	10	6	5	6	62	
B-1	0	0	10	10	10	0	10	10	50	
B-2	0	0	5	5	10	3	10	10	43	39
B-3	0	0	0	10	10	3	10	0	33	
B-4	0	0	0	5	5	10	5	3	28	
C-1	4	10	10	10	10	10	5	10	69	
C-2	5	10	5	10	10	10	5	6	61	63
C-3	4	10	10	10	10	6	5	3	58	
C-4	5	10	5	10	10	6	10	6	62	
D-1	0	0	5	0	10	6	10	6	37	
D-2	0	0	5	0	10	0	10	3	28	31
D-3	0	0	5	0	10	3	10	3	31	
D-4	0	0	0	0	10	3	10	6	29	
E-1	6	10	10	10	5	10	5	6	62	
E-2	8	10	10	10	0	10	10	3	51	53
E-3	6	10	10	10	5	6	5	0	52	
E-4	8	10	10	10	0	6	10	3	47	
F-1	3	10	10	10	5	6	5	6	55	
F-2	6	10	10	10	0	3	10	3	42	51
F-3	3	10	10	10	5	6	5	6	55	
F-4	6	10	10	10	0	3	5	6	50	

TEST FOR SERVICEABILITY PERFORMANCE

Crack Bridging Test for Specimens with High Temperature Treatment

The aim of the tests are to check for the bending crack bridging capability for waterproof membranes systems that have been subjected to heat during placing of hot asphalt concrete.

The waterproof membrane systems were coated on concrete prism specimens (15x15x55cm) with pre-cracks introduced on the concrete surfaces. Heat treatment was then introduced to simulate the effects of heating due to placing of hot asphalt concrete. The amount of heat energy measured from the asphalt concrete paving tests were applied here. Based on the thermal hysteresis results, heat energy of $100°C×100$min. was applied to the test specimens. The test method was based on the procedure specified by Hanshin Expressway Public Corporation[2](Japan). In order to visually inspect for cracks occurring in the waterproof membrane system throughout the tests, the surface where the layer of waterproof membrane was coated was set facing upwards, with the tensile region being on the top surface. The specimens were subjected to loading at three points (center loading). Crack formation in the waterproof membrane system and also its progression was monitored with the help of crack gauges placed across the cracks in the concrete specimen, while the midspan deflection was measured using electronic deflection meters and the load was measured using a load cell. Formation of cavities or cracks on the waterproof membrane were checked through visual inspection.

Results and Discussions

Comparisons of the crack bridging capability before and after heating and also the period of the tests (in October and December) are shown in Figs.1 and 2, respectively. The results obtainable from both figures can be summarized as:
① The crack bridging capacities for post cure type of epoxy based waterproof membrane (type E membrane) increase after heating.
② The crack bridging capacities for room temperature cure type epoxy based waterproof membrane (type C membrane) show a tendency to decrease after heat treatment.
③ No cracking was observed in the Hot-melt type rubber based waterproof membrane (type A membrane) after heating and also during the different test periods. It can be concluded that an increase in crack bridging capability can be expected for this type of membrane.
④ The crack bridging capacities for post cure type and room temperature cure type show a tendency to decrease for tests carried out in December (room temperature: $13°C$), when compared to results of tests carried out in October (room temperature: $27°C$).

ANALYSIS OF SHEAR AND PEELING RESISTANCE

The main aim of this analysis is to clarify certain mechanical properties that are considered to be necessary for application of

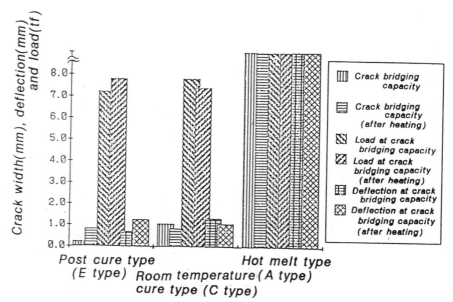

Fig. 1 Results of crack bridging capability test
(Comparison between before and after heating)

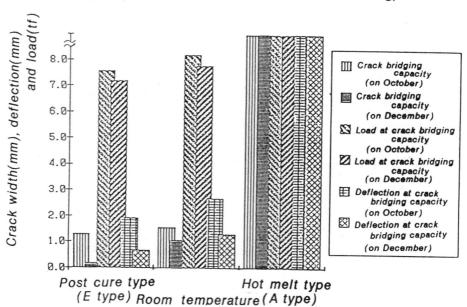

Fig. 2 Results of crack bridging capability test
(Comparison between the test results on Oct. and the test results on Dec.)

polymers to waterproof membranes for protection of concrete bridge decks. The mechanical properties that will be considered here are the effects of normal(peeling) stresses on the membranes under heavy moving traffic load and also the effects of shear stresses during sudden traffic movements such as acceleration and braking.

The analytical model is as shown in Fig.3. The finite element method (elastic analysis) is applied here. The waterproof membrane and asphalt concrete layer is assumed to be on a rigid deck (RC deck slab) with a composition such as; total depth: 7cm, asphalt concrete layer: 6.8cm, membrane layer: 2mm and 5mm, width: 50cm. The rear tires of a heavy vehicle with a total weight of 9tf. is assumed to act directly on the asphalt concrete layer. The asphalt concrete and membrane layers are considered to be visco-elastic bodies and since the modulus of elasticity varies with temperature, the modulus of elasticity for asphalt concrete, E_a and modulus of elasticity for resin, E_r are taken as parameters in the analysis.

Results and Discussions

Typical examples of the relation between the maximum shear stress, maximum normal stress and asphalt concrete modulus of elasticity, E_a are shown in Figs.4 and 5. The modulus of elasticity for the resin membrane layer, E_r was altered from 100 to 100000kgf/cm^2 while the thickness was set at 2mm. The main results derived from both figures can be summed as:
① The maximum shear stress decreases with an increase in the value of E_a and this trend becomes more significant as the value of E_r becomes smaller. The same trend is also noticeable for the maximum normal stress.
② The specifications provided by the Public Works Research Institute of the Japan Ministry of Construction[3] require waterproof membranes to have a maximum shear stress of above 8.0kgf/cm^2 at -10°C and a level above 1.5kgf/cm^2 at 20°C. Calculations for resins that satisfy the above specifications show that all the resins satisfy the calculated values at 0°C, but some resins that do not satisfy the calculated value of 3.7kgf/cm^2 at 20°C cannot be considered to be appropriate for use as waterproof membranes.

From the analytical results, mathematical functions for the maximum shear stress and maximum normal stress in terms of the moduli of elasticity for asphalt concrete and resin were calculated as shown in Eqs.(1) to (3). The equations can be employed to predict the maximum shear stress and maximum normal stress for different resin combinations.

$$S_{max} = [0.117 \cdot (\log_{10} E_r) - 0.558] \cdot E_a + [0.375 \cdot (\log_{10} E_r) + 2.018] \qquad (1)$$

$$H_{max} = \begin{cases} 0.660 & ; E_r \geqq 5000 \text{kgf/cm}^2 \qquad (2) \\ [0.033 \cdot (\log_{10} E_r) - 0.143] \cdot E_a + [0.140 \cdot (\log_{10} E_r) + 0.214] \\ \qquad\qquad ; E_r \leqq 5000 \text{kgf/cm}^2 \qquad (3) \end{cases}$$

where, S_{max} : maximum shear stress (kgf/cm^2),
H_{max} : maximum normal (peeling) stress (kgf/cm^2),

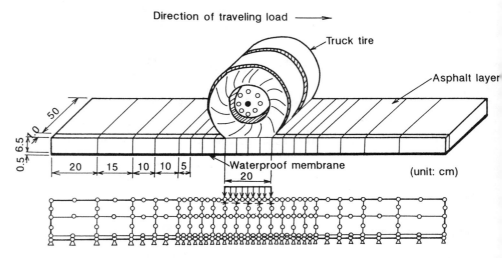

Fig. 3 Model for analytical study (finite element meshes)

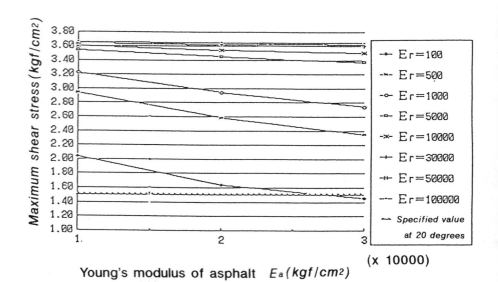

Young's modulus of asphalt $E_a(kgf/cm^2)$

Fig. 4   Analytical results of maximum shear stress

E_a : asphalt concrete modulus of elasticity (kgf/cm^2),
E_r : resin modulus of elasticity (kgf/cm^2).

PROPERTIES REQUIRED FOR WATERPROOF MEMBRANE MATERIALS

An integrated evaluation can be carried out based on all the results mentioned above. Properties which are considered to be necessary and also the most suitable performance capabilities together with the degree of satisfaction from each resin are compiled in Table 9. Three new items (construction period, mixing and curing conditions, workability during placing of asphalt concrete) are introduced as additional properties required. These properties are considered to be necessary for speedy construction and also in producing a uniform quality for coating of membrane. Comparing the degree of satisfaction of the resins tested, the Hot-melt type is the most suitable for application as waterproof membrane. The suggested properties to be improved for each resin are also included in Table 9. The improvements suggested can be used as a guideline in development of new materials for waterproof membrane. For example, the crack bridging capability can be improved by decreasing the modulus of elasticity at room temperature and also by decreasing the glass transition temperature below -10°C.

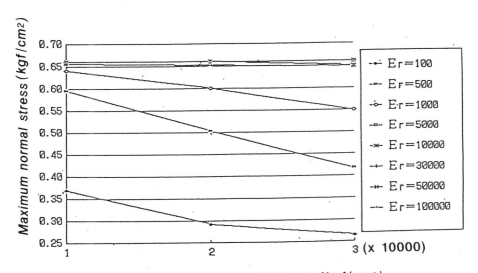

Fig. 5   Analytical results of maximum normal stress

Table 9 Summarize the required and improving properties for waterproof membrane material

	Required properties	Most suitable performance	Specified values*	Degree of satisfaction of each waterproof membrane system						Suggested properties to be improved
				A	B	C	D	E	F	
In construction and in placing of asphalt concrete	Construction period	Complete construction 2 or 3 hours before placing of asphalt concrete.	——	O	—	—	O	O	O	
	Mixing and curing conditions	Two component and room temperature cure type.	——	—	—	—	O	O	O	
	Workability during placing of asphalt	Contain no weakness into waterproof membrane under track tire and caterpillar.	——	O	O	O	—	—	—	
	Bond(adhesive) strength	Strong.	More than 12kgf/cm²(-10°C) More than 6kgf/cm²(20°C)	O	—	—	—	O	O	Give large penetration depth of polymer into asphalt. Give no discolor during placing of hot asphalt concrete.
	Interception of water	Zero water leakage.	Less than 0.5ml of water leakage	O	—	O	—	—	—	Give no starve joint and no discolor during placing of hot asphalt concrete.
In service	Peeling(isolation) resistance (Bond strength)	No isolation under repeated normal stress by heavy traveling load.	More than 12kgf/cm²(-10°C) More than 6kgf/cm²(20°C)	—	—	—	—	—	—	Give as small as possible Young's modulus in room temperature range.
	Shear resistance	No sliding under repeated shear stress by suddenly braking and accelerating.	More than 8kgf/cm²(-10°C) More than 1.5kgf/cm²(20°C)	—	—	—	—	—	—	Give as small as possible Young's modulus in room temperature range.
	Crack bridging capacity	As large as possible in room temperature range. And also, no change of crack bridging capacity before and after heating.	——	O	—	—	—	—	—	Give as small as possible Young's modulus and small change of it in room temperature range. Reduce the glass transition temperature, TG to low temperature region (less than -10°C). Give large elongation and shear deformation to the polymer.

Note*: Specified values at present in Japan.

CONCLUSIONS

The mechanical properties required during construction and also the necessary properties during service-life for polymers to be successfully applied to concrete bridge decks are evaluated through experiments and also analysis. An integrated evaluation is proposed based on available results. The main results of this study can be summarized as:

The requirements assigned in the construction specifications are considered together with requirements derived from this study in order to assign improvements necessary for each material and also to set out new requirements for future application. The relationship between the required mechanical properties and material properties are clarified and used for proposing an integrated evaluation method.

Considering the required properties and also the degree of satisfaction provided by waterproof membrane type A~F, it can be concluded that the Hot-melt type resin satisfies the requirements most and can be assumed as the best choice for application to waterproof membrane.

Within the scope of this study, the most suitable material would be the Hot-melt type of resin, and in order to ensure speedy construction, the temperature should be near room temperature as possible. Moreover, there should be little discoloration and not much saturation into asphalt during placing of asphalt concrete. From the material property point of view, materials with a low Young's modulus at room temperature and a low glass transition temperature range (below $-10°C$) would be ideal.

REFERENCES

1   HANSHIN EXPRESSWAY PUBLIC CORPORATION. Study on durability of waterproof membranes for RC decks of highway bridges (in Japanese), March 1987.

2   HANSHIN EXPRESSWAY PUBLIC CORPORATION. Test procedure for materials used for repairing concrete (in Japanese), April 1987.

3   MAEDA T. Crack bridging capacity for deck lining and repair materials (in Japanese), Undergraduate Thesis (Kobe University, Japan), March 1988.

FIELD SERVICE LIFE PERFORMANCE OF DEEP POLYMER IMPREGNATION AS A BRIDGE DECK CORROSION PROTECTION METHOD

R E Weyers

Virginia Polytechnic Institute & State University,

P D Cady

Pennsylvania State University,

M Henry

Virginia Polytechnic Institute & State University,

United States of America

ABSTRACT. Two sound concrete bridge decks were impregnated with methyl methacrylate to a depth of 3 to 4 inches and polymerized insitu. Both bridges were in sound condition with no surface spalling at the time of impregnation. After 15 and 5 years, respectively, the small scale and full scale field trial sections and control sections were evaluated for their corrosion abatement performance. The evaluation program included delamination survey, chloride content, corrosion potentials, petrographic analysis of core sections, and corrosion rate measurements. The small scale field trial has undergone dramatic differences in performance between the control and impregnated section with spalling and very high corrosion rates occurring in the control section. The control section for the full scale field trial shows higher chloride contents and corrosion rates than the impregnated sections.

Keywords: Bridge decks, Polymer impregnation, Linear polarization.

Dr. Richard E. Weyers is an Associate Professor of Civil Engineering at the Virginia Polytechnic Institute and State University, Blacksburg, VA, USA. He is Associate Director of the Civil Engineering Structures and Materials Research Laboratory. His current research work includes: maintenance and rehabilitation of concrete bridges, and service life prediction of concrete structures.

Dr. Philip D. Cady is a Professor of Civil Engineering at The Pennsylvania State University, University Park, PA, USA. He is the Distinguished Alumni Professor of Engineering. His current research work includes: assessment of the physical condition of concrete bridges, and cost-effective maintenance and rehabilitation of concrete bridges.

Mr. Mark Henry is an undergraduate student in Civil Engineering at the Virginia Polytechnic Institute and State University, Blacksburg, VA, USA.

INTRODUCTION

The early deterioration of reinforced concrete bridge decks was first recognized by highway agencies in the United States in the late 1950's and early 1960's. Spalling was identified as the major contributing factor to the early deterioration of concrete bridge decks [1, 2]. High chloride contents of the concrete as a result of winter maintenance activities was related to corrosion of the reinforcing steel and subsequent spalling of the concrete. This accelerated rate of deterioration of reinforced concrete bridge decks parallels the increased use of road salts which were used to implement a "bare pavement policy". In 1955 deicing salt usage in the United States was about 1 million tons [3]. By 1970, deicer salt usage had increased to over 9 million tons [3].

By the early 1970's, highway agencies had begun to realize the magnitude of financial exposure represented by the United States bridge deck problems. The Federal Highway Administration in 1973 estimated the cost to repair bridge decks at $70 million per year [4]. By 1986, the unfunded liability to correct corrosion-induced deterioration of bridges was estimated at $20 billion and to be increasing at a rate of $500 million annually [4].

In 1972, the National Cooperative Highway Research Program sponsored a research project with an objective to develop the deep polymer impregnation process for bridge decks. The deep impregnation process consists of drying the concrete, impregnating the concrete with a monomer to a depth of 4 inches in order to encapsulate the upper rebar mat, and polymerizing the monomer insitu. The deep impregnation progress should abate or arrest the corrosion of the rebar by replacing the corrosion cell electrolyte (concrete pore water) with a dielectric polymer, immobilizing the existing chloride and reducing the ingress of further chlorides, water and oxygen. The research project culminated in the deep impregnation of a test section 3.5 feet by 11.5 feet of bridge deck in Bethlehem, Pennsylvania. In 1985, the Pennsylvania Department of Transportation sponsored a research project to demonstrate the technical and economical feasibility of full scale impregnation of bridge decks. Approximately one-half, 60 feet by 44 feet, of a 131 feet long center span of a bridge deck in Boalsburg, Pennsylvania was impregnated to a depth of about 4 inches using the grooving technique in May-June of 1985.

This paper presents the results of the field corrosion performance of the two depth polymer impregnated concrete bridge decks in Pennsylvania. The field corrosion performance investigation, sponsored by the Strategic Highway Research Program, consisted of a visual inspection and delamination survey, chloride contamination levels, corrosion potentials, petrographic analysis of drilled concrete cores, and corrosion rate measurements.

BOALSBURG BRIDGE DECK

Background

The bridge is a three-span multigrider bridge, simply supported with steel plate girders. The end spans are 42 feet and 38 feet and the center span is 131 feet. The deck width, curb to curb, is 44 feet, consisting of two 12-foot traffic lanes and two 10-foot aprons. The concrete deck was placed in April 1972 using permanent steel forms. It is of composite design. The main reinforcement consists of No. 5 bars on 6-inch centers in the transverse direction, top and bottom. The longitudinal steel consists of No. 4 bars 12-inches on center in the top and No. 5 bars on 9-inch centers in the bottom. The design deck thickness is 8-inches, with a 2-inch minimum concrete cover over the top reinforcing steel.

In March, 1983 a visual inspection of the deck showed the deck to be in excellent condition. The only deterioration observed was a series of shallow spalls about 0.5 inches deep immediately adjacent to the expansion dam cover plate at the east end of the center span. The deterioration appeared to be a result of poor construction practices. A cover depth survey showed the mean cover depth to be 2.86 inches with a range of 2.3 to 3.3 inches and a standard deviation of 0.22 inch. The mean value of the copper copper-sulfate (CSE) half-cell measurements performed in March, 1983 was -0.176 volts with a standard deviation of -0.028 volts. Therefore, the probability was less than 10 percent that corrosion cells exist in about 80 percent of the deck, and the remaining 20 percent of the area showed potentials in the questionable zone over two years prior to impregnation (impregnated June, 1985). Chloride sampling and analyses demonstrated that less than 0.005 percent of the reinforcing steel had a chance of being critically contaminated with chlorides in March of 1983. No delaminations associated with the corrosion of the reinforcing steel were discovered in March, 1983.

A section, approximately one-half of the center span, 60-feet long by 44-feet wide was impregnated using the grooving technique with a monomer (methyl methacrylate) and polymerized insitu in June, 1985 [5]. The grooves were backfilled with latex modified mortar. The typical depth of impregnation was about 3.5 inches. Drying shrinkage cracking was observed in both the impregnated and control section cores. The observed cracks were fine and generally of shallow depth (less than 0.50 inches). The frequency of cracking was not significantly different for the control and the impregnated section.

However, the cracks in the impregnated section were generally deeper with 32 percent of the cracks in the impregnated area deeper than 0.50 inch versus

Visual Inspection and Delaminations Survey

In March, 1989 a visual inspection and delamination survey showed that the deck was in excellent condition approximately 4 years after impregnation and 6 years after the post-impregnation condition survey. The only visual evidence of concrete deterioration was a few spalls along the east expansion dam in the control section that were observed in the March, 1983 survey and are related to poor construction practices. In addition, no delamination planes were detected in the impregnated nor in the control section.

Chloride Contamination Levels

In March, 1983 powdered samples for chloride analyses were taken at mean depths of 0.25, 0.75, 1.50, and 2.5 inches in the aprons, right wheel path, and between-wheel-path locations in both the to be impregnated and the control areas. Since the chloride contents of bridge decks are generally highly variable and the highest levels of contamination are generally in the wheel paths, five powdered samples were taken in the right wheel path in both the impregnated and control sections in March, 1989. Sample mean depths were 0.5, 1.0, 1.5, 2.0, 2.75, and 3.63 inches.

The difference between the right wheel path chloride contents for the sample years 1989 and 1983 for the impregnated and control would be a measure of the effectiveness of the impregnation process to exclude chloride ions, see Table 1. As illustrated the difference in the control section chloride content (1989-83) is higher for all depths. Because of a significant decrease in permeability normally associated with polymer impregnation, there should be little to no increase in chloride content in the impregnated section. Whereas, Table 1 shows that the chloride did increase in the impregnated section. However, it needs to be pointed out that the chloride samples taken in March, 1983 were some 2.25 years before the section was impregnated in June, 1985.

Table 1. Difference in Average Chloride Content in Right Wheel Path

Depth in.	Chloride Content, lb/cy Difference Between 1989-83		Difference Between Con.-Imp.	Percent Excluded
	Control	Impregnated		
0.5	3.5	1.8	1.7	48
1.0	3.5	0.1	3.5	100
1.5	1.9	0.5	1.4	73
2.5	0.5	0.4	0.1	20

A measure of the effectiveness of the impregnation process to exclude chlorides would be the percent chloride excluded (difference between the control and impregnation relative to the control section). As presented in Table 1, the impregnation process has excluded a significantly high percentage of the chlorides for all the depths except the depth of 2.5 where the exclusion is only 20 percent. Again, it needs to be pointed out that even though the percent exclusion is low the actual chloride increase is low and that the deck was exposed to chlorides 2.25 years after the sampling and before being impregnated.

In addition to the effectiveness of the impregnation process to exclude chlorides, the percent of reinforcing steel presently in critically contaminated concrete is also of interest. The March, 1989 chloride contamination levels for the impregnated and control sections are approximately equal. Assuming 1.2 lb. chloride per cubic yard of concrete as the corrosion threshold level, 0.002 percent of the reinforcing steel in the wheel paths have a chance of being in critically contaminated concrete for the given cover depth distribution (previously discussed in the background section).

Corrosion Potentials

Corrosion half cell potentials were measured with a CSE in March of 1983 and 1989. Table 2 presents the mean, standard deviation and number of readings for both the March 1983 and 1989 surveys.
As shown in Table 2, the mean corrosion potential for both the control and impregnated sections are approximately equal in both 1983 and 1989 and both have slightly increased in the six year period. For the control section in 1989, approximately 30 percent of the area surveyed is in the uncertain corrosion activity zone (-200 to -350 millivolts). Whereas, 40 percent of the impregnated section is in the uncertain corrosion activity zone. There were no readings with a greater than 90 percent probability of active corrosion (greater than -350 millivolts).

Table 2. Boalsburg Bridge Deck Half-Cell Potential Readings

Statistical Parameter	Potential (MV)			
	Impregnated Section		Control Section	
	1989	1983	1989	1983
Mean, $\bar{x}$	192	173	200	176
Standard deviation, σ	19	5	26	28
Number observations, n	24	9	28	72

Petrographic Analysis

Four inch diameter cores were drilled with a water cooled diamond bit for petrographic analysis, two in the impregnated and two in the control section. The depth of impregnation was determined to be about 3 to 4 inches which agrees with previous findings [5].

The concrete volume composition values are in the range of typical construction grade concretes. The coarse aggregate is a crushed limestone. The fine aggregate is a highly siliceous natural gravel sand. Both the fine and coarse aggregate are good quality aggregates. However, the quality of the cement paste is poor, showing considerable evidence of excessive mixing water. In addition to drying shrinkage cracks, there is excessive near-surface porosity and bleeding channels and high porosity and large irregularly-shaped voids adjacent to coarse aggregate particles. The drying shrinkage/thermal and non-specific cracks are numerous, though minor, and are equally distributed between the impregnated and the control areas.

Two of the four cores contained rebar, one in the impregnated and one in the control area. The core from the impregnated area showed no corrosion. The core in the non-impregnated showed heavy corrosion deposits from a supporting chair. The core is located immediately adjacent to a sliding plate expansion joint at the east end of the span. This is the area where spalling has occurred for some time due to poor construction practices. The corrosion of the chair was most likely pre-existing and is to be considered a special case. The inordinately deep cover of the reinforcing steel appears to have prevented the corrosion of the reinforcing steel as of March 1989, in spite of the poor quality concrete.

Corrosion Rate

In March 1990, one year after the visual, chloride, and corrosion potential surveys, 10 corrosion rate measurements were taken in both the control and impregnated areas. The corrosion rate measurements were made with a 3LP device. The device is based on the linear polarization resistance technique with changes in cathodic polarization currents measured at changes in potentials of 0, 4, 8, and 12 millivolts. The values used for the anodic and cathodic tafel slopes are 150 and 250 mV/decade. Table 3 presents the results of the corrosion rate measurements and the cover depth of the reinforcing steel at the test locations. As shown, for approximately the same cover depth, the control section was corroding at a rate of about 2.5 times the impregnated section. Also, note that there is no correlation between the rate of corrosion and the corrosion potentials taken during the corrosion rate measurements. Thus, it appears that the polymer impregnation is providing an increased level of corrosion protection for the Boalsburg deck.

Table 3. Boalsburg Bridge Deck Corrosion Rate Measurements

Location	Cover Depth (in.)	Corrosion Potentials (mV,CSE)	Corrosion Rate (mA/sq.ft.)	(mpy)
Impregnated Section				
1	2.375	236	0.94	0.49
2	2.375	244	1.63	0.79
3	2.375	241	1.57	0.77
4	2.500	298	1.79	0.87
5	2.375	232	2.01	0.98
6	2.250	216	1.41	0.68
7	2.375	224	1.55	0.75
8	2.500	199	1.11	0.54
9	2.250	239	1.28	0.62
10	2.250	211	1.00	0.49
x	2.363	234	1.43	0.70
Control Section				
1	2.375	218	2.80	1.37
2	2.375	207	4.15	2.03
3	2.375	202	2.91	1.42
4	2.625	212	3.76	1.84
5	2.375	206	2.58	1.26
6	2.375	208	3.89	1.90
7	2.375	219	5.65	2.76
8	2.250	199	2.97	1.45
9	2.750	203	3.73	1.83
10	2.750	197	4.13	2.02
x	2.46	207	3.66	1.79

Bethlehem Bridge Deck

Background

The dual-lane bridge carries Pennsylvania Route 378, the spur route linking Bethlehem to U.S. Route 22 to I-78, over Union Boulevard. In March, 1975 a test section at the south end of the bridge, 3.5 ft. by 11.5 ft., was deep impregnated from the surface using the pressure method. The bridge was 8 years old at the time of the impregnation. The wheel path areas were deeply rutted.   The chloride content at the depth of the reinforcing steel in the impregnated area exceeded the corrosion threshold level but the deck was sound, no spalled or patched areas. Details of the deep impregnation with methyl methacrylate and in situ polymerization of the field test installation are presented elsewhere [6].

A visual examination of the bridge in December, 1983, revealed some obvious differences in the performance between the deck in general and the deep impregnated test area. The visual performance difference between the test area and the remaining deck area initiated an investigation into identifying the cause of the difference in the visual corrosion performance. The investigation performed in February, 1984 consisted of a delamination survey, corrosion potential measurements, chloride content analysis, and a microscopic analysis of drilled concrete cores. The results of the February, 1984 study have been reported elsewhere [7]. A follow-up investigation performed in March, 1989 consisted of a visual inspection and delamination survey, corrosion potentials and microscopic analysis of drilled concrete cores. The performance of the deep polymer impregnated corrosion abatement test section follows.

Visual Inspection and Delaminations

A visual inspection and delamination survey of an area about 14 feet by 37 feet encompassing the impregnated test area was performed at the south end of the northbound traffic lane. Figure 1 presents the observed spalled, patched spalled and delamination zones in relationship to the impregnated area. The patched spalled area encroaching on the southeast corner of impregnated area is not the result of the corrosion of rebar within the impregnated area. The patching material used to repair a large spall in this section of the investigation area was merely parged over the sound impregnated area. Of the non-impregnated section, approximately 20 percent had deteriorated from the corrosion of the reinforcing steel. Whereas, the impregnated area remains sound, no patched spalls, spalls or delaminations within the impregnated area even though the surrounding area is severely deteriorated.

Figure 1. Bethlehem bridge deck spalled and delamination areas.

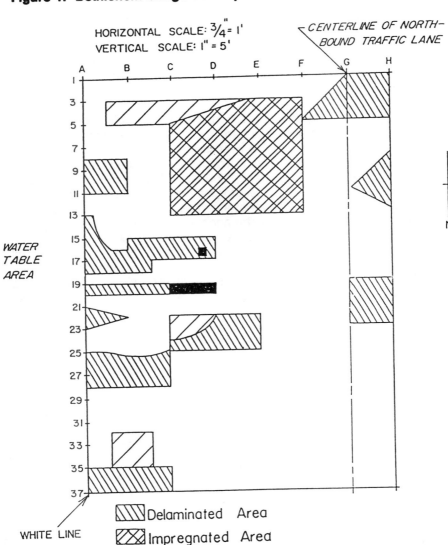

Figure 1.   Bethlehem bridge deck spalled and delamination areas.

Concrete Cover Depth and Chloride Content

A pachometer was used to determine the depth of concrete cover over the reinforcing steel. Forty measurements were taken, 20 within and 20 outside of the impregnated area. The mean depth of cover is 1.45 in. with a standard deviation of 0.40 in. and a range of 1.00 in. to 2.00 in. Thus, the depth of cover is relatively small which would account for the premature deterioration of the deck concrete.

During chloride sampling the top 0.25 in. sample was discarded and four samples were taken at depth increments of 0.50 with the fifth sample increment being 1.00 in. Chloride content samples were taken in the left wheel path at ten locations, five within and outside of the impregnated area. The mean depth of the samples was 0.50, 1.00, 1.50, 2.00 and 2.75 in. Table 4 presents the average chloride content for the wheel path areas as function of the depth for the 1984 and 1989 samplings. The chloride content at the depth of the reinforcing steel was above the corrosion threshold value in 1975 [6] and there was not a significant difference between the chloride contents of the impregnated area and the rest of the deck as a whole at the 95 percent confidence level [7]. Thus, any difference in the chloride contents between the impregnated section and the non-impregnated section should be a measure of the efficacy of impregnation to reduce the rate of chloride diffusion into concrete.

Table 4. Difference in Average Chloride Content
in Right Wheel Path for Bethlehem Bridge.

Depth in.	Chloride Content, lb/cy Difference Between 1989-84		Difference Between Con-Imp	Percent Excluded
	Control	Impregnated		
0.5	0.2	3.0	-2.8	--
1.0	-0.1	1.5	-1.6	--
1.5	2.6	1.1	1.5	58
2.0	1.9	0.7	1.2	63
2.7	3.0	1.0	2.0	67

As presented in Table 4, chloride contents in both the impregnated and control area appeared to have continued to increase during the five year time period between 1989 and 1984. The difference in chloride contents between the measurements indicate that more chloride penetrated the impregnated concrete than the control concrete for the top one inch of concrete. However for the depths of 1.50 to 2.75 inches the impregnated concrete excluded an average 1.5 lbs. chloride/cy of concrete, see Table 4. The average percent exclusion which would be a measure of the effectiveness to prevent further chloride penetration is about 60 percent for the depths of 1.50 inches and below. It is interesting to note that the chloride content at the depth of the reinforcing steel in the impregnated area remained above the threshold level of 1.2 lbs. chloride/cy for 14 years without spalling or delaminating.

Corrosion Potentials

Corrosion half cell potentials were measured with a CSE. Table 5 presents the mean, standard deviation, number of measurements, the percent more negative than 350 millivolts and the percent less negative than 200 millivolts. As shown the non-impregnated area potentials have slightly increased in the percent more negative than 350 millivolts. Thus indicating that the percent greater with a greater than 90 percent probability that reinforcing steel corrosion is occurring has increased 6 percent.

Table 5. Mean, Standard Deviation and Number
of Corrosion Half Cell Potentials

Time, Year	x Millivolts	x Millivolts	No.	More Negative Than -350 %	Less Negative Than -200 %
Impregnated Area					
1984	260	60	5	0	0
1989	350	90	18	50	0
Non-Impregnated Area					
1984	340	110	34	41	9
1989	340	110	134	47	12

Also the percent with a greater than 90 percent probability that no reinforcing is corroding has slightly increased by 3 percent. However, there is no change in the mean or the standard deviation. Thus, for all practical purposes there appears to have been little to no change in the corrosion potentials in the non-impregnated area. However, for the impregnated area the corrosion potentials have increased by a negative 90 millivolts from a negative of 260 millivolts to a negative 350 millivolts and the standard deviation has increased by a negative 30 millivolts.   Also, the percent more negative than negative 350 millivolts indicating a greater than 90 percent probability that corrosion is occurring has increased from zero to 50 percent. Thus, the potentials appear to indicate that the corrosion activity in the impregnated area has increased during the five year period from 1984 to 1989.

Microscopic Analysis

Four inch diameter cores were drilled with a water cooled diamond drill bit. In general, the cores were about five inches long. Vertical sections were cut, polished and examined with the aid of a petrographic microscope.

The coarse aggregate is a blast furnace slag and the fine aggregate a natural sand. The cement paste appeared to be of excellent quality. The depth of impregnation is approximately 3 inches which agrees with previous results [6].

The primary interest in the examination was the corrosion of the reinforcing steel since the purpose of deep impregnation is to prevent/arrest the corrosion process.   Table 6 presents the summary of the corrosion performance examinations. The microscopic examination of polished and fractured section revealed details of the corrosion state of the deck. With the exception of the Conn. core, corrosion products were observed around the reinforcing steel. Cores A and D from the non-impregnated area showed evidence of slight corrosion. All four cores from the impregnated area showed evidence of corroding reinforcing steel.   However, the corrosion products were impregnated with polymer. Thus, the deep impregnation process successfully arrested active corrosion cells in all four of the cores. Figure 2 presents a microphotograph of impregnated core J showing a vertical crack caused by the corrosion of the rebar. Figure 3 is a microphotograph of a fractured surface of core J showing the impregnated corrosion products. Thus, the microscopic investigation presents direct evidence that the deep polymer impregnation process will arrest the corrosion of steel in concrete. These results verify the previous results of deep polymer impregnation arresting the corrosion of steel in concrete [7].

Table 6. Microscopical Examination of Polished Vertical Sections from Cores from the Bethlehem Bridge Deck.

| Core No. | Impreg. Area | Cracking Observed | | | | Cover in | Corr. of Rebar |
		Drying Shrink	Micro Shrink	Subsi-dence	Rebar Corr.		
A	No	No	No	No	No	2.2	Slight
D	No	No	No	No	No	1.3	Slight
Conn	No	Few	No	No	No	1.4	No
F	Yes	Two	No	Yes	No	1.5	Y/Arrested
H	Yes	Few	No	No	No	1.4	S1/Arrested
I	Yes	(a)	No	(a)	(a)	1.2	Y/Arrested
J	Yes	No	Minor	No	Yes	1.4	Y/Arrested

(a)   Two cracks originating from rebar -- cannot discern if shrinkage, subsidence, or rebar corrosion related, but definitely pre-exist impregnation.

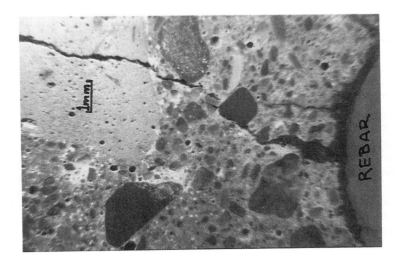

Figure 2. Core J, Bethlehem Bridge
Rebar corrosion and associated vertical crack

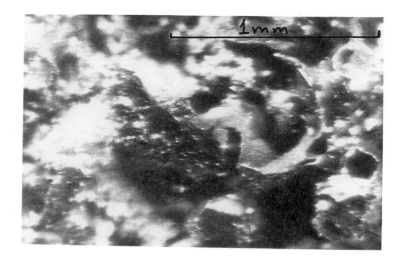

Figure 3. Core J, Bethlehem Bridge
Polymer impregnated corrosion products--fractured section.

Corrosion Rate

In March 1990, one year after the previously described investigation, 8 corrosion rate measurements were taken in both the control and impregnated areas shown in Figure 1. The measurements in the control area were taken in sound areas as were the measurements in the impregnated area. Because no delaminations were detected in the impregnated area. Table 7 presents the results of the 3LP corrosion rate measurements. As shown in Table 7, there is no relationship between the corrosion potentials and rate of corrosion. Also, the rate of corrosion in the impregnated section is significantly less, 15 times slower, than the rate of corrosion in the control section. Thus, it is reasonable to expect that the impregnated section would not be damaged by corrosion within 10 to 15 years whereas the control would expect damage in 2 to 10 years [8].

Table 7. Bethlehem Bridge Deck Corrosion Rate Measurement

Location	Impregnated Area Corrosion Potential (mV,CSE)	Corrosion Rate (mpy)	Control Area Corrosion Potential (mV,CSE)	Corrosion Rate (mpy)
1	475	0.31	354	1.38
2	422	0.17	507	3.12
3	288	0.11	398	2.17
4	280	0.07	280	1.27
5	310	0.07	325	1.49
6	312	0.05	329	1.41
7	342	0.08	403	1.98
8	440	0.15	404	2.77
x	358	0.13	375	1.95

CONCLUSION

The major finding of the investigations is that deep polymer impregnation will prevent corrosion damage of reinforced concrete bridge decks for a period of at least 25 to 30 years after corrosion initiation.

ACKNOWLEDGEMENTS

In helping to gather the recent data that formed the basis for this paper, the authors are especially indebted to Messrs. Gary Hoffman, Robert Donovan, and Joseph O'Melia of the Pennsylvania Department of Transportation and Dr. John P. Broomfield of the Strategic Highway Research Program (SHRP).

The research report here was financed by SHRP. The opinions, findings, and conclusions are those of the authors and not necessarily those of the sponsoring agencies.

REFERENCES

1.   PORTLAND CEMENT ASSOCIATION, "Durability of Concrete Bridge Decks", Report 5, Skokie, Illinois, 1969, pp 46.

2.   PORTLAND CEMENT ASSOCIATION, "Durability of Concrete Bridge Decks", Final report, Skokie, Illinois, 1970, pp 35.

412   Weyers, Cady, Henry

3.   DALLAIRE, G., "Designing Bridges That Won't Deteriorate", Civil
     Engineering, August 1973, Volume 43, Number 8, pp 43-48.

4.   AASHTO, FHWA, TRB and NCHRP, Strategic Highway Research
     Program-Research Plans", Final report, Washington, D.C., May 1986,
     pp TRA 4.1-4.60.

5.   WEYERS, R. E., and CADY, P.D. "Deep Impregnation of Concrete
     Bridge Decks", Transportation Research Record 1184, Transportation
     Research Board, 1988, pp. 41-49.

6.   MASON, J. A., CHEN,W.F., VANDERHOFF, J. W., MEHTA, H. C.,
     CADY, P. D., KLINE, D.E. and BLANKENHORN, P.R., "Use of
     Polymers in Highway Concrete," NCHRP Report 190, Transportation
     Research Board, 1978, pp 77.

7.   CADY, P. D. and WEYERS, R. E., "Field Performance of Deep
     Polymer Impregnations," Journal of Transportation Engineering,
     American Society of Civil Engineers, Vol. 113, No. 1, January 1987,
     pp. 1-15.

8.   CLEAR, K. C., "Measuring Rate of Corrosion of Steel in Field
     Concrete Structures," Transportation Research Record 1264, 1990,
     pp 12-17.

WATERPROOFING OF BRIDGE DECKS USING SPRAYED LIQUID MEMBRANE

J Batchelor

D J Helliwell

Stirling Lloyd Ltd.,

United Kingdom

ABSTRACT. Conventional waterproofing techniques rely on the use of preformed sheets based upon bituminous materials. These techniques have not been successful and maintenance costs unacceptable. The main limitations of sheet materials are poor adhesion to the deck, thermoplastic nature, problems of tailoring, failure of joints and labour intensive application.

In recent years, spray applied membranes have been developed which cure rapidly to give a tough, flexible, seamless coating. The main advantages of these systems are excellent adhesion to the substrate and surfacing, resistance to high temperature, long durability, tolerance of texture and detail and speed of application. The development of acrylic systems which can be applied down to -10°C is discussed in detail. This material has been used successfully since 1975 on over 500 bridges without a single reported failure.

Keywords: Waterproofing, Concrete, Bridge Decks, Sprayed Systems, Polymers, Acrylics.

Dr J Batchelor is the Technical Director of Stirling Lloyd Limited and was formerly Director of IPTME, Loughborough University of Technology (1985-1989). He was previously with British Rail Research from 1970-1985, where his interest in civil engineering applications of polymers was developed.

D J Helliwell has been Development Manager of Stirling Lloyd Limited since 1985. From 1969-1985 he was a Senior Research Scientist at British Rail Research, with specific responsibility for civil engineering applications of polymers.

INTRODUCTION

The deterioration of bridge decks as a result of defective waterproofing has become a major problem [1]. In concrete decks, water seeps
through hairline cracks and pores and attacks the steel reinforcement.
This process is exacerbated by chloride ions from road salts and by
carbon dioxide which reduce the alkalinity of the protective concrete
and hence encourage corrosion. As the steel rusts, it expands and
creates high tensile stresses which subsequently fracture the concrete.
Further cracking and spalling takes place due to the freeze – thaw
action of water entering these cracks [2].

Several methods of preventing this deterioration have been used,
including coating of the rebars, incorporating integral waterproofers
and the application of an external membrane. Coating of the reinforcing steel is not common in the UK, although with the current
awareness of the causes and effects of concrete deterioration it may
gain more widespread acceptance, probably as extra insurance with a
membrane system.

Integral waterproofers can seal the pores in the concrete but will not
prevent water penetrating through shrinkage cracks. Similar problems
will arise when silanes are used as water repellents.

External waterproofing membranes have been the most widely used and
most have been developed from the primitive use of pitch as a sealing
medium. Refinements have led to the manufacture of a wide variety of
modified bituminous sheets which are preformed off site. They are
formed into a membrane by bonding to the deck and jointing with a
bituminous adhesive either by hot pour method or in the form of self-
adhesive sheets.

The above technique can give variable results and is very dependent on
the skill of the operator. The multitude of joints between the sheets
is a potential source of failure and achieving a good bond to the deck
can be a problem with uneven surfaces. Other potential application
problems occur with geometrical changes in plane and the increase in
stiffness of the bituminous sheets in cold weather making them
difficult to handle.

In the late 1960s British Rail Civil Engineers identified defective
waterproofing as a major cause of the degradation of railway structures.
This group laid down stringent requirements which a new waterproofing
system must achieve. The overall objective was to produce a continuous
seamless coating which could accommodate any geometric variation in the
deck. The key criteria specified were:

1.   Long and effective life.
2.   Easy and quick to apply.
3.   Applicable all the year round.
4.   Capable of carrying load after 1 hour.
5.   Tough enough to resist random site damage.
6.   Bridge shrinkage cracks in concrete decks over a wide temperature

range.
7. High bond strength to deck.
8. Easily repairable.

DEVELOPMENT

British Rail Research were asked to develop a new system based on the Civil Engineers' requirements. The research team examined a number of potential polymer options in the laboratory and it was soon apparent that only liquid applied systems would meet the engineers' requirements. It would be a completely seamless membrane with intimate adhesion to the substrate. The experimental programme considered the following properties as being particularly significant:

Physical properties of membrane
Adhesion of membrane to substrate
Crack bridging capability on concrete decks
Application properties

They focused on two liquid systems. One was based on methyl methacrylate resins and the other on polyurethane. Other polymers in both liquid and sheet form had been examined but were rejected.

Subsequent site trials over a number of years showed that the methacrylate system met the criteria and was more "user friendly" under site conditions. This system has been used extensively in British Rail since 1974, including all the over bridges on the Selby diversion to the East Coast Main Line [3].

A further development and testing programme followed, which was run in conjunction with the UK Department of Transport. A primary aim of the programme was to modify the membrane to accommodate the particular requirements for road bridge protection. This included chemical modification, enabling the membrane to withstand elevated surfacing temperatures. The programme was a success and the liquid applied system (Eliminator) became the first sprayable membrane to be approved by the Department of Transport.

The acrylic system developed is a two-part solventless liquid material which can be sprayed onto the deck to give a tough, flexible, seamless membrane which cures quickly to accept traffic after 1 hour at 20°C. Although the acrylic membrane is the key element in the waterproofing system, there are other components which are essential. Thus a range of primers have been developed to ensure that we achieve a good bond between the deck and membrane. Similarly a compatible tack coat is available which promotes adhesion between the membrane and surfacing.

EXPERIMENTAL

Physical Properties

The physical properties of the cured acrylic membrane are summarised in Table 1. It is clear that the membrane has excellent properties over a range of temperatures and will satisfy the requirements for use in this application.

TABLE 1   Physical Properties of Acrylic Membrane

	Test Temp (°C)	Tensile Strength (N/mm^2) BS903 pt A2	Elongation at break (%) BS903 pt A2	Tear Strength (N/mm) BS903 pt A3	Wt Change on Ageing (%)
Before	20	7.5	140	25	–
Ageing	–10	15.0	100		
After heat ageing at 70°C for 28 days	20	7.5	140	25	–0.1
	–10	15.0	100		
After ageing in water 60°C for 28 days	20	7.0	150	24	+0.3
	–10	14.0	100		

Resistance to Substrate Cracking

The membrane must be able to withstand hairline cracks in a concrete deck. It is measured by the crack bridging technique [4], which is now part of the Department of Transport specification for bridge deck waterproofing systems. Typical results for the acrylic membrane are given in Table 2. The material is sufficiently flexible at –20°C to bridge a crack up to 0.5mm wide.

TABLE 2   Results of Crack Bridging Tests

Temperature (°C)	Crack bridging capability (mm)
23	⟩2
0	⟩2
−20	0.5

Adhesion to the Substrate

In early work on liquid systems [5], a peel test was used to assess the adhesion of the membrane to the substrate. However, this technique is not suitable for site measurements and a portable method using an Elcometer Adhesion Tester has been developed. Typical values obtained from site measurements are given in Table 3.

TABLE 3   Site Adhesion Test Results

Site	Adhesion Strength (N/mm²)	Mode of Failure
Tees Viaduct at location (52–51N)	1.2	Cohesive failure of concrete
Severn Bridge concrete deck at location B304	1.0	Cohesive failure of concrete
Severn Bridge steel deck at location B309	⟩6	No failure

The bond of the acrylic membrane to concrete is invariably limited by the cohesive strength of the concrete and is over ten times the bond strength of any sheet system. The best figure achieved for sheets is 0.1N/mm². TRRL in the UK has calculated that a minimum figure of 0.8N/mm² is necessary if the membrane is to resist the shear forces imposed by the surfacing roller. The acrylic membrane easily satisfies this requirement.

Adhesion to Surfacing

A tack coat must be applied to the membrane to ensure a good bond is obtained to the surfacing. It is a hot melt adhesive based on acrylic resins. When the hot surfacing is laid, the tack coat melts and, as it cools, membrane and surface are intimately bonded. The adhesion strength of the bond is determined in shear and tension, using methods illustrated in Figures 1 and 2. Typical results are given in Table 4. Excellent adhesion is achieved between the membrane and surfacing in both shear and tension.

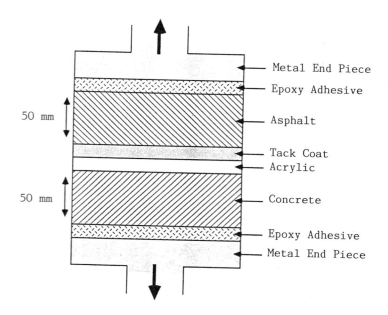

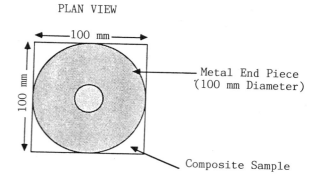

PLAN VIEW

Figure 1: Tensile bond strength test

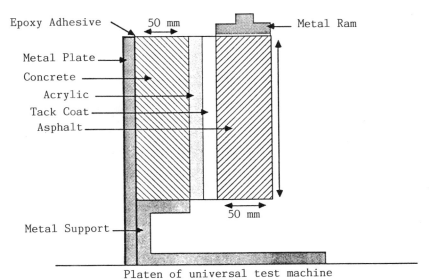

Figure 2: Shear bond strength test

TABLE 4   Adhesion of tack coat to asphaltic concrete

Sample	Shear Strength (N/mm²)	Tensile Strength (N/mm²)
DF9001/2001	1.34	5.60 (asphalt failure)
DF9001/2002	2.10	4.10 (asphalt failure)

Note: Bitumen content of asphalt = 5.5%

Thermal Analysis

It is essential that the membrane is resistant to the hot surfacing
even when mastic asphalt at 240°C is applied. Thermogravimetric
analysis indicated that there is no change in the membrane up to this
temperature. This has been confirmed in experiments carried out at
TRRL where the acrylic membrane was not punctured by the aggregate
during surfacing.

Viscosity

Acrylic resins have a low viscosity and are not affected significantly by temperature. In contrast, polyurethanes tend to have much higher viscosities which are more dependent on temperature.

Cure

The chemistry of the acrylic and polyurethane systems are quite different which strongly influences their handling on site. The cure mechanism of acrylic resins is not too critical to the proportions of the two components. The polymer is formed from monomer in the presence of catalyst and accelerator. In the early stages of the reaction, the viscosity remains approximately constant before it suddenly increases and cures (Figure 3). The length of the induction period can be controlled by adjusting the formulation.

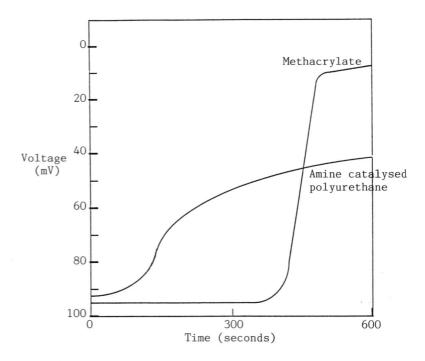

Figure 3:   Typical Vibrating Needle Curemeter traces
           (low voltage equates to high viscosity)

Conversely, polyurethane cures by a process which produces a steady increase in viscosity (Figure 3). It is also essential that the proportioning of the two components is critical and this factor must be reflected in the design of spray equipment.

The viscosity and curing characteristics of acrylics are a major advantage in site applications. The acrylic is easier to handle in the spray machine and, if required, the equipment can be remote from the job to allow up to 60m between mixer and gun.

APPLICATION

Substrate Preparation

This operation is one of the most important elements of any coating or waterproofing operation. The surface must be clean, dry and dust free before commencing any other operation. It is necessary to remove laitence, etc., by shot blasting or scabbling.

Priming

It is necessary to apply primer to the deck to improve adhesion and seal the surface to prevent pinholing. A number of primers are available depending on the substrate and weather conditions.

Membrane Application

Plural component spray equipment which meters and mixes the two components in the correct proportions has been developed. We use a system based on Graco pumps. Because of the very short pot lives of modern polyurethane systems, the two components are fed directly to the spray gun where mixing takes place.

The membrane is applied after the primer has dried or cured. For DTp applications with a U4 finish, 2 coats of nominal wet film thickness of 1.5mm are applied. The coats are colour coded to aid visual inspection and the thickness is checked by the operatives using wet film thickness gauges.

Due to the speed of cure, the second coat can be applied about 20 minutes after the first coat, depending on ambient temperature. The material is thixotropic to ensure that we achieve a minimum coverage of 2mm on the peaks of the substrate.

For concrete decks which do not have a U4 finish, the coverage rate will increase from a nominal 3.4 kg/m^2. This can be estimated by the 'Sand Patch Method'[6] which measures the depth of the undulations in the deck. For steel decks, it is usual to apply one coat of nominal thickness 2mm. Polyurethanes are always applied in one coat, partly due to the difficulty of achieving a good intercoat adhesion.

Tack Coat

The tack is applied by brush, roller or spray after the membrane has
cured (about 30 minutes at 20°C). The surfacing can be applied
directly to the tack coat after about one hour depending on ambient
temperature.

ADVANTAGES OF SPRAYED MEMBRANES

(i)   An excellent bond to the substrate is achieved and water is not
        trapped between the membrane and the deck.

(ii)   Resistant to elevated temperatures.

(iii)   They have excellent ageing characteristics.

(iv)   Good adhesion to surfacing.

(v)   Resistant to all the usual on-site abuses.

(vi)   They can cope with varying surface textures, but are
        thixotropic enough not to run off peaks. The seamless system
        has no vulnerable joints and vertical and horizontal details
        are protected by a continuous membrane.

(vii)   Liquid applied systems are more expensive on a first cost basis
        than traditional sheet materials. However, they are cost
        effective if we take into account the total costs of
        application and maintenance throughout the life of a bridge as
        a result of:

        -   speed and ease of application reduces possession times and
            keeps disruption to a minimum.

        -   ease of effective jointing and repair of accidental damage
            reduces time delays and cost of this work.

        -   membrane can accept traffic quickly after application which
            eases traffic management problems and facilitates contract
            progression.

In addition to the general advantages of liquid systems there are a
number of additional advantages specific to acrylic membranes.

(a)   Sprayed membranes can be applied quickly, even on abnormal
        shapes where tailoring of sheet materials would be time-
        consuming. As an example, the total acrylic system was applied
        to 2100m^2 of the Tamar Bridge (Cornwall : UK) in under 7 hours.
        Normal application rates are between 1000 and 1500m^2 per day.

(b)   The acrylic system can be applied all year round from -10°C to
        +50°C and under high humidity. Only actual rain or snowfall
        will prevent spraying. It can be applied in a wider range of
        climatic conditions than other systems.

(c)   Repair of accidental damage and ease of effective jointing
        reduces time delays and cost of this work.

(d)   The materials are not waterproofing membranes until they have
been applied and bonded to the deck. Suitable quality
assurance procedures have been established for the acrylic
system. Adhesion tests and wet film thickness are checked
regularly. In addition, film samples are produced on site for
subsequent physical testing. Thus quality assurance extends
through to the finished product.

These types of tests are not carried out during the application
of polyurethanes or sheet systems. This is due to the
difficulty of repairing any adhesion test holes.

CONCLUSIONS

The sprayed acrylic system has been used successfully on over 500
bridges without any single reported failure. The advantages of liquid
applied systems have been discussed in detail and it is clear that they
will be the bridge deck waterproofing of the future. They are being
used increasingly throughout the world as engineers appreciate the
longer lasting effective protection offered and the resultant large
savings in long term maintenance costs.

REFERENCES

1   Technical Report Number 11, 18th World Road Congress, Brussels,
Belgium, September 1987, pp32.

2   DALLAIRE G. Designing bridge decks that won't deteriorate.
Civil Engineering, ASCE, Volume 43, Number 8, 1973, pp43–48.

3   BASTIN R D. Selby Diversion of the East Coast Main Line,
Part 3, Bridges, Proc. Inst. Civ. Engrs., Part 1, Volume 74,
1983, pp719–747.

4   MACDONALD M D. Waterproofing concrete bridge decks : materials
and methods. TRRL Laboratory Report 636, 1974, pp14.

5   BATCHELOR J, ROBINSON M and LAMBROPOULOS V L. Adhesion of
Polyurethane Rubbers to Concrete. Adhesion 1, edited by
K W Allen. Applied Science, London, 1977, pp53–62.

6   Der Bundesminister für Verkehr, ZTV–BEL–B87. Part 3, 1987,
Dichtungschicht aus Flüssigkunstoff, Dortmund, pp23.

REPAIR OF SALT-DAMAGED CONCRETE STRUCTURES

T Kamijoh

Sho-bond Construction Co. Ltd.,

Japan

ABSTRACT. By being subjected to various deteriorative actions caused by the environmental conditions under which they have been built and use conditions, concrete structures develop all kinds of deteriorative phenomena. Of the various deteriorative actions, corrosion of the reinforcing bars caused by chlorides is discussed in this report, taking up a representative case of repair conducted for a concrete structure suffering from salt injury. The structure which was the object of repair consisted of a PC bridge built over the sea. Roughly 17 years had passed since it was completed.
Prior to repair, a detailed survey was conducted to obtain an overall understanding of the damaged state. The survey consisted of an appearance survey on deterioration such as cracking and exfoliation of concrete, and an internal deterioration survey including an examination of the chloride content in concrete and the state of rebar corrosion.
As for the repair method, the coating method which boasts results which make it the most popularly used repair method in Japan, was adopted. This method combines rust—inhibiting treatment with sectional repair and coating of the concrete surface. Execution was completed in roughly two months, during which time, strict execution control was effected based on the control items set up for each step of work.

Keywords: Concrete, Chlorides, Reinforcing bars, Corrosion, Deterioration, Repair, Deterioration diagnosis, Sectional reconstruction, Coating, Painting, Rust inhibition.

Tatsuyuki KAMIJO is Chief of the Eastern Japan Technical Center, Sho-Bond Construction Co., Ltd., Japan. He majored in polymer chemistry, but is currently engaged in research work concerning the deterioration of concrete structures and the setting up of countermeasures. He is also applying himself to consultant activities.

INTRODUCTION

Of the repairs carried out for concrete structures suffering damage
due to the corrosion of reinforcing bars, a representative type of
repair is that executed for concrete deterioration originating from
neutralization. Repair of another type of deterioration that has, in
fact, become a social issue in recent years because of leading to
premature concrete deterioration, concerns that caused by salt injury.

Rebar corrosion caused by neutralization, in most cases, is found to
occur in localized defective portions of the concrete as in such
places as where the concrete cover is not sufficiently thick, or in
honeycomb areas. For this reason, repairs are ordinarily localized.
By contrast, in salt injury repairs,damage is frequently found to
cover a wide area since the speed at which deterioration advances, as
compared with that of neutralization, is significantly faster. In
addition, locations which, at the present stage, may appear to be
quite sound can, in the span of only several years, be found damaged.
Salt injury repairs, therefore, essentially turn out to be large-scale
repairs covering a wide area. Therefore, the quality of repair
execution has a substantial influence on the effects of repair,
thereby contributing in no small measure to determining the durability
of the structure.

Taking up a model case of repair carried out for a bridge, this paper
introduces repair execution in outline, while pointing out matters
that call for special care and attention.

OUTLINE OF REPAIRED STRUCTURE

Figure 1: Locational environment of PC bridge

The structure in question consists of a PC bridge (Figure 1) built over the sea in the offing of Kashima, Ibaragi Prefecture. Roughly 17 years have passed since the bridge was completed.
Figure 2 shows the schematic configuration of the structure.

Overall view

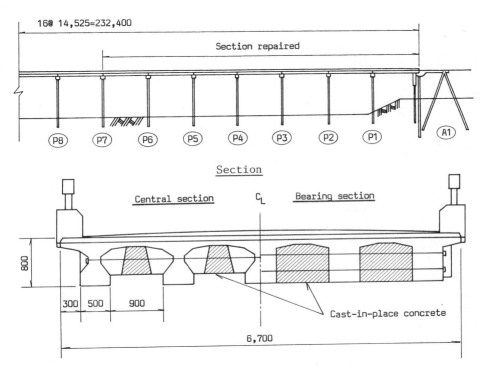

Figure 2: Configuration of repaired structure

REPAIR PROCEDURES

The repair procedures taken for the present structure from inspection up to completion of execution are as shown in Figure 3.
Generally, orders are placed separately for survey works (including repair design) and for repair work. however, for detailed surveys, where the erection of scaffoldings will be required, survey works are frequently incorporated in the repairs work as "Preliminary survey for repair" as shown in Figure 3. In such cases, since repair design and budgeting are performed based on the results obtained from the primary survey (appearance inspection), it will be necessary to review the repair design again after completing the detailed survey.

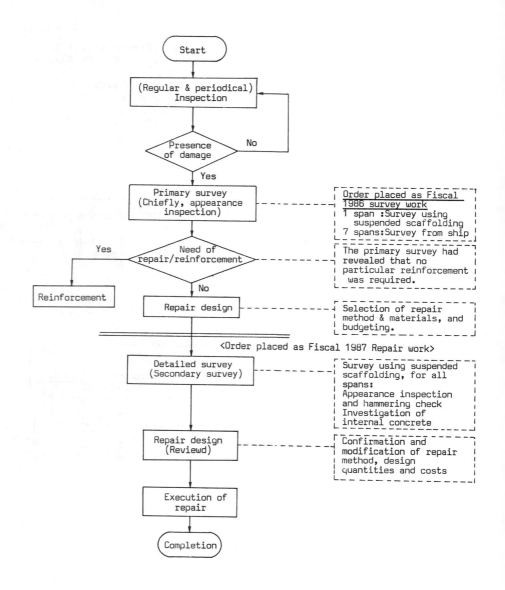

Figure 3: Repair procedures

DETAILED SURVEY

Survey Method

Table 1 shows the method adopted for conducting the detailed survey.

Table 1: Detailed survey method

Item	Objective and details
(1) Damaged quantity survey	With scaffoldings erected, a detailed survey is conducted to obtain a correct and full understanding of the range and quantities of damage.
(2) Neutralization depth measurement	The depth of concrete neutralization is measured, and its relevancy to bar corrosion is studied.
(3) Schmidt hammer test	The drop in the concrete strength caused by deterioration is examined.
(4) Chloride content survey	To obtain a full understanding of the ingress of chlorides, the chloride concentration in the concrete is measured for representative areas, and its relevancy to bar corrosion is studied.
(5) Investigation of rebar corrosion	The corroded state of the rebars inside the concrete is examined both for the damaged portions and sound portions as judged from the appearance inspection.
(6) Measurement of cover thickness	to obtain a full understanding of the chloride quantity at the locations of the reinforcing bars for the overall area, the thickness of the cover is measured at various locations, and its relevancy to rebar corrosion is studied.
(7) PC cable survey	The state of PC cable corrosion is investigated to confirm that no problem exists in terms of strength.

Results of Survey

Investigation of damaged quantities Between the findings obtained from the primary survey (conducted in 1986, using suspended scaffoldings for Span No. 1 and from a boat for other spans) and those from the detailed survey (conducted in 1987, using suspended scaffoldings for all spans), such variances as shown in figure 4 and Table 2 were found to occur in terms of the damaged range and quantities.

These results are believed to suggest that (a) in the primary survey which centers on appearance inspection, it is difficult to grasp the damaged range and quantities accurately, and (b) when a long interval

exists between the primary survey and detailed survey, it will be
necessary to take into account the increase in the damaged quantities
caused by the progress of deterioration.

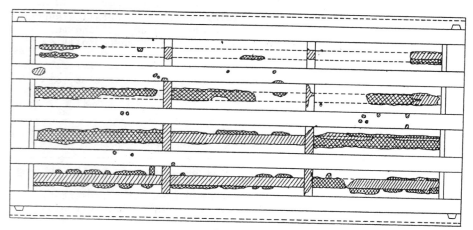

Legend: ▨:Damaged section found by primary survey

▨:Damaged section found by detailed survey

Figure 4: Development drawings of damage found by primary
survey and detailed survey (Example)

Table 2: Comparison of damaged quantities found by primary
survey and detailed survey

Span	Classification of deteriorated state	Unit	Damaged quantities			Remark
			Primary survey(A)	Detailed survey(B)	Ratio B/A	
1	Loosening, exfoliation	m²	20.4	27.2	1.3	A:1986 Suspended scaffolding
	Cracks	m	8.3	11.3	1.4	B:1987
	Rusty liquid, rebars exposed	places	91	179	2.0	Suspended scaffolding
2-8	Loosening, exfoliation	m²	45.8	85.1	1.9	A:1986 Visual inspection from boat
	Cracks	m	7.5	10.6	1.4	B:1987 Suspended scafolding
	Rusty liquid, rebars exposed	places	12	1254	105	

Concrete deterioration survey The depth of concrete neutralization
was very little, not extending as far as where the reinforcing bars
are located. The concrete strength was found to be quite sufficient
as well.
Meanwhile, a notable increase was found in the chloride content in
concrete. As may be seen from Figure 5, rebar corrosion occurred at
locations where the chloride concentration exceeded 0.04wt% (soluble
chlorine quantity). Furthermore, the occurrence of such corrosion
showed a tendency of concentrating in areas where the cover thickness
was less than 4cm as shown in Figure 6.
It was assumed that in the case of the present structure, the
chlorides contained in seawater splashes had permeated into the
concrete, increasing the chloride concentration level at the surface
course section. And this, in areas where the thickness of the cover
was relatively small, had eventual led to corroding the reinforcing
bars located in those areas.
Meanwhile, as a result of the investigating the PC cables, it was
confirmed that the sheath tubes as well as the PC wires remained
sound, and that no problem existed in terms of yield strength.

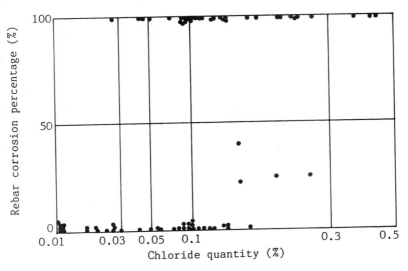

Figure 5: Relationship between chloride quantity
(soluble chlorine) and rebar corrosion rate

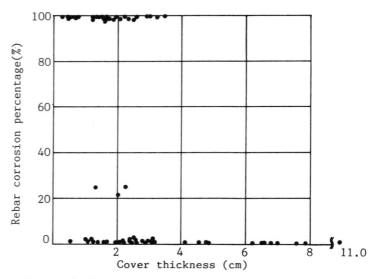

Figure 6: Relationship between cover thickness and
rebar corrosion percentage

REPAIR DESIGN (REVIEW)

Repair Method

As a result of reviewing the repair design with the results of the
detailed survey taken into account, it was decided that the coating
method (Figure 7) and repair method (Table 3) as originally planned
will be adopted.

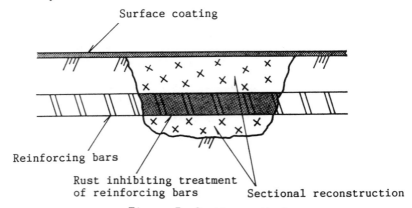

Figure 7: Coating method

Table 3: Repair method

Repair method	Objective	Countermeasure
(1)Rust-inhibiting rebar treatment	To inhibit further corrosion of rusted rebars	After exposing the rebars and removing all rust by blasting, rust-inhibiting paint is applied.
(2)Sectional reconstruction	To reconstruct the sectional configuration of areas where portions of concrete are missing, by performing, chipping, etc.	Sectional reconstruction (patching) material is filled in.
(3)Coating of concrete surface	To inhibit the progress of rebar corrosion by reducing the ingress of chlorides, oxygen and moisture	The concrete surface is coated using a material having interceptibility.
(4)Replacement and additional supply of rebars	To compensate for the sectional reduction of rusted rebars	Rebars showing significant corrosion are replaced, or additional rebars are supplied.

Design Quantities

Presuming that the damaged range and quantities obtained from the primary survey will lack in accuracy, design quantities provided with a certain degree of allowance was decided from the very beginning. However, since the detailed survey revealed values even greater than those provided with the above allowance, the design quantities were modified by a roughly 10% increase.

EXECUTION

Execution Procedures

In salt-injury repair works, the quality of execution is closely associated with the effects of repair. In view of this, in the present project, workers' education was carried out prior to execution to have them recognize (1) the objective and significance of the repair works, and (2) the need to perform carefully and proper execution at all stages of work. Of the eight spans, one was repaired ahead of the others to serve as test execution. Execution control and quality control were carried out with particular carefulness. Figure 8 shows the execution procedures taken for the present project.

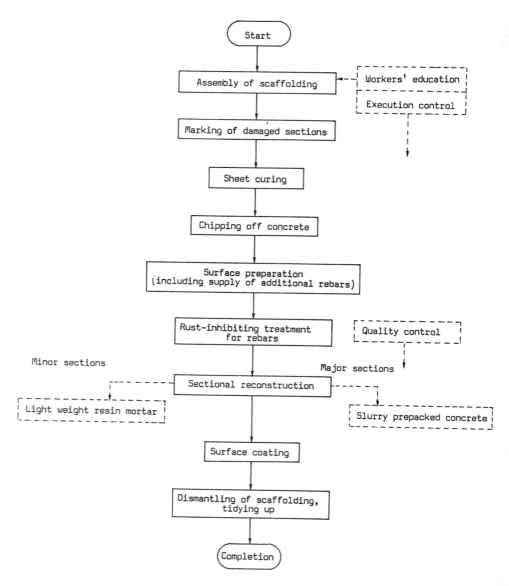

Figure 8 Execution procedures

Erection of Scaffolding, Sheet Curing, and Marking

After erecting the scaffolding, sheet curing was performed as a measure to protect the concrete surface from seawater splashes, dust and seawater pollution. Then, abnormal locations (showing loosening,

exfoliation, rusty liquid, cracking, etc.) on the concrete surface
were marked off with chalk.

Chipping off Concrete

Using such means as a pick and chisel, concrete of the marked off
locations were chipped off until the corroded rebars were exposed. To
confirm the positions of the rebars, the design drawings and a metal
sensor were used.
Further, to enable blast treatment of the rebars and application of
rust-inhibiting paint, the concrete in the periphery of corroded
rebars were chipped off to the extent as shown in Figure 9.

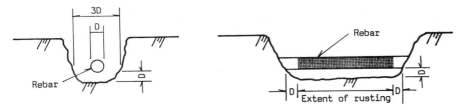

Figure 9: Extent to which concrete is chipped off
when rebars have corroded

Surface Preparation and Supply of Additional Rebars

Rust on the reinforcing bars and depositions 8chlorides, oil and
grease, dust, etc.) on the concrete surface were removed by means of
sandblasting and a blow brush. The extent of rust removal was set to
that where the texture of the reinforcing bars becomes exposed. If
the chloride quantity deposited on the concrete surface is large,
there is risk of the adhesive strength dropping. Therefore,
measurements were taken on the deposited chloride quantity on
completion of surface preparation to confirm that the values did not
exceed the control value ($1000mg/m^2$).
For those locations showing significant sectional losses due to rebar
corrosion, stiffening bars were added to supplement the loss.

Applications of Rust-inhibiting Paint to Reinforcing Bars

Following rust removal by blasting, epoxy resin paint was applied to
the surfaces of exposed reinforcing bars (Figure 10). In
consideration of the marine environment and the fact that reinforcing
bars regain their active state once they have undergone blast
treatment making them more susceptible to rusting, rust-inhibiting
painting was carried out as soonest as possible (that is, within the
same day).

Figure 10: Rust-inhibiting painting of exposed reinforcing bars

Sectional Reconstruction

Locations where portions of concrete were missing (sections exfoliated
due to deteriorations, and those from which concrete has been chipped
off) were restored using sectional reconstruction (patching) material.
There are various kinds of sectional reconstruction (patching)
materials. As shown in Table 4, two kinds of materials were used
separately in this case so as to match the volume (extent and depth)
of the section from which concrete was missing. Tables 5 and 6 show
the general physical properties of the sectional reconstruction
materials used.

Table 4: Selective use of sectional reconstruction
materials volume of missing section

Volume of missing section		
Classification	Range & depth (approximate.)	Sectional reconstruction materials
(1) Minor missing section	Less than $0.5m^2$ or less than 2cm	Light-weight epoxy resin mortar
(2) Major missing section	Other than above	Polymer cement slurry prepacked concrete

Table 5: General physical properties of light-weight epoxy resin mortar

Test item	Testing method	Unit	Specified value
Specific gravity	JIS K 7112	–	0.70±0.10
Flexural strength	JIS K 7203	kg/cm^2	100, or more
Compressive yield strength	JIS K 7208	kg/cm^2	280, or more
Adhesive strength	BRI system*	kg/cm^2	20, or more

Testing conditions: 20°C, 7 days
* Building Research Institute System

Table 6: General physical properties of slurry prepacked concrete

	Polymer cement-based prepacked concrete			
	Unit	Test value	Specified value	Testing method
Specific gravity	–	2.30	2.30±0.1	Underwater displacement method
Compressive strength	kg/cm^2	326	250, or more	JIS A 1108
Compressive Young's modulus	kg/cm^2	2.30x10^5	–	Strain gauge method
Flexural strength	kg/cm^2	63	–	JIS A 1106
Flexural adhesive strength	kg/cm^2	*46	–	JIS A 1106

*Cement concrete failure 100%

Surface Coating

The surface coating process adopted for the present repair project, in which a rubber-based resin paint was used is shown in Table 7. The key points in execution control effected for each process are described below.

Table 7: Painting process

Process	Material used	Standard application quantity (kg/cm^2)	Means of application	Coating interval (Approx. 20°C)
Priming	Epoxy resin primer	0.15	Roller, brush	2 hrs. – 1 day
Puttying	Epoxy resin putty	0.40	Putty, pallet	1 day – 5 days
Intermediate coat (1)	Thick-film	0.75	Rubber trowel, Rubber pallet	1 day – 5 days
Finishing coat (1)	Polyurethane resin paint	0.12	Roller, brush	1 day – 2 days
Finishing coat (2)	Polyurethane resin paint	0.12	Roller, brush	–

Pretreatment for painting Pretreatment for painting was performed to ensure satisfactory application of the intermediate coat.

Puttying, in particular, serves to prevent such film defects as pinholes from occurring in the intermediate coat by filling in all honeycombs on the concrete surface as well as leveling all unlevels portions.

Prior to executing pretreatment, measurements of the moisture quantity on the concrete surface were taken. Also, to confirm that the deposited chloride quantity was kept within the range of the control values (Moisture quantity, No more than 10%, Deposited chloride quantity* No more than 100 mg/m^2) measurements were taken before starting each painting process.

Intermediate coat and finishing coat The intermediate coating process is a process by which a film is formed to intercept the rebar corroding factors (Chlorides, oxygen and moisture) which would otherwise permeate into the concrete. Accordingly, since it is essential that the specified film thickness is secured, film thickness control was effected by taking measurements at 2 locations per span: a film thickness gauge was employed.

The finishing coat is applied for the purpose of improving the weather resistance and durability of the intermediate film (figure 11). Due to the thin film thickness of a single film applied, defects are liable to generate. For this reason, film control was effected for the finishing coat by applying it in two coats.

Furthermore, during painting work, temperature and humidity measurements were taken twice a day to ensure that application was carried out under humidity conditions of no less than 85%RH.

Figure 11: Completion of execution

CONCLUSION

Recently, with interest toward the maintenance and repair of concrete structures mounting higher than ever before, the development of materials, methods and execution technologies that would assure effective and durable repair is anxiously awaited. It is the earnest wish of the author that vigorous studies on such repair technologies will be undertaken in the future by the public agencies and offices concerned.

PROTECTION AGAINST CORROSION OF REINFORCING BARS USING HIGH-STRENGTH CONCRETE

H Seki

Waseda University,

K Fujii

Japan Press Concrete Co. Ltd.,

Japan

ABSTRACT. The paper mainly discussed the relationship between crack width and corrosion of bars. High-strength concrete was examined so as to prevent corrosion of bars. Accelerated test of duration from 1 week to 2 years was adopted. Main test results were as follows : 1) corroded area showed the quantitative tendency with corroded length, 2) high-strength member had clear relationship between crack and corrosion compared to normal strength concrete members, 3) there was no clear tendency between crack width and corrosion, but corrosion tended to become deeper and larger over the range of crack width 0.2mm ~ 0.25mm, and 4) high-strength concrete contained less chloride compared to normal strength concrete.

Keywords: Corrosion, Crack width, High-strength concrete, Press concrete, Vaccum concrete, Chloride ion, Permeability

Dr. Hiroshi Seki is Professor of Concrete Technology at the Department of Civil Engineering, Waseda University, Tokyo, Japan. He has been active in research work and in JSCE (Japan Society of Civil Engineers), especially durability of concrete. His research works include corrosion of bars, protection of concrete, estimation of service life of structures, under-water concreting, new materials for concrete and rigidity of concrete members.

Mr. Kentaro Fujii is Chief of Technical Committee at Japan Association of Concrete of Sheet Pile, and Head of Research Cent er and Director at Japan Press Concrete Comp. Ltd, Tokyo, Japan. He has been actively engaged in research works and technical developments regarding high-strength concrete, slag concrete and fiber concrete.

INTRODUCTION

Reinforced concrete has generally high durability. Under the conditions that structures are exposed to severe sea conditions, reinforcing bars tend to corrode due to immersion of chloride ion into concrete. One effective countermeasure toward deterioration is to use low-permeable concrete.

The paper discussed about the effect of high-strength concrete toward durability of concrete members. Concrete was produced with normal method, high-pressed technique (press concrete) and vacuum method (vacuum concrete). Regarding corrosion of bars, accelerated testing method was adopted.

The relation between corrosion of bars and crack width was mainly examined, because corrosion might be more severe at the reinforced concrete members. Furthermore, evaluation of corrosion data and penetration of chloride ion into concrete were discussed.

TESTING METHODS

Outline of Test

Test was divided into three series, that is, Series I to Series III. Series I was carried out so as to make clear the relation between accelerated time (1 week to 2 months) and corrosion. Series II aimed at obtaining the following relationship, such as the relation between crack width and corrosion, and the relation between production method of concrete (high-strength concrete, normal concrete) and corrosion. Accelerated time was set for one month.

Main purpose of Series III was to obtain the data regarding the decrease of tensile strength of corroded bars, immersion of chloride ion into concrete and so forth. Accelerated time ranged from 2 weeks to 2 years.

Fabrication of Specimens

Normal portland cement was used. Aggregates were river sand and river gravel, and maximum size of gravel was 20 mm. Specific gravity, absorption and F.M. for sand and gravel were 2.16~2.63 and 2.64~2.65, 0.96~1.21 and 0.55~0.66, and 2.80~3.15 and 6.67~6.79, respectively. Main reinforcements had diameter of 5.5 mm, and mechanical properties were 1,180~1,250 MPa for yield point, 1,490~1,590 MPa for tensile strength, and 9~10% for elongation, respectively.

Fig. 1 indicates the shape and dimension of specimens for Series
II and III. Most of the specimens had the cross-section of
70×150 mm, and cover for main bars (φ5.5 mm) was 7mm. Mix
proportion of concrete was shown in Tab. 1.

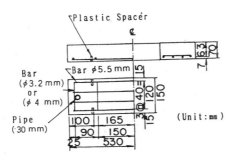

Fig. 1   Dimensions of Specimens (Series II and III)

Tab. 1   Mix Proportions of Concrete

Series	Kind of Concrete	Slump (cm)	V. B. (sec)	W/C (%)	s/a (%)	C (kg/m³)	S (kg/m³)
I	High-strength (Press)	1～4	10	36	40	420	736
II	High-strength (Press, Vaccum)	2	10	36	40	420	735
	Normal	8	—	55	43	314	808
III	High-strength (Press)	2. 5	10	37	39	420	713
	Normal	8	—	55	43	318	811

High-strength concrete was produced with high-pressed technique
(press concrete) and vaccum method (vaccum concrete). After
casting concrete into mold, specimens were cured in steam
condition (100℃, 3 hrs). Tab. 3 shows test results of
compression strength.

Tab. 2 Compression Strength of Concrete

Series	Kind of Concrete	Standard Curing (28 days)	Steam Curing Indoor (28 days)	Steam Curing → Indoor → Accelerated Test	
I	High-strength (Press)	60	70	83	(195~227 days)
II	High-strength (Press)	59	79	78	(87~122 days)
	High-strength (Vaccum)	59	64	60	(91~126 days)
	Normal	38	29	36	(89~124 days)
III	High-strength (Press)	53	74	105	(83~89 days)
	Normal	34	29	37	(84~88 days)

(N/mm²)

Testing Methods

Tests were carried out mainly according to the following process, ① load was applied, ② specimens were exposed to accelerating test in the salt-water spray apparatus, ③ concrete specimens were broken and corrosion of bars was measured, and ④ tensile test of corroded bars was done.

Loading system applied was shown in Fig. 2. Accelerated test[1].[2] adopted in this paper was the following conditions,
Salt water : 3% of salt (by weight)
No. of cycles : 2 cycles/day, Temperature : 60℃ (const.)
Salt water spray : 3 hrs./cycle, Drying : 40% R.H.

Corrosion of bars was measured with the following items, that is, corroded area, corroded length, depth of corrosion and loss of weight due to corrosion. Corroded parts were removed by way of solution of 2 anmoniun citrate. Chloride content in concrete was measured with the tilting method of silver nitrate[3]

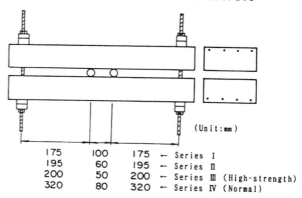

(Unit:mm)

175	100	175	← Series I
195	60	195	← Series II
200	50	200	← Series III (High-strength)
320	80	320	← Series IV (Normal)

Fig. 2 Loading Method

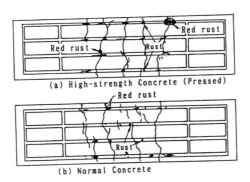

(a) High-strength Concrete (Pressed)

(b) Normal Concrete

Fig.3 Crack Pattern and Corrosion of Bars

EVALUATION OF CORROSION

Corrosion of Bars Embedded in Concrete

Fig.3 shows the flexural crack pattern at the stage of applied load and the visual state of bar corrosion just after specimens being broken.

It was likely that corrosion occurred mainly at the cracked section of specimens, and concrete surface of some specimens were stained with red rust. Corrosion pattern depended on quality of concrete. Though corrosion generally extended along longitudinal direction of bars for normal concrete, corrosion was relatively limited near cracked section for high-strength concrete.

Reciprocal Relation Regarding Amount of Corrosion.

Among corroded area, corroded length, depth of corrosion and loss of weight due to corrosion, corroded area had high correlation with corroded length. Fig.4 shows this tendency. The coefficient of correlation between loss of weight due to corrosion and tensile strength was approximately 0.6. As indicated in Fig.5, tensile strength qualitatively decreased according to increase of loss of weight. Futhermore, high-strength concrete had generally less corrosion loss of weight compared with normal concrete.

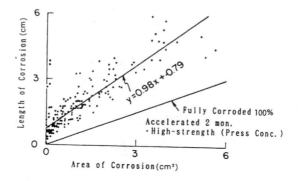

Fig.4   Relation between Area and
Length of Corrosion

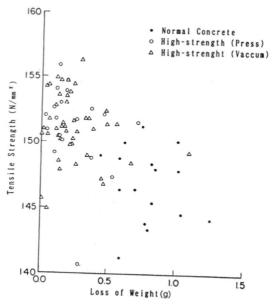

Fig.5   Change of Tensile Strength due to
Corrosion Loss of Weigth

CRACK AND CORROSION OF BARS

Occurrence of Corrosion

Fig. 6 indicates the ratio of corrosion occured (called
"occurrence ratio of corrosion" in this paper) in every 0.05 mm
of crack width. At the stage of 0.3 mm of crack width,
occurrence ratio ranged from 70% to 100%, and almost all of the
bars corroded at the cracked section providing crack width was
over 0.5 mm. Occurrence ratio depended on the kind of concrete,
and occurrence ratio of corrosion of high-strength concrete
always exceeded that of normal concrete under the same crack
width. This might be reasoned with the characteristics of crack
phenomenon.

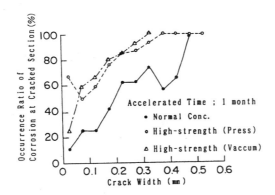

Fig. 6 Occurrence of Corrosion
at Cracked Section

Crack Width and Corrosion

The coefficient of correlation between corroded area and
flexural crack width ranged 0.2 to 0.3, and it was impossible to
obtain the clear relation between two items. Then, corroded area
was evaluated within smaller range of crack width, as was shown
in Fig. 7. Due to Fig. 7, it is likely that distributation pattern
of corroded area was clear distinction at the boundary of crack
width of approximately 0.2 ~ 0.25 mm. This tendency was also
recognized regarding the relation between crack width and
corroded depth.

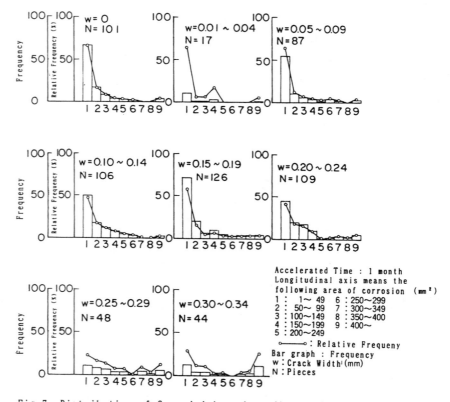

Fig. 7 Distribution of Corroded Area According as Crack Width

PENETRATION OF CHLORIDE INTO CONCRETE

In Series III, some concrete specimens had no flexural crack.
Though bars in high-strength concrete had no corrosion, bars in
normal concrete showed slight corrosion. Chloride ion,
accumulated in concrete due to spray of salt water, was measured.
Fig.8 indicated the test results of chloride ion. It was made
clear that high-strength concrete had less chloride and low
permeability compared to normal concrete.

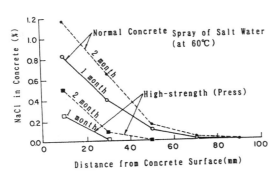

Fig.8 Salt Penetration into Concrete (Series III)

CONCLUSION

Followings might become clear based on the experimental works
done in this papar.

(1) Though corrosion generally extended along longitudinal
     direction of bars for normal concrete, it was likely that
     corrosion was limited near cracked section for high-strength
     concrete.
(2) Regarding correlation of corrosion data, it was observed
     clear relationship between corroded area and corroded length.
(3) At the stage of 0.3 mm of crack width, occurrence ratio of
     corrosion ranged from 70% to 100%, and almost all of the
     bars corroded over crack width of 0.5 mm.
(4) Relation between crack width and corroded area might have
     boundary at the crack width of 0.2~0.25 mm.
(5) Penetration of chloride ion highly depended on quality of
     concrete, and high-strength concrete had low permeability.

REFERENCES

1. KISHITANI K. et al Experimental Study on Drying Accelerated
   Test for Anti-corrosion Admixtures of Concrete.
   Cemend and Concrete, No. 408, Feb. 1981

2. OKADA K. et al Corrosion of Reinforcing Bars Embedded in
   Concrete Under Accelerated Test of Hot Water.
   Materials, Vol. 26, No. 290, Nov. 1977

3. CORROSION COMMITTEE OF JAPAN CONCRETE INSTITUTE
   Chemical Analysis of Chloride in Concrete. Dec. 1983

PROTECTION OF
STRUCTURAL CONCRETE

Session B

Chairmen

Dr J Larralde

Department of Civil Engineering,

Drexel University,

USA

Professor H R Sasse

Director,

Institut fur Bauforschung,

West Germany

FIELD APPLICATIONS OF CONCRETE SURFACE IMPREGNATION

W G Smoak

U.S. Bureau of Reclamation,

United States of America

Abstract. The Bureau of Reclamation, U.S. Department of the
Interior, treated the first full-size highway bridge deck with
polymer surface impregnation in 1974 and subsequently provided
technical assistance to the Federal Highway Administration, U.S.
Department of Transportation, on surface impregnation demonstration
projects on eight additional highway bridge projects. In 1982-83,
the roadway over Grand Coulee Dam, Washington, was treated and
protected by surface impregnation, a project of some 19,600 yd²
(14,980 m²) of treated surface. This paper uses the case study
approach to discuss the polymer surface impregnation processes,
costs, problems, and benefits achieved on these projects.

Keywords: Concrete Bridge Decks, Polymer, Monomer, Polymer
Impregnated Concrete, Surface Impregnation.

W. Glenn Smoak is a senior Research Civil Engineer with the
Research and Laboratory Services Division, Bureau of Reclamation,
Denver, Colorado. He has been active in research and development
of concrete polymer materials since 1969. He currently is the
Principal Investigator of Reclamation's Concrete Materials research
program.

INTRODUCTION

As a result of the increased emphasis placed upon highway safety in the late 1960's, the Federal and State highway organizations in the United States adopted a bare-pavement policy during the winter months. Implementation of this policy greatly increased the utilization of chloride deicing salts on highways and highway bridge decks. A significant increase in the rate of reinforced concrete bridge deck deterioration coincided with the increased use of deicing salts. Following numerous investigations conducted to determine the cause of this deterioration, it was found that chloride-ion-induced corrosion of the reinforcing steel caused subsequent spalling, delamination, and pot hole formation in the upper zones of the concrete decks.

Repair of chloride-induced deterioration of concrete highway bridge decks has become a major portion of highway maintenance budgets in all states affected by subfreezing winter weather. Other types of concrete structures are also affected. Corrosion of concrete reinforcing steel caused by chloride contamination is not limited to northern structures. Concrete structures located in the coastal states also experience this of type corrosion as a result of windblown sea spray and of course, direct contact with seawater. Reinforced concrete parking garages are also affected by chloride-induced corrosion. Automobiles carrying salt contaminated ice and water enter these structures and repeatedly deposit chlorides on the entry/exit ramps and at the designated parking spaces. Very high concentrations of chloride can thus occur in short time periods.

The Bureau of Reclamation began investigation and development of concrete polymer materials in 1966. These materials exhibited increased resistance to freeze-thaw deterioration, reduced permeability and greatly increased durability. It was believed that these materials might offer a deterrent to the deterioration caused by chloride intrusion. A cooperative research program sponsored by the FHWA (Federal Highway Administration) was performed at the Reclamation facilities to determine if it was feasible to increase the durability of concrete bridge decks by impregnating the concrete surface to a depth of 1/2 to 1 inch (12 to 25 mm) with acrylic polymer (1). This initial study indicated that it was technically feasible and a full-scale research effort was performed to develop the materials, equipment and process technology necessary to accomplish surface impregnation on concrete highway bridge decks (2). Subsequently, surface impregnation was applied and demonstrated on highway bridge decks in eight states. The largest application of the technology was the impregnation of the entire roadway and part of the pumping plant parking area of Grand Coulee Dam. This involved treating approximately 20,000 yd^2 (15,000 m^2) of concrete surface.

This paper is a brief summary of these field applications. It should be noted, however, that a final quantitative evaluation of the polymer surface impregnation process has not been performed, and thus final conclusions on the effectiveness of the process cannot be made at this time.

SURFACE IMPREGNATION PROCESS AND MATERIALS

The polymer impregnation or surface impregnation process consists of four basic steps:

1. Preparation of the concrete surface to remove oils, contaminants, and shallow areas of deteriorated concrete.

2. Drying with heat to remove moisture to the desired depth of polymer impregnation.

3. Application of a MMA (methyl methacrylate) based monomer system to the concrete.

4. Thermal-catalytic polymerization of the monomer system in the concrete.

Surface Preparation

The surface of the concrete to be impregnated must be clean and free of debris. Contaminants such as motor oil, tire rubber, paint or curing compounds must be removed using techniques such as sand or shot blasting if adequate penetration of the monomer system is to occur. Environmental concerns limit the use of dry sand blasting but self contained shot blasting or high pressure water blasting equipment are very effective. If the concrete surface has been damaged by scaling, carbonation or freeze-thaw action, the deteriorated material must be removed to sound concrete.

Drying

It is essential to dry the concrete to remove moisture from the zone to be impregnated. Concrete, even in geographical locations of very low ambient humidity, contains sufficient moisture to reduce the impregnation rate to levels impractical for economical application.

A number of forced drying techniques have been used with the surface impregnation process. The most common technique is to use hot forced air or radiant infrared heaters to raise the concrete surface temperature to 250 °F (121 °C) and maintain that temperature for a period of 6 to 8 hours. This technique normally

454 Smoak

requires construction of an insulated enclosure over the area to be
dried.

Monomer System

The impregnation monomer system is composed of 95 percent by mass
MMA (methyl methacrylate) and 5 percent by mass TMPTMA
(trimetholpropane trimethacrylate). Just prior to application, a
polymerization "catalyst" AMVN [2,2-azobis-(2,4-
dimethylvaleronitrile)], is added to the system at a rate of 0.5
percent by mass of monomer system.

Monomer Application

After the concrete has been dried, it is allowed to cool to a
temperature not exceeding 100 °F (38 °C). A layer of dry sand is
then spread over the concrete surface to retain the monomer system
in place while it impregnates the concrete surface. Best results
have been obtained with a sand layer 1/4 to 1/2 inch (6 to 13 mm)
thick. Sand gradation is important. If the sand particles are too
fine, the monomer system will be retained in the sand and not
penetrate the concrete. If the sand particles are too coarse, the
monomer will quickly drain off the surface. The ideal gradation
will pass a No. 16 (1.18 mm) sieve with not more than 5 percent
passing a No. 100 (150 μm) sieve. If dry sand is not available,
sand containing moisture can be applied prior to the drying step
whereby it will dry with the concrete.

The monomer system is applied to the sand in a single application
at a rate of about 0.8 lb/ft² (3.9 kg/m²) and allowed to soak into
the concrete for a period of about 6 hours. The monomer system is
very volatile and must be covered with a polyethylene or mylar film
during the impregnation and subsequent polymerization steps to
prevent evaporation.

Polymerization

The word "polymerization", as used in this paper, is defined as the
chemical reaction that converts the low viscosity liquid MMA-
TMPTMA monomer system into a hard, glassy, solid or polymer. The
monomer system applied to the concrete contains a polymerization
"catalyst" that initiates the polymerization reaction at elevated
temperatures. It is thus necessary to reheat the concrete surface
to a temperature of 165 to 185 °F (74 to 85 °C) and maintain that
temperature for about 5 hours to ensure complete polymerization of
the monomer system in the concrete.

FIELD APPLICATIONS OF SURFACE IMPREGNATION

The surface impregnation process was initially developed to provide concrete highway bridge decks with protection from the intrusion of chloride deicing salts. The first full-size bridge deck treated with this process was located in Denver, Colorado. The treatment process was tested and evaluated on the Denver bridge and then refined and used on a small bridge located near Sandpoint, Idaho. The next application of surface impregnation was on a much larger bridge located 3 mi (4.8 km) east of Forest Glen, California, on State Highway No. 36. This application demonstrated the use of a gas-fired, hot forced air heating system for drying the concrete and will be discussed in detail.

California Bridge

The California bridge is 211 feet (64.3 m) long by 34 feet (10.4 m) wide with an 8-inch (203-mm) thick concrete deck. The deck has 2 inches (50 mm) of concrete cover over the top mat of reinforcing steel. This newly constructed bridge required no surface preparation prior to surface impregnation. The bridge carries a high volume of large truck traffic due to its location in an area of heavy timber operations. The bridge is shown in figure 1.

Figure 1: Surface impregnated bridge near Forest Glen, California

Because of the large size of the bridge, treatment was accomplished in three sections of roughly equal length. The drying enclosure was 70 feet (21.3 m) long and covered the entire width of the bridge. Enclosure height was 22 inch (560 mm). Heat was supplied by four 500,000-Btu (527-MJ) gas-fired, forced air heaters. Two 70-feet (21.3-m) long air distribution ducts were installed beneath the enclosure to distribute the hot air uniformly over the concrete surface. A heater was installed on each end of each duct. The enclosure was constructed of steel supports with a commercially available prefabricated aluminum cover. Fiberglass roll insulation was placed over the aluminum cover to retain heat and the entire enclosure was covered with black polyethylene plastic for protection from rain. This enclosure is shown installed on the bridge in figure 2.

Figure 2: Heating enclosure installed on California bridge

Prior to erecting the enclosure, thermocouples were attached to the concrete surface and at depths of 1 and 2 inches (25 and 51 mm) to monitor and control concrete treatment temperatures.

This heating system proved under-sized for this application and 12 hours of operation were required to reach the target concrete surface temperature of 250 °F (121° C). This temperature was then maintained for 24 hours. The concrete deck was then allowed to cool naturally until the concrete at the 2-inch (51 mm) depth was 86 °F (30 °C). A 3/8-inch (10-mm) thick layer of dry sand was then spread over the cooled concrete surface.

The monomer system was composed of 95 percent MMA and 5 percent TMPTMA and contained 1 percent AMVN, all by mass. The monomer was applied to the sand layer in two applications 3 hours apart. The first application was at a rate of 1 lb/ft² (4.9 kg/m²) and the second application was at a rate of 0.4 lb/ft² (1.9 kg/m²). The surface of the sand was then covered with polyethylene plastic film to reduce monomer evaporation and left for 18 hours. During this time period the heating enclosure and heat system were reinstalled over the treated surface in preparation for the polymerization cycle.

Polymerization of the impregnated monomer was accomplished by reheating the concrete to 165 °F (74 °C) and maintaining that temperature for 12 hours.

The remaining two 70-foot (21.3-m) long sections of the bridge were treated similarly.

Figure 3:   Polymer penetration appears darker in cores from California bridge.

Polymer penetration depths, as determined in cores taken from the deck, were 1 to 2-1/2 inches (25 to 63 mm). Figure 3 shows the polymer penetration as a darker zone in a core from the deck. Penetrations this deep exceeded the requirements for durability and indicated that the process times and monomer application quantities could be reduced.

Other Bridges

Following completion of the bridge in California, the hot air drying system was used to surface impregnate a bridge in West Virginia, and for some precast bridge members in the State of Virginia. A novel system using electric infrared radiant heaters was developed and used to dry and polymerize seven 125-foot (38.1-m) long prestressed concrete bridge girders that were surface impregnated in a precasting yard in Yakima, Washington, and then transported to the bridge location near Santa Rosa, Washington for construction. Figure 4 is a drawing of the prestressed decked girder and heating enclosure used in that treatment.

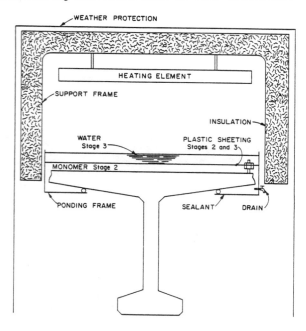

STAGE 1. Deck drying cycle
          Cool deck
STAGE 2. Monomer soak cycle
STAGE 3. Polymerization heat cycle

Figure 4: Heating enclosure for precast decked girder bridge.

Simultaneously with this work, the FHWA was developing and evaluating an electric resistance heating blanket system for another type of bridge heating application. These blankets showed potential for use in the surface impregnation process and were used for a full-scale bridge treatment in the State of Maine. The blankets provided a tight cover over the monomer saturated sand and greatly reduced monomer evaporation during the impregnation and polymerization cycles. The resulting excess monomer bonded the sand to the concrete surface. Removal of this bonded sand was extremely difficult and costly.

Georgia Bridge

A highway bridge on Star Route 20 over Interstate 575 near Canton, Georgia had been constructed with areas of insufficient cover over the top layer of reinforcing steel in the deck. The defective area measured 12,000 ft² (1,115 m²). This bridge was chosen for surface impregnation to protect the shallow steel from corrosion. The Georgia Department of Transportation directed and performed this work with technical assistance from the author and FHWA personnel. (3)

The heating enclosure for this project utilized twenty-five 75,000-Btu (79-MJ) enclosed flame, gas-fired infrared heaters. The enclosure was constructed in five sections, mounted on wheels, which, when bolted together, covered a 75-by 20-feet (22.9-by 6.1-m) concrete area. Eight separate set-ups of the heating enclosure were required to surface impregnate all the defective areas of the bridge. Figure 5 shows the heating enclosure in place on the bridge.

The infrared heaters proved to be an effective heat source but forced ventilation of the enclosure with the portable heaters seen in figure 5 was required during the polymerization step to prevent the accumulation of flammable monomer vapors. In this application, the impregnation sand was applied to the deck prior to the drying step and left in place until polymerization was complete. Mylar film was used to cover the sand to reduce monomer evaporation during the impregnation and polymerization steps. This material was found more heat resistant than the polyethylene film used previously. A monomer application rate of 1.1 lb/ft² (5.4 kg/m²) was initially used, but found to be excessive, resulting in sand bonding to the concrete deck. The application rate was reduced to 0.8 lb/ft² (3.9 kg/m²) with good results, and polymer penetrations averaged about 3/4 inch (20 mm). Drying, cooling, impregnation, and polymerization times and temperatures were roughly equivalent to those used on the California bridge.

Figure 5:   Infrared heating enclosure on Georgia bridge.

Grand Coulee Dam Roadway

Construction of Grand Coulee Dam was completed in 1942. The
spillway area of the main dam is shown in figure 6. The two-lane
dam roadway is over 1 mile (1.6 km) long and consists of an 18-
inch (457-mm) thick layer of roadway concrete placed directly over
the mass concrete of the dam. The roadway has been in use
continuously for over 45 years, experiencing heavy wheeled traffic
and exposed to the severe winter conditions of central Washington
State. Freeze-thaw conditions during this long exposure and
possibly carbonation have resulted in surface deterioration,
shallow spalling and cracking on the roadway surface. This damage
does not affect the safety of the dam but unless arrested would
ultimately lead to replacement of the roadway. The roadway
contains numerous hatches, access ports, electrical service outlets
and the embedded rails for the dam gantry cranes. The clearance
between the crane wheels and the roadway concrete is less than 3/4
inch (19 mm). These features, particularly the close clearance of
the crane wheels, eliminated the use of a conventional asphalt
overlay as a repair technique and a construction contract was
awarded to surface impregnate the roadway and an adjacent parking
lot as a means of permanent repair. Figure 7 shows the dam roadway
looking toward the right abutment.

Figure 6: Spillway area of Grand Coulee Dam.

Figure 7: Grand Coulee Dam roadway. The entire roadway surface, from curb to curb, was surface impregnated.

Concrete surface preparation consisted of sandblasting to remove oils, other surface contamination, and the shallow areas of deteriorated concrete from the surface of the roadway.

Ample electrical power was available from the dam's generating facility and the contractor elected to use electric infrared radiant heaters to supply heat to the drying-polymerization enclosure. The enclosure was 15 feet (4.6 m) wide by 260 feet (79.2 m) long and consisted of a structural steel channel base which was then covered by 22 heating panel units. Each panel was 15 feet (4.6 m) long and 12 feet (3.6 m) wide and contained nine electric infrared radiant heaters rated at 7.5-kW each. The enclosure thus had a total power rating of 1,485-kW. Temperature control was by means of an interval timer for each heat panel. This timer controlled the percentage of time all the heaters in a panel received electric power during a 30-second interval. That is, the nine heaters in a panel were all on or off simultaneously for a selected percentage of each 30-second time interval during a heat cycle. A forced air ventilation system with monomer vapor sensors was installed in the enclosure to prevent development of flammable and unsafe vapor concentrations. Figure 8 shows one end of the heating enclosure.

Figure 8: Electric infrared heating enclosure on Grand Coulee Dam.

Concrete temperatures were monitored by thermocouples epoxied to the surface and at depths of 1 inch (25 mm) below the surface. These temperatures were continuously recorded on multipoint analog and digital recorders and provided a real time record of the process temperatures.

Reduced traffic was maintained over the roadway during surface impregnation by limiting the operations to one lane of travel. The contractor elected to perform the work on a 24-hour-a-day basis except for special holidays or events that required full access to the dam roadway and suspension of work. Fifty setups of the enclosure were required to treat the entire roadway. The work began in May and was completed in September.

A typical process cycle began with spreading a 1/2 inch (12 mm) thick layer of sand on the concrete surface within the enclosure base. The heat panels were positioned over the base and the electrical connections made. The drying cycle normally consisted of 6 to 8 hours of heat-up time and 8 hours of heat maintenance at 250 °F (121 °C). The concrete was then allowed to cool for 12 to 36 hours, depending upon ambient conditions, until the temperature at 1 inch (25 mm) below the surface reached 100 °F (38 °C). The monomer system composed of 95 percent MMA and 5 percent TMPTMA with 0.5 to 1 percent AMVN, all by mass, was applied at a rate of 0.4 lb/ft^2 (2.0 kg/m^2). (The MMA-TMPTMA monomer system was obtained premixed from the manufacturer so that only the proper quantity of AMVN had to be added and mixed at the job-site.) The monomer was applied using a hand-held spraybar attached to the monomer tank truck with a flexible supply hose, figure 9. Mylar film was placed over the saturated sand immediately following monomer application to reduce monomer evaporation. The combination of highly efficient heaters and mylar film so reduced the polymerization heat up time and monomer evaporation that a reduction in monomer application rate from previous projects was required. During the 6-hour impregnation step, the heat panels were replaced and the enclosure was prepared for the polymerization step. The polymerization step required 5 hours at 165 °F (74 °C). Reheating to this temperature only required about 2 to 3 hours, for a total reheat polymerization cycle of 7 or 8 hours duration. This process yielded consistent polymer penetrations of 1/2 inch (12 mm). The contractor became highly efficient in this operation. In the absence of inclement weather, the complete impregnation process could be completed in about 48 hours. Only a minimum work crew was on hand during the cooling cycle.

Figure 9:   Monomer application to the dried concrete surface at
Grand Coulee Dam.

BENEFITS, LIMITATIONS, AND COSTS OF SURFACE IMPREGNATION

Benefits

Polymer impregnation is a very effective method of improving the
durability of concrete to intrusion of corrosive materials such as
deicing salts and to freeze-thaw deterioration. It also results in
significant improvement in abrasion resistance (4). The surface
impregnation process is adaptable for use on precast or cast-in-
place concrete and can be performed with very simple equipment.
The projects reported in this paper, with one exception, were
performed by private contractors with only a minimum of initial
instruction, and in every instance they made improvements in
process equipment and procedures over those used on previous
projects.

Limitations

The use of the surface impregnation process as a means of
protecting concrete is subject to limitations. The inherent nature
of the monomer soaking process makes application on other than flat

horizontal surfaces difficult. Even the superelevation present on some of the bridge decks presented problems with monomer drainage on the projects reported in this paper.

Safety considerations also limit the use of this process. The MMA-TMPTMA monomer system is flammable and has a moderately high vapor pressure. The flash point of the monomer is 55 °F (13 °C) as measured by the ASTM D1310 (tag) method. The monomer and catalyst and their vapors are also toxic. Storage, handling, and use of these materials must be done with consideration of these potential hazards.

The heat application necessary to remove moisture from the concrete zone to be impregnated accelerates formation of the shallow (1 to 3 inches [25 to 76 mm] deep) hairline pattern cracks that ultimately occur on concrete with the passage of time. This type of cracking, caused by drying shrinkage, occurred during the drying heat cycle on every newly constructed concrete surface treated with the surface impregnation process, with the exception of the prestressed girders of the Santa Rosa bridge. A series of laboratory tests were performed to determine if these cracks were detrimental to the long-term corrosion resistance of the treated concrete surfaces (1,2). Although some of the cracks extended below the depth of the top level of reinforcing steel, no instance of corrosion of reinforcing steel occurred either in the laboratory tests or subsequently in the long-term exposure of treated structures in the field. It is recommended, however, that, if the immediate formation of these pattern cracks is objectionable on newly constructed concrete then, the surface impregnation process should be reserved for use on older concrete structures which have already experienced this type of cracking.

Costs of Surface Impregnation

The records of the costs associated with the surface impregnation of each of the projects include, in most instances, the costs of additional research, development, and expediency items that should not be encountered in the performance of surface impregnation if the process were a standard construction practice. As such, the unit costs of the surface impregnation projects varied from a low of $30.00/yd^2 ($36.00/m^2) to a high of $110.00/yd^2 ($140.00/m^2).

The Grand Coulee Dam roadway project, however, was performed by a private contractor who was awarded the construction contract under competitive bidding procedures. This project, therefore, has the most reliable cost data available to date. These data are shown in the following table and include contractor profit:

Costs of Grand Coulee Roadway Project

Area Treated	No. of Setups	Costs Total	Unit
19,600 yd²	50+	$830,500	$42.37
14,980 m²	50+	$830,500	$55.44

CONCLUSIONS AND RECOMMENDATIONS ON FUTURE USE OF SURFACE IMPREGNATION

The author has visited and inspected most of the structures that were subjected to surface impregnation in order to monitor and evaluate their performance over the 15-year period since the treatments were first performed. In no instance has damage or deterioration been detected on the treated concrete surfaces as a result of chloride intrusion, abrasion, or freeze-thaw exposure.

The bridges in Colorado, California, and Idaho have now been covered with a thin asphalt overlay system known as "chip seal". This work was done by local highway maintenance crews, unaware of the significance of the treated bridges, as part of standard highway maintenance programs. The bridges were treated in this manner simply as a matter of convenience to the highway crews, not because of deterioration or damage to the bridge decks. This is unfortunate because it terminated the long-term evaluation of the surface impregnation process on these bridges. It should be recognized, however, that during the 8 to 13 year period between impregnation and chip seal application, no damage due to freeze-thaw deterioration or corrosion of the reinforcing steel was noted on these bridge decks.

The appearance of the prefabricated, prestressed bridge located in Santa Rosa, Washington, remains as constructed after 10 years' exposure and use. Even the small beads of polymer left on the concrete surface from the impregnation process are still visible. The nonskid texture broomed into the riding surface remains sharp and well-defined.

Grand Coulee Dam is inspected frequently by Reclamation personnel located at the damsite. Additionally, the author has visually inspected the roadway three times since impregnation was completed. The deterioration of the roadway surface appears to have been arrested and there has been no further damage to the concrete surface during the intervening 7 years.

Although the field performance of the surface impregnation process has been successful since first application 15 years ago, the future use of the process in the United States is uncertain at this time. The costs of the process are high, though not as high as the costs of removing and replacing a deteriorating concrete structure. Other concrete sealing compounds such as the alkoxy silanes, alkoxy siloxanes and the metallic stearates have entered the marketplace and provide surface protection benefits similar to polymer impregnation. These materials, however, do not offer the improvement in abrasion resistance, depth of penetration, or the degree of long-term durability protection of polymer impregnation. They are particularly susceptible to removal from the concrete as a result of the very shallow penetration depths achieved by their recommended application techniques.

The FHWA has set a moratorium on additional bridge deck polymer impregnation treatments until the bridges that have been treated with the process have been fully evaluated after long-term exposure. Conversely, the Bureau of Reclamation includes the process in its standard procedures for repair of concrete (5), for use in instances where the permanent protection of treated concrete must be assured.

Polymer surface impregnation is much superior to any "penetrating" sealant treatment and should thus be considered as a means of protecting concrete in those site specific instances where the benefits of assured long-term or permanent protection outweigh the costs of the treatment. These costs could be significantly reduced by the development of more economical methods of removing moisture from the concrete zone to be impregnated.

REFERENCES

1. Smoak, W. G., Development and Field Evaluation of a Technique for Polymer Impregnation of New Concrete Bridge Deck Surfaces. Federal Highway Administration, FHWA-RD-75-72, June 1975.

2. Smoak, W. G., Implementation of Polymer Impregnation as a Bridge Deck Sealant. Bureau of Reclamation, U.S. Department of the Interior, REC-ERC-84-11, September 1984.

3. Neill, C. G, Jr., Polymer Impregnation as a Bridge Deck Sealant. Georgia Department of Transportation, Office of Materials and Research, Report No. FHWA/GA/001, February 1982.

4. DePuy, G. W., Dikeou, J. T., et al, Concrete-Polymer Materials, Final Report, USBR-BNL-OSW Cooperative Program, Bureau of Reclamation, U.S. Department of the Interior, REC-ERC-76-10, August 1976.

5. Standard Technical Procedures for the Repair of Concrete. Bureau of Reclamation, U.S. Department of the Interior, M-47, Revised draft 1990.

A FIELD STUDY ON DURABILITY OF R C BUILDINGS EXPOSED TO A MARINE ENVIRONMENT

T Oshiro

University of the Ryukyus,

S Tanikawa

Toagosei Chemical Industry Co. Ltd.,

Japan

ABSTRACT. Corrosion of steel reinforcements in concrete is one of the major problems with respect to the durability of reinforced concrete structures, and chloride ions are considered to be the major causes of premature corrosion of steel reinforcements.
To study the process of material deterioration of reinforced concrete structures, an experimental building was constructed in 1984 and has been exposed for five years and three months (5 1/4 years) to marine environment under sub-tropical weather of Okinawa, Japan. Field tests were performed to this building to determine concentration and amount of chloride ions in concrete, variation in the half-cell potentials with relation to time, corroded areas on steel reinforcements, and bond strength of the coating being used. Based on the test results, the coating effects for protecting reinforced concrete structures are clearly identified.

Keywords: Field Study, Durability of RC Building, Marine Environment, Chloride Ions, Half-Cell Potentials, Corrosion, Acrylic Rubber Coating.

Dr. Takeshi Oshiro, a member of ACI and JCI, is a Professor of Architectural Engineering at the University of the Ryukyus, Okinawa, Japan. He received his Ph.D. in 1970 from University of Delaware, and has been active in the research of durability of reinforced concrete structures.

Mr. Shin Tanikawa, a member of ACI and JCI, is a research engineer at the research laboratory of Toagosei Chemical Industry Co. Ltd. He received his M.S. in 1975 from Nagoya Institute of Technology, Nagoya, Japan and has been active in projects dealing with application of surface coatings.

INTRODUCTION

Corrosion of steel reinforcements in concrete is one of the major problems with respect to the durability of reinforced concrete structures. Chloride ions are considered to be the major causes of premature corrosion of steel reinforcements in concrete. Dissolved chloride ions may penetrate unprotected hardened concrete exposed to marine environment. Also, chloride salts may originate in aggregates dredged from sea.

As a method of excluding external sources of chloride ions from concrete, waterproof membranes have been used extensively to minimize the ingress of chloride ions into concrete. There is a growing tendency to use highly elastic coatings as the membrane in Japan.

The objective of this paper is to study the process of material deterioration on an experimental building constructed in 1984, which has been exposed for five years and three months (5 1/4 years) to marine environment under sub-tropical weather of Okinawa, Japan. Field tests were performed to this building to determine concentration and amount of chloride ions in concrete, variation in the half-cell potentials, corroded areas on steel reinforcements, bond strength of the coating to concrete, and protective effects of the coating. [1]~[3]

OUTLINES OF EXPERIMENT

A reinforced concrete experimental building was constructed at seashore of Bise-Cape, Motobu-cho, Okinawa Prefecture, Japan, in 1984 shown in Figs. 1 and 2. The three sides of this building have been exposed to salt breeze and splashed with sea water during heavy winds under high temperature and humidity which are the typical characteristics of sub-tropical weather.

Two kinds of concrete mix were used for this building; one containing non-chloride ions referred to concrete with no salt, and the other

Figure 1: General view of experimental building

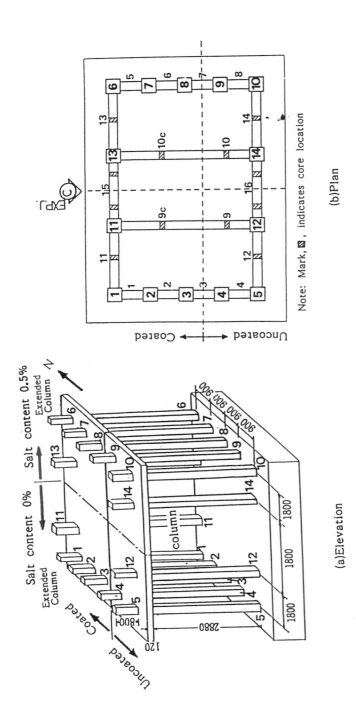

(b)Plan

Note: Mark, ▨ , indicates core location

(a)Elevation

Figure 2: Dimensions, test conditions and element numbers

containing 0.5% sodium chloride (NaCl) by weight of concrete referred to concrete with salt. The most common concrete design was made with a water to cement ratio of 0.63, a slump of 180mm, and concrete strength of 21 N/mm². The thickness of concrete cover was 40mm for columns and beams and 30mm for slabs.

Half of the building on sea side was coated with acrylic rubber type coating, and another half being left as uncoated. This coating is described in JIS A 6021 (Roof coatings for water proofing), consisting of primer, base coat and top coat. The acrylic rubber contains 54% of acrylic polymer by weight and the rest comprises inorganic filler and pigment. The acrylic polymer consists of rubber like, bridged 2-ethylhexyl acrylate. Total thickness of the coating becomes 1.1mm and this coating is considered to meet the requirements for ideal waterproofing; good bond to substrate, and low permeability to chlorides and moisture under service conditions, especially temperature extremes, crack bridging and aging.

RESULTS AND DISCUSSIONS

Concentration of Chloride Content

For core samples from beams and columns, water-soluble chloride content of concrete was measured by a potential difference titration method using a chloride selective electrode, which is expressed in terms of NaCl content (%) by weight of concrete.

Concentrations of chloride content of samples from beams with non-salt are shown in Fig. 3, where the results for 2 1/4 and 5 1/4 years' exposure (measured in 1986 and 1989) are compared. Location of coated beams 11 and 15 is northern sea side, where-as uncoated beams 12 and 16 are located on southern land side. Surface chloride content of coated beam 15 presents the result of salt penetration for three months before coating and tends to be leveled. Both uncoated surfaces of this building that have been exposed to sea and land show a large amount of penetrated chlorides, especially on sea side due to influences of marine environment. Comparing the chloride content of coated and uncoated beams, it is apparent that the coating has a great effects on resistance to chloride penetration for longer periods.

Two cores were obtained from beam with non-salt at east-west direction, and the resulting concentrations of chloride content are shown in the same figure. The eastern uncoated surface of beam 9 shows greater amount of chloride content compared to western uncoated surface, where-as both coated surfaces of beam 9C show a little except chloride content that penetrated before coating.

Concentrations of chloride content on beams with salt are shown in Fig. 4. Both uncoated surfaces on sea and land sides of beams 10 and 14 present a great penetration of chlorides from external sources, where-as coated surfaces of beam 10C and 13 show a little increase in chloride content due to low permeability of the coating to chloride penetration. But, some increases in chlorides on coated surfaces occur since chlorides inside tend to move to surfaces in case of high chloride content.

The results obtained from coated and uncoated beams with salt show that high amount of chloride content tends to increase on surface portions where steel reinforcements are embedded, which makes it difficult to

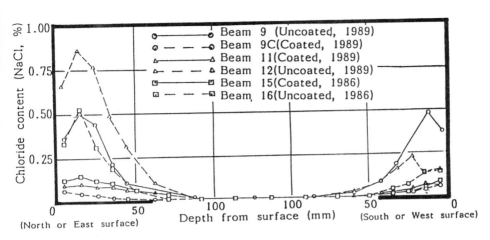

Figure 3: Concentration of chloride content
(Beams with non-salt)

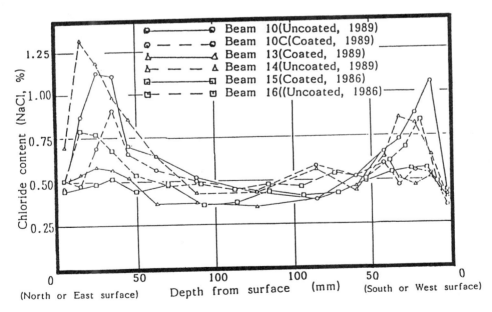

Figure 4: Concentration of chloride content
(Beams with salt)

prevent corrosion of steel in concrete.
Concentrations of chloride content for coated and uncoated columns with non-salt are shown in Fig. 5. Uncoated columns 4(side 1-3) and 12(side 2-4) for 5 1/4 years' exposure show that, on northern side 1 and eastern side 2, a great chloride concentrations appear corresponding to marine environment. As examples of coated columns, extended column 11(side 1-3, side 2-4) for 4 1/4 years' exposure and column 1(side 1) for 1 1/4 years' exposure are shown in the same figure. Comparing these columns, no increase in chloride content for 3 years is observed, presenting the coating effects on chloride penetration.
Concentrations of chloride content for columns with salt become almost the same as beams with salt shown in Fig. 5, and uncoated columns present a great ingress of chloride from external sources.

Half-cell Potentials

The electrical potentials were measured by the use of a copper-copper sulfate half-cell(CSE) and a high impedance volt-meter for 5 years since 1984 to 1989, and the results are shown in Figs. 6 - 11.
Initial values measured in 1984 on beams with non-salt are compared to the results after five years in Fig. 6. Results on coated beam 11 show practically no variation for the periods. But, values of uncoated beam 12 on northern sea side drop to more negative values about 100 mV, whereas ones on the southern land side show only a slight variation.
The beam running on north-south has two portions, coated on north and uncoated on south, and the results are shown in Fig. 7. Resulting values on coated portion show no variations during the period, but uncoated portion drops to more negative on both eastern and western sides.
The potential variation with relation to time for five years are also shown for columns with non-salt 4 and 5 in Figs. 8 and 9, and with salt 10 in Fig. 10. Columns 4 and 5 show that potential shifts to negative every year indicate more undesirable of corrosive environment. Center portion of column 5 was repaired by prepacked concrete after 3 years' exposure and the potentials shift to noble one year later shown as 4 years' exposure.
Column with salt 10 in Fig. 10 shows the initial potential range of -400--450 mV and the range shifts to -450--500 mV after 3 years, showing a slight variation with relation to time. Since increasing growth of cracks on surfaces was observed, it was repaired totally by prepacked concrete where polymer cement mortar was grouted. One year later, the potentials became about -200mV, shifting to noble and a slight variation about 40 mV in a year is observed two years after repair.
To determine potential differences with respect to side exposure, the potentials for each side are shown in Fig. 11 for uncoated column with non-salt 4, where the potential values are results of 4 years' exposure. The potentials on side 1 facing the sea present the most negative, followed by sides 2 and 4. The side 3 facing the land presents only a slight drop to negative from the initial values.
Based on the results, it becomes evident that the potentials vary according to influence of external environment, which also correspond to corrosion of steel reinforcements in concrete.

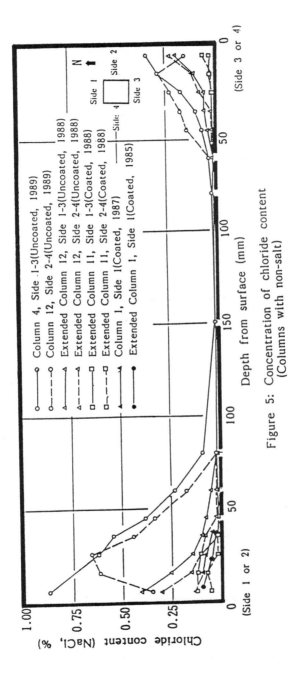

Figure 5: Concentration of chloride content
(Columns with non-salt)

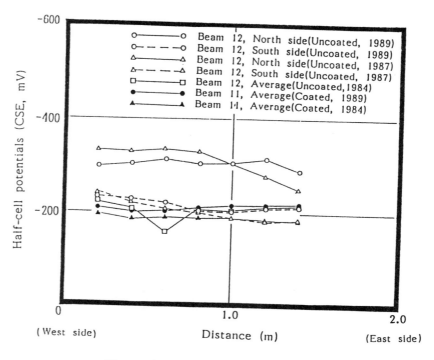

Figure 6: Results of half-cell potentials
(Beams with non-salt)

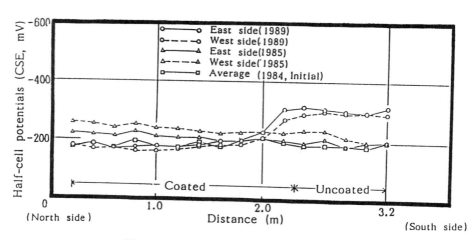

Figure 7: Results of half-cell potentials
(Beam with non-salt, 9)

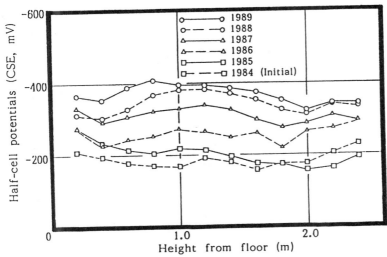

Figure 8: Results of half-cell potentials
(Uncoated column with non-salt, 4)

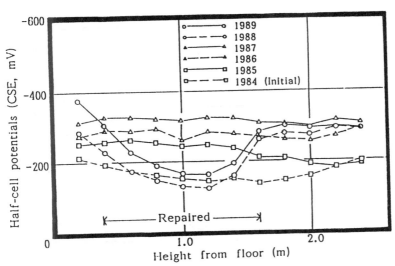

Figure 9: Results of half-cell potentials
(Uncoated column with non-salt, 5)

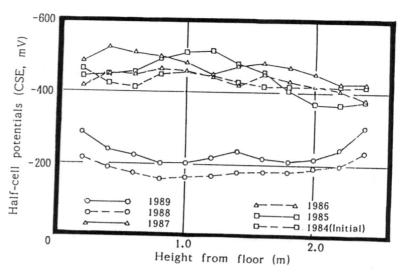

Figure 10: Results of half-cell potentials
(Uncoated column with salt, 10)

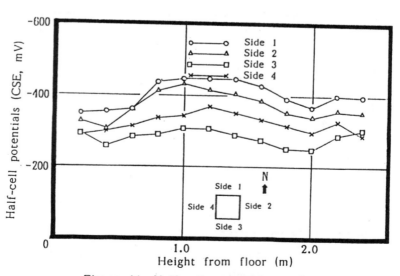

Figure 11: Half-cell potentials on four sides
(Uncoated column with non-salt, 4, 1989)

Corroded Area Ratio

Corroded area ratio is defined as ratio of corroded area to total area of
a reinforcement. Traces of corroded area of samples are measured by an
automatic planimeter, and the results are shown in Fig.12 and Table 1
for beams and columns. On coated beams with non-salt 9C and 11, only
small spots of corrosion are observed, presenting 0% corroded area ratio.
Uncoated beam with non-salt 9 presents 28% on east side, where-as west
side presents 0%. Similarly, uncoated beam 12 presents 25% on sea side
and 0% on land side.
Beams with salt show a great corrosion of steel reinforcements, 60-70%
on coated portions and 86% on uncoated portions. Therefore, the effect
of the coating is not considered referring to the corroded area ratios in
case of high chloride content, but the effects might be evaluated by
crack appearance instead. Coated portions show a small crack of max. 1
mm width, where-as, spallings of concrete cover appear on uncoated
portions.
Corrosion area ratios were measured for columns and the results are
shown in Table 1. Coated portions with non-salt show only small spots of
corrosion, but uncoated portions show corrosion which varies with sides
corresponding to marine environment.
For coated columns with salt, corroded areas of column 6 and extended
beam 13 were measured in 1988 and the results are shown in Table 1,
where effects of coating appear with respect to the ratios. Uncoated
columns with salt show heavy deterioration of concrete.

Bond Test

To determine bond strength of the coating to the substrate, bond test
was performed at 14 points on beams and columns after 5 1/4 years' exposure according to JIS A 6910-1988. An average of all measured values
becomes 1.7 N/mm², where the measured values vary from min. 1.2

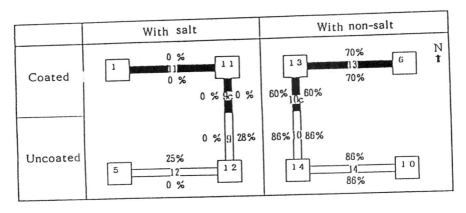

Figure 12: Corroded area ratios (Beams)

Table-1 Corroded area ratios (Columns)

Column No.	Side	Corroded area ratios (%)[1]	Chloride content (%)[2]	Column No.	Side	Corroded area ratios (%)[3]	Chloride content (%)[2]
Column 1	1	1	0.134	Extended Beam 11	1	2	0.013
	2	2	0.143		2	0	0.014
	3	1	0.059		3	0	0.015
	4	2	0.060		4	0	0.014
Column 5	1	12	0.880	Extended Beam 12	1	67	0.092
	2	12	0.437		2	43	0.075
	3	1	0.127		3	8	0.049
	4	3	0.168		4	34	0.16
Column 6	1	1	0.590	Extended Beam 13	1	23	0.508
	2	2	0.497		2	20	0.506
	3	6	0.468		3	5	0.491
	4	0	0.496		4	16	0.524
Column 10	1	100	0.694	Extended Beam 14	1	97	0.582
	2	100	0.408		2	94	0.473
	3	100	0.583		3	65	0.374
	4	100	0.645		4	67	0.650

Note: 1)measured in 1987   2)chloride content at location of reinforcement   3)measured in 1988

N/mm^2 to max. 2.4 N/mm^2, and all the values meet the specification of 0.7 N/mm^2 without losing bond strength for the longer periods.

CONCLUSIONS

(1) Concrete portions with non-salt at the experimental building present no increase in chloride content if protected by the acrylic rubber coating which has low permeability to chlorides. Uncoated portions show a great ingress of chlorides, especially on surfaces facing to sea, and the necessity of providing a proper protective measure at an early age is shown.
Besides ingress of chlorides from external sources for uncoated portions, great concentrations of chlorides appear at the location of steel reinforcements due to movement of chlorides from inside to surfaces in case of high chloride content.

(2) Half-cell potentials vary sligtly on coated portions for 5 years, where-as, on uncoated portions, the potentials drop greatly to negative with relation to time corresponding to corrosion of steel rein-forcements.
Half-cell potentials at repaired portions shift to noble greatly, present-ing improvement of corrosion environment.

(3) The acrylic rubber coating used for this test shows a great resistance to chloride penetration, and also prevents ingress of water and oxgen. Therefore, the protection of steel reinforcements in concrete from corrosion is verified referring to corroded area ratios and half-cell potentials, especially on concrete portions with non-salt.

REFERENCES

1   Oshiro, T., Tanikawa, S. and Nagai, K. : Field Exposure Test for Deterioration of RC Structure, Trans. of the Japan Concrete Institute, Vol. 8, 1986, pp 161-168.

2   Oshiro, T., Horizono, Y., Tanikawa, S. and Nagai, K. : Experimental and Analytical Studies on Penetration of Chloride Ions into Concrete, Trans. of the Japan Concrete Institute, Vol. 9, 1987, pp 155-162.

3   Oshiro, T. and Tanikawa, S. : Effect of Surface Coatings on the Durability of Concrete Exposed to Marine Environment, Concrete in Marine Environment, American Concrete Institute SP-109, 1988, pp 179-198.

THE BEHAVIOUR OF SEVERAL TYPES OF SURFACE COATINGS SUBJECTED TO AGGRESSIVE WATER BUT NOT EXPOSED TO DIRECT SUNLIGHT

C Oosthuizen

P A Naude

A G van Heerden

Department of Water Affairs,

South Africa

ABSTRACT. The Department of Water Affairs owns, operates and maintains some 130 dams listed in the ICOLD world register of dams. Engineers in the Department are often faced with the task of prescribing methods and/or products to be used either for repair-work to deteriorated or damaged concrete surfaces or for the protection of new surfaces. This is not an easy task owing to budgetary constraints and/or biased advice of experts. A testing program was therefore launched by the Department of Water Affairs in which suppliers of surface coatings were invited to apply their product(s) on test strips in the river outlet tunnel of the Theewaterskloof Dam. These products were then evaluated in terms of their relative resistance to the abrasive and aggressive water flowing in the outlet tunnel. The results of these tests should however be used with caution as they are site specific and based on subjective judgement and not controlled laboratory measurements.

Keywords: Abrasive Water, Aggressive water, Surface Coating, Concrete Surface Protection, Behaviour, Cost.

Dr Chris Oosthuizen is Deputy Chief Engineer: Civil Design in charge of Dam Safety at the Department of Water Affairs, South Africa. He has for a number of years been actively involved in all facets of dam engineering, in particular the behaviour of and risk based safety evaluation of dams.

Mr Piet A Naude is Control Industrial Technician: Dam Safety in charge of all Instrumentation in the Department of Water Affairs, South Africa. He is project leader for the evaluation of various coating systems in the Theewaterskloof Tunnel.

Mr A George van Heerden is a Senior Industrial Technician (Civil) who participated as the Department of Water Affairs' observer during application and evaluation of the various coating systems.

BACKGROUND

After a boom in dam construction during the past three decades a new phase or era in dam engineering is emerging in South Africa, namely that of maintenance and rehabilitation. The cost of remedial repairs is already beginning to represent a significant portion of the Department's budget. During the regular dam safety inspections performed by dam safety experts in the department, it was noticed that some of the products applied to protect or repair concrete were performing well under certain conditions but not under slightly different conditions. To ensure that mistakes of the past are not repeated in future and to provide a basis for evaluating concrete protection products, a test program has been initiated by the Department which involves the evaluation of a variety of products currently used to protect concrete surfaces.

INTRODUCTION

The Theewaterskloof Dam, designed, constructed and operated by the Department of Water Affairs, was completed in 1980. Three years after completion, the operations engineer on site reported deterioration of the tunnel lining (Naude et al:1985). The 155 metre long outlet tunnel measures 3,75 m x 3,75 m and is designed as a free flow tunnel. The floor of the tunnel is permanently submerged owing to leaking gate-valves while the roof of the tunnel is for all practical purposes always dry. The flow profile of the water in the tunnel can be distinguished clearly by the degree of deterioration of the lining which varies from severe at the floor to negligible at the roof.

The river outlet tunnel of the Theewaterskloof Dam was chosen as test site for various reasons. Firstly, deterioration of the concrete lining had already taken place to a degree where a protective coating would have to be applied in the near future. Secondly, several kilometres of tunnel used for inter basin water transfer in the upper region of the dam are subjected to similar conditions and would also require similar attention in the future.

TEST CONDITIONS AND SURFACE PREPARATION

A uniformly weathered section mid-way in the tunnel was chosen as test area comprising two identical test sites. At each of the two test sites, panels of 500 mm wide and 2 metre high, were marked on the left and right hand side of the tunnel, i.e. four panels per product. The test panels on the left side were sand-blasted and those on the right were cleaned with a water-jet (compressed air and water) in order to evaluate both methods of surface preparation.

At both test sections the flow is confined within the first 2 metres from the floor and flow velocities during the test period varied between 10 and 12 metres per second. The releases for 326 days of operation during the test period amounted to 644million cubic metres.

The chemical properties of the water downstream of the dam are
determined on a regular basis. The pH of 109 samples taken during
the test period varied between 4 and 5 during releases but increased
to between 6 and 6,9 when, except for the leaking slab gate valves,
no releases were made . The Electrical Conductivity (EC) varied
between 6 and 28,8 milli-Siemen/metre at 25 degrees Celsius, the
Total Dissolved Salts (TDS) between 26 and 141 ppm and the Total
Alkalinity (TAL) as $CaCO_3$ between 0,2 and 21,8.

The concrete mix used for the sides and roof of the river outlet
tunnel contained "1 part Cape Eland Portland Cement, 2,09 parts sand
and 2,97 parts 20mm stone by weight, i.e. 344 Kg Cement per cubic
metre. The water/cement ratio was 0,55 and the average 28 day
strength 30 MPa". The stone was described as "crushed water worn
Granite consisting of less than 10% Quartzitic Sandstone of which
some particles are slightly weathered", the sand as "Quartzitic pit
sand with a Fineness Modulus (fm) of 2,2." (VAN DER WALT:1990)

Abrasion is caused by solids in the water. The actual size of the
solids carried by the relatively high velocity water has not yet been
determined. An indication of the abrasion is deduced from the
behaviour of three 8 mm stainless steel studs mounted on each test
panel. These studs were drilled and epoxied into the lining to
protrude 30 mm with the lowest stud 100 mm above the floor level.

EVALUATION OF COATING SYSTEMS

The test panels were inspected visually and photographed at regular
intervals. On the 18th of January 1990 the surface coatings were
inspected by the authors accompanied by an independent Professional
Engineer from the Portland Cement Institute (PCI). The second and
third authors were directly involved with the application and regular
evaluation of the products. As they could relate product numbers to
suppliers and costs (and may have developed some biases) they were
excluded from this evaluation. Both the first author, Chris
Oosthuizen, and Alex Peters from the PCI conducted their evaluations
independently. Each panel was closely inspected for extent of damage
from lack of adhesion, peeling, disintegration and wear from abrasive
solids in the water. Their findings were then grouped into five
categories ranging from "excellent" to "very poor". The results are
summarised in table 1.

The products tested, their suppliers and panel numbers as well as
notes on their application are presented as an addendum (VAN HEERDEN
et al:1986).

The panels on the left hand side of the tunnel were clearly subjected
to more abrasion than those on the right as all the studs on the left
hand side were either ripped out of the lining or were bent, in some
instances up to 90 degrees. As far as preparation methods are
concerned, it has been deduced that both methods are adequate.

Table 1: Subjective evaluation and rating of the surface coatings

Visual Evaluation	Coating reference number in order of preference	
	Alex Peters (PCI)	Chris Oosthuizen (DWA)
EXCELLENT	10	10
GOOD	6B, 6C, 7, 8, 11, 28AT, 28A	6B, 6C, 8, 11, 7, 28AT, 28A
FAIR	12C, 12B, 29A, 27, 29C, 29B, 19	29A, 12C, 12B, 29B, 18A, 27, 29C, 19, 24
POOR	32D, 22, 33, 24, 34, 40A, 23, 18A, 32B, 32C, 12A, 18B, 32A	32D, 22, 12A, 18B, 32C, 40A, 23, 32B, 33, 34, 32A
VERY POOR	40B, 6A, 30, 15 no trace of coat visible	40B, 15, 30, 6A no trace of coat visible

It is significant that both adjudicators classified the same products under the headings "excellent" and "good". Although not in the same order of preference the same products appear in the top ten products on both lists. Table 2 gives a weighed priority list of these products and their cost.

Table 2: Comparison between weighed rating of behaviour and cost.

RATING (behaviour)	PANEL No	COST in SA Rand	RATING (cost)
1	10	106.75	10
2	6B	8.60	5
3	6C	28.40	8
4	8	12.95	6
5	7	26.00	7
6	11	42.95	9
7	28AT	7.08	4
8	28A	4.30	3
9	12C	No longer available in SA	
10	29A	3.80	2
11	27	2.62	1

With the exception of products 6B, 7 and 8 the cost are in direct relation to their efficacy. The cost and skills needed for the application of the products is not reflected in the costs. However an attempt has been made with the description of the products in the addendum to reflect the relative effort needed. Cognisance must be taken of the fact that these results are only valid for conditions similar to those at the Theewaterskloof Dam river outlet tunnel.

CONCLUSIONS

The danger of this type of subjective evaluation is that conclusions on the efficacy of products may be drawn, which are strictly speaking not scientifically based. It must be stressed that these tests are being carried out under specific conditions and furthermore several of the test products are being applied under conditions not recommended by the manufacturers.

Product No 10 performed extremely well at the river outlet tunnel of Theewaterskloof Dam and is recommended for use wherever protection is of prime importance and where cost and ease of application are of secondary importance. From an economic standpoint product No 6B is the authors choice. It is one of the lower cost products, easily applied and has been rated second best after four years in operation.

The implementation of regular inspections has proven invaluable for early identification of potential problem areas thereby allowing time for full scale trials on repair products and/or methods.

The next phase in the test program involves the exposure of a selected number of these products to other extremes often experienced, i.e. high levels of UV and Ozone as well as wet and dry cycles. Some of the products will require an additional protection layer under these conditions. This is scheduled for later this year.

ACKNOWLEDGEMENTS

The authors wish to express their gratitude to all the suppliers who not only gave permission to publish the results of the tests before the tests commenced, but supplied and applied their products free of charge;the PCI and Alex Peters in particular for his contribution; members of the Subdirectorates Dam Safety and Data Management for their assistance and the Department of Water Affairs for permission to publish this paper. The findings, opinions, conclusions and recommendations presented in this report are those of the authors and may not necessarily reflect the views of the Department.

REFERENCES

1   NAUDE P A, OOSTHUIZEN C, CROUCAMP W S. Betonhersteltoetse in Theewaterskloofuitlaattonnel. Department of Water Affairs. Report 1/85, February 1985, pp 1-6.

2   VAN DER WALT M A. Personal communication. March 1990.

3   VAN HEERDEN A G, NAUDE P A, OOSTHUIZEN C. Notes on Concrete Renovation Tests at the Theewaterskloof Dam Outlet Tunnel. Report 1/86, August 1986, pp

APPENDIX: List of supplier, product, number and notes on application.

SUPPLIER	PRODUCT	NO.	NOTES ON APPLICATION
A.B.E.	DURAPROOF	6A	1. Mixed $5\frac{1}{2}$ to 1 by vol; with water brought from Cape Town. 2. Brush off surface dust. 3. First coat applied to satura=tion 2nd coat after 24hrs also saturated. 4. Can be spray applied - airless 5. Very easily applied. 6. When dry looks like wet concrete.
	ABECOTE 352	6B	1. Modified 352 used. 2. Applied to normal concrete surface. 3. Applied by brush - roller or airless spray, can be used. 4. Black. 5. Easy to apply.
	BARRESKIM/BARREFLEX	6C	1. Wet concrete surface thoroughly with water. 2. Apply BARRESKIM with block=brush thereafter trowel with grooved trowel to obtain uniform thickness then smooth down with ordinary trowel. 3. Apply 1st & 2nd coat BARREFLEX after 24 & 48 hrs resp; using BARRESKIM method of appli=cation, except surface not to be wetted again BARRESKIM-Grey. BARREFLEX- White. 4. Pot-life both systems ±30 min. depending temperature. 5. Applied fairly easily, but time consuming.
AE & CI	AMERCOAT 2200/2203	28A	1. Apply sealer/primer - AMERCOAT 2200 - diluted 20% with dam water to normal concrete surface. 2. Apply AMERCOAT 2203 as topcoat after 24 hrs. 3. Apply 2nd topcoat 2203 after 24 hrs.

List of supplier, product, number and notes on application (continue)

SUPPLIER	PRODUCT	NO.	NOTES ON APPLICATION
AE & CI	AMERCOAT 2200/2203	28A	4. Applied by brush - easily applied. 5. Colour light grey.
	AMERCOAT 2200/2202/ 2203	28	1. Apply sealer/primer - AMERCOAT 2200 - diluted 20% with dam water to normal concrete surface. 2. After 24 hrs apply AMERCOAT 2202 with a trowel to primed surface - leave to dry for 24 hrs. 3. Apply 1st coat AMERCOAT 2203 to trowelled surface and leave to dry for ±24 hrs. 4. Apply 2nd coat AMERCOAT 2203. 5. AMERCOAT 2200 & 2203 with brush - can be airless/conven=tional sprayed rolled or brush applied - 6. Colour light grey.
ARCOTEX	TENAXON 70 Z U/S only	12A	1. Applied directly to concrete. 2. Applied by short brush - fair=ly difficult to apply - brush or trowel only. 3. Grey.
	TENAXON 70 Z U/S & D/S	12B	1. Applied directly to concrete. 2. Easily applied by brush - can be sprayed or rolled on - 3. White.
	TENAXON 70 Z U/S & D/S	12C	1. Applied directly to concrete. 2. Easily applied by brush - can be sprayed or rolled on - 3. Black.
BURMAH ADHESIVES	HEY'DI POWDER No.1 & POWDER X	23	1. Wet concrete surface thoroughly. 2. Apply Powder No.1 in liquid form - mixed with water - onto surface with a block brush. 3. Apply Powder X by rubbing onto wet Powder No.1 surface.

List of supplier, product, number and notes on application (continue)

SUPPLIER	PRODUCT	NO.	NOTES ON APPLICATION
BURMAH ADHESIVES	HEY'DI POWDER No.1 & POWDER X	23	4. Apply sealer. 5. Reapply last coat Powder No.1 by brush, block brush - easily applied. 6. Cement colour.
EMAQ	CONCRETE PROOFER	24	1. Single pack bakelite con= taining epoxy. 2. 1st coat onto normal concrete surface and left to touch dry. 3. 2nd coat applied. 4. Applied by brush - can also be rolled or airless sprayed on. 5. D/S R.H.S. panel wetted with water before application as a test. 6. Colour clear.
EMBECON	VERTIPATCH	18A	1. Prime concrete surface with diluted "vertipatch" - fairly easily applied using block brush - 2. Plaster on "vertipatch" ± 2 mm thick. 3. Plaster application similar to normal cement mortar plaster. 4. Grey.
	SEALCOAT	18B	1. Dampen concrete surface with water. 2. Apply "SEALCOAT" with block brush. 3. One coat applied. 4. Fairly easily applied. 5. Grey.
FORBES WATER= PROOFING	CARBOFOL MEMBRANE	26	1. Hot Bituminous primer/adhesive applied to concrete surface. 2. CARBOFOL rolled onto heated primer/adhesive surface and pressed down with narrow paint roller to ensure adhesion and remove air bubbles.

List of supplier, product, number and notes on application (continue)

SUPPLIER	PRODUCT	NO.	NOTES ON APPLICATION
FORBES WATER= PROOFING	CARBOFOL MEMBRANE	26	3. Joins hot welded with special tool. 4. The start & ± 2 m D/S strapped down with special metal straps that are anchored to the concrete - metal strap red on photo. 5. Time consuming to apply.
FOSROCK	CITAL	7	1. Applied on brush wetted left side and normal concrete right side panel. 2. Easily applied with roller. 3. Apply 2nd coat immediately after 1st coat. 4. Can be airless sprayed rolled or brushed. 5. Colour light blue.
	NITOMOTAR E L	8	1. Epoxy mortar. 2. Apply directly onto cleaned concrete surface with trowel. 3. Fairly difficult to apply - like most epoxy mortars - 4. Colour sand. 5. Surface can be moist.
HELPSEAL IT	H203	32A	1. Apply directly to concrete surface. 2. U/S R.H.S. received 2nd coat H203 within 24 hrs. 3. U/S L.H.S. problem with 1st coat due to wet concrete surface at ± bottom ½m of panel caused by splash of water dropping from roof drain hole - system would not adhere properly. 4. Easily applied - applied by brush - can be brushed/rolled or sprayed. 5. D/S only one coat applied.

List of supplier, product, number and notes on application (continue)

SUPPLIER	PRODUCT	NO.	NOTES ON APPLICATION
HELPSEAL IT	CLEAR SEALER COAT FOR DAMS	32B	1. Apply directly onto concrete surface. 2. After 3 days (5 hrs normally) H112 applied as final coat with brush. 3. L.H.S. U/S coated with grey H112, moisture caused coating problem at ± bottom 0,5 m of this panel. 4. Easily applied - can be rolled brushed or sprayed on.
	H112	32C	1. Applied directly onto concrete surface. 2. 1st coat with brush 2nd coat with roller to get thickness. 3. U/S grey D/S white.
IVORY CHEMICALS	EPOXY MORTAR 329	10	1. Prime surface with 318/20 - before dry, cast clean graded silica sand onto primed surface to give a sandpaper effect - 2. After the 318/20 has dried and within 24 hrs the epoxy mortar 329 is applied by trowel - as with all epoxy mortars trowelling is fairly difficult especially when it starts to set - 3. Colour grey.
	EPOXY 301	11	1. Prime surface with diluted 301 - applied to normal surface - 2. Leave to dry-coat within 24 hrs - 3. After 301 primer has dried apply 301 epoxy mastic with brush - airless spray or roller can be used. 4. Easily applied. 5. Colour white.

List of supplier, product, number and notes on application (continue)

SUPPLIER	PRODUCT	NO.	NOTES ON APPLICATION
MARLEY	DEEPSEAL	15	1. Wet concrete surface thoroughly - wait for free water to disappear - 2. Apply one coat Deepseal with block brush. 3. Easily applied. 4. Cement colour.
PLASCON	WATERBASED EPOXY SEALER	29A	1. Dilute with 20% water - milky colour - apply to surface with a brush - surface normal - 2. Drying time min : 24 hrs. 3. Easily applied. 4. When dry apply E.P.D. 100.
	EPOXY TAR FINISHING COAT E.P.D. 100	(29A)	1. Applied by brush on sealer surface. 2. Easily applied 3. Colour black.
	WET BLAST PRIMER (CODE L41153)	29B	1. Applied to normal concrete surface 2. Brush applied 3. Easily applied 4. Colour grey 5. Finishing coat E.P.D 100 as above in 1.
	PURE ACRYLIC EMULSION (CODE CXO 24)	29C	1. Dilute with 10% water 2. Brush applied to surface - surface normal - 3. Minimum drying time 12 hrs 4. Very easily applied - can be rolled or sprayed - 5. Colour blue. FINISH COAT: 1. One coat undiluted CXO 24 2. Very easily applied by brush 3. Min : 48 hrs drying time 4. Colour blue

List of supplier, product, number and notes on application (continue)

SUPPLIER	PRODUCT	NO.	NOTES ON APPLICATION
PROSTRUCT	EPOXY RESIN HIGH BUILD 632	19	1. Applied directly onto concrete surface 2. Easily applied by brush - airless spray or roller can be used 3. Colour fawn
VANDEX	VANDEX SUPER: (DARK GREY)	40A	1. Wet concrete surface thoroughly 2. Mix with water to a creamy consistency (2 parts/1 part water by vol.) 3. Apply with block brush
	SEMREP: (LIGHT GREY)	40B	1. Application the same as for VANDEX SUPER.
VIALIT	BITUMEN EMULSION	27	1. Apply primer - leave to dry (tacky) - applied by brush - 2. Final coat applied by brush 3. Easily applied 4. Black
WATERPROOF GUARANTEE	DURAPROOF	30	1. Silicone resin 2. Spray on application - air spray 3. Sprayed directly onto surface 4. The drier the surface the deeper will be the penetration 5. Very easily applied 6. Colourless
ZENDON	ZENLAY 345	22	1. Surface must be damp but not saturated 2. Applied directly to surface with brush 3. 2nd coat applied within 1 hr 4. Easily applied 5. Brushed or rolled on 6. White

PROTECTION OF CONCRETE IN WATER-COOLING TOWER STRUCTURES

R A Khartanovich

D N Lazovsky

Novopolotsk Polytechnical Institute,

USSR

ABSTRACT. Constant effect of the aggressive fluid and gaseous medium, systematic supermoistening-drying, freezing-thawing cause reduction of durability of the water-cooling tower structures as compared with the design figures. The secondary concrete protection often is not effective. The paper describes the problems of the primary concrete protection, based on the concrete structural theory. It includes dependence on quality enabling to use the model"concrete composition-structure -properties" to solve the technological problems: well . grounded comparison of concrete compositions, statistics accumulation of laboratory and production data and producing on this basis thick, corrosion resistant concrete.

Keywords: Water-Cooling Tower Structures, Cement Paste, Concrete Structural Theory, Aggressive Medium, Foliation, Density, Water Absorption

Roald A. Khartanovich is Reader at the Department of Reinforced Concrete and Stone Structures, Polytechnical Institute, Novopolotsk, BSSR, USSR. He is actively engaged in inspecting structures and buildings and working out recommendations on increasing durability of the structures by means of improving material structure by adding some additives into concrete.

Dmitry N Lazovsky is a Senior Lecturer at the Department of Reinforced Concrete and Stone Structures, Polytechnical Institute, Novopolotsk, BSSR, USSR. He is engaged in inspecting structures and buildings and has undertaken studies into research into detachable reinforced concrete structures and their improvement.

INTRODUCTION

At actual inspection of reinforced concrete water cooling
tower structures used for cooling industrial waters after
20 years of operation in heavy conditions it was found
that constant attack of aggressive fluid and gaseous
medium, systematic supermoistening-drying, freezing-thawing resulted in reducing durability of these structures
as compared with the design figures (1).This is applied
for example, to the foundation and wall reinforced concrete structures of the water accumulating basin.
Aggressiveness of technological waters, washing the
reinforced concrete structures, is characterized by
content of sulphates (0,6-0.9 g/l), chlorides (4-6.3 mg/l)
pH fluctuation (6-9), suspended particles (14-40 mg/l)
and specific impurities of hydrocarbons. The water
temperature fluctuates from 17 to $45^{o}C$, the ambient air
temperature changes from $25^{o}C$ to $-30^{o}C$.

RESULTS OF INSPECTION

During the inspection some defects and damages were found
among which the most substantial were: foliation and
destruction of the gluing insulation (fibreglass based
on epoxy glue) and the plastering insulation (compound
based on epoxy resin), concrete and reinforcement
corrosion. The structures, located in the zones with a
variable level of the technological waters, i.e., in
specially heavy operating conditions, have suffered the
most substantial damages. The structures having lower
concrete density, when in equal conditions, were found
to have suffered the heaviest damages.

As a result of these inspections durability of the
gluing and plastering anticorrosive covering was found
to be low. Alternative moistening-drying and
temperature fluctuation and also contraction occurred
in these coverings, having substantial thickness (2-5 mm)
in the first hand resulted in local foliation from
concrete and after accumulation brought to complete
destruction of the insulation. This damaging of the
insulation resulted in its turn in chemical and physical
corrosion of concrete and reinforcement. The walls
of the water accumulating basin, located in extremely
heavy conditions of the variable water level and
freezing-thawing, are liable to these types of corrosions
There were no substantial damages in the structures
but as a result of deterioration of water drainage
condition physical-mechanical characteristics of bottom
grounds were lowered.

ANALYSIS

As a consequence of the chemical reactions of interaction of aggressive medium with hydration products some not-well soluable salts are accumulating in pores and capillary of concrete. When concrete dries the salts get crystallized, giving strain to the pore walls, resulting in mechanical destruction of the concrete structure. Besides, during the alternative moistening-drying process there are some physical processes taking place washing away soluable components and mass transfer processes are increasing. Also freezing-thawing process accelerates corrosive processes, accompanied by internal strain in concrete at freezing of porous water, resulting in increased numbre and opening width of microcracks, concentrating strains in their mouths. Besides, changing from minus to plus temperature strains, appeared at the alternative freezing-thawing process of concrete, weaken adhesion between granules of coarse aggregate and cement-sandy stone due to the difference of the temperature expansion factors of these materials (2). The above mentioned multiple repeated power loads, applied to the pore walls and capillary, cause fatigue, weaken the links in the structural grid of the material and facilitate formation of microcracks in the cement stone, reducing more and more strength and resistance of the concrete.

Besides the above mentioned factors of mechanical physical structure destruction when designing durable concretes it is necessary to consider diffusive penetration of aggresive substances, and also possibility to wash away calcium hydrate $Ca(OH)_2$ from the protective layer. It is characteristic for the aggresive medium in our case to have ions SO_4^{--} in the industrial refrigerating water, causing formation of sulphoamominate corrosion. During the process of washing away calcium hydrate from the protective layer it was observed that the concrete strength and density lowered and also its alkalinity, characterized by hydrogen value of the porous fluid. Reduction of the cement stone alkalinity was proved by treating the samples taken from the concrete with the alcohol solution of phenol phthalein. Reducing hydrogen pH to 11.8 resulted in stopping passivation of the reinforcement in the concrete accompanied by intensive corrosion of the reinforcement, destruction and foliation of the protective layer.

RECOMENDATIONS

For working out recomendations how to reinforce the damaged structures we have applied the wellknown

method (3), allowing to evaluate expecting durability of
the structures with the given thickness of the protecting
layer, density of concrete and mineralogical composition
of cement or to prolong its durability. The method is
based on defining quantity of SO_3 bonded with cement and
CaO carried away and based on assumption that reduction
of concrete durability begins with the quantity of SO_3
bonded with cement to be above 2% at $Q_{SO_3} = 10.5\%$ the
concrete durability reduces 2 times, but at $Q_{SO_3} = 14\%$
complete destruction of concrete is taking place. As a
concrete density factor was taken its water absorption
W_{max}. The required water absorption at average structure
durability was defined using the formula

$$W_{max} = \frac{0.24 h_z + n_{C_3A} \cdot C_s \,(1 - e^{-0.05\sqrt{\gamma t}})}{1 - e^{-0.05\sqrt{\gamma t}}}, \text{ where}$$

t - durability
h_z - protection layer thickness (cm)
n_{C_3A} - C_3A content in cement (g/g of cement)

C_s - SO_4^{--} ion content in aggressive medium (g/l)

γ - factor, taking into consideration acceleration of
   destruction processes under effect of the
   alternative freezing/thawing process, which is
   equal according to the attained data to 1.13 (1)
It must be noted that this method can be used to solve
the opposite task; definition of protection layer
thickness at given concrete water absorption.

So the subject of more detailed study during designing
concrete having higher protective features such as
water absorption, penetration and diffusion. In the
first hand these features are defining durability of
concrete, make it possible to predict structures
durability, to work out measures against corrosive
destruction of the structures. Because concrete
durability and resistance are ensured at high quality of
the basic materials and advanced technology it is worth
while to put into life the principle of predicting
penetration features of concrete.

To our mind the concrete structural theory is meeting
these requirements to the greater extent(4). It
includes dependence on quality enabling to use the model
"concrete composition-structure-properties" to solve
technological problems: well-grounded comparison of
concrete compositions, statistical accumulation of

laboratory and production data and producing on this basis concrete with the specified properties including corrosion resisting concrete. The theory is based on the assumption that the matrix, formed during cement hydration, is "a primary framework" exerting decisive influence on the cement stone structure. The mixture composition at the time of transforming the cement paste into cement stone at the final setting of cement predetermines the concrete properties. The concrete structure is changing in time depending on the conditions of further hardening, but it is an inherited function of the volumetric concentration of the cement paste in concrete - C_v and its true water-cement ratio - W/C. Under the true water cement ratio is meant such a ratio W/C, at which the concrete mixture has the same mobility and setting time, as the pure cement paste has.

Without changing the essence of the concrete structural theory, we suggested to simplify that part of the model, which is characterized by the true water cement ratio index (5). In place of the true water cement ratio is assumed as an actual water cement ratio in concrete - W_α/C at which the mobility and setting time are not connected with the hypothetical cement paste. And here with the principle to consider aggregate water requirements is kept. For this aim some mixture making water is considered to be used for wetting sand and crushed stone. It is assumed that:

$$W_\alpha = W - S \cdot F_s - R_m \cdot F_{zm} , \qquad \text{where}$$

F_s and F_{zm} - water absorption factors for fine and large -sized aggregates

S, R_m - consumption of sand and crushed stone

The volumetric cement paste concentration in concrete at the initial period mixture making can be expressed as a sum of cement and water volumes:

$$C_v = \frac{C}{\rho_c} + W_\alpha$$

It is reasonable to introduce the second graphic criterion to consider in complex water requirements for aggregates and for cement as well, which is the quotient obtained when the actual water cement ratio is divided by the normal thickness of the cement paste - $F_{N\,thk}$

$$X_\alpha = (\frac{W_\alpha}{C})/ F_{N\,thk} \qquad , \text{ then}$$

$$W_a = X_a \cdot C \cdot F_{N\,thk} \quad , \quad \text{and}$$

$$C_v = \frac{C}{\rho_c} + X_a \cdot C \cdot F_{N\,thk} = C\left(\frac{1}{\rho_c} + X_a \cdot F_{N\,thk}\right) \ .$$

The volumetric concentration of aggregate can be expressed as:

$$A_v = 1 - C_v \ .$$

The main advantage of the described model that it is clear and informative as compared ,for example, with the model being solved using the conventional method of the absolute volumes. Really C_v quantitatively characterizes relative content of the cement paste in concrete, X_a = aggregate and cement water content. The design model considers all the major physical properties of the concrete mixture components: density (ρ_c, ρ_s, ρ_{tm}) water absorption (F_s, F_{tm}), normal thickness of the cement paste ($F_{N\,thk}$). The relation between the fine and large-sized aggregate can be expressed as follows;

$$S = \frac{A_v \cdot \rho_{tm} - R_m (1 + \rho_{tm} \cdot F_{tm})}{\rho_{tm}(1 + \rho_s F_s)} \rho_s \ .$$

The indicated criteria clearly characterize the basic concrete mixture composition, the initial concrete structure, the matrix or framework, on which the hydrating cement system is built. The value X_a,actual cement paste water content and consequently fully indicates the assumed concrete density, because the main source of porosity is excess of water in the initial concrete mixture. The water content in the cement paste is characterized by the limiting indices (4)at X_a=0.876 relatively full watering of the cement particles is taking place, at $X_a = 1$ the cement paste contains optimum quantity of water, ensuring complete hydration of cement, at $X_a = 1.65$ the critical water content is reached, at which the cement paste begins to foliate.

Having formalized in such a way the indices of the initial concrete mixture composition more favourable conditions were attained to use mathematical models for solving technological problems, production of thick and durable concretes, ensuring reliable primary protection of the structures, included. To solve the above problems using the methods mathematical experiment planning relation between water absorption - W_{max} ,average pore size index - $\bar{\Lambda}z$ and the pore size homogeneity index - α and water content - X_α and volumetric concentration of the cement paste in concrete - C_v. The concrete porosity indices were defined with kinetics of water absorption using known solutions of the absorption equations by polycapillary material of wetting liquid - water. These solutions were modified to some extent in comparison with the norms (5). The pore size homogeneity index was defined using the expression:

$$\alpha = \ln \left[\frac{\ln \left(1 - \frac{W_{2.72}}{W_{max}}\right)}{\ln \left(1 - \frac{W_1}{W_{max}}\right)} \right] \quad ,$$

and the average pore size index was defined using the expression:

$$\bar{\Lambda}_z = \frac{\ln \left[\ln \left(\frac{W_{max}}{W_{max}} - W_{2.72} \right) \right]}{\alpha} - 1 \quad ,$$

where W_{max}, $W_{2.72}$, W_1 - maximum concrete water absorption after 2.72 hours and 1 hour respectively.

Studying the compositions while solving the problems of producing durable concrete for cooling water towers degree of water absorption $X_\alpha = X_1$, and volumetric concentration of the cement paste $C_v = X_2$ were assumed as the factors defining variability of the technological parameters. The plan of the experiments provided changing the factors on 3 levels. The properties (W_{max}, $\bar{\Lambda}_z$, α) were studied on 7x7x7 cm cube samples at the age of 28 days, 3 twin-samples for each concrete composition. The obtained results were processed using methods of mathematical statistics resulting in algebraic equations of dependence in which water absorption, pore size homogeneity index and average pore size index are related to the criteria, characterizing the initial concrete composition. The equations, having particular features, are not given here. But the detailed analysis of these algebraic equations enabled us to carry out optimization of the compositions, that is to say, to find such a final solution for the initial composition, which ensures production of the set concrete having technical properties not lower than required ones at minimum consumption of cement.

The essence of this operation is clearly represented when the isolines were considered. For example, when plotting

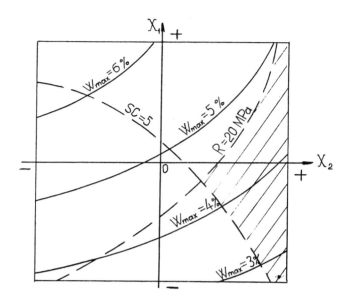

Figure 1: Isolines R, SC, W_{max}

the isolines W_{max} , R, SC (Figure 1) the range of the
values X_1 and X_2 is clearly represented, in which the
required values of the concrete water absorption $W_{max} \leqslant$
$\leqslant 5\%$, durability R $\geqslant 20$ MPa, mobility SC $\geqslant 5$ cm are
executed.

When using the above methods for predicting concrete
properties including those used for production of thick
and durable concrete in the industrial conditions,
additional criteria can be suggested, graphical
representation of which will be useful for making
additional nomographs. So at constant consumption rate
of cement (C = const) we have (Figure 2)

$$C_v = \frac{C}{\rho_c} + X_a \cdot tg\alpha ,$$

where
$tg\alpha = C \cdot F_{N thk}$ may be called as a factor effecting
cement consumption.

At constant degree of water saturation X_a = const) we
get (Figure 3)

$$C_v = C(\frac{1}{\rho_c} + X_a \cdot F_{N thk}) = C \cdot tg\beta ,$$

where $tg\beta = (\frac{1}{\rho_c} + X_a \cdot F_{N thk}) = C_v / C$ and may be

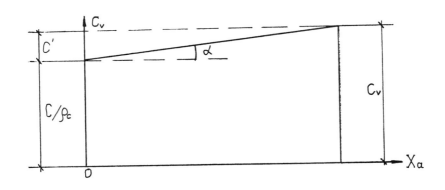

Figure 2: Graph of dependence $C_v = f(X_a)$

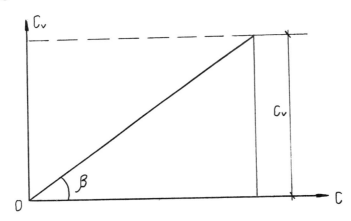

Figure 3: Graph of dependence $C_v = f(C)$

water content factor.

At constant volumetric concentration of the cement paste in concrete (C_v = const) we get (Figure 4)

$$X_a = \frac{C_v}{C \cdot F_{N\,thk}} - \frac{1}{\rho_c \cdot F_{N\,thk}} = \frac{C_v}{tg\alpha} - F_{qc}$$

where

$$F_{qc} = \frac{1}{\rho_c \cdot F_{N\,thk}}$$ and may be called a constant of cement quality.

When accumulating statistical data nomographs can be made for operative selection of concrete compositions,

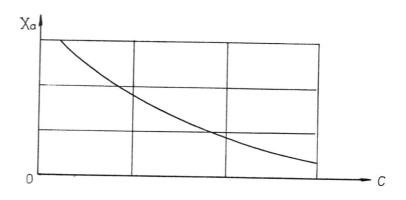

Figure 4: Graph of dependence $X_a = f(C)$.

having predetermined properties for optimizing and
correcting these compositions with partial changing of
the initial materials or their properties and for
reducing volume of the control tests.

CONCLUSIONS

Based on the experience of operating the water cooling
tower structures the causes of corrosion destructions of
the concrete and reinforcement have been analysed. Some
suggestions on designing protection of concrete and
increasing durability of such structures have been
considered. Designing of the primary protection and
production of concrete with the specified thickness,
water absorption without additional cost have been
considered. New criteria in the concrete structural
theory have been suggested and possibility of their
practical application in construction when selecting
concrete compositions and their optimization.

REFERENCE

1   KHARTANOVICH R A, LAZOVSKY D N, SOROKIN A V.   Actual
    inspection of the ventilating water towers.
    Industrial Construction, USSR, 1989, pp 16-17

2   MOSKVIN V M, KAPKIN N M, MAZUR B M, PODVALNY A M.
    Durability of concrete and reinforcement of negative
    temperature. Stroiisdat, USSR, 1967.

3   Manual for providing durability of the reinforced

concrete structures during reconstruction and renewal at the plants in the ferrous metallurgy. Stroiisdat, USSR, 1982.

4   BAZHENOV Y M, GORCHAKOV G I, ALIMOV L A, VORONIN V V. Production of concrete with the specified properties. Stroiisdat, USSR, 1978.

5   KHARTANOVICH R A.   Structural characteristics and properties of concrete with admixture of ethylene production waste.   Dissertation for bachelorhood. Rostov-on-Don, 1985.

AGGRESSIVE UNDERGROUND ENVIRONMENTS: FACTORS AFFECTING DURABILITY OF STRUCTURES AND SPECIFICATION OF APPROPRIATE PROTECTIVE SYSTEMS

C R Ecob

E S King

Mott MacDonald Group

United Kingdom

ABSTRACT : Achieving long term durability of reinforced concrete buried structures, constructed in seawater environments can be considerably more difficult than that normally associated with above ground structures. The design life requirements and severity of conditions are dependent on the type of use, i.e. railway or road tunnel, sewage outfall, etc. and the factors likely to affect durability vary. Particular risks include sulphate attack, carbonation, chloride induced and stray current corrosion, bacterial attack and contaminated ground. There is evidence of an increasing number of durability failures.

It is becoming possible to predict the durability of protected reinforced concrete structures by a calculation of the time to initiation of corrosion and it's rate. The time to loss of serviceability in aggressive environments can often be less than the required life and in such cases the specification of additional protective measures as construction is therefore considered to be necessary.

Various approaches to enhancing durability are considered, including the use of surface coatings to concrete, cathodic protection, admixtures and an innovative use of fusion bonded epoxy reinforcement coatings. The development work and application of such techniques to actual structures is discussed.

Keywords : Durability, underground structures, protective measures.

Chris R. Ecob : is an Associate in the Special Services Division of Mott MacDonald. He is currently advising on durability aspects on the Channel Tunnel, Great Belt Tunnel in Denmark and for undersea outfall tunnels and pipes in the USA and Hong Kong.

Elizabeth S. King : is a Senior Engineer in the Special Services Division of Mott MacDonald. She is involved in a number of projects looking at durability of new structures and investigation and repair of existing structures.

INTRODUCTION

The increasing demands of society for an improved environmentally friendly infrastructure, is pushing construction underground. Often this necessitates construction in below sea environments and in estuarine or reclaimed land. Such structures range from basements for office/industrial complexes, long-term storage facilities for hazardous wastes to large tunnels constructed for transportation. It is very important for such structures to be durable owing to the difficulties associated with inspection, maintenance and repair and consequently clients require that structures have long service lives with minimal requirements for maintenance.

Where the soil or groundwater surrounding the structure is benign, (i.e. not containing substances aggressive to concrete or other construction materials) durability problems can be less severe than for above ground structures as reinforced concrete benefits from steady damp conditions.   However, it is often found that exposure conditions underground are more aggressive than those generally considered in existing structural codes of practice.

The durability of structures can be affected by both external exposure conditions and the internal environment.   The most commonly encountered external factors are levels of sulphate and sodium chloride, organic matter for bacterial attack where there is contaminated reclaimed ground, low pH, salt water and the multitude of less common chemicals that are aggressive to concrete and metals. As a result of groundwater pressures developing around a buried structure, aggressive substances are driven in by the pressure differential as well as diffusion and may concentrate on the inside as a result of internal dyring.   The most significant internal factors are the temperature and relative humidity (RH) with other aggressive substances that are contained in the structure during operation, e.g. hydrogen sulphide (H_2S) in sewers or concentrated groundwater in drains.

Requirements for watertightness vary enormously between structures used for varying purposes, e.g. railway tunnels can tolerate a significantly higher water inflow than a basement of a building containing sensitive computer equipment. It can be considered that the basis for "watertight" reinforced concrete structure is good quality design and construction of the concrete and jointing arrangements.   The principal factors in producing watertight concrete are the four 'C's' – Constituents of the mix, Cover, Compaction and Curing (ref. 1).   Although achieving the above is a good start in producing durable concrete structures, a number of other performance characteristics need to be satisfied, and are discussed in this paper.

The significance of loss of durability of underground structures varies according to the form and use of the particular structure.   The principal consequences are likely to be as follows:-

(i)   Risks to users or equipment from spalling concrete or significant water inflows.

(ii)   Risks of reducing the operational capacity by reduction of the structure gauge (particularly in tunnels where excessive lining deformation could occur) or due to the additional space requirements of remedial measures.

(iii)   Costs of unexpected remedial works, both to the structure itself, but also services such as electricity, ventilation, and equipment located in the structure.

(iv)   Loss of operation loss and convenience to users. Particularly in the case of transportation tunnels.

FACTORS INFLUENCING DURABILITY

The principal factors affecting durability performance are illustrated by figure 1 and are listed below.

1)   Chloride Induced Corrosion of Reinforcing Steel

Structures located undersea or close to estuaries rivers will usually be surrounded by chloride laden groundwater, (NaCl). The ingress of chloride ions into the concrete can have the effect of depassivating the protective layer around reinforcing steel formed by the highly alkaline cement paste, if they are permitted to reach the reinforcement. Once the passivation layer is broken down, corrosion cells can be set up, and corrosion in the form of pitting can occur, (ref. 2). Case history studies of structures subject to saline groundwater have revealed that chloride ions can migrate up to 100mm into good quality OPC concrete of specified 40 N/mm^2 strength in less than 10 years, leading to severe reinforcement corrosion, (ref. 3).

The principal mechanisms by which chlorides pass through and concentrate in the concrete are as follows:-

(i)   Hydraulic Pressure Gradients through thickness of wall.

The flow of groundwater through the wall is governed by the permeability of concrete and results in a continuous inflow of chlorides. The permeability of concrete will be several orders of magnitude lower than the surrounding ground, and it can be assumed that the concrete wall/lining will carry virtually all the hydrostatic pressure head.

(ii)   Ionic Diffusion.

Chloride ions penetrate the concrete by movement through the water in the capillary pores. The rate that they move is related to:

- their concentration gradient.
- the size, number and orientation of capillary pores which can be dependent on the mix constituents of the concrete and their proportions and the curing procedures.
- the cement chemistry (i.e. the C_3A content which has the ability to bind chlorides).

(iii)   Capillary and Drying Effects.

Water movement through the concrete structure will be influenced by the internal environment. As groundwater passes through the concrete from the saturated outer face towards the inner face, a point will be reached where the water supply rate is equal to the evaporation rate into the structure. At this point all soluble salts contained in the water, will be deposited into the pores of the concrete. As a result, previously innocuous concentrations of chlorides and other salts can be increased to levels that can attack the reinforcing steel or lead to salt crystallisation damage.

(iv)   Wetting/Drying Effects.

Wetting/drying effects can be readily observed in structures containing construction joints or other features subject to leakage. Depending on the environmental conditions within the structure, the wet/dry interface can move having the effect of increasing the concentration of salts. During the wetting

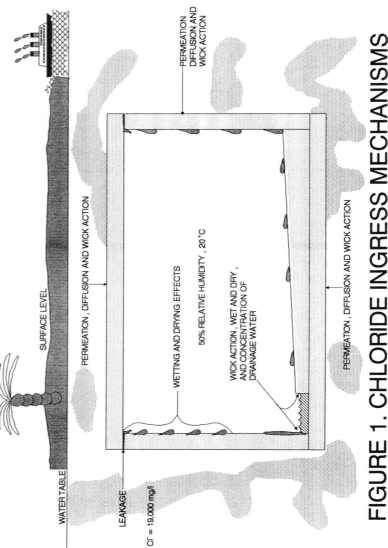

FIGURE 1. CHLORIDE INGRESS MECHANISMS

stage, groundwater will be absorbed into the outer surface of the concrete, and during drying, the water will evaporate leaving behind any salts in the concrete. Concentration of salts increases with subsequent cycles of wetting and drying.

The reinforcing steel at greatest risk from corrosion is that on the internal faces of the structure or adjacent to joints (precast segmental tunnel linings), although external reinforcement could be subject to corrosion due to the "hollow-leg" phenonmena which has been observed on oil rig legs. This occurs when corrosion cells are established with the external concrete saturated with chloride ions acting as the anode being driven by a large, oxygen rich cathode at the inside face of the structure.

There is debate regarding the concentration of chlorides required to initiate corrosion although 0.4% Cl^- or 0.2% Cl^- by weight of cementitious material (OPC or SRPC respectively) are currently referred to.

2)   Reinforcement Corrosion due to Carbonation

Carbon dioxide present in the structure forms carbonic acid is the presence of moisture, which then reacts with the hydration products of cement, particularly calcium hydroxide $(Ca(OH)_2)$ which then reacts to form calcium carbonate.

$$Ca(OH)_2 + CO_2 \rightarrow CaCO_3 + H_2O$$

If carbonation is permitted to reach the level of reinforcement, it reduces the highly alkaline passivation layer around the reinforcement leaving it susceptible to corrosion.

The extent that carbonation occurs depends on the concentration of CO_2, which could be as high as 3% in railway tunnels operating diesel locomotives, the permeability of the cement paste, the moisture content of the concrete and the ambient temperature and relative humidity (maximum carbonation when RH is between 50-90%). Although risk of carbonation is low in good quality concretes (40 N/mm² and above), carbonation can still occur down cracks to reach the level of reinforcement in quite short periods.

3)   Alkali Aggregate Reaction (AAR)

This occurs when reactive silica in the aggregate reacts with alkalis in the cement to form an alkali silicate gel which attracts water by absorption or osmosis and tends to increase in volume. As a result of the gel being confined by the surrounding cement paste, internal pressures occur which eventually lead to expansion, cracking and disruption of the cement paste and map-cracking of the concrete.

Factors influencing the progress of AAR include the nature and porosity of the aggregate, the alkali content of the cement and other mix ingredients. In addition where structures are subjected to ingress of groundwater, alkali's (e.g. sodium) from external sources can increase the available alkali content to trigger the reaction, within the concrete, (ref. 4).

Due to the above factors it is recommended that alkali aggregate reaction is controlled by specifying non-reactive aggregate, rather than trying to limit alkali levels. However, current U.K. simple rules for selecting non-reactive aggregates are not in themselves not rigorous enough for contract specification and quality control on major projects with a long design life.

4)   Sulphate Attack of Concrete

In solution, sulphates found in groundwater, e.g. sodium, calcium or magnesium sulphates, react with the calcium hydroxide $(CaOH)_2$ and the hydrate C_3A in the cement paste to form gypsum and ettringite. These products occupy greater volumes than the compounds they replace, so that expansion and disruption of the hardened concrete takes place. When the source of sulphates is in seawater, the potential harmful effects are diminished owing to the fact that gypsum and ettringite are more soluble in chloride solution than in water, and can therefore be more easily leached out. This results in reduced mechanical damage, although there is a slow increase in porosity and therefore decrease in strength, (ref. 5).

Following the guidelines for mix design recommended in BRE Digest 250 (ref. 6) has proven satisfactory for varying groundwater sulphate concentration although where sulphates are present in seawater environment, a C_3A level for cement that limits risk of chloride related corrosion is recommended. The standard for Maritime Structures (BS6349) recommends in a recent amendment that a C_3A level of between 4 and 10% should be used.

5)   Aggressive Chemical Attack

The form of attack can originate from several sources.

(i)   Acidic Groundwater.

       The hydration products of portland cements require an alkaline environment to retain their stability, and if anything alters the pH balance in the pore water, the stability of the cement paste will be affected. If the groundwater is acidic, it can decompose the hydration products, dissolving the lime (CaO) with continued leaching, eventually leaving a residue of incoherent hydrated silica, iron oxide and alumina. Deterioration of unprotected concrete will normally occur with pH levels of less than 6.

(ii)   Aggressive substances origination from inside structure.

       A wide range of chemicals, particularly acids and some hydrocarbons are aggressive to concrete. The risk of spillage is particularly prevalent in underground storage facilities or railway tunnels where spillage can occur from trains.

       If concrete surfaces are likely to be affected by aggressive chemicals it is usual to apply protective coatings that are compatible with the environmental and operating requirements of the structure concerned.

6)   Bacteriological Attack

Deterioration of concrete and metals can occur as a consequence of bacterial activity. This is particularly so when groundwater contains ferrous iron or reduced sulphur compounds. Both aerobic and anaerobic bacteria co-exist in the ground, their prospective dominance dependent on the availability of oxygen. Although in general, anaerobic bacteria will be dominant with underground structures, aerobic conditions can occur as a result of:-

       -    oxygen diffusion from inside the completed structure through construction joints.
       -    disturbance of the ground due to construction activities.
       -    draw down of oxygen containing groundwater.

The principal reactions that can occur include:

(a) Anaerobic bacterial action in clays containing organic material, breaking down the proteins and hydrocarbons to release hydrogen sulphide and methane.

(b) Anaerobic bacterial action, reducing sulphate within the soil to form iron sulphides which in turn react to form hydrogen sulphide and methane.

(c) Aerobic bacterial action, oxidising organic and inorganic sulphur compounds, this sulphates and sulphides to form sulphates and free sulphuric acid.

(d) Anaerobic bacterial action, oxidising hydrogen sulphide (produced by aerobic bacteria in reactions (a) and (b)) in the presence of water to form sulphuric acid. This reaction is observed and forms the principal deterioration factor in concrete lined sewers.

If bacterial attack is a potential risk, then the specification of protective membranes/coatings should be considered in a manner similar to those for chemical attack.

7) Mechanical/Structural Damage

Long term durability is significantly influenced by the design and construction procedures adopted for the structure. Factors where durability can be influenced include:-

-   structures, particularly tunnel linings, should be designed to act in compression as much as possible, it must be ensured that loadings occurring in service do not result in excessive cracking in the tension faces.

-   jointing arrangements should be so designed to either make them watertight or to channel water to drainage systems.

-   construction procedures should not exert undue stresses in the permanent structure, e.g. in tunnel construction loadings from Tunnel Boring Machines.

-   use of concrete mix constituents and proportions and casting procedures, e.g. vibration and curing that limit thermal and shrinkage cracking and provide a dense homogenous concrete.

Cracks within concrete even when as narrow as 0.1mm can be subjected to preferential reinforcement corrosion due to carbonation fronts and chloride ingress down the crack.

SPECIFICATION TO ENHANCE DURABILITY PERFORMANCE

A large number of options exist to improve the durability performance of new underground structures, that can be used singly or in combination. Their adoption for a particular situation depends on several factors including, ease of application, effectiveness in relation to risk of damage occuring during construction, whether the system can be applied retroactively and also the cost benefit and ease of and requirements for maintenance.

Potential Approaches in Improving Durability Performance Include:-

1)   High Quality Reinforced Concrete.

The principal factors in producing durable reinforced concrete structures are:-

-       establishing optimum cover to reinforcement.
-       selection and proportioning of concrete constituents.
    -       cementitious materials
    -       aggregate porosity and grading
    -       water/cementitious material ratio
    -       admixtures.
-       vibration techniques to provide optimum compaction.
-       strength gain requirements particularly with precast units or walls requiring early backfilling.
-       curing methods to optimise hydration and minimise drying shrinkage cracking and micro-cracking.
-       cooling requirements to prevent thermal cracking.

The use of cement replacement materials such as Pulverised Fly Ash (PFA) and Ground Granulated Blast furnace slag (GGBFS) with very low w/c ratios (0.35 - 0.40) is known to provide concrete of very low permeability and high ionic diffusion resistance. Diffusion Coefficients as low as 500×10^{-15} m^2/sec are now achievable with current concrete when GGBFS or PFA are included in the mix in approximate proportions of 70% and 30% of cement respectively.

Testing methods that can measure extremely low diffusion (D) and permeability (K) coefficients of concrete and are suitable for normal site quality control testing, have recently been developed (ref. 7).

2)   Protective Surface Coatings.

A wide range of coating generic types are currently available to provide protection against chloride and sulphate ingress, carbonation acid and other aggresive chemical attack, freeze thaw and abrasion. Numerous factors should be taken into account in selecting suitable coatings including:-

-       Location of application - inside or outside of structure, precast or insitu.
-       Resistance to alkalis, acids, chlorides, sulphates, and other aggresive substances.
-       Permeability to water vapour and water.
-       Bond to substrate.
-       Toughness against impact.
-       Flamability and emmision of toxic fumes.
-       Surface preparation and application requirements.
-       Durability and associated maintenance costs.
-       Ease of maintenance re-application.

The generic types and the suitability of application for underground structures can be generally summarised as below:-

(a)   Surface Coatings.

These are viscous fluids that form a film on the surface of concrete and provide a protection as a result of the qualities of the film itself. They are often extended or filled to provide greater applied thickness.

Examples include : epoxies, polyurethanes, vinyls, acrylics and bitumens.

The selection of a particular generic type is dependent on such factors as protection provided, applicability and flammibility characteristics.

(b)   Sealers.

These are viscous fluids which both penetrate concrete and form a thin film on its surface. They are sometimes used as primers to coatings.

Examples include : epoxies, polyurethanes and acrylics.

They are unlikely to provide suitable protection in the presence of strong chemical attack risk.

(c)   Penetrant – Pore Liners.

Low viscosity fluids that penetrate several millimeters into the concrete. They react with the substrate and line the pores of the concrete. These hydrophobic materials work by repelling water and restrict access of substrates in solution, e.g. chloride ion and carbon dioxide. They are not suitable for use in saturated conditions or for protection against chemical attack.

Examples include : silanes, siloxanes and silicones.

3)   Fusion Bonded Epoxy Coated Reinforcement.

The application of FBECR is essentially a controlled factory operation. Cleaning is required by decontamination and then blast cleaning with steel grit. The next step in the process is to uniformly heat the bar to a temperature in the region of 230°C. Once heated the bar passes through a powder coating machine and the coating is allowed to cure, usually by reheating.

To date the principal technique for coating application has been by electrostatic spraying onto straight bars, which are subsequently bent to the shape required after the coating has cured.

As an alternative the bars could be coated by fluidised bed dipping (FBD) following bending and fabrication (ref. 8,9). The advantages of this technique are:

–   complex reinforcement cages that benefit from pre-fabrication, such as tunnel lining segment reinforcement, can be coated.

–   the bending of coated bars is avoided and thus any potentially weak areas of coating on the outside of bends.

–   a thicker coating can be provided which would enhance the chemical and electrical resistance of the steel.

Therefore a thicker and more durable coating can be specified with FBD than is the norm for electrostatically sprayed reinforcement. With all epoxy coated steel some reduction in pullout resistance, an increase in crack widths and greater deflections must be considered in structural design. This can normally be achieved by small modifications to the design details.

4)   Cathodic Protection.

Cathodic protection (CP) is most effectively employed on reinforced concrete where substantial areas of the concrete may become laden with chloride ions and large scale repairs or complete replacement are unacceptably disruptive. For effective application cathodic protection needs a uniform electrolyte condition in and around the concrete. Buried structures especially pipelines and the extension of immersed tube tunnels are then naturally suitable for C.P. The interior conditions of trafficked tunnels with strong gradients in electrical characteristics from wet to dry areas are not so amenable to C.P.

The types of anode systems presently available for applied current cathodic protection are as follows:

-    metal oxide coated titanium mesh laid in a cementitious overlay
-    conductive polymer wire laid in a cementitious overlay
-    conductive coating
-    sprayed metallic coating
-    distributed wire system
-    conductive asphalt

The most appropriate anode for a particular structure would depend on the nature of the installation. Guidance on specification for cathodic protection system can be found in a number of publications (ref. 10).

5)   Special Concrete Admixtures.

(a)   Corrosion Inhibitors.

Corrosion inhibitors are available which are able to protect the reinforcement by developing a competing oxidiation reaction between nitrite ions and the steel resulting in a regeneration and/or thickening of the passive film. Such products may not be suitable for structures subjected to chloride ions and differential pressure.

(b)   Permeability Reducers.

The principal components of permeability reducers are ammounium stearate and emulsified asphaltic globules. Ammonium stearate reacts with calcium hydroxide to produce calcium stearate which is precipitated on the inner surfaces of capillaries in the pore structure to form an impermeable coating on the surface. Electrical fields created by the stearate have the effect of repelling water molecules instead of allowing the surface tension of the capillary surfaces to attract water. The asphaltic globules remain in the capillary pores, and when water under pressure enters the capillaries, the globules come together and coalesce into a type of asphaltic plug.

CONCLUSIONS

It can be seen that the estimation of life of a concrete buried structure is a complex matter because of the large number of factors involved that influence its durability. The relevant factors must be assessed individually for a particular structure and the measure to be adopted for enhancing the durability selected accordingly from the many options available.

REFERENCES

1. SOMERVILLE, G., "The Design Life of Structures", Proceedings of Institution of Structural Engineers, Volume 64A Number 9, September 1986.

2. PAGE, C.L., TREADWAY, K., Aspects of Electrochemistry of Steel in Concrete. Nature Magazine, 13 May 1982, Volume 297, page 109-115.

3. BAEKLAGSUNDERSOGELSE I STOREBAELTSOMRADET, 1979, (Chloridtraenging i. beton), Institutet for Metallaere – Denmarks Tekniske Hojskole.

4. Concrete Society Technical Report, No. 30 – Alkali Silica Reaction – Minimising the risk of damage to concrete, Guidance Notes and Model Specification Clauses, October 1987, Section 7.

5. A.M. NEVILLE., Properties of Concrete. 3rd Edition 1983, page 452.

6. BUILDING RESEARCH ESTABLISHMENT. Digest 250, Concrete in Sulphate – Bearing Soils and Groundwater, June 1981.

7. WOOD, J.G.M., WILSON, J.R., LEEK, D.S., "Improved Testing for Chloride Ingress Resistance of Concretes and Relation of Results to Calculated Behaviour", 3rd International Conference on Deterioration and Repair of Reinforced Concrete in the Arabian Gulf, Bahrain, October 1989.

8. ECOB, C.R., KING, E.S., ROSTAM, S., VINCENTSEN, L., "Epoxy Coated Reinforcement Cages in Precast Concrete Segmental Tunnel Linings – Pre-production Trials and Manufacturing Procedures", XIth International Congress on Prestressed Concrete Technology FIP 90, Hamburg, June 1990.

9. ECOB, C.R., KING, E.S., ROSTAM, S., "Epoxy Coated Reinforcement Cages in Precast Concrete Segmental Tunnel Linings – Durability, 3rd International Symposium on Corrosion of Reinforcement in Concrete Construction", The Society of Chemical Industry, The Belfrey, Warwickshire, May 1990.

10. Concrete Society Technical Report No. 36. Cathodic Protection of Reinforced Concrete, 1989.

INVESTIGATION OF DETERIORATION OF CONCRETE IN A CHEMICAL PLANT

T P Ganesan

P Kalyanasundaram

R Ambalavanan

S K Sharma

Indian Institute of Technology,

India

ABSTRACT. The paper reports the investigation of corrosion damages of structural concrete in an ammonium chloride plant building in the coastal area of North Madras, India. Ammonium chloride can react with the free lime present in the concrete members in a base exchange process to form ammonium hydroxide and calcium chloride. This reaction causes porosity in concrete due to volatilization of ammonia and leaching of calcium chloride; pH value of concrete also gets reduced, thus producing an environment conducive to corrosion. The details of investigation and the remedial measures adopted are described in the paper.

Keywords: Base Exchange, Concrete, Corrosion of Steel, Guniting, Leaching, pH, Porosity, Remedial Measures.

Dr T P Ganesan is Professor and Head of Civil Engineering Department, Indian Institute of Technology (I.I.T), Madras. He has over 35 years' experience in teaching, research and professional work in the areas of structural engineering, model analysis and design, prefabrication, housing and energy conservation.

Dr P Kalyanasundaram is Professor and Head, Building Technology Division at Civil Engineering Department, I.I.T, Madras. He has over 35 years' experience in teaching, research and consultancy in building materials, concrete technology, construction management and maintenance.

Mr R Ambalavanan is Senior Scientific Officer in the Department of Civil Engineering, I.I.T, Madras. He has over 20 years of professional experience including teaching and research in concrete technology and masonry.

Mr S K Sharma is Executive Engineer, Central Public Works Department, Government of India and is currently doing research on the deterioration of reinforced concrete in adverse environments at I.I.T, Madras.

INTRODUCTION

In reinforced concrete the external environment most often affects the concrete and through it, the reinforcement (1,2). When reinforced concrete shows some signs of distress such as cracking, spalling or rust staining, it becomes necessary to carry out investigations to find out how serious it is, what has caused it and how it can be set right (3). This paper discusses one such investigation recently undertaken to study the corrosion damage in the structural members of a factory building housing ammonium chloride plant in the coastal area in North Madras. The scope of the work included attempts to evolve suitable remedial measures to make the building serviceable for a further period of 5 to 10 years.

SITE SURVEY

The chemical reactor plant building is a reinforced concrete framed structure of five floors with a water tank located at the fifth floor (Figure 1). At the time of inspection, a large number of the reinforced concrete components of the reinforced concrete components of the building including the staircases had already suffered corrosion damage to varying degrees in about seven years of service exposure to chemically aggressive environment contaminated with ammonium chloride, vapours of ammonia and hydrochloric acid. It could be ascertained on a rough estimate that about 50% of the structural elements were badly affected by corrosion damage.

The production process being carried out in the building was the manufacture of ammonium chloride from the chemical combination of the raw materials hydrochloric acid and ammonia. In some areas around the processing plant, overflow and leakage of the chemical solutions were noted and there was no proper provisions for drainage and overflow of the effluents. The floors were also exposed to the action of chemicals carried by wind and rain water since the construction was of open frame type. Also the effluents overflowing from around the plant could be carried by the rain water to various locations in the floor. Some of the open drainage channels were seen to be heavily incrusted with the chemicals. The columns located in the immediate vicinity of open channels carrying chemicals and rain water were heavily damaged by corrosion. No proper slopes or isolation arrangements were provided for adequate drainage of rain water and effluent water separately in any of the floors.

INVESTIGATIONS

The role of thorough investigation and systematic site surveys for analysis of deterioration in concrete structures and for arriving at the protective measures, cannot be overemphasized. It is of interest to note that the concrete research group at Queen Mary College, University of London is carrying out a major four-year study for evaluating procedures and techniques for inspecting, appraising and field testing various types of concrete structures (4). In site surveys, the structure's history and environment should be studied first. The results of previous studies, if any, should also be consulted. Visual and photographic survey should be carried out and supplemented by non-destructive tests. The detailed survey should establish concrete covers and corrosion state of embedded steel. Concrete and steel samples should be obtained to make a comprehensive study.

In this investigation, an attempt was made to establish the cause of damage. Keeping in view the objective of the repair viz., and restoring the structure as nearly as possible to the original condition, areas requiring repair were identified and the protective measures were evolved based on the needs, feasibility and cost of repair.

All the components of the building were critically inspected and in particular the locations of the damaged areas, to ascertain the physical condition of the concrete and its chemical nature with respect to alkalinity and presence of soluble chlorides. Detailed survey and mapping of the corroded members were carried out for assessing the exact extent of corrosion, establishing the causes of deterioration and arriving at suitable measures. Damaged concrete was removed at selected locations to establish the depth of concrete cover that has suffered significant loss of alkalinity. The condition of embedded reinforcements were assessed at different locations in affected portions as well as in apparently undamaged concrete. The details of the observations are briefly outlined below.

Cracks in Concrete

Longitudinal cracks in concrete along the reinforcements were generally 0.1 to 0.5 mm wide in columns and 0.1 to 0.2 mm in beams. Large cracks upto 15 mm wide in a few columns and upto 20 mm in a few beams were also noted in a few cases. The affected ceilings showed hair cracks and rust stains in some panels but no severe damage was noted. Some ring beams were severely corroded with spalling of concrete cover and corrosion of exposed reinforcement (Figure 2).

Figure 1 : A general view of the ammonium chloride plant building

Figure 2 : A corroded ring beam

Alkalinity Tests

It has been established that the depth of carbonation in concrete can be assessed visually by means of a spray test (5). In this test an indicator solution such as phenolpthalein is sprayed on to a freshly broken surface of concrete and those areas on the concrete which turn pink in colour indicate alkaline concrete with pH more than 9.5 or 10. It was found from the present study that the spray test was very useful in identifying areas where the pH of concrete has been lowered significantly by heavy chloride intrusion. The spray test was used as a rapid method for identifying the damaged areas and the depth of concrete cover that had been affected in these areas. Samples of concrete were taken both from damaged and undamaged areas and tested in laboratory. The soluble chloride was found to be as high as 28000 ppm, and pH value between 8 and 9.5 for the damaged concrete.

Extent of Rusting of Steel

In case of members which exhibited signs of corrosion cracking along the direction of the main reinforcing steel, the extents of corrosion were ascertained by direct measurement of the reduction of diameter of the steel rods using calipers, after breaking open the concrete cover. Addition of make up reinforcement was suggested upto 25%, 50% and full area of cross-section depending upon the extent of corrosion and strength tests on steel specimens taken out of the corroded members.

Assessment of Strength of Concrete

A Schmidt Rebound Hammer (ELE make, UK) was used for estimating in-situ strength of concrete before and after repair work in a few members. Average value of 25 acceptable impact values was considered for strength assessment. The strength was assessed to be about 15 N/mm^2 in case of deteriorated concrete and 20 N/mm^2 in case of sound concrete. The original designed strength of concrete was 20 N/mm^2 at the time of construction.

Foundations

Foundations of concrete appeared to be in good condition. This was confirmed by the fact that foundation of the worst affected column opened out upto the pile cap level was found to be in very good condition.

DISCUSSION

Chemical Attack

While hydro-chloric acid vapours present at plant site can directly result in acid attack on the concrete, the ammonium chloride can react with free lime in the concrete in a base exchange process to form ammonium hydroxide and calcium chloride. Ammonia from the hydroxide volatilizes leaving the pores of the concrete cover filled with water only (6). Thus the pH value of the concrete cover can get reduced very much, resulting in corrosive environments around the embedded reinforcement. In addition, the presence of calcium chloride formed from the reaction with ammonium chloride increases the corrosivity of the environment. Calcium chloride can also be leached out to some extent, by rain and waste water leaving the concrete cover porous and susceptible to further attack by the harmful environment. The cracking caused by corroding

reinforcement   aggravates the problem further   and   exposes
the reinforcement to direct chemical attack.

RECOMMENDATIONS FOR REMEDIAL MEASURES

Classification of Structural Members

Based   on detailed surveys and investigations as   mentioned
above, the various structural members affected by corrosion
of reinforcement and spalling of concrete were   enumerated
and classified according to the level of the treatment
needed, for repair and rehabilitation as follows:

  i)   Members requiring only surface treatment (10%   of
        total   number of columns, 60% of beams and 40% of
        panels required this treatment).

  ii)   Members which need treatment by chipping upto a
        depth of 50 mm but without any replacement of
        reinforcement (about 50% of total number of
        columns, 30% of beams and 40% of panels required
        this treatment).

  iii) Members which require to be chipped and provided
        with additional reinforcement to replace corroded
        steel rods (about 40% of columns, 10% of beams
        and 20% of panels as well as all the staircases
        fell in this category).

and iv) Members which require to be completely demolished
        and replaced (seven columns and two ring beams
        required this treatment).

At the   outset, suitable guidelines were   worked out for
ensuring the safety of the structure and components during
the repair work.

Guniting

Guniting is regarded as the principal method of repair and
strengthening of concrete. It is generally considered the
most satisfactory and economical method (7,8). The process
adopted is outlined below.

The surface to be gunited is chipped (Figure 3) upto the
required depth. The loose rust products were removed by
wire brushing and sand blasting. Additional longitudinal
and transverse reinforcement was welded to the existing
steel rods to make up for the lost area of cross-section
due to corrosion. Concrete surface was thoroughly washed

by water jets to remove chlorides, if any, and was allowed to dry for one day. Before guniting, the old concrete surface was coated with an epoxy formulation for enhancing bonding with the new concrete (9).

Guniting was done with cement: sand mix of 1:3 to bring the structural members to the original dimensions using grillwork in convenient stages. The amount of rebound of the cement mortar varies with the position of the work, air pressure, angle of nozzle, water and cement content, size and grading of sand, amount of reinforcement and thickness of layer being gunited. Air pressure was maintained at 0.75 N/mm^2. The rebound was kept within 15% during execution.

The strength of the gunited surface was assessed using a rebound hammer and was found to be about 15 N/mm^2 and 20 N/mm^2 after one week and one month respectively. Guniting work in progress is shown in Figure 4.

Figure 3: Chipping of concrete from a column is in progress

Figure 4: Guniting is in progress (after sand-blasting of concrete and reinforcing steel)

Surface Coating

When the plant building was constructed initially, coal tar epoxy had been used as a surface coating to concrete between ground level and the plinth level. The steel reinforcement behind concrete protected by coal tar epoxy was seen to be unaffected by corrosion, which thus indicated the protective potential of the coating layer. The critical factors to be considered for selecting a suitable protective coating, are the binder system used and the dry film thickness obtained in addition to other parameters like flexibility, adhesion, abrasion and chemical resistance (10). Further, a barrier coating should withstand swelling, dissolving, cracking and embrittlement when exposed to chemicals. Keeping these factors in view, specifications of the commonly available protective systems were studied and a suitable coal tar epoxy system was recommended for protection of the concrete surfaces.

Additional Safeguards

Apart from the above measures, the following recommendations were made :

1. Spillages of ammonium chloride and raw materials should be stopped by plugging all leakages and sealing joints. If any spillages or overflow are contemplated in particular locations, the relevant area of the floor around such locations is to be provided with adequate non-absorbent treatment and quick drainage arrangements.

2. The floors are to be given proper slopes for drainage and given a mastic lining around the processing areas and openings in floors.

3. Separate drainage arrangements are to be provided for drainage of rainwater and flow of effluents.

4. The existing service water reservoir on the terrace is to be replaced with a lighter synthetic plastic tank to relieve the load on the columns.

CONCLUSIONS

The corrosion damage was attributed to the leaching of calcium chloride formed due to the reaction between free lime in concrete and ammonium chloride. The porous concrete having low alkalinity resulted in severe corrosion distress to reinforcement and consequent longitudinal cracking in concrete in the affected locations.

In all, about 50% of structural elements required only surface treatment, 30% required nominal treatment upto 50mm depth but thorough treatment was found necessary in the remaining 20% of total number of structural members. Two staircases required intensive treatment. Foundation appeared to be in good condition in general. Selective remedial measures as discussed in the paper were adopted for making the building serviceable in the adverse environment.

REFERENCES

1   MOSKVIN V et al. Concrete and Reinforced Concrete Deterioration and Protection. Mir Publishers, Moscow, 1980, pp 264.

2   MUNRO E V. Concrete Durability in a free market system. Concrete International, October 1986 (Reproduced in the Bulletin of ACI - Maharashtra India Chapter, July-September 1989, pp 5).

3   PULLAR - STRECKER P. Corrosion Damaged Concrete: Assessment and Repair. Construction Industry Research and Information Association, London, 1987, pp 1-48.

4   EDITORIAL. Protecting Concrete from Deterioration. Indian Concrete Journal, October 1987, pp 272.

5   ROBERTS M H. Carbonation of Concrete made with Dense Natural Aggregates. Building Research Establishment, Information Paper Number IP 6/87, April 1981, pp 1.

6   BICZOK I. Concrete Corrosion: Concrete Protection. Akademiai Kiado, Budapest, 1972, pp 301.

7   PERKINS P H. The Deterioration and Repair of Concrete Structures. Concrete, U.K, March 1980, pp 35.

8   ACI JOURNAL COMMITTEE 201. Guide to Durable Concrete. Title Number 74-53, December 1977, pp 605.

9   HINDUSTAN CIBA-GEIGY LIMITED. Araldite for Bonding New to Old Concrete. 1982, pp 1-4.

10 BHAKRE V S et al. High Build Surface Coatings for concrete. Transactions of the Society for Advancement of Electrochemical Science and Technology, India, Volume 23, Number 2-3, April-September 1988, pp 191.

USE OF EPOXY CONRETE OVERLAYS FOR INDUSTRIAL FLOORS

J A Peter

N P Rajamane

S Selvi Rajan

V S Parameswaran

Structural Engineering Research Centre,

India

ABSTRACT. Industrial floors are subjected to severe abrasion and wear-and-tear due to movement of heavy loads. Epoxy concretes are not only highly resistant to chemicals but also are capable of resisting shear stresses, developed across the interface between them and the base concrete, caused due to variations in loads and temperature.

Since epoxy resins are very expensive, it is necessary to use them judiciously and evolve optimum mix proportions that possess the required bond and shear strength and resistant to abrasion and chemicals. Detailed investigations were carried out at the Structural Engineering Research Centre (SERC), Madras, India, on different formulations of epoxy concretes using flyash and silica flour as filler materials. An experimental epoxy concrete overlay measuring 2.5 m x 2.5 m was cast to evaluate its long-term performance under various degrees of loads, temperature change, etc. Tests were carried out to determine the effect of strong and corrosive chemicals on them. The paper outlines the details of the investigations and tests carried out on the overlay and summarises the conclusions drawn from the study. Recommendations on the use of epoxy concrete overlays for industrial applications are also included at the end of the paper.

Keywords : Epoxy Resin, Epoxy Mortar, Epoxy Concrete, Overlay, Abrasion, Acid/Alkali Attack, Shrinkage, Mix Proportion, Flyash, Silica.

Mrs.Annie Peter, Mr.Rajamane, and Mrs.Selvi Rajan are Scientists at the Concrete Composites Laboratory (CCL) of the Structural Engineering Research Centre, Madras, India. Their research interests are in the area of special concretes. Mr. Parameswaran is the Deputy Director and Head of CCL and has wide R&D experience in reinforced, prestressed, and special concretes.

INTRODUCTION

Industrial floors are the highly stressed structural elements of a building. These floors have to withstand considerable amount of wear and tear due to heavy movements of loads. Epoxies are ideal binding material for bonding together a variety of inert fillers such as aggregates, silica flour, flyash, etc. Properties of epoxy such as high mechanical strengths, superior adhesive power and ability to withstand attack by a wide ranging chemicals, have contributed to the development of chemical and abrasion-resistant overlays involving their use as a chief binding material.

Overlays are generally classified into three categories based on the procedure adopted in laying them (Fig.1). They are (i) unbonded overlay (ii) partially bonded overlay and (iii) bonded overlay.

Unbonded overlays are overlays that are placed on top of the existing concrete surface, by applying a separation course in between them. Partially bonded overlays are those without the separation course and have a minimum thickness of 125 mm. An overlay of varying thicknesses from 25 mm to the desired design thickness that can be bonded monolithically to the base concrete to form a combined thickness equal to a single slab of adequate design for the planned loading is called a bonded overlay.

In an unbonded overlay, the underlying cracks in the base concrete are not normally reflected but the thickness of the overlay has to be more and sometimes steel reinforcement has to be provided to attain a particular load-carrying capacity. Partially bonded overlays usually reflect the underlying cracks and the joints in the base concrete and in the overlay should be matched while placing the overlay. Epoxies are used in the form of bonded overlay since they bind effectively on to the existing concrete floors. They are in use for airfield pavements and for highway pavements since 1956 (1).

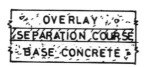

UNBONDED OVERLAY

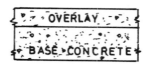

PARTIALLY BONDED OVERLAY

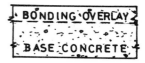

BONDED OVERLAY

Figure 1: Types of Overlay

Owing to their highly beneficial properties such as low shrinkage, resistance to chemical attack, good abrasion resistance, and superior mechanical bonding to the hardened surface of concrete, epoxy bonded overlays are ideally suitable for industrial floors.

LABORATORY TESTS ON OVERLAYS

Abrasion Resistance

An accelerated abrasion test was performed on epoxy concrete and epoxy mortar specimens following the recommendations of IS:1237-1980[2] to study the quality of the surface matrix which mainly governs the abrasion resistance. The epoxy resin employed for this investigation was a modified type of DGEBA called 'Epoxite'. Properties of these resins are furnished in Table 1.

The Indian Standard defines an average wear in terms of the average abraded thickness, 't', which is calculated using a formula:

$$t = (\text{Weight change})/(\text{density} \times \text{surface area}).$$

Table 2 and Fig.2 provide the abrasion test results on epoxy mortar and epoxy concrete specimens.

From Table 2, it can be noticed that mere increase in the quantity of resin would not substantially improve the abrasion resistance and that particles of larger size in the matrix enhance the abrasion resistance. Indian Standard recommends an average wear of 2 mm and 3.5 mm for general purpose and heavy-duty floors, respectively. Thus, epoxy mortars made with flyash and sand were found to be suitable for all general purpose flooring. For heavy-duty flooring, epoxy concretes containing aggregates of maximum size of 5 mm were found to be adequate as the wear in such concrete was only 1.65mm for wet specimens and 1.52 mm for dry specimens.

Table 1   Details of Epoxy Resins

Epoxite system	Mixing Ratio (parts by weight)	Viscosity at 25°C mPa.s	Density at 25°C g/ml
EMG 182	100	680-1300	1.11 - 1.16
EH 220	45	3800-4800	0.96 - 0.99
EH 230	15	15000-21000	-

Table 2 Results from Abrasion Tests

Designa-tion	Mix proportion Flyash:Sand:Aggregate	Resin Quantity % by weight	Loss in thickness	
			Observed (mm)	Theoretical (mm)
A	1:6	12	1.93	2.37
B	1:8	12	2.46	2.58
C	1:9	12	2.17	2.51
D	1:6	15	2.79	2.46
E	1:8	15	3.50	3.65
F	1:9	15	2.57	2.60
G	1:3:4.7	15	1.53	1.52
H	1:3:4.7	15	1.75	1.65

Mix A,B,C,G,H contained fine aggregates passing through 1.18mm sieve;

Mix D,E,F had Ennore Standard sand as fine aggregates;
Mix H was tested in Wet condition;

Durability

Four different chemicals were selected as attacking agents to evaluate the resistance of epoxy mortars to acids and alkalis of different concentrations at room temperature (30°C - 35°C). Disc specimens of size 50 mm diameter by 25 mm height, using 1:6, 1:8, and 1:9 mix ratio by weight of flyash and sand with 12% epoxy resin, were used. For the comparative studies, cement mortar specimens containing

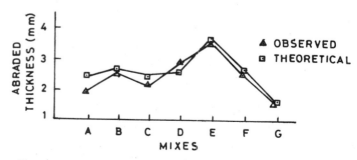

Figure 2 : Abraded thickness Versus Different
Mix ratio

cement and sand in the proportion of 1:2 and water/cement ratio
of 0.40 were also subjected to the attack of the same chemicals.
The specimens were immersed in the test solutions and were periodically weighed as per the recommendations of ASTM(3). Table 3 and
Fig.3 present the observations made after and during 200 days of
immersion in chemicals.

Low Shrinkage

To study the effect of variations in service temperatures with
reference to shrinkage, thermal compatibility between epoxy concrete
and conventional concrete, and the bond between the overlay and
the concrete substrate, concrete slab specimens of size 300 mm x
300 mm x 75 mm were cast. As per ASTM specifications (4), a 12.5mm
thick epoxy concrete overlay was laid on precast concrete slabs.
The specimens were then subjected to alternate heat and cold cycles,
each cycle consisting of heating of the specimens to 60°C and cooling
to a temperature of -10°C. Typical variations of strains in the
specimens for test cycles is shown in Fig.4.

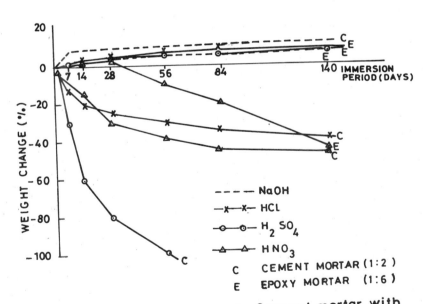

Figure 3: Deterioration of Cement mortar with respect to Epoxy mortar

Table 3   Chemical Resistance of Epoxy Mortar Specimens

Test solutions*	Percentage change in weight after 200 days of immersion				Remarks
	1:6	1:8	1:9	Control	
10% NaOH	8.6	9.4	7.9	11.3	Due to deposition of salt, increase in weight was noticed.
15% NaOH	7.8	10.8	9.5	11.6	
Saturated NaOH	7.9	8.4	7.9	9.6	
10% HCl	8.8	9.8	10.2	-39.4	Physically there was no change in the appearance of epoxy mortar specimen.
15% HCl	8.7	9.9	9.7	-20.6	
Conc.HCl	9.4	10.8	9.6	-24.2	
10% H_2SO_4	10.1	13.2	6.9	-100.0	Cement mortar (control) specimens were destroyed in 47 days
40% H_2SO_4	7.0	11.2	11.9	-100.0	
Conc. H_2SO_4	5.0	10.2	5.1	-100.0	
10% HNO_3	-27.5	-35.7	-68.3	-26.7	Epoxy mortar specimens were destroyed in concentrated solution in 11 days
25% HNO_3	-91.5	-95.2	-54.6	-25.7	
Conc.HNO_3	-100.00	-100.00	-100.00	-21.2	

* Test solutions were changed every 30 days or earlier when the solutions became too much contaminated with the products released from test specimens.

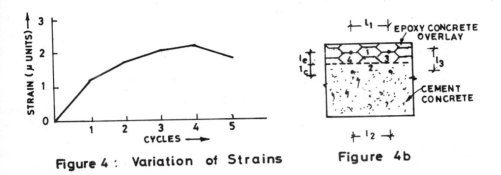

Figure 4 : Variation of Strains   Figure 4b

Referring to Fig.4(b), let ϵ_1 be the strain in epoxy concrete overlay, ϵ_2 be the strain in cement concrete base, and ϵ_3 and ϵ_4 be the strains across the joint. The separation at the joint is theoretically calculated as:

The displacement of epoxy concrete across 3 or 4 $= l_1 \times \epsilon_1$

The displacement of cement concrete across 3 or 4 $= l_2 \times \epsilon_2$

$\therefore$ The (calculated) total displacement across 3 or 4

$$= (l_1 \epsilon_1) + (l_2 \epsilon_2)$$

The observed displacement across 3 or 4 $= \epsilon_3 (l_e + l_c)$ or $\epsilon_4 (l_e + l_c)$

Hence the separation across the joint line between 1 and 2 is:

$$(l_1 \epsilon_1 + l_2 \epsilon_2) - \left[\epsilon_3 (l_e + l_c)\right] \text{ or } (l_1 \epsilon_1 + l_2 \epsilon_2) - \left[\epsilon_4 (l_e + l_c)\right]$$

It may be noted that the base concrete and the overlay are assumed to be isotropic.

The calculated separation at bond line between epoxy concrete and cement concrete was in the range of 16×10^{-5} mm to 184×10^{-5} mm for the test conditions adopted . Physically no crack was observed at or near the bond line. Thus, the epoxy concrete overlays were found to possess low shrinkage and adequate thermal compatibility with cement concrete.

CHARACTERISTICS OF EPOXY RESIN

The epoxy resin used in the investigations has the composition as shown in Fig.5. Its physical and mechanical properties are given in Table 5.

DESIGN OF MIX PROPORTION FOR CASTING OF EXPERIMENTAL OVERLAY

As epoxies are very expensive resins, proper mix design and application techniques are essential for achieving maximum economy in their use. For this, experiments were conducted to determine the gradation and quantity of different fillers so that epoxy mortars could be prepared with minimum quantity of resin.

Figures 6 and 7 show the sieve analysis of fine fillers (flyash, cement and silica) and coarse fillers (of size between 2.4 mm and 4.8 mm, and 12.5 mm), respectively.

Specific gravities of cement, flyash, silica flour, sand and 12.5mm size aggregate were determined using the pycnometer method that is described in Indian Standards[6] and their values are 3.10, 2.24, 2.66, 2.65, and 2.92, respectively. As the specific gravities of the fillers are known, the total volume of voids between the particles is calculated for different proportions of flyash and sand mixture; silica flour and cement mixture; and sand, flyash and aggregate mixture. Based on voids analysis of dry mixtures of these fillers, the optimum quantity of resin was found to be 12% by weight of fillers, taking the need for proper workability also into consideration. Epoxy mortars with flyash to sand ratios

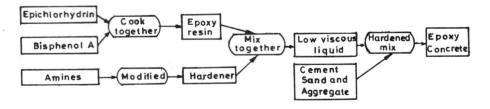

Figure 5: The Making of Epoxy Concrete

Table 5   Characteristics of Epoxy Resin System

Tests	Observed value	Recommended value [5]
Gel time (min.)	30	30
Water absorption 24 hr. (%)	0	1.5
Filler content (phr)	0.56	0.5 to 300
Volatile content (%)	2.07	3.0
Bond strength (N/mm^2)	34.70	30.0
Compressive strength (N/mm^2)	197.00	-
Tensile strength (N/mm^2)	55.70	-
Flexural strength (N/mm^2)	89.60	-
Impact strength (N/mm^2)	2.30	-

of 1:9, 1:8, 1:7, and 1:6 with a resin quantity at 12% by weight of fillers were cast and their compressive strengths were 36.16, 46.46, 56.34 and 79.86 N/mm^2, respectively. The void content of mix containing aggregates, in the ratio of 1:3:5 by weight of flyash: sand:aggregate, was 30%. Mixes containing more aggregates than that in the above ratio were having increased void ratio. Epoxy concrete specimens with ratios of flyash:sand:aggregate of 1:3:2, 1:3:3, 1:3:4, 1:3:5, and 1:3:6 were cast and the optimum mix ratios from the experiments were 1:3:3, 1:3:4, and 1:3:5 with compressive strengths of 58.73, 66.67, and 63.23 N/mm^2,respectively.

EXPERIMENTAL OVERLAY

A portion of the workshop floor of the Structural Engineering Research Centre, Madras, which was built more than 15 years back, was used for the study. An area of about 2.5 m x 2.5 m of the floor where heavy loads were expected to move, was chosen for the experimental overlay. The compressive strength of this concrete flooring was found to be 30 N/mm^2 from Schmidts hammer tests. The top portion of the concrete floor was chipped off to a depth of 55 mm to expose the hard concrete inside. The bottom and the side faces of the trench were cleaned with a wire brush and compressed air was used to blow off all the loose particles. An epoxy resin recommended for bonding hardened concrete to fresh concrete was applied before placing a 40 mm thick M30 grade concrete mix in the trench. The top surface of the concrete was rendered rough to obtain better mechanical bond between this concrete and the epoxy concrete overlay to be laid on top of it. After 28 days of curing in moisture, an epoxy concrete overlay was laid in four sections. The geometry of each section is given in Fig.8. Flyash:sand:aggregate:epoxy resin was 1:3:4.7:1.3 which possessed a compressive strength equivalent

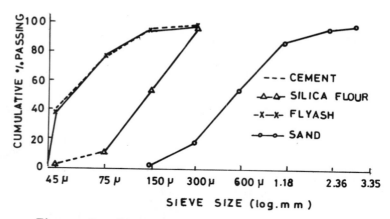

Figure 6: Sieve Analysis of Fine fillers

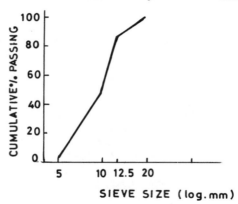

Figure 7: Sieve Analysis of 12.5 mm
Aggregate

to the base concrete. The required thickness of the overlay was 15 mm and hence, aggregates of size less than 5 mm were only used. Before laying the epoxy concrete over the base concrete, a prime coat of epoxy resin recommended by the manufacturers, was applied.

A companion polyester resin based concrete overlay was also laid beside the experimental epoxy concrete overlay to study the effect of shrinkage. The shrinkage strains were measured periodically from the mechanical strain gauge pellets pasted on the top surfaces of the overlays. Due to large amount of shrinkage in the polyester based resin concrete, prominent separation was observed between the overlay and the surrounding old concrete, after a month, which was not the case in epoxy concrete overlay.

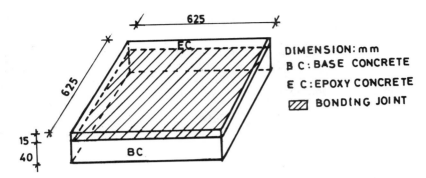

DIMENSION: mm
B C: BASE CONCRETE
E C: EPOXY CONCRETE
BONDING JOINT

Figure 8 : Geometry of the Experimental Overlay

Epoxy mortars having a mix ratio of 1:6 by weight of calcium carbonate and fine sand (passing through 1.18 mm sieve) and with a 12% resin content have a marble like appearance. A thin layer of such mortars can be used on top of epoxy concrete overlays for improving aesthetic appearance of the floor. This technique was used for the experimental overlay also.

RECOMMENDATIONS AND CONCLUSIONS

Epoxy concrete overlays can be utilised for industrial floors since its wear is well within the limits recommended in standard text books and codal specifications.

From the results of chemical attack tests, it is concluded that epoxy concretes are quite capable of resisting acid/alkali attack to a satisfactory extent except for concentrated nitric acid. As can be seen from Table 3, cement mortar could not resist even 10% concentration of sulphuric acid. The superior resisting property of epoxy mortars to cement mortar has been proved even by other researchers[7,8] .

From the mix proportioning of several mixes, it is concluded that for epoxy concrete overlays, the maximum size of aggregates may be taken as 1/3 to 1/5 of its thickness.

Epoxy concrete overlays possess low shrinkage compared to other resin based systems. They have also good mechanical bond to base concrete.

From the cost analysis, it is seen that epoxy resin alone accounts to more than 80-90% of the cost of the overlay. Therefore, it is essential to carefully design the mix before its use in any major work so that minimum possible quantity of resin is utilised in achieving the required workability and strength.

540   Peter, Rajamane, Rajan, Parameswaran

REFERENCES

1   AMERICAN CONCRETE INSTITUTE COMMITTEE 325. Design of Concrete Overlays for Pavements. ACI Manual of Concrete Practice Part I, Materials and Properties of Concrete Construction Practices and Inspection Pavements, 1968, pp 75-84.

2   INDIAN STANDARDS INSTITUTION IS 1237:1980. Indian Standard Specification for Cement Concrete Flooring Tiles, India, pp 12-14.

3   AMERICAN SOCIETY FOR TESTING AND MATERIALS. ASTM C267-77. Standard Test Method for Chemical Resistance of Mortars. 1982. Annual Book of ASTM Standards Part 16, pp 162-166.

4   AMERICAN SOCIETY FOR TESTING AND MATERIALS, ASTM C884-78. Standard Test Method for Thermal Compatibility Between Concrete and an Epoxy Concrete Overlay. 1982 Annual Book of ASTM Standards, Part 14, pp 582-584.

5   AMERICAN SOCIETY FOR TESTING AND MATERIALS. ASTM C881-78. Standard Specification for Epoxy-resin Base Bonding Systems for Concrete. 1982 Annual Book of ASTM, Part 14, pp 571-575.

6   INDIAN STANDARDS INSTITUTION IS 2386:1963. Methods of Test for Aggregate for Concrete, Part III, India, pp 3-14.

7   YOSHIHIKO OHAMA,KATSUNDRI DEMURA, TAKAYUKI OGI. Mix Proportioning and Properties of Epoxy Modified Mortars. Proceedings of the International Symposium on Brittle Matrix Composites (2). Poland 1988, pp 516-525.

8   EPOXIES WITH CONCRETE, ACI Special Publication, SP-21, American Concrete Institute, 1968, pp 145.

EXTERIOR WALLS IN EXTREME CLIMATE

A Rusinov

Railway Engineering Institute

USSR

ABSTRACT. Maintenance of the exterior of concrete buildings in the maritime winter climate is characterised by their low durability. This paper identifies the essential cause of the concrete deterioration and proposes a constructive remedy which makes it possible to lower the corrosive effect of atmospheric exposure. It has been proved that the main cause is the negative osmotic pressure caused by the ingress of atmospheric salts into the pore system of the concrete. The cause of the negative osmotic pressure is the change of osmotic diffusional flux direction with time. This flux arises because of the passage of atmospheric salt solution through the semi-permeable membrane of cement gel. The membrane permits the passage of salt ions but retains the water molecules. Once the salts have accumulated beyond the membrane, balancing of the solution concentrations on both sides of the membrane occurs, stimulating a flux in the opposite direction. The process continues until the gel is completely saturated with chloride ions, causing self excited oscillations in the concrete structure and ultimately its destruction. The discovered effect has been called Rusinov osmotic oscillator. Reduction of the effect is achieved by arranging in the wall, at the freezing depth, a continuous layer of funnels with their wide mouth towards the outside surface.

Keywords: Concrete, Osmotic Pressure, Oscillator, Chyophase, Funnels, Thermally Active Layer.

Cand Sc (Tech) Rusinov Alexander. Is the Manager of the Scientific Laboratory at the Railway Engineering Institute in Khabarowsk. This current research work includes the use of the laws of thermodynamics with regard to durability problems in building structures.

INTRODUCTION

The onset of the destruction mechanism of exterior concrete walls in maritime winter climatic conditions is characterised by the following stages: the appearance of damp spots on the interior surface, peeling of the exterior surface, the appearance of cracks on the exterior surface, water saturation and finally spalling of the exterior concrete surface. Failure of the heat insulation properties begin at the stage of water saturation. The cause of failure is the layering of the wall at the freezing depth. A network of microcracks develop in the longitudinal cross-section at the freezing depth; later these micocracks develop into a cavity. Thus, the wall thickness decreases, it's moisture permeability increases and the heat insulation characteristics deteriorate until failure. Surplus moisture inside the exterior layer of the wall forms a cryophase volume as the exterior temperatures drop, resulting in critical expansion stresses in the concrete structure. The existence of atmospheric salts in the pore system aggravates the situation because of the appearance of osmotic pressures which increase the expansive stresses; as the exterior temperatures drop the effect increases. Destructional self-excited oscillation of the concrete matrix in the process of its saturation with atmospheric salt solutions is considered to be the prerequisite for the development of these defects. Cement paste gel allows the passage of salt ions but retains water molecules in a process of capillary infiltration. This gives rise to solutions with high and low concentrations on both sides of the semi-permeable gel membrane. The process of capillary infiltration during the balancing of the concentrations can be observed by following the penetration of chloride ions in the opposite direction. Migrational fluxes of the salt solutions through the semi-permeable membrane in forward and reverse directions give rise to negative forces in the concrete matrix.

This phenomenon has been called the Rusinov osmotic oscillator and has been corroborated both theoretically and experimentally. The lessening of the effect of this phenomenon can be guaranteed by arranging intraporous moisture concentrators inside the wall body. A layer of funnels with their wide mouths towards the outside surface is arranged at the seasonal freezing depth. These funnels are made from moisture proof material. An artificial pressure differential occurs in the concrete pores within the limits between the wide and narrow tunnel mouths which causes stimulation of the migrational flux and its transference deep into the concrete structure, out of the freezing zone. This process balances the solution concentrations in winter maritime climatic conditions and increases the necessary maintenance repair period by a factor of 1.47 times.

CONCRETE DESTRUCTION

The deterioration of concrete exterior walls in winter maritime climatic conditions is a serious problem in maritime regions with very cold winters and temperatures always below 5°C. Numerous studies were carried out to determine whether the main cause of deterioration was osmotic or crystallisational pressure. Further studies have shown that these factors are consecutive factors in the same process. Formation of crystalline-hydrate nucleii within the concrete structure pore system is difficult for a variety of reasons including: pore size, existence of infra pore vapour pressure, the strength of the pore walls, chemical cement formation, its exothermia, etc. It has been proved that the development of crystal nucleii is possible, if the pore space volume and evaporation processes are great enough. Such conditions arise in the process of microcrack networks forming cavities in the interior wall layer at the freezing depth. Microcrack networks and cavities form at the beginning of the deterioration cycle during the saturation of the concrete with atmospheric salt solution by the forces of capillary infiltration.

Osmotic Oscillator

The saturation of exterior concrete walls with atmospheric salt solution occurs during the initial stage by capillary infiltration, a process which ends with the balancing of the intrapore pressures. Penetration of the salt solutions occurs both through the pores and through the moist cement hydrate gel. It is known that the gel allows the passage of salt ions, but retains water molecules, acting like a semi-permeable membrane. This causes an imbalance in the intrapore pressure, which results in a diffusional balancing of the solution concentrations formed on either side of the semi-permeable membrane. The passage of the salts solution through the semi-permeable membrane in both directions, gives rise to both compressive and tensile stresses within the concrete matrix, resulting in the formation of microcrack networks and cavities.

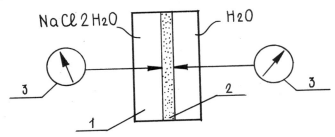

Figure 1: The diagram of the installation for appreciation of Rusinov osmotic oscillator effect. 1. a hermetic box, 2. a cement-sand mortar plate, 3. indicator.

Self excited oscillations of the concrete matrix during the differential transport of the salt solutions were called the Rustinov osmotic oscillation. Evaluation of the effect of the osmotic oscillator was carried out experimentally. The apparatus (shown diagrammatically in Figure 1) consists of a hermetic box, hermetically divided into two equal volumes by a cement-sand mortar plate with indicators set on both sides of the plate. The indicators were capable of measuring deflections of up to 0.002 mm. One volume was filled with water, and the other with an aqueous solution of sodium chloride (20% concentration). The box was sealed hermetically and readings of the changes in deflection were taken every hour for 5 days and subsequently four times a day. The constancy of the temperature and humidity of the room air was also monitored. The plot of the deflection of the left and right plate surfaces an its central axis is shown in Figure 2.

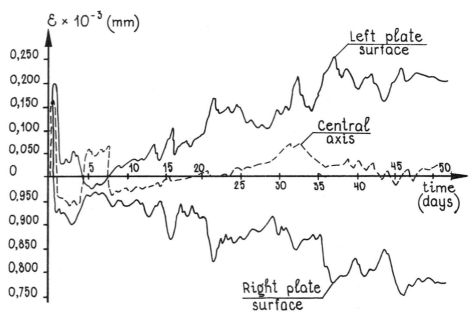

Figure 2: The plot of the self-excited oscillations

A thermo-dynamical description of the self excited oscillation process, which can be explained by analysis of Want-Goff and Corrence equations, is given in a previously published paper by this author (5). It follows from the analysis, that it is possible to arrest the process, if the solution concentrations within the concrete layer between the saturated surface and the capillary infiltration depth are balanced. In practice a balancing process is taking place during the entire deterioration period, which can be explained in terms of the high chemical potential of the atmospheric

salt medium and a limited drying of the gel in the concrete pores.
Damping of the self excited oscillations is explained by the
deposition of salts on the pore walls, egress of salts to the
surface by evaporation, and the dewatering effect of the thermo-
diffusional flux of the moisture from the interior of the wall to
the external surface.

Subzero Temperatures

Subzero temperatures have a catalytic effect on the osmotic
oscillation phenomena, which can be attributed to the osmotic nature
of the process and the cryoscopic disintegration of the intraporous
salt solution. It is known that as the temperature drops weak
solutions disintegrate, evolving the water cryophase (ice), and
increasing the concentration of the remaining solution. It is also
connected with an increase in the amount of chloride ions
transported bidirectionally across the semi-permeable cement hydrate
gel, causing an increase in the frequency of the negative
oscillation cycle. This explains the rapid drop of concrete
strength in a saline medium accompanying a drop in temperature.
With further reduction of temperature the evolved water cryophase
experiences an anomalous increase in volume resulting in
considerable mutual stress with the pore walls. Thus a constant
background of the tensioned stresses coupled with the negative
stresses of osmotic oscillator, destroy the matrix of the concrete
wall exterior layer. Obviously the most durable structures are well
matured, with developed pore systems and dry cement hydrate gel,
however this is technologically difficult to achieve.

THE PROTECTION OF CONCRETE

The concrete of exterior walls exposed to winter maritime climate
conditions has to be protected from both the Rusinov osmotic
oscillator and the pressure of the cryophase volume in the
intraporous system. It is known that while walls of buildings are
dampened by atmospheres mineralised moisture, the concentration from
the surface to the interior of the wall follows a decreasing
distribution curve.

An approximate distribution curve of the solution concentration at
the freezing depth of the wall is shown in Figure 3. Obviously the
motive force for the migrational flux of moisture in the gradient of
concentration across the wall thickness. It is possible to halt the
flux by reducing the gradient through balancing of the solutions
concentrations. Figure 4 shows a section of the continuous layer of
tunnels with their wide mouths towards the outside wall surface at
the freezing depth.

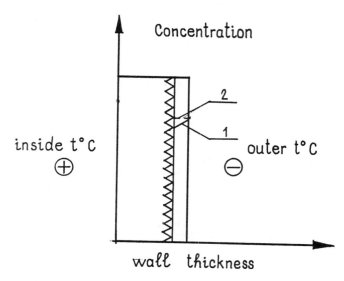

Figure 3:   The distribution of the intraporous solution
            concentration from the surface deep into the
            walls: without the funnels 1. and with the
            funnels 2.

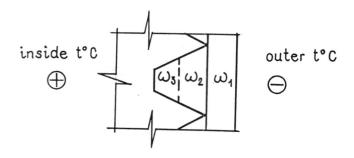

Figure 4: The distribution of the intraporous moisture volumes in
          the exterior layer W1, by the wide mouth of the funnel
          W2 and by the narrow mouth

The concrete inside the volume of the tunnel has a diminishing pore
character in relation to the exterior wall layer, and the
intraporous moisture is distributed in a similar manner. As is
known, the level of osmotic pressure and that of the moisture volume
are inversely related, and therefore the pressure of the migrational
flux will increase from the exterior layer to deep into the wall.
In conditions of low temperature this process is accelerated and it
is possible to move the moisture from the exterior layer into the
warm zone of the wall, ie towards the narrow end of the funnels.

Simultaneous with the migration of moisture, the solution concentrations are being balanced, the gradient is lowered and the cryophase volume in the freezing layer diminished. The S.V. Alexandrovsky method of predicting exterior wall durability (4) was applied to determine the effect of the usage of the funnels. It predicted a rise in wall durability of 1.47 times for Far East conditions. Experimental studies have proved that the cryophase volume in the freezing layer reduces by a factor of 2, whilst the wall surface heat transfer resistance is increased by 30%.

Field of Application

The application of the intraporous moisture concentrator of a continuous layer of funnels, is hot only limited to maritime climatic and natural salt conditions, but includes urbanised environments.

The influence of such environments on exterior wall durability has been found to be extremely unfavourable by a number of investigations. Some research has been done into the peculiarities of the urbanised environment in terms of chemical reagent. The results of these investigations led to the proposal of the protection of the existing structures with the help of a rolled matrix of the funnels and subsequent re-rendering.

REFERENCES

1 ALEXANDROVSKY S V. Temperature and humidity exposure calculation of the concrete and reinforced concrete constructions. Moscow, 1973, pp 442.

2 MOSKVIN V M, KAPKIN M M, MAZUR B M and PODVALNY A M. Concrete and reinforced concrete resistance to the subzero temperature. Moscow, 1967, pp 863.

3 RUSINOV A V. Thermodynamical analysis of parameters of calculation of exterior panel wall durability for the maritime conditions. Intercollege collection of scientific works. Khabarovsk Railway Engineering Institute, 1987, pp 56-69.

4 RUSINOV A V. Prediction of constructive design of wall panels on the principles of the method of durability calculation. Intercollege Collection of Scientific Works. Khabarovsk Railway Engineering Institute 1987, pp 59-65.

5 RUSKINOV A V. The phenomenon of the osmotic oscillator by absorption of the atmospheric salt solutions by surface layers of exterior building walls. Collection of scientivic works. Khabarovsk Railway Engineering Institute, 1988, pp 59-63

Theme 4

PROTECTION THROUGH DESIGN

Session A

Chairmen

Professor P C Kreijger

Department of Civil Engineering,

Eindhoven University of Technology,

The Netherlands

Professor K Mahmood

Department of Civil Engineering,

King Abdulaziz University,

Saudi Arabia

ASPECTS OF DESIGN FOR DURABILITY AT THE BRITISH LIBRARY

P J Ryalls

R Cather

A Stevens

Ove Arup & Partners,

United Kingdom

ABSTRACT: The design of the new British Library has taken account of an assumed 'extended life' of 500 years. Various structural elements are in locations where future maintenance and repair would be impossible. Achievement of the 'extended life' for concrete elements has been approached by consideration of concreting materials, mix design and curing.

Keywords: Concrete, Durability, Long life, Mix design, Pfa, Blast furnace slag, Curing.

Peter Ryalls has been involved in the design of universities, shopping centres, hospital and office developments, both in the UK and overseas. They have included tall buildings incorporating the use of dense and lightweight concrete, pre-stressed and pre-cast concrete, structural steel and deep basement construction. He has been responsible for the design of British Library since 1976.

Bob Cather is part of an advisory team within Arup Research & Development providing specialist advice to designers on the properties and performance of construction materials and particularly on cement based materials. He is a Technical Director of the Ove Arup Partnership.

Anthony Stevens has specialised in the design and analysis or large, complex civil and structural works, geotechnics, prestressed concrete and structural steelwork. He has wide experience of many building types both in the UK and overseas. He is a Director of Ove Arup & Partners, responsible for a number of large developments, including the British Library.

INTRODUCTION

The British Library is intended to have an 'extended life'.

Put simply, once the 28 km of basement bookshelves are filled with 22+ million books it seems unlikely that they will be moved to a new home for the 'foreseeable future'.

What does this mean in numerical terms?

Nobody has answered this question - leaving the design engineer to answer it for himself.

Looking around at other buildings for guidance - the nearest building of some antiquity is St Pauls Cathedral - some 300+ years old. Nobody seems to be suggesting that it has come to the end of its useful life - and indeed maintenance work continues to prolong its life.

Using this as a guide it seemed to the authors that it would not be unreasonable to design elements exposed to attack for a design life of 500 years. The particular elements chosen for such 'extended life span' were those where inspection and remedial measures would in future be effectively impossible, namely:

1)   Secant pile perimeter walls to the basements.

2)   1.8m diameter bored piles underreamed by 4.3m.

3)   Precast concrete boot lintols supporting superstructure brick cladding.

LOCATION

The building is located in central London, has significant buildings on the East, South, and West side. St Pancras Station and Chambers a listed building, is located some 20m to the East, well within the depth of the excavation.

The running tunnels of the London Underground District Line pass east-west, just below street level, immediately to the south of the south perimeter wall.

The running tunnels of the London Underground Victoria line pass east-west across the site in the Central Area, at a depth of some 18m below street level, and some 3m to the north of the northern-most extremity of the deep basements.

THE BASEMENTS

There are two different types of basement adjacent to each other.

In the 'South Area' some 11 million books will be stored in 5 levels of basements, extending some 25m below street level, each basement covering an area of 2¼ acres. It is surrounded on all four sides by secant walls 30m deep.

Immediately to the north lies the 'Central Area'. This area is only 2 basements deep and serves as the location for the mechanical services plant. The perimeter secant walls on the East, West and South are approximately 20m deep.

The South Area

There are 5 levels of basements, each 4.85m floor to floor. A void, heave space, generally 1.1m in depth, but extending to 2.2m to form an internal continuous walkway, is located below the lowest basement level.

The basements are polygonal in plan being some 130m max. width E-W and approximately 100m N-S.

Each basement has an area of approximately 2¼ acres. The volume of clay to be removed was some 250,000m³. This volume constituted the largest known civilian excavation in London Clay.

The construction method used in this area was the 'top down' method, in order to minimise ground movements of the adjacent buildings, in particular that of St Pancras Station.

The secant pile walls in this area are 1180mm diameter and 30m long. They are thought to be the longest secant piles to have been used.

The Central Area

Because the tunnels of the London Underground Victoria Line pass across the site, the basements are restricted in depth, the use of pile foundations was thus precluded. Five reinforced concrete rafts were used, three at Basement 1 Level and two at Basement 2.

The use of r.c. raft foundations required the construction to be made in 'open cut', with excavations up to 15m in depth.

The east and west perimeter walls are formed of secant pile construction, 1180m dia. extending in depth to 20m. In order to permit the excavation to proceed without the need for internal propping, 3 rows of temporary ground anchors (loads up to 70 tonnes) were used to support the piles during the construction process.

Permanent propping of the walls is provided by the basement slabs

acting as permanent struts E-W across the site. The ground anchors
were then de-stressed.

SECANT PILE WALL

Structural Design

The piles extend through the London Clay and into the underlying
Woolwich Clay Beds. A perched water table exists 2m below street
level and 60% hydrostatic pressure extends well beyond the toe level
of the formed pile.

The clay contains sulphates to Classes 2 and 3. A layer of wet
carbonaceous material also containing sulphates, occurs some 7m
below the horizon of the Woolwich and Reading Beds. It was
therefore decided to adopt Class 3 provisions throughout.

Layers of water bearing sand were known to occur in the Central
Area.

Analysis showed that for the most part, during excavation, inward
movement of the pile walls would result in single curvature
deformation with compression of the external face. Inward bowing of
the piles due to inward ground movements amounted to some 12mm in
30m.

Reinforcement consisted of universal beams installed in each female
pile (see Fig.1). Each steel beam was designed to take both its own
load, and that of one adjacent male pile. The adjacent male pile
was only lightly reinforced to resist shrinkage cracking.

The resistance moment was calculated on the steel section only. No
contribution was assumed from the concrete. In choosing the section
some allowance was made for future rusting to occur.

Concrete

The concrete was specified so as to produce a dense, impermeable,
workable concrete. Such concrete should satisfy both the
requirements of resistance to sulphate attack Class 3, and that of
longevity. The general approach was to use code of practice and BRE
guidance but with great emphasis on implementation.

Density and impermeability were to be achieved by use of a 'low
heat' cementitious component. This would reduce temperature rise,
and subsequent shrinkage cracking and porosity.

A 'low heat' cementitious material was to be produced by reducing
the amount of C_3A in the cementitious fraction.

C_3A is the material which contributes most to early generation of heat of hydration. Therefore elimination of as much of this material as reasonably possible would reduce the early temperature rise in the mix, which would thus reduce porosity and thermal cracking in the immature concrete.

Reduction of C_3A was achieved by two means:

1) Specifying cement replacement materials, thus reducing at a stroke, the amount of O.P.C. in the mix, and with it, the C_3A contained therein.

   Cement replacement materials specified were:

   either

   min. 70% Portland Blast Furnace Slag (Cemsave) + 30% max. O.P.C.

   or

   max. 40% Pozzolan (Boral Pozzolanic) +min. 60% O.P.C.

2) Specifying a maximum content (7%) of C_3A in the cement.

Concrete Mixes Used

In order to facilitate the balance between low early strength for pile cutting and the later structural and durability requirements it was deemed to be adequate to specify the required minimum strength of $30N/mm^2$, to be achieved at 56 days instead of the usual 28 days. Various modifications to the original mix were used during the contract in attempts to reduce this early strength gain. This culminated in the use of:

1) 40mm aggregate which permitted the cement content to be reduced.

2) Increasing the Cemsave fraction from 70% to 85%, thus reducing the portland cement fraction further.

3) Use of the plasticiser to improve workability without the addition of water.

4) Acceptance of a characteristic strength at 56 days of $29N/mm^2$, evidence having previously shown that further strength development of this actual mix at 90 days would carry the strength beyond $30N/mm^2$.

Some measurements of insitu temperatures were carried to help assess the effects of these changes.

1.8m DIA. BORED PILES - UNDERREAMED UP TO 4.3m DIA.

119 piles 12m long were constructed at the bottom of a 19m long shaft, 2.2m diameter and temporarily lined with a 2m dia. 'Armco' casing. The empty casing would serve to permit the future installation of the 19m long steel column complete with brackets, prior to commencement of the 'top-down' construction.

Ground conditions and mix specification were the same as for the secant wall, but the majority of the piles were case using pfa rather than blast furnace slag.

It was assumed that the concrete will be attacked by sulphate conditions to a depth of 150mm during a period of say 500 years, thus reducing a pile to 1500mm diameter. The pile is therefore designed as a 1500mm diameter pile and reinforced accordingly (see Fig.2).

PRE-CAST CONCRETE BOOT LINTOLS

The vertical cladding to the superstructure is brick and is chosen to blend with brickwork of the adjacent St Pancras Station.

Brickwork panels are either storey height, or where there are windows, act as infill panels between windows above and below a given floor. Brickwork extends downwards approximately one metre below the floor soffit, and is continuously supported on a precast concrete boot lintol. Columns are at 7.8m c/c. Precast boot lintols are 7.8m long and separated from each other by a gap. Each boot lintol is hung from the floor at two points, each 1/5 span from the support (see Fig.3). The connection is made by interlocking looped bars from unit and slab, cast into a rebate formed in the floor slab (see Fig.4). Some lintol supports extend upwards and are connected at the top to form a window cill.

The external wall is of cavity construction, the cavity being the space left between the outer skin of brickwork and the inner skin of lightweight blockwork (see Fig. 5).

Once the brickwork is built on the boot lintol, it will not, in future, be possible to inspect the condition of the concrete boot at the point of brick support within the cavity.

The external skin of brickwork being porous, the cavity is expected to be damp. On the east, north and west faces the conditions within the cavity are considered to be permanently adverse.

It was therefore decided that:

1)   Reinforcement in the boot itself would need to be stainless
      steel
      (Grade 316).

2)   The specification required the production of a dense,
      impermeable concrete, in order to achieve maximum durability.

The intended permeability was to be achieved by specifying:

1)   Cement content O.P.C. to BS12   max. 420 Kg/m^3
                                             min. 330 Kg/m^3

2)   Maximum water/cement ratio   0.35

3)   Mixer to be Forced Action Paddle Mixer to BS1305.

4)   7 day curing under controlled conditions of 100% humidity and
      temperature 20°C   ±2°C.

Work was carried out in two contracts separated in time.

Trial tests on one contract established a mean 28 day design
strength
of   57.5 N/mm^2
and a characteristic strength of   42.5 N/mm^2

Works results for the duration of one contract produced average 28
day results of   59 N/mm^2
with standard deviation varying from 2.8 to 4.2 N/mm^2.

In lieu of a 'mist curing room', permission was given to shrink wrap
units immediately after de-moulding with introduction of free water
to maintain the humidity. The shrink wrapping was maintained in
position during transportation to site.

CLIENT
The Office of Arts & Libraries

USER
The British Library

PROJECT MANAGERS
Property Services Agency,
Department of Civil Projects

ARCHITECT
Colin St John Wilson & Partners

CIVIL & STRUCTURAL ENGINEERS
Ove Arup & Partners

SERVICE ENGINEERS
Steensen Varming Mulcahy & Partners

QUANTITY SURVEYORS
Davis Langdon & Everest

BUILDING SURVEYORS
Baxter Payne & Lepper, inc. Britton Poole & Burns

CONSTRUCTION MANAGERS
Laing Management Contracting Limited

CONTRACTORS

Secant Pile Walls	Lilley Construction Co. Ltd.
Underreamed Bored Bearing Piles	Piggot Foundations Ltd.
Precast Concrete Boot Lintols	Malling Concrete Ltd. Crendon Concrete Ltd.

SECANT PILE WALL MIXES

The mix proportions of the first and last mixes are given below:

Cementitious Content (Kg/m3)	Cemsave: Cement Ratio	W/C Ratio	Plasticiser	Max. Aggregate size (mm)	Design Slump (mm)	7 Days AV	7 Days F_{cu}	28 Days AV	28 Days F_{cu}	56 Days AV	56 Days F_{cu}	90 Days AV	90 Days F_{cu}
360	70:30	0.47	-	20	150	19	15½	35	24	42	29	45	33
340	85:15	0.48	Cormix P1	40	150	15	10	31½	25	36	30	40	34

CONCRETE MIXES USED IN BORED 1.8m DIA. PILES

Cementitious Content Kg/m3	Cement Replacement Material	Replacement Cement Ratio	W/C Ratio	Plasticiser	Design Slump mm	Crushing Results Cube 7 Day AV	7 Day F_{cu}	28 Day AV	28 Day F_{cu}	56 Day AV	56 Day F_{cu}
425	Cemsave	70:30	0.43	Cormix P2	150	20.5	16	39.2	32.3	41	34
430	Pozzolan	25:75	0.43	-	150	37	31.7	48.3	43.6	56	50

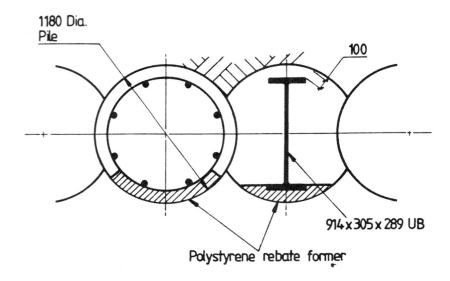

Figure 1: Typical Plan on Secant Pile Wall

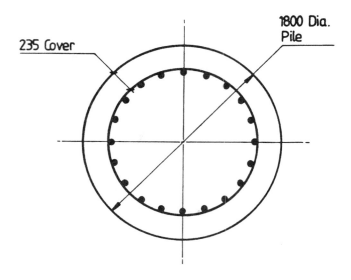

Figure 2: Plan on Typical 1.8m diameter Bored Pile

Figure 3: Elevation on Building
showing typical precast unit
stitched to structure @ 1/5
span points

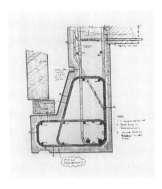

Figure 4: Typical boot
lintol reinforcement
detail

Figure 5: End view on precast concrete unit
and cladding brickwork

IMPORTANCE OF DETAILING IN
R C STRUCTURES

M I Soliman

Ain Shams University,

Egypt

ABSTRACT. One of the most important pre-requests of reinforced
concrete, is the bond between the reinforcement and the concrete.
This phenomenon is important for the continuous cooperation of these
two materials in any reinforced concrete element. To assure this
property the steel must be embedded in concrete to a certain length.
To translate this required developed length to the construction
engineer, the designer must supply a full detail of reinforcement
for the different concrete elements, specially at the junction of
two concrete elements, This study is divided into two parts. In the
first part the behaviour of RC corners subjected to opening
movements was studied, and in the second part, the effect of
confinement of concrete, by the transverse reinforcement of stirrups
was investigated. The results of this study were combined with
other available information to formulate some recommendations for
the designer and researcher for the importance of the detailing and
the confinement in the life time and serviceability of RC
structures.

Keywords: Reinforced Concrete, Steel stress, Bond, Splice,
             Frames, Joints, Anchorage, Confinement, Deformations
             and Cracks.

Prof Dr M I Soliman is an Assistant Professor at the Department of
Civil Engineering, Ain Shams University, Cairo, Egypt. He received
his BSc degree in Civil Engineering from Ain Shams University in
1969, his MSc degree from the same University in 1972. He received
a M Eng degree from McGill University, Canada in 1975 and his PhD in
1979 from the same University.

INTRODUCTION

There are many details for reinforced concrete joints, and corners.
Few studies have been made on frame joints to define the joint
detail that could satisfy the strength requirements, cracking,
ductility, in addition to the simplicity of construction. The most
critical type of joints are those subjected to opening moments,
which produce tensile stresses in the inside surface of the corner.
The reinforcement layout of the frame joints, should meet certain
basic requirements. The joint should resist a moment of at least
the failure moment of the section adjacent to the joint. In this
case, a flexural failure may occur outside the joint, which then
should not limit the load bearing capacity of the frame. If the
first requirement is not fulfilled, then the joint should have the
necessary ductility, and rotational capacity, so that the
redistribution of forces within the structure, can proceed without
the occurrence of brittle failures of the joints. The present
research work is divided into two parts. In the first part of the
presented work, the effect of the following factors on the general
deformational behaviour of the reinforced concrete joints and
corners are studied. These factors area as follows:

1) The effect of using different types of details of reinforcement in
   the frame joints.
2) The effect of adding special inclined steel bars at the tension
   side of these joints.
3) The effect of using different ratios of the main steel
   reinforcement in the joint behaviour.
4) Proposing the ideal detail of reinforcement, and the ratio of the
   main steel, which provide the maximum strength, and ductility for
   this type of joint.
5) The effect of the joint confinement by stirrups.

The stirrups, which are normally placed in R C elements, confine the
concrete in these elements. Accordingly improves the bond
resistance by increasing the characteristic strength of concrete in
these regions. Therefore, in the second part of this study, the
problem of splicing steel reinforcement is studied. Splices are
necessary because of the reinforcing bars are produced normally in
standard lengths. In general, bars are cut to a shorter length, and
lapped, mostly, at the location of minimum critical bending moments.
The forces in the spliced bars are transmitted, as normal, to the
concrete by bond. Most national building codes disregarded the
effect of the lateral confinement of concrete on the behaviour of
the structural element, although they all contain clauses limiting
the sizes and spacing of ties. It appears that most code
recommendations are based on the assumption that the main function
of the lateral reinforcement is to prevent the longitudinal
reinforcement from buckling, and to resist the shear stresses
existing in these elements. The purpose of the present work is to
explain the effect of the confinement of concrete by stirrups, in

reducing the probability of crack formation, at the location of a splice for rounded smooth tension bars. Also to study the effect of confinement in reducing the length of this splice due to the increase of the concrete strength at this location.

EXPERIMENTAL WORK

The experimental work of the present study was divided into two main parts. In the first part six triangular reinforced concrete frames of the same concrete dimensions were tested. In the second part a set of nine reinforced concrete simple beams were tested with a splice at the middle section of the beam.

Description of the Triangular Frames

The overall dimensions of the six tested triangular frames were 1.30m wide by 1m high. The cross section of the included members were 0.15m x 0.25m while that of the upper joint was 0.15m x 0.35m. The difference between the tested six frames were the details of reinforcement, the presence, and the amount of the additional included bars, the stirrup distribution inside the joint, and the percentage of the main reinforcement. Figure 1 shows the detail reinforcement of each frame.

Reinforcement used in manufacturing of these frames was of mild steel of tensile strength 240N/mm², while the characteristic strength of concrete after 28 days was about 30N/mm².

Description of the Simple Beams

The overall dimensions of the tested beams were kept constant with a breadth of 0.1m and an overall depth of 0.4m and span 2m. Each beam had a splice at the middle section of it. The ratio of the transverse reinforcement at the location of the splice were varied, Figure 2. These beams were cast of concrete of characteristic strength equal to 22.5N/mm². Plain reinforcement bars made of mild steel were used in construction of these beams. The beams were designed to prevent shear failure.

Test Procedure

The pre-described specimens were tested using an incremental static loading procedure. The loading force was provided by a hydraulic jack of 20 ton capacity and 0.05 ton accuracy. An incremental load of about 10% of the ultimate load was applied to the tested specimens during a period of about 60 seconds. This load was kept constant for about 30 minutes during each increment while readings were being taken.

For the six frames, a single concentrated load was applied at the top of the frame, while the nine beams were tested under two concentrated loads, symmetrically placed on the top of these beams.

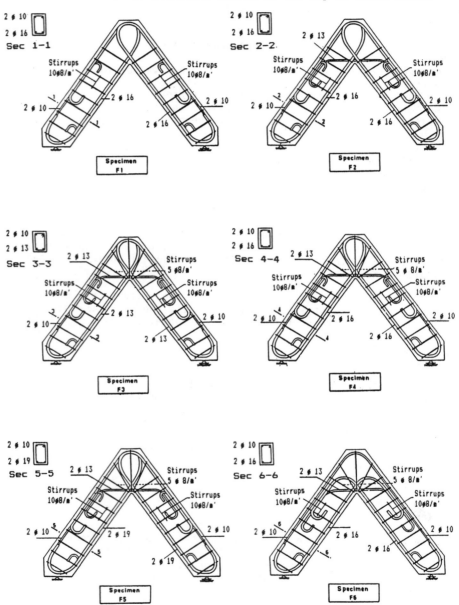

Figure 1: Reinforcement of the Triangular Frames

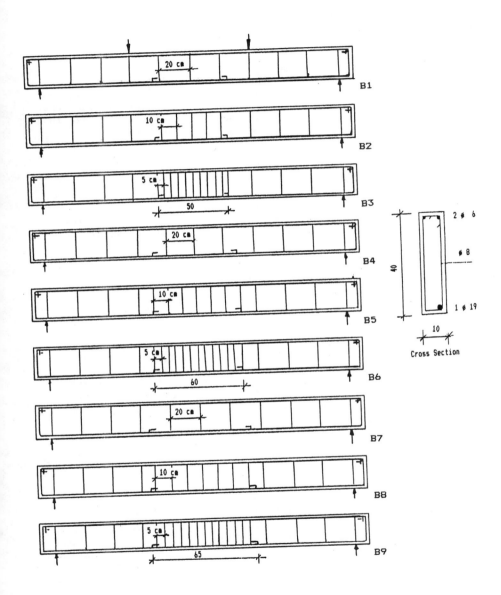

Figure 2: Reinforcement of the tested beams

ANALYSIS OF THE EXPERIMENTAL RESULTS

The experimental results of crack width and propagation, strains and deflections were recorded at different stages of loading from zero up to the failure load.

Results of the Triangular Frames

For the six tested frames, the cracks appeared first at the tension side of the joint section. These cracks developed towards the upper surface of the frame joint, (the compression side). At further load increments, other cracks appeared at the inclined beams of the frames.

Cracking and Crack Patterns

Figure 3 indicates the crack patterns of the tested frames at failure. Also, Table 1, shows a comparison for the cracking loads, and the crack widths, at a load level equal to 85% of the ultimate load.

From the observations of the crack patterns of the tested frames, the following remarks can be concluded:

a) For a joint with a reinforcement detail, as frame F1, the cracking load level was very low compared with the cracking load of the other tested frames.
b) The crack pattern of frame F2, shows that, the presence of the inclined bars at the tension side of the joint, increases the cracking load level of this frame.
c) Confinement of the frame joint by means of stirrups, increases the cracking load. The comparison of the crack patterns of frames F2 and F4 showed that, the presence of the stirrups increases the cracking load of these frames.
d) Increasing the percentage of the main reinforcement improves the cracking behaviour of these frames.
e) The presence of the cross reinforcement in the joint, (frame F6), decreases the cracking load of these frames.

TABLE 1: Cracking Loads, Crack Widths and Failure Loads

Frame Number	F1	F2	F3	F4	F5	F6
Cracking Load (kN)	5.0	20.0	30.0	40.0	50.0	30.0
Crack width (mm) at 85% Ult. load	Failed	0.85	0.71	0.55	0.49	0.79
Failure Load	80.0	147.0	120.0	165.0	203.0	132.0

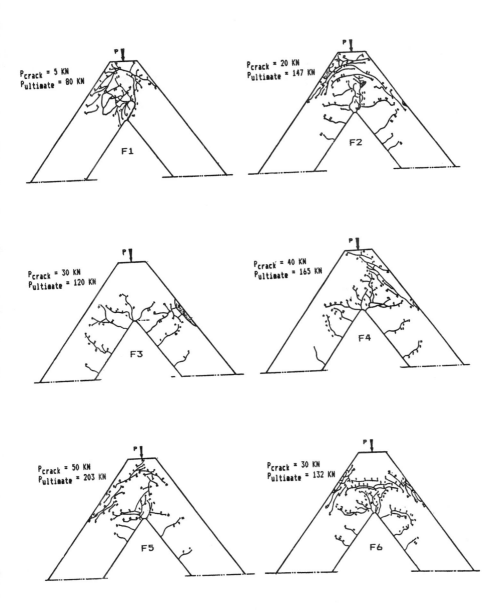

Figure 3: Crack Pattern of the Triangular Frames

Deflections and Horizontal Movement of the Roller Support of the
Triangular Frames

By studying the results of deflections at mid span and the
horizontal movement of the roller support respectively, from zero up
to the failure load, the following remarks can be concluded:

a) The displacements of the tested frámes without inclined bars, or
   stirrups inside the joint were higher than those of the other
   frames.
b) Using the inclined bar reinforcement, and joint stirrups
   increases the frame stiffness, and consequently reduce the
   resulting displacements.
c) Comparing the displacements obtained from the tested frames F3, F4
   and F5, showed that by increasing the ratio of the main
   reinforcement of these frames decreases the resulting
   displacements.
d) The cross detail of the main reinforcement decreases the frame
   stiffness. However the loop detail improves the deformational
   behaviour of these frames.

Results of the R C Beams

The results of the tested beams are presented in Figures 4 and 5,
which show the beams deflections at different load levels. At the
working stages of loading, these figures show that the variation in
the transverse reinforcement has no effect on the value of the
midspan defection. The deflections of the tested beams ranges
between 0.94 to 1.88mm as shown in the figures, which is about
1/1000 of the span length. However, at the failure load, it has
been noticed that by reducing the splice length, and the transverse
reinforcement, the midspan deflection increases.

From the crack pattern of the tested beams it was noticed that all
beams started to crack within a load level of 20 - 30 kN in the
following manner:

1) Beams with splice length of 50cm

The tested beams cracked first at both ends of the splice on the
nearer face of the U hook of the bar. The increase of the
transverse reinforcement along the splice length, indicated an
improvement in the splice behaviour and strength. The longitudinal
cracks parallel to the reinforcing bars in the region of the splice
were formed in the bottom face of all the tested beams. In beam B1,
(20cm stirrups spacing), these longitudinal cracks were continuous
along the splice length, while in the other two beams, (B2 and B3),
these cracks appeared only at both ends of these splices.

2) Beams with a splice length of 60cm

These beams started to crack as the beams of group (1). A complete crack appeared on the bottom face of beam B4, along the splice length. For beam B5, (10cm stirrup spacing), a longitudinal bottom crack has occurred at both ends of the splice, while for beam B6 (5cm stirrup spacing) no longitudinal cracks occurred. These cracks spread and propagated along the sides of the beam B6, however, these cracks were less than those of the other two beams.

3) Beams with a splice length of 65cm

Beams B7 and B8 (20 and 10cm stirrups spacing respectively) cracked at both ends of the splice. While the cracks of beam B9 (5cm stirrup spacing) started under the applied two concentrated loads on the beam, then, a second crack occurred at the ends of the splice of the beam.

At the failure of beam B7, a longitudinal bottom crack started to appear. The propagation and extension of these cracks along the beam decreases at the per cent of the transverse reinforcement increases. This indicates an increase in the strength of the splice, and consequently a lower crack width occurs.

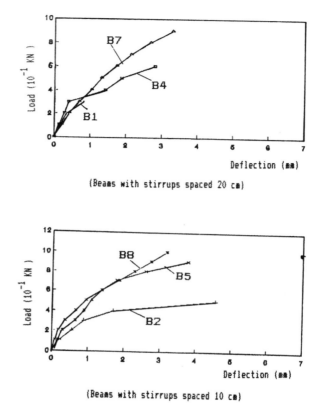

(Beams with stirrups spaced 20 cm)

(Beams with stirrups spaced 10 cm)

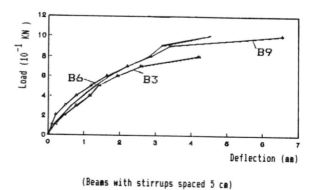

(Beams with stirrups spaced 5 cm)

Figure 4: Load Deflection Curves for the Tested Beams

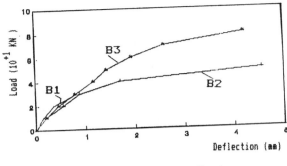

(Beams with splice length 50 cm)

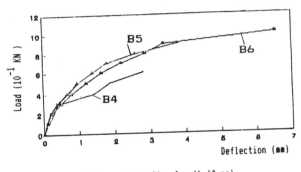

(Beams with splice length 60 cm)

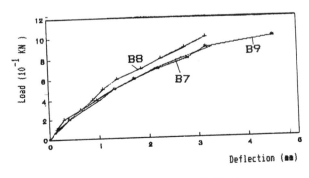

(Beams with splice length 65 cm)

Figure 5: Load Deflection Curves for the Tested Beams

CONCLUSIONS

1) The actual ultimate capacity of the frame joints of the tested triangular frames, were lower than their designed values, due to the resulting stress concentration at the joint region. Studying the general deformational behaviour of the six tested frames, the following may be recommended:

   a) The main reinforcing bars, have to be looped into the joint region, as far as the concrete cover allows, and then brought back into the same cross-section to the level of the included bars.

   b) It is recommended to use inclined steel bars at the tension side of the joint section, with a value not less than one half of the area of the main reinforcing bars. These inclined bars increase the joint capacity, and also improve the deformational behaviour of these joints.

   c) The main reinforcement of these joints should be designed for the moments and the normal forces of the adjacent sections of these joints. The effect of the loop reinforcement in these regions, and the effect of the inclined bars are to be ignored during the design.

   d) In sharp edged corners, there is a stress concentration factor of about two, and this causes a bond failure in these corners. Then it is recommended to use deformed bars, with small diameters, and increasing the reinforcement ratio of these types of joints.

   e) Confinement of the joint be means of stirrups, increases the load carrying capacity of these joints by about 10-15%.

   f) The increasing of the reinforcement ratio of these joints, decreases the ratio of the actual carrying capacity of the joint to the designed capacity. Therefore, it is recommended to limit the reinforcement ratio below 1.2%

2) Increasing the splice length of the tested R C beams, and increasing the transverse reinforcement, has no effect on the deflection of the beams within the working load level. However, they have a significant effect on reducing the deflection near the ultimate stages.

3) Increasing the confinement of concrete at the splice region, prevents the occurrence of the longitudinal cracks along the splice length.

4) Increasing the confinement of concrete in the region of the splice, increases its strength and accordingly, the splice length can be reduced.

ACKNOWLEDGMENT

This paper constitutes a part of the MSc Thesis of Eng. T M Hakim from the Civil Engineering Department, Ain Shams University, December, 1989. This paper also constitutes a part of the MSc Thesis of Eng. M M Awad - at the Civil Engineering Department, Ain Shams University, (under preparation).

REFERENCES

1   NILSSON H E and ANDERS LOSBRG. Design of Reinforced Concrete Corners and Joints Subjected to Bending Moments.

2   HAKIM T. Behaviour of R C Corners Subjected to Opening Moments and Normal Forces. MSc Thesis, Faculty of Engineering, Ain Shams University, Cairo, December 1989.

3   MAYFIELDO, KONG, BENSSISON and DAVIS. Corner Joints Details in Structural Lightweight Concrete. ACI Journal, May 1971.

4   SKETTRUP E, STRABO J, ANDERSEN H and NIELSEN T. Concrete Frame Corners. ACI Journal, December 1984.

5   DAVID M F ORR. Lap Splicing of Deformed Reinforcing Bars. ACI Journal, November 1976, pp 622-627.

6   ORANGUN C O, JERSA J O and BREEN J E. A Re-evaluation of Test Data on Development Length and Splices. ACI Journal, March 1977, pp 114-122.

7   GHALI K N, SOLIMAN M I and AWAD MM. Effect of Confinement of Concrete on the Region of Splice. El Azhar Engineering First Conference, Cairo, Egypt, December 1989.

8   GHALI K N, SOLIMAN M I and HAKIM T M. Ultimate Limit Moment of R C Corners Subjected to Opening Moments and Normal Forces. El Azhar Engineering First Conference, Cairo, Egypt, December 1989.

THE EFFECTS OF REINFORCEMENT DETAILING ON THE PROTECTION OF REINFORCED CONCRETE BEAM/COLUMN CONNECTIONS

R H Scott

P A T Gill

University of Durham,

United Kingdom

ABSTRACT. This paper describes laboratory tests on fifteen reinforced concrete external beam/column connections. Both the beam and the column reinforcement were internally strain gauged with, typically, a total of 230 strain gauges being installed in each specimen. The tests produced very detailed information concerning strain and bond stress distributions in the connection zone for each of the three detailing arrangements investigated - bending the beam tension steel down into the column, bending it up into the column, or bending it back into the beam to form a 'U' bar. Bending the beam tension steel down into the column was best able to transfer load by bond at all stages in the load history of the specimens. The 'U' bar detail was satisfactory, whilst the results indicated that bending the beam tension steel up into the column should be avoided.

Keywords: Reinforced Concrete Connections, Strain Gauges, Reinforcement Details, Reinforcement Strain Distributions, Bond Stresses.

Dr R H Scott is a Lecturer in the School of Engineering and Applied Science at the University of Durham. His research interests are concerned with the behaviour of reinforced concrete structures, in particular the detailed measurement of reinforcement strain and bond stress distributions.

Dr P A T Gill is a Senior Lecturer in the School of Engineering and Applied Science at the University of Durham. His research interests are concerned with dynamics and vibrations, materials and experimental stress analysis.

INTRODUCTION

The performance of the connections between beams and columns in reinforced concrete structures has a significant effect on the behaviour of these structures as a whole.   Many factors affect the performance of these behaviourally complex regions, amongst them being the detailing of the reinforcement in the connection zone itself.   This is particularly important in external connections where the beam is present on one side of the joint only.   Many tests have been conducted to formulate design requirements, the results of such work leading to the rules and guidelines which appear in the various national codes and standards.   Nevertheless, the understanding of connection behaviour is still incomplete, particularly with regard to the detailed distributions of strain and bond stress along the reinforcement in the connection zone itself.   These parameters influence the development of cracks in the connection zone and hence the degree of protection that the reinforcement receives from the concrete.   An improved understanding of strain and bond stress distributions would thus seem likely to assist in selecting reinforcement layouts most suited to producing a connection that was both structurally efficient and durable.   This paper seeks to contribute to this understanding by describing work which made detailed measurements of reinforcement strains and bond stresses in external beam/column connections using internally strain gauged reinforcing rods.   The authors addressed this problem by making detailed measurements of reinforcement strains using a technique for strain gauging the reinforcement which they had developed during earlier work (1).   This paper describes the work that was undertaken and the results obtained.

SPECIMEN DETAILS

Each specimen had a beam 850mm long and 110mm wide framing at mid-height into one side of a column 1700mm high.   Column cross-sections were all 150 x 150mm, but two beam depths were used - 210mm (twelve specimens) and 300mm (three specimens).   Both the beams and the columns were loaded, the former downwards at a point 100mm from the free-end;   the tension steel was thus at the top of the beam's cross-section.

Columns were reinforced with four 16mm diameter high yield rods (Torbar) and 6mm diameter mild steel links at 150mm centres, which included a link at the mid-height of the beam.   Tension reinforcement in the beams was a pair of 12mm or 16mm diameter high yield rods. Three arrangements of this reinforcement were studied - bending it down into the column, bending it up into the column, or bending it into a 'U' to form the beam compression reinforcement also.   With the first two details, 12mm diameter straight high yield rods were used for the beam compression reinforcement.   In the connection zone the beam reinforcement fitted between the column reinforcement, but the two sets of rods touched where they became adjacent.   Shear reinforcement in the beams was 6mm diameter mild steel links at 100mm centres. Parameters studied were thus beam depth, beam steel percentage and beam detailing arrangement. In addition two column loads were used - 275kN which produced around 500 microstrain in compression in the column reinforcement, and 50kN which gave about 100 microstrain. The higher load, used for eleven of the fifteen specimens tested, produced column strains representative of those expected in a full size structure under normal working loads. The low column load was used to investigate the effects of increased rotation of the connection zone, an effect also achieved in the three specimens with the increased beam depth. Specimen details are given in Table 1, the suffix 'L' to a specimen number indicating a low column load specimen.

Rods on one side of each specimen were machined to permit the installation of electric resistance strain gauges for strain measurement. The technique, which has been used by the authors in a number of investigations (1-4), involved milling two reinforcing rods down to a half round and then machining a 4mm wide by 2mm deep longitudinal groove in each to accommodate the strain gauges (gauge length 3mm) and their wiring. The two halves of the rod were then bonded together with an epoxy resin which also filled any remaining spaces in the duct, giving the outward appearance of a normal reinforcing rod, but with the lead wires coming out at the ends. A three wire, common dummy installation was used which, because of the severe space limitations within the duct, used very fine lead wires. Typically 230 strain gauges were installed in the beam and column reinforcement, including the bent beam rods. The gauging layouts were designed to give both an overall picture of a specimen's behaviour plus very detailed information in the connection zone itself. The minimum gauge spacing was 12.5mm.

Concrete for the specimens used 10mm aggregate, with an aggregate/cement ratio of 5.5 and a water/cement ratio of 0.6. Specimens were cast vertically in two stages - first the bottom column and beam together, followed the next day by the top column. A 50mm high kicker above the connection zone, cast with the first pour, assisted with setting the formwork for the top column. Cubes and cylinders were cast with each specimen for the determination of the concrete's compressive and indirect tensile strength respectively.

TEST PROCEDURE

The column was loaded first to its full load in increments of 25kN, and this load held whilst the beam was loaded in 1kN increments until failure occurred. At each load stage a full set of strain gauge readings was recorded using a computer controlled data acquisition system. This also logged the applied loads, column shear forces, and the load carried by a prop provided at the beam end to control sidesway. Where possible, increments of deflection were applied to the beam after joint failure had occurred. Strains at the surface of the concrete were measured at selected load stages using a 200mm Demec gauge with a grillage of studs on both the beam and column faces.

TABLE 1 : Specimen Details

Specimen	Beam Depth (mm)	Diameter of Beam Tension Reinforcement (mm)	% Beam Tension Reinforcement	Detailing Type (see below)
C1	210	12	1.0	A
C1A	210	12	1.0	A
C1AL	210	12	1.0	A
C2	210	12	1.0	B
C3	210	12	1.0	C
C3L	210	12	1.0	C
C4	210	16	1.9	A
C4A	210	16	1.9	A
C4AL	210	16	1.9	A
C5	210	16	1.9	B
C6	210	16	1.9	C
C6L	210	16	1.9	C
C7	300	16	1.3	A
C8	300	16	1.3	B
C9	300	16	1.3	C

A: Beam reinforcement bent DOWN into column
B: Beam reinforcement bent UP into column
C: Beam reinforcement bent into a 'U' bar.

RESULTS

The strain gauging technique produced very detailed information regarding the strain distributions in the beam and column reinforcement. Typical strain distributions along the beam tension reinforcement are shown in Figures 1-3 to illustrate the effect of detailing arrangement and reinforcement percentage. The rods have been 'straightened' for ease of interpretation with a key being provided with each figure to show its context within the specimen. The loads indicated are those applied to the beam; the column load is always 275kN.

Loading the column produced compressive strains in the column reinforcement and in the vertical legs of the beam tension reinforcement - around 100 microstrain for the 50kN column load and 500 microstrain for the 275kN column load. The Poisson's ratio effect caused by the column shortening led to small tensile strains being observed in the bottom beam rods where they crossed the connection zone. Loading the beam soon eliminated these and caused a steady increase in tension in the top beam reinforcement and compression in the bottom beam reinforcement. The distributions of tension in the top beam reinforcement exhibited a series of peaks as cracks developed in the beam. A progressive movement through the connection zone of the point of zero strain was observed at each load stage as an increasing length of each beam rod became tensile. Cracks also developed in the column both above and below the connection zone in specimens with the low column load.

Diagonal cracks in the connection zone were first observed at the loads and strains indicated in Table 2, the latter being the maximum strains in the beam tension reinforcement either at, or very close to, the face of the column. Table 2 indicates that diagonal cracks in the low column load specimens generally occurred at lower beam loads and strains than in the companion specimens with the high column load. In contrast, strains at the end of the first bend in the tension reinforcement (the only bend when the rods were bent up or down) when diagonal cracking occurred (points C,C,E, in Figs 1-3 respectively) were similar, thus bond stresses around the first bend in the tension reinforcement were generally higher in the high column load specimens than in those with the low column load.

Average bond stresses when joint cracking occurred are tabulated in Table 3, both around the first bend and, where sensible calculations could be made, along the vertical leg of the reinforcement also, although these latter bond stresses were developed over short distances only. It was observed that, up to the onset of joint cracking, most of the load transfer from the beam tension reinforcement into the concrete had taken place once the end of the first bend was reached. The contribution of the vertical leg of the reinforcement to this process was small, thus load transfer by bond was largely independent of the type of detailing arrangement used.

TABLE 2 :　Summary of Loads and Strains for Initial Joint
Cracking and Specimen Failure

Specimen	Initial Joint Cracking		Specimen Failure	
	Beam Load	Max. tensile reinf. strain (microstrain)	Beam Load (kN)	Max tensile reinf. strain (microstrain)
C1	17.1	1821	26.2	>23000
C1A	19.0	2628	26.8	>17000
C1AL	11.1	1208	22.0	3749
C2	18.9	2547	21.5	5782
C3	19.9	2442	25.9	7129
C3L	12.5	1271	21.6	3640
C4	18.8	1190	29.7	2125
C4A	18.8	1109	31.8	2335
C4AL	12.1	764	28.3	1796
C5	9.7	616	13.8	1182
C6	14.8	899	21.8	1304
C6L	14.8	967	26.0	1678
C7	22.2	859	32.0	1424
C8	26.2	1174	27.3	1483
C9	22.9	1092	27.9	1460

After diagonal cracking occurred, the relative performances of the different reinforcement details became more marked. Table 2 lists the beam load, and maximum beam tension strains at which specimen failure occurred. It is convenient to deal with the high column load specimens first.

With specimens C1 and C1A, which had 1.0% tension reinforcement bent down into the column, an increasing length of the vertical leg of the beam tension reinforcement became tensile with each load increment until, at the end of the test, the whole length of these rods was in tension even though the adjacent column rods were still in compression. Failure was due to a plastic hinge forming in the beam at the face of the column. This detail exhibited considerable ductility with a significant increase in load between joint cracking and specimen failure. With specimen C2, which had the beam tension reinforcement bent up into the column, loads for joint cracking and specimen failure were much closer. There was a sudden propagation of tensile strain along the vertical leg of the beam reinforcement just prior to failure, causing the behaviour of this specimen to be more brittle. In contrast, the behaviour of specimen C3, with the 'U' bar, was more like that of C1 and C1A with a plastic hinge forming at the column face. Tensile strains extended around both bends of the 'U' and back out into the beam by the time failure load was reached. C3 was, however, not so ductile as C1 and C1A since peak tensile strains were in the order of 7000 microstrain - values of over 23 000 and 17 000 microstrain being recorded in C1 and C1A respectively.

Specimens with 1.9% beam tension reinforcement exhibited more extensive cracking in the connection zone. Failure was governed by the capacity of the connection zone, which was consistent with the findings of Taylor (5). Strains in the beam tension reinforcement were lower than those in the specimens with 1.0% beam tension steel. This was particularly marked in specimens C4 and C4A which had beam reinforcement bent down into the column although, as with C1 and C1A, tensile strains had again reached the end of the vertical leg of the beam rods by the end of the tests. The beam tension reinforcement in C5 and C6 was still behaving elastically when failure occurred.

Specimens C7, C8 and C9 were all brittle, C8 with the beam tension reinforcement bent up into the column particularly so as joint failure occurred at the next load stage after diagonal cracking was first observed. Strains in the beam tension reinforcement in all three specimens were still elastic when joint failure occurred.

Specimens with the low column load had reduced beam loads and strains at failure than their high column load counterparts (with the exception of specimen C6L). This was particularly marked with specimen C1AL. Cracks in the columns and extensive cracking in the connection zone prevented a full plastic hinge being developed at the column face. Consequently, this specimen was unable to develop the high strains recorded in C1 and C1A. The low column load thus led to a reduction in the degree of ductility which this type of specimen could develop. This was less marked with C4AL compared with C4A

since the failure of C4A was already governed by the performance of
the   connection   itself   -   the   low   column   load   of C4AL led to an
emphasizing of this effect rather than a radical reduction in
ductility.   C6L, which had the T16 'U' bar behaved very similarly to
C6,   although both the beam load and the maximum strain at failure
were slightly higher.

Average bond stresses at failure are listed in Table 3,   both around
the first bend and along the vertical leg of the reinforcement. Some
values could not be calculated due to reinforcement strains being too
non-linear.   Bending the beam reinforcement down into the column
produced a detail which was able to adjust to the changing conditions
in the connection zone.   Load transfer by bond continued along the
vertical leg of the reinforcement after bond around the bend had been
largely lost either as a result of high steel strains - specimens C1
and C1A - or extensive joint cracking - specimens C1AL,   C4,   C4A and
C4AL.   This   reinforcement detail did not limit the performance of
these specimens under ultimate loads.   In contrast,   bending the
reinforcement up into the column did limit a specimen's behaviour
since compressive stresses developed in the beam steel's vertical leg
as   a   result of column loading hindered the subsequent spread of
tension due to beam loading. This resulted in load transfer from the
beam reinforcement having to occur almost entirely around the bend at
all load stages making it difficult for this detail to adjust readily
to changing conditions in the connection zone. This was reflected in
an inferior specimen performance.   With the 'U' bar detail load
transfer by bond could occur along the short straight leg and around
both bends of the 'U' but was then opposed by the compressive
stresses induced in the steel by flexure of the beam. This is likely
to have limited the performance of specimens having this detail.
Nevertheless, the 'U' bar detail was much more satisfactory than that
of bending the reinforcement up, although the detail of bending the
reinforcement down was clearly superior to it.

An interesting tendency was for tensile strains to develop in the
normally compressive bottom beam reinforcement as failure was
approached.   These strains were first observed within the column
width, but later sometimes moved out into the beam itself.   The
effect was most pronounced in those specimens having the beam tension
reinforcement bent up into the column, but was also observed in
specimens having the other details.   It was believed to be caused by
the propagation of a diagonal joint crack across the line of this
reinforcement,   and strains could be quite substantial with peak
values of around 1300 microstrain being recorded.   Thus, at the end
of a test, it was common for both the top and bottom beam rods to be
in tension over a short distance close to the column. This was not
the result of sidesway since the load on the prop at the end of the
beam was always compressive.

CONCLUSIONS

The strain gauging technique gave detailed information concerning strain and bond stress distributions along the reinforcement in fifteen reinforced concrete beam/column connection specimens. Up to the onset of diagonal cracking in the connection zone, load transfer from the beam reinforcement into the column was by bond stresses developed around the first bend in the beam tension rods, for all three detailing arrangements investigated. After diagonal cracks formed in the connection zone, the detail of bending the beam rods down into the column was best able to adjust to bond degredation in the connection zone by its ability to develop bond stresses along its vertical leg. The detail of bending the beam rods up into the column was unable to do this, which led to brittle behaviour as a result. The 'U' bar detail performed more satisfactorily, but transfer of load by bond was limited both by the shortness of its vertical leg, and by compressive stresses produced by beam flexure.

It is recommended that the detail of bending the beam tension reinforcement down into the column be used wherever possible when detailing connections in reinforced concrete structures. The 'U' bar detail is the most satisfactory alternative, but cannot produce quite the same degree of ductility. Bending the beam reinforcement up into the column should be avoided due to its inability to develop a satisfactory mechanism for load transfer by bond under ultimate load conditions.

ACKNOWLEDGEMENTS

The technical support of Mr S.P. Wilkinson is gratefully acknowledged. The SERC is thanked for its financial support.

REFERENCES

1   SCOTT, R H , GILL, P A T. Short-term Distributions of Strain and Bond Stress Along Tension Reinforcement. The Structural Engineer. Volume 65B, Number 2, June 1987, pp 39-43, 46.

2. SCOTT, R H., GILL P A T. Time Dependent Distributions of Strain and Bond Stress Along Tension Reinforcement, to be published in Proc. ICE (June 1990).

3. JUDGE, R C B, SCOTT, R H, GILL, P A T. Force Transfer in Compression Lap Joints in Reinforced Concrete. Magazine of Concrete Research, Volume 41, Number 146, March 1989, pp 27-31.

4. JUDGE, R C B, SCOTT, R H, GILL P A T. Strain and Bond Stress Distributions in Tension Lap Joints in Reinforced Concrete. To be published in Magazine of Concrete Research (March 1990).

5. TAYLOR, H P J. The Behaviour of In-Situ Concrete Beam-Column Joints. C & CA Technical Report 42.492, May 1974.

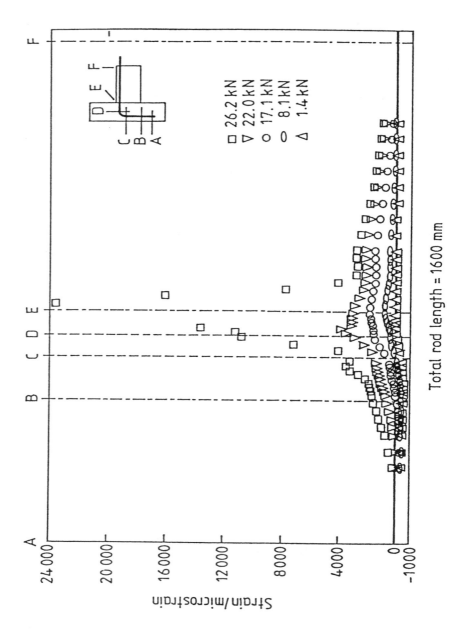

Fig. 1: Specimen C1: top beam rod

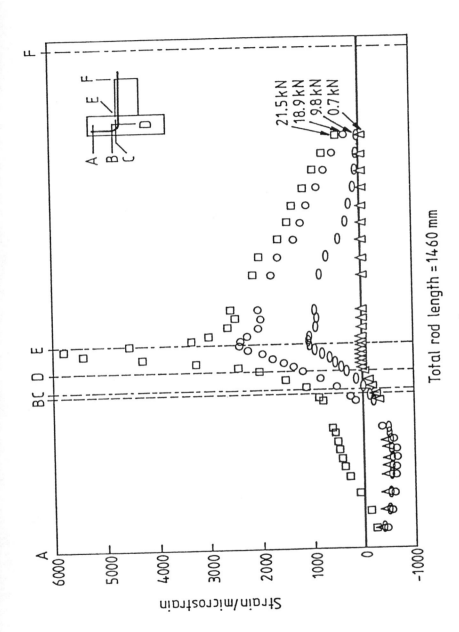

Fig. 2: Specimen C2 : top beam rod

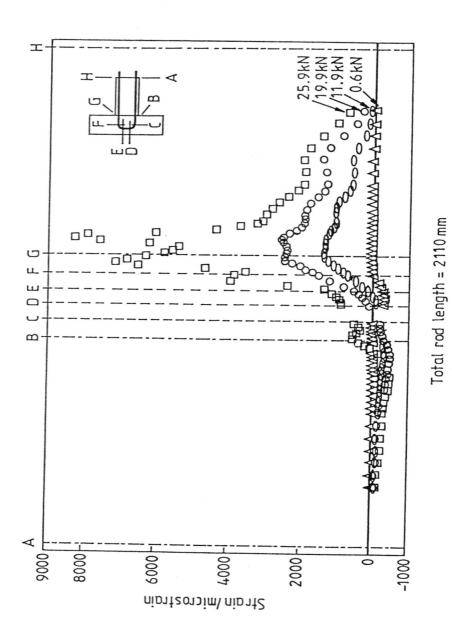

Fig. 3: Specimen C3: 'U' bar

TABLE 3 : Bond Stresses (N/mm^2) at Initial Joint Cracking and Specimen Failure

Specimen	Initial Joint Cracking		Specimen Failure	
	Bend	Vertical Leg	Bend	Vertical Leg
C1	7.0	2.7	-	4.0
C1A	5.7	-	-	3.3
C1AL	2.0	-	-	3.6
C2	6.3	-	6.6	1.4
C3	7.7	3.0	-	5.6
C3L	3.8	3.2	-	3.9
C4	5.1	2.4	-	4.3
C4A	5.3	2.8	-	4.2
C4AL	2.8	2.8	-	2.5
C5	2.8	-	6.8	-
C6	3.9	-	6.7	-
C6L	4.4	-	9.6	-
C7	3.4	-	-	-
C8	5.0	-	-	-
C9	4.6	1.9	7.5	2.6

TENSION STIFFENING, CRACK SPACINGS, CRACK WIDTHS AND BOND-SLIP

G Gunther

G Mehlhorn

University of Kassel,

West Germany

ABSTRACT. The durability of reinforced concrete structures depends essentially on crack widths and deformations. The protection against corrosion of reinforcement is lost with too large crack widths, and the stability is endangered. At too large deformation increases owing to cyclic and long-term loads considerable use restriction of the structure may result. In the contribution a summary of the results of about 200 experiments will be presented. The experimental investigations were carried out in two steps. With "global" tests especially the deformation and crack width depending on load duration and number of repeated loads were determined. The influence of the stress state effected by concrete shrinkage on the deformation increase was considered. Besides the tension and flexion tests also biaxial compression-tension tests with reinforced concrete specimens were carried out. The "local" investigations with centrically reinforced tension specimens were to determine the local relative displacements between steel and concrete depending on bond stress. With these investigations it is possible to explain the material behavior determined in the "global" investigations more exactly. Further detailed information can be taken from [1] as an addition to the test results presented in this paper.

Keywords: Tension Stiffening, Shrinkage, Transverse Compression, Prestressed, Cyclic Loading, Crack Spacings, Crack Widths, Local Bond, Axial and Radial Displacements.

Dr.-Ing. Gerd Günther is assistant in the Division of Reinforced Concrete at the University of Kassel. He is active in the research of material behavior of reinforced and prestressed concrete.

Prof. Dr.-Ing. Gerhard Mehlhorn is Professor of Civil Engineering at the University of Kassel. His main research activities are in the field of FEM applications for the analysis of reinforced and prestressed concrete constructions. Furthermore he is engaged in the experimental investigation of concrete material behavior.

TENSION STIFFENING

Centrically and Symmetrically Reinforced Tension Specimens

Tension specimens without concrete shrinkage

Fig. 1 shows the principal relation between average strain over the cracks and external force related to the steel cross sectional area. In these experiments the load was cyclic or on long-term.

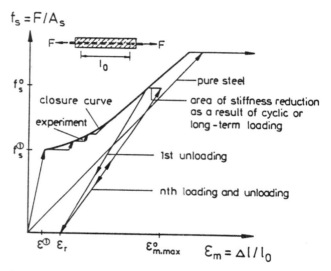

Figure 1: Relation between average strain over the cracks and external force related to pure steel cross sectional area

For centrically and symmetrically reinforced specimens loaded in tension good correlation was found for initial loading with the following idealization:

$$\varepsilon_m = \frac{f_s}{E_s} \sqrt{1 - \left(\frac{f_s^\oplus}{f_s}\right)^2}$$

$$(1)$$

The steel stress in the crack cross sectional area immediately after the first crack could be determined as follows:

$$f_s^\oplus = \frac{f_t}{A_s} (A_{c,n} + n \cdot A_s) \approx \frac{f_t}{\rho} .$$

$$(2)$$

The largest average strain at top load which stabilizes after a certain amount of repeated loads can be approximated to:

$$\varepsilon^o_{m,max} = \frac{f_s}{E_s} \sqrt{1 - \left(\frac{f^o_s}{f^o_s}\right)^5} \qquad (3)$$

These deformations appear in the short-term test already after 5000 repeated loads. They correspond approximately to those appearing after a six months long-term load. After six months no significant increases of deformations are to be expected at a long-term load of $f_{s,l-t} \geq 235$ N/mm^2 (compare Fig. 2).

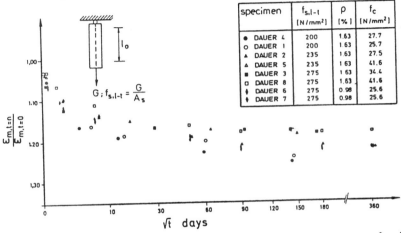

specimen	$f_{s,l-t}$ [N/mm^2]	ρ [%]	f_c [N/mm^2]
● DAUER 4	200	1.63	27.7
○ DAUER 1	200	1.63	25.7
▲ DAUER 2	235	1.63	27.5
△ DAUER 5	235	1.63	41.6
■ DAUER 3	275	1.63	34.4
□ DAUER 8	275	1.63	41.6
◆ DAUER 6	275	0.98	25.6
◆ DAUER 7	275	0.98	25.6

Figure 2: Development of average strain due to long-term load

At complete unloading a residual strain of

$$\varepsilon_r = 0,2 \cdot 10^{-3} \qquad (4)$$

remains in the specimen because the cracks do not close completely (compare Fig. 3). The residual strain is independent of the amount of repeated loads, the reinforcement percentage, and the top load.

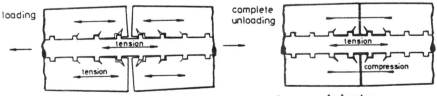

Figure 3: Basic deformation and stress behavior
of reinforcing bar and concrete

Thus, besides the cross sectional values and the modulus of elasticity of steel only the concrete tensile strength must be known to analyze Tension Stiffening.

The average related tensile strength for concrete at an age of 28 days depending on the eccentricity e/h could be determined by:

$$f_t = (0{,}25 - 0{,}6 \tfrac{e}{h}) \cdot \sqrt[3]{f_c^2} \quad ; \qquad f_t, f_c \quad [N/mm^2] \;. \tag{5}$$

The strength deviation is between $0{,}7\, f_t$ and $1{,}3\, f_t$.

The largest marginal strains owing to 85 % of the cracking load can be calculated with

$$\varepsilon_{max}^{\mathbb{O}} = (0{,}25 \cdot f_t^e + 0{,}25) \cdot 10^{-4} \quad ; \qquad f_t^e \quad [N/mm^2] \tag{6}$$

and

$$f_t^e = f_t + \frac{F \cdot e}{S} \;. \tag{7}$$

Influence of concrete shrinkage

Is the shrinkage influence approximated with

$$f_{s,sh}^{\mathbb{O}} = f_s^{\mathbb{O}} - \varepsilon_{s,sh} \cdot E_s \tag{8}$$

and

$$\varepsilon_{s,sh} = \frac{\varepsilon_{c,sh}}{1 + n \cdot \rho} \tag{9}$$

in order to determine the steel stress at the first crack, the average strain related to initial loading can also be calculated according to Eq. (1) applying $f_{s,sh}^{\mathbb{O}}$:

$$\varepsilon_{m,sh} = \frac{f_s}{E_s} \sqrt{1 - \left(\frac{f_{s,sh}^{\mathbb{O}}}{f_s}\right)^2} \;. \tag{10}$$

The largest average strain increase at top load results as for the specimens not pre-loaded by shrinkage to:

$$\Delta\varepsilon_{m,max}^o = \frac{f_s^o}{E_s} \left(\sqrt{1 - \left(\frac{f_s^{\mathbb{O}}}{f_s^o}\right)^5} - \sqrt{1 - \left(\frac{f_s^{\mathbb{O}}}{f_s^o}\right)^2} \right) \;. \tag{11}$$

At complete unloading of the specimen a residual strain of

$$\varepsilon_r = 0{,}2 \cdot 10^{-3} + \varepsilon_{m,sh}^o + \Delta\varepsilon_{m,max}^o - \varepsilon_s^o \tag{12}$$

remains under the condition

$$\varepsilon_{m,sh}^o + \Delta\varepsilon_{m,max}^o - \varepsilon_s^o \geq 0 \;.$$

The compression induced in the reinforcing steel by shrinkage of concrete can also effect larger average strains of the reinforcing specimen than those analyzed for pure steel (compare Fig. 4).

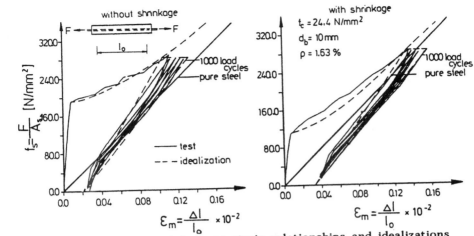

Figure 4: Experimental stress-strain relationships and idealizations of specimens without and with shrinkage

Influence of externally applied transverse compression

The principal relationships between average strain over the cracks and external tension force related to the steel cross sectional area which derived from the tests are shown in Fig. 5.

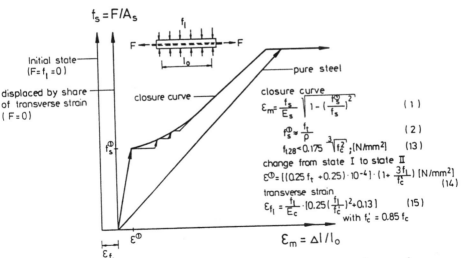

Figure 5: Relationship between average strain over the cracks and external force related to the steel cross sectional area dependent on lateral pressure

The influence of lateral pressure is registered by displacement of the origin of the coordinate system by the amount of lateral strain.

Prestressed, and Centrically Reinforced Tension Specimens

The idealizations for monotonically increasing and cyclic loadings can be transferred to specimens prestressed in a prestressing bed when referring to the steel stress and the average strain at which the concrete is free of stresses (see Fig. 6).

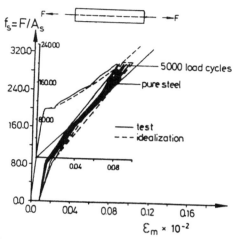

Figure 6: Stress-strain relationship of a prestressed specimen

Eccentrically Reinforced Tension Specimens

The stress-strain behavior of eccentrically reinforced tension specimens can also be described with Eqs. (1), (3), and (4). However, the change from state I to state II leads to a larger strain compared with the centrically reinforced tension specimens:

$$\varepsilon^{\textcircled{1}} = 2 \cdot (0,25 \, f_t + 0,25) \cdot 10^{-4} \, ; \, f_t \, [\text{N/mm}^2] \, . \tag{16}$$

The steel stress at the first crack can be calculated approximately by

$$f_s^{\textcircled{1}} \approx f_t \cdot \frac{A_{cef}}{A_s} \tag{17}$$

using the effective zone of reinforcement stated by Gergely et al. [2]

$$A_{cef} = 2 \cdot b \, (h - d) \tag{18}$$

and to the centrical tensile strength of concrete at an age of 28 days

$$f_t = 0,25 \sqrt[3]{f_c^2} \, ; \, f_t, \, f_c \, [\text{N/mm}^2] \, . \tag{19}$$

Reinforced Concrete Beams

In Fig. 7 the measured stress-strain relation and the stress-strain relation analyzed with equilibrium considerations in the crack cross section neglecting the concrete tensile stress are compared. The comparison shows the influence of the concrete tensile stress in the crack cross section on the amount of the steel stress. For equal strains, calculated stresses are much larger than measured stresses. Owing to the complex material behavior of concrete under tension it is not possible to calculate Tension Stiffening accurately using the external moment.

In contrast to the tension specimens, the steel stresses in the crack increase under repeated loads with unchanged top load.

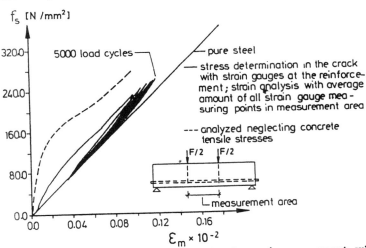

Figure 7: Steel stress-strain behavior of a beam (measurement with strain gauges at the reinforcement or analyzed neglecting the concrete tension in the crack cross section)

CRACK SPACINGS AND CRACK WIDTHS

According to the CEB/FIP requirements [3] the following approximation of the experimental crack spacings resulted (compare Table 1). This approximation is sufficient for the draft for ribbed bars or mats.

$$s_m = 2 (c + k \cdot a) + 0,1 \frac{d_b}{\rho} \quad \text{with} \tag{20}$$

c = smallest concrete cover of reinforcement at pure tension
= distance between lower side of reinforcing bar and tension edge at eccentric tension and at flexure
k = 0,1 for ribbed bars
a = reinforcing bar distance
d_b = diameter of reinforcing bar
ρ = $\dfrac{A_s}{A_{cef}}$

TABLE 1 Crack spacings

Specimen		d_b [mm]	c [mm]	a [mm]	reinforcement	number of specimens	crack spacing			
							experiments		calculated	
							s_m	s_{max}	s_m	s_{max}
reinforced concrete panels		10x8.5	10	50	ribbed bars with transverse bars	2	134	185	105	178
					ribbed bars	5	136	200		
					ribbed mats	2	107	150		
		10x5.5	11	50	ribbed bars with transverse bars	2	126	200	147	249
					ribbed bars	2	103	200		
					ribbed mats	1	102	125		
		5x16.0	42	100	ribbed bars	2	140	250	183	311
centrically reinforced tension specimens		1x10.0	30	70	ribbed bars	10	128	210	135	230
		1x10.0	40	90		2	127	220	200	340
		1x16.0	32	80		3	115	160	131	223
		1x16.0	48	112		2	183	240	219	372
		1x16.0	64	144		2	333	390	322	547
		1x22.0	44	110		20	174	310	180	306
		1x22.0	66	154		27	290	450	300	511
		1x22.0	88	198		16	359	630	442	752
		1x28.0	56	140		2	200	300	229	390
		1x28.0	84	196		2	333	450	382	650
		1x28.0	112	252		2	417	530	563	957
excentrically reinforced tension specimens		1x22.0	44	110	ribbed bars	8	156	280	180	306
		1x22.0	66	154		4	285	425	300	511
reinforced concrete beams		1x22.0	44	150	ribbed bars	3	244	300	214	363
		2x22.0	44	75	ribbed bars	2	162	250	151	256
		3x22.0	44	54	ribbed bars	2	113	185	131	223

As the deformations of a reinforced concrete specimen loaded in tension mainly occur in the cracks the average crack width can be determined by:

$$w_m = \varepsilon_m \cdot s_m \cdot \tag{21}$$

The analysis of the average strain is calculated as mentioned above. The largest crack width can be approximated by:

$$w_{max} = 1,7 \cdot w_m \cdot \tag{22}$$

LOCAL BOND

Measuring System for the Study of the Local Axial and Radial Displacements Between Concrete and Reinforcing Steel

The measuring system mentioned above allows among other things to measure local axial and radial displacements between reinforcing steel and concrete very accurately without any bond impairment. For the measurement of relative displacements between steel and concrete a permanent magnet ist attached to the reinforcing steel and a Hall-sensor is fastened to the concrete at a certain distance to the magnet (see Fig. 8).

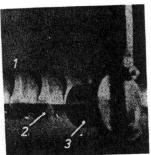

1   reinforcing bar
2   permanent magnet fastened to the reinforcing bar
3   cylinder with Hall-sensor attached to the concrete

Figure 8: Arrangement of measuring elements for the measurement of local relative displacements between steel and concrete

The magnetic induction which affects the Hall-sensor changes when the distance between the measuring elements is altered. Thus, the induction change can be used as unit for the change in distance. Depending on the used magnets and Hall-sensors relative displacements of 0,0005 mm at a distance of 20 mm between the measuring system elements can be registered.

Bond Stress - Axial Displacement Relations

The local bond investigations carried out so far had the following results: Fig. 9 shows that with increasing concrete strength and decreasing distance to the middle of the test specimen bond becomes stiffer.

598   Gunther, Mehlhorn

At equal distance to the edge of the test specimen the bond stiffness becomes smaller at increasing length of the specimen. Already after ten repeated loads the maximum bond stress at the largest external load decreases by 10 to 40 % compared with the initial loading. At equal concrete strength no significant differences dependent on the relative rib area were determined. The crack formation at initial loading has no significant importance on the bond behavior compared with the influence of the concrete strength.

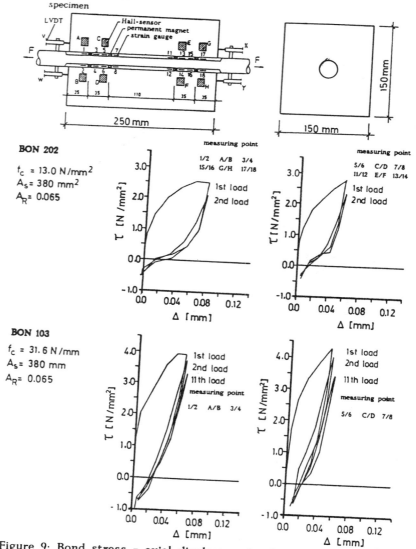

Figure 9: Bond stress – axial displacement relations dependent on concrete strength and the distance to the center of the specimen

Local Radial Displacements Between Steel and Concrete

In Fig. 10 the axial and radial displacements over the length of a test specimen for initial loading are shown. The course of the axial displacements was assumed to be approximately linear. With short specimens the realistic course of the axial displacements differs only insignificantly from the linearized approximation. The radial displacements are considerably smaller than the axial displacements. The radial displacements increase with the decrease of the distance to the edge of the specimen and with increasing length of the specimen. This explains the poor bond stiffness of the marginal areas compared with the inner areas. It also clarifies the bad transfer of bond strength in the longer specimens compared with the shorter ones at equal distance of the measuring points to the edge of the specimen.

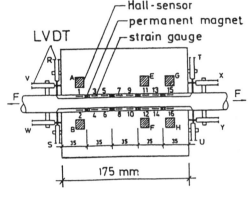

Axial displacements   Radial displacements

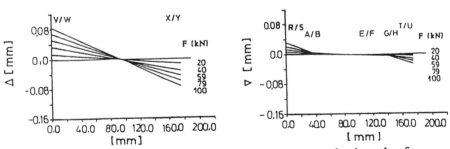

Figure 10: Axial and radial displacements over the length of a specimen for initial loading

REFERENCES

1   Günther, Gerd: Verbundverhalten zwischen Stahl und Beton unter monoton steigender, schwellender und lang andauernder Belastung. Dissertation, Kassel, 1989 (Summary, notations, and legends to figures both in German and in English).

2   Gergely, Peter, L.A. Lutz: Maximum Crack Width in Reinforced Concrete
     Flexural Members. At: Causes, Mechanisms, and Control of Cracking
     in Concrete, SP-20, American Concrete Institute, Detroit, 1968, pp. 87–117.
3   CEB/FIP-Mustervorschrift für Tragwerke aus Stahlbeton und Spann-
     beton. Deutscher Ausschuß für Stahlbeton, 1978.

NOTATIONS

Subscripts

s   steel
c   concrete
m   average
max   maximum
sh   shrinkage
l-t   long-term load

Superscripts

①   at the first crack
o   upper limit of cyclic load

Cross sectional properties

e   eccentricity of force
h   total height
b   width
d   height from center of gravity
       of tension reinforcement to
       compression edge
c   concrete cover
d_b reinforcing bar diameter
A_s steel cross sectional area
$A_{c,n}$ area of concrete (net)
A_{cef} effective area of concrete
ρ reinforcement percentage

Material properties

E_s modulus of elasticity of steel
E_c modulus of elasticity of concrete
$n = E_s/E_c$ modular ratio
f_c concrete cube strength
f_c^{ς} concrete cylinder strength
f_t direct tensile strength of
       concrete
f_t^e strength at eccentric tension

Stresses

f   stress
τ bond stress

Strains

ε strain
ε_m average strain over the
       cracks
ε_r residual strain

Lengths

Δ relative axial
       displacement between
       steel and concrete
∇ relative radial dis-
       placement between
       steel and concrete
Δl change in length
l_o initial length
s   crack spacing
w   crack width

Miscellaneous notations

t   days
S   section modulus
Δ increment
F   force
f_l lateral pressure
A_R relative rib area of bar

BOND STRESS-SLIP BEHAVIOUR OF CLASS C FRP REBARS

J Larralde

Drexel University,

United States of America

ABSTRACT A newly developed type of reinforcement bar for Portland Cement concrete, Fiberglass Reinforced Plastic (FRP) rebar, can be used as reinforcement in concrete which will be exposed to marine or corrosive environments. FRP rebars are corrosion resistant and have a high tensile strength. To fully utilize the strength of FRP rebars, however, they have to be adequately anchored in the concrete. FRP rebars do not have the characteristic surface deformations of steel rebars thus their bond in concrete is different and needs to be defined. Little experimental information is available about the bond of FRP rebars. In this paper, the results of bond tests of FRP rebars are presented.

Keywords: Bond, Corrosion, Concrete, Composites, Reinforcement, Protection, Construction Materials, Fiberglass, Plastics.

Dr. J. Larralde is Assistant Professor of Civil Engineering at the Department of Civil and Architectural Engineering, Drexel University, Philadelphia, PA, USA. His research areas are in materials for civil engineering, highway engineering, and highway materials. His current research includes uses of reinforced plastics in the form of concrete reinforcement and structural shapes, and in nondestructive testing of concrete.

INTRODUCTION

Fiberglass Reinforced Plastic (FRP) rebars can be used advantageously as concrete reinforcement when concrete will be exposed to corrosive environments. FRP are corrosion free, non-electromagnetic, and have a very high strength-to-weight ratio. To fully utilize the high tensile strength of FRP rebars in reinforced concrete and to avoid extreme cracking or even failure, the bond between the rebar and the concrete has to be large enough for the rebar to develop its ultimate tensile strength. Because tensile strength, modulus of elasticity, and surface deformation pattern of FRP rebars are different from those of steel rebars, the design guidelines for anchorage of steel cannot be used directly in the anchorage design of FRP rebars. Moreover, only little information about the behavior of FRP rebars in concrete is available, in particular, experimental information about the bond and anchorage of FRP rebars is needed.

FRP rebars are produced commercially in the United States in diameters from 0.25 in. (6.4 mm) to 1.5 in. (38.1 mm) and in lengths of 20 ft (5.08 m) or longer. FRP rebars are manufactured from vinyl ester resin reinforced with continuous C glass roving. The rebars are formed through a pultrusion process which produces rods with a relatively smooth surface and with constant diameter. To provide bond with concrete, a single strand spiral is wrapped around the rod before the resin sets. The resulting rebar has a spiral indentation which produces an irregular surface envelope (Fig. 1).

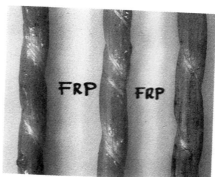

Fig. 1: Photograph of 1/2-in. (12.7 mm) FRP rebar

From preliminary tests conducted by the author, consisting of FRP rebars embedded in concrete cylinders and subjected to a concentric pullout force, average nominal bond strengths of 1,400 psi (9.65 MPa) and 925 psi (6.38 MPa) were obtained for FRP rebars with nominal diameters of 3/8 in. (9.5 mm) and 5/8 in. (15.9 mm) respectively (Larralde and Silva[1]). Values of loaded-end slip at the onset of failure ranged from 153 mils (3.9 mm) to 336 mils (8.5 mm) for the 3/8 in. (9.5 mm) rebar, and from 51 mils (1.3 mm) to 114 mils (2.9 mm) for the 5/8 in. (15.9 mm) rebar. The average compressive strength of concrete was 3,456 psi (23.8 MPa) and embedment lengths were 3 in (76.2 mm) and 6 in. (152.4 mm). The state of stresses in the concrete of pullout tests, however, does not reflect the stress conditions of reinforced concrete elements subject to bending. The concrete in pullout tests is under compressive stresses, increasing the bond strength that otherwise exists in flexural members where the concrete surrounding the rebars is under tension.

In this paper, the results of bond-slip tests conducted in small concrete beams with FRP rebars with different embedment lengths are reported.

EXPERIMENTAL PROCEDURE

Specimens consisting of concrete beams with one FRP rebar embedded at a fixed depth and with controlled length of embedment were prepared. The beams had a 0.5 in. (12.7 mm) by 3 in. (76.2 mm) notch at the midspan (Fig. 2)

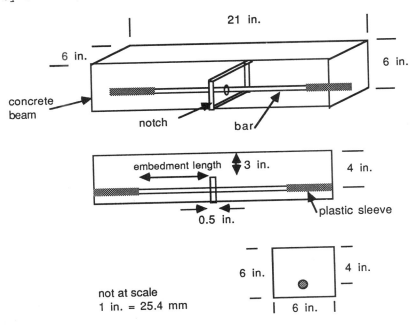

Fig. 2: Concrete beam specimens

Concrete was made using a 1/2-in.(12.7 mm) maximum size, crushed-limestone coarse aggregate, river sand, cement type III, and water. Concrete was mixed with the following proportions: 2.0:1.0:0.6:0.3 by weight of coarse aggregate, fine aggregate, cement, and water respectively. The beam specimens were cast in steel molds in three layers and vibrated for 3 minutes on a vibrating table. Six 3 in. (76.2 mm) by 6 in. (152.4 mm) companion specimens were cast simultaneously with the beams to determine the compressive strength of concrete. The notch was formed by placing a 0.5 in. (12.7 mm) by 3 in. (76.2 mm) by 6 in. (152.4 mm) plywood form at the midspan. The plywood was coated with oil before concrete was cast. After casting the concrete, all specimens were cured for 20 days at 100% humidity and 73 °F +/- 3 °F (23 °C +/- 1.6 °C) and were dried for 1 day at room temperature prior to testing.

FRP rebars are manufactured through a pultrusion process which produces rods with relatively smooth surface. To provide bond strength with concrete, a spiral is wrapped around the rebar to produce regularly spaced indentations in the surface of the rebar. The rebar then has a circular cross section with variable diameter.

The FRP rebars used in the tests were 21 in. (533.4 mm) long with nominal diameters of 0.25 in. (6.4 mm) and 0.5 in. (12.7 mm). Three specimens form each diameter were prepared to determine the ultimate tensile strength and modulus of elasticity. To provide the desired embedment length, the ends of the rebars were covered with plastic sleeves which prevented contact with concrete. The range of values of diameter and pitch of the spiral for the tested rebars are given in the following table.

TABLE 1. Diameter and spiral pitch of tested rebars.

bar nominal diameter	diameter range (in.)	pitch range (in.)
0.25 in.	0.26 - 0.28	0.71 - 0.80
0.50 in.	0.52 - 0.54	1.08 - 1.20
1 in. = 25.4 mm		

Twelve beam specimens were prepared with two different bar diameters and three different embedment lengths. Two beams were cast for each combination of embedment length and diameter. A list of the specimens is given in Table 2.

TABLE 2 List of beam specimens

specimen	bar diameter (in.)	embedment length (in.)
#2-5-1	0.25	5.0
#2-5-2	0.25	5.0
#2-7-1	0.25	7.0
#2-7-2	0.25	7.0
#2-9-1	0.25	9.0
#2-9-2	0.25	9.0
#4-5-1	0.50	5.0
#4-5-2	0.50	5.0
#4-7-1	0.50	7.0
#4-7-2	0.50	7.0
#4-9-1	0.50	9.0
#4-9-2	0.50	9.0

1 in. = 25.4 mm

The beam specimens were tested under one-point bending with the load applied gradually at increments of 200 lb ((890 N). The notch opening, mid-span deflection, and slip of bar at both free ends were measured at the end of each load increment (Fig. 3).

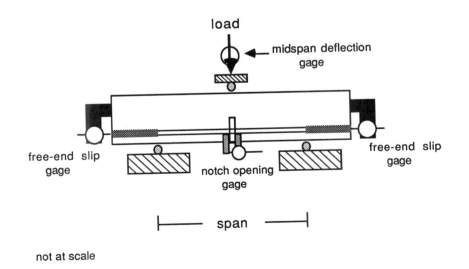

Fig. 3: sketch of testing setup

RESULTS

A summary of the results is given in Table 3. The results are reported in terms of ultimate load, midspan deflection, notch opening, and free-end slip of bar.

TABLE 3   Summary of test results

specimen	span (in.)	load[1] (lb)	deflection[2] (mil)	notch opening[3] (mil)	left slip[4] (mil)	right slip[5] (mil)
#2-5-1	10.5	7420	197	250	55	39
#2-5-2	10.5	7020	189	240	57	68
#2-7-1	14.5	4900	230	270	76	12
#2-7-2	14.5	4850	270	314	41	42
#2-9-1	18.5	4080	345	280	47	31
#2-9-2	18.5	3850	272	252	20	43
#4-5-1	10.5	11340	103	116	28	30
#4-5-2	10.5	10800	109	107	19	31
#4-7-1	14.5	9000	134	108	13	21
#4-7-2	14.5	10580	160	131	22	26
#4-9-1	18.5	8000	183	114	12	15
#4-9-2	18.5	8800	239	156	23	22

Compressive strength of concrete = 3,100 psi
Tensile strength of rebars = 98,600 psi
Tensile modulus of elasticity of rebars = 6.6x10^6 psi

[1] load at failure
[2] deflection at failure
[3] notch opening at failure
[4] slip at left free end of bar
[5] slip at right free end of bar
1 in. = 25.4 mm, 1 mil = 0.0254 mm, 1 lb = 4.5 N, 1 psi = 6.896 kPa

Table 4 shows the results in terms of ultimate bending moment, ultimate nominal normal stress, ultimate nominal bond stress, and slip at the loaded and free ends of the rebar. The normal and bond stresses were calculated from the equilibrium conditions at the section of maximum bending moment and assuming cracked section. The slip at the loaded end of the bar is the average of the slip at both faces of the notch. It was calculated as the notch opening minus the deformation due to normal stresses in the rebar. The slip at the free end of the rebars is the average of the values of slip at both ends.

TABLE 4 Test results

specimen	moment (lb-in.)	normal stress (psi)	nominal bond (psi)	loaded-end slip (mil)	free-end slip (mil)	
#2-5-1	19477	105815	1322	121	47	*
#2-5-2	18427	99733	1246	116	63	*
#2-7-1	17762	95906	856	131	44	*
#2-7-2	17581	94867	847	153	42	*
#2-9-1	18870	102290	710	136	39	*
#2-9-2	17806	96158	667	122	31	*
#4-5-1	29767	41876	1046	56	15	**
#4-5-2	28350	39667	991	51	25	**
#4-7-1	32625	46408	828	52	17	**
#4-7-2	38352	55843	997	63	24	**
#4-9-1	37000	53570	744	54	13	**
#4-9-2	40700	59857	831	75	22	**

* indicates that bar failed in tension

** indicates that bar did not fail in tension

moment = maximum bending moment at midspan

normal stress = nominal normal stress in rebar at failure $(4P/\pi d_b^2)$

nominal bond = nominal bond stress in rebar at failure $(P/\pi d_b l_{db})$

P = calculated normal force in rebar at failure

d_b = nominal bar diameter

l_{db} = embedment length

1 lb-in. = 0.011 N-m, 1 psi = 6.896 kPa, 1 mil = 0.0254 mm

Test results are shown in Fig. 4   through Fig. 9. In these figures results for only one specimen of each case are presented in terms of the deflection and notch opening as a function of bending moment; and the amount of free-end slip at both ends of rebar versus bending moment.

As can be seen in Table 4, in all specimens with 0.25 in. (6.4 mm) bar diameter the rebar failed in tension while all specimen with 0.5 in. (12.7 mm) bar diameter the failure occurred as a result of slip of the rebar.

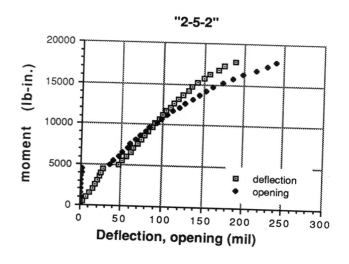

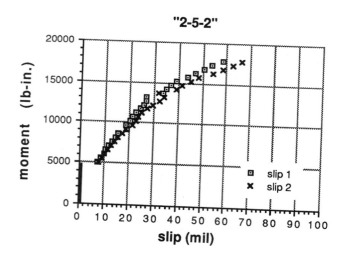

Fig. 4: Deflection, notch opening, and free-end slip of rebar with 0.25 in. (6.4 mm) diameter and 5 in. (127 mm) embedment length (specimen 2-5-2).

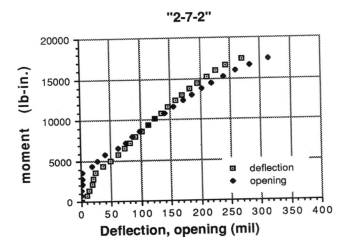

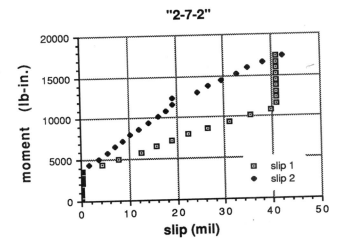

Fig. 5: Deflection, notch opening, and free-end slip of rebar with 0.25 in. (6.4 mm) diameter and 7 in. (178 mm) embedment length (specimen 2-7-2).

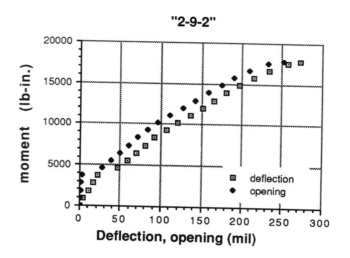

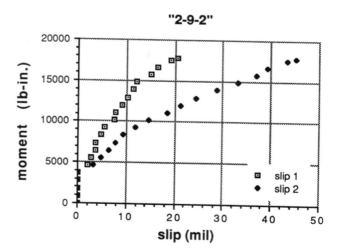

Fig. 6: Deflection, notch opening, and free-end slip of rebar with 0.25 in. (6.4 mm) diameter and 9 in. (229 mm) embedment length (specimen 2-9-2).

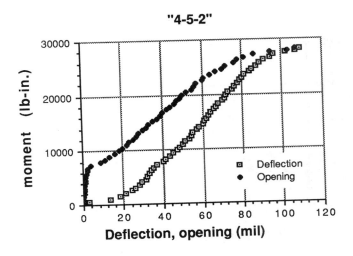

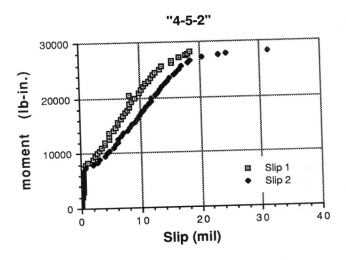

Fig. 7: Deflection, notch opening, and free-end slip of rebar with 0.5 in. (12.7 mm) diameter and 5 in. (127 mm) embedment length (specimen 4-5-2).

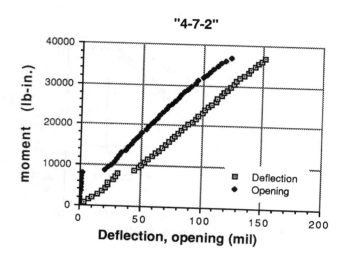

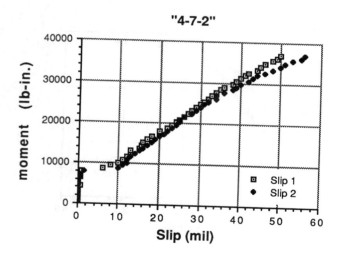

Fig. 8: Deflection, notch opening, and free-end slip of rebar with 0.5 in. (12.7 mm) diameter and 7 in. (178 mm) embedment length (specimen 4-7-2).

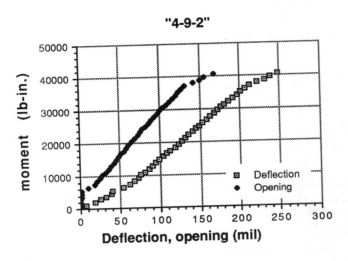

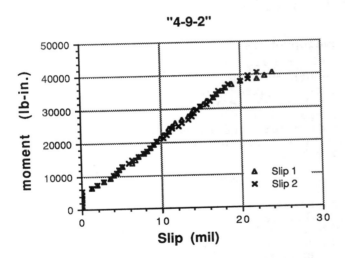

Fig. 9: Deflection, notch opening, and free-end slip of rebar with 0.5 in. (12.7 mm) diameter and 9 in. (229 mm) embedment length (specimen 4-9-2).

In all cases, the rebars have slipped at failure. The average free-end slip was lower for longer embedment length for both bar diameters. However, the amount of loaded-end slip was higher for longer embedment lengths. That is, for both bar diameters, the increase in embedment length effectively increased the anchorage in concrete.

In all cases of bar diameter equal to 0.25 in. (6.4 mm), the bars failed in tension and the bar was able to develop all the tensile force necessary to produce ultimate moment capacity. The moment capacity calculated from the average tensile strength of the rebars and average compressive strength of concrete is 18,586 lb-in. (204.4 N-m) and the average moment capacity of the tested beams was 18,320 lb-in. (201.5 N-m) with a coefficient of variation of 4%.

For the specimens with rebars with 0.5 in. (12.7 mm) diameter, the ultimate moment increased as the embedment length increased. The ultimate moment varied from 39,667 lb-in. (436.3 N-m) to 40,700 lb-in. (447.7 N-m). Embedment lengths of 5 in. (127 mm), 7 in. (178 mm), and 9 in. (228.6 mm) produced moment capacities of approximately 44%, 54%, and 60%, respectively, of the ultimate moment (65,586 lb-in. (721 N-m)) calculated from the rebar and concrete strengths.

The basic developmet length (l_{db}) required by ACI-318-89 (2) for steel rebars is $0.04A_b f_y / (f'c)^{0.5}$. If the development length for FRP rebars were calculated using the ACI-318 code, the required development lengths for number 2 and number 4 rebars would be 4 in. (101.6 mm) and 14 in. (355.6 mm) respectively. In the specimens with 0.25 in. (6.4 mm) bar diameter, the embedment lengths were greater than required by ACI-318 while for the 0.5 in. (12.7 mm) diameter, the embedment lengths were shorter.

CONCLUSIONS

FRP rebars can be used advantageously as reinforcement for concrete that will be exposed to corrosive environments. FRP rebars are corrosion resistant and have a very high strength-to-weight ratio. Nevertheless, to fully utilize FRP rebars the anchorage and bond with concrete has to be adequate enough for the rebar to develop all its tensile strength. FRP rebars have surface deformation different from those of steel rebars. As such the design guidelines for bond and anchorage of steel rebars cannot be used on FRP rebars.

In this paper, the results of bond tests on concrete beams reinforced with FRP rebars are reported. From the tests, it is found that for embedment lengths equal or slightly greater than the length required by ACI-318 for steel rebars, bond is large enough to develop the tensile strength, in spite of the fact that slip was observed at the end of the embedded length. For embedment lengths shorter than required by ACI-318, bond was not enough to develop the tensile strength of the rebar. Also, at the onset of failure, shorter embedment lengths resulted on smaller amounts of slip at both the free end and the loaded end of the rebars.

Although the required embedment lengths found from the tests reported herein, are approximately equal to those required by ACI-318, more research is necessary to accurately define the length of embedment of FRP rebars required to develop the utlimate tensile strength.

REFERENCES

1   LARRALDE, J. AND SILVA, R.Bond stress-slip relationships in concrete. Submitted to ASCE Materials Engineering Congress, Denver, Colorado, USA, 1990.

2   ACI Building Code Requirements for Reinforced Concrete (ACI Committee 318). American Concrete Institute, Detroit, MI., USA 1989

A COMPARISON OF BOND CHARACTERISTICS BETWEEN MILD STEEL AND TYPE 3CR12 CORROSION RESISTING STEEL REINFORCING BAR

G J van den Berg

P van der Merwe

M W Pretorius

Rand Afrikaans University

South Africa

ABSTRACT. Type 3CR12 corrosion resisting steel is a 12% chromium containing steel recently been developed by the specialty steel producing company, Middelburg Steel and Alloys. Type 3CR12 steel may prove to be a cost effective material of construction in applications where corrosion is an important consideration. A comparative study is made of the bond characteristics between Type 3CR12 stainless steel and mild steel used as reinforcing bar. Only plain round bars were used in the experiments. A series of pull—out tests with lengths that varied from 200 mm up to 2000 mm were carried out where the reinforcing bars were embedded in concrete. It was concluded that Type 3CR12 corrosion resisting steel bar can successfully be used as reinforcing bar in structural concrete and that the bond characteristics compare well with that of mild steel bar.

Keywords: 3CR12, Corrosion Resisting, Chromium, Bond, Mild Steel, Stainless Steel, Reinforcing Bar, Pull Out Tests, Cost Effective.

Dr. Gerhardus Johannes van den Berg is Associate Professor, Chairman of Civil Engineering and Chairman of the Materials Laboratory in the Faculty of Engineering at the Rand Afrikaans University, Republic of South Africa. He has been active in the Chromium Steels Research Group at the Rand Afrikaans University for a number of years by means of research and publishing of papers.

Dr. Pieter van der Merwe is Professor of Mechanical Engineering and Dean of the Faculty of Engineering at the Rand Afrikaans University, Republic of South Africa. He is the project leader of the Chromium Steels Research Group at the Rand Afrikaans University.

Marthinus Wessel Pretorius is a Research Assistant in the Materials Laboratory of the Faculty of Engineering at the Rand Afrikaans University, Republic of South Africa.

INTRODUCTION

Today one of the major concerns for engineers is the failure of reinforced concrete to meet the design life of structures. Probably the most worrying aspect is the problems encountered with cracking and spalling of concrete due to the corroding of reinforcement.

According to a report by the Agricultural Department of the United States of America 26% of the 468 095 rural bridges were classified as structurally deficient and 20% were classified as functionally obsolete. Structurally deficient does not mean that the bridges are unsafe. The structurally deficient bridges are either closed or restricted to lighter vehicles only because of deteriorated structural components. The average age of all United States rural bridges was found to be 36,6 years. The National Association of Counties reports in a survey that three quarters of all bridges need to be repaired or replaced at a cost of billion dollars. This report did not take into account the bridges under State or Federal jurisdiction.

Unprotected reinforcing steel can corrode in certain environments because moisture and oxygen penetrate into the concrete through cracks and pores. This problem is accelerated in coastal areas by penetration of chloride ions from the marine environment. The ultimate effect of this penetration of moisture and chloride ions is corrosion of the reinforcing bar and thus spalling of the concrete.

In 1984 it was reported by the Federal Highway Authority[1,2] in the United States that it has on record more than 160 000 bridges in distress due to the corrosion of reinforcing bar. This has prompted large scale research into the methods of protection for new, refurbished and existing bridges. Among the techniques tried is the use of stainless steel as reinforcing bar.

STAINLESS STEEL REINFORCEMENT

Stainless steels offer an excellent alternative where reinforcing steel is subject to corrosion. Stainless steels has established itself as a corrosion resistant construction material, with a wide usage in many industries where the environmental aggressiveness is beyond any circumstances envisaged in construction. Even in the most severe construction applications stainless steel will equal or exceed the life of adjacent constructional materials. The yield strength of stainless steels can be as high as 835 MPa for prestress applications and 460 MPa for reinforcing bar.[2]

Classification of Stainless Steels

A wide range of stainless steels, which are iron based alloys containing at least 11% chromium with an upper limit of 30% for practical considerations, are utilized, especially for their resistance to corrosion, in a wide range of environments. Chromium is not the only alloying element which is used to produce the different types and grades of stainless steels. To enhance the corrosion resistance, and to resist more aggressive corrosion conditions, the chromium contents are increased and additional alloying elements are added, mainly nickel and molybdenum. Other elements, such as carbon, manganese, silicon, copper, titanium, niobium, nitrogen, sulphur, selenium and aluminium may also be used, not only to increase the corrosion resistance and heat resistance, but also to influence the crystal structure, the mechanical properties and hence the formability, machinability and weldability.[3],[4]

The classification of the different types of stainless steels are based on the crystal structure which are developed within the steel, due to both the chemical composition and thermal treatment. Although stainless steels are classified into austenitic, ferritic, martensitic, duplex and precipitation hardening stainless steels, only the first two will be discussed as they are the two which are commonly used as reinforcing bar in concrete.

Austenitic Stainless Steels

The American Iron and Steel Institute (AISI) Type 200– and 300–series are austenitic stainless steels. The two types that are used in reinforced concrete are AISI Type 304 and 316. The formation and stabilization of the austenitic crystal structure, over a wide range of chromium contents and temperatures, is promoted by the amount of nickel (6–20%) in the austenitic stainless steels. In the 200–series some of the nickel is replaced by manganese in the ratio of two parts of manganese for each part of nickel. The austenitic stainless steels have a high level of corrosion resistance in a wide variety of aggressive conditions, as well as a good high temperature strength, a high resistance against scaling at high temperatures and excellent toughness and ductility down to very low (cryogenic) temperatures.

Stainless steel Type 304 is commonly available and is used in a wide range of applications. Although stainless steel Type 304 is less corrosion resistant than Type 316 it has a wider field of application as it is more price competitive. Stainless steel Type 304 has a corrosion resistance in industrial areas where there is a combination of moisture, carbonaceous and other pollutants.

Ferritic Stainless Steels

Some of the AISI Type 400 – series are ferritic stainless steels. Type 3CR12 corrosion resisting steel, a 12% chromium steel, is a modified AISI Type 409 steel, developed by the specialty steel producing company, Middelburg Steel and Alloys, to overcome the weldability problems of plain chromium ferritic stainless steels. Type 3CR12 steel has sufficient chromium to impact a useful, cost effective level of corrosion resistance.

The corrosion resistance of Type 3CR12 steel has been tested in numerous applications involving dilute acid, base and salt solutions.[5] The corrosion rate in many of these applications is low enough to consider use of the material as an alternative to coated mild steels.

Melville et al[6] reported on the results of immersion tests in sea water and gold mine water on Type 3CR12 steel (pickled and scaled), Corten Steel, mild steel and galvanized steel. It was found that pickled Type 3CR12 steel and galvanized steel showed virtually zero corrosion and was far superior to Corten and mild steels. The scaled Type 3CR12 steel showed superficial corrosion of the scaled surface, but that this did not penetrate to any significant depth. The scaled Type 3CR12 steel was however still far superior to the Corten and mild steels. In the same publication the results of certain accelerated corrosion tests were discussed. These tests included salt spray tests (5% sodium chloride solution), CASS and Corrodcote tests and were performed on the same materials mentioned above, as well as Type 430, 304 and 316 stainless steels.

It was concluded that in mildly corrosive conditions there was little difference between the corrosion of pickled Type 3CR12 steel and that of Type 430 and 304 stainless steels, but that the pickled Type 3CR12 steel showed itself to be far superior to the other materials tested. Type 316 stainless steel did not corrode at all in these environments.

From the discussion above it can be concluded that Type 3CR12 corrosion resisting steel can be a substitute for the austenitic stainless steels in mildly to severe corrosive conditions. Corrosion of Type 3CR12 steel will not take place under these condition and breaking and spalling of concrete will thus not take place. The ferritic stainless steel Type 3CR12 is more cost effective than the austenitic stainless steels because of its lower chromium and nickel contents. Although the austenitic stainless steels are commonly used as reinforcing bar in concrete, Type 3CR12 or a 12% ferritic type stainless steel should be considered as an alternative for reinforcing bar.

MECHANICAL PROPERTIES

Although Type 3CR12 corrosion resisting steel is not commonly used in reinforced concrete some plain round bar has been rolled by Middelburg Steel and Alloys. Some deformed round bar will also be rolled soon.

Type 3CR12 and mild steel round bar was obtained. The Type 3CR12 steel round bar was descaled while the mild steel bar was scaled bar used as reinforcing bar in concrete.

Uniaxial tension test specimens were prepared in accordance with the dimensions outlined by the ASTM Standard A370–77[7] and BS18[8]. No specification exists for the preparation of compression test specimens and a similar method described by Parks[8] was used.

TABLE 1. MECHANICAL PROPERTIES

	3CR12 STEEL		MILD STEEL	
	LT	LC	LT	LC
Elastic Modulus, E(GPa)	192,7	191,6	215,8	211,0
Yield Strength, F_y(MPa)	381,3	370,2	306,2	304,2
Proportional Limit, P_p(MPa)	227,4	220,5	300,8	289,2
Ultimate Strength, F_u(MPa)	464,0	–	456,0	–
Elongation (%)	34,1	–	47	–

LT = Longitudinal Tension
LC = Longitudinal Compression

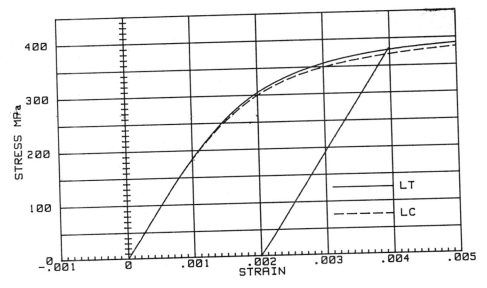

FIGURE 1   STRESS STRAIN CURVE FOR TYPE 3CR12 STEEL

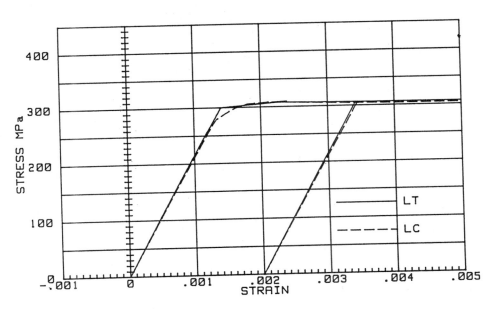

FIGURE 2   STRESS-STRAIN CURVE FOR MILD STEEL

Tensile and compression tests were conducted generally in accordance with the procedures outlined by ASTM Standard, A370–77.[7] All specimens were tested using an Instron 1195 universal testing machine. Average strain was measured using two strain gauges mounted on either side of the specimen in a full bridge configuration with temperature compensation. Compression test specimens were mounted in a specially manufactured compression test fixture which prevents overall buckling of the test specimen.

The mechanical properties of the two steels are given in Table 1 for longitudinal tension and compression. Figures 1 and 2 show the stress strain curves for the two steels in tension and in compression.

EXPERIMENTAL PROGRAM

Preparation of Test Specimens

Ten specimens each for mild steel and Type 3CR12 corrosion resisting steel, with bond lengths that varied from 200 mm up to 2000 mm and 16 mm diameter were prepared for pull–out tests. The lengths of the beams were 100 mm longer that the bond lengths that were investigated. A piece of paper was wrapped around this 100 mm. This was done to reduce the influence of the disturbed area that forms close to the bearing plate. The bar to be tested extends beyond the two sides of the specimen with the load applied to the longer end. The mild steel bars tested were without loose mill scale, entirely free from rust and carefully degreased. The Type 3CR12 steel bars were descaled bars and carefully degreased. The beams were cast horizontally to ensure that a good compaction is obtained. For an unknown reason the concrete received from a ready mix company was 10 MPa instead of 25 MPa that was ordered. As this was a comparative study the tests were carried out on these specimens. Six test cubes were taken. The test specimen is shown in Figure 3.

Testing Procedure

The specimen is placed vertically under a bearing plate provided with a hole for the reinforcing bar to be pulled through. The reinforcing bar is gripped by hydraulic grips. The pull–out tests were carried out in a 250 kN hydraulic Instron universal testing machine. The layout of the testing device is shown in Figure 1. The procedure to follow for the pull–out tests is described in Reference 10 and 11. The load was applied to the top end of the reinforcing bar in the pull–out test until failure occurred either by yielding of the steel or excessive slip between the bar and concrete was attained.

Results

Bond stresses were calculated from Equation 1.

$$F_b \quad = \frac{F_s A_s}{\Sigma_0 L} = \frac{P}{\Sigma_0} \tag{1}$$

where

TESTING MACHINE

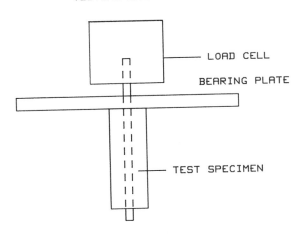

FIGURE 3   TEST SPECIMEN IN TESTING MACHINE

F_b = bond stress
F_{bs} = bond stress in stainless steel reinforcing bar
F_{bm} = bond stress in mild steel reinforcing bar
F_s = stress in reinforcing bar
A_s = cross–sectional area of reinforcing bar
Σ_0 = nominal perimeter of reinforcing bar
L = bond length
P = tensile force applied to reinforcing bar
P_s = tensile force applied to stainless steel reinforcing bar
P_m = tensile force applied to mild steel reinforcing bar

The results of the pull out tests on Type 3CR12 corrosion resisting steel and mild steel are given in Table 2 and Figures 4 to 5. In this table a comparison is made of the bond stresses between the two types of steel.

Because a difference was found in the performance of Type 3CR12 steel and mild steel the roughness of each steel was determined with an electronic profile projector. The mean roughness of the mild steel was 5,0 microns while the Type 3CR12 corrosion resisting steel's roughness was 2,0 microns. The mild steel reinforcing bar was also descaled and the roughness was then found to be 5,0 microns.

Discussion of Results

From Table 2 it can be seen that the mild steel gave higher bond strengths than the Type 3CR12 steel. One reason for this difference in strength can be attributed to the difference in roughness that was found between the descaled Type 3CR12 steel and the scaled mild steel bar. The scale on the mild steel bar did not contribute to the roughness on the surface.

TABLE 2. COMPARISSON OF MILD STEEL AND TYPE 3CR12 STEEL BOND STRENGTHS

Length (mm)	MILD STEEL			TYPE 3CR12 STEEL			$\dfrac{F_{bm}}{F_{bs}}$
	P_m (kN)	F_{bm} (MPa)	F_{sm} (MPa)	P_s (kN)	F_{bs} (MPa)	F_{ss} (MPa)	
200	4,6	0,45	22,3	2,9	0,29	14,7	1,55
400	11,5	0,56	55,7	10,9	0,55	55,2	1,55
600	19,5	0,64	94,5	15,3	0,51	77,5	1,25
800	36,5	0,90	176,9	31,7	0,80	160,7	1,13
1000	43,1	0,85	208,9	23,7	0,48	120,1	1,77
1200	62,7	1,03	303,8	34,2	0,57	173,3	1,81
1400	64,2	0,90	311,1	46,1	0,66	233,6	1,36
1600	64,1	0,79	310,6	44,8	0,56	227,1	1,41
1800	68,7	0,75	332,9	42,6	0,48	215,9	1,56
2000	68,8	0,68	333,4	49,1	0,49	248,9	1,39
Mean							1,43
Coefficient of Variation							0,18

L = Length of reinforcing bar
P_m = tensile force in mild steel reinforcing bar
P_s = tensile force in stainless steel reinforcing bar
F_{bm} = bond stress for mild steel reinforcing bar
F_{bs} = bond stress for stainless steel reinforcing bar
F_{sm} = tensile stress in mild steel reinforcing bar
F_{ss} = tensile stress in stainless steel reinforcing bar
Diameter of mild steel = 16,21 mm
Diameter of 3CR12 steel = 15,85 mm

It can be seen in Table 2 and from Figures 4 to 5 that because of the low rough-ness of the Type 3CR12 steel bars it could not develop its full strength. The maximum tensile stress in the bar was 248,9 MPa which is lower than the yield strength.

The stress–strain curves of Type 3CR12 steel differ from that of mild steel. From Figures 1 and 2 it can be seen that Type 3CR12 steel yields gradually under load. The effect of this gradual yielding, which mean that these type of steels will have higher strains at lower stresses, on the ultimate bond strength should be examined more carefully. It is possible that this earlier yielding can also contribute to the lower strength of the type 3CR12 steel bar.

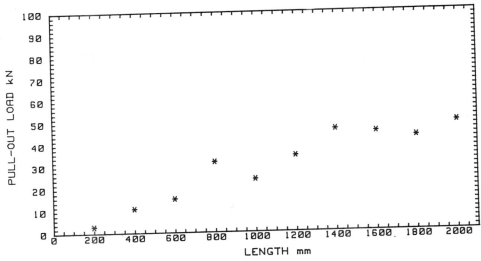

FIGURE 4   PULL-OUT LOAD VERSUS LENGTH FOR TYPE 3CR12 STEEL

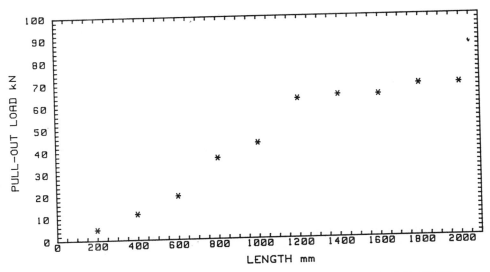

FIGURE 5   PULL-OUT LOAD VERSUS LENGTH FOR MILD STEEL

CONCLUSION

It can be concluded in this study that although the bond strength of Type 3CR12 corrosion resisting steel is lower than that of mild steel it can be successfully used as reinforcing bar in structural concrete.

ACKNOWLEDGEMENTS

The authors would like to acknowledge the financial assistance provided by Chromium Centre.

REFERENCES

1.   Federal Highway Administration, Bridge Devision, Office of Engineering, 400 Seventh Street SW, Washington DC.

2.   Haynes, J M, Stainless Steel Reinforcement. Civil Engineering. August 1984.

3.   Van den Berg, G J, The Torsional Flexural Buckling Strength of Cold–Formed Stainless Steel Columns. D. Eng Thesis. Rand Afrikaans University. 1988.

4.   Van den Merwe, P, Development of Design Criteria for Ferritic Stainless Steel Cold–Formed Structural Members and Connections. Ph.D. Thesis. University of Missouri–Rolla. 1987

5.   Thomas, C R, Hofman, J P, Metallurgy of a 12% Chromium Steel. Specialty Steels and Hard Metals. Ed. Comins, N R, and Clark, J B. Pergaman Press 1983.

6.   Melvill, M L, Mahony, C S, Hofman, J P, Dewar, K, The Development of a Chromium Containing Corrosion Resisting Steel. Paper Read at the Third South African Corrosion Conference. March 1980. Johannesburg.

7.   American Society for Testing and Materials. Standard Methods and Definitions for Mechanical Testing of Steel Products. ASTM A370–77. Annual Book of ASTM Standards. 1981.

8.   British Standards BS18. Part 2. Methods for Tensile Testing of Metal. 1971.

9.   Parks, M B, Yu, W W, Design of Automotive Structural Components Using High Strength Sheet Steel. Civil Engineering Study 83–3. University of Missouri–Rolla. August 1983.

10.   International Union of Testing and Research Laboratories for Materials and Structures. RILEM/CEB/FIP Recommendation RC6. Bond Testing Reinforcing Steel 2. Pull Out Tests.

11.   Carrasquello, P M, Pullout Tests on Straight Deformed Bars Embedded in Superplasticized Concrete. ACI Materials Journal. Title No. 85–M11 March–April 1988.

GALVANIC CATHODIC PROTECTION FOR REINFORCED CONCRETE

S Mita
K Ikawa

Nakagawa Corrosion Protecting Co. Ltd.,

Japan

ABSTRACT. Today, problem of a reinforcing bar (rebar) corrosion in salt-contaminated concrete has been reported. So, various researches for the corrosion protection of the rebar have been carried out. The cathodic protection which is one of the countermeasures is recognized as highly effective method, and we can find many applied examples to bridges and garages in the U.S.A. and Canada.[1][2][3] But, the examples of the galvanic cathodic protection system are few. The galvanic cathodic protection system has many advantages such as maintenance-free, no requirement of power source, and etc. Electrical resistivity of salt-contaminated concrete is not so high. Therefore, it is possible to apply the galvanic cathodic protection system using zinc sheets for protecting reinforced concrete under marine environments.

Keywords: Galvanic Cathodic Protection System, Impressed Current Cathodic Protection System, Salt-Induced Damage, Rebar Corrosion, Backfill, Corrosion Potential, Polarization, Shift Value.

Shunichiro Mita is a research fellow at the institute of Nakagawa Corrosion Protecting Co.,Ltd., JAPAN, where he has undertaken studies into cathodic protection for reinforced concrete. At present, he is interested in evaluation for impedance characteristics of the rebar under cathodic protection and development of reference electrode embedded in concrete.

Kazuhiro Ikawa is a research fellow at the institute of Nakagawa Corrosion Protecting Co.,Ltd., JAPAN. He is actively engaged in research into cathodic protection for reinforced concrete. His current research work includes: study on impressed current cathodic protection system, standerdization of cathodic protection for reinforced concrete.

INTRODUCTION

A typical form of the early deterioration of reinforced concrete, which has become a considerable social problem in Japan, is salt-induced damage. Although cathodic protection is being hailed as an effective countermeasure, in Japan it is still only in the experimental stage.

Cathodic protection has advantages as follows: ①It is possible to apply to newly installed as well as existing structure. ②Even if salt is contained in concrete, it is possible to prevent corrosion. ③Though structure to be protected is situated under severe corrosive condition, durability of the effect is expected. ④Though steel embedded in concrete is already corroded, it is possible to prevent corrosion.

Cathodic protection can be applied using an impressed current or a galvanic cathodic protection system, the former of which has been widely used in the U.S.A. and Canada. In reality, however, reinforced concrete in an environment beeing liable to suffer salt-induced damage is often under conditions more suited for the use of the galvanic cathodic protection system. The system can be applied as satisfactory measure against salt-induced damage when the features of the galvanic anode are assured.

Hence, development of the galvanic cathodic protection system was performed by following processes for the purpose of facilitating the practical application. And, concerning this study, zinc sheet is used for the material of the galvanic anode because this material has many good results for the galvanic anode under various conditions in the past. ①Development of backfill.(Material to assure the performance of an anode) ②Study on evaluation criteria for cathodic protection. ③Application of galvanic cathodic protection to a test piece. ④Application of galvanic cathodic protection to actual structure. As a result of these tests, the galvanic cathodic protection system is a reasonable cathodic protection for rebars embedded in concrete.

This paper describes fundamental studies and experimental test results performed under actual environments.

PRINCIPLE OF CATHODIC PROTECTION

The rebar embedded in concrete becomes passive state to be almost free from corrosion due to the high alkalinity of concrete. However, when salt permeates and reaches a specific concentration in the concrete, the passive film on the rebar is broken down, resulting in corrosion. The part broken the film has more negative potential than the other parts and thus, a current flows from this corroded part to the non-corroded parts. In other words, this corroded part serves as an anode and the non-corroded parts as a cathode. When a corrosion current flows from the anodic part, the current first flows into the cathodic parts, reducing the potential at the cathodic parts. When the anode and cathode are balanced with the same potential, no current will flow, thereby preventing corrosion. This protective method is called cathodic protection.

The galvanic cathodic protection system, used in our experimental tests, provides a direct current for protection of corrosion through connecting the metal to be protected with a metal of a less noble potential, based on the principle of a galvanic cell utilizing the potential difference between two metals. Fig.1 shows a schematic diagram of the galvanic cathodic protection system.

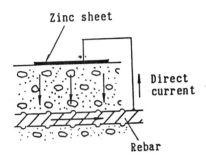

Fig.1   Galvanic cathodic protection system

DEVELOPMENT OF BACKFILL(MATERIAL TO ASSURE THE PERFORMANCE OF AN ANODE)

Considering the available data, workability and cost, various types of bentonite-based materials were tested for selection of an optimal backfill.

Testing

A backfill testing cell was prepared, as shown in Fig.2, to perform anodic electrolysis under a constant current of 10 $\mu A/cm^2$ for evaluation of the stability of varying anode potential with time. And, the testing conditions are listed in Table 1.

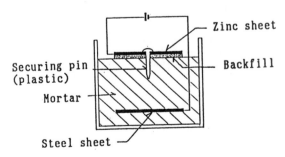

- Mortar : Water-cement ratio→ 0.5
  Salt concentration→ 10 kg/m^3 in terms of NaCl
- Steel sheet dimensions : $100 \times 70 \times 0.5$ mm
- Covering thickness : 70 mm

Fig.2   Backfill testing cell

Table 1   Testing conditions for backfill selection

Back-fill	Materials	Classi-fica-tion	Weight ratio	Exposure conditions
A	Bentonite, Gypsum, $MgSO_4$	A1	7:2:1	Water is supplemented when necessary(Sealed)
		A2	7:2:1	Water is supplemented only at the start of testing
B	Bentonite, Gypsum, $MgSO_4$ Silica gel	B1 B2 B3	7:2:1:1 7:2:1:2 7:2:1:4	
C	Bentonite, Gypsum, $MgSO_4$ $CaCl_2$	C1 C2 C3	7:2:1:1 7:2:1:2 7:2:1:4	Sealed for the first 40 days, after which it was exposed to air.
A	Bentonite, Gypsum, $MgSO_4$	A3	7:2:1	Water is supplemented only at the start of testing
D	Bentonite, Gypsum, $MgSO_4$ $MgCl_2$	D1 D2 D3 D4 D5	7:2:1:1 7:2:1:2 7:2:1:3 7:2:1:4 7:2:1:5	Exposed to air.

Test Results

Effects of moisture retention

Fig.3 shows the varying anode potentials with the passage of time (days) for three different compositions of backfill A : A1, A2, and A3. (vs.SCE : Saturated Calomel Electrode.) As shown in the figure, the retention of moisture at the boundary between the concrete and anode is required for a stable, less noble potential of the anode.

Selection of optimal backfill

Among the B, C, and D groups, backfills exhibiting a stable, less noble potential over a long time were selected. For those selected backfills, plus A2, the varying anode potentials with time were measured and are indicated in Fig.4. Using the figure, the optimal backfill was selected which had excellent water-absorbing and water-retention features to maintain moisture at the boundary between the concrete and anode. This was composed of a mixture of bentonite, gypsum, $MgSO_4$, and $MgCl_2$ in the ratio 7:2:1:3-5 by weight.

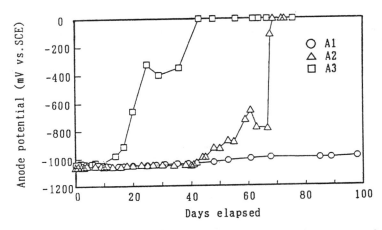

Fig.3   Varying anode potentials over the course of time (Days)
(Effect of moisture-retention)

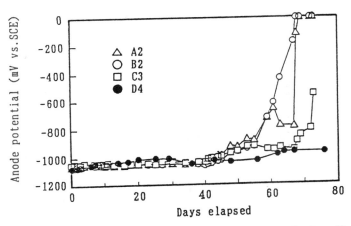

Fig.4   Varying anode potentials over the course of time (Days)
(Backfill selection testing)

STUDY ON EVALUATION CRITERIA FOR CATHODIC PROTECTION

As a criterion for evaluating the effect of cathodic protection on
rebars embedded in concrete, the -100 mV shift[4], measured from the
corrosion potential, can be applied. In reality, however, this is
somewhat unreliable, due to inaccurate correspondence with the corrosion
rate. Hence, a study was conducted to confirm a standard for the
evaluation of cathodic protection.

Testing

Controlled potential testing was performed with varying concentrations
of NaCl, ranging from zero to 30 kg/m^3, using sat.Ca(OH)$_2$ aq. as a
simulation of concrete solution. In conjunction with this, the immersion
test without applying current was performed for the quantitative
analysis of all iron ions in the solution, and also the determination of
the weight loss of the test piece to estimate the corrosion rate. A
schematic diagram of the experimental testing system is shown in Fig.5.

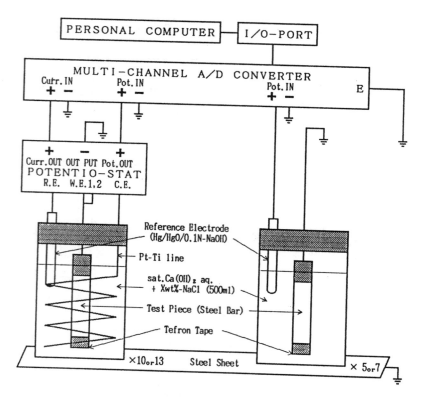

Fig.5   Schematic diagram of controlled potential testing system

Results

Fig.6 shows a graph of potential against corrosion current density,
converted from the estimated corrosion rates for varying salt
concentrations. For a corrosion rate of zero, the corresponding current
density is assumed to be 0.01 mA/m^2 for convenience.

Considering the absolute values of potential data in Fig.6, -600 mV vs. SCE may be an evaluation criterion of effective cathodic protection for all possible environmental conditions. However, due to the fact that the corrosion potential varies with different salt concentrations and also in consideration of the liquid junction potential in concrete[5], control of the absolute value-based potential may be difficult to perform, making it unsatisfactory as an evaluation criterion for cathodic protection.

Therefore, studies were conducted in relation to shift values based on the corrosion potential. Fig.7 shows the relationship between the shift values for average corrosion potentials of an immersed test piece under a stable condition, and the corrosion rates estimated for varying salt concentrations. For a corrosion rate of zero, the corresponding shift value is assumed to be 0.01 mdd(mg/dm²/day), for convenience.

Fig.7 indicates that 50-100 mV cathodic polarization from the corrosion potential for every salt concentration results in a corrosion rate of zero. In other words, this result reveals that the commonly used -100 mV shift from the corrosion potential is a reasonable evaluation criterion for cathodic protection. Therefore, the -100 mV shift from the corrosion potential was used in our experimental tests as the evaluation criterion for cathodic protection.

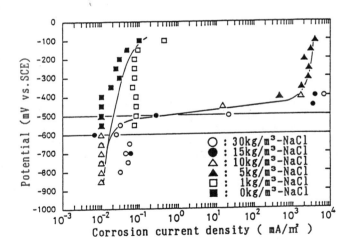

Fig.6   Anode polarization curve estimated from corrosion rates for varying salt concentrations

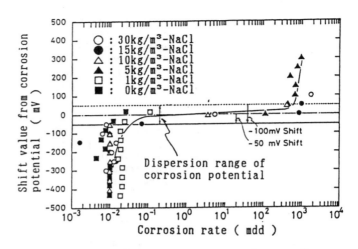

Fig.7   Shift values and corrosion rates for varying
        salt concentrations

APPLICATION OF GALVANIC CATHODIC PROTECTION TO A TEST PIECE

Galvanic cathodic protection was applied to a test piece to evaluate the
degree of protection.

Testing

The test piece, as shown in Fig.8, simulates reinforced concrete, with
the underside exposed to salt-induced damage. The test piece was
installed on a steel frame located one(1)-meter above ground outdoors.
On its underside, cathodic protection testing was performed using a zinc
sheet. The zinc sheet was placed on the surface of the concrete through
the backfill. In Japan, the reason for testing the underside is based on
the fact that sea water spray and/or splashed sea-salt particles often
adhere to the underside of structures, such as piers and road bridges,
causing salt-induced damage.

Results

Figs.9 and 10 show the generated current densities from the zinc anode,
and the variations in the anode and rebar potential against the course
of time, respectively. In addition, Table 2 shows the degree of
polarization of the rebar potential.

Regarding the test piece, polarizations of -100 mV or more were observed
for all rebars, with the result that the zinc anode system is effective
against salt-induced damage.

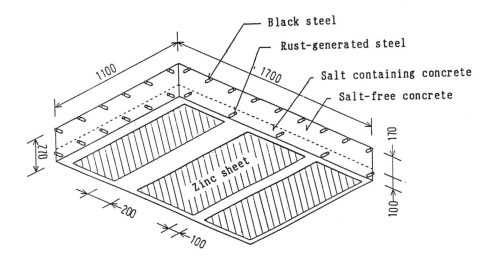

Fig.8   Test piece

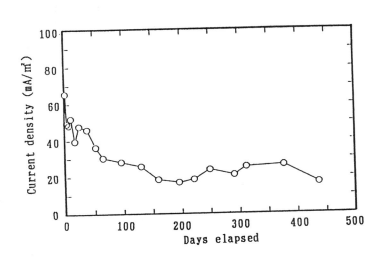

Fig.9   Generated current density variation at anode over
the course of time (Days)

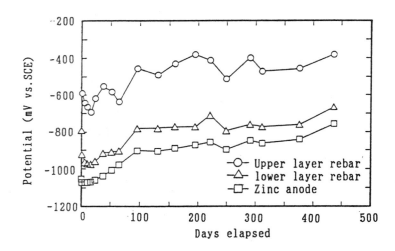

Fig.10   Variations in zinc anode and rebar potentials
over the course of time (Days)

Table 2   Degree of polarization of the rebar potential

Days elapsed	Position of rebar	ON potential (mV vs.SCE)	Potential 4 hours after protective current is turned off (mV vs.SCE)	Cathodic Polarization (mV)
130	Upper Layer	-484	-149	335
	Lower Layer	-767	-383	404
291	Upper Layer	-392	-151	241
	Lower Layer	-747	-306	441
376	Upper Layer	-447	-195	252
	Lower Layer	-746	-341	405

APPLICATION OF GALVANIC CATHODIC PROTECTION TO ACTUAL STRUCTURES

Based on the various results, the galvanic cathodic protection system
has been experimentally applied to actual structures.

Testing

Table 3 lists the tested structures and Fig.11 shows the typical anode
configurations used in the tests. At a part of area to be protected on
each structure, the potential of the rebar and the flowing current were
set to be easily monitored and measured at a specified interval. (The
measurement for structure D is in progress.)

Table 3   Actual structures tested

Structures		Construction date	Application date of CP	Area subjected to CP (surface area of concrete)
A	Wharf(Newly installed)	2. '88	2. '88	14 ㎡
B	Quaywall (Existing)	'69	1. '89	478 ㎡
C	Quaywall (Existing)	'56	3. '89	52 ㎡
D	Road Bridge (Existing)	2. '72	1. '90	17 ㎡

* CP : Cathodic Protection

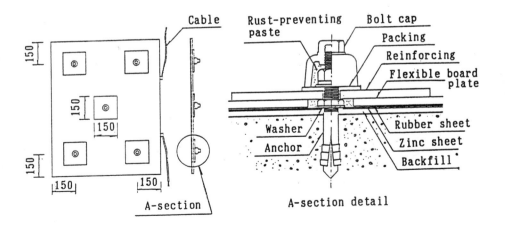

Fig.11   Galvanic anode configurations

Results

Figs.12 and 13 show anode current densities varying and rebar polarization varying with the passage of time (months), respectively. Just like the test piece, all structures were provided with a satisfactory protective current with a rebar polarization of -100 mV or more, and thus, they are estimated to have been subjected to effective cathodic protection. (Measurements were made in relation to the Hg/HgO reference electrode embedded along the rebar.) The varying anode current with different seasons may be attributed to different climates and other varying environmental conditions.

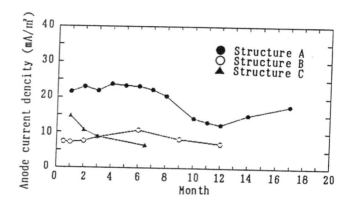

Fig.12   Variations of anode current density over the course of time (Months)

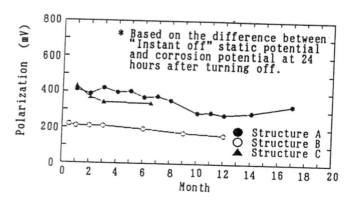

Fig.13   Variations of polarization over the course of time (Months)

SUMMARY

The most commonly used cathodic protection method for steel materials
exposed to sea water is the galvanic cathodic protection system. However,
in the U.S.A., for reinforced concrete exposed to the atmosphere, main
structural parts are mostly subjected to the impressed current cathodic
protection system, with only a few galvanic cathodic protection systems
among the hundreds of cathodic protection systems employed. This may be
based on the concept that the concrete in the actual environment has a
relatively high resistivity, making it difficult to supply a sufficient
protective current, and thus, the galvanic cathodic protection system is
not appropriate.

On the contrary, our series of tests has revealed that the galvanic
cathodic protection system is a reasonable cathodic protection for
rebars embedded in concrete. This may be explained in terms of the use
of backfill with excellent moisture-absorbing and moisture-retention
capabilities, and the fact that the undersides of structures such as
slabs and beams of piled pier, mostly subjected to cathodic protection,
tend to get wet in Japan.

By monitoring the rebar potential, protective current density and
resistance between the rebar and the galvanic anode, investigation of
the protective effect and durability of the galvanic cathodic protection
systems at the above mentioned locations will be continued.

REFERENCES

1   SCHUTT W R,   Steel-in-Concrete Cathodic Protection Results of a
     10-year Experience.   Corrosion '85, Boston, 1985, Paper No.267.

2   SCHELL H C,   Evaluating the Performance of Cathodic Protection
     System on Reinforced Concrete Bridge Substructures.   Corrosion '85,
     Boston, 1985, Paper No.263.

3   IKAWA K, KATAWAKI K,   Cathodic Protection Experiment for Rebar
     Embedded in Concrete of Actual Road Bridge.   17th Japan Road
     Conference, Tokyo, 1987, pp 832-833.

4   APOSTOLOS J A,   Cathodic Protection Using a Metallic-Sprayed Anode.
     Conference on Cathodic Protection of reinforced Concrete Bridge
     Decks, San Antonio, 1985, Paper No.16.

5   MOCHIZUKI N et al,   Evaluation of Rebar Corrosion Based on
     Corrosion Potential.   Proceedings of Fushoku-Boshoku '87, Tokyo,
     1987, pp 25-28.

CATHODIC PROTECTION OF REINFORCED CONCRETE BRIDGES

M McKenzie

Transport and Road Research Laboratory,

United Kingdom

ABSTRACT. Cathodic protection (CP) is under investigation as a repair option for concrete bridges deteriorating as a result of chloride induced corrosion of the reinforcement. An outline of the special requirements for bridges is given along with experience of the technique in North America. Trials being carried out by the Department of Transport are described with particular reference to research underway at the Transport and Road Research Laboratory. This demonstrated that cathodic protection can significantly reduce corrosion in chloride contaminated reinforced concrete and that the levels of CP used did not significantly affect the steel-to-concrete bond strength. The use of potential decay as a technique for routine monitoring of CP installations is discussed. Factors to be considered when costing CP against other options are given.

Keywords: Cathodic Protection, Concrete, Bridges, Corrosion, Chlorides, Repair, Department of Transport, Transport and Road Research Laboratory, Monitoring, Potential Decay, Bond Strength, Costs.

Malcolm McKenzie is a Senior Scientific Officer in the Bridges Division of the Transport and Road Research Laboratory and deals with corrosion problems on highway bridges. In addition to cathodic protection his research interests include fusion bonded epoxy coated reinforcement, weathering steel and bridge enclosure.

INTRODUCTION

Concrete bridges can suffer considerable deterioration as a result of chloride induced corrosion of the reinforcement. Apart from loss of strength as a direct result of loss in section of the reinforcement, expansive corrosion products can cause the concrete to crack and spall. In the USA and Canada cathodic protection (CP) has become an accepted method of treating structures and is now under investigation for use in the UK. CP is particularly attractive as it should prevent further corrosion of the steelwork even when chloride contamination continues.

BASIS OF CATHODIC PROTECTION

Steel in concrete is in contact with an electrolyte of high alkalinity, typically of pH 12.5 to 13.5. This leads to the formation of protective oxide films on the steel surface which render the steel passive so that it does not corrode. However, this passivity can be destroyed by the action of aggressive anions such as chlorides or due to reduction in alkalinity by the action of acidic gases such as CO_2. Corrosion can then commence.

The corrosion mechanism is electrochemical. Metal ions pass into solution at corroding anodic areas liberating electrons into the metal. These electrons are consumed at non-corroding cathodic areas in reactions such as the reduction of oxygen and water to form hydroxyl ions. The electrochemical circuit is completed by the pore water electrolyte.

Cathodic protection uses an external anode to apply a current to the reinforcement so that all the steel is rendered a cathode and therefore does not corrode. The protective current can be provided by using an external anode of a more reactive metal (sacrificial anode cathodic protection) or by applying a current from a low voltage DC supply (impressed current cathodic protection). The latter method is generally used for bridges. Current is usually supplied from the mains using a transformer/rectifier to provide the DC supply.

APPLICATION OF CP TO BRIDGES

CP is a well established method of protecting pipelines buried in soils and structures in the sea. For CP to be effective the metallic component to be protected must be continuous and in an electrolyte of high enough conductivity to conduct the current from the external anode to the steel. Whereas saturated soils and seawater are good conductors of electricity the conductivity of concrete is lower and can be very variable. Ensuring that sufficient current reaches all the reinforcement to be protected is not straightforward. A practical consequence of this is that the external anode has to cover the entire concrete surface to minimise the electrical pathway to the steel. Instead of applying CP from anodes in an electrolyte SURROUNDING the structure, as is the case with buried or immersed structures, CP is applied from an anode in contact with an electrolyte, the concrete, which is PART OF THE STRUCTURE. Hence reactions between the external anode and the electrolyte are of far more importance.

Typical current requirements of bridge CP systems are of the order of 10 to 20 mA/m^2 of steel surface and operating voltages tend to be less than 12 V. Control of CP systems needs to take into account the changing conductivity of concrete as weather conditions alter. The two main approaches are either to establish a high enough potential difference between the external anode and the reinforcement such that sufficient current is always being supplied, or to use a constant current supply.

CP is not a panacea for corroding concrete bridges. It is not recommended where reinforcement corrosion has occurred solely as a result of carbonation of concrete. The resistivity of carbonated concrete is likely to be higher and this might make it difficult to impress sufficient current to the reinforcement. In any case carbonation only affects limited areas of the structure where the reinforcement cover is low, and there are satisfactory proprietary repair systems already available. CP should not normally be applied to prestressed concrete. There is a possibility of hydrogen evolution during CP and this could embrittle the high strength steels used in prestressing. Concern has also been expressed that CP might affect the steel-to-concrete bond strength and exacerbate alkali aggregate reaction. It should also be borne in mind that CP will at best do no more than stop further deterioration. It will not regenerate lost steel.

The success of a CP system is assessed ultimately on the basis of the maintenance requirements of the structure after the application of the system. It is not possible to examine the reinforcement directly to see if corrosion has been stopped. This is a major difference between the use of CP on concrete and more traditional applications. Pipes can be dug up; divers can be sent down to inspect oil rigs. Such procedures may be expensive but are at least possible. For reinforced concrete the steel cannot be inspected directly without removing the concrete cover which is part of the structure. These restrictions render the choice of techniques to monitor performance on a routine basis particularly important.

NORTH AMERICAN EXPERIENCE

The majority of information on the use of CP on reinforced concrete bridges comes from the USA and Canada. In the late 1960s a major maintenance problem arose with bridge decks because of reinforcement corrosion caused by deicing salt. The use of unprotected concrete as a riding surface accelerated salt penetration. CP was first used as a repair method for decks in North America about 15 years ago.

A number of different anode systems were used during the development of CP for decks. Initially conductive asphalt was laid on top of the deck. Dead load problems associated with this type of system were eliminated in a second generation system of slotted anodes. Developments of this system were necessary to overcome acid production at the anodes and to determine optimum slot spacing. Other systems that have been tried include conductive polymer rope systems and wire mesh anodes in concrete overlays.

Although CP of bridge decks is now considered by the majority of North American engineers as a proven technology, CP of substructures is still in the experimental stage with experience over only about 6 years. Anode systems tried include conductive paints, sprayed metal coatings and conductive mesh systems in concrete overlays. Problems that have arisen include poor durability of paint systems and debonding in overlay systems.

A fuller description of North American experience with CP is given elsewhere (1).

UK EXPERIENCE

In the UK, the use of waterproofing membranes under an asphaltic wearing course has led to a much lower incidence of corrosion in bridge decks than in the USA and CP has generally not been considered appropriate. The major UK interest is in the use of CP on substructures. Hence the majority of North American experience is of limited value. The Department of Transport has set up two investigations into the use of CP for bridge substructures; a major practical comparative study on elements of the Midland Links motorway viaduct and a direct assessment of the effectiveness of CP at the Transport and Road Research Laboratory (TRRL).

Midland Links Trials

Four different CP systems have been installed by different contractors on four similar pier tops and their performance is being monitored over a period of years. This will provide information on the relative costs of different systems, their long term durability and effectiveness, their maintenance requirements and costs and their practicability for use on a routine basis. The anode systems involved include two different conductive paint coatings, a conductive polymer in an overlay and a discrete anode system of conductive tiles with connecting titanium strips. No information has yet been published from these trials. The relative performance of the various systems is clearly a matter of great commercial sensitivity and technical performance also requires evaluation over a longer period before meaningful conclusions can be reported. The trial is to be extended to look at two further anode systems- sprayed zinc and a wire mesh with an overlay.

TRRL Research

TRRL has set up a trial in which a number of reinforced concrete test specimens have been exposed to salt to encourage corrosion. Some of the specimens are cathodically protected and others are not. Each specimen (Fig 1) is L shaped in section with a height of 850 mm and volume of approximately 0.5 m^3. The reinforcement consists of a cage of 16 mm diameter mild steel bar and includes monitoring bars which are electrically isolated inside the specimen but connected to the rest of the reinforcement outside the specimen via a resistor. Reference cells have been embedded within the concrete at specific positions on each isolated bar. The purpose of the isolated bars and associated reference cells is to allow measurements of current flow onto each bar and the change in steel potential produced. CP is applied by impressed current through conductive coatings on one vertical and one horizontal face. CP is under voltage control at nominally constant levels of 2 V in one set of specimens and 4 V in another.

The effect of CP on the steel-to-concrete bond strength was assessed using sets of cylindrical test specimens each 500 mm long and 150 mm in diameter. Three mild steel bars ran the length of each specimen, one through the centre with the other two in the same plane on either side. The same levels of CP used in the main experiment (control, 2 V and 4 V were applied to sets of bond strength specimens.

In both types of specimen the steel was initially uncorroded and the concrete of poor quality so that applied chlorides would cause corrosion of unprotected steel.

Both main and bond strength specimens were installed at an outdoor test site at TRRL in January 1985 and the CP systems switched on in April 1985.

Some specimens within each set had salt additions to the concrete mix. All specimens had salt applied externally on a weekly basis except for one set of control bond strength specimens which were exposed with the rest but not salted either in the concrete mix or afterwards.

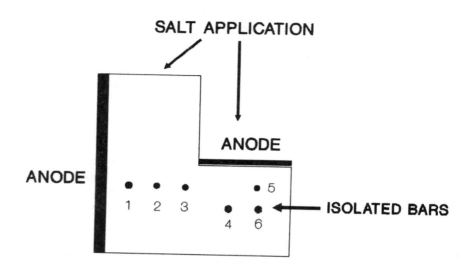

FIG 1 Cross section of test specimen

Effectiveness of CP

Groups of specimens are being broken open at intervals so that the reinforcement can be examined directly for the extent of corrosion in comparison with unprotected controls. Two demolitions have been carried out so far, one after 18 months and the other after 32 months exposure. Results from the trial (2) have clearly indicated that CP is effective in reducing if not eliminating the corrosion that occurs in reinforcement exposed to chlorides. Control specimens which had not been cathodically protected were heavily stained with rust on the concrete surface and extensive areas of corrosion were apparent on the reinforcement when the specimens were broken up. The cathodically protected specimens were generally free of external rust staining and reinforcement corrosion was much reduced in comparison with the control specimens. The corrosion that did occur in the cathodically protected specimens might not have resulted entirely from a lack in efficacy of CP. Some initial rusting might have occurred in the 2 months before the CP system was switched on. Also, deterioration of the conductive coating anodes that occurred periodically during the trial reduced current supply and that might have allowed corrosion in some areas until the anodes were renewed.

Effect of CP on steel-to-concrete bond strength

The bond strength specimens were removed from exposure after 30 months and pull-out tests performed by clamping the two outer bars and pulling the centre one. Ultimate bond stress was calculated from the maximum load recorded during each test and the surface area of the bar over which it was acting. There was no indication that CP had significantly affected the ultimate bond stress (2). However, scatter in the results for each set of specimens was considerable (Fig 2).

Further work

The original trial started with uncorroded steel in the test specimens and was really an assessment of the ability of CP to prevent corrosion as chloride levels increase. A further stage in the experiment has now been started in which CP is being applied to the same design of specimen but with corrosion already

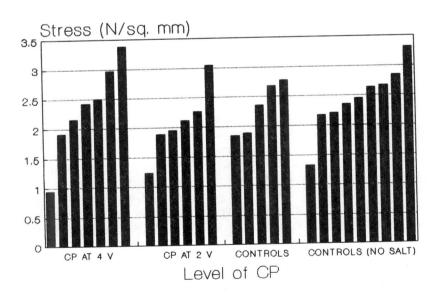

FIG 2 Effect of CP on ultimate bond stress

well established.

Thirteen specimens were exposed at the TRRL site with weekly salting to encourage corrosion. After 3 years, 2 of the specimens were demolished to assess the degree of corrosion prior to CP being applied. Five specimens were then protected using an impressed current applied through a sprayed zinc anode on all faces of the specimen except the base. [NOTE Zinc can also be used as a sacrificial anode for CP but this was not the case here]. The remaining 6 specimens are acting as controls. Five of these were also zinc sprayed but this was not powered or connected to the reinforcement. This was to eliminate any effect of the coating alone on comparisons and to provide information on the effect of current on anode durability.

In this second trial CP is under constant current control with a target current density of 10 mA/m^2 of steel surface. It had been found in the first trial that the use of voltage control led to large changes in the current applied to the specimens as the conductivity of the concrete changed with the weather.

The CP system was energised in February 1989 and, as in the first trial, effectiveness will be assessed by breaking up the specimens and examining the reinforcement directly for the extent of the corrosion in relation to unprotected controls. The first examination will be carried out after the CP system has been running for 2 years. Monitoring of the current spread using the isolated bars has shown that although the total current flowing into the concrete can be controlled current distribution can still vary considerably within a specimen (Fig 3).

MONITORING CRITERIA

A number of electrical criteria have been proposed for routine monitoring of CP installations but at the moment the trend is to use potential decay.

As CP is applied the potential of the steel reinforcement is driven more negative with respect to a reference cell. This leads to a reduction in corrosion rate so that the change in potential gives some indication of the effectiveness of CP. The difficulty lies in deciding what potential change is needed for a specific reduction in corrosion rate. Once a CP installation is operating these potential changes can be measured by monitoring the reverse process - the decay of potential towards more positive values when CP is interrupted. This involves measuring the steel potential relative to a reference cell immediately after the CP system has been turned off and remeasuring the potential after 4 hours during which the CP current remains off. A potential change of at least 100 mV is considered to indicate adequate protection. There is still some

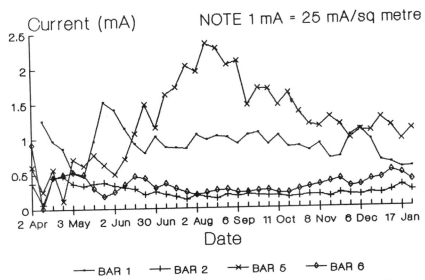

FIG 3 Currents to isolated bars in a specimen under constant current control

discussion regarding the 4 hour period with some authorities recommending ⸱ longer periods if 100 mV is not achieved in 4 hours. This criterion has the advantage of being straightforward to carry out and, because only differences in potential are involved, not too dependent on the accuracy of the reference cell. However there is little theoretical or practical evidence to demonstrate just how reliable it is.

Potential decays measured during the first TRRL trial met this criterion except at times of reduced current when the anode had deteriorated. The fact that this trial demonstrated the effectiveness of CP in at least reducing corrosion gives some support to the use of this criterion. The potential decays measured showed a correlation with the current flowing just before the CP system was turned off although variability in the decays associated with a particular current were considerable (Fig 4) and the relationship varied throughout the experiment (2). It was not possible to relate the limited and isolated corrosion which did develop on some of the protected specimens to the measured potential decays and confirm 100 mV in 4 hours as adequate. Further work is now underway in which specimens are being cathodically protected under laboratory conditions designed to give a range of potential decays in a controlled corrosive environment. Examination of the extent of corrosion after demolition of the samples should provide valuable evidence on the relevance of the potential decay criterion.

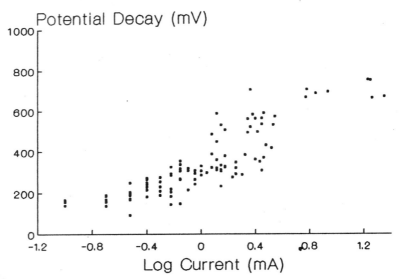

FIG 4 Potential decay against log current on isolated bars in all specimens

COSTS OF CP INSTALLATIONS

At the moment it is very difficult to compare CP with other repair options because most of the relevant information comes from abroad and long term performance of anode systems is largely unknown. It is important to take into account the total system costs including pre-application repair costs. These are likely to be lower than conventional repairs as there should be no need to remove significant amounts of chloride contaminated but otherwise sound concrete, or to provide temporary structure support during repairs. Installation costs for substructures based on American experience range from £30 to £100/m² (3). The main running cost will be in upkeep of the anode systems. Assuming the CP system is effective in preventing further deterioration of the structure, attention is transferred from upkeep of the structure to upkeep of the anode system. More information is needed on the durability and maintenance costs of the various anode systems on offer so that CP can be assessed against other options to produce the most economic repair strategy.

SUMMARY

In the UK, CP is being investigated mainly as a repair option for substructures deteriorating as a result of contamination from chloride deicing salts. Research has shown that it can reduce the rate of corrosion but more information is needed on the extent of the reduction and monitoring criteria. To compare CP with other repair options more information is also required on the long term durability and maintenance costs of anode systems.

REFERENCES

1. TASKER J, HUMPHREY M, McANOY R and MONTGOMERY R. Coatings for concrete and cathodic protection. Report of a mission sponsored by the Institution of Civil Engineers and supported by the Department of Trade and Industry, Thomas Telford Ltd, London 1989.

2. McKENZIE M and CHESS P M. The effectiveness of cathodic protection of reinforced concrete. UK Corrosion 88, Brighton 1988.

3. CHESS P M. Guidelines for the selection of cathodic protection systems. Cathodic protection of concrete structures - the way ahead. Seminar held at the Institution of Structural Engineers 7 July 1988.

ACKNOWLEDGEMENTS

The work described in this paper forms part of the programme of the Transport and Road Research Laboratory and the paper is published by permission of the Director.

POLYMERS IN CONCRETE: A PROTECTIVE MEASURE

D Kruger

Rand Afrikaans University,

D Penhall

Patrick Draper and Associates,

South Africa

ABSTRACT. The use of polymers in concrete as a protective measure against various means and forms of deterioration is investigated and discussed in this paper. Attention is given to the design of new concrete as well as the use of polymers in the rehabilitation of existing concrete.

Looking at impregnation with a monomer and subsequent polymerisation, mixing of monomers or polymers into standard Portland cement mixes and the use of a polymer as the only binder in a concrete mix, it is clear that a wide range of solutions is available against concrete deterioration problems when considering polymers in concrete.

Attention is given to the mechanism of protection when using a polymer in the concrete matrics, as well as the disadvantages in using this material.

To illustrate the use of polymers in the rehabilitation of existing corroded concrete, some case studies are presented which represent typical concrete corrosion problems in South Africa.

Keywords: Polymer Concrete, Polymer Cement Concrete, Polymer Impregnated Concrete, Permeability, Voids, Micro Reinforcing, Durability, Rehabilitation, Repair.

Mr Deon Kruger is a senior lecturer in Civil Engineering at the Rand Afrikaans University in Johannesburg. He leads the polymer concrete research program at this university and has presented several papers at international conferences on this subject. His current research work includes: polymer concrete precast pipes and other precast products, shrinkage of polymer concrete and the development of special polymer concrete mixes.

Mr Derek Penhall is a partner in the consulting practice of Patrick Draper and Associates based in Durban. He is a civil engineer who has specialized in the field of corrosion control and in particular in concrete in aggressive environments. The practice is multi–disciplinary having an industrial chemist and a chemical engineer as the other partners and is specifically structured toward industrial maintenance.

INTRODUCTION

It is well known and documented that ordinary Portland cement concrete (OPC concrete) is subject to attack by various chemicals. In practice most concrete is used together with reinforcing steel as the composite product Reinforced Concrete. Attack by aggressive media such as sulfates, chlorides or acids and soft water (demineralised water) is predominantly classed into three categories:

1.   Surface attack e.g. by acidic media or soft water

2.   Penetrative attack e.g. by sulphates

3.   Penetrative attack e.g. by chlorides

The first two having influence on the concrete portion predominantly while the latter having no effect on the concrete but influencing the corrosion of the reinforcing steel.

All three reactions lead to various forms of decay to the reinforced concrete and subsequent loss of engineering properties.

Since high density and low permeability of the concrete matrix retard the influx of deleterious elements, it is important that these two aspects are addressed when considering high durability.

Excess water over that theoretically required for the chemical cement hydration reaction is necessary to ensure the workability required for proper placing and compaction of the concrete. During the curing process this free water evaporates leaving a micro system of interconnecting capillaries and voids.

By introducing a polymeric system with the Portland cement binder, a composite material is formed in which the polymer lines and seals the capillaries and micro cracks effectively reducing the porosity and making the material less permeable. Figures 1 and 2 illustrate this mechanism.

In addition to reducing the permeability, the polymer also tends to form a protective layer over the crystalline cement portion which is otherwise susceptible to surface attack. These modifications to the material matrix limit the access of the aggressive elements and significantly increases the durability of the material under this form of attack.

PROPERTIES OF POLYMERS IN CONCRETE

Polymers can be introduced into concrete in various ways. This gives rise to three main categories of polymers in concrete:

Polymer impregnated concrete
Polymer cement concrete
Polymer concrete

Polymer impregnated concrete is a previously hardened Portland cement concrete which is impregnated with a monomer or a partially polymerised monomer to controllable depths of the material. Only after impregnation polymerisation is induced by means of radiation, heat or a retarded catalyst. This impregnation process ensures that most of the capillary and interconnecting compaction pores are sealed with a resulting impermeability. Another advantage of this method is that the polymeric chains which are formed, act as a micro—reinforcing system and nominally increases the tensile strength of the material. This in itself is a contributing factor to increased durability as it results in less tensile cracking which would normally promote the inflow of aggressive liquids.

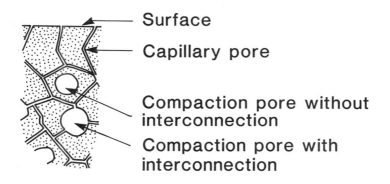

Surface

Capillary pore

Compaction pore without interconnection

Compaction pore with interconnection

Figure 1: Unmodified concrete with interconnecting voids

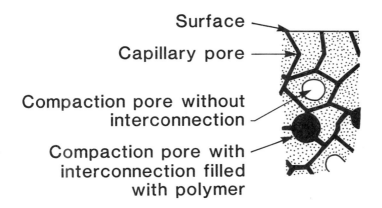

Surface

Capillary pore

Compaction pore without interconnection

Compaction pore with interconnection filled with polymer

Figure 2: Polymer modified concrete with interconnecting voids sealed

The impregnation of conventional concrete that is being damaged by ingress of aggressive elements can be undertaken and excellent results are achievable. Epoxies of low viscosity are frequently used in practice to impart additional tensile and cohesive properties to the surface of concrete or plasters suffering from weathering failure, often brought about by wetting and drying cycles. Methyl methacrylates, being normally less vicious than epoxies, are also frequently used for this purpose. Solvent free epoxies, however, tend to be less moisture sensitive and although often less penetrative, are more reliable under outdoor site conditions.

Polymer cement concrete and polymer concrete can be described as a polymeric system introduced into the wet aggregate mix either as an addition to the cement binder or as a total replacement of the Portland cement. Complete polymerisation is again ensured during or just after the setting of the concrete.

Polymer cement concrete is fast gaining popularity as a profiling product where thin sections require additional adhesive and cohesive properties from the cement matrix. Traditionally styrene butadine rubbers were used but these dispersions have been shown to absorb up to 12% moisture in comparison to the current acrylic suspensions which generally have water absorptions of less than 8%[1].

Normally polymer cement concrete and polymer concrete have a refined matrix using small aggregates, the course component seldom exceeding 13 mm. As with polymer impregnated concrete, the polymer fills and seals voids while in addition tends to coat the components and generally provides additional tensile strength.

Polymer cement concrete is cost effective in the sense that an inexpensive cement base material is modified with a relatively small amount of a more expensive polymer resulting in greatly improved properties. In order of chronological use the typical polymers used to date include polyvinyl acetate (PVA), polyvinyl propionate (PVP), styrene butadine copolymer (SBR) and acrylate copolymer (AC).

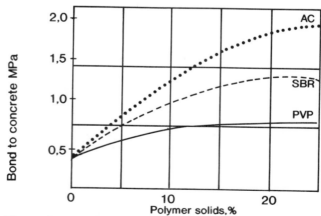

Figure 3: The relationship between bond to concrete and polymer solid contents of various polymer types(Burger)

The latest acrylate copolymers are emulsified dispersions with relatively low water absorption, lower viscosity and good resistance against re-emulsification.

Research has shown that increased bond strength appears to have a direct relationship corresponding to decreased water absorption of the polymer at the interface[1]. This is shown in Figure 3 for three types of polymer at varying concentrations. It follows logically that the expansion of the polymer cement concrete is also directly proportional to the degree of water absorption by the polymer.

Polymer concrete is classed as a concrete matrix where the cement binder is totally replaced by a polymeric substitute. Some of the more common polymers used in polymer concretes are polyester, epoxy, furan and methyl methacrylate

As the durability of concrete is more often than not derived from the performance of the binder component, it stands to reason that the performance of polymer concrete varies considerably from that having a cementicious binder. This is further illustrated by the following stress–strain graph which compares these properties for conventional concrete to that of two types of polymer concrete.[2].

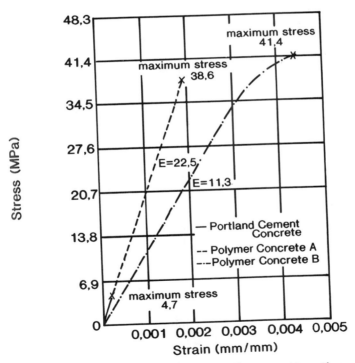

Figure 4: Stress–strain curves illustrating the difference between that of two types of polymer concrete and that of OPC concrete (Burleson)

Of interest is a comparison of the average properties of Portland cement concrete with that of the more common polymer concretes. Table 1 provides such a comparison and is derived from various publications in order to give an overview of average properties.

Table 1:   A comparison of average values for some properties of polymer concretes and Portland cement concrete.

Product	Compressive Strength MPa	Bond Tensile Strength MPa	Coefficient of Thermal Expansion	Modulus of Elasticity MPa	Water Vapour diffusion resistance coefficient	Carbon Dioxide diffusion resistance coefficient	Flexural strength MPa
Polymer Concrete Epoxy	70	20	45.10^{-6}	9 000	25 000	30.10^6	30
Polymer Concrete Polyester	60	15	50.10^{-6}	7 000	20 000	20.10^6	25
Polymer Concrete Methyl Methacrylate	95	15	28.10^{-6}	9 000	17 000	25.10^6	30
Polymer Cement Concrete Acrylic Emulsion Fine Aggregate	35	2.5	13.10^{-6}	15 000	800	10 000	11
Polymer Cement Concrete Acrylic Emulsion Coarse Aggregate	45	2,5	12.10^{-6}	20 000	350	5 000	10
Portland Cement Concrete Good Quality	40	4	11.10^{-6}	40 000	75	450	5
Portland Cement Plaster	45	2	12.10^{-6}	35 000	70	350	6

THE ADVANTAGES OF USING POLYMER IN CONCRETE

The use of a polymer in concrete has many advantages. These advantages vary with the type of polymer used in the concrete as well as with the method of introducing the polymer into the material.

In general some advantages achieved in using a polymer in concrete include the following:

corrosion resistance
higher durability under chemical attack
faster curing and high early strength development
increased final tensile strength characteristics
improved aesthetics
better water retaining properties
increased bonding to other materials

A practical advantage of a polymer modified concrete is that products could be manufactured in thinner sections compared to using OPC concrete due to its reduced permeability and higher strength characteristics. This gives rise to the product being less bulky and easier to handle and install. For this reason, as well as for aesthetics, precast polymer concrete panels used for building cladding are becoming more and more popular. Especially in the more aggressive coastal conditions it has a major advantage above conventional concrete cladding.[3] An additional advantage is that the building support structure can also be of lighter design.

The reduction in weight of products while retaining strength and corrosion resistance is a major advantage in the mining industry. Drainage and fluid transportation channels which traditionally have been manufactured from OPC are difficult to handle and transport in the confines of mining tunnels and stopes. The use of asbestos and pure polymer pipes are often avoided for health and fine hazard reasons and the polymer cement concrete pipe provides a lightweight yet strong and highly durable alternative. Especially in the acid groundwater conditions encountered in the South African gold mines, the durability of ordinary Portland cement reinforced channels and pipes is low. Polymer modified concrete pipes are being investigated as a possible replacement for OPC pipes.

In addition to the use of a polymer to enhance the corrosion resistance of a Portland cement concrete, it can also improve the wear and cavitation resistance of concrete linings such as is found in hydropower structures. The upper St. Anthony Falls Lock, located on the Mississippi river in Minnesota is an example of a hydro–structure repaired with a polymer modified concrete to prevent abrasion erosion.[4] Figure 5 compares the erosion resistance of conventional concrete with that of a concrete impregnated with a methyl methacrylate monomer.[5]

Polymer concretes, in particular concrete modified with methylmethacrylate will give the same strength at 5 °C as at 25 °C although over a longer period. This is not necessarily true for Portland cement concrete and therefore a major advantage where cold environment work is unavoidable. Methyl methacrylate will polymerise at temperatures well below freezing.

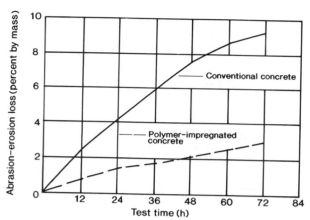

Figure 5: By impregnation with a methyl methacrylate monomer the erosion resistance of concrete can be improved dramatically.

(US Army Corps of Engineers, 1978)

Using different polymeric systems as modifiers to ordinary concrete, a wide range of improved properties can be achieved. Figure 6 compares the carbonation resistance of a polymer modified concrete to that of an ordinary concrete.[6] The polymeric systems used in this test were polyacrylic ester (PAE) and styrene butadiene (SBR). It is seen that even with an addition of 10% PAE (percentage of cement content) a dramatic improvement in carbonation is obtained. With a 20% SBR addition, results are even more favourable.

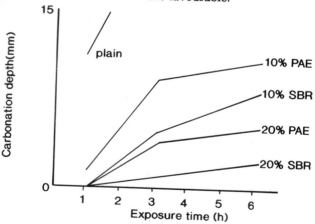

Figure 6: Accelerated test using CO_2 under pressure shows improved carbonation resistance of concrete with polymer additions(%polymer on cement)(The Concrete Society, 1987)

Freeze—thaw durability is also improved with the use of a polymer modifier. It was reported at the 5th International Congress on Polymers in concrete held in Brighton (UK) in 1987 that using a 5% SBR addition to form a polymer cement concrete, a total loss of dynamic modulus of elasticity which occurs at an unmodified concrete after about 70 freeze/thaw cycles, is prevented. When using a polyester concrete made with unsaturated resin as binder, no loss in dynamic modulus of elasticity is experienced.

Considering some disadvantages in using a polymer in concrete as a component to protect the concrete or to increase its durability, it is found that material costs are relatively high. The rapid curing rates and excellent properties obtained from polymer concrete and polymer cement concrete often make the use of these polymers in concrete cost effective though. Polymer impregnation is the least practiced version of polymers in concrete since special and expensive equipment is required to ensure adequate and uniform impregnation.

As the polymerisation of a monomer involves complex chemistry, care must be taken to select a system that is both cost effective and compatible with the other materials in the matrix. Creep, elongation and bonding properties vary between systems and care must be taken to match these to that of the substrates on which it is to be used. Moisture also negatively affects the curing of some polymers and must be taken into account during the design to ensure long term durability.

Polymer concrete's fire resistance is related to the polymer used. In general they tend to have a low fire resistance and are often combustible. As seen in Table 1 the coefficient of expansion of an epoxy polymer concrete is significant and this aspect alone has caused many compatibility failures when used with conventional Portland cement concrete, despite the improved interface bond strength.

Most polymer cement concrete, polymer concrete and polymer impregnated concretes are relatively expensive, highly sophisticated and sensitive to abuse during the curing phase. Experienced people are therefore a prerequisite for the successful application on site.

CASE STUDIES

Because of the cost of polymer concrete (up to 20 times that of OPC concrete) it is usually only used by itself under special circumstances at this stage. Polymer concrete and polymer cement concrete are however frequently used as repair material. Especially in the flooring industry, polymer concrete has gained general acceptance as a protective coating.

Various polymer modified lining systems exist on the market which can be used to cover the exposed concrete surface in order to protect it from a corrosive environment. Although these systems perform well and make use of polymers and fine fill material, it cannot necessarily be called a polymer concrete. It is rather a polymer modified lining or coating system. However, when such a lining system contains aggregates with or without Portland cement in order to use it as a thick layering material for repair of damaged concrete structures it would be classed as a polymer concrete.

Case One

Indian Ocean Fertilizer, a manufacturer of phosphoric acid used in the fertilizer industry and situated on the Natal north coast of South Africa needed to rehabilitate part of a concrete cooling tower and yet maintain production with the remaining cells of the tower in operation.

The problem

1. The repair was to be done in as quick a time as possible.

2. The repaired area was to be overlined with an acid resistant liner of vinylester picking up the existing tower liner.

3. The environment had a relative humidity in excess of 95% and was contaminated with hydrofluoric acid (only 904 stainless steel survives).

The solution

An aromatic amine solvent free epoxy was selected which together with selected graded silica sand and 6 mm stone was mixed to form a voidfree saturated epoxy polymer concrete. The polymer and curing agent ensured that the concrete was not sensitive to moisture and had high chemical resistance to the acid contaminated micro environment. No significant shrinkage was expected. The casts were made in pours of 0,35 m³ each and the linings were reapplied over the polymer concrete within 24 hours and the tower recommissioned.

Case Two

The upgrading of road structures to carry abnormal heavy loads from the coast to the Transvaal was necessary when Escom were building large electrical power stations in the Transvaal. A simply supported bridge in the Drakensberg area was upgraded by effectively creating a continuous deck. This required stressing between piers once temporary props had been installed. A concrete had then to be cast to encase the stress bars and link deck and piers. No shrinkage could be tolerated. The work was done during winter so cold weather concreting was unavoidable. The contractor fixed the shutters between adjacent piers and below the existing slab and filled the void created with river boulders to act as plumbs. A solvent free epoxy polymer concrete was now poured into this void in pours of 1,3 m³ each until the volume was totally filled. A monolithic pier—deck system was thus created while the structure remained operational.

Case Three

Re—nosing and repair of expansion joints in cold rooms were done during 1987 for Natal Co—operative Dairies. Floor sections and expansion joints in the large freezer rooms had been damaged by heavy wheel loads and reprofiling was required while the cold rooms remained in operation.

The contractor cordoned off the area to prevent contamination of foodstuffs and the work was put to hand over a weekend using a methyl methacrylate polymer concrete at sub zero temperatures. The floor was opened again to traffic the following day and the repair proved to be most successful.

Case Four

The Natal Provincial Administration required to repair a bridge joint nosing during 1987. This bridge was on a main arterial route into Umlazi and was heavily trafficed. The contractor was only permitted to occupy one lane at a time and then only between peak periods. The break out and repair was done within 2 hours using a methyl methacrylate polymer concrete.

Case Five

A laundry in the Umgeni area of Natal had suffered severe flood damage during the 1988 flood which caused the Umgeni river to flood its banks and the laundry building. Some differential settlement had occurred and floor slabs had to be reprofiled. The floor was scabbed and repaired with a 15 mm layer of acrylic modified polymer cement concrete formulated to be self–levelling. The high abrasion resistance coupled with improved modulus of elasticity and bond to the original slab have shown ideal to withstand impact, heat and the wear of general traffic.

CONCLUSION

Polymers can be used as a protective measure in concrete in any one of three methods. These are

- Polymer impregnated concrete
- Polymer cement concrete
- Polymer concrete

In addition to the above a polymer can be used in a thin lining or coating system to protect underlying material.

Polymers in concrete must satisfy many requirements including:

improved chemical resistance
reduced shrinkage
improved abrasion resistance
higher bond strength
improved workability

in comparison with conventional Portland cement concrete. To achieve this special polymers are researched, refined and modified so as to be compatible with the alkali Portland cement binder or with the cement interface. The polymer must have no corrosive effect on steel reinforcing and must have polymerisation or strength–gain times that closely match that of Portland cements in the case of polymer cement concrete.

Obviously, by varying the relative quantity of polymer modifier added, the range of modified properties that can be achieved becomes extensive. Equally the type of and method of addition of the polymer significantly affects the performance of the end product. This versatility has therefore established polymers in concrete as a field of unequalled opportunity with the potential to for materials in the civil engineering field what the microchip has done for electronics.

REFERENCES

1   BURGER, T A. Sika Product Tehcnology Management — Concrete and mortar. Research and Development. Sika AG Zurich, 1987.

2   BARLESON, J D ET AL. An investigation of the Properties of Polymer Concrete SP 58–1, pp 1 – 19.

3   PRUGINSKI, R C. Study of commercial Development in Precast Polymer Concrete SP 58–5, 1978, pp 75 – 84.

4   SCANLON, J M. Application of Polymer Concrete ACI Publication SP69. Applications of Concrete Polymer Materials in Hydrotechnical Construction, 1981, pp 45 – 62.

5   Publication. US Army Corps of Engineers (CE) 1978 Survey 1978.

6   The Concrete Society. Polymer Concrete: Revised Technical Report no 9 London, 1987.

PERFORMANCE PROPERTIES OF POLYMER COMPOSITES

L Y Lavrega

Byelorussian Polytechnical Institute,

USSR

ABSTRACT. Reliable corrosion protection of concrete using polymer coatings may be ensured not only by their chemical resistance with respect to aggressive media, but also by their impermeability, excellent adhesion to the base and sufficient tensile strength during their long-term life in the aggressive environment. Permeability of the polymer coatings is closely associated with their porous structure and may be decreased by adding admixtures - adhesion promoters, which form silicone bonds on division boundaries. Porous structure and physico-mechanical properties of alkylresorcinol epoxy compositions was studied. It was found that their maintenance properties influenced by temperature gradients, water and organic acid media (solutions of acetic or lactic acids) increased with admixture in optimum quantities of the proper silicone adhesion promoters. Regulation of the porous structure of the polymer compositions and the required adhesion on the division boundaries may considerably increase service life of the polymer coatings.

Keywords: Polymer coating, Concrete, Alkylresorcinol epoxy binder, Porous Structure, Diffusion Permeability, Silicone Adhesion Promoters, Corrosion Resistance

Dr Lavrega L.Y. is an Assistant Professor at the Department of Concrete and Reinforeced Concrete, the Byelorussian Polytechnic Institute, Minsk, USSR. Her current research work includes: durability of polymer concrete, corrosion resistance of polymer coatings applied to cement concrete and reinforcement, resistance of polymer coatings to various aggressive media. She has gained experience and skill in this field.

INTRODUCTION

The use of concrete and reinforced concrete structures in agro-industrial and processing industries invariably entails protection of the above structures from corrosion. The presence of organic acids (lactic, acetic, citric and other acids), detergent solutions including soda, salt, chlorinated lime etc. in the industrial waste waters causes destruction of concrete due to formation of easy-soluble or weak compounds. Furthermore, considering mechanical factors, such as abrasion and shock loads, microorganisms, mould, fungi, temperature gradients, it is impossible to slow down diffusion processes (1) by the formed layer of corrosion products. The most expedient way would be to exclude the corroded cement binder from concrete and substitute it for a polymer or polymer-silicate binder. This would cause, however, certain problems of economic and constructional nature. High shrinkage, creepage, thermal expansion, tendency to aging of the polymer compositions require very careful approach to their application as a construction material. In addition, the polymers are rather expensive and scarce.

All this necessitated the use of protective polymer coatings characterized by chemical stability in the aggressive environment, sufficient adhesion to the substrate as well as strength and elasticity. One of the most important qualities of the polymer coatings is their density and impermeability with respect to the concrete-destroying aggressive media.

At present polymer coatings of various thickness (from several microns to 7-10 mm and more, depending on their application) have been widely used as protective coverings. It is very economic indeed to use polymer materials. The calculations show that the expenditures on repairs to concrete structures as well as downtime of equipment greatly exceed the cost of polymer coatings.

The choice of polymer materials for anticorrosive protection of buildings in food and agricultural industries largely depends upon hygienic properties of polymers. The tests of alkylresorcinol composites in imitating media (distilled water, 0,3% lemon acid, 5% NaCl solution) to determine qualitative content of volatile toxic substances which migrate into the simulated media while exposing extracts for 1, 2 and 10 days showed that the content of harmful volatile substances (epichlorohydrin) did not exceed standard limits.

Properties of the polymer compositions vary over a wide

range, the main criterion for their prediction being the content of compositions. Since properties of the polymer compositions are inadequately influenced by the components (3), there comes a necessity for determining a number of other criteria which would stipulate physico-mechanical and maintenance properties of the polymer compositions. For many non-organic materials such a criterion will be their porous structure (4). Furthermore, it may be assumed that this point of view holds true for the polymer compositions, such as mastics, polymer mortars and concretes, too. Thus an increase in porosity by 10% may weaken compression strength of polymer concrete by 50% and its elastic modulus by 25% (3).

Experience of using protective coatings in the aggressive environment has shown that durability, in some cases, may be determined not by the binder chemical stability (which is very high for a number of polymers), but by the structural permeability of the coating and its compatibility with the base.

Tight contact between the organic polymer and hydrophilic material of the substrate does not yet ensure formation of a corrosion-resistant material. Neither this can be achieved through direct chemical bonding, since the organic polymer with stable covalent bonds and the mineral with ionic bonds are rather heterogeneous materials. The study of interfaces in the polymer composites (5) revealed that sufficient adhesion between these heterogeneous materials may be acquired by the use of a third material, ie intermmediate layer either between the matrix and filler, or between the polymer coating and substrate. These materials were called adhesion promoters.

This paper deals with the questions of forming porous structures in the polymer compositions subject to their content and preparation technology, role of adhesion promoters and influence of the above mentioned factors upon physico-mechanical and maintenance properties of the polymer coatings.

FORMATION OF POROUS STRUCTURES IN POLYMER COMPOSITIONS

Porosity (P) of polymer mortars and polymer concretes is made up of two components: porosity of intergranular hollows in the aggregate (P_j) and a binder (P_s), i.e.

$$P = P_j + P_s \quad (3).$$

The first component is wholly determined by the grain

content of the aggregate, compaction of the composition and binder consumption. As for the other component, it is commonly accepted that its insignificantly changing value (within 5%) may be neglected.

Nevertheless, if the porosity is of constant value, the character of the porous structure may seriously affect the properties of a polymer composition as well as its impermeability and durability.

The goal of our researh was studying porous structure of the binder subject to the type, degree of filling and dispersity ($S_{sp.}$) of the filler, preparation techniques and exposure to aggressive media. Dianic and alkylresorcinol epoxy resins were used as binders. Polyolefins, industrial oils, plasticizer ISK-11 and silicone monomer ES-32 served as modifiers. Various aggregates were used: fine quartz sand, silica gel, siliceous fly ash, limestone, ochre and titanium phosphate.

Complex porosimetric analysis of the polymer compositions was conducted using a high-pressure mercury porosimeter MPa P2000 ('Carlo Erba Instrumentazione', Milano), a device for detecting macropores of 0,6 mm radius and a computer with automatic read-off. The results were corrected due to the stress-strain behavior of the specimens.

Microstructure of the polymer compositions was studied simultaneously using a scanning electron microscope 'Tesla BC-301' and 'Stereoscan' ('Cambridge Scientific Instruments Ltd.'). The investigations were carried out at the Slovacian Academy of Science under the guidance of Dr J.Jambor.

The experiments showed that the porosity value of the polymer compositions and its structure depend upon many factors, while total porosity and its character alter with time. A similar effect was produced by the aggressive media. A number of interesting mechanisms was revealed:
- compounds containing optimum quantities of a modifier were characterized by minimum porosity as to physico-mechanical properties;
- porosity of specimens on pure non-modified resin base decreased with time, while the porosity of modified specimens increased in all respects (micro- and macro-porosity, medium pores radii, specific pore surface). It is characteristic that the compounds having optimum quantity of modifier suffered much less from structure deterioration.

Boiling for 36 hours increases porosity of the polymer compositions containing modifiers. Characteristically, the porosity of specimens based on pure resin and having optimum quantity of modifier decreases (Fig. 1).

Reactionless diluents (acetone) usually increase total porosity of compositions and volume of medium and macropores in comparison with non-diluted compounds and those with diluents capable of forming chemical bonds with epoxy resins. Very interesting results have been obtained through modifying the mixtures with polyolefins and silicone monomer ES-32. The latter does not influence the porosity values much (when the quantity is up to 10%), but highly increases consolidation of the polymer compositions after boiling within 36 hours. Thus, the volume of micropores rad. up to 7500 Nm is reduced by 24%, pores rad. up to 0,1 mm by 23%. The medium radius of micropores decreases from 10,91 to 9,62 Nm and pores rad. up to 0,1 mm from 12,12 to 11,13 Nm; specific pores surface decreases from 8,34 m^2g^{-1} to 6,56 m^2g^{-1}. These results indirectly confirm the fact that the silicone monomers can form chemical bonds on the division boundaries, what will be discussed later.

Loading of compositions with fillers may somewhat reduce total porosity. However, as the studies have shown, its character wholly depends on specific surface of the filler. As the filler dispersity grows, the midium radius of micropores (micropores median) considerably increases in comparison with a filler-free composition, while the use of fine sand as a filler (Ssp.= 0,15m^2g) increases pores size , micropores and macropores median, approximately, 6 times as much against filler-free composition and 58-100 times as much exceeds this value in mixtures loaded with highly dispersive fillers. Characteristically, total porosity value is approximately identical for all compositions. It reveals insufficiency of this characteristic for talking about density or impermeability of the polymer coatings (Table 1).

The fillers dispersity only slightly affects the strength of the polymer compositions. For this reason, a group of scientists considered it inexpedient to increase specific surfaces of the fillers to more than 0,15 m^2/g (7). However, the study of porous structure of the polymer compositions refutes the above stated point of view. Moreover, the higher dispersity of the filler, the more effective phisico-mechanical properties of the composition may be obtained, while consumption of the filler will be reduced.

While analysing the experimental data, we can divide the

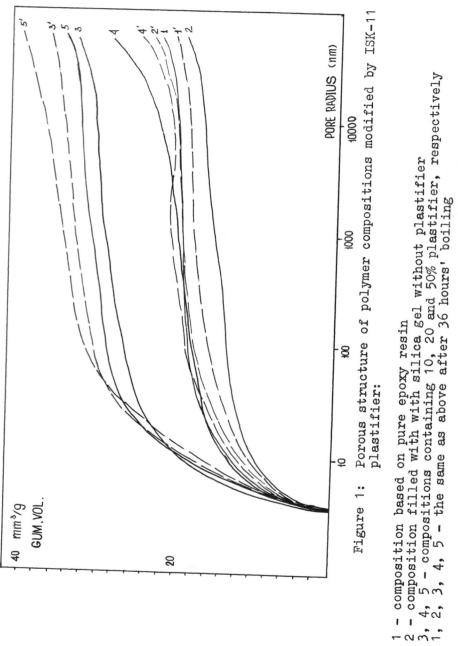

Figure 1:   Porous structure of polymer compositions modified by ISK-11
            plastifier:

1 - composition based on pure epoxy resin
2 - composition filled with with silica gel without plastifier
3, 4, 5 - compositions containing 10, 20 and 50% plastifier, respectively
1, 2, 3, 4, 5 - the same as above after 36 hours' boiling

TABLE 1   Structural and physico-mechanical properties of polymer compositions

No.	Fillers Type	Dispersity, m²g⁻¹	Strength, MPa	Total porosity, %	Micropores rad. 3,73-7500 nm Vol. nm³g⁻¹	%	Median radius, nm	Pores rad. ≤0,6 mm Vol. nm³g⁻¹	%	Median radius, μm	Specific surface of pores, m²g⁻¹
1.	-	-	128,8	4,6	13,14	1,67	10,2	20,32	2,59	119,1	2,911
2.	-	-	127,75	6,1	12,852	1,62	8,8	19,4	2	192,2	3,015
3.	Sand	0,15	98,27	4,9	18,47	2,8	47,7	29,24	4,5	1124,6	2,343
4.	Limestone	1,3	91,45	5,8	23,38	3,1	12,1	43,23	5,8	19,1	3,692
5.	Amorphous silice	5,37	70,66	4,6	17,87	2,42	8,1	28,48	3,82	10,4	4,762
6.	Silica gel, SiO2	30	49,85	6,6	32,86	4,4	30,6	45,03	6,09	601,3	4,274
7.	Titanium phosphate	125	77,22	6,5	17,58	2,24	11,0	24,64	3,22	22,3	3,707

pores formed in the polymer binder in 2 groups:
- "polymerizational" micropores, the form and dimension of which depend on the character of microblocks formed during hardening, on their conformational properties (micellar, folded, globular etc.) as well as super-molecular structure of the polymer compositions. The presumable value of the polymerizational micropores will be below 1 μm (1000 nm);
- "technological" pores rad.>1 μm. This group includes the pores formed as a result of evaporation of chemically unbound diluents, diffusion of reactionless plasticizers, air intake during preparation of mixtures and insufficient compaction of the mixture.

The less is the pores median, the more hardened the composition and the higher strength and density of the material. This may be proved by comparing strength values of the polymer compositions of one group with the median radius of the micropores.

Formation of a super-molecular structure in the polymer composition is closely associated with the filler's dispersity and degree of filling. As is well known, the fillers may slow down crystallization, form colloidal super-structures within the super-molecular structure, modify the structure of the boundary layers, etc. At the same time, the character of microporosity and its value may, to some extent, reflect stability and perfection of the super-molecular structure in the polymer composition. Thus, we can assert that it is possible to judge of density and impermeability of the polymer compositions in the aggressive environment using the median radius of micropores. This is well illustrated in the micro-photographs of the polymer compositions. However, the size of this little article is not enough for these photographs to be demonstrated.

Testing of specimens under aggressive conditions has shown that both water and acid resistance of the polymer binders are closely connected with the character of porosity (Table 2).

Porosity of the binder containing quartz sand (compound No.3) increases in water. This compound is characterized by maximum water absorption and minimum water- and acid resistance.

Penetration of the aggressive medium into the specimens filled with titanium phosphate (comp. No.7) decreases their total porosity. Water resistance of these specimens is higher than that of compounds No.3 and 4, whereby their acid resistance is slightly lower. Thus, not only dispersity of the filler, but also its chemical composition affect protection properties of the polymer coatings.

TABLE 2   Properties of polymer binders
after 25 days of exposure to
aggressive conditions

No. of compound	5% CH_3COOH solution		Distilled Water		
	Resistance factor	Variations of total porosity	Resist. factor	Total porosity,%	Water absorption,%
3	0,92	− 0,39	0,91	+0,08	0,70
4	0,95	− 0,57	0,95	−0,14	0,12
5	0,99	− 0,40	0,99	−0,02	0,03
7	0,87	− 1,08	0,98	−1,53	0,05

INFLUENCE OF ADHESION PROMOTERS ON
PROPERTIES OF POLYMER COMPOUNDS

The study showed that the fillers exposed to the water or other aggressive media, while strengthening polymer compositions in general, may worsen the properties of coatings due to weak contact zones. A similar picture is observed in the "coating-base" contact zone. The expediency of using admixtures (apprets) in this case has already been mentioned. As stated by a number of authors, the most effective admixtures are silanes (monomer silicons), which are compatible with almost all organic polymers. They may be added either directly into the polymer (integral mixing method), or used for curing substrate. In the first case they reach the substrate by way of migration in the process of standard mixing or during use.

General formula for the industrial silanes is $x_3Si(CH_2)nY$, where $0 \leqslant n \leqslant 3$ and x is the silicon atomic group being hydrolized, and Y is the organic-functional group ensuring compatibility with the given resin.

If interacted with water, the x-groups are hydrolized to form silanols:

$$x_3Si(CH_2)_nY + H_2O \rightarrow (HO)_3Si(CH_2)_nY + 3H.$$

Exposure to diluted acids (acetic, lactic, pH=4) facilitates hydrolization of silanes and formation of silantriols. Thus, a part of SiX_3 molecule or its reaction product may ensure adhesion adhesion to the composite's non-organic component by means of covalent or hydrogen bonds. The appret must contain at least one reactive group, which interacts with resin during hardening. As a result, different groups of one and the same molecule bond with different composite components, the activation energy on the interface reaching 50 to 100 Cal/mole. Characteristically, the above mentioned bonds appear in the presence of water, whose inevitable diffusion in the protective layer of the polymer coating may be employed as a positive factor (8).

The influence of a number of silicone adhesion promoters on physico-mechanical properties and corrosion resistance of alkylresorcinol compositions was studied. Water, 5% acetic and lactic acid solutions (pH=2,7 and 2,43) were used as aggressive media. The type of the silicone appret depended upon the character of hydrolitically unstable substitute x in SiX_3 group and reactive R-group which ensured interaction with the polymer matrix (Table 3).

TABLE 3   Properties of adhesion promoters

N N	Compound	Reactive bond	Hydrolisis conditions	Adsorption limit, Mmole/g	Hydrophobic effect on hardened cement W,%	Θ,gr	h,mm
1.	ES-32	Si -OC$_2$H$_5$	Basic, acid	0,141	0,36	125	1,8
2.	113-63	Si -OC$_2$H$_5$	Basic, acid	0,047	0,47	126	2,4
3.	MVDHS	Si - Cl	Neutral, basic,acid	0,176	0,6	105	1,9

Notes: W% is the humidity of the specimens after 3 days' exposure to water; θ is the moistening angle, grad; h is the depth of unmoistening layer.

In addition, a highly dispersive amorphous silice – butasil (pH 4,8; filling mass – 20 to 40 g/l; S_{sp}. – 1800 m/g) was used as an adhesion promoter. Ground quartz sand (S_{sp}.=0,18 m/g) and Portland cement 400 (S_{sp}.=0,32 m/g) served as aggregates.

The investigations were carried out using 2x2x2 cm cubes and 4x4x16 cm beams made of polymer mastics and cement-sand mortar (1:3) coated with 0,6 mm polymer mastic. The cement-sand mortar had the following physico-mechanical properties: compression **strength** (R_c) = 31,4 MPa, bending strength (R_b) = 7,05 MPa, tensile strength (R_t) = 2,57 MPa, water absorption 5.33%, dynamic elastic modulus $29x10^3$ MPa. The specimens were subjected to cyclic heating up to 60°C under aggressive conditions within 4 to 6 hours followed by exposure to atmosphere within 18 to 20 hours. Outward appearance, weight, bending strength, compression strength, tensile strength and elasticity modulus were determined upon 7, 15, 30 and 60 test cycles.

Cyclic tests in water of polymer mastics containing adhesion promoters showed that irrespective of the type of an adhesion promoter there is some optimum quantity of it in the polymer composition. This becomes evident from variations in bending strength of the polymer compositions that take place upon 30 test cycles in water (Figure 2). Butasil affects the alkylresorcinol compounds just as silicone apprets do. As is seen from Figure 2, it is inexpedient to admix large quantities of silane since those oily, gel-like and powder-like polymer products, which appear on the division boundaries, cannot facilitate strengthening of the polymer composition.

After 30 test cycles in water and organic acid solutions the colour of coatings did not change. The coatings were solid and had no peels. Nearer to 60 test cycles in water there appeared bubbles and blisters (compounds No.3 and No.4).

Specimens of compounds No.1 and No.2 increased their strength in water at the early stage of testing. Then there was a decrease in strength, but still it remained 5 to 20% higher than the initial. Specimens' weight did not change much in contrast to the standard.

Elastic modulus of specimens covered with mastics of compounds No.1 and No.2 within 30 test cycles stayed

almost the same. When covered with mastic of compounds
No.3 and No.4, it decreased by 10% and 15%, respectively.
60 test cycles showed that solidity of the coatings in
all the media as well as colour were maintained only
when mastic of compound No.1 was used, while compound
No.2 in acid solutions suffered blisters and cracks.
The same was earlier observed for compound No.3. Weight
of specimens and their strength decreased. Elasticity
modulus of compound No.1 remained almost the same, but
decreased by 7% and 15% for compounds No.2 and 3,
respectively.

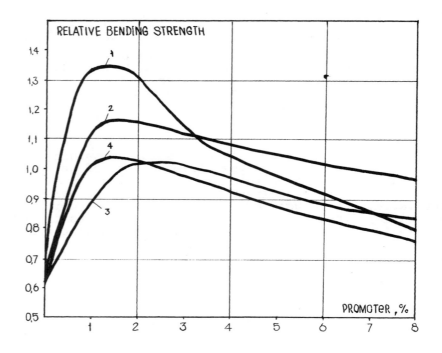

Figure 2: Water resistance of alkylresorcinol compositions via
amount of adhesion promoter:

1 - MVDHS;   2 - ES-32;   3 - 113-63;   4 - butasil

Experiments showed that adhesion promoters improve performance of the polymer compositions and protective coatings on their base in a number of the most frequent aggressive media.

The character of the hydrolitically unstable SiX_3 group as well as its ability to form bonds on division boundary (subject to pH of the midium) may influence the performance of the polymer compositions and coatings.

Modification of the polymer compositions with 1 to 10% adhesion promoters shall not considerably affect porosity, although there is a tendency towards its increase with enlarged quantity of silane. However, boiling of specimens for 36 hours may cause a decrease in micropores volume by 24-49% (23-40% for pores rad.up to 0,1 mm, half as much again for pores rad.median). Specific surface of pores is also decreased by 20-60%. The minimum microporosity corresponds to that of compounds containing an optimum quantity of adhesion promoters.

Changes in the porous structure of the polymer compositions modified by adhesion promoters (ES-32) as well as natural changes in strength and elasticity confirm E.Pludeman's theory of reversible destructions and reductions of tense bonds between the appret and substrate in the presence of water. Silanes provide relaxation of tension at the interface without deterioration of adhesion. This is very important when the coatings are used under alternately acting aggressive conditions and temperature drops (8, 9).

PREPARATION METHODS AND PROPERTIES
OF POLYMER COMPOSITIONS

The investigations and practical data have shown that quality of multicomponent and highly viscous materials may be ensured only through accurate mixing and dispersing. The dispersing systems employed (VNIISMI system) provide good homogeneity, mobility and solidity of compounds (6).

Tests of porous structures of the dispersed polymer compositions demonstrated a decrease in total porosity by 20%. Volume of micropores and pores rad. ≤ 0.6 mm decreased half as much, and specific surface of pores decreased half as much again, from 5,428 m^2g^{-1} to 3,707 m^2g^{-1}. As for the strength of the specimens, it increased half as much again or even twice as much. All these data may be explained by possible alteration of conformational properties of macromolecula, higher supermolecular structure of the polymer composition and activation of chemical interaction between the components.

Diffusive permeability of the dispersed mixtures decreases respectively due to alteration of porous structure (10).

CONCLUSIONS

Improvement of maintenance properties of the polymer compositions involves several factors: degree of filling. type of modifier, preparation methods, etc. In any case, however, all these properties affect porous structure of the polymer compositions and nature of adhesion bonds on division boundaries that cause penetration of aggressive media into the depth of coating and resistibility to any distructive conditions.

REFERENCES

1   MOSKVIN V M.   Korrozia i Zashchita Zhelezobetonnych Konstrukcij. Beton i Zhelezobeton, number 3, 1976, pp 2-3.

2   RUBETSKAYA T V., SHNEIDEROVA V V., LUBARSKAYA G V. Korrozia Betona v Kislych Aggressivnych Sredach i Metod Ozenki Zashchitnych Svoistv Lakokrasochnych Pokrytij. Trudy Instituta (NIIZHB), Volume 19 "Povyshenie Korrozionnoj Stoikosti Betona i Zhelezobetonnych Konstruktzij", M. Strojizdat, 1975, pp 17-20.

3   LECH CZARNECKI. Betony Żywiczne. Arkady. Warszawa, 1980, pp 180.

4   JAROMIR JAMBOR.   Porosita, Pórová Štruktura a Pevnost Çementovych Kompozita. Stavebuicky Časopis, Vol. 33, c 9, Véda, Bratislava, 1985, pp 743-765.

5   PLUDEMAN E.   Poverchnosty Razdela v Polimernych Kompozitach. Moskva, Izd. "Mir", 1978, pp 1-290.

6   LAVREGA L Y., NOVIK A F.   Izgotovlenie Antikorrozionnych Polimernych Kompozitsij s Uluchshennymi Svoistvami. Stroiteljnye Materialy, Number 5, 1986, pp 20.

7   SOLOMATOV V I., KNIPPENBERG A K.   Issledovanje Struktury i Svoistv Poliefirnogo Polimerbetona.

Stroitelstvo i Arhitektura, No. 6, 1977.

8   LAVREGA L Y. Povyshenie Dolgovechnosty Polimernych Pokrytij. Ctroitelnye Materialy, No.6, 1986, pp 15-16.

9   LAVREGA L Y., BORISLAVSKAYA I V., BAJZA A I., UNCHIK S Y. Povyshenie Dolgovechnosty Betona pri Vozdejstvii Organicheskych Kislych Sred. Beton i Zhelezobeton, No.3, 1989, pp 20-22.

10 LAVREGA L Y., NOVIK A F. Poristost i Porovaya Struktura Polimernych Pokrytij Betona i Ich Vlijanie na Zaschitnye Funktsii. Sbornik "Nowe Zozwiçzania orar Wdroienia z Zakresu Postç Pu Technicznego w Budownictwie". Politechnika Hubelska, 1989, pp 243-250.

PERMEABILITY OF CONCRETE AS A METHOD FOR CALCULATING THE LIFE EXPECTANCY OF REINFORCED CONCRETE COMPONENTS

J Bilcik

Slovak Technical University, Bratislava

Czechoslovakia

ABSTRACT. The close relationship between the specific permeability coefficient and the specific diffusion coefficient simplifies the assesment of the structure of concrete and its ability to resist various external agressive agents. Carbonation is presupossed to act as a factor harmful to reinforced concrete and this period of the life expectancy can be theoretically calculated in dependance of the permeability coefficient. Permeability of concrete determines also the transportation of oxygen required for corrosion after the initiation of corrosion. The design problem is to ensure that the time from construction to the occurence of unacceptable corrosion damage is greater than or equal to the design life of the structure. Laboratory results are reported in order to evaluate the influence of different surface coatings.

Keywords: Permeability, Diffusion, Carbonation, Corrosion, Life Expectancy, Coatings.

Juraj Bilčík is a Senior Lecturer at the Department of Concrete Structures and Bridges, Slovak Technical University, Bratislava, Czechoslovakia. His current research work includes all aspects of concrete durability and, in particular, corrosion of reinforcement.

INTRODUCTION

Insufficient durability of concrete structures has become a serious problem. One of the most important parameters influencing the durability of concrete, in particular that portion which provides protection to reinforcement, is its permeability. Permeability of concrete determines the ease with which gases, liquids and dissolved deleterious substances such as carbon dioxide or oxygen or chloride ions, can penetrate the concrete.

The process of converting the products of hydratation of cement to calcium carbonate by means of carbon dioxide is called carbonation. Carbonation is in general not harmful to unreinforced concrete. But it does have a damaging effect on reinforced concrete when the carbonation front extends to the reinforcing steel. When this happens the passive layer that covers the steel – preventing futher reaction so that the steel remains unaltered – is destroyed. That is why the carbonation front in concrete should not, during the service life of the structure, extend to the reinforcement.

If the corrosion process has started the rate of corrosion is still dependent on the suply of oxygen. The permeability of concrete is a major factor affecting the service life of reinforced concrete components.

EFFECT OF PERMEABILITY ON CARBONATION

The most important factors in carbonation are the concentration of the carbon dioxide and the diffusion characteristic of the concrete. Since the concentration of CO_2 in the atmosphere is approximately constant, it is a factor that can be disregarded.

Permeability of Concrete

One of the easiest and fastes techniques for measuring permeability of mortars and concretes directly is the differential pressure technique using a inert gas. For determing the gas permeability of concrete – the gas flow rate through the material – Nischer[1] apllied the following expressions:

$$Q = K \cdot \frac{A}{\eta \cdot L} \cdot \frac{/P_e - P_a / \cdot /P_e + P_a/}{2p} \quad . \quad /1/$$

With the aid of equation /1/ the specific permeability coefficient

$$K = \frac{Q \cdot \eta \cdot L}{A} \cdot \frac{2p}{/P_e - P_a//P_e + P_a/} \quad , \quad /2/$$

where
   Q = rate of flow /m³. s⁻¹/
   K = specific permeability coefficient /m²/
   A = cross-sectional area of the specimen /m²/
   p = pressure, at which the gas flow rate is measured /bar/
       /1 bar = 10 N.m⁻²/
   η = viscosity of the gas /N.s.m⁻²/
   L = thickness of the specimen /m/
   p_e, p_a = entry and exit pressure of the gas, absolute /bar/.

Relationship between Permeability and Diffusion

For the penetration of gases such as CO_2 or O_2 into a concrete
structural only a difference in partial pressure will have to be
considered. The propagation of moleculas or ions in consequence of
such differences in concentration is called diffusion. The specific
permeability coefficient K and the specific diffusion coefficient D
of concrete with regard to oxygen are connected by a relationship as
indicated by Lawrence [2] is here presented in Figure 1.

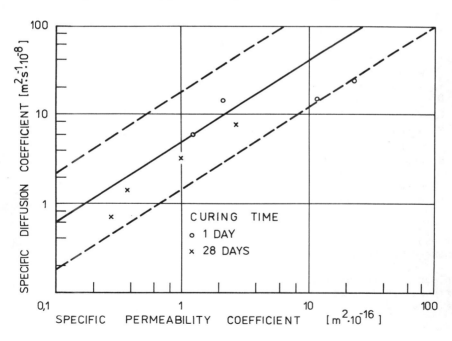

Figure 1: Relationship between specific permeability
          coefficient and specific diffusion
          coefficient for concrete [2]

Since the determination of diffusion coefficient is difficult and
time consuming, it appears to be adventageous to determine the
permeability coefficient as an auxiliary parameter. The close
relationship between the permeability and diffusion coefficient
simplifies the assessment of the structure of concrete and its
ability to resist various external aggressive agents.

Calculation of Carbonation Progress

The law of $\sqrt{t}$ frequently used in literature to describe the advance
of the carbonation is not sufficient for quantitative assessment of
the carbonation process. The increase of the diffusion resistance
from the concrete exterior and a diffusion of $Ca/OH/_2$ from the non
carbonated interior towards the carbonated marginal zone of the
concrete lead to a permanetly increasing deviation from the law of
$\sqrt{t}$ as the duration of carbonation increases. Using these two
variables a law of carbonation was derived by Schiessl [3] which
describes very well the actual advance of the carbonation:

$$x_\infty = \frac{D . /C_1 - C_2 /}{b} \qquad\qquad /3/$$

$$t = \frac{a}{b} \left[x + x_\infty . \ln /1 - \frac{x}{x_\infty} / \right] , \quad /4/$$

where

x = actual depth of carbonation /m/
x_∞ = final depth of carbonation /m/
D = diffusion coefficient /m^2. s^{-1}/
C_1 = concentration of CO_2 in the atmosphere /$g.m^{-3}$/
C_2 = concentration of CO_2 in concrete at the carbonation front
/$g.m^{-3}$/

a = quantity of CO_2 required for the carbonation of 1 m^3 of
concrete /$g.m^{-3}$/
b = retardation factor /$g.m^{-2}.s^{-1}$/; b ≐ 1,2 . 10^6 $g.m^{-2}$ .s^{-1}
t = carbonation time /years/.

Permeability and its Role in Predicting Structural Life of Reinforced Concrete Components

The service life of a concrete structure with regard to reinforcement
corrosion is devided into an initiation stage and propagation stage,
see Figure 2.

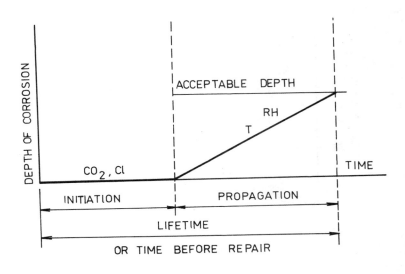

Figure 2: Schematic sketch of steel corrosion
sequence in concrete [4]

If carbonation is presupposed to act as the most harmful factor to
concrete at the first stage, then the length of initiation period can
be theoretically calculated: In analogy with the diffusion coefficient
we obtain flow curves of carbonation for various permeability
coefficients calculated from equation /3/ and /4/. With this curves
it becomes possible to assess the length of time it takes for the
carbonation to reach the reinforcement if the depth of concrete cover
is known.

The same procedure can be used in principle for calculating the
expected length of initiation period of concrete components treated
with appropriate carbonation retarders.

In an experimental program [5] the permeability coefficient K was
determined for concrete with different types of coatings. The
permeability coefficients are shown in Table 1.

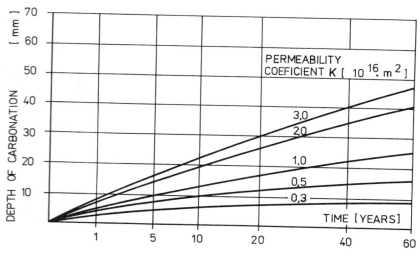

Figure 3: Calculated flow curves of carbonation
of various grades of concrete

TABLE 1   Permeability coefficient

	CONCRETE WITHOUT COATING K [m².10¹⁶]	ART OF COATING	CONCRETE WITH COATING K'[m².10¹⁶]	$\frac{K}{K'}$
A	2,0	SOKRAT 2804 - acrylic emulsion	0,7	2,8
B	2,0	UNIAKRYL - diss. acrylic resin	1,17	1,7
C	2,0	DISKOLOR - emulsion paint	1,4	1,4
D	2,0	LUKOFOB-water repelent solution on silicone resin basis	2,8	0,7

Each value in table 1 represents an average of 5 individual measurements. In Figure 4 are shown the assumed curves of carbonation of
concretes with and without coatings.

When corrosion has been initiated, the rate of attack is determined
both by the rate of the anode and cathode reactions. The cathode
process, which consumes oxygen, can be a limiting factor only in
that case when the diffusion coefficient falls below a value of
$0,3.10^{-8} m^2.s^{-1}$ [4] .This limitation takes place at relative humidities
in excess of about 90 % for low W/C ratios, while the value becomes
about 95 % for high W/C values. If uncertainty prevails concerning
the moisture state the following values can be assumed:
- indoors = no corrosion
- concrete constantly saturated with water = no corrosion
- CO_2 initiated corrosion   approx. 50 μm/year
- Cl initiated corrosion   approx. 200 μm/year.
The maximum attack is about 5 - 10 times larger than the above mean
attacks.

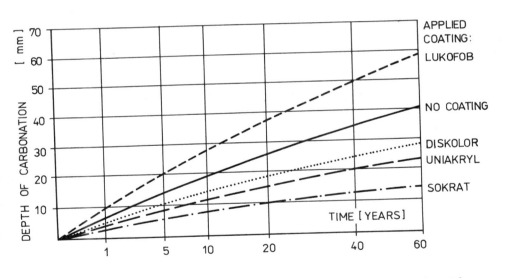

Figure 4: Calculated flow curves of carbonation
of concrete with and without coatings

CONCLUSIONS

The close relationship between the permeability and diffusion
coefficient simplifies to assess the life expectancy of reinforced
concrete components.

The length of the initiation period is very sensitive to the permeability coefficient and can be theoretically calculated with the aid of Eq. /3/ and /4/.

The laboratory results indicate that the permeability test can be used for comparing the effectiveness of coatings to resist various external aggressive agents. Experimental results show that water--repellent treatments /serie D/ slightly increased the permeability coefficient, while the other applied coatings affected a decreasing of the permeability coefficient.

REFERENCES

1   NISHER P. Improving the durability of structures – concrete – technological influencing factors, Betonwerk + Fertigteil – Technik, Number 5, 1987, pp 341–351.

2   LAWRENCE C D. Transport of Oxygen through Concrete. The British Ceramic Society Meeting, Chemistry and Chemically – Related Properties of Cement, Imperial College, London April 1984.

3   SCHIESSL P. Zur Frage der zulässigen Rissbreite und der erforderlichen Betondeckung im Stahlbetonbau unter besonderer Berücksichtigung der Karbonatisierung des Betons, W. Ernst und Sohn, Berlin 1976. pp 43–47.

4   TUUTTI K. Corrosion of steel in concrete, Swedish Cement and Concrete Resarch Institute, Stockholm 1982. pp 18 and 80.

5   BILČÍK J. Analysis of demages of reinforced concrete constructions and possibilities of their protection /in Slovac/, Slovac Technical University; Bratislava 1988, pp 25–37.

MECHANISMS OF WATER ABSORPTION BY CONCRETE

M Emerson

Transport and Road Research Laboratory,

United Kingdom

ABSTRACT. Data from a literature survey on the absorption of water by concrete are presented, together with the main results from some experimental work on water absorption. The literature survey highlights a diversity of approaches, in both the understanding of the basic concepts and the mathematical solutions. The capillarity approach for unsaturated conditions and the permeability approach for saturated conditions are discussed and compared. The concept of sorptivity associated with water flow in unsaturated concrete is also discussed, together with the variety of definitions found. The experimental work is associated with the measurement of the absorption of water by samples of concrete of different thicknesses. Absorption through both cut and as cast surfaces is investigated. The results are presented in terms of weights of water absorbed and distances of penetration. It is suggested that the sorptivity of cover concrete will give a better idea of the rate of water absorption than its permeability.

Keywords: Concrete, Water Absorption, Permeability, Sorptivity, Durability.

Mary Emerson is a Senior Scientific Officer at the Scottish Branch of the Transport and Road Research Laboratory, Craigshill West, Livingston, West Lothian, UK.

INTRODUCTION

One of the topics at present dominating the literature, thinking and finances associated with the maintenance of concrete bridges is the corrosion of reinforcement caused by the ingress of chloride contaminated water. In an uncontaminated state, the highly alkaline portland cement used in concrete encourages the formation of an oxide film on the surface of steel reinforcing bars. In theory, this film protects the reinforcement from corrosion. In practice, water contaminated with chlorides from de-icing salts used in winter maintenance, or from a marine environment, sometimes penetrates the concrete cover to the reinforcement. The protective oxide film is then destroyed, and the reinforcement corrodes.

Reports of bridges suffering from the problem of chloride contamination are numerous[1]. Remedial measures can be expensive, are sometimes very difficult to carry out, and frequently have no guarantee of long-term success. It is also possible that the presence of repair materials can alter the path of electrolytic corrosion currents within the surrounding concrete and actively encourage corrosion in previously sound areas.

Perhaps the most certain way to eliminate chloride contamination is to prevent contaminated water from getting into the concrete. In order to prevent ingress, it is first necessary to understand the mechanisms by which it occurs. Historically, permeability has been regarded as the property of concrete associated with the ease of passage of water. However, a literature survey on the absorption of water by concrete revealed that more recently the concept of sorptivity, a property associated with capillary effects, had begun to find favour.

The Paper describes some of the more important findings from the literature survey and from some experimental work on water absorption.

PERMEABILITY, POROSITY AND CAPILLARITY

In order that the reader may understand the terminology used in the literature survey, brief descriptions of permeability, porosity and capillarity, are given below. For the purposes of this Paper, the descriptions relate specifically to the movement of water through concrete.

Permeability is a property of concrete which is associated with the ease with which water passes through it. It is defined by D'Arcy's law which, when related to concrete, states that the steady flow of water through saturated concrete is proportional to the product of the cross-sectional area through which the flow occurs and the imposed hydraulic head, divided by the length of the flow path. The constant of proportionality is called the permeability of the concrete, or the coefficient of permeability.

Porosity is also a property of concrete. For the purposes of this

Paper it is defined as the ratio of the volume of the pores to the volume of the concrete.

In capillarity the force which drives the water results from the difference in pressure between the two sides of the menisci formed by water in the pores of the concrete. Thus capillary forces will not be present in fully saturated concrete and movement of water will cease because there is no longer a driving force.

It therefore follows that if water is brought into contact with unsaturated concrete, it will, even in the absence of an imposed force, be absorbed into the concrete by capillary action. If water is to move through saturated concrete, then an externally applied force, such as a head of water, is required.

Historically, the permeability of concrete was the criterion by which its durability was judged, low permeability being indicative of high resistance to the passage of water. Indeed, many researchers still regard permeability as the most important durability related property. This may be true for permanently submerged concrete, which is both saturated and subjected to a head of water, but it is difficult to see how it relates directly to concrete which is exposed to the environment, such as that in a bridge. In these latter conditions, the outer layers of the concrete are not likely to be permanently saturated, nor, unless in a tidal zone, are they subjected to a head of water. They will in fact be in a fluctuating wet/dry situation according to the prevailing weather. In these circumstances capillarity rather than permeability will control the passage of water.

THE LITERATURE SURVEY

According to a recent Department of Transport Report[1], the majority of damage to concrete bridges is caused by penetration of contaminated water through the cover to the reinforcement. Thus, when considering water ingress, the main interest must lie in the speed and distance of penetration of the water, and the mechanisms by which it penetrates.

In order to investigate the current situation regarding these aspects of water absorption, a literature survey was carried out. A full description of the survey may be found elsewhere[2]; a summary of some of the main findings is given below.

A number of researchers reported the results of measurements of water absorption by concrete. Unfortunately, there were considerable differences in the way they presented their results, and in the units used. For example, Fagerlund[3] discusses his results in terms of the increase in weight gain with time of small samples of concrete absorbing water from wet sponges. Ho & Lewis[4,5,6] sprayed water on to their samples and measured the distance of penetration of the water by visual inspection after breaking open the samples at various absorption times. Kelham[7] sealed all but one surface of his samples, suspended them from a load measuring beam, and then submerged

them in a tank of water. Other researchers used similar approaches,
e.g. Hall[8,9], Gummerson et al[10,11], Bamforth et al[12].

One common theme did emerge from the literature mentioned above. It
was that in the early stages the absorption process was governed by
the square root of time ($\sqrt{t}$). A linear relationship was observed
between $\sqrt{t}$ and the weight gain of a sample, or the distance/depth/
height of penetration of the water, or the volume of the absorbed
water, or the volume of the absorbed water per unit area of the
absorbing surface.

Here, however, the common theme ended. The slope of the line was in
general described by researchers as the sorptivity of the concrete,
irrespective of which of the above relationships was used. This
introduces a confusion of units. Other researchers, for example
Valenta[13], describe the slope of the line as the mean value of the
coefficient of permeability of the concrete. Examples of some of the
mathematical equations used to describe the water absorption process
are given in Table 1 below.

Table 1

Examples of mathematical equations

Fagerlund[3]

$t = mz^2$, where:

t = time,
m = resistance to water penetration,
z = depth of penetration.

Ho and Lewis[4]

$i = St^{1/2}$, where:

i = volume of water absorbed per
unit area of inflow surface,
S = sorptivity,
t = time.

Hall[8]

$i = St^{1/2}$, where:

i = cumulative absorption,
S = sorptivity,
t = time.

Bamforth et al[12]

(a) $x = \frac{r}{2}\left(\frac{P_o t}{\eta}\right)^{1/2}$, where:

x = distance travelled,
r = radius (mean) of capillaries,
P_o = driving pressure,
t = time,
η = viscosity of water.

(b) $K = \frac{r^2 \nu \rho g}{8\eta}$, where:

K = permeability,
r = pore radius,
ν = porosity,
ρ = density,
g = acceleration due
to gravity,
η = viscosity.

(c) $x = \sqrt{\frac{2kT(P_1 - P_2)}{\nu \rho g}}$, where:

x = penetration
distance,
k = permeability,
T = penetration time,
$(P_1 - P_2)$ = pressure
difference,
ν = porosity,
ρ = density,
g = acceleration due
to gravity.

Vuorinen[14]

$x = (2Kht)^{1/2}$, where:

x = depth of penetration,
K = coefficient of
permeability,
h = hydraulic head,
t = time.

These examples illustrate clearly the diversity of theoretical approaches to the absorption process. While each individual equation is no doubt mathematically correct, collectively they confuse and complicate the situation. The main reason for this is that some of the equations have been derived by applying D'Arcy's permeability law to unsaturated conditions, while others have been derived from the capillary approach. The distinction between saturated and unsaturated conditions must be made clear, for the forces/pressures necessary to move water through concrete, and the rates at which it moves are quite different.

In the same way that permeability is associated with the steady flow of water in saturated concrete, sorptivity can be associated with the unsteady flow of water in unsaturated concrete. Here the similarity ends. By definition, saturated concrete has only one moisture condition. In contrast unsaturated concrete can have a moisture condition varying from dry to nearly saturated. It must therefore be expected that the sorptivity will depend on the initial moisture state of the concrete. Hall, and Gummerson et al, emphasise this point, and also make clear the difficulties associated with the determination of moisture contents of concrete. They define sorptivity as the slope of the line obtained from a plot of volume of water absorbed per unit area of suction surface against the square root of the absorption time.

LABORATORY EXPERIMENTS

Although wide ranging work on the subject of water absorption by concrete has been carried out, no clear picture of a unified approach to the theory, experimentation, or units of measurement emerged from the literature survey. It was also evident that the majority of the experimental work had been carried out on relatively thin samples of concrete, whereas in practice concrete cover is an integral part of a much greater volume of concrete.

Another factor requiring clarification is the state of the suction surface of the concrete. For example, Fagerlund[3] states that the 'suction surface should normally be sawn and be free from carbonation and other contaminations'. Other researchers used suction surfaces as they were cast, i.e. with the outer skin of the concrete intact. Some researchers make no comment about the state of the suction surface.

In an attempt to understand the situation regarding the ingress of water, and to investigate the effect of sample size, state of suction surface, and extended periods of water absorption, some experimental work was carried out. A full description of this work may be found elsewhere[2]. Relevant details are summarised below.

Experimental Procedure

Samples from different batches of the same mix of concrete were cut from 508mm x 102mm x 102mm ordinary portland cement concrete beams which had been cast 6 years earlier and then stored under cover in an

unheated environment. Seven different thicknesses were cut, namely
25mm, 50mm, 75mm, 100mm, 150mm, 200mm, and 250mm. Six samples of each
thickness were tested, three of which absorbed water through a cut
surface and three of which absorbed water through an as cast surface.
The surface through which water was absorbed is referred to as the
suction surface.

After being cut from the beams with a water-cooled diamond saw, the
samples were left to dry until they achieved equilibrium with the
laboratory environment, i.e. their weights increased or decreased
slightly in accordance with the variations in temperature and humidity
of the air in the laboratory. The four sides of each sample were then
coated with two layers of an epoxy resin. This was to prevent
evaporation losses from the sides of the samples when they were
absorbing water, and to ensure that the water flow was one-dimensio-
nal. The bottom and top surfaces of the samples were not coated, thus
allowing water to be absorbed and evaporation to take place.

Each sample was then weighed and placed with its suction surface
resting on synthetic foam saturated with water contained in a tray.
The top of the foam was clear of the water so that the sides of the
samples were above the water level. The trays were placed under
polythene covers to restrict evaporation losses from the tops of the
samples.

After being placed on the wet foam, the samples were periodically
removed and weighed. Prior to each weighing the suction surface was
wiped with a damp sponge to remove excess water. This operation, and
the actual weighing, were carried out as quickly as was feasible in
order that the interruptions to the absorption process were as brief
as possible.

Results

In general there was little difference between the results of
replicate tests, and the data from one set of samples with cut suction
surfaces and another with as cast suction surfaces are sufficient to
represent the results.

Plots of weight gain for the first 9 hours of absorption time are
shown in Fig. 1a, and for the first 60 days in Fig. 1b. For the
samples with the cut suction surfaces, it can be seen from Fig. 1a
that, with the exception of the 25mm sample, the rate of weight gain
of all the samples is very similar. This indicates that the rate of
water absorption is not affected by sample thickness. The weight gain
of the 25mm thick sample was much the same as the others for the first
2 hours, but after this the rate of water absorption decreased. It is
suggested that this is because after approximately 2 hours the easily
accessible pores have been filled, and the water then takes longer to
penetrate the less accessible pores.

Comparison of the data in Fig. 1a for both cut and as cast suction surfaces shows that there is a significant reduction in the rate at which water is absorbed through an as cast suction surface. The 25mm sample with the as cast suction surface is still gaining weight at the same rate as the thicker samples.

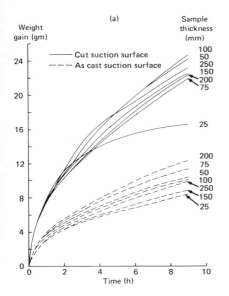
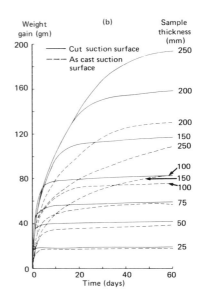

Fig.1 Weight gain v time : cut and as cast suction surfaces

On the extended timescale of Fig. 1b it can be seen that the weight gains of most of the samples eventually begin to level off. The weight gains of the samples with the cut suction surfaces increase in the order of increasing sample thickness, and, at saturation, are proportional to the sample thickness. The weight gains of the samples with the as cast suction surfaces do not follow such an obvious pattern.

The lack of a simple pattern in the results from the samples with the as cast suction surfaces makes analysis more difficult. However, the data from these experiments show that the rate at which water is absorbed is clearly significantly less than the rate at which it is absorbed through a cut suction surface. It should be remembered that the surface of a bridge in service will be an as cast surface.

Sorptivity

The work discussed in the literature survey had indicated that a linear relationship existed between weight gain and the square root of the absorption time. The data in Figs. 1a and 1b were accordingly plotted in this fashion. The results are shown in Figs. 2a and 2b. The linear relationship is well illustrated in both Figures for the

samples with the cut suction surfaces. It is not so evident on the
extended timescale for the samples with the as cast suction surfaces.

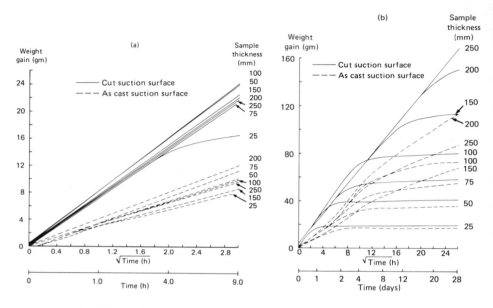

Fig.2 Weight gain v square root of time : cut and as cast suction surfaces

The term 'sorptivity' was mentioned in the literature survey, where it
had been used by various authors to define a variety of relationships
between water absorbed and the square root of the absorption time.
Hall, and Gummerson et al, defined sorptivity as the slope of the line
obtained from a plot of the volume of water absorbed per unit area of
suction surface and the square root of the absorption time. This
definition has two advantages. The first is that it involves an
easily measurable quantity - the volume of the water absorbed. (Other
definitions involved the distance of penetration of the water, which
can only be determined by breaking open the concrete samples.) The
second is that the quantity of absorbed water becomes independent of
the area of the suction surface. For these reasons, the Hall, and
Gummerson et al, definition has been adopted.

The mean sorptivity value (taken from the results for the first 9
hours) of the samples with the cut suction surfaces used in the experimental work was $0.72 \text{mm/h}^{0.5}$. The mean 'apparent' sorptivity of the
samples with the as cast suction surfaces (again taken over the first
9 hours) was $0.32 \text{mm/h}^{0.5}$. Unfortunately none of the results relating
to sorptivity found during the literature survey were directly
comparable. However, work published recently by Hall and Yau[15]
lists sorptivities for a variety of concrete mixes. Values given for
the concretes nearest in composition to that used in the experimental
work described in this Paper are $0.73 \text{mm/h}^{0.5}$ and $0.93 \text{mm/h}^{0.5}$.

Of equal interest to actual values of sorptivity is the fact that the sorptivity of samples of concrete absorbing water through as cast suction surfaces is less than one half of the sorptivity *of samples of the same concrete* absorbing water through cut suction surfaces. This difference is so significant that it is suggested that the composition of the skin of concrete warrants considerably more attention than it has received to date[16,17,18]. It is also of interest to note that the skin of the concrete is not always a barrier to water, because Kreijger[16] reports increased rates of water absorption through as cast surfaces.

Comparison of experimental and theoretical conditions

The porosity of the concrete can be determined from the saturated weight gains shown in Fig. 1b. Thus the distance of penetration of the water at any instant may be calculated from the corresponding measured weight gain (assuming that the water penetrates as a well defined front). The results, for the 250mm sample with a cut suction surface, are shown in Fig. 3. In ideal conditions, the weight gain would remain proportional to the square root of time until saturation was achieved[3]. In these circumstances distances of penetration, and theoretical weight gains can be calculated, as shown in Fig. 3. For ease of comparison, measured weight gains from Fig. 1b are also given.

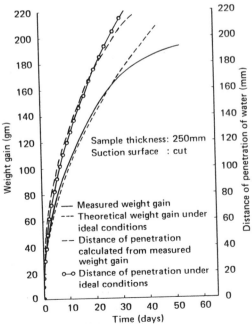

Fig.3 Comparison of measured and theoretical weight
gains and distances of penetration of water

It can be seen from Fig. 3 that the measured and calculated weight gains begin to diverge when the water has penetrated to approximately three quarters of the full height of the sample. Similar relationships exist for the other six thicknesses of sample.

Rates of water absorption/flow

Consider next the rates of absorption. It can be seen from Fig. 1a that in the first hour of the absorption process, the weight of water absorbed by the 25mm sample with the cut suction surface was 8gm, which is equivalent to an average rate of flow of $2.2 \times 10^{-9} m^3/s$. Classical permeability theory would suggest that a head of water in excess of 1600m was required to produce this flow through good quality concrete (with an assumed permeability of $10^{-12} m/s$). Clearly then permeability is playing no significant part in the mechanism of water absorption observed in these experiments.

CONCLUSIONS

1. The literature survey highlights a diversity of approaches to the theoretical and experimental aspects of the absorption of water by concrete. The main reasons for this are a lack of distinction between saturated and unsaturated flow, and a lack of adherence to the classical concept of permeability.

2. Except in saturated conditions, capillary forces will govern the process of water absorption, and it is the sorptivity of cover concrete which will control the rate of ingress of water, not its permeability. The concept of sorptivity warrants further investigation.

3. The effect of the skin of the concrete also warrants further investigation. In the experiments described in this Paper, its presence reduced considerably the rate of water absorption. Kreijger reports an increase in the rate of water absorption through the skin.

ACKNOWLEDGEMENT

The work described in this Paper forms part of the programme of the Transport and Road Research Laboratory and the Paper is published by permission of the Director.

REFERENCES

1.   WALLBANK E J. The performance of concrete in bridges: a survey of 200 highway bridges. A report prepared by G Maunsell & Partners for the Department of Transport. HMSO, London, April 1989.

2.   EMERSON, MARY. The absorption of water by concrete: the state
     of the art. TRRL Working Paper No. WP/SB/1/90. Transport and
     Road Research Laboratory, Scottish Branch, Livingston, April
     1990.

3.   FAGERLUND G. On the capillarity of concrete. Nordic concrete
     research. 1982. Oslo. Publication No. 1. December.

4.   HO D W S, LEWIS R K. Water penetration into concrete - a measure
     of quality as affected by material composition and environment.
     Symposium on concrete. 1983. Perth (Australia), 20 - 21
     October.

5.   HO D W S, LEWIS R K. The water sorptivity of concretes: the
     influence of constituents under continuous curing. Durability of
     Building Materials. 1987. Vol. 4, pp 241 - 252.

6.   HO D W S, LEWIS R K. Concrete quality after one year of outdoor
     exposure. Durability of Building Materials. 1987. Vol. 5, pp 1
     - 11.

7.   KELHAM S. A water absorption test for concrete. Magazine of
     Concrete Research. 1988. Vol. 40, No. 143.

8.   HALL C. Water movement in porous building materials - I.
     Building and Environment. 1977. Vol. 12, pp 117 - 125.

9.   HALL C. Water movement in porous building materials - IV.
     Building and Environment. 1981. Vol. 16, No. 3, pp 201 - 207.

10. GUMMERSON R J, HALL C, HOFF W D. Water movement in porous
     building materials - II. Building and Environment. 1980. Vol.
     15, pp 101 - 108.

11. GUMMERSON R J, HALL C, HOFF W D. Capillary water transport in
     masonry structures: building construction applications of
     D'Arcy's law. Construction Papers. 1980. Vol. 1, No. 1.

12. BAMFORTH P B, POCOCK D C, ROBERY P C. The sorptivity of
     concrete. Conference: Our World in Concrete and Structures.
     1985. Singapore, 27 - 28 August.

13. VALENTA O. From the 2nd RILEM Symposium: Durability of
     concrete: in Prague. Materiaux et Constructions. 1970. Vol.
     3, No. 17, pp 333 - 345.

14. VUORINEN J. Applications of diffusion theory to permeability
     tests on concrete, Part 1: depth of water penetration into
     concrete and coefficient of permeability. Magazine of Concrete
     Research. September, 1985. Vol. 37, No. 132.

15. HALL C, YAU M H R. Water movement in porous building materials -
     IX. The water absorption and sorptivity of concretes. Building

and Environment. 1987. Vol. 22, No. 1, pp 77 - 82.

16. KREIJGER P C. The skin of concrete: composition and properties.
     Materiaux et Constructions. April 1984.

17. KREIJGER P C. The 'skins' of concrete - research needs.
     Magazine of Concrete Research. September 1987. Vol. 39, No.
     140.

18. SENBETTA E, SCHOLER C F. Absorptivity, a measure of curing
     quality as related to durability of concrete surfaces. Paper
     from a Conference held in Maryland, USA, 1981. pp 153 - 159.

PROTECTION THROUGH DESIGN

Session B

Chairmen

Dr W G Smoak

Division of Research and Laboratory Services,

United States Department of the Interior,

USA

Prof P Spinelli

Department of Civil Engineering,

University of Florence,

Italy

IMPROVING THE PERFORMANCE OF CONCRETE IN TRUNK ROAD BRIDGES IN SCOTLAND

R R Johnstone

Scottish Development Department,

United Kingdom

ABSTRACT.   Concrete elements in trunk road and motorway bridges throughout Scotland have been found to be deteriorating. The commonest problems are reinforcement corrosion and spalling, cracking and staining of concrete.   This deterioration can be caused by a number of physical, chemical and environmental factors, acting alone or in combination.   Chlorides from de-icing salt are the principal cause of concrete deterioration in trunk road bridges in Scotland. This Paper describes the measures that are being taken to secure the life of existing bridges and to achieve greater durability and better performance in new construction.

Keywords: Bridges (Structures), Concrete, Deterioration, Reinforcement Corrosion, Chlorides, Durability, Inspection, Maintenance, Design Requirements.

Raymund R Johnstone is a Senior Engineer in the Scottish Development Department Roads Directorate, New St Andrew's House, Edinburgh.

INTRODUCTION

The Scottish Development Department's Roads Directorate manages and maintains the trunk road all-purpose and motorway network on behalf of the Secretary of State for Scotland.

The Department is a major bridge owner, being responsible for some 1545 bridges with spans greater than 3 metres on the trunk road network. Of these, approximately 1300 are of reinforced concrete, prestressed concrete or composite concrete/steel construction and as many as three quarters of these have been built since 1961.

Reinforced and prestressed concrete are economic and durable forms of construction for bridges provided they are properly designed, constructed and maintained. Sadly, in recent years deterioration has been found in many of the Department's bridges, sometimes at an early stage in their design life. The most serious problem is reinforcement corrosion arising from contamination by chlorides from de-icing salt. Although an immediate threat to safety is unlikely, if undetected or if action is not taken promptly when discovered then concrete deterioration can threaten the serviceability of structures and greatly reduce their design life.

This paper describes how it is hoped to achieve greater durability in highway structures using the experience gained from inspection and maintenance of a variety of existing trunk road bridges in Scotland and by adoption, where appropriate, of the recommendations contained in recent research reports, in particular the report prepared for the Department of Transport by G Maunsell and Partners.[1]

DETERIORATION OF CONCRETE IN TRUNK ROAD BRIDGES IN SCOTLAND

The Common Problems

The commonest problems found in the concrete elements of trunk road bridges in Scotland are:

-   Corrosion of reinforcement and prestressing tendons

-   Cracking, spalling and rust staining of concrete, usually associated with reinforcement corrosion

-   Scaling of concrete

-   Disintegration of the top surfaces of decks under the asphalt surfacing

-   Water staining and leaching

The most serious and widespread problem is reinforcement corrosion. Corroding reinforcement has been found in all the different concrete elements of the Department's bridges. There are two types of corrosion that occur in reinforced and prestressed concrete; general and localised.(2) General corrosion is an expansive process which leads to cracking, spalling and rust staining of the concrete. These signs of general corrosion are usually visible long before structural weakening has occurred. Of greater concern is localised or pitting corrosion which can cause a serious loss of section from a reinforcing bar without any evidence appearing on the surface of the concrete. Consequently, it is more difficult to detect. This type of corrosion has been found in bridge decks and in the top sections of supports below deck joints.

Concrete cracking is common and can be due to a variety of reasons other than reinforcement corrosion such as structural stresses, thermal movement and shrinkage. Cracking can accelerate existing deteriorative processes and initiate new ones.

Scaling is another type of surface deterioration found on some bridges. It reduces cover thus increasing the risk of reinforcement corrosion. The worst cases of scaling have been found on parapet upstands.

It is of great concern that the top surfaces of the decks on some of the Department's bridges have been found to be disintegrating. This problem is not usually detected during routine inspection and may only be discovered when the surfacing has been removed as part of carriageway resurfacing operations or when the asphalt surfacing breaks up as a consequence of the disintegration of the concrete.

Leakage of water, often containing chlorides, through movement joints and through cracks and construction joints in concrete decks where waterproofing is absent or has failed is a major problem on a number of bridges. It can cause unsightly water runs or staining on soffits, abutments and piers and can leach out lime compounds from hardened concrete which can cause efflorescence, incrustation and the formation of stalactites. Apart from its adverse affect on the appearance 'of structures', leakage can promote reinforcement corrosion by contaminating the concrete with chlorides.

The Common Causes

The common causes of concrete deterioration in the Department's bridges are:

- Chloride attack

- Frost damage

- Inadequate specification

- Design and detailing faults

- Poor workmanship and construction

Most deterioration has been caused by chlorides penetrating the concrete and breaking down the passive oxide film, which normally forms on the surface of the reinforcing bars in the alkaline environment of the surrounding concrete and initiating progressive corrosion. De-icing salt used during the winter months is the main source of chlorides in the concrete of bridges. Contamination is caused by the wetting of concrete surfaces by salt spray thrown up by passing traffic and by leakage of salt laden water at defective deck joints and malfunctioning drainage systems and into concrete decks where waterproofing is absent or has failed. Sea water and wind-borne chlorides at coastal sites are another source of chloride contamination.

Saturated concrete is prone to damage by frost action especially when high chloride levels are present. Scaling of parapet edge beams and the disintegration of top surfaces of concrete bridge decks where waterproofing is absent or defective are generally attributed to the alternate cycles of freezing and thawing associated with frost action. This disruption of the concrete makes it easier for chlorides to reach the reinforcement and for corrosion to occur.

Previous specifications for materials and workmanship were more concerned with providing satisfactory performance in terms of the strength of concrete structures. Unfortunately the need to provide for the durability of the structures was not sufficiently recognised. Consequently, lesser standards were specified for many of the properties considered to provide durability, for example, cement content and water/cement ratio, and there is no doubt that this has had an adverse effect on the performance of concrete in bridges.

Poor design and detailing have caused or contributed to concrete durability problems on some bridges. Inadequate design for structural, thermal and shrinkage stresses has resulted in excessive concrete cracking in some structures which has led to deterioration of the concrete. In many bridges little or no provision has been made for inspection and maintenance in areas such as bearing shelves and deck ends and this has resulted in concrete deterioriation.

Bad workmanship and construction permitted by inadequate supervision have had an adverse effect on concrete durability in a number of structures. Two common faults in particular have contributed to concrete deterioration in trunk road bridges; low cover and porous concrete. Inadequate cover and porous concrete reduce the protection normally given by the outer layer of concrete making it easier for the ingress of harmful agents, principally chlorides, which cause corrosion.

The above factors, acting alone or in combination, are the main causes of concrete deterioration in the Department's bridges.

The Maunsell Report - Its Relevance to Scotland

In August 1986, the Department of Transport appointed consulting engineers G Maunsell and Partners to undertake a survey of some of the concrete elements of a random, but representative sample, of the trunk road bridges in England, to obtain data on their condition so that a reasonable forecast of bridge maintenance requirements could be made. The study involved the visual inspection and limited testing of 200 trunk road and motorway bridges. The findings, together with recommendations for future bridge maintenance needs, a review of current design standards and specifications and recommendations for future research, were published in April 1989.(1)

The test results for chloride concentration and half-cell potential values were used to classify the bridges into good, fair and poor condition. Of the random sample of 200 bridges studied, 41 were found to be in 'poor' condition, meaning that significant repairs were required. Only 59 bridges were classed as 'good', the remaining 100 being classed as 'fair'. The study confirmed that chloride contamination was the main cause of deterioration in bridges.

The study's findings and recommendations are very relevant to bridges in Scotland. The commonest defects found by the researchers in England are the same as those found in bridges in Scotland, and the principal cause is the same; chloride contamination from de-icing salt. This is not surprising as most trunk road and motorway bridges in Scotland and England are of the same types, constructed to the same specification and standards and are of a similar age to those investigated. In fact, the potential for corrosion may be even greater in Scotland as the survey found that the bridges investigated in the North of England were visually in worse condition than those in the South and contained higher chloride levels. This was attributed to the Northern bridges being subjected to more salt than the Southern ones because of the more severe climate, although this was not verified by comparing salting rates.

NEW CONSTRUCTION

Much can be done and has already been done to improve the durability of new bridges in Scotland.

Departmental Documents

The Department's requirements for design, materials and workmanship for bridge construction are set down in comprehensive design standards and in the Specification for Highway Works.

Design standards and specifications are continually reviewed and updated taking account of feedback on bridge performance, from inspections and maintenance works and from new information being produced by organisations such as the Government's main research body for roads and bridges, the Transport and Road Research Laboratory. Further changes are likely as some of the recommendations of the Maunsell Report are implemented.

Current requirements for the design of concrete bridges are given in BS 5400:Part 4(3) and related technical memoranda. In addition to the ultimate limit state requirements for the strength of structures, comprehensive requirements for the serviceability limit state are given. The design of a structure should comply with these requirements to ensure that it does not suffer local damage which would shorten its intended life or would incur excessive maintenance costs. In particular maximum permissible crack widths and minimum cover requirements are given for different exposure conditions.

The Specification for Highway Works(4) published in 1986 and subsequent amendments contain the requirements for materials and workmanship for bridges. It was drawn up with durability as a principal aim and incorporates a number of important measures which were not in the previous version including:

- Permitting the use of composite cements for concrete

- Specifying the use of air entrained concrete wherever the concrete is liable to freezing when saturated

- Specifying minimum cement content

- Specifying maximum water/cement ratios

Improving Durability and Reducing Maintenance Problems by Good Detailing

Many existing bridges were designed on the assumption that concrete was a maintenance-free material. The shortcomings of this approach have become very apparent and the Department in its technical approval role has been particularly keen to ensure that the knowledge gained from inspections and maintenance is fed back into the design process to avoid repeating past mistakes.

Many of the common problems relate to deck joint, waterproofing and drainage inadequacies.

Leaking deck joints have been a major cause of problems.   The problems are very much worse in areas of bridges where inspection and maintenance is difficult, for example, on bridges of cantilever and suspended span construction.   In some cases, it has been necessary to lift the entire suspended span to carry out inspection and repairs to the areas of the half joints which were not visible. Consequently, this form of construction has not been approved by the Department for some years.

The design of continuous structures has been encouraged to minimise the number of deck joints and thus improve durability. For bridges with spans under 10 metres, it is recommended that designers should endeavour to dispense with all joints, and should adopt simply supported spans without curtain walls, box culverts or portal frames for the design of these smaller structures.

Where joints have to be used, the designer should assume that leaks will occur, and a system of positive drainage should be provided.   In recent years the Department has recommended that wherever possible abutment galleries should be provided under bridge joints to facilitate the inspection and maintenance of joints, bearings, curtain walls and deck ends, and to collect and remove the inevitable leakage.

The waterproofing of bridge decks became mandatory for trunk road and motorway bridges in Scotland in 1971.   Waterproofing systems are applied over bridge decks to prevent water containing chlorides penetrating the concrete substrate and corroding the embedded steel reinforcement.   Sadly there is evidence that many of the waterproofing systems applied have not performed effectively in service.   Recently spray applied acrylic and urethane systems have been coming into use and trials by the Transport and Road Research Laboratory indicate that these may offer advantages over some of the earlier waterproofing systems.(5)

Structures are often located at or near the top or bottom of vertical curves on road alignment.   Consequently, the decks on many bridges have inadequate falls and do not shed water effectively resulting in pools of standing water forming.   It is important when designing structures that adequate falls and drainage are provided to ensure that chloride contaminated water is removed as effectively as possible.

Other Measures to Improve the Performance of Concrete

Experience in other countries has shown that impregnation of concrete surfaces with silane provides effective resistance to penetration by water containing chlorides from de-icing salt and in marine environments.(6)   It is therefore proposed to treat all vulnerable concrete surface on new concrete structures before they come into service.   The Specification for Highway Works will be amended to include silane impregnation.

Epoxy coated reinforcement has been widely used in Canada and the USA for a number of years and appears to have been successful in protecting steel from the corrosion which has caused so many of the problems in concrete structures in the UK.(7)

It has recently been approved by the Department for use on a trial basis in two bridges under construction in very aggressive environments.   Epoxy coated reinforcement is more expensive than normal reinforcement but if it proves to be successful in preventing corrosion then it may be adopted more often in future.

INSPECTION AND PREVENTATIVE MAINTENANCE

Various measures are proposed to arrest the deterioration of concrete in existing bridges and to improve inspection and maintenance procedures.

Improving Inspection Procedures

Periodic inspections are an essential part of the bridge maintenance process.   They provide a record of the condition of bridges and identify defects and also provide a valuable feedback of information to enable design and construction practices to be improved.

The Department's procedures for inspection and reporting on trunk road and motorway bridges are set down in Technical Memorandum SB1/78.(8)   General inspections are required at intervals not exceeding 2 years and the more rigorous Principal Inspections, which involve close examination of all parts of the structure, at intervals not exceeding 6 years.

General and Principal Inspections have up till now been limited to visual examination of the structure.   However, visual inspection alone cannot usually detect areas of potential corrosion or areas where corrosion is occurring without any visible evidence on the surface of the concrete.

In recognition of this it is the Department's intention to implement the recommendation contained in the Maunsell Report that the scope of Principal Inspections should be extended to include a limited amount of testing.   This should enable those areas requiring further, more detailed investigation to be identified.

Preventative Maintenance

A high priority is being given to preventative maintenance because of the serious damage being caused to reinforced and prestressed concrete structures early in their design life by chlorides from de-icing salts and marine environments. Substantially increased expenditure is planned over the next few years on measures to combat the deterioration occurring in the concrete elements of highway structures.

Preventative maintenance involves a major programme of silane impregnation of vulnerable concrete surfaces, where suitable, on highway structures already in service. Bridge deck joints are also to be given a high priority for inspection and, if defective, repair or replacement.

It is proposed to protect all concrete surfaces on existing bridges which are exposed to chloride contaminated spray or leakage by impregnating them with silane, unless chloride levels and half-cell potential measurements indicate that there is a high probability of corrosion having commenced. However, as a first priority, those structures which have been in service for not more than 6 winters are to be impregnated as soon as practicable without prior testing to determine the likelihood of corrosion having started. It is estimated that 150 bridges on trunk roads and motorways in Scotland will qualify for this immediate treatment.

The evidence of chloride contamination due to leaking joints shows how important it is to keep all joints as watertight as possible. Every effort should be made to repair leaking joints or replace if necessary.

In cases where it is not possible to seal the joint or to provide some other means for removing salt-laden water, then the bearing shelves and drainage systems should be regularly maintained.

CONCLUSION

.The repair of deteriorated concrete in highway structures is expensive, can be technically difficult and can cause great disruption to traffic. The methods and materials for the repair of concrete structures form a major topic which I have not attempted to cover in this Paper.

Now that we have identified the major causes of deterioration in concrete structures, in particular the chlorides from de-icing salt, the developments described in this Paper and the advances in concrete repair technology give cause for optimism that structures currently being designed and constructed will have greater durability and that we shall be able to extend the lives of existing structures.

ACKNOWLEDGEMENT

The Author would like to thank Mr J Dawson, Director, Scottish Development Department Roads Directorate, for his permission to write this Paper.

REFERENCES

1. WALLBANK, E J. The Performance of Concrete in Bridges. A Survey of 200 Highway Bridges. A report prepared for the Department of Transport by G Maunsell & Partners, HMSO, London, 1989.

2. VASSIE, P R. Reinforcement corrosion and the durability of concrete bridges. Proceedings, Institution of Civil Engineers, Part 1, 76, August 1984, pp 715-716.

3. BRITISH STANDARDS INSTITUTION. BS 5400 Steel, concrete and composite bridges - Part 4:1984, Code of practice for design of concrete bridges, London, 1984.

4. SCOTTISH DEVELOPMENT DEPARTMENT. Specification for Highway Works. HMSO, London, August 1986, and Appendix L, March 1988.

5. PRICE, A R. A field trial of waterproofing systems for concrete bridge decks. TRRL Research Report 185, Transport and Road Research Laboratory, Crowthorne, 1989.

6. WALLBANK, E J. The Performance of Concrete in Bridges. A Survey of 200 Highway Bridges. A Report prepared for the Department of Transport by G Maunsell & Partners, HMSO, London, 1989, p 63.

7. INSTITUTION OF CIVIL ENGINEERS. Coatings for concrete and cathodic protection. Report of a mission sponsored by the ICE and supported by the Department of Trade and Industry, Thomas Telford, London, 1989, pp 26-29.

8. SCOTTISH DEVELOPMENT DEPARTMENT. Technical Memorandum (Bridges) SB1/78, The Inspection of Highway Structures, SDD, Edinburgh, July 1978.

DETAIL OF STRUCTURAL DESIGN FOR PREVENTING FROST DAMAGE OF CONCRETE USED IN ROAD BRIDGES

T Fujiwara

K Katabira

H Kawamura

Iwate University,

Japan

ABSTRACT. Concrete structures in the northern regions of Japan have a higher risk of deterioration due to freezing and thawing coming from the cold climate and heavy snows. The laboratory evaluation of resistance of concrete to freezing and thawing is studied actively in Japan, but there is still much left to be studied hereafter about the deterioration of concrete structures exposed in the field. A survey of the deterioration of concrete used in road bridges was made in Iwate Prefecture, a district in northeastern Japan. The number of bridges examined in this survey was 382. Many of the bridges were in greater or lesser degree damaged by freezing and thawing. The degree of deterioration differed according to the different parts of the bridges. Although the main cause of deterioration is presumed to be the result of poor construction, it can be pointed out that the lack of consideration for design of these structures adds significantly to their deterioration. In this paper, some designing methods for preventing frost damages of concrete are listed, based on the results of the survey.

Keywords: Frost Damage, Concrete, Road Bridges, Surveys, Structural Design, Detail, Preventing Method

Dr Tadashi Fujiwara is Associate Professor at The Department of Civil Engineering, Iwate University, Morioka, Japan. He has made researches on the durability of concrete to freezing and thawing actions, the mechanism of drying shrinkage of concrete and the effect of aggregate on the properties of concrete.

Mr Kunishige Katabira is Research Assistant at The Department of Civil Engineering, Iwate University, Morioka, Japan, where he has undertaken experiments and surveys on the durability of concrete.

Mr Hiroji Kawamura is President of RC Structure Design Co., Morioka, Japan. His concern is to design the durable concrete structures.

INTRODUCTION

Concrete structures in the northern regions of Japan have a higher risk of deterioration due to freezing and thawing coming from the cold climate and heavy snows.

Many researches on the deterioration have been done in our country so, that the mechanisms which cause the deterioration have become clear to a certain extent and some methods of preventing the deterioration are proposed. These researches resulted in recomendations for producing durable concrete such as the standard specification of the Japan Society of Civil Engineers. The specification recommends that air entrainment agent is sure to be used in concrete located in regions with severe climatic conditions, and that the maximum water-cement ratio is limited for different types of construction and exposure conditions.

In spite of these specifications, quite a number of concrete structures are damaged by freezing and thawing action. There is little doubt that the deterioration is still one of the major problems of concrete industry in the northern regions of Japan.

In order to establish methods for producing durable concrete, it seems to be necessary to make a detailed investigation on the deterioration of exposed concrete structures in the field, besides laboratory experiments. Such investigation will make clear the actual reasons for deterioration and there is the possibility of finding new methods of preventing deterioration. However, there are very few such investigations in our country.

In this investigation, we examined the deterioration of concrete used in road bridges and tried to understand the actual situation of the damages. Through this investigation, we found many points which should be considered in the designing of bridges in order to prevent deterioration of concrete due to freezing and thawing. The structural details of each part of the bridge are listed in this paper.

OBJECTS OF INVESTIGATION

Objects of this investigation were 382 road bridges in Iwate, a prefecture in Northeast Japan, as shown in Fig.1. There are two north-south mountain ranges, with a plain between and seaside in the east. The mountain range in the west is known as region of heavy snow, the range in the east as an area of extreme cold. Considering these geographic and climatic conditions, this prefecture seems to be suitable for investigation of frost damage of concrete.

Before starting the investigation, we tried to obtain data about design and construction of each bridge. Unfortunately these data were not preserved any more in the institution concerned, except for general

Fig. 1   Observed region

Table 1   State of Deterioration

degree	state	depth
1	cracking	-
2	light scale	~ 5mm
3	medium scale	5 ~ 20mm
4	heavy scale	20 ~ 50mm
5	severe scale	50mm ~

Table 2   Area of deterioration

degree	area	
a	partially	~ 5%
b	conspicuously	5 ~ 15%
c	broadly	15% ~

Table 3   Classification of deterioration

	area		
state	a	b	c
1	light	light	medium
2	light	light	medium
3	light	medium	heavy
4	medium	heavy	heavy
5	medium	heavy	heavy

Table 4   Damage rate

part of bridge		damage rate (%)			
		light	medium	heavy	total
newel post	exposed concrete	8.0	5.3	6.1	19.4
	finishing with mortar	20.6	13.8	9.2	43.6
	total	12.4	8.2	7.1	27.7
handrail		10.3	10.3	25.6	46.2
curb		14.1	8.3	11.1	33.5
slab		1.8	2.1	3.9	7.8
beam	reinforced concrete	6.4	4.3	4.3	15.0
	prestressed concrete	5.1	1.7	1.7	8.5
	total	5.7	2.8	2.8	11.3
abutment		6.5	7.1	6.0	19.6
pier	jutting out	15.3	11.9	3.4	30.6
	within bridge width	7.0	5.3	2.6	14.9
	total	9.8	7.5	2.9	20.2

drawings of the bridges. Because of the lack of basic data such as materials used and mix proportion, which could have explained existing deteriorations caused by freezing and thawing, the investigation had to be restricted to visual observation on the spot.

OBSERVED DETERIORATION

Many of the observed bridges showed signs of damage due to freezing and thawing. Some of them seem to have been repaired because of such damage. Altogether we can say that preventing deterioration due to freezing and thawing is very important in this prefecture.

According to the visual observations, the state of deterioration was classified as shown in Table 1 and the area of deterioration as shown in Table 2. The degree of deterioration is estimated as shown in Table 3 in accordance with the state and area of deterioration. Table 4 shows the damage rate of each part of bridge. The damage rate represents the ratio between the number of parts with deterioration caused by freezing and thawing and the total number of observed parts. Characteristic deterioration of the structural parts of the bridges was as follows.

The total damage rate of the newel post amounts to 28%. The newel post is structurally not an important part of the bridge, but at the "face" of the bridge the occurrence of deterioration gives an unfavorable impression to the users of the bridge. The newel post showed various forms of deterioration. In Fig.2, there is a sample of D-line crack, its characteristic being running parallel to the edge. The cracks develop into scaling and end in disintegration. This photograph shows the process very clearly. In the case of finishing with mortar, there occurs vertical and lateral cracking on the mortar as shown in Fig.3. In many cases the bond to the concrete is lost and the mortar falls. As is shown in Table 4, the damage rate of mortar finishing is considerably larger than that of exposed concrete.

The damage rate of handrail was 46%, it is the part most susceptible to damage. As a part that attracts the attention of the users, it gives a feeling of uneasiness in the case of conspicuous damage. As is shown in Fig.4, there is severe damage of the upper portion of the handrail.

The damage rate of the curb was 34%, and as with the handrails the damage was very severe. As is shown in Fig.5, there is considerable damage at places, which frequently are in contact with water. Although this part is structurally not so important, it must be feared that deterioration of the curb extends to the slab and beam.

The damage rate of the slab and beam, the most important structural parts of the bridge, was 8% and 11% respectively, relatively low as could be expected. The casting of the concrete was presumably done more consciously and there is less contact with water compared with other parts. But, as is shown in Fig.8 ~ 11, there are also extreme cases of damage, such as the concrete peeling off and the reinforcement being exposed. If the deterioration has reached that state, appropriate repair works are necessary in order to guaranty the function of the bridge.

The damage rate of the abutment and pier was 20%. As Fig.16 shows, the part, where snow heaps up, melts and flows down, is susceptible to deterioration. If the damage has reached the shoe of the beam, the overall function of the bridge is endangered.

DETAIL OF STRUCTURAL DESIGN FOR PREVENTING FROST DAMAGE

The resistance of concrete structures to freezing and thawing is basically determined by the quality of the concrete itself, but in many cases deterioration could be avoided by careful consideration in design. Damages due to freezing and thawing can only occur where there is contact with water. In other words, it can be expected that deterioration is prevented by appropriate design which cuts off the contact with water.

From this standpoint, we found 63 points which should be considered in the designing of the bridges in order to prevent deterioration of concrete. Main points among these are listed in Table 5, which contains the state and cause of deterioration together with the detail of structural design for preventing frost damage.

Table 5 Detail of structural design for preventing frost damage

1) Newel post

a)	state	D-line cracking. Cracks can develop into scaling at edges. See Fig. 2.
	cause	Stress concentration due to freezing and thawing at edges.
	detail	Adequate beveling to prevent stress concentration at edges.
b)	state	Peeling of surface mortar. See Fig. 3.
	cause	Water permeates into the boundary between surface mortar and concrete through cracks of mortar. Bond between both materials is lost by expansion of freezing water.
	detail	Mortar finishing is susceptible to frost damage. See Table 4. Exposed concrete is more durable.

2) Handrail

a)	state	Degradation of upper portion. See Fig. 4.
	cause	Snow piles up on the top of handrail and the degree of water content within concrete at this portion increases.
	detail	To avoid shapes which snow is apt to pile on the top. Round shapes are better.

b)	state	Deterioration of post. Post is made of concrete and railing is made of metal.
	cause	Water permeates into concrete at the contact point between concrete post and metal railing. Frozen water causes damage of concrete.
	detail	To avoid using concrete and metal in combination.

3) Curb

a)	state	Severe damage all over curb.
	cause	Curb is susceptible to climatic conditions such as melting snow and solar radiation.
	detail	To cover overall curb with an impermeable material such as metal.
b)	state	Deterioration of upper face of curb.
	cause	Thickness of pavement is increased by overlay, so that the height from the bridge face to the upper face of curb decreases.
	detail	To keep appropriate height, h in Fig. 17, considering future overlay.
c)	state	Deterioration which developed from cracking.
	cause	Cracking is caused by drying shrinkage. Water permeates through the cracks and frozen water causes deterioration of curb concrete.
	detail	To increase longitudinal reinforcing bar for preventing such cracks. See Fig. 18. To equip contraction joint with adequate space.
d)	state	Considerable damage near drainage. See Fig. 5.
	cause	Melting snow gathers at drainage and increases the degree of saturation of concrete surrounding the drainage because of drain blockage.
	detail	Structure of drainage which can easily drain water. Large opening of screen is better.
e)	state	Cracking in curb concrete at the foot of post of handrail. See Fig. 6
	cause	Swelling of water which enters from the joint of handrail into the post of handrail and freezes. If the joint is placed exactly over the post, the risk of damage is heigher.
	detail	To avoid joint being placed exactly over post. To equip drainage as shown in Fig. 19. To adopt a type of post which water does not permeate, such as H.
f)	state	Deterioration of side of curb. See Fig. 7.
	cause	Snow heaps upon the pipe installed along the side of bridge.
	detail	To inatall pipe under bridge.

4) Slab

a)	state	Damage of under face of overhanging. See Fig. 8.
	cause	Water flows from bridge face to slab because of inadequate water drip and shape of curb and slab.
	detail	Proper water drip and shape of curb and slab. See Fig. 17. The gradient θ is desirable to be as small as possible.
b)	state	Deteraioration which developed from cracking.
	cause	Cracking is caused by overload.
	detail	Design to prevent the occurence of large structural cracking.
c)	state	Deterioration of under face of overhanging.
	cause	Water permeates from base course of sidewalk to under face of slab through structural cracking.
	detail	To prevent occurence of structural cracking. To equip drainage as shown in Fig. 20.
d)	state	Deterioration at the part of expansion joint. See Fig. 9
	cause	Water is apt to flow down at the part of joint and to permeate into the space between metal and concrete.
	detail	To equip drainage as shown in Fig. 21.

5) Beam

a)	state	Exposure of reinforcing bar at the side of outside beam. See Fig. 10.
	cause	Because of the obtuse angle between slab and beam, water is apt to flow from slab to beam. See Fig. 22.
	detail	Proper water drip. The intersection θ_1, θ_2 in Fig. 22 should be a right angle.
b)	state	Deterioration at side of outside beam.
	cause	Overhanging of slab is small so that water flows on to the beam. See Fig. 22.
	detail	Proper water drip. It is desirable that the distance ℓ in Fig. 22 is large.
c)	state	Deterioration at overhanging of compound beam.
	cause	Structure which water flows from bridge face to beam. See Fig. 22.
	detail	Proper water drip. Close to a right angle of θ_3 in Fig. 22 is desirable.
d)	state	Deterioration at side and under face of prestressed concrete beam.
	cause	Inadequate shape of curb. Water falls along side curb reaching beam. See Fig. 22.
	detail	Shape of curb which does not allow water to reach beam. It is desirable that θ_4 in Fig. 22 is a right angle.
e)	state	Deterioration at side of outside beam.
	cause	Water wets concrete surrounding drainage pipe because the drainage is equiped near beam.
	detail	To pay attention the position of drainage in the case that water is directly drained from bridge face to under bridge.

f)	state	Deterioration of beam. See Fig. 11.
	cause	Water comes to beam through structural cracking of pavement and slab.
	detail	Design to prevent the occurence of large structural cracking of pavement and slab.
g)	state	Deterioration at hinge of Gerber beam.
	cause	Water flows through space of hinge. Water content within concrete at hinge is large.
	detail	To cover concrete at hinge with an impermeable material.
h)	state	Deterioration at side and under face of outside beam.
	cause	During snow removal snow is piled up around beam, causing saturation of the concrete.
	detail	Covering the beam with a protective substance such as metal is effective if snow removal is necessary.

6) Abutment

a)	state	Scaling at bridge seat, edges and corners. See Fig. 12
	cause	Snow piles on the bridge seat and melting snow supplies water to the concrete. Stress concentration occurs at the edges and corners of abutment.
	detail	To slop the bridge seat for drainage. Proper beveling of the edges and corners. To cover the bridge seat with a protective material.
b)	state	Deterioration at bridge seat of abutment. See Fig. 13.
	cause	Water flows down through the space of expansion joint and wets the concrete of bridge seat.
	detail	To devise a method in order to prevent water from flowing through expansion joint. See Fig. 21.
c)	state	Deterioration at edge between side and front of abutment.
	cause	Because of the acute angle between the side and front of abutment, stress concentration occurs at the edge.
	detail	To avoid the acute angle. Proper beveling in the case that the acute angle is necessary.
d)	state	Deterioration at front of abutment. See Fig. 14.
	cause	Water from drainage pipe flows down directly to the front of abutment.
	detail	To consider the position of drainage. To bend drainage pipe toward the opposite direction of abutment if drainage is inevitably placed right above abutment.
e)	state	Deterioration at construction joint. See fig. 15.
	cause	Construction joint is generally a weak point of abutment. Water which comes directly to this part causes deterioration of the concrete. See Fig. 23.
	detail	To avoid shapes which water comes directly to the construction joint of abutment. Perpendicular abutment is desirable.
f)	state	Deterioration which developed from cracking.
	cause	Abutment is generally a large-sized concrete structure. Thermal stress causes the cracking.
	detail	To control thermal cracking.

7) Pier

a)	state	Scaling at top of pier.
	cause	Snow accumulates on the top of pier causing saturation of the concrete.
	detail	To design the pier so that the top is sloped and snow does not accumulate. To cover the pier with a protective material.
b)	state	Deterioration near top of pier. See Fig. 16.
	cause	Pier extends out from the bridge and snow accumulates on the top the pier. See Table 4.
	detail	To avoid a structure in which the pier extends out from the bridge. To cover the pier with a protective material if the jutting out is necessary.
c)	state	Deterioration at lower part of pier.
	cause	Repetition of freezing and thawing due to fluctuation of water level. The degree of saturation is always high at the part.
	detail	To cover the part with a protective material in the case of a quite number of the repetition.

Fig. 2

Fig. 3

Fig. 4

Fig. 5

Fig. 6

Fig. 7

Fig. 8

Fig. 9

Fig. 10

Fig. 11

Fig. 12

Fig. 13

Fig. 14

Fig. 15

Fig. 16

Fig. 17   Shape of curb and slab

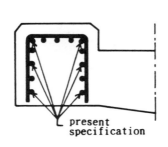

present
specification

Fig. 18   Reinforcement of curb

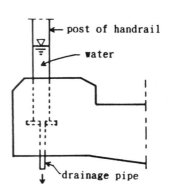

post of handrail

water

drainage pipe

Fig. 19   Drainage from
handrail post

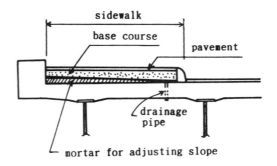

sidewalk

base course

pavement

drainage
pipe

mortar for adjusting slope

Fig. 20   Drainage from base course of sidewalk

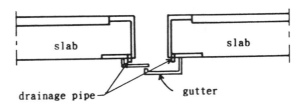

slab

slab

drainage pipe

gutter

Fig. 21   Drainage at expansion joint

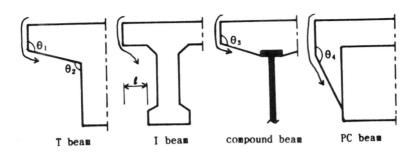

Fig. 22   Shape of beam

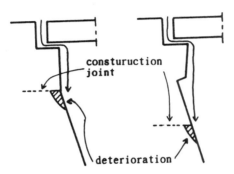

Fig. 23   Damage of abutment

CONCLUSION

In spite of many specifications for preventing deterioration due to freezing and thawing, quite a number of concrete structures are damaged by freezing and thawing. The deterioration seems to be mainly caused by poor construction, but in many cases deterioration could be avoided by careful considerations to questions of design of concrete.

In this investigation, 382 road bridges were examind and many points which should be considered in the designing of bridges were found. It is expected that the details of structural design proposed here could be helpful in preventing deterioration due to freezing and thawing.

OXYGEN PERMEABILITY OF STRUCTURAL LIGHTWEIGHT AGGREGATE CONCRETE

B Ben-Othman

N R Buénfeld

Imperial College, London,

United Kingdom

ABSTRACT. An investigation of the oxygen permeability of structural lightweight aggregate concrete is reported. Concretes produced from sintered pulverised fuel ash (Lytag) and pelletised expanded blastfurnace slag (Pellite) coarse aggregates were compared with basalt concrete of identical volumetric composition. The mix variables also included free water to cement ratio and Lytag aggregate volume fraction. Test samples were cured for 28 days either in a sealed condition or at 100% RH. Samples were then dried to comparable moisture conditions prior to testing. Three drying procedures (85% RH at room temperature, 65% RH at 30°C and oven drying at 50°C) were employed.

Both Pellite and Lytag concretes were found to have similar or lower permeabilities than equivalent basalt concretes. The lower permeability of lightweight aggregate concretes was most marked in the sealed cured samples, particularly those dried at 85% RH. The contributions of a number of mechanisms that may be responsible for reducing the permeability of the material are discussed.

Keywords: Permeability, Oxygen, Concrete, Lightweight, Aggregate, Curing, Drying, Interface, Self-desiccation, microcracking.

Mr Bechir Ben-Othman is a research assistant at the Department of Civil Engineering, City University, London, UK. He is also writing up his PhD thesis on transport processes through structural lightweight aggregate concrete.

Dr Nick R Buenfeld is a lecturer at the Department of Civil Engineering, Imperial College of Science, Technology and Medicine, London, UK. His research centres on the measurement of transport processes in cementitious materials. Current projects include a study of admixtures that reduce chloride penetration into concrete, investigations into the durability of high strength lightweight aggregate concrete and concrete repair materials and development of cementitious systems to stabilize hazardous waste materials.

INTRODUCTION

Since its modern development, lightweight aggregate concrete (LWAC) has found applications in a variety of constructions including buildings, bridges and marine structures. A review of the literature on the transport properties of LWAC which are of relevance to its durability shows that, with the exception of carbonation, little data is available on transport processes. The available data indicate that although lightweight aggregates (LWAs) are very likely to be highly permeable, the transport properties of LWAC are generally comparable to those of normal weight concrete (NWC) [1]. A recent investigation [2] showed that LWAC has a much lower permeability than NWC of similar strength. For instance, it has been found that the air permeability of sealed Lytag concretes was about 30 times lower than that of gravel concrete of similar compressive strength [2]. It remains, however, that little is known of the effect of such factors as mix proportions, aggregate type or curing on most transport properties of LWAC. Besides, the relative importance of factors which are thought to favourably affect the transport properties of LWAC is not clear. This highlights the need for further research into the transport properties of LWAC, in order to provide a better insight into its potential durability.

The work reported forms part of a research program into a range of transport properties of LWAC and aims principally to assess the effect of coarse lightweight aggregates on oxygen permeability of structural OPC concrete. Accordingly, the main approach adopted is to compare the properties of LWA/sand mixes with those of equivalent NWC mixes of identical volumetric composition. This approach may have the added advantage of allowing some comparisons between normal weight and lightweight mixes of similar strength if a number of mixes are accordingly designed. In the light of earlier tests which suggested that the relative performance of lightweight aggregate concrete is affected by the type of curing regime and by the conditions of drying [1], these factors are futher investigated in the work reported here.

MATERIALS

Ordinary Portland Cement, natural sand fines (zone M) and sintered pulverised fuel ash (Lytag) and pelletised expanded blastfurnace slag (Pellite) coarse lightweight aggregates were used in this investigation. Basalt coarse aggregate of the same maximum aggregate size and similar grading to the lightweight aggregates was used in the equivalent normal weight aggregate mixes. Some of the physical properties of the aggregates are given in Table 1. In addition to the type of coarse aggregate, the following mix variables were considered: two free water/cement ratios (W/C, 0.39 and 0.49), two moisture conditions of Lytag aggregate (as-received and presoaked for 24 hours) and 4 different levels of volume fraction of Lytag aggregate (0, 32, 42 and 52%). Altogether ten mixes (A to J) corresponding to the combinations shown in Table 2 were investigated, these were cast in alphabetical order.

TABLE 1   Some physical properties of coarse aggregates

Property		Lytag	Pellite	Basalt
Relative density on an oven dry basis (kg/m^3)		1440-1450	1650	2870
Absorption (% dry weight)	30 min	12.1-13.0	13.6	
	24 hrs	14.5-15.5	14.1	1.4

TABLE 2   Mix proportions, workability and compressive strength

Mix ref.	COARSE AGGREGATE		W/C (free)	C kg/m^3	S/C**	Slump (mm)	Comp. strength (N/mm^2)
	type*	% Volume fraction					
A	L2.ar	42	0.49	360	1.94	65	45
C	B2	42	0.49	360	1.94	25	57
I	P	42	0.49	360	1.94	50	38
B	M	0	0.39	780	1.39	130+	61
H	L2.ar	32	0.39	530	1.41	5	53
E	L2.ar	42	0.39	450	1.39	135	60
D	L2.ar	52	0.39	370	1.39	5	53
J	L2.P	42	0.39	450	1.39	15	60
G	B2	42	0.39	450	1.39	20	72
F	P	42	0.39	450	1.39	40	44

```
*    L2.ar  Lytag, as-received moisture condition.
     L2.p   Lytag, presoaked for 24 hours.
     P      Pellite
     B2     Basalt
     M      No coarse aggregate (mortar)

**   Sand/Cement ratio

+    Flow test (BS 4551:1980)
```

Mixing was carried out in a horizontal pan mixer and the following mixing procedure was followed for all concrete mixes. The coarse aggregate was mixed with about a third of the total water for approximately one minute, the sand was then added and mixed for about 30 seconds. The cement and the remaining water were then added and mixing was continued for a further three minutes. For the lightweight aggregates used in the as received moisture condition, an absorption equal to the 30 min value was allowed for in the calculations of total water. Presoaked Lytag aggregates were drained for about 20 min and the total mixing water was calculated taking into account the excess moisture content of the aggregate. The mix proportions and properties of fresh concrete are given in Table 2.

Oxygen permeability test samples were cast to have 101.6mm diameter and 50mm nominal thickness. Immediately after casting, specimens were stored under wet hessian and polythene sheet where they were cured overnight; after demoulding the following day they were weighed and cured at room temperature for another 27 days. Most of the samples were sealed cured in order to simulate curing under an efficient curing membrane; the remaining samples were cured at 100% RH to study the effect of the availability of an external supply of water during curing (Table 3). Compressive strength was measured at 28 days on 101.6mm cubes which were cured in water at 20°C, the results are given in Table 2.

OXYGEN PERMEABILITY APPARATUS

Figure 1: View of the oxygen permeability apparatus.

(showing a: permeability cell, b: hydraulic jack, c: Pump, d: Bubble flow meters in parallel, oxygen supply, pressure regulator and pressure gauges)

The apparatus used for the measurement of the oxygen permeability of concrete is illustrated in Fig. 1. The design of the permeability cell was based on that used by Mills [3]. The cell consists of two brass plates, a silicone rubber ring-seal and a mild steel containing ring. Using a hydraulic jack and a loading frame, a vertical load is applied through the top cover onto the silicone rubber ring-seal. The latter is compressed and forced inwards against the sample, thus providing a seal against the curved surface of the sample.

Steady state flow was measured, using bubble flow meters mounted in parallel, either at 5 increasing levels of gauge pressure (0.5, 1.0, 1.5, 2.0 and 2.5 bars) when the relation between flow rate and pressure was investigated or at only one pressure (1.5 bars) to obtain a permeability value.

The repeatability of the measurement of the flow rate of oxygen with the permeability apparatus was evaluated on a basalt concrete sample cut from a cylinder used in an earlier investigation. The sample, which was dried at room temperature at 85% for three months, was repeatedly tested at 5 increasing pressures levels for nine successive assembly/diassembly cycles of the permeability cell. The coefficient of variation was typically 5% and was considered satisfactory.

The oxygen permeability coefficient K was calculated from:

$$K = \frac{2P_2 VL\eta}{A(P_1{}^2 - P_2{}^2)}$$

where P_1 = absolute applied pressure = 2.5 bars (0.25 N/mm^2)
$\quad\quad P_2$ = absolute outlet pressure= 1 bar (0.1 N/mm^2)
$\quad\quad V$ = outlet volume rate of flow
$\quad\quad \eta$ = viscosity of oxygen (= 2.02 x 10^{-5} Ns/m^2 at 20°C)
$\quad\quad L$ = length of specimen
$\quad\quad A$ = cross-section area of specimen

CONDITIONING OF TEST SAMPLES

Three drying regimes corresponding to three levels of saturation were used: 85% RH at room temperature, 65% RH at 30°C and oven drying at 50°C. Details of the test programme are given in Table 3.

TABLE 3 Oxygen permeability test Program

Curing	Drying	Mixes
27d 100% RH	65% RH at 30°C	W/C=0.39 only
27d Sealed	65% RH at 30°C	All
27d Sealed	85% RH at room temperature	B,E,F,G and J
27d Sealed	Oven drying at 50°C	B,E,F,G and J

The bulk of the testing was done on samples dried at 65% RH, 30°C in a conditioning cabinet. For drying at 85% RH, test samples were stored above a saturated potassium chloride solution in contact with an excess of solids at ambient temperature in cabinet desiccators. For both drying at 85% RH at room temperature and at 65% RH at 30°C, samples were stored in the presence of self-indicating soda lime to prevent carbonation. Samples were tested after drying for one year +/- 2 weeks.

The samples to be oven dried were first stored in laboratory air for 2 days after the end of the curing period. During that time, two coats of Rezex ES2 (Cormix Ltd.) resin were applied to the curved surface of test samples in order to achieve near unidirectional drying conditions. Test samples were then dried over silica gel in the oven at 50°C for 8 weeks, then cooled to room temperature and tested. Preliminary tests indicated that this would allow all mixes to dry to constant weight (<0.05% weight loss in 24 hrs); this was later confirmed.

RESULTS

The mean oxygen permeability coefficients of three replicates ranged from 0.466×10^{-16} to 1.78×10^{-16} m^2 for the mortar and basalt mixes and from 0.0360×10^{-16} to 1.76×10^{-16} m^2 for LWACs. Standard deviations were within the range $(0.008-0.293) \times 10^{-16}$ m^2, the corresponding coefficients of variation were from 2.1 to 30.0% for the mortar and basalt concretes and from 4.0 to 42.5% for LWACs.

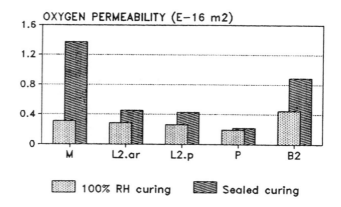

Figure 2: Effect of curing and coarse aggregate type
(volume fraction = 0.42 in concrete) on oxygen
permeability (65% RH at 30°C; W/C = 0.39)

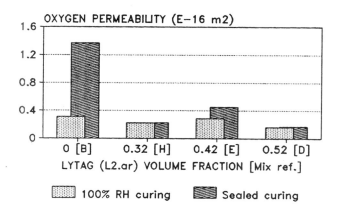

Figure 3: Effect of curing and Lytag volume fraction on
          oxygen permeability (65% RH at 30°C; W/C = 0.39)

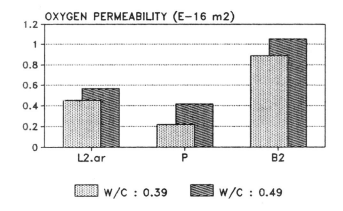

Figure 4: Effect of W/C and coarse aggregate type
          (volume fraction = 0.42 in concrete) on oxygen
          permeability (65% RH at 30°C; sealed curing)

The experimental results are given in Figures 2-5 which show the
effect of the various variables investigated on oxygen permeability.
For all the combinations of test variables which were included in the
test programme, it is found that lightweight aggregate concretes have
equal or lower permeability compared with both basalt concrete of
identical volumetric composition, but of higher compressive strength,
and mortar of equal W/C and S/C.

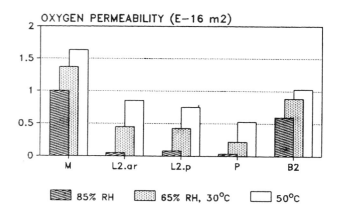

Figure 5: Effect of drying regime and coarse aggregate type
(volume fraction = 0.42 in concrete) on oxygen
permeability (65% RH at 30°C; sealed curing)

For sealed curing, lightweight concretes are much less permeable than
equivalent mortar and normal weight concrete (Fig. 2 and 3). For
both curing regimes, the data show that an increase in the aggregate
volume fraction in the range 32-52% does not necessarily increase
oxygen permeability (Fig. 3). A change of the curing regime from 100%
RH to sealed curing is found to result in about two and four fold
increases in oxygen permeability for basalt concrete and mortar
respectively. A relatively smaller effect of the curing regime is
observed with both Lytag mixes with a coarse aggregate volume fraction
of 42%. Whereas in the Lytag (L2.ar) mixes with a coarse aggregate
volume fraction of 32 and 52% and the Pellite concrete, the type of
curing had no significant effect.

Figure 4 shows the effect of aggregate type and W/C ratio on the
oxygen permeability of concrete samples which were sealed cured prior
to drying at 65% RH, 30°C. It is worth noting, that lightweight mixes
having a W/C of 0.49 (C= 360 kg/m³, compressive strength= 45 and 38
N/mm² for the Lytag and Pellite concretes respectively) have a lower
permeability than the basalt mix with a W/C of 0.39 (C= 450 kg/m³,
compressive strength= 57 N/mm²). Statistical analysis of the data
shown in Fig. 4 indicates that the effect of W/C ratio is not quite
significant at the 5% level. For NWC, the influence of W/C on oxygen
permeability observed in this work is in agreement with that reported
by other workers for oxygen [4] and air permeability [5, 6] of dry
normal weight concretes of W/C lower or equal to 0.5.

Changes in the drying procedure are found to result in relatively
small changes in the oxygen permeability of basalt concrete and mortar
despite considerable differences in the amount of moisture lost on
drying (Fig. 5). Greater changes are observed with lightweight

concretes, particularly as the drying regime is changed from 85% RH at room temperature to 65% RH at 30°C (Fig. 5). The ratios of the permeability coefficients of LWACs to that of equivalent basalt concrete or mortar is found to be lowest for drying at 85% RH where LWACs are about an order of magnitude (7-27 times) less permeable than basalt concrete or mortar (Fig. 5).

Pellite concretes are found to have lower permeability coefficients than equivalent Lytag concretes (Fig. 2, 4 and 5). Whilst the moisture condition of Lytag aggregate prior to mixing is found to have no significant influence on oxygen permeability (Fig. 2 and 5).

DISCUSSION

Lytag and Pellite aggregates are highly porous and absorptive and are therefore expected to be highly permeable in comparison with cement mortar. Yet, lightweight aggregate concretes are found to have equal or lower permeability compared to mortar of equal W/C and S/C and to basalt concrete of identical volumetric composition. This suggests that LWA reduces the permeability of the surrounding mortar. The good interfacial behaviour between LWAs and cement paste [7, 8] could be partly responsible for the discrepancy between LWAC and NWC [2, 9]. However, flow paths through lightweight aggregate particles are still likely to exist and would be expected to result in greater permeability for LWAC in comparison with equivalent NWC. As will be discussed in the following sections, the results of this work point to the existence of other favourable effects of LWA particles on mechanisms influencing the permeability of the surrounding mortar, namely hydration and microcracking.

Effect of aggregate on hydration and shrinkage of surrounding paste

For basalt concrete and mortar, at W/C =0.39, sealed curing is found to result in oxygen permeabilities 2 and 4 times the values measured when curing at 100% RH. Whereas for LWACs, a significant increase (of about 50%) is observed with only two of the five mixes. This suggests that in LWAC, self-desiccation effects are less marked than in mortar and NWC and are probably of minor influence. A similar conclusion was reached from a number of investigations of the compressive and axisymmetric compression/tension strengths of Lytag and limestone concretes [10]. In order to provide an indication of the potential of Lytag and Pellite to oppose self-desiccation, the amount of water lost at room temperature at about 97% RH (above a saturated solution of potassium sulphate in contact with an excess of solids) was measured for Lytag and Pellite aggregate samples which were first oven dried at 105°C then presoaked in water for either 30 minutes or 24 hours. The water lost was over 5% of the oven dry weight and ranged between 78% and 92% of the amount absorbed; this may be linked to the relatively coarse pore structure of these aggregates [8, 11]. It shows that the water absorbed by lightweight aggregates prior to or during mixing constitutes a vast reserve which is readily lost even at very high humidities and which would therefore maintain the cement matrix of sealed cured LWAC very near saturation. Hydration would therefore continue more favourably in LWAC than in equivalent NWC or mortar [2, 10]. Further evidence for this is that lightweight concretes showed negligible moisture gain during curing at 100% RH whilst mortar and basalt concrete gained significant amounts of water. The water in the LWA particles is also thought to allow hydration to continue more favourably in LWAC than in NWC during drying out [8, 10].

Self-desiccation is also accompanied by chemical shrinkage which in turn can cause microcracking in NWC [12]. Little if any chemical shrinkage would be expected in sealed cured LWAC. In fact, it has been reported that partially sealed Lytag and Solite concretes showed expansion while limestone concrete showed shrinkage [13].

Effect of aggregate on microcracking of cement paste

For drying at 65% RH and 30°C, LWACs were found to be of similar or lower permeability than equivalent basalt concrete, even for curing at 100% RH. This suggests that, besides the favourable influences of LWA particles on the coarse aggregate/cement interfacial zone and on the hydration of the surrounding mortar, another mechanism is contributing to the reduced permeability of LWACs. This is likely to be a reduction in microcracking in the mortar matrix of LWACs in comparison with that of NWC. Indeed, in contrast to normal weight aggregates whose elastic moduli are much larger than hardened cement mortar, LWAs have similar elastic moduli to mature mortar [14]. For a given differential deformation between the coarse aggregate and the mortar, the self-induced stresses would therefore be lower in LWAC than in NWC [15]. Accordingly, LWAC would be less prone to microcracking caused by differential deformation between aggregates and cement paste during curing and drying; this should be further enhanced by the generally spherical shape of LWAs which may help reduce stress concentrations [2] and the good bond between LWAs and cement paste. Evidence for the greater resistance to microcracking of LWAC could be inferred from the observations on LWAC cores taken from structures which showed little evidence of microcracking even at magnifications of 200x despite long-term exposure to severe conditions [7]. Whilst in laboratory samples of NWC, microcracks are evident at lower magnifications [16]. Hornain and Regourd[17] also found that at the macro level the crack density in a NWC was about 50% greater than that in expanded shale lightweight concrete. However, using higher magnifications, the crack densities at the level of the mortar were found to be similar for both concretes [17]. The data for the various drying procedures for sealed cured samples also suggests that microcracking is influencing the results. Indeed, the much lower permeability coefficients of LWACs by comparison with equivalent mortar or NWC after drying to 85% RH seem to indicate much less microcracking in the former. Changes in the drying regime are found to result in rather small changes in the oxygen permeability of basalt concrete and mortar; similar, though not directly comparable, trends have been reported [4, 18]. This suggests that for NWC and mortar most of the detrimental effect of microcracking on permeability occurred before the specimen reached equilibrium at 85% RH. On the contrary, the LWACs are found to be far more sensitive to the drying regimes adopted in this work, suggesting that most of the microcracking takes place at relative humidities of less than 85% RH.

CONCLUSIONS

For the combinations of curing regime (sealed or 100% RH curing), drying procedure (85% RH at room temperature, 65% RH at 30°C and oven drying at 50°C) and mix proportions adopted in this work, the oxygen permeability coefficients of LWACs were found to be similar or lower than those of basalt concrete of identical volumetric composition and mortar of equal W/C and S/C. The lower permeability of LWACs was most marked for sealed cured samples, particularly for drying at 85% RH. These observations point to the beneficial effects of LWA particles on hydration and microcracking of the surrounding mortar. These effects together with the good interfacial behaviour between LWAs and cement paste are favourable to the durability of LWAC.

Lower permeability coefficients were observed for Pellite concrete than for Lytag concrete of identical volumetric composition, this could be partly due to the relatively sealed surface and disconnected pore structure of Pellite aggregate by comparison with Lytag aggregates [11]. The moisture condition of Lytag aggregate and its volumetric composition had little effect on oxygen permeability.

REFERENCES

1   BEN-OTHMAN B. Transport processes in structural lightweight aggregate concrete. PhD Thesis to be submitted to the University of London, 1990.

2   BAMFORTH P B. The properties of high strength lightweight concrete. Concrete, Vol. 21, No. 4, April 1987,pp. 8-9.

3   MILLS R. Mass transfer of gas and water through concrete. In: Scanlon J M. ed. Concrete Durability — Katharine and Bryant Mather International conference, ACI SP-100, 1987, Vol. 1, pp. 621-644.

4   GOHRING C and REIDEL W. Gas permeability of concrete and its influence on the corrosion resistance in salt solution. Proceedings of RILEM/IUPAC International Symposium on Pore structure and properties of materials, Prague, 1973, pp. F95-118.

5   DHIR R K, HEWLETT P C and CHAN Y N. Near surface characteristics of concrete: intrinsic permeability. Magazine of Concrete Research, Vol. 41, No. 147, June 1989, pp. 87-97.

6   PERRATON D, AITCIN D C and VEZINA D. Permeability of silica fume concrete. In: WHITING D. and WALITT A. ed. Permeability of

concrete. ACI SP-108, 1989, pp. 63-84.

7 HOLM T A, BREMNER T W and NEWMAN J B. Lightweight aggregate concrete subject to severe weathering. Concrete International – design and construction, Vol.6, No. 6, June 1984, pp. 49-54.

8 SWAMY, R.N. and LAMBERT, G.H. (1984) Microstructure of Lytag aggregate. The International Journal of Cement Composites and Lightweight Concrete, Vol.3, No. 4, pp. 273-82.

9 DHIR, R K. Durability potential of lightweight aggregate concrete. Concrete, Vol.21, No. 4, April 1987, p. 10.

10 LYDON F D. An outline comparison between high strength concretes containing normal weight or lightweight aggregates. FIP notes, 1988 , No. 4, pp. 5-8.

11 HANSON J M. The thermal conductivity properties of lightweight expanded slag aggregates and concrete. Tarmac Pellite Ltd, Reprint from the Journal of Chartered Institute of Building Services, Jan 1982, 2 pp.

12 CHATTERJI S. Probable mechanisms of crack formation at early ages of concretes: a literature survey. Cement and Concrete Research, Vol. 12, No. 3, 1982, pp. 371-376.

13 LYDON F D. Contribution to discussion of paper by LITTLE M E R. The use of lightweight aggregates for structural concrete. The Structural Engineer, Vol. 57A, No. 5, May 1979, pp. 158-166.

14 BREMNER T W and HOLM T A. Elastic compatibility and the behaviour of concrete. ACI Journal, Vol. 83, March-April 1986, pp. 244-250.

15 WITTMANN F H. Private communication.

16 HSU T T C, SLATE F O, STURMAN G M and WINTER G. Microcracking of plain concrete and the shape of the stress strain curve. Journal of the American Concrete Institute. Vol. 60, February 1963, pp. 209-223.

17 HORNAIN H and REGOURD M. Microcracking of concrete. Proc. 8th Intenational Congress on the Chemistry of cement. Rio de Janeiro, 1986, Vol. V, pp. 53-59. (In french).

18 DAY R L, JOSHI B W, LANGAN B W and WARD M A. In Proc. 5th International Ash Symp. , Orlando, 1985, Vol. 2, pp. 811-821. Quoted in: HOOTON R D. What is needed in a permeability test for evaluation of concrete quality. In Pore structure and permeability of cementitious materials, ed. Roberts L R and Skalny J P, Materials Research Society Symposium Proceedings, Vol. 137, Boston, 1988, pp. 141-149.

CRITICAL CONCRETE COVER THICKNESS IN CORROSION PROTECTION OF REINFORCEMENT

W H Chen

National Central University,

Republic of China

ABSTRACT. The shape of the crack is an important factor in assessments of the effectiveness of the concrete cover for corrosion protection of embedded reinforcement. The four factors, i.e., cover thickness, level of steel stress, compressive strength of concrete and bar diameter are studied with respect to the shape of the crack, based on the data from the test of 216 uniaxial tension specimens. It has been found that the cleavage length of cover at the steel-concrete interface under loading is more than 170 times and 650 times larger than the concrete surface crack width under loading and unloading, respectively. The increase of cover thickness has the tendency to increase the cleavage length of cover and the concrete surface crack width under loading and unloading. The statistical expressions of these relationships are presented. Using the proposed expressions and considering the barried-effect of cover against steel corrosion, a new concept of critical concrete cover thickness for various corrosive environment are suggested.

Keywords: Critical cover thickness, Corrosion, Crack width, Cleavage failure, Durability, Internal crack, Reinforcing steels, Reinforced concrete, Spalling, Structural design.

Dr Wen Hsiung Chen is Professor of Civil Engineering, National Central University, Taiwan, Republic of China. He received his Dr. degree from Waseda University, Tokyo, Japan in 1974. Dr. Chen has been engaged in teaching, research and consultation work on various projects in Japan, Canada and United States, and has published numerous research papers. His main research interests are in the field of cracking behavior and durability studies of reinforced concrete structures. He is a member of the revising committee on construction code of the Chinese Institute of Civil and Hydrolic Engineering of ROC.

INTRODUCTION

Corrosion of reinforcing steel embedded in concrete structures is becoming more severe because of the aggravation of environment due to
industrial pollution; shortage of suitable materials, and skilled man-
power in construction practices; inadequate specification, upgrading of
the allowable steel stress, and using ultimate load procedures in
design. This problem has received increasing attention because of its
widespread occurrence in reinforced concrete structures and the high
cost of repairs.

Several methods of protecting embedded steel against corrosion have been
suggested by numerous researchers [1] [2] [3]. The approaches can be
classified into three main categories: (A) to utilize the inherent
protective characteristics afforded by the portland cement concrete,
(B) to exclude the corrosion-inducing substance away from the concrete
surface, and (C) to shield the reinforcement directly. In the first
case, the protective properties of the portland cement concrete may be
ascribed to its: (a) high alkalinity, (b) high impermeability, and (c)
high electrical resistivity. However, these properties can provide
effective corrosion protection only if the reinforced concrete structures are properly designed and constructed. That is: (1) an adequate
concrete cover thickness over the embedded steel should be designed,
(2) wide cracks and large cleavage of concrete cover should be
eliminated by the proper design, and (3) good quality impervious
concrete and good workmanship should be followed throughout the entire
construction operation. In this paper, only the case of (1) and (2)
are being discussed.

RESEARCH SIGNIFICANCE

The role of concrete cover thickness in the corrosion protection is
complicated. The effectiveness of corrosion protection increase with
the increase of the thickness of the concrete cover because of increasing the length of the path through which corrosion-inducing substances
must travel before reaching the steel. Conversely, the effectiveness
of corrosion protection decrease with the increase of the thickness of
the cover because of inducing the wider cleavage of concrete cover from
the steel surface to make greater access to corrosion-inducing
substances and thus accelerate their attack. This paper introduces
the concept of critical concrete cover thickness, which imply to
satisfy both the requirement of the minimum concrete cover thickness
and the maximum permissible crack width. The presented experssions
include the effect of concrete cover thickness on cracking pattern.

TEST PROGRAM

The 216 uniaxial specimens simulating a segment of the tensile portion
of a reinforced beam between two successive tensile cracks were tested
to gain information on the relationships of the crack widths to the
variable. The crack width measurements identified by the Wc and Wo are

the measured crack width on the concrete surface under loading and
unloading, respectively (Fig. 1). The crack width measurements
identified by the Ws are the cleaved length of the concrete cover
measured at the steel-concrete interface under loading. The variables
considered were level of steel stress fs (Table 1), concrete strength
fc', concrete clear cover thickness Dn, and bar diameter Db.

The test specimens shown in Fig. 2 were the prism of square cross
section, 500 mm. long, with a reinforcing bar embedded along the
longitudinal axis of the specimen. Four series of concrete mixtures
were proportioned. The average compressive strength fc', obtained
from 150 x 300 mm standard cylinders at 28 days was 19.8, 24.1, 35.2
and 37.8 N/mm², respectively. Ordinary Portland Cement (Taiwan Suini)
was used in making concrete. The Dah Hann River sand and graval were
used in all of the mixes. The maximum size of the course aggregate
was 20 mm. The fineness modulus of the sand was: (a) 2.77 for series
19.8; and (b) 2.75 for series 24.1, 35.2 and 37.8, respectively.
The No. 5 (16 mm), No. 6 (19 mm) and No. 8 (25 mm) bars had yield strength
of: (a) 314, 304 and 314 N/mm² for series 19.8 and 35.2; and (b) 314,
397 and 364 N/mm² for series 24.1 and 37.8, respectively, were used.

TEST PROCEDURE

The specimens were tested by applying a force to the ends of the bar by

TABLE 1   Variables considered in the experimental program

fc'	Db	b	Dn	Load. (kN)			
N/mm²	mm	mm	mm	fs=157	fs=216	fs=275	(N/mm²)
	16	60	22.05	31.15	42.83	54.51	
	19	60	21.00	39.90	54.87	69.83	
19.8	25	60	17.75	73.93	101.66	129.39	
	16	80	32.05	31.15	42.83	54.51	
&	19	80	31.00	39.90	54.87	69.83	
	25	80	27.75	73.93	101.66	129.39	
35.2	16	100	42.05	31.15	42.83	54.51	
	19	100	41.00	39.90	54.87	69.83	
	25	100	37.75	73.93	101.66	129.39	
	16	60	22.05	31.19	42.86	54.53	
	19	60	21.00	39.91	54.92	69.83	
24.1	25	80	27.30	81.99	112.68	143.38	
	16	80	32.05	31.19	42.86	54.53	
&	19	80	31.00	39.91	54.92	69.83	
	25	100	37.30	81.99	112.68	143.38	
37.8	16	100	42.05	31.19	42.86	54.53	
	19	100	41.00	39.91	54.92	69.83	
	25	120	47.30	81.99	112.68	143.38	

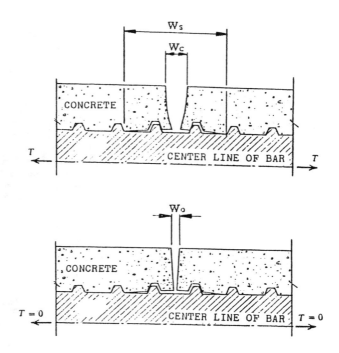

Fig. 1: Schematic diagrm of cleavage of concrete
cover and definitions of crack widths

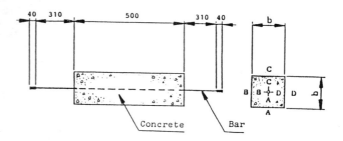

Fig. 2: Specimen and locations of crack width
measured (mm)

a specially fabricated loading frame to allow loading in tension by hooking the bar. Measurements during the test included the crack widths Wc, and Wo. When Wc was measured, the 1% phenolphthalein alcohol solution was injected from the opening of crack into the concrete. After Wo was measured, the specimen was removed from the frame and inserted in a compression testing machine in the horizontal position, so that compression was applied uniformly along two opposite generators. Pads were inserted between the compression platens of the machine and the specimen. The pads and longitudinal axis of the specimen should be lay on the central vertical plane of compression testing machine. After loading, the specimen will spilt into halves along that plane, the length to be dyed red with the phenolphthalein alkaline reaction on the plane at the steel surface was measured as the length of Ws.

TEST RESULTS AND ANALYSIS

1. Based on the test results of 108 specimens, the average ratio of the cleavage length of concrete cover under loading Ws to the crack width on the surface of concrete under loading Wc for these specimens was 175.55 with a coefficient of variation of 54.3% and a standard deviation of 95.42. It was observed that Ws increased in a nonlinear relationship with increasing Wc. A regression equation in form of

$$Ws = 1.81 + 193\ Wc - 149\ Wc^2 \quad \dots\dots\dots\dots\dots\dots\dots (1)$$

was found as best fit for the data.

2. Based on the test results of 216 specimens, the average ratio of the crack width on the surface of concrete under loading Wc to the crack width on the surface of concrete under unloading Wo was 4.88 with a coefficient of variation of 58.9% and a standard deviation of 2.88. It was observed that Wc increased with increasing Wo. The relationship between Wc and Wo can be expressed as follows

$$Wc = 0.08 + 2.68\ Wo - 5.03\ Wo^2 \quad \dots\dots\dots\dots\dots\dots (2)$$

3. The average ratio of Ws to Wo was 658.75 with a coefficient of variation of 55.7% and a standard deviation of 367.96. It was observed that Ws increased with increasing Wo. The relationship between Ws and Wo can be expressed as

$$Ws = 1.57 + 740\ Wo - 2410\ Wo^2 \quad \dots\dots\dots\dots\dots\dots (3)$$

4. It was observed that the more the depth of concrete cover Dn, the wider the Wc, Ws and Wo. The correlations between them can be express by the following equations

$$Wc = 0.0217 + 0.00578\ Dn \quad \dots\dots\dots\dots\dots\dots\dots\dots (4)$$
$$Ws = 4.98 + 0.969\ Dn \quad \dots\dots\dots\dots\dots\dots\dots\dots\dots\dots (5)$$
$$Wo = 0.0159 + 0.00108\ Dn \quad \dots\dots\dots\dots\dots\dots\dots\dots (6)$$

5. For a given concrete cover thickness, Wc, Ws and Wo increase with increase of steel stress fs, as shown in Figs. 3, 4 and 5. The relationships between Wc and Dn for various level of steel stress can be approximated for engineering purposed as follows

$$Wc = 0.0107 + 0.00461\ Dn \quad \text{for } fs = 157\ N/mm^2 \quad \dots\dots\dots (7)$$

$$Wc = 0.0146 + 0.00614 \text{ Dn} \quad \text{for fs} = 216 \text{ N/mm}^2 \dots\dots\dots(8)$$
$$Wc = 0.0442 + 0.00637 \text{ Dn} \quad \text{for fs} = 275 \text{ N/mm}^2 \dots\dots\dots(9)$$

6. As the bar diameter Db increases, the Wc, Ws and Wo have the tendency to become larger. However, the bar diameter Db exhibits a substantially smallar influence on the Wc, Ws and Wo. The correlations between Db and the Wc, Ws and Wo are give by

$$Wc = 0.132 + 0.00301 \text{ Db} \dots\dots\dots\dots\dots\dots(10)$$
$$Ws = 18.1 + 0.737 \text{ Db} \dots\dots\dots\dots\dots\dots(11)$$
$$Wo = 0.0482 - 0.00000670 \text{ Db} \dots\dots\dots\dots\dots(12)$$

7. The crack widths Wc, Ws and Wo were found to increase with increases in the steel stress for a bar of a given size, as shown in Figs. 6, 7 and 8.

CRITICAL COVER THICKNESS

In spite of considerable experimentation the exact relationships between crack width, the thickness of concrete cover and steel corrosion are not fully understood [4] [5]. However, from the investigation described herin, the idea of critical concrete cover thickness could be established based on the following considerations.

1. The concrete cover can provide effective corrosion protection if the concrete is high quality, dense, impervious and durable. The effectiveness of the protection of steel well be reduced or even rendered ineffective if one of the following conditions may exist along the embedded steel. Accordingly, the corrosion of steel can be categorised into three main groups; termed as follows:

Group 1. When a crack opens up from surface of concrete to steel and provides access to external corrosive agents. Thus this lead to steel corrosion.

Group 2. When certain external corrosive agents, such as chloride iorn, carbon dioxide, oxygen and water, penetrate through the concrete cover to the surface of steel and lead to steel corrosion.

Group 3. Others types of steel corrosion such as stray current corrosion. These types of corrosion are not appreciably affected by the crack width or concrete cover thickness, and will not be considered further in this paper.

2. Field data show that the corrosion of reinforcing steel is apparently somewhat greater when embedded in the cracked concrete than when embedded in the uncracked concrete. After the concrete cover spalls, the steel corroded more rapidly and severely than the steel embedded in the cracked concrete. This fact indicates that there is a critical crack width Wcr, as defined by the maximum crack width that may not lead to corrosion by type of group 1.

The tolerable crack width for reinforced concrete structures is affected by many parameters, the most important of them being : (1)

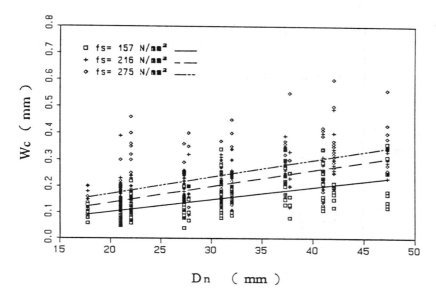

Fig. 3: Wc-Dn-Fs relationships

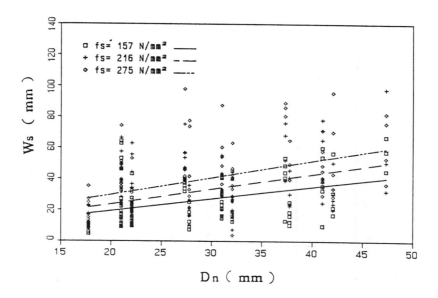

Fig. 4: Ws-Dn-Fs relationships

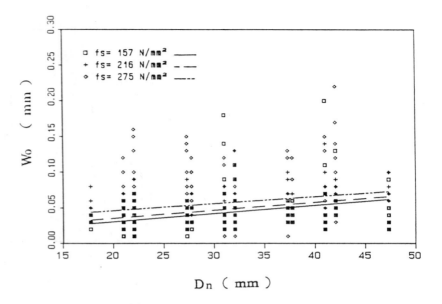

Fig. 5: Wo-Dn-Fs relationships

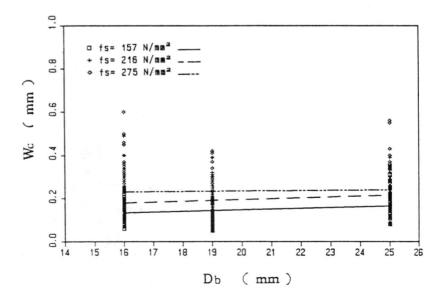

Fig. 6: Wc-Db-Fs relationships

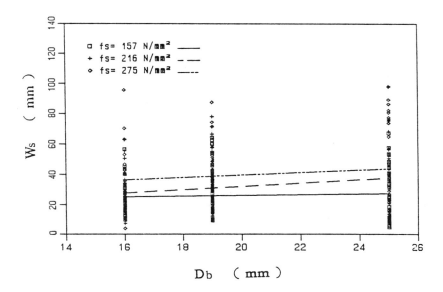

Fig. 7: Ws-Db-Fs relationships

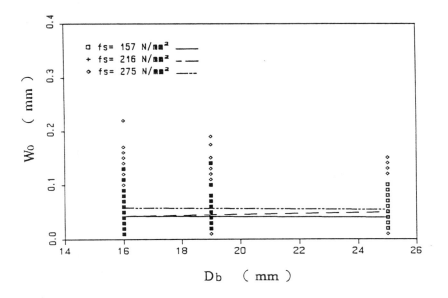

Fig. 8: Wo-Db-Fs relationships

aesthetic and appearance requirements, (2) functional requirements,
(3) the environmental factors, (4) the exposure conditions, (5) the
shape of crack (e.g., the crack width at concrete surface and the bar
surface, orientation, crack length and spacing), (6) cover thickness
and (7) the type of loading (e.g., live loads or sustained loads).
However, for practical application the crack control criteria in the
current ACI 318-83 Code [6] may be used as the critical crack width
as the surface of the concrete for given types of environment.

$$Z = fs \sqrt[3]{dcA} \quad \cdots \cdot (ACI\ 318-83\ Eq.\ 10-4)\ldots\ldots\ldots\ldots\ldots (13)$$

in which

$$Z = \frac{W}{11.0\ \beta\ x\ 10^{-6}} = \frac{W}{13.21\ x\ 10^{-6}} \quad \cdots\cdots\cdots\cdots\cdots\cdots (14)$$

Where W is the maximum width of crack, in mm, fs is the steel stress
at the load for which the crack width is to be determind, in N/mm ,
dc is the thickness of concrete cover measured from tension face to
center of bar closest to that face, in mm, A is the concrete area
surrounding one bar, equal to total effective tension area of concrete
surrounding reinforcement and having same centroid, divided by number
of bars, mm^2, β is the ratio of distances from tension face and from
steel centroid to neutral axis, 1.2 may be used for beams.

These equations were obtained from the research work of Gergley and
Lutz, together with simplifying approximations adpoted by ACI committee
318. The ACI committee 318 and 350 [7] specify that Z shall not
exceed 30.6 kN/mm (175 kips/in) for interior exposure, 25.4 (145) for
exterior exposure and 20.1 (115) for normal sanitary exposure for
corrosion protection. These limits correspond to the tolerable crack
widths of 0.41, 0.33 and 0.27 mm, respectively. These tolerable crack
widths may be considered as the critical crack width for the purposes
of design.

3. A thick cover has greater mechanical strength against abrasion,
impacts and spalling. Further more, a thick cover has the better
barried effect on the progress of penetration of carbon dioxide,
chloride or sulfid ions, moisture and oxygen. In general, up to a
certain degree, there is a tendency that the thicker the concrete
cover the higher the corrosion protecting capacity. However, for
practical application the minimum concrete cover thickness Dmin can
be determined considering the environment and exposure of the
structure, quality of concrete used.

4. The data analysis from this investigation indicate that increasing
the concrete cover Dn increases the crack width Wc and Ws, and hence
decreased the effectiveness for corrosion protection. This tendency
is inconsistent with the tendency of the thicker the concrete cover
the higher the corrosion protecting capacity as previously mentioned.
However, there is an optimum solution for these two alternative
conditions, defined as the critical cover thickness Dcr, to satisfy
the corrosion protection requirements. This concept of the critical
cover thickness is hown schematically in Fig. 9.

The line AB in Fig. 9 represents the critical crack width Wcr, for example Z = 25.4 KN/mm (145 kips/in), W = 0.33 mm for interior member of normal exposure by ACI in Code 10.6. The crack width under this line no corrosion of group 1 can occur in the vicinity of the crack.

The inclined line BE, representing the maximun crack width Wc, is derived from the regression study of the experimental results of this research. The line BE intersects line AB at point B. From point B, project a vertical line down until it intersects the horizonal axis, giving point C.

The significance of the point C is that the concrete cover thickness beyond this point the cracks are wide enough to allow group 1 reinforcing bar corrosion if the concrete cracking occurs. This cover thickness C is defined as the critical concrete cover thickness. Point D is the minimum concrete cover thickness obtained as described above. The cover thickness between F and D, type of group 2 corrosion may occur whether crack exists or not. The cover thickness between C and D, corrosion may not occur whether crack exists or not. Therefore the practical cover thickness should be chosen from this range. The critical concrete cover thickness for various corrosive environment are shown in Table 2.

TABLE 2   Critical cover thickness (mm)

Exoposure condition	Tolerable crack width W mm	Cleavage length Ws mm	Steel stress N/mm^2		
			157	216	275
Interior exposure	0.41	55.89	87	64	57
Exterior exposure	0.34	50.21	71	53	46
Normal sanitary exposure	0.27	43.06	56	42	35

DISCUSSION

1. Test results show that the average ratio of the cleavage length of concrete cover Ws to the crack width at the concrete surface under loading Wc is 176. This mean that the barrier effect of the concrete cover for steel corrosion is destroyed by cracking of the concrete more than 170 times wider than the crack width measured at the surface of the concrete. This cleavage failure of concrete cover should be considered as an important factor for assessment of durable structures as well as the maximum crack width at the surface of the concrete for given types of environment.

2. The average ratio of the crack width on the surface of concrete under loading Wc to unloading Wo is 5. This ratio will change in practical cases by the ratio of live load to dead load of structures.

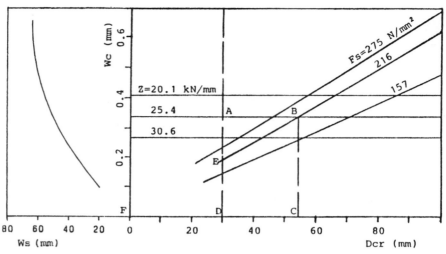

Fig. 9: Schematic description of critical concrete cover thickness (mm)

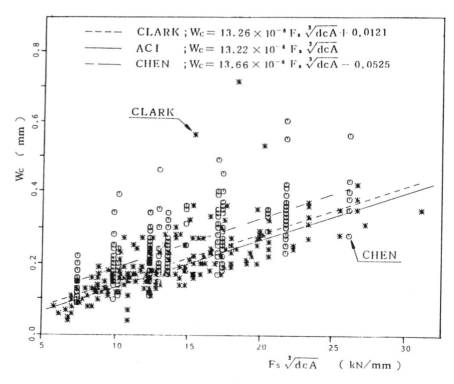

Fig. 10: Comparison of Author's results with existing test results

However, the cracks attributed to live loads applied for short periods may not be as serious as cracks due to dead load, since the cracks due to live load may be expected to close or at least decrease in width upon removal of the load. Therefore, the crack width Wo is equally important to the crack width Wc in the assessment of aesthetic requirement of reinforced concrete structures.

3. The cleavage length of concrete cover is more 650 times wider than the residual crack width. This implies the corrosion will take place widely on the certain length along the bar both side of the crack rather than only in the room of the opening of the crack. As the result, the spalling of the concrete cover due to rust of rebars will take place. Consequently more sever corrosion will be commenced and loss of reinforcement area may be induced. The ratio Ws/Wo is unneglibigle facotr for control of steel corrosion in concrete structures.

4. Statistical study based on the present experimental results shows that there are large variations in crack widths, even in careful laboratory controlled works. However, the results obtained are acceptable for engineering purpose as cracking is inherently a random behavior subject to a large degree of scatter up to 50 percent or more [8] [9] [10].

5. Numerous studies show that crack width does not scale form model to prototype, particularly if the steel-concrete interface bond interlock is not possible to scale truly [8]. In order to examine the size effect of the tested specimens, the comparison with A.P. Clack's full size beams test data [11] were made, as shown in Fig. 10. It is noted that no significant size effect was observed.

CONCLUSIONS

1. The internal crack shape of reinforced concrete members is quite different from that observed on the surface of the concrete. In the range 157 to 275 N/mm² fs, 2.2 to 4.7 cm Dn, 19.8 to 37.8 N/mm² fc' and 16 to 25 mm Db, the cleavage length of concrete cover is more than 170 times larger than the crack width on the surface of concrete under loading, and also more than 650 times larger than the crack width on the surface under unloading, the crack width on the surface under loading is 5 times larger than that of unloading. These facts should be considered for the design of crack width controll against steel corrosion.

2. The importance of the factors affecting the cleavage of concrete cover is the steel stress, the cover thickness, the concrete compression strength and the bar diameter in that order.

3. It is not an effective approach to distribute the reinforcement for crack control using a large number of smaller bars rather than a few large bars.

4. On the standpoint of corrosion protection of reinforcing steel, it is desirable to provide the concrete cover for reinforcement to be equal to the critical concrete cover thickness as proposed by this paper. Otherwise, the reduction of steel stresses at service loads should be considered.

REFERENCES

1. AMERICAN CONCRETE INSTITUTE COMMITTEE 222. Corrosion of metals in concrete. Journal of the American Concrete Institute, Volume 82, Number 1, January/February 1985, pp 3-32.

2. AMERICAN CONCRETE INSTITUTE COMMITTEE 201. Guide to durable concrete. Journal of the American Concrete Institute, Proceedings V. 74, No. 12, December 1977, pp 573-609.

3. AMERICAN CONCRETE INSTITUTE COMMITTEE 201. Durability of concrete in service. Journal of the American Concrete Institute, Proceedings V. 59, No. 12, December 1962, pp 1771-1820.

4. GERGELY P and Lutz A L. Maximum crack width in reinforced concrete flexural members. ACI Special Publication, SP-20, American Concrete Institute, Detroit, 1968, pp 87-117.

5. BEEBY A W. Cracking, cover and corrosion of reinforcement. Concrete International, February 1983, pp 35-40.

6. AMERICAN CONCRETE INSTITUTE COMMITTEE 318. Building Code Requirements for Reinforced Concrete (ACI 318-83). American Concrete Institute, Detroit, 1983, pp 813-38.

7. AMERICAN CONCRETE INSTITUTE COMMITTEE 350. Concrete Sanitary Engineering Structures. Journal of the American Concrete Institute Proceedings V. 74, No. 6, June 1977, pp 235.

8. AMERICAN CONCRETE INSTITUTE COMMITTEE 224. Control of cracking in concrete structures. Journal of the American Concrete Institute, V. 69, No. 12, December 1972, pp 717-753.

9. FERGUSON P M, BREEN J E and JIRSA J O. Reinforced Concrete Foundamentals, 5th ed., John Wily & Sons, New York, 1988, pp 102.

10. Nilson A H and Winter G. Design of Concrete Structures, 10th ed., McGraw-Hill, New York, 1982, pp 199.

11. CLACK A P. Cracking in Reinforced Concrete Flexural Members. Journal of the American Concrete Institute, Proceedings V. 52, No. 8, April 1956, pp 851-862.

PROTECTING CONCRETE AGAINST CHLORIDE INGRESS BY LATEX ADDITIONS

H Justnes
B A Oye

Cement and Concrete Research Institute,

S P Dennington

Borregaard Industries Ltd.,

Norway

ABSTRACT. A latex has been developed that effectively blocks chloride from entering mortar or concrete that has been modified by this polymer. The necessary dosage for adequate protection against chloride ingress is found to be dependent on the w/c-ratio and the content of entrained air. A 10 % addition of the polymer has proven to be sufficient even at high air contents ($\approx$ 10 vol%) and moderate w/c-ratios (0.55). The effectiveness of this new polymer, compared with three commercially available polymers, is probably due to the functional monomers incorporated in the chain. Thus, since the functional monomers are capable of forming chemical bonds with the cement minerals, the polymer-cement complex will be insensitive to moisture once it is formed. Other polymers are believed to partial reemulsify when the PCC's are immersed in water. One commercial available polymer, containing copolymerized vinylchloride, has been proven to produce free chloride during hydrolysis in the alkaline interior of the mortar. The protection of concrete by the new latex is also superior to that provided by condensed silica fume (10% replacement).

Keywords: Chloride, Ingress, Concrete, Mortar, Latex, PCC, Protection

Dr Harald Justnes is a Research Engineer at the Cement and Concrete Research Institute at the Foundation for Scientific and Industrial Research (SINTEF) in Trondheim, Norway. Although educated at the Department of Inorganic Chemistry at the Norwegian Institute of Technology (NTH), his main field is Concrete Polymer Composites. He is an active member in both RILEM TC 105-CPC and 113-CPT. He also has a general interest in the micro (and nano) structure of concrete.

Dr Bjarte A Øye is a Research Engineer at the Cement and Concrete Research Institute at SINTEF in Trondheim, Norway, where he recently completed his PhD study with the thesis: "Repair Systems for Concrete - Polymer Cement Mortars. An Evaluation of Some Polymer Systems".

Simon P Dennington B.Sc. is the Laboratory Manager for the Polymer Emulsions Section of Borregaard Industries Ltd, N-1701 Sarpsborg, Norway, where the experimental latex was synthesized.

INTRODUCTION

Along with the neutralization of concrete (for example, by the carbon dioxide in air), ingress of chloride is the major cause of deterioration of reinforced concrete. The source of chloride may either be from the components of the concrete (e.g accelerators like $CaCl_2$) or from the service environment of the structure (exposure to sea water, de-icing salts etc). The deterioration mechanism is well known as the initiation of corrosion, its progress with the formation of expanding corrosion products leading to, eventually, the formation of cracks that leave the way open for further attack.

Reinforced concrete structures in a chloride contaminated environment may be protected in many ways. However, the cheapest method at the stage of necessary repair is usually chosen, rather than evaluating the cost/efficiency-ratio of the alternative measures available. The most cost/effective measure is probably to make a high quality concrete. That is, a concrete with a low w/c-ratio (<0.45), proper post-curing and adequate rebar coverage (>50mm). In the tidal or splash zone, the concrete may be further protected by the addition of a hydrophobizing agent (e.g. latex) or surface treatment with water-repellants (e.g. silanes) or polymer coatings (e.g. epoxies).

Special measures, like cathodic protection (impressed current or sacrificial anodes), may also be chosen in order to prevent corrosion of the rebar, either during planning of the structure or after damage has occurred. For instance, some of the concrete of the oil rigs in the North Sea has a rather high chloride content at the depth of the rebars, but still corrosion is not a problem. The reason might be that the rebars are connected to the metal structure of the oil platform, which in turn is protected by sacrificial anodes.

Chemicals inhibiting corrosion may also be added to the concrete, but the efficiency of these may be doubtful in a chloride contaminated environment. And as is well known, the high pH-level (≈13) of the pore water in concrete creates a passivation layer on the rebar. However, this is broken down by the action of chloride.

Perhaps the most common repair system when corrosion damage of reinforced concrete is observed is: i) removal of damaged concrete to behind the corroded rebar, ii) cleaning the cavity, iii) protecting the exposed rebar (e.g. corrosion inhibitor), iv) treating the cavity with an adhesive and v) filling the hole with a compatible mortar (often modified with a latex). Finally the surface should be coated.

Since the use of latex is a common denominator for many of the protective measures for concrete in an aggressive environment, the present investigation was undertaken in order to quantify the effect of the polymer phase on chloride ingress. This was done both for cementitious mortars up to high dosages of polymer (which may be used as protective layers or repair mortars) and concrete with small dosages of polymer. Due to the high cost of refined products such as polymer, additions of more than 5 % to the bulk concrete may be difficult to justify.

EXPERIMENTAL

Polymer Cement Mortars

The investigation was based on two OPC mortars with water/cement ratios (w/c) of 0.40 and 0.55, respectively. The sand was dried and graded before use (1452 kg/m^3 made up of equal amounts of the fractions 0-0.5, 0.5-2 and 2-4 mm), and limestone filler (96.8 kg $CaCO_3$/m^3) was added as additional fine aggregate. The four tested latices ($\approx$ 50% polymer) consisted of;

Polymer 1 - the commercial terpolymer vinyl acetate/vinyl chloride/
            ethylene
Polymer 2 - the commercial copolymer vinyl acetate/Versatic ester
            (Veova)
Polymer 3 - the commercial copolymer styrene/butadiene rubber (SBR)
Polymer 4 - the experimental terpolymer n-butyl acrylate/methyl meth-
            acrylate/functional monomer (as a "chemical anchor").

The synthesis of latices with polymers containing functional monomers that may act as "chemical anchors" towards cementitious minerals, is described by Justnes and Dennington[1].

The water in the latices was subtracted from the total mixing water in the PCC's. The mortars were all blended in a 50 1 SKG1 forced action mixer, and the latices were the last component added. The basic compositions of the reference mortars were modified by partial replacement of the cementitious paste with polymer solids, and the composition and quantity of the aggregate phase were not altered at any point in the test. The binder volume was held constant regardless of the w/c ratio and degree of polymer modification. As Table 1 indicates, the cement and water content of the polymer modified mortar (PCC) compositions were reduced in accordance to eq's (1) (V - volume) and (2):

$$V_{cement} + V_{water} + V_{polymer} = \text{Constant (420 1 per m}^3\text{)} \qquad (1)$$

$$w/c = \text{Constant (0.40 or 0.55)} \qquad (2)$$

TABLE 1 Composition of basic and polymer-modified mortars.

POLYMER CONTENT OF BINDER (vol%)	BINDER COMPOSITION (kg/m^3) Total binder vol = 420 1/m^3					
	w/c = 0.40			w/c = 0.55		
	CEMENT	POLYM.	WATER	CEMENT	POLYM.	WATER
0	585.4	0	234.1	484.0	0	266.2
5	556.1	21.0	222.4	460.0	21.0	252.9
10	526.7	42.0	210.7	435.5	42.0	239.6
15	497.4	63.0	199.0	411.3	63.0	226.3
20	468.1	84.0	187.3	387.1	84.0	213.0

The above basis for comparison is in contradiction to most literature[2] which tends to base comparisons on the concept of constant
flow of the fresh mix, but should be more precise in order to evaluate
the effect of the polymer phase itself rather than the effect of the
lower w/c-ratio usually obtained for latex modified mortar and concrete.

A reference mortar with a 10% replacement of cement (c) by condensed
silica fume (s) and w/(c+s) = 0.55 was also made in order to compare
the effect of silica fume with the effects of the polymers.

In addition to the resistance towards chloride ingress, all mortars
were characterized with respect to rheological and mechanical properties.

The rheological properties of the fresh mix were measured as slump
(analogous to ASTM C143, but cone of 12 cm height), flow (ASTM C124-
39), air content and density (according to ASTM C231).

The hydration of the PCC's involved three different curing schemes,
chosen to monitor the effects of the polymer on the mortar properties
at different moisture conditions: I (28 days dry), II (14 days dry/7
wet/7 dry) and III (21 days dry/7 wet), with a temperature of approx.
20°C throughout the curing period. All samples were left in the moulds
covered with wet burlap for two days. This was done to ensure good
initial cement curing conditions. The dry period corresponded to 50 %
R.H. The compressive and flexural strength were tested for all mortars
on RILEM prisms (dimension 40x40x160 mm^3) after 28 days hardening
according to the three curing schemes.

Polymer Cement Concrete

All the concrete mixes had a constant w/c-ratio of 0.45, since both
the moisture content of the sand (3.1 %) and the water in the latices
(50 %) was subtracted from the mixing water. The composition of the
concrete was 400 kg OPC, 875 kg sand (0-8 mm) and 875 kg crushed
gravel (8-16 mm) per m^3. Concrete containing latex generally received
an addition of 1.0 kg/m^3 of the super-plasticizer Mighty 150 and 0.6
kg/m^3 of the defoaming agent Foammaster ENA-224. However, the concrete
with 5 % of polymer 4 needed 0.9 kg/m^3 of the defoaming agent in order
to get a comparable amount of entrained air as the rest of the series.
A reference concrete with entrained air was also made by adding 0.15
kg/m^3 of an air-entraining agent (Rescon L). All concrete reference
mixes (without latex) received 1.5 instead of 1.0 kg/m^3 of the super-
plasticising agent. Note that the stated polymer dosage in the concrete series (5 and 3 weight% of cement) was in addition to the cement, leading to a slightly increased binder content. However, the
quality of the cement paste remained the same (i.e. constant w/c).

The rheological properties of the concretes were monitored analogously
to those of the mortars (except for the use of a 30 cm slump cone).

The concrete was cured according to the following scheme for 90 days: 1 day in the mould, 6 days in water, 40 days at 50% R.H., 7 days submerged in water and 36 days at 50% R.H. The reason for the last period in water, was to obtain a more realistic curing schedule. Micro cracks formed by moisture cycling may, for example, ease the access of chlorides to the rebar.

In addition to their rheological properties, all concretes were characterized by their compressive strengths after 1, 3, 7, 28 and 90 days age, measured on 10x10x10 cm^3 cubes according to ISO-Standard 4012-1978.

Chloride Resistance Test

The chloride ingress test was carried out on cylinders with diameter 10 cm and height 20 cm, containing a reinforced steel rod of diameter 10 mm embedded in the centre of the cylinder with the lower end 5 cm from the bottom. The cylinders were cured for 12 weeks: 2 days in the covered moulds and the remaining time at 50% R.H. and 20 °C.

This test is similar to a method used by the Federal Highway Administration, USA[3]. The method is an accelerated impressed current test which measures time to failure of reinforced concrete specimens. The chloride source was natural seawater surrounding the mortar cylinder (Fig.1). A constant direct current field of 5 volt was maintained between the embedded steel (positive) and a steel sheet (negative) immersed in the seawater. The electrical current in the circuit was recorded, and when the chloride gets through to the rebar, an increase in current will be observed (Fig. 2). This is taken as the prime criterion for chloride penetration. Time to failure is defined as time corresponding to the observation of cracks, which is taken as a secondary criterion (expansive corrosion products have been formed). In addition, brown coloured liquid may be observed.

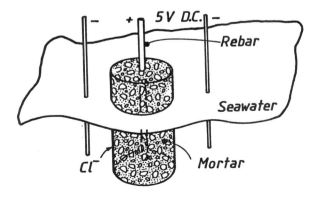

Figure 1: Experimental set-up for the chloride resistance test.

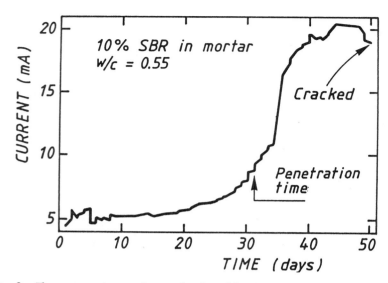

Figure 2: The current vs time relationship in the chloride ingress
test, exemplified by a PCC containing 10 % of polymer 3.

RESULTS AND DISCUSSIONS

Polymer Cement Mortars

The rheological properties of the investigated mortars are given in
Table 2. Note that most of the latices increase the slump and flow of
the mortars substantially, with the greatest effect for polymer 3
(SBR). In addition to the effect of surfactants in the latices and the
ball-bearing effect of polymer particles, the slump and flow of the
mortars also increase with increasing air content. The measured air
volumes were gauged just after the completion of mixing. However, the
handling of the mix, such as moulding and vibrating, may affect the
initial air content (unstable air). This was checked by calculating
the air content of the cured mortars on the basis of data from a
capillary suction test[4]. As seen from the table, these values are in
general lower (or insignificantly higher) than the corresponding
volumes in the fresh mix, as expected.

The mechanical properties of the mortars are shown in Table 3 as the
compressive and flexural strength determined after 28 days of hardening according to the three curing schemes (C.S.). It is difficult to
compare the mechanical strengths of the mortars due to the varying
contents of entrained air. However, as a rule of thumb; the mechanical
strength is reduced by about 5% for each extra vol% of entrained air.

TABLE 2   Slump, flow, air content and densities of the mortars.

POLY-MER	DOS-AGE	w/c = 0.40				w/c = 0.55			
		Slump (cm)	Flow (%)	Air[a] (vol%)	Density (kg/m^3)	Slump (cm)	Flow (%)	Air[a] (vol%)	Density (kg/m^3)
REF.	0	2.1[b]	90[b]	4.7/3.6[b]	2.248[b]	7.0	170	3.6/3.4	2.194
REF.[c]	0	-	-	-	-	2.5	105	5.2/5.7	2.213
No 1	5	1.4	85	4.7/4.0	2.255	9.3	>225	3.7/4.3	2.226
-"-	10	6.7	155	4.9/4.5	2.240	10.1	>225	3.7/4.8	2.226
-"-	15	6.9	125	4.9/4.2	2.233	10.3	>225	4.1/4.7	2.205
-"-	20	5.0	80	5.1/5.1	2.201	9.5	170	5.5/5.1	2.157
No 2	5	1.5	90	10.7/8.2	2.133	6.8	155	12.2/10.1	1.981
-"-	10	2.3	100	11.3/8.0	2.015	9.1	180	12.8/14.6	1.959
-"-	15	7.4	135	15.3/11.4	1.831	9.0	180	15.5/16.2	1.840
-"-	20	8.0	140	14.3/14.0	1.877	10.2	210	19.0/19.2	1.801
No 3	5	6.3	160	9.4/8.2	2.105	10.0	>225	11.6/11.5	2.004
-"-	10	8.5	>225	12.2/9.4	1.965	10.1	>225	13.7/13.4	1.860
-"-	15	9.6	>225	14.7/13.4	1.895	-	-	-	-
-"-	20	11.6	>225	15.6/17.2	1.807	-	-	-	-
No 4	5	0.4	50	11.6/8.0	2.042	8.0	180	7.6/7.0	2.112
-"-	10	3.8	75	10.3/10.8	2.022	8.0	165	8.4/9.1	2.059
-"-	15	7.5	140	8.9/11.5	2.084	10.7	>225	8.2/7.9	2.069
-"-	20	9.8	>225	11.1/14.5	1.991	11.2	>225	6.9/8.0	2.064

a: Measured/calculated air content.
b: Containing 0.3% of the cement weight of Mighty 150 superplasticizer
c: Reference mortar with 10% condensed silica fume.

The series with polymer 1 have similar air-contents to the plain mortars. The tendency for this series at w/c = 0.40 is for the compressive
strength to decrease with increasing polymer content, while the flexural strength is significantly improved. Note that with only 5 vol%
addition of polymer 1 at w/c = 0.40, the compressive strength is unaltered, while the flexural strength is greatly improved.

The mortar with 10% polymer 3 (SBR) at w/c = 0.55 was also made with
an addition of a defoaming agent, which resulted in an air content
slightly lower than the control. It can be seen from Table 3 for this
mixture that; the compressive strength is insignificantly lowered, and
the flexural strength is only slightly improved after curing scheme I
and not improved at all after C.S. II and III. This observation may be
explained by a partial reemulsification of polymer 3, with a direct
effect on flexural strength after C.S. III and indirectly after C.S.
II (the polymer is not able to prevent microcracking during dry/wet
cycles).

The results from the chloride ingress test are given in Table 4. The
first result one should notice is the effect of the w/c-ratio for the
cement mortars without polymer additions. The reduction in w/c-ratio
from 0.55 to 0.40 increased the penetration time for chloride by a

factor of 35! The reason is that a mortar with w/c = 0.40 has a low content of segmented capillary pores ($\approx$ 0 at 100% hydration), while w/c = 0.55 leads to a continuous capillary pore system. This illustrates the importance of using high quality (i.e. low w/c) concrete in aggressive environments, and the importance of using constant w/c-ratio when PCC's are to be compared. Remember, a low w/c-ratio may be obtained by cheaper means (small dosages of plasticizing agents) than by the addition of polymers.

TABLE 3 Compressive and flexural strengths (MPa) after 28 days curing

POLY-MER	DOS-AGE	w/c	COMPRESSIVE STRENGTH (MPa)			FLEXURAL STRENGTH (MPa)			AIR (Vol %)
			C.S. I	C.S. II	C.S.III	C.S. I	C.S. II	C.S.III	
REF.	0	.40	57±3	62±2	53±3	8.0±.4	5.6±1.	7.8±.8	3.6
No 1	5	.40	61±1	63±3	52±1	12.9±.3	10.1±2.	10.0±.2	4.0
-"-	10	-"-	57±1	56±1	48±1	14.0±.3	9.9±.1	10.0±.1	4.5
-"-	15	-"-	46±3	53±1	43±6	11.6±2.	9.9±.3	9.6±1.	4.2
-"-	20	-"-	51±1	49±1	42±1	13.0±.2	11.2±.7	9.9±.3	5.1
No 2	5	.40	46±1	42±2	41±1	8.7±.4	5.5±.6	6.9±.3	8.2
-"-	10	-"-	43±2	42±3	36±1	9.8±.4	4.8±.6	6.3±.3	8.0
-"-	15	-"-	35±1	31±1	23±3	8.7±.4	4.9±.2	5.4±.5	11.4
-"-	20	-"-	28±3	21±1	16±2	8.2±.3	5.2±.3	4.4±.2	14.0
No 3	5	.40	43±1	44±1	40±1	8.6±.4	6.6±.1	7.2±.2	8.2
-"-	10	-"-	35±1	34±1	26±1	8.1±.5	7.1±.6	5.8±.2	9.4
-"-	15	-"-	29±1	29±1	24±1	8.5±.3	7.0±.4	6.1±.0	13.4
-"-	20	-"-	24±2	26±1	17±1	7.6±.6	7.2±.6	4.6±.3	17.2
No 4	5	.40	41±1	40±2	33±2	9.1±.3	6.7±1.	7.9±.3	8.0
-"-	10	-"-	41±1	39±1	29±2	10.7±.6	7.8±.1	7.4±.6	10.8
-"-	15	-"-	38±1	33±1	27±1	11.3±.0	8.5±.4	7.3±.2	11.5
-"-	20	-"-	32±1	27±0	22±1	10.3±.5	7.7±1.	6.6±.3	14.5
REF.	0	.55	38±3	43±4	35±2	7.8±.8	6.4±.6	6.5±.6	3.4
REF.[a]	0	.55	50±2	57±2	51±2	8.2±.3	6.2±.5	8.7±.3	5.7
No 1	5	.55	44±1	44±1	36±1	9.4±.3	5.8±.2	6.2±.4	4.3
-"-	10	-"-	43±1	40±0	32±1	10.9±.0	7.3±.3	6.2±.1	4.8
-"-	15	-"-	47±1	42±2	35±1	12.0±.5	8.7±1.	7.2±.4	4.7
-"-	20	-"-	44±1	39±2	32±2	12.0±.4	9.5±.6	7.4±.3	5.1
No 2	5	.55	29±1	24±1	18±1	7.3±.6	3.9±.6	4.2±.1	10.1
-"-	10	-"-	26±1	21±2	15±1	7.4±.7	3.5±.3	3.3±.2	14.6
-"-	15	-"-	22±1	17±0	11±1	7.6±.2	3.6±.3	2.9±.1	16.2
-"-	20	-"-	19±1	12±1	8±1	6.7±.3	3.5±.5	2.4±.1	19.2
No 3	5	.55	27±1	29±1	20±1	6.4±.5	5.2±.1	4.6±.1	11.5
-"-	10	-"-	21±1	23±0	15±1	6.3±.3	5.2±.2	3.8±.1	13.4
-"-[b]	10	-"-	38±1	40±0	32±1	8.9±.6	6.3±.3	5.9±.1	3.0
No 4	5	.55	41±1	40±1	29±1	9.1±.3	5.4±.1	5.9±.1	7.0
-"-	10	-"-	41±1	34±1	26±1	10.7±.6	6.9±.6	6.3±.5	9.1
-"-	15	-"-	38±1	29±0	24±1	11.3±.0	7.9±.3	6.1±.2	7.9
-"-	20	-"-	32±1	25±0	21±0	10.3±.5	7.7±.5	5.4±.1	8.0

a: with 10% condensed silica fume. b: with defoaming agent.

TABLE 4   Time before chloride penetration (current rise) and visible corrosion for PCC cylinders subjected to the chloride penetration test.

POLYMER	POLYMER CONTENT (Vol%)	w/c - 0.4 Chloride penetr. (days)	w/c - 0.4 Visible corr. (days)	w/c - 0.55 Chloride penetr. (days)	w/c - 0.55 Visible corr. (days)
REFERENCE	0	103±47	118±57	3± 1	14±18
REFERENCE[a]	0	-	-	16± 0	16± 0
NUMBER 1	5	175±35	181±46	24± 2	27± 1
-"-	10	59± 7	68± 8	12± 3	21± 3
-"-	15	31± 6	32±13	23± 0	24± 1
-"-	20	35± 9	41±15	25± 0	32± 3
NUMBER 2	5	91[b]	98[b]	31± 0	46± 0
-"-	10	185[b]	190[b]	32± 0	86±14
-"-	15	110± 0	188± 1	35± 0	39± 6
-"-	20	118± 0	154± 0	32± 0	43± 0
NUMBER 3	5	125± 0	148± 0	30± 0	43± 0
-"-	10	118±28	139±40	26± 1	43± 0
-"-	15	103± 7	129±13	41± 0[c]	52± 3[c]
-"-	20	90± 5	118±10	-	-
NUMBER 4	5	351± 3	353± 1	35± 5	42±13
-"-	10	> 400[d]	> 400[d]	> 365[e]	> 365[e]
-"-	15	> 400[d]	> 400[d]	> 365[e]	> 365[e]
-"-	20	> 400[d]	> 400[d]	> 280[b]	> 298[b]

a: with 10% condensed silica fume
b: Only one out of two parallels failed within one year
c: 10% of polymer 3 (SBR) with defoaming agent
d: The test is still in progress after more than 1 year
e: The test was stopped after 1 yr, even though mortars did not fail

Even though the replacement of cement with 10 % condensed silica fume increased the mechanical strength considerably at a constant w/(c+s) = 0.55, the effect on the chloride penetration time was less than it was with the commercial latices (table 4), and in particular with the experimental latex. Note that compressive strength is not necessarily synonymous to a long service life in an aggressive environment for a concrete. Therefore, the revised Norwegian Standard specifies w/c-ratio rather than strength in relation to service environment.

Another feature that can be observed from the chloride penetration results, is that even though the three commercial latices (polymers 1-3) prolonged the penetration time for chloride by a factor of 10 at w/c = 0.55, they did not lead to any significant improvement (with the exception of 5 % polymer 1 and 10% polymer 2) compared to the control at w/c - 0.40. In other words, the performance of an unmodified mortar with w/c = 0.40 may be equally good in a chloride contaminated environment.

A puzzling observation was that polymer 1 performed <u>worse</u> with increasing polymer dosage in the chloride ingress test. Thus, prisms
(40x40x160 mm^3) of mortars with increasing dosages of polymer 1 were
stored in <u>fresh</u> water for 2 months before the pore water was pressed
from the mortars. The pore water was then analysed for free chloride
ions. The results showed a concentration of 0.280, 0.323, 0.397 and
0.649 g chloride/litre for PCC's with 5, 10, 15 and 20 vol% polymer 1,
respectively. These measurements lead to the conclusion that polymerized vinylchloride (VC) undergoes hydrolysis (Eq. 3 or 4) in the
alkaline interior (pH $\approx$ 13) of a cementitious mortar:

$$-(CH_2-CH)- + OH^- = -(CH=CH)- + H_2O + Cl^- \tag{3}$$

$$\quad\quad\quad VC \quad Cl \quad\quad\quad\quad E \quad\quad\quad\quad\quad\quad\quad\quad\quad E - \text{Ethylene unit}$$

$$-(CH_2-CH)- + OH^- = -(CH_2-CH)- + Cl^- \tag{4}$$

$$\quad\quad\quad VC \quad Cl \quad\quad\quad\quad VA \quad OH \quad\quad\quad VA - \text{Vinyl alcohol unit}$$

The results in Table 4 reveal that the experimental polymer 4 vastly
improves the chloride resistance of the mortars, regardless of the
w/c-ratio. For w/c - 0.55, the experiment was terminated after one
year (> 100 fold increase in penetration time whith polymer dosage $\geq$
10 vol%), while the experiment continues for w/c - 0.40 (> a projected
4 times increase in penetration time for chlorides). A dosage of 10
vol% (8 % of the cement weight at w/c - 0.40 and 10 % at w/c - 0.55)
of polymer 4 seems to be sufficient for adequate chloride protection.
The effectiveness of the experimental latex compared with that of the
three tested commercial latices, is attributed to the effect of the
"chemical anchor" in the polymer chain leading to cement-polymer
complexes insensitive to moisture[1].

Polymer Cement Concrete

The rheological properties of concrete with and without latex addition
are given in Table 5. The increase in workability (slump and flow) due
to latex additions is evident, remembering that the concrete without
latex contained 50% extra super-plasticizing agent. In particular
polymer 3 (SBR) increases the workability effectively.

The compressive strengths are listed for all the concretes at different ages in Table 6. The results reveal that both polymers lower the
compressive strength somewhat, with the most pronounced effect for
polymer 3 (SBR). The relative compressive strengths indicate a certain
retardation at the early age.

The results from the chloride penetration test are listed in Table 7.
First of all, it seems as though the additional entrained air (+1.8
vol%) leads to a somewhat shorter penetration time for chlorides in
this test. Secondly, the penetration time for concrete with polymer 3
compared to the reference, reveals that this polymer (SBR) gave no
extra protective effect against chloride to the rebar at this water/-
cement-ratio (0.45). Thirdly, the experimental latex prolongs the

penetration time for chlorides significantly, and particularly at dosages of 5% (> twice the time).

Cost-efficient use of PCC

Since polymers in general are about 10 times as expensive as cement, and since a 5 - 10 % dosage of the new latex seems to be necessary in order to secure proper chloride protection, the following scenarios may be utilized to decrease the costs:

i) Latex is easily mixed with ordinary concrete, and no special equipment is required. Since latex based PCC is compatible with CC, different sections of, for instance, a marine structure may be made of either PCC or CC. Those parts which are in the tidal or splash zone would be the main areas to be constructed of PCC. With only exposed parts of the structure containing polymers, the investment can pay off in increased service life before necessary repair/maintenance.

ii) Since most latices lead to excellent adhesion between the PCC and a CC substrate, the second scenario would be to shotcrete the exposed areas of a structure of cement concrete with a latex modified mortar.

Both these courses of action occur at the stage of construction, and this is recommended since the measures then are most effective and least costly. There is enough experience around to pin-point the weak parts of a structure in, for instance, a marine construction. Patch-work repair after corrosion damage has been observed is more or less of an esthetic nature only (i.e. a further slight delay of the process) and very expensive (work intensive).

TABLE 5   Rheological properties of fresh concrete

POLYMER	DOSAGE	Slump (cm)	Flow (%)	Density (kg/m^3)	Air (vol%)
REF	-[a]	12.0	135.0	2.343	3.8
-"-	-	12.0	120.0	2.364	2.4
No 3	5 %	23.0	180.0	2.355	1.8
-"-	3 %	20.0	150.0	2.374	2.0
No 4	5 %[b]	5.5	92.5	2.301	4.3
-"-	5 %	6.5	100.0	2.358	2.5
-"-	3 %	3.0	-	2.365	2.2

a: with air-entraining agent
b: without defoaming agent

TABLE 6 Density and compressive strength of the PCC's

POLYMER/ DOSAGE	Age (d)	Density at demoulding (g/dm^3)	Density at testing (g/dm^3)	Compressive strength (MPa)	Relative Compr. strength (% of ref.)
REF.[a]	1	2395±9	-	25.6±0.2	-
	3	2409±16	2429±15	34.4±0.4	-
	7	2407±25	2428±26	37.5±0.2	-
	28	2400±15	2382±16	50.5±0.3	-
	90	2393±25	2369±25	56.7±2.9	-
REF.	1	-	2401±11	28.0±0.8	-
	3	2405±8	2423±8	35.6±0.9	-
	7	2401±17	2409±15	39.5±0.2	-
	28	2399±2	2384±2	52.2±0.9	-
	90	2385±22	2363±19	58.7±0.3	-
No 3/5%	1	-	2367±5	18.7±0.5	66.8
	3	2358±4	2373±5	26.8±0.4	75.3
	7	2361±9	2373±6	30.2±0.1	76.5
	28	2351±22	2337±24	40.4±0.4	77.4
	90	2359±6	2332±6	48.1±0.7	81.9
No 3/3%	1	2401±7	-	24.7±0.1	88.2
	3	2385±2	2391±9	31.6±0.2	88.8
	7	2381±4	2397±4	35.3±0.3	89.4
	28	2406±26	2388±27	46.7±0.3	89.5
	90	2414±7	2387±5	53.3±0.4	90.8
No 4/5%[b]	1	-	2358±4	18.2±0.3	71.1
	3	2367±9	2379±8	27.4±1.0	79.7
	7	2355±12	2373±12	31.8±0.3	84.8
	28	2333±5	2307±6	50.6±0.1	100.2
	90	2310±17	2283±18	56.7±1.0	100.0
No 4/5%	1	-	2395±27	23.9±0.6	85.4
	3	2369±13	2383±14	30.5±0.6	85.7
	7	2372±17	2386±16	34.0±0.2	86.1
	28	2401±5	2388±4	48.0±0.1	92.0
	90	2381±1	2356±3	55.4±0.1	94.4
No 4/3%	1	-	2389±10	21.8±0.7	77.9
	3	2389±7	2404±9	31.4±0.3	88.2
	7	2399±2	2417±4	36.0±0.3	91.1
	28	2400±9	2387±10	50.4±0.4	96.6
	90	2401±6	2377±7	59.8±2.8	101.9

a: with air-entraining agent
b: without defoaming agent

TABLE 7 Time (days) for chloride penetration through concrete (10 cm) with and without latex additions

POLYMER	DOSAGE	Chloride penetration (current increase)	Visible corrosion (cracks)
REF.	0%[a]	44±8	49±8
REF.	0%	58±0[d]	63±0[d]
No 3	5%	59±1	59±2
No 3	3%	58±0	59±2
No 4	5%[b]	??[e]	123±9
No 4	5%	> 150	> 150
No 4	3%	> 83[c]	> 83[c]

a: with air-entraining agent
b: without defoaming agent
c: one parallel failed after 70 days, the second after 84 days
d: only two parallels instead of the usual three
e: the current vs time curve has not been analysed yet

CONCLUSIONS

A latex has been developed that effectively blocks chloride from entering mortar or concrete that has been modified by its incorporation. A 10 % addition of this polymer at w/c – 0.55 and a 5 % addition at w/c – 0.40 has been proven to give adequate protection towards chloride ingress. The test results for unmodified mortars at w/c = 0.40 and 0.55, illustrate the importance of using a concrete with low w/c-ratio in aggressive environments. The three tested commercial latices gave no extra protection against concrete at a w/c-ratio of 0.40. A 10 % replacement of cement with condenced silica fume, gave only a minor improvement of chloride resistance at w/c+s – 0.55.

REFERENCES

1 JUSTNES, H, DENNINGTON, S P. Designing Latex for Cement and Concrete. Nordic Concrete Research. Publ. No. 7 (1988) pp. 188-206.

2 OHAMA, Y. Polymer Modified Mortars and Concrete. Chapter 7 in Concrete Admixtures Handbook. Noyes Publ.(1984) NJ, USA.

3 BROWN, R P, KESSLER, R J. An Accelerated Laboratory Method for Corrosion Testing of Reinforced Concrete Using Impressed Current. Report No. 206. Florida Dept. Transportation Office of Materials and Research. October 1978.

4 ØYE, B A. Repair Systems for Concrete - Polymer Cement Mortars. An Evaluation of some Polymer Systems. PhD Thesis No. 54 (1989). the Norwegian Institute of Technology. N-7034 Trondheim, 194 pp.

CRACK CONTROL IN MASONRY STRUCTURES

A H Hosny

M I Soliman

H A Fouad

Ain Shams University,

Egypt

ABSTRACT. When a concrete block masonry wall is subjected to an in-plane concentrated load, high local stresses are developed in the region beneath that load. These stresses can cause spalling, or splitting of the masonry. In the present paper, two different techniques are proposed to reduce the stress gradients and concentration within these walls, and accordingly increasing their load carrying capacity, and the cracking load. These methods are by adding a reinforced concrete bond beam at the top of the wall, or by grouting the two cells beneath the concentrated load. In the experimental phase of the present work, eight wall panels were tested under a single centric concentrated load at the top of the wall. In the theoretical phase of the present work, a layered finite element program was developed, which is capable of predicting the non-linear behaviour of masonry structures. The finite element results were combined with those obtained experimentally, to formulate some recommendations for the design, strengthening and repair of this type of structures.

Keywords: Masonry, Concrete Block, Concentrated Loads, Grouting, Bond Beams, Finite Element, Plain Stress, Joint Element.

Prof. Dr. A.H. Hosny is a Professor at the Department. of Civil Eng., Ain Shams Univ., Cairo, Egypt. He received his B.Sc. degree in civil eng. from Cairo Univ. in 1950, his M.Sc. deg. from the same Univ. in 1954. He got his Ph.D. from Leeds Univ., U.K. in 1957.

Prof. Dr. M.I. Soliman is an Assistant Professor at the Department. of Civil Eng., Ain Shams Univ., Cairo, Egypt. He received his B.Sc. degree in civil eng. from Ain Shams Univ., Cairo in 1969, his M.Sc. deg. from the same Univ. in 1972. He got a M. Eng. deg. from McGill Univ., Canada in 1975 and his Ph.D. in 1979 from the same university.

Eng. H.A. Fouad is an Assistant Lecturer at the Department of Civil Eng., Ain Shams Univ., Cairo, Egypt. He received his B.Sc. degree in civil eng. from Ain Shams Univ., Cairo in 1987, his M.Sc. deg. from the same Univ. in 1990.

INTRODUCTION

Masonry is one of man's oldest building materials. Ancient Egyptians were the first to build extensively in stone masonry, eventually becoming so skilled in cutting and fitting individual stones that no joining material was necessary between units. Stability was the only structural criterion that had to be satisfied in this construction. However, the first study on masonries was performed by Coulomb (1776). In his calculation for the bearing capacity of arches, developed different concepts, and methods which two centuries later, formed the basis for the limit analysis [3].

Masonry is a composite material. The main types of units in use today are either stone or man made units. Within these main classes, there is an infinite variation in size, shape, degree of perforation, in addition to the basic variable of material strength. Masonry is hand-made with prefabricated units. The usual material for joining the units together nowadays, is a portland cement—sand mortar. The mortar is also subjected to variation in compressive strength, the strength of the adhesion of the mortar to the units, and the efficiency with which it is placed. Generally, the basic philosophy of masonry construction using prefabricated units of sizes appropriate to man's physical abilities and conditions, appears to be a sensible and, hence, rational construction method.

Masonry is the new challenge in the area of research work. There are so many variables in it which to tie it down is not easy. Thus, this is the additional challenge to the researchers.

The compressive strength of masonry is the most important parameter in the design of masonry structures. Recent design provisions for concrete masonry [1,2] list values for the specified compressive strength, which are based on the compressive strength of the block, and the type of mortar.

The tensile strength of masonry is also an important parameter in the behaviour of structural masonry elements such as shear walls. A possible failure mode of masonry shear walls is diagonal tension failure, which is mainly governed by the tensile strength characteristics of the combined masonry materials. Numerous studies on the diagonal tensile strength of masonry have been done, however most of it for brick masonry [4].

Masonry walls can be subjected to an in-plan concentrated load from a beam, lintel or prestressing anchorage. Current design rules for predicting the strength of walls subjected to concentrated loads are approximate and semi-empirical, due to lack of research in this area.

Therefore, studies of the behaviour of concrete block masonry wall structures under concentrated loads is essential for both the researchers and the designers of masonry structures.

EXPERIMENTAL WORK

In the experimental phase of the present study, eight wall panels were tested under compression using a single concentrated load at the top of the wall panels. These wall panels were divided into two groups as follows :

Group (A) : Consisted of four walls. All walls were without bond beams. The only parameter was in the position of the grouted cells.

Group (B) : Consisted of four walls similar to those of group (A). The only difference was in adding a reinforced concrete bond beam at the top of these walls.

Material Characteristics

Concrete blocks

Autoclaved masonry units were used in this investigation. The properties of the blocks were based on five test repetitions for each property. Three types of blocks were used. Block type (A) is the flat ended block with two pear shaped cores. Block type (B) is a half block (splitter block). Block type (C) is a bond beam or lintel unit used for the bond beam on top of the wall panels. The blocks were tested in compression to determine their compressive strength as well as their modulus of elasticity and poisson's ratio. Table [1] gives a summary of the properties of the blocks used in the present study.

TABLE 1 Characteristics of the used blocks

Block type	Shape	Dimensions(cm) L	W	H	Weight (Kg)	Area (cm²) Gross	Net
A		39	19	19	18.200	741	430
B		19	19	19	10.500	361	244
C		19	19	19	5.600	361	172

Mortar

The mix proportions of the used mortar were 1 : 2.8 parts by weight of portland cement and sand. Batching was controlled by weight. The water-cement ratio was kept 0.6. The sand was sieved to meet the specifications of ASTM C-144. The mortars were thrown out after a ½-hr. period to avoid variations resulting from retempering. The control specimens were tested under axial compression according to ASTM C-476 at approximately the same age as the corresponding wall.

Grout

A coarse grout mix (ASTM C476) was used in filling the cores and the reinforced concrete bond beams at the top of the walls. The mix proportions were 1 : 2.25 : 1.90 parts by volume and (1) : (2.50) : (2.20) parts by weight of portland cement, sand and gravel. Batching was controlled by weight. The water cement ratio was kept 0.60, which gave a slump value of 210 mm, assuring a fluid grout that could be poured in the cores without separation of its components. The control specimens used were block-molded prisms using paper as a porous separator so that water could be absorbed by the blocks. The prism dimensions were 95 mm x 95 mm x 190 mm. The control specimens were tested under axial compression at the time of testing corresponding assemblages.

Reinforcement

Plain rounded bars of mild steel were used for the reinforcement of the bond beams. Young's modulus of the steel reinforcement was found to be 2100 t/cm^2, and its yield strength was found to be 2450 kg/cm^2.

Fabrication Of The Wall Panels

An experienced mason laid the blocks using flush mortar joints. Excess mortar around the joints was removed from the cores before the grout was poured two days later. Between the age of 28 to 40 days, the wall panels were tested under compression.

Test Specimens

Eight wall panels of dimensions 1.20 x 1.00 ms and a nominal thickness of 0.20 ms. were build using the blocks, mortar, grout and reinforcement described previously. All wall panels were of the single-wythe construction. Table [2] shows the description of the wall panels tested in this experimental program. Also Figure [1] shows the shape and overall dimensions of the wall panels.

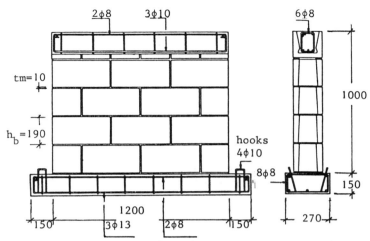

Figure 1: Shape and Dimensions of the Wall Panels.

TABLE 2   Details Of The Wall Panels

Group	Wall	Grouted Cells	Bond Beam
A	W1	. . . . . .	No
	W2	. . ▪ ▪ . .	No
	W3	. ▪ . . ▪ .	No
	W4	▪ ▪ ▪ ▪ ▪ ▪	No
B	W5	. . . . . .	Yes
	W6	. . ▪ ▪ . .	Yes
	W7	. ▪ . . ▪ .	Yes
	W8	▪ ▪ ▪ ▪ ▪ ▪	Yes

where : ▪ = The position of the grouted cells.

Test Setup

The loading was applied by a hydraulic machine of 2500 KN capacity and 10 KN accuracy. The machine had a rigid bed which acted as a support at the wall bottom.

Measurements

The strains were measured using a mechanical strain gauge of 0.002 mm accuracy and 5 mm capacity. The gauge length was kept constant at about 200 mm. The vertical and lateral deformations of the wall panels were measured using dial gauges of 0.01 mm accuracy and 10 mm capacity.

THEORETICAL WORK

A finite element model is developed which treats the blocks and the joints separately. It also allows for the nonlinear deformational characteristics, and progressive failure of both blocks and joints. The load is applied incrementally, thus allowing the behaviour of the wall to be traced from low load levels, up to the failure loads.

Proposed Finite Element Model

In the present study, two different types of elements were used to represent the different properties of masonry. The first is a layered plain stress rectangular element with two degrees of freedom at each node, which was used to represent the concrete block, grout and the bond beam. Accordingly, an eight D.O.F. rectangular element was obtained, as shown in figure [2].

The second type was a joint element, used to represent the vertical and horizontal mortar joints. The technique of adding the stiffness of the joint element to the total structure stiffness, was first developed by Ngo and Scordellis [10], in modeling reinforced concrete, and has been adopted to the field of rock mechanics by several investigators [6,12]. The joint element was assumed to be capable of deforming in the x and y directions only, figure [3], i.e. in the normal and shear directions.

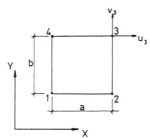

Figure 2: The eight D.O.F. Plain Stress Rectangular Element.

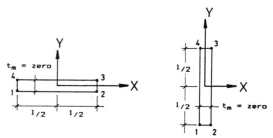

a) Horizontal joint element.   b) Vertical joint element.

Figure 3: The Mortar Joint Element.

Strength Envelopes

For the plain stress rectangular element, a biaxial strength envelope was used as shown in figure [4]. This envelops is divided into four regions which depend on the state of stress as represented by the stress ratio.

For the joint element, two separate sub-criteria were used. For joint slip failure in the shear-compression stress field, coulomb's theory of internal friction [9], can reasonably predict the joint strength of masonry assemblages, figure [5]. While for the shear-tension stress field, the modified Cowan theory [11], which was proposed to express the failure of the brittle materials in the shear-tension stress field, was adopted for this criterion.

The Finite Element Meshes

An irregular mesh was chosen with a fine mesh beneath the concentrated load, in order to determine the distribution of the stress concentration in this region, figure [6].

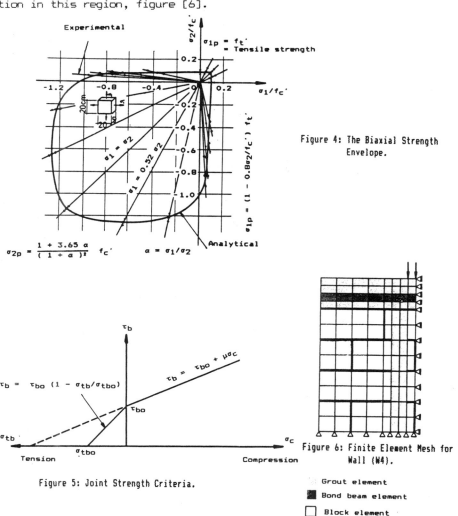

Figure 4: The Biaxial Strength Envelope.

Figure 5: Joint Strength Criteria.

Figure 6: Finite Element Mesh for Wall (W4).

Grout element
Bond beam element
Block element
Joint element

COMPARISON BETWEEN EXPERIMENTAL AND THEORETICAL RESULTS

A comparison is performed between the observed experimental behaviour of the tested walls, and those obtained theoretically from the proposed finite element analysis.

Table [3] shows the experimental cracking and failure loads, and the theoretical failure loads. A very good correlation between the experimental and theoretical results were observed.

TABLE 3   Cracking and Failure Loads

Wall	Experimental		Theoretical
	Cracking Load (KN)	Failure Load (KN)	Failure Load Lies Between (KN)
W1	325.0	395.0	375.0 — 400.0
W2	433.0	652.0	650.0 — 700.0
W3	416.0	631.0	650.0 — 700.0
W4	522.0	853.0	850.0 — 900.0
W5	332.0	402.0	375.0 — 400.0
W6	433.0	670.0	650.0 — 700.0
W7	430.0	662.0	650.0 — 700.0
W8	531.0	934.0	850.0 — 900.0

The experimental and theoretical crack patterns were compared in figure [7] for different walls. An excellent correlation was observed between the experimental and the theoretical crack patterns in the upper three courses of the walls. Due to the assumed boundary conditions in the F.E.M., movement was prevented at the base of the wall. This assumption prevented the formation of the cracks in the lower coarse, while in the experimental crack patterns, the cracks propagated till the base of the wall.

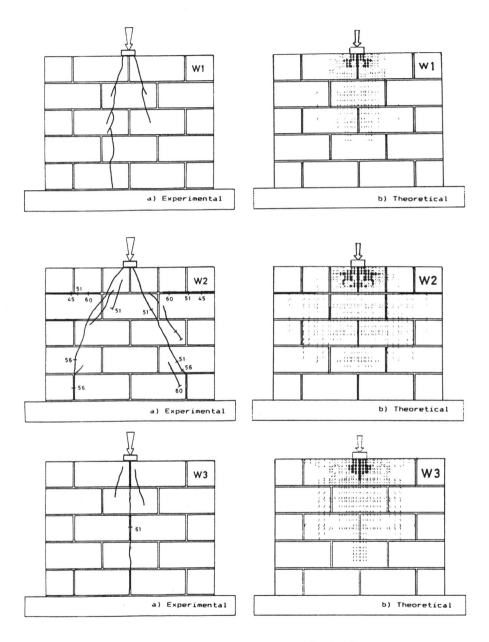

Figure 7: Experimental and Theoretical Crack Patterns.

/ Cracked Gaussian point
• Crushed Gaussian point
▬ Debonded mortar joint

CONCLUSIONS

1 Grouting the two cores beneath the load, had similar effects as fully grouting the wall. This is due to that the two cells beneath the load bears most of it. The advantages of grouting these two cores, on the behaviour of these type of walls are as follows :

a) Increasing the cracking load by about 20 to 30% for walls with and without bond beams, respectively.

b) Increasing the failure load by about 30 to 45%, and about 60 to 70%, for walls with and without bond beams, respectively.

c) For walls without bond beams, grouting the cores beneath the load increases the cracking load, (the cracking load is about 65% of the failure load). In this case, also a gradual failure occurs, compared to the ungrouted walls where the ratio between the cracking and failure loads was about 82%.

d) The vertical strains in the face shells of the walls decreases since the grout cores beneath the load bears most of it. The grouting also decreases the wall deflections under the load due to the increase of the stiffness of these walls.

2 Grouting the two cells beneath the concentrated load is quite enough and sufficient to increase the load carrying capacity of the wall. Therefore, there is no need for fully grouting this wall, nor grouting the two cores next to the loaded central cores.

3 High local stresses were developed directly beneath the concentrated load. These stresses were very high in ungrouted walls without bond beams. However, these stresses reduced by about 55 to 60% by the presence of bond beams.

4 In all the tested walls, the resulting stress concentration took place only within the upper course of these walls, and just beneath the load, which is about twice the width of the bearing plate. Then these stresses were rapidly distributed through the height of the wall.

5 Horizontal compressive strains were produced just under the concentrated load, in all the tested walls at all load levels. This is due to the effect of the bearing plate which confined the zone beneath it. This confinement, is localized in the upper part of the top course only, with a depth of about ½ the width of the bearing plate. Beneath this zone, tensile stresses were developed with a maximum value occurred at the lower third part of the top course, and along the vertical line beneath the concentrated load.

6 Providing a reinforced concrete bond beam at the top of the masonry walls, had the following advantages :

a) An increase in the cracking load by about 20 to 30%..

b) An increases in the failure load by about 65 to 70% for ungrouted walls, and about 35 to 40% for grouted walls with at least two grout cores under the load.

c) A cracking load of about 55 to 70% of the failure load for all walls, and hence prevent the sudden failure.

d) A reduction in the deflection values beneath the load due to the increase in the walls stiffness. It also results in a better vertical stress distribution, and also decreases the horizontal tensile stresses within the wall.

7 Due to the presence of the bond beam, the load disperses before reaching the hollow masonry. Accordingly, increasing the depth of the bond beam increases the dispersion of the load. Accordingly, it is recommended to select the depth of the bond beam not more than 3 to 5 times the thickness of the floor slab, since it has no significant effect beyond this limit

8 Grouting the two cells beneath the load, or adding a reinforced concrete bond beam at the top of the wall, had approximately the same effect on the behaviour of the wall. Therefore either techniques could be used to increase the load carrying capacity of these masonry walls. They also change the mode of failure of these types of walls to a gradual failure, and prevents the occurrence of sudden failure.

REFERENCES

1   AMERICAN CONCRETE INSTITUTE COMMITTEE 531. Commentary on Building Code Requirements for Concrete Masonry Structures. Journal of the American Concrete Institute, Volume 75, Number 9, September 78, pp. 460-498.

2   AMERICAN CONCRETE INSTITUTE COMMITTEE 531. Proposed ACI Standards: Building Code Requirements for Concrete Masonry Structures. Journal of the American Concrete Institute, Volume 75, Number 8, August 78, pp. 384-403.

3   BIOLZI, L. Evaluation of Compressive Strength of Masonry Walls by Limit Analysis. Journal of Structural Division, ASCE, Volume 14, Number 10, October 88, pp. 2179-2189.

4   DRYSDALE, R.G., HAMID A.A. Tension Failure Criteria for Plain Concrete Masonry. Journal of Structural Division, ASCE, Volume 110, Number 2, February 84, pp. 228-244.

5   FOUAD, H.A. Behaviour of Partially Grouted Concrete Block Masonry Walls Under Concentrated Loads. M.Sc. Thesis, Civil Eng. Dept., Ain Shams University, Cairo, Egypt, February 1990.

6   GOODMAN, R.E., TAYLOR, R.L. BREKKE, T.L. A Model for the Mechanics of Jointed Rock. Journal of the Soil Mechanics and Foundations Division, ASCE, Volume 94, Number SM3, Proc. Paper 5936, May 1968, pp. 637-659.

7   HOSNY, A.H., SOLIMAN, M.I., HAMDY, K.A. Behaviour of Concrete Block Masonry Walls Under Concentrated Loads. 8th International Brick/Block Masonry Conference, Ireland, 1988.

8   HOSNY, A.H., SOLIMAN, M.I., RAHMAN, A.A. Behaviour of Reinforced Concrete Block Masonry Walls Under Concentrated Loads. International Conference on Economic Design, University of London, June 1989.

9   NADAI, A. Theory of Flow and Fracture of Solids. Volume 1, McGraw-Hill Book Co., New York, 1950.

10 NGO, D. ,SCORDELIS, A.C. Finite Element Analysis of Reinforced Concrete Beams. Journal of the American Concrete Institute, Volume 64, Number 3, March 1967, pp. 152-163.

11 Zia, P. Torsional Strength of Prestressed Concrete Members. Magazine of Concrete Research, Number 14, 1953, pp. 75-86.

12 Zienkiewicz, O.C., et al. Analysis of Non Linear Problems in Rock Mechanics with Particular Reference to Jointed Rock Systems. Proceedings of the Second Congress of the International Society for Rock Mechanics, Belgrade, Yugoslavia, 1970, pp. 501-509.

THE RESEARCH OF EXTERNAL WALLS: DESTRUCTION CAUSES IN THE CONSTRUCTION WITH MAINTAINED MICROCLIMATE

I L Nikolaeva

N L Gavrilov-Krjamichev

V V Ovtcharov

Zaporozhe Industrial Institute,

USSR

ABSTRACT. The protective coating of wall panel lightweight concrete in buildings with constant microclimate (temperature,humidity), which is needed to create necessary conditions for production process, often fails. The temperature-humidity conditions of the light aggregate concrete walls exploitation were studied taking into account the climate influence. It was ascertained that the main walls destruction cause is the condensate formation, owing to distinations of the heat conductivity qualities of the walls elements (window frames, window pier). Moisture, accumulating in the light aggregate concrete panel pores under the alternating twenty-four hours temperature conditions, breaks the concrete panel core and exfoliates the protective coating. The variants of the wall panel lightweight conrete protection from the destruction are suggested.

Keywords: Wall Panel, Lightweigt Concrete, Microclimate, Temperature - Humidity Conditions, Fail, Exploitation, Analysis, Protection.

Mrs Irina L Nicolaeva is a scientific collaborator. Her interests lie in the field of construction structures reliability extention of the structures made of reinforced concrete. She is the author of a number of articles and author´s certificates.

Dr-Ing Nicholai L Gavrilov-Krjamichev is the head of the laboratory. He is the specialist in the field of the mechanics of destruction and author of more than 30 articles and author´s certificates.

Mr Vitaly V Ovtcharov is a scientific collaborator. He carries on researches in the field of methods of protection of the reinforced construction structures from the atmospheric exposure.

INTRODUCTION

Wall panel lightweight concrete and aluminium window frames were used in 70-80-ties while constructing industrial buildings with maintained microclimate. During the use of such buildings one could see destruction of wall panel lightweight concrete.

The task to find the causes of wall panel lightweight concrete fail and to develop recommendations on the protection of panel walls from fail was set.

A number of high-rise buildings in the South-East of the Ukraine were examined. The following things were done: wall panel lightweight concrete was examined; the conditions of the exploitation of wall panels were studied; the temperature-humidity conditions of the use of wall panels were researched; the calculations of the temperature of the dew-point were made.

The examined buildings had window frames of about 38 - 48% from the total area of the walls.

CHARACTER OF DESTRUCTION

The following distinctions of wall panel lightweight concrete destruction were found during the investigation of the walls.

The main area of destruction was a continuous belt 0.15-0.2m wide with downward local sections 0.2-0.3 m long, located under the window opening and lengthways the joints between the panels. Sometimes the place of destruction reached 40-48 mm depth and the reinforcement of the panel was exposed (1).

THE CONDITIONS OF EXPLOITATION

The formation of the strong condensate on the window cross casement and partitions between the windows in the period of the outside air temperature lowering down to -10 $^\circ$C and lower was seen while exploring the conditions of wall panels exploitation. In some rooms glass partitions lengthways outside walls at a distance of 0.5 m were installed to maintain temperature and humidity. Under these conditions the exploitation of the outside walls improved. But in winter, when the temperature of the outside air was lower then -13°C the condensate on the window frames and partitions appeared.

accumulated moisture, under the alternating twenty-four hours temperature condition crystallizes and breaks the wall panel lightweight concrete. In consequence of it, the superficial protection of the exterior walls (facing, stucco, etc.) does not yield desired results, since the moisture accumulating process takes place at the protective covering boundary and, as a result of it separation and rapid destruction of the latter. The destruction process proceeds most actively in winter – spring period.

CONCLUSIONS AND SUGGESTIONS

The main cause of the wall panel destru ction is the accumulation of the moisture in the wall panel lightweight concrete pores in the heating season of building exploitation. The moisture accumulates in the wall panel lightweight concrete pores owing to the condensate formation on the window frames and partitions between the windows.

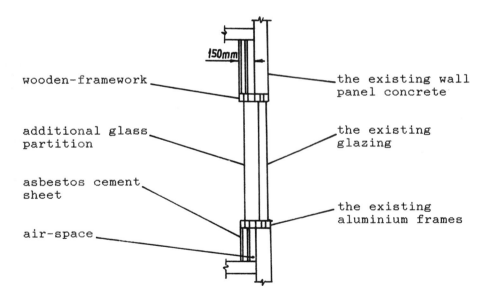

Figure 1: First variant protection

The emerging aerodynamic processes intensify the process of distruction of the wall panels.

The intensive diffusion process favours the penetration of the moisture into the lightweight concrete pores in winter season.

One should mention, that the emission of the ammonia
during the production process favoured the emergence of
the condensate in many rooms. Livings vapour seal was
used for outside walls protection. But, owing to the
constant bondensate formation, the vapour seal in a short
period of time failed.

The repeated repairs of the protective outside walls
covering sharply increased the costs of the exploitation
of the buildings. For instance, current expenditures on
the repairs of one 9-storey building during 3-year-period
of exploitation were equal to more than 43% of its
initial cost.

TEMPERATURE HUMIDITY CONDITION

Automatic control of ventilaition and airconditioning was
used to maintain microclimate. Nevertheless, local
temperature changes were $\pm 7^\circ C$ and humidity fluctuations
made $\pm$ 10%. In autumn-winter period the surface
temperature of the aluminium window frames and partitions
between the windows inside the building goes down lower,
than the dew-point of the air in the room and it causes
the condensate formation. Aerodynamic processes,
appearing during building exploitation, influence the
temperature humidity condition and, accordingly, the
outside walls condition. They are: positive pressure,
caused by the plenum-exhaust ventilation; the inleakage
of the outside air and outflow of the indoor air, caused
by the porosity of the walls; the convection of the air
inside the building during the heating period.

In addition-in winter period, there is a water vapour
diffusion in buildings with heating through exterior
walls from the inside to outside surface (2). All the
above mentioned processes favour the water vapour
penetration into the wall panel lightweight concrete
pores.

DESTRUCTION MECHANISM

The study of the dynamics of defect formation revealed,
that the wall panel lightweight concrete destructions
maximize in winter-spring period. This maximum
correlates with the alternating twenty-four hours
temperature fluctuations of the atmospheric air.

The moisture, pene trating through badly protected
sections of the wall panel lightweight concrete surface (
joints, panel faces, breakdowns of vapour seal)
accumulates in the wall panel lightweight concrete pores
and penetrates to the exterior surface of the walls. The

Taking into account the above-stated it should be admitted, that the principal measures of the wall panel lightweight concrete protection lie in the elimination of the condensate. It was taken into account while developing the variants of protection.

First Variant

The lining of the wall panel lightweight concrete with asbestos cement sheets on the wooden framework with an air- space; the installation of additional wooden window frames; obligatory thorough hermetic sealing (Figure 1).

Second Variant

The installation of a glass hermetic partition with a metal framework. The connection of the air-space automatic ventilation system between the glass partition and the exterior wall to the general automatic ventilation sustem. The temperature humidity condition of the air-space should lie in the range of: temperature 18-16°C; humidity 40-45% (Figure 2).

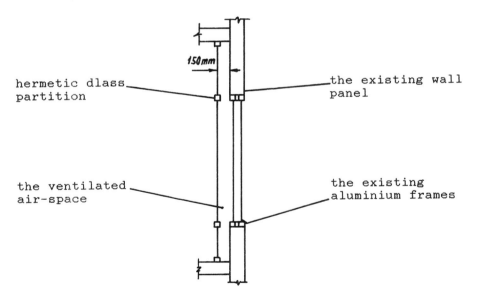

Figure 2: Second variant protection

REFERENCES

1   Luzhin O V , Zlochevsky A B , Gotbunov I A , Volohov V A. Inspection and Testing of Civil Engineering Structures Contents. Stroiizdat, Moscow, 1987, pp 262.

2   Iljinsky V M. Construction Thermal Physics (fence sructures and microclimate of the buildings). Vysshaja shkola, Moscow, 1974, pp 320.

MECHANICAL PROPERTIES OF CORRODED STEEL PILES PARTIALLY COMPOSED BY REINFORCED CONCRETE

O Kiyomiya

H Yokota

T Fukute

Port and Harbour Research Institute,

Japan

ABSTRACT. A steel pile under marine environments may have been damaged greatly by corrosion during long term service. Damage is concentrated at the splash zone and causes lack of strength. Repair is required to such a damaged pile. Reinforced concrete covering is one of the effective repair techniques. Although this technique has been applied to field work, its design method has not been established. Flexural loading tests on concrete covered steel pipe and sheet piles have been undertaken to investigate their mechanical properties and to confirm the applicability of the tentative design method. It was made clear that the arrangement of shear connectors has greatly affected the effectiveness of this repair technique.

Keywords: Repair, Composite Structure, Steel Pipe Pile, Steel Sheet Pile, Loading Test, Structural Analysis.

Dr Osamu Kiyomiya is a Chief of Structural Mechanics Laboratory, Port and Harbour Research Institute, Ministry of Transport, Yokosuka, Japan. He got his Dr Eng degree at Tokyo Institute of Technology in 1984. His recent research work includes: structural analysis and design, earthquake resistant design, repair and maintenance.

Hiroshi Yokota is a Senior Research Engineer, Port and Harbour Research Institute, Ministry of Transport, Yokosuka, Japan. He got his MSc degree at Tokyo Institute of Technology in 1980. He is actively engaged in research into composite structures and various aspects of structural problems.

Dr Tsutomu Fukute is a Chief of Materials Laboratory, Port and Harbour Research Institute, Ministry of Transport, Yokosuka, Japan. He got his Dr Eng degree at Nagoya University in 1985. He has actively undertaken research into durability of concrete and pavement design.

INTRODUCTION

A steel pile is one of the principal materials for port and harbour structures. Since about 1955, steel piles have been used rather frequently for facilities as piers and mooring quays in Japan. Corrosion of steel is a serious problem where is in directly contact with sea water. Although those piles employed corrosion prevention procedures such as cathodic protection and had extra thickness against presumed loss, development of corrosion is far remarkable than expectation particularly in the zone between high and low water levels from the result of corrosion surveys. Corrosion causes poor performance under service load and inadequate ultimate strength.

Such damaged structures may have to be demolished or rebuilt, or locally repaired and strengthened. The first alternative is not a desirable answer because it needs lots of costs and time, but the second is more attractive. Most of the damaged facilities can keep their functions sufficiently after being repaired and strengthened.

The reinforced concrete covering technique, where reinforced concrete covers over the damaged part of steel piles, is effective and has considerable potential because it is simple, quick, and cheap and, more attractively, it makes corrosion rate slow. Although the technique has been used extensively, many problems particularly their strength and durability are remaining. The repaired part of a damaged pile becomes a composite structure and then stress transfer has to be surely made by certain shear connectors.

The mechanical property of repaired piles with the reinforced concrete covering, which is a partially composed structure with steel and concrete, are investigated experimentally. In particular, its covered range and the mechanism of shear resistance of studs in thinly covered concrete are examined. This paper presents the ultimate strength, the mode of failure, and structural details of reinforced concrete covered steel piles. Furthermore, those experimental results are compared of calculated ones from the tentative design manual [1] which is followed by discussion of its applicability.

STATE OF CORROSION ON STEEL PILES

Steel pipe piles and sheet piles are used for the foundation of piers and are used in front of piers respectively as illustrated in Figure 1. Corrosion investigation on those piles has been made since 1967 mainly by the Port and Harbour Research Institute. The results [2,3] point out that corrosion patterns and corrosion amounts depend on the location of piers. Figure 2, for example, shows the corrosion amount of a steel sheet pile after 15-year service. The corrosion is serious in the zone between high and low water levels and this phenomenon is often observed. Within the zone, corrosion rate is beyond the expectation and, in some severe cases, it reaches about 1 mm per a year. On the other hand, the corrosion rate is very low where always

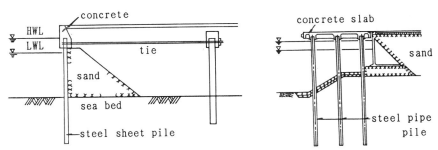

Figure 1 Typical structures of piers

submerged in sea water and in sea mud, that is, less than 0.2 mm per a year. Thus, corrosion tends to concentrate in the zone between high and low water levels and may cause serious problems there.

OUTLINE OF REPAIR WORK

The purpose of repair is to restore the lost strength of piles and to prevent further progress of corrosion. When stress induced by design external forces exceeds the limited value, or when the stress is feared to do so in future at a thin part of piles, repair should be considered.

Figure 3 shows the outline of the reinforced concrete covering method for repair. At first as surface preparation, rust and sticking organisms are removed from a pile. Holes are patched by steel plates. Studs are welded as shear connectors to above and below the area where repair should be made. Studs of 12 through 16 mm in diameter are often used. Where plate thickness is less than 5 mm, however, studs should not be supplied. Longitudinal reinforcing bars and stirrups are arranged around a steel pipe pile or in front of a steel sheet pile. Steel and reinforced concrete are combined mechanically by studs. Then concrete placed within shuttering. The thickness of concrete cover is generally 150 through 250 mm and the cover depth to embedded bars is required to be more than 70 mm from the viewpoint of durability. The range of concrete cover is usually on the part from just below the slab to the

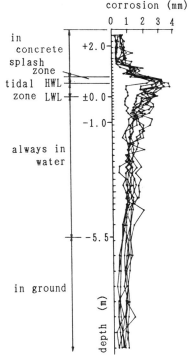

Figure 2 Corrosion amount of a sheet pile

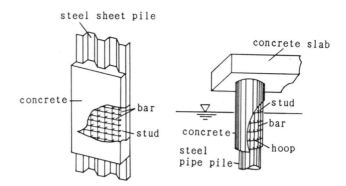

Figure 3   Reinforced concrete covering techniques

level of LWL-2 m, where LWL stands for the low water level. The
composite part carries external loads by both steel and concrete.

DESIGN REQUIREMENT OF COMPOSITE PILES

Ultimate Strength

Flexural moments and axial forces are induced in piles by horizontal
external forces due to earth pressures, earthquakes, ship mooring, and
so on. The resisting flexural moment of a composite pipe which is
repaired by the covering method is calculated on the assumption that
additional reinforcement is regarded as a pipe of uniform thickness
with the same cross sectional area and a parent steel pipe and
reinforced concrete are fully composed. The stiffness of the pipe
should be neglected at the part where studs are not provided or its
thickness is less than 5 mm.

Ultimate strength of the composite sheet pile is calculated by the
method similar to that on normal reinforced concrete structures on the
assumption that the parent steel and the concrete are fully composed.
Here, the thickness of the sheet pile is neglected because it is
considerably smaller than the neutral axis depth.

Ultimate shear force is obtained by the tied-arch resisting mechanism.

Width of Crack

Sufficient durability is required for port and harbour structures made
of reinforced concrete. It is important to control the occurrence of
cracks for preventing corrosion of embedded reinforcement. The width
of crack under marine environments due to flexural moments is usually
limited to 0.0035c, where c stands for the cover depth, and the cover
depth should be more than 70 mm.

Strength of Studs

Studs are usually arranged in longitudinal and transverse directions
with required intervals. The minimum plate thickness required for
welding studs is 5 mm or one third of the diameter of studs. Thus,
studs shall not be attached to the part where corrosion is quite heavy.
An allowable shearing force per one stud is specified in various
standards, which has been investigated through push-off tests and
fatigue tests. These values depend on the compressive strength or the
casting direction of concrete, the size of studs, the arrangement of
reinforcement, and kinds of external forces. The allowable force of a
stud in marine repair work should be decided likewise.

LOADING TESTS OF COMPOSITE PIPE PILES

Since flexural moments act on repaired piles mainly, flexural loading
tests [4] on seven composite beams with different structural details
have been undertaken.

Test Specimen

Details of the test beam are shown in Figure 4. Their overall length
is identical of 3.5 m. The outer diameter and the thickness of the
parent steel pipe is 318.5 mm and 6.9 mm respectively and its yield
strength is 370 MPa. Table 1 presents the structural modifications.
SP-1 is a reinforced concrete pipe beam and SP-7 is a steel pipe beam
without covering. Both the beams are made for investigating mechanical
properties of respective materials only. The others are partially
composite beams with one pair of virgin steel pipes and reinforced
concrete. The thickness of the covering concrete is 140 mm. Within
500 mm of the midspan of the beam, steel is not provided for the
simulation of thinning due to corrosion. The covering regions and
numbers of studs are altered as described in Table 1. The conventional
design method requires 24 studs. The covering concrete is reinforced
by ten deformed bars of 16 mm in diameter at equal intervals. Hoops of
10 mm dia deformed bars are arranged at intervals of 150 mm. The yield

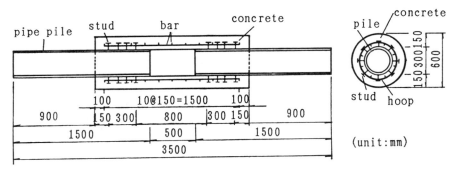

Figure 4   Details of SP-4

Table 1 List of SP beams

No	covering range	number of studs (one side)	length of pipe
SP-1	3.5 m	–	–
SP-2	3.5 m	32 (8 x 4)	1.5 m x 2
SP-3	2.7 m	32 (8 x 4)	1.5 m x 2
SP-4	1.7 m	32 (8 x 4)	1.5 m x 2
SP-5	1.7 m	24 (8 x 3)	1.5 m x 2
SP-6	1.7 m	12 (4 x 3)	1.5 m x 2
SP-7	–	–	3.5 m x 1

strength of these bars is approximately 360 MPa. The headed stud is 16
mm in diameter and 75 mm in height. Longitudinal bars are arranged
just under the head of studs and hoops are on the side surface of the
studs. The cover depth to the bars is about 80 mm. Concrete is cast
keeping the test beam in vertical position which is the same manner as
field work. The maximum size of the aggregate is 10 mm. The design
strength of the concrete is 24 MPa.

Test Procedure

Figure 5 shows the test set-up. Symmetrical loads are statically
applied to the beam. The distance between the supports is 3 m and that
between the loading points is 0.85 m. Purpose made supports and loading apparatus whose faces are one quarter circle so as to contact with
the beam at right angle with thin rubber sheets. Loads are applied by
a manually operated hydraulic jack with monotonous increase in 20 kN.

An applied load and deflection are measured by a load cell and
displacement transducers respectively. Strains of steel and concrete
are measured by strain gauges, and widths of crack are measured by the
contact type gauge. Its measuring distance is 100 mm. Measurement of
crack widths is made along the lower edge of the beam. Development of
cracks is observed
visually and sketched.

Test Results

The test results are
summarized in Table 2.
Here, the ultimate load
is the maximum applied
load and the cracking
load is the value when a
flexural crack is initi-
ated. The ultimate load
of SP-2 hardly showed
difference with that of
SP-1 in spite of the
absence of the steel
pipe at the centre. As

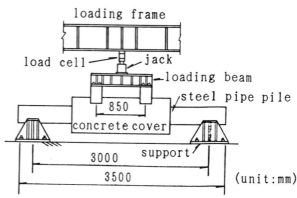

Figure 5 Test set-up for SP beams

Table 2   Results of bending loading test

No	cracking load (kN)	0.2 mm cracking load (kN)	yielding load (kN)	ultimate load (kN)
SP-1	99.2	116.8	289.8	530.8
SP-2	77.6	116.8	260.5	534.0
SP-3	48.9	138.4	237.6	466.8
SP-4	79.6	116.8	197.1	424.9
SP-5	98.5	118.8	197.1	432.2
SP-6	41.1	118.8	198.5	362.3
SP-7	–	–	*254.8	344.8

*) yield of the steel pipe

the covering range decreased, the ultimate load, etc. also decreased. Although the numbers of studs were different, the ultimate loads were almost the same between SP-4 and SP-5. When the number of studs is 24 (SP-6), however, approximately 16% of the ultimate load reduced.

Figure 6 shows crack formation in SP-4. When the load reaches 118 kN, axial cracks at the position about 45 degrees from the upper end of the concrete developed and the concrete separated from the steel pipe. Along with the development of flexural

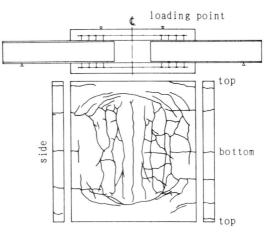

Figure 6   Crack formation (SP-4)

cracks, shear cracks were observed. When the load exceeded 392 kN, shear cracks were remarkable near the loading points. Finally, crushing of concrete occurred between the loading points. Destruction of steel including studs was not observed. Similar to SP-4, shear failure was dominant in SP-5 and SP-6.

Figure 7 shows the distribution of strains of the steel pipe, the reinforcing bar and the concrete in SP-3. In the right half of the beam, the concrete was covering over the steel pipe completely, and strain of concrete was larger than that of the pipe. Therefore, resistance against external force was made mainly by the covering concrete. In the left half of the beam, on the other hand, the concrete was covering just one part of the pipe, and both the pipe and the concrete resisted against the loads. At point b in this figure, strain of the pipe was large, but that of the concrete was very small. At point c, however, strain of the pipe was small and that of concrete was large. It is found that the pipe and the concrete were composed by studs and the transmission of forces was relatively smooth there. The

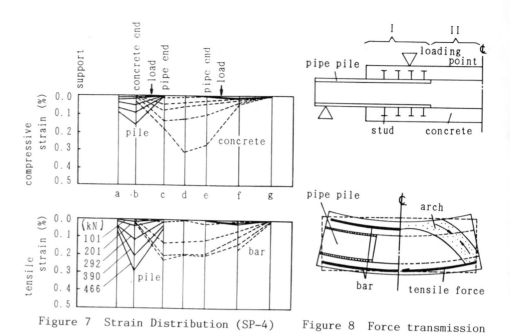

Figure 7 Strain Distribution (SP-4)   Figure 8 Force transmission

calculated yielding load, 204 kN, showed fairly good agreement with the
test result as indicated in Table 2. The calculated ultimate load, 297
kN, was, however, considerably smaller than the test results. The
reason is that the beams could resist the loads with the arch-resisting
mechanism after they collapsed. Figure 8 illustrates the aspect of
shear failure, which may be useful to explain the mechanism. The shear
failure occurred at zone I, but did not at zone II. Since zone II was
a composite structure, its stiffness was about twice as large as that
of zone II. Hence, zone II was supposed to be a fixed point for zone
I. That is, zone I was regarded as a beam with fixed supports.

LOADING TESTS OF COMPOSITE SHEET PILES

Loading tests [5] on eight specimens have been undertaken.

Test Specimen

The overall length and width of the test beam are 5.0 m and 1.6 m
respectively. Figure 9 shows the details of SS-1. Four virgin steel
sheet piles whose yield strength is 450 MPa are assembled to fabricate
the test beam. Its central area of 3.0 m is covered with reinforced
concrete. Deformed bars of 16 mm in diameter are arranged along the
axial direction of the beam at intervals of 150 mm, while those of 13
mm in diameter are arranged transversely at intervals of 130 mm. Their
strength was about 360 MPa. The beams are modified their details

according to Table 3. For SS-1 to SS-4, studs of 16 mm in diameter and 75 mm in length are used as shear connectors. They are arranged in 2 lines at intervals of 600 mm for SS-1 through SS-3 and in 1 line at intervals of 1200 mm for SS-4. For SS-5 and SS-6, J-shaped bars of about 230 mm in length and stirrups made of deformed bars of 13 mm in diameter are arranged respectively. In SS-5 and SS-6, the shear connectors are densely arranged near the end of the concrete. The test beam is strengthened by steel at its loading point and supports to prevent widenning or twisting. Initial stresses of 175 MPa and 290 MPa are introduced to SS-2 and SS-3 respectively before concrete casting. The initial stress simulates actual condition of a pipe which is induced fairly large stress due to external forces on corroded parts. The maximum size of aggregate of the concrete is 25 mm. Compressive strength of the concrete ranged from 21 to 30 MPa at the tests.

Test Procedure

Figure 10 shows the test set-up. The load is applied from the sheet pile side symmetrically and the distance between supports is 4.0 m.The loading programme and the instrumentation are the same as those for the pipe pile test.

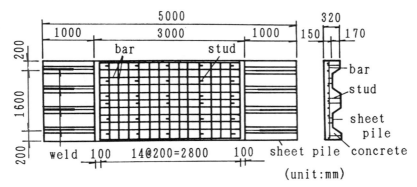

Figure 9   Details of SS-1

Table 3   List of test specimen

No	shear connector	initial stress
SS-1	40 headed studs	no
SS-2	40 headed studs	176 MPa
SS-3	40 headed studs	294 MPa
SS-4	15 headed studs	no
SS-5	36 headed studs 36 J-shaped bars	no
SS-6	72 headed studs 45 stirrups	no
SS-7	sheet pile only	no
SS-8	reinforced concrete	no

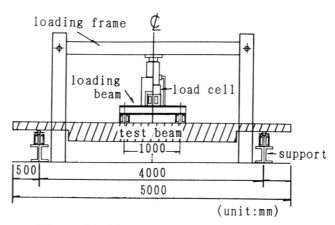

Figure 10   Test set-up for SS beams

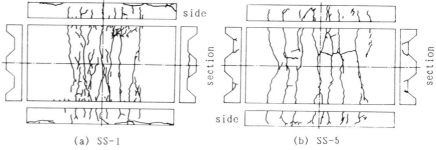

(a) SS-1   (b) SS-5

Figure 11   Crack formation of SS-1 and SS-5

Test Results

Crack formations of SS-1 and SS-5 are sketched in Figure 11. In SS-1, a crack was initiated near the midspan at the load of 39 kN and then the number of cracks increased. When the load reached 157 kN, cracks occurred horizontally at the end of the concrete. This was caused by shear failure of the concrete around the studs. When the load reached 314 kN, the concrete was crushed in conical shape and the sheet pile and the concrete separated from its end. Subsequently, the beam collapsed quickly and buckling of the pile finally occurred. The process of collapse of SS-2 to SS-4 was similar to that of SS-1. In SS-5, a crack was initiated at the load of 90 kN, and then flexural cracks were formed. Finally, shear failure of the concrete around the studs was observed. However, collapse at the end of the concrete like SS-1 was not observed. The ultimate load of SS-5 was the largest among the beams. In SS-6, flexural cracks developed like the other beams, but finally, shear failure of the concrete occurred at the stirrup. In all the beams, destruction of the studs and the bars did not occur. The calculated ultimate moment of the beam is 120 kNm, which is 307 kN

in terms of load. It can be
said that SS-1 through SS-6
have higher yield strengths
than this value. SS-1 through
SS-4, however, did not have
sufficient added strengths
because of destruction caused
by pull-out shearing. Hence,
mechanical properties of these
beams were not preferable.
Therefore, use of long studs or
stirrups should be necessary.
Strength characteristics of the
beams with initial stresses
showed no remarkable difference
with those without them.

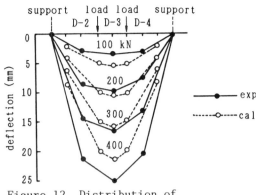

Figure 12   Distribution of
deflection (SS-4)

Figure 12 shows the distribution of deflection along the axis of SS-4.
While the load was small, the distribution line bent at D-2 and D-4 in
this figure where the flexural stiffness changed. As the load was
increasing, the deflected curvature of the concrete part also became
large because of the decrement in the stiffness due to cracks. The
calculated values on the assumption of a fully composite beam are also
shown in this figure. In this calculation, the joint efficiency [6] of
the sheet pile was determined to be 0.83 which was the test result from
SS-7. The test and calculated deflections showed fairly good agreement
until steel yielded.

Figures 13 and 14 show the slip and the separation between the sheet
pile and the concrete at its end. The slip was about 1 mm at the load
of 294 kN, which was not so large. Separation began when the load was
small and reached 1.0 mm at the load of 294 kN. The separation has to
be paid attention, because it should cause further corrosion of steel.

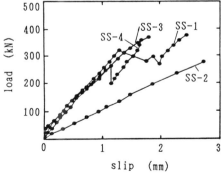

Figure 13 Slip between the
sheet pile and the concrete

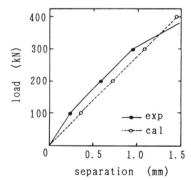

Figure 14   Separation of the concrete from the sheet pile (SS-4)

CONCLUSIONS

The number of studs greatly affected ultimate loads and structural characteristics. When specified number of studs from the current design criteria is used, the composition of steel and concrete is made sufficiently. As increasing the load, shear cracks developed and finally caused flexural-shear failure as the tied-arch resisting mechanism. Predicted ultimate strength based on this mechanism was almost the same as the test results.

It was necessary to provide sufficient reinforcement to the end part of the covered concrete. If shear connectors were not fixed tightly, concrete around them near the end of the concrete collapsed by pull-out forces and the overall structure rapidly deteriorated. Slip between the parent pile and the concrete was not so much, but separation was found at small loads. When long studs or stirrups are supplied there, such phenomena were not observed.

The predicted ultimate strength of the partially composite structures was the safe side compared of the test results. For sheet piles, it was necessary to consider the joint efficiency, which was about 0.8 based on the test results. This should be decided according to actual situation of sheet piles and the condition of surrounding ground.

REFERENCES

1   COASTAL DEVELOPMENT INSTITUTE OF TECHNOLOGY. Manual for repair of steel structures in port and harbours. Tokyo, March 1986.

2   ABE M, YOKOI T, OTUKI N, and YAMAMOTO K. Corrosion Survey Compilation of Steel Structures. Technical Note of the Port and Harbour Research Institute, No 628, September 1988, pp 1-261.

3   KIYOMIYA O, CHIBA T, YOKOTA H and ABE M. Corrosion Survey and Residual Strength of Steel Pipe Piles for Pier. Technical Note of the Port and Harbour Research Institute, No 593, September 1987, pp 1-29.

4   KIYOMIYA O, CHIBA T and YOKOI T. Mechanical Properties of Repaired Steel Pipe Pile Covered by Reinforced Concrete. Report of the Port and Harbour Research Institute, Vol 27, No 1, March 1988, pp 125-173.

5   KIYOMIYA O, NOGUCHI T and YOKOTA H. Mechanical Properties of Repaired Steel Sheet Piles. Report of the Port and Harbour Research Institute, Vol 28, No 3, March 1990, pp 181-232.

6   SHIRAISHI M. Theoretical analysis of U-type sheet pile wall on shearing resistance force of interlocking joints and sectional properties. Proceedings of the Japan Society of Civil Engineers, No 385/VI-7, September 1987, pp 49-58.

STUDY OF THE PROTECTION OF PRESTRESSED CONIC ANCHORAGES OF ALUMINA CEMENT CONCRETE

H B Sun

Zhejiang Institute of Technology

Republic of China

ABSTRACT. In post-tensioned structures, high alumina cement (HAC) concrete was adopted to anchor the ends of the prestressed bars in China because of its high strength and quick hardening.

HAC concrete suffers from a slow decomposition reaction. This conversion results in a decrease in strength especially in hot wet conditions.

Based on experiments, however, it is revealed that the strength of HAC concrete covered and coated with rigid polyurethane(PU) foam and other coatings is significantly higher than those that are bared. The PU foam has an extremely low thermal conductivity. The PU coatings combined with others exhibit excellent resistance to water vapour, water and chemical attacks. These coverings and coatings insulate influence of the bad environment, so that the rate of conversions is decreased and the minimum strength may be higher.

On the basis of pull-out bond tests, it is also indicated that HAC concrete conic anchorages surrounded by the block are forced in three dimensions when the prestressed bars are drawn, consequently, the bond strength between the prestressed bars and the surrounding HAC concrete within the female cone will increase. The increased bond strength partially compensates HAC concrete in end anchorages for a loss of conver-'sion. Lining and coating with rigid PU foam and others combined are suitable for important HAC concrete structures which suffer from hot, wet and chemical attack conditions.

Keywords: Anchorage, High Alumina Cement (HAC)., Conversion, Bond Strength, Polyurethane (PU) Foam, Slide, Restrained bar.

Huan Bin Sun is a Prof. at the Department of Civil Engineering, ZheJiang Institute of Technology, HangZhou, ZheJiang Province, the People's Republic of China, where he is a former head of the department and has been involved in teaching, research and design work. He has published numerous papers and was awared several prizes for achievements in scientific research in China.

INTRODUCTION

In the late fifties prestressed concrete was initially used in China. The prestressed reinforcement was steel rods, but after a few years low-alloy deformed steel bars in sizes 10 and 12 mm diameter with ultimate strength of 750 N/mm^2 were available and several wedge-type metal end anchors were adopted to anchor the prestressed bars. Some afterwards, high alumina cement (HAC) concrete with fine aggregate was used instead of metal anchorages in order to save metal and machining. The processes are as follow: At the ends of the concrete member there are female cones (Figure 1,(a) & (b)). Between the end of the concrete and the jack, there is a rack (Figure 1,(b)), used as a tool, the prestressed bars are wedged around the inner side of the cone and are stretched by the jack. When the required tension is reached, the prestressed bars are fixed on the rack and the jack is released. High-strength fine HAC concrete is then poured into the cone. After the concrete quickly hardens, the prestressed bars are cut and the pre-stress is stransferred to the HAC concrete anchorage and the rack is moved. In this way, the turnover rate of the rack is raised. This measure was widely used in the early 1970s.

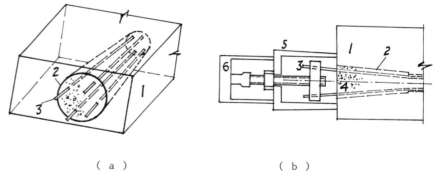

(a)   (b)

1 --- The end of the concrete member   2 --- Female cone
3 --- Prestressed bars   4 --- HAC concrete
5 --- Rack   6 --- Jack

Figure 1: The end block of a prestressed member

The collapse of the roof over a swimming pool in Britain, it alerted the designer to potential hazards to buildings in which HAC had been used. End anchorages are the most important part of a prestressed member. The immediate problems have been brought to the fore with existing HAC concrete anchorages mainly used in the prestressed members of industrial buildings, such as roof girders and trusses,crane girders, trussed girder. Sometimes these members are in an environment with higher temperatures and humidity where HAC concrete suffers from a

conversion reaction and loses strength. Appraisal of its inner condition and appropriate preventive measures became necessary and a series
of research programmes were started.

THE PRINCIPLES OF CONVERSIONS

The most important reactions during the hardening of the HAC concrete
are the formation of monocalcium aluminate decahydrate CAH_{10}, but they
are metastable and convert to tricalcium aluminate hexahydrate C_3AH_6 and
gibbsite AH_3 which are more stable, the slow decomposition is:
$$3CAH_{10} \rightleftharpoons C_3AH_6 + 2AH_3 + 18H$$
This reaction liberates water and results in a reduction in the volume
of the solid and an increase in the porosity so the concrete will lose
strength and durability is reduced. The rate at which conversion and
loss of strength takes place is affected mainly by changes in temperature and moisture levels[1]. The higher the temperature and moisture,
the faster the rate of conversions[2]. On the contrary, the rate of
conversions will be slower, and the minimum strength higher when
temperature and moisture are lower.

PROTECTION TESTS AND THEIR RESULTS

Preparation of Samples

HAC with strength of 40 N/mm^2 was produced by ZhengZou No. 503 Factory.
Chemical content is shown in Table 1.

TABLE 1   Chemical content of HAC

SiO_2	AL_2O_3	Fe_2O_3	CaO	MgO	TiO_2	loss on ignition
4.66	57.51	1.33	28.5	0.4	3.0	1.37

Concrete mix and test specimens

Granite aggregate of 10 mm maximum size togather with silica sand was
used. Cube strength of 35 N/mm^2 and water-cement ratio of 0.38 were
adopted. Five groups, each of which consists of three cubes of 100 mm
concrete samples. Four groups were cured at room temperature 20 ± 1°C
with humidity of less than 65 % RH for more than one month.   One grounp
was cured for only one day.

Protection of HAC concrete

Two groups were covered with rigid moulded polyurethane (PU) foam
completely enclosing them while the others were exposed to the air.

The constituent of PU is shown in Table 2.

TABLE 2   The constituent of PU foam

Polymeric MDI
Polyether
Monofluorotrichloromethane (R11)
Ethylenediamine
Polyetherpolysiloxanes
Other addtional

The thickness of foam around was 120 mm, except for 150 mm at the
bottom. After demoulding multiple layers of PU protective coatings
alternating with coal tar and other anti-ageing materials were used to
resist water vapour and water formed by condensation.

Protection Tests

Two thermocouples and two humidometers were glued on opposite sides of
one of the cubes of the group 4 samples. It was set in a small curing
room. The room temperature was controlled by electric rings and the
humidity kept at about 95% RH by necessary adjustments to water surface.
The heating curve1 (Figure 2) was adopted. The curve is the upper
envelop of the curves which were measured with thermocouples from an
inner point standing about100 mm from the surface on which the sun
shone of a crane girder in the open air. Each measurements took
place over a few days in different cities in the height of summer.
These measured inner points represant the situations of HAC anchorages in bad conditions. Curve 2 shows the mean value of the two
thermocouples and the humidity indicated relatively no change from the
oringinal. It is shown that it is very efficient to insulate heat and
damp by means of covering and coating with PU.

At a nearby pool where slag, from a blast furnace, is scoured by water,
two groups of PU covered cubes and one group of bare cubes were placed
in a steel mill. All these cubes togather with a thermometer and a humi-

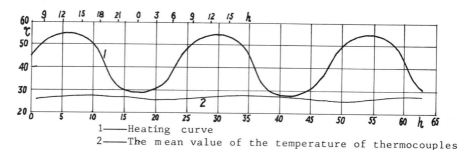

1——Heating curve
2——The mean value of the temperature of thermocouples

Figure 2: Curves of heating and thermocouples

dometer were put in a reinforced cage and kept out of sun and rain. Year in year out, the temperature and humidity was kept at a high level. The temperature fluctuated from $34°C - 62°C$ and the appropriate relative humidity was 95–98% . From March, 1978 to October, 1988, the cubes underwent a long process of conversions. The results are shown in Table 3.

TABLE 3   The results of cube strength

CONDITIONS	COMPRESSIVE STRENGTH OF CUBES N/mm²				
	IN ROOM AT 1 DAY	IN ROOM AT 28 DAYS	COVERED WITH PU AT 10 YEARS		HOT & DAMP AT 10 YEARS
GROUP	1	2	3	4	5
STRENGTH	24.4 25.3 23.1	36.1 34.9 34.6	23.5 17.8 25.1	29.6 26.7 20.1	14.3 10.2 12.6
MEAN VALUE	24.3	35.2	23.8		12.4
PERCENTAGE OF 1 DAY %	100	145	98		51

Results

It is known from Table 3 that the bare cubes, in group 5, in a hot and damp environment, lost about half of their strength, while the closed cubes, group 3 and 4, only lost approximate 2% of strength. It follows that if cubes are completely cut off from a hot and damp environment, the rate of conversion will be very slow, and the minimum strength may be higher than that of exposed ones, also it will be better for the HAC concrete anchorage to be insulated from hot and wet conditions.

STRENGTH EXPERIMENTS

On Specimens Of End Anchorages ––– Pull Out Bond Test

Aggregate, sand and HAC are completely the same as protection tests indicated. Low-alloy deformed steel bar of 12 mm diameter was drawn at room temperature. The ultimate strength f_{pu}= 750 N/mm² and a stress of 550 N/mm² was used in the jacking force. The details of block samples are shown in Table 4.

TABLE 4   Details of block samples

BLOCK CODE	COMPRESSIVE STRENGTH OF BLOCKS AND FEMALE CONES N/mm²				
	B1R	B1F	B1	B4	B6
NUMBER OF PRESTRESSED BARS	1	1	1	4	6
PORTLAND CEMENT BLOCK	14 12 17	---	> 30	>30	>30
HAC BLOCK	---	13 15 16	---	---	---
HAC FEMALE CONE	---	---	19 13 15 17 15 12	17 10 15 11 16 13	15 18 12 17 13 17
MEAN VALUE	14.3	14.6	15.2	13.7	15.3

The pull—out test was made on a universal testing machine. Micrometers were set on the two ends of the block to measure the slip of the bar.

For a single bar in a non—restrained condition

The dimension of an HAC concrete block with an average cube strength of 14.6 N/mm^2 numbered B1F was 150×150×400. It is a pretensioned member. A prestressed bar was placed at the center of the cross section 150×150, the experiment results are shown in Figure 4.

For restrained bars

The block shown in Figure 3 is made of Portland cement. It had been reinforced in order to resist the inner expansion force. After the block was hardened, 1—12 mm dia. bar/ 4—12 mm dia. bars / 6—12 mm dia.bars (see Table 4) were put into the cone and stretched, then HAC concrete was carefully poured into the female cone, being vibrated with a vibrator. After the HAC concrete was hardened,

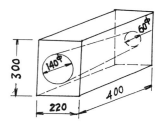

Figure 3: Dimensions of the block & female cone

the prestressed bars were cut. When experimenting, the stretched force, the slide between the prestressed bars and the HAC concrete, and also the slide between the HAC concrete anchorage and the block were measured. The results are shown in Figure 4 and Figure 5.

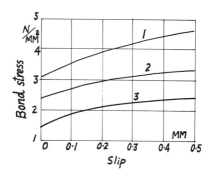
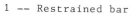
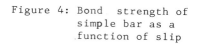

1 -- Restrained bar
2 -- Bar in prestressed concrete
3 -- Bar in reinforced concrete

Figure 4: Bond strength of simple bar as a function of slip

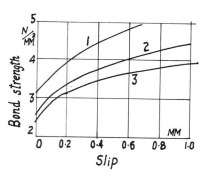

1 -- 1—12 mm dia. bar restrained
2 -- 4—12 mm dia. bars restrained
3 -- 6—12 mm dia. bars restrained

Figure 5: Bond strength of restrained bars as a function of slip

Results and discussions

It is indicated in Figure 4 that the bond strength between the steel
bar and HAC concrete of the prestressed members (B1F), is a bit higher
than that of those non-prestressed ones, because the prestressed
bar is compressed by the surrounding concrete. When it is cut,the bond
strength in the female cone (B1), is still higher than that in B1F, because the former is forced in three dimensions (Figure 6). The prestressed bar was stretched while the slide between the HAC concrete
anchorage and the block around was begun.

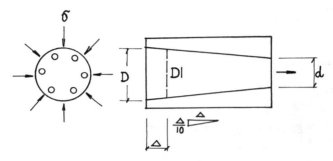

d --- Diameter of small hole
D --- Diameter of great hole
D1 --- Diameter of great hole after drawing back

Figure 6: HAC concrete of female cone was forced in three dimensions

This transverse force may be of considerable value:

$$D1 = D - 2\, \frac{\triangle}{10} = D - \frac{\triangle}{5} \quad,$$

Circle of great hole $C = \pi D$,
Circle of great hole after drawing back $C1 = \pi (D1)$,

$$\frac{C - C1}{C} = \frac{\pi D - \pi(D - \frac{\triangle}{5})}{\pi D} = \frac{\triangle}{5D} \quad,$$

$$\sigma = E \cdot \frac{\triangle}{5D} = 2\times 10^{4} \cdot \frac{\triangle}{5\times 140} = 28.6 \triangle \; \text{N/mm}^{3} \;.$$

For example, if the slide value $\triangle = 0.3$ mm is assumed, then the compressive stress will be: $\sigma = 28.6 \triangle = 28.6 \times 0.3 \doteq 8.5$ N/mm^{2}, the more the
slide value and the steeper the conic slope, the greater the transverse
compressive stress. The stress increases the bond strength between
prestressed bars and the surrounding HAC concrete. The increased fact-
or was suggested in some research papers[4], as:

$$K = [\, 1 + \mu(\, 1 - \mu\,) \cdot f_{y}/ f_{c}^{'}\,] \quad,$$

here, μ — The ratio of the steel area to concrete,
      f_{y} — Yeild strength of steel bar,
      $f_{c}^{'}$ — Ultimate compressive strength in concrete . In this
experiment, K will be greater than 1.5 .   However, when the amount of
bars is added, the bond strength will be a slight decrease as shown in

Figure 7, the slopes of the curves from bottom to top gradually steepen. It will be seen that the degree in which the bond strength is influenced by the cube strength of HAC concrete is gradually decreased as the bond strength develops.

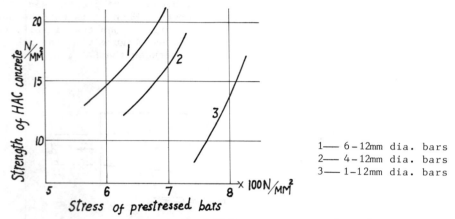

1—— 6-12mm dia. bars
2—— 4-12mm dia. bars
3—— 1-12mm dia. bars

Figure 7: Strength of HAC concrete as a function
of stress of prestressed bars with
relation to the number of bars

On Triangular Trusses

The external form of the trusses was completely the same except for the duct, one was grouted while the other was not. The strength of these two trusses was well designed everywhere, except the end block represented the real service conditions. It is the hope that the collapse would occur at the HAC concrete anchorage which cube strength of 15 N/mm^2 was adopted. The slides of the anchorages of the two trusses were approximate the same. The slides of the anchorages of the one truss are shown in Figure 8. The details of the trusses are shown in Figure 9. When the service loads were loaded on the trusses, slides of HAC concrete anchorages were just about 0.05 mm. It is to be consided that the bond in the whole block was in good conditions. It has even been proved that a truss with low HAC concrete strength in

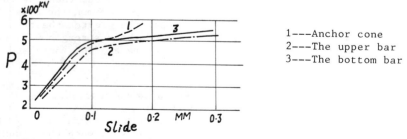

1---Anchor cone
2---The upper bar
3---The bottom bar

Figure 8: The slides of the bars & anchorage of the truss

anchorages works safely, if it is brought to a strength not lower than 15 N/mm^2. At the ultimate stage, some fine cracks radiating in shape[3] were found at the surface of the anchor regions. It is a warning signal for a strengthening examination.

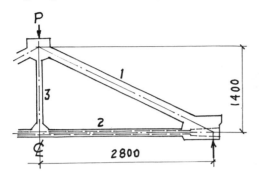

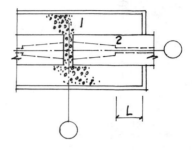

1---200×240, 4-14 mm dia. bars
2---200×150, 4-10 mm dia. bars
  +6-12 mm dia. bars
3---200×100, 4-10 mm dia. bars
Figure 9: Dimensions of truss
          and its reinforcement

1---Rigid PU foam
2---Protected structures

Figure 10: Protection of the
HAC concrete structures
moulded & foamed

PROTECTION AND STRENGTHENING

Polyurethane foam covered with coatings is an ideal material for protecting HAC concrete from hot, wet and chemical attack, because PU foam has an extremely low thermal conductivity and the PU coatings combined with others possess a combination of properties that are not attainable with any other coating system. It exhibits excellent resistance to water vapour, water and chemical attacks. Lining with PU foam is suitable for any HAC concrete structures which are needed to be protected. Pouring is better than bonding. A mould was built, then the reaction mixture was poured into the mould and foamed (Figure 10). The cover length L should have sufficient distance from the anchorages' end to avoid heat bridge. Some structures have been bonded with polyvinychloride (PVC) foam. They maybe not so effective as PU.

Strengthening was only used for the HAC concrete that suffered from a serious conversion reaction, and a substandard quality of HAC concrete. Also the cube strength was lower than 15 N/mm^2 and its W/C was above 0.4 . The HAC concrete shows to have not enough reliability even if they are kept in bad conditions. The following scheme is suggested (Figure 11). The main points are to reduce the service stress an HAC concrete end anchorage carried. Post-tensioned system is adopted. It is preferable to use strands instead of bars. If steel bars are used, a combination of electrical and machenical cross stressing will be efficient in some existing cases.

1--- Steel bars for strengthening
2--- To be strengthened
       structure

Figure 11: A scheme of strengthening

CONCLUSIONS

HAC concerte bared in a hot and damp environment lost a considerable amount of strength, while concrete covered and coated with rigid polyurethane (PU) foam and PU coatings combined with others lost a bit of strength, because PU foam has an extremely low thermal conductivity to insulate heat and PU coatings combined with others exhibit high water, solvent and chemical resistance. PU series is an ideal material for protecting HAC concrete from hot, wet and chemical attacks.

End anchorages shaped female cone are pressed in three dimensions when the prestressed bars are drawn, consequently, the bond strength between the prestressed bar and the surrounding HAC concrete is increased . The increased bond strength will partially compensate HAC concrete in end anchorages for a loss of conversions. HAC concrete must be strengthened when it is kept in bad conditions and is poor in quality, and the cube strength is lower than 15 N/mm^2.

REFERENCES

1   COLLINS R J and GUTT W. Research on Long – term Properties of HAC Concrete. Magazine of Concrete Research, Volume 40, Number 145, December 1988, pp 195 – 208.

2   MIDGLEY N G and MIDGLEY A. The conversion of high alumina cement. Magazine of Concrete Research, Volume 27, Number 91, June 1975, pp 59 – 77.

3   LEONHARDT F. Prestressed Concrete Design and Construction. 1964.

4   THE NINTH INSTITUTE OF THE MINISTRY OF MACHINE BUILDING. The experiment of concrete in steel tube. Shanghai, 1974 (In Chinese).

EXTERNAL PRESTRESSING : AN AID FOR THE PROTECTION AND REPAIRING OF REINFORCED CONCRETE STRUCTURES

P Spinelli

S Morano

University of Florence,

Italy

ABSTRACT: The prestressing with unbonded tendons, externally disposed with respect to reinforced concrete sections, is a tecnique spreading in some countries for the construction of prestressed concrete bridges as well as for the repairing and upgrading of existing reinforced concrete ones. For the new construction the use of external prestressing leads to more durable works; for the repairing or upgrading of existing structures the external prestressing can be adopted in the case of lack of prestress or deterioration of steel rebars.
The behaviour of prestressed concrete structures with unbonded reinforcement shows an equivalence in comparison to the traditional prestressed concrete structures, as long as the section remains entirely reacting, whilst a dangerous drop in safety margins can verify from the decompression of sections up to the ultimate limit state. The present work has the aim to underline the benefits of the presence of even a modest quantity of passive reinforcement in the behaviour of unbonded reinforced concrete sections. This can be typically the case of the use of unbonded technique for the consolidation of existing structures.

Key words ; Prestressed concrete, external prestressing, unbonded tendons, durability, structural reparing

Paolo Spinelli: born in 1950, formely lecturer in Bridge Construction, now associate professor in Advanced Theory of Structures at the Faculty of Engineering, University of Florence.
Fields of interest: wind and earthquake actions on structures, tensile structures, prestressed concrete, monitoring of monumental structures.

Salvatore Morano: born in 1963, graduate at the Faculty of Engineering, University of Florence, with a thesis entitled "Strutture precompresse con cavi non aderenti" (Structures prestressed by external tendons).
Fields of interest: prestressed concrete structures, new bridge construction technologies.

INTRODUCTION

The protection of concrete is a major problem for concrete and prestressed concrete structures. Particularly referring to prestressed concrete two important sides of the problem appear:
- the first is to prevent concrete deterioration and structure durability for new applications;
- the second is to face concrete (and consequently prestressed steel) deterioration and to repair and restore the prestressed concrete structures.
In both these applications the use of unbonded external tendons shows widely its validity.
In the new bridges the strands, externally disposed with respect to the concrete, are deviated on the supports and in the span through transversal beams or deviation blocks.
The disposing of the strands outside the reinforcing bars reduces reinforcement congestion and therefore it is easier to guarantee proprer reinforcement covers and to achieve more accurate concrete casting with a consequent advantage of the concrete quality and hence of the structure durability . Besides all that this offers the chance of substituting the prestressing strands which allows to have efficient prestressing reinforcement during the whole life of the structure.
In the repaired or upgraded structures with external prestressing, the tendons can be either straight or deviated along the span, and are connected directly to the existing structure or to transversal beams. The use of additional prestressing is necessary for the reparation of prestressed concrete structures when a reduced efficiency of reinforcing steel is experienced or because of increased loads or deterioration in actual (non-prestressed) reinforcement.
In both cases the technology proposed is a valid alternative to other reparation techniques (like beton plaques, for example), and has an intrinsic validity expecially when the reinforcement has to be added in high quantity or is difficult to connect to the concrete.
These technological advantages have on the other hand some disadvantages as far as the statical behaviour is concerned, because the external prestressing has a different behaviour in comparison to the traditional (internal bonded) one, often associated with some risks.
This work has the aim to illustrate and quantify these differences and disadvantages and to demonstrate that they are however rather modest if even a little amount of bonded (prestressed or non-prestressed) reinforcements is present.

STATICAL BEHAVIOUR

The worse statical behaviour of externally prestressed structures is due to a double cause:
1) the different displacement of the concrete section and the prestressing tendons: under load the actual eccentricity of external tendons is reduced;
2) under overload effect, the beam sections more stressed show a higher increase in tendon action for bonded tendons than for external ones (as its deformation can only be uniform between two deviation blocks or even for the whole beam length).
As a matter of fact, the effects of the first cause have little significance and can be neglected if the deviation blocks are sufficient and well disposed.
The second cause constitutes the really big difference between the two techniques. It is evident at its most after the cracking of concrete and therefore it highly affects the ultimate behaviour of the beam. In fact, when the most stressed section cracks, it is

in such section that the bonded reinforcement has a high and concentrated increase in deformation and hence of tension, whilst if the strand is unbonded the force in the reinforcement, is substantially constant. It follows that the increase in bending moment is absorbed in the bonded reinforcement through the increase both of tension resultant and of lever arm. In the unbonded beams, in absence of bonded reinforcement (prestressed or non-prestressed) the stress resultant does not change and the only way to face the increase in external moment is the increase in internal lever arm and this happens with a reduction of compressed zone with a consequent quick increase in the concrete compressions. This leads to the beam fragile rupture due to concrete crushing at an ultimate moment level which is lower than in the case of bonded reinforcement. As an example, the statical behaviour of a prestressed beam with rectangular cross-section both in the case of a bonded and an unbonded reinforcement is shown. This behaviour is obtained through a numerical simulation carried out by an ad hoc written computer programme.

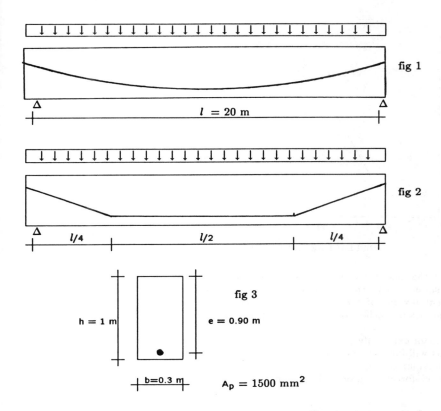

fig 1

$l = 20$ m

fig 2

$l/4$ $l/2$ $l/4$

fig 3

$h = 1$ m $e = 0.90$ m

$b=0.3$ m $A_p = 1500$ mm^2

The mechanical properties assumed for the materials are the following:

concrete
- modulus of elasticity, $E_c = 33700$ N mm^{-2}
- design compression strengh $f_{cd} = 15.4$ N m^{-2}

- design tension strenght, $f_{ctd} = 0$
prestressing reinforcement
  - design tension strenght, $f_{pd} = 1565$ N m^{-2}
  - initial prestressing force, $N_{pi} = 1500$ KN

The bonded tendon has the same cross-section A_p and internal force of the external one and with a parabolic lay-out with eccentricities on the supports and at middle span equal to the external prestressing strand. For both beams the uniform load which leads to cracking of the middle span section is 17 KN/m.

LOAD- DISPLACEMENT AT MIDDLE SPAN

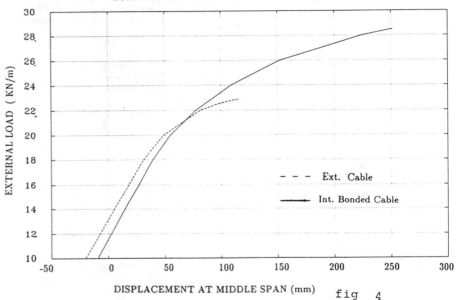

fig 4

Looking at the load-displacement curves one can notice that the behaviour of the two beams is similar up to the decompression load. The small difference is due mainly to the different lay-out of the prestressing strand. At the ultimate state the externally prestressed beam reaches an ultimate load and displacement lower than the bonded one.
If in the beam externally prestressed a non-prestressed (bonded) reinforcement, A_s, is added, this will increase its tension when the section cracks. This of course induces a better behaviour at rupture. Fig. no. 7 shows the different behaviour of the beams externally reinforced with or without non-prestressed (bonded) reinforcement.

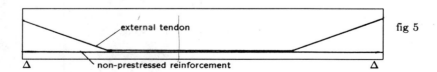

fig 5

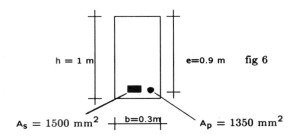

h = 1 m   e=0.9 m   fig 6

$A_s = 1500$ mm^2 b=0.3m $A_p = 1350$ mm^2

steel (non-prestressed reinforcement)
- design tension strenght, $f_{sd} = 383$ KN mm^{-2}

The tendon section is reduced in comparison to the precedent exemplification (fig. no. 5,6) in order to obtain a behaviour in service similar to the two precedent beams.

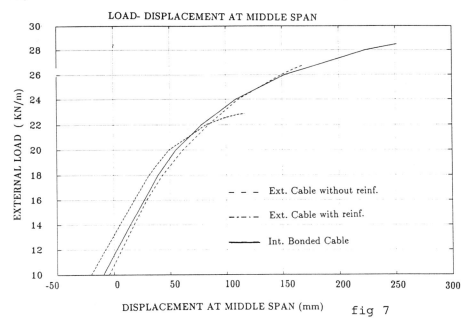

LOAD- DISPLACEMENT AT MIDDLE SPAN

– – – Ext. Cable without reinf.

–..–.– Ext. Cable with reinf.

—— Int. Bonded Cable

DISPLACEMENT AT MIDDLE SPAN (mm)   fig 7

Synthetically:

	ultimate load	ultimate displacement	increase in tendon action at ultimate state
bonded	28.5 KN/m	252 mm	+ 44 %
ext. no reinf.	22.9 KN/m	116 mm	+ 16 %
ext. with reinf.	26.8 KN/m	169 mm	+ 18 %

One can see how, in presence of non-prestressed reinforcement, the ultimate load together with the ductility of externally reinforced beam is increased.

ULTIMATE LOAD PREDICTION OF UNBONDED BEAMS

As far as prestressed bonded beams are concerned, the evaluation of the ultimate load can be done by evaluating, for the most stressed section, the ultimate load and by obtaining the load causing it. This is possible because the hypothesis of aderence and bond in every section leads to this fact: the deformation of prestressed (and non prestressed) reinforcement depends only on the actual deformation if the examinated section and hence the ultimate moment is an intrinsic characteristic of the section itself.

In the prestressed concrecte beams with unbonded tendons the deformation of the tendon depends on the deformed shape of the whole beam and therefore, even if neglecting the eccentricity variation of the tendon under load, as the actual force in the tendon at the rupture is not known, it is impossible to know the actual ultimate load whitout knowing the deformed shape of the structure.

COMPARISON BETWEEN THE ULTIMATE MOMENT FOR PRESTRESSED CONCRETE SECTION WITH BONDED OR UNBONDED TENDONS.

It is known that the lack of bond between cable and concrete leads to a reduction of the value of the ultimate moment. In order to quantify this reduction we analyse two rectangular sections, equal in geometry and reinforcements (prestressed and not prestressed), one with bonded tendons, and the other with unbonded ones. The initial prestressing force is equal for the two sections. By indicating the ultimate moment for the section with bonded tendons with $M_{du}{}^{bo}$ and the one for section with unbonded ones with $M_{du}{}^{un}$, we studied the parameter Γ (called comparison factor) in order to try to identify the parametres that mostly influence its value.

$$\Gamma = \frac{M_{du}{}^{bo}}{M_{du}{}^{un}}$$

To do so a value α is assigned to the ratio between the force in the unbonded tendon at the ultimate state and the initial one:

$$\dot{\alpha} = \frac{N_{pd}{}^{un}}{N_{pi}}$$

$N_{pi} = $ initial prestressing force

$N_{pd}{}^{bo} = $ ultimate force for the bonded tendon

$N_{pd}{}^{un} = $ ultimate force for the unbonded tendon

Hence it is possibly to obtain the resistance domain both for bonded sections and unbonded ones, i.e. the value of the rupture moment as a function of initial prestressing force. If the initial tension in the tendon is fixed, the initial force is obviously proportional to the prestressing reinforcement section A_p.

obviously proportional to the prestressing reinforcement section A_p.

Showing the results, to reach more generality, it is preferred to refer to the adimensional quantities:

$$m_{du}{}^{un} = \frac{M_{du}{}^{un}}{f_{cd} \cdot b \cdot e^2} \ , \qquad m_{du}{}^{bo} = \frac{M_{du}{}^{bo}}{f_{cd} \cdot b \cdot e^2} \ , \qquad n_{pi} = \frac{N_{pi}}{f_{cd} \cdot b \cdot e}$$

where the geometrical dimensions e and b are indicated in the fig 8 below.

RESISTANCE DOMAIN AND BEHAVIOUR OF THE COMPARISON FACTOR

As an example we show the results obtained for a section with different non-prestressed reinforcement percentage. The data for the investigated section are resumed in table 1

Table 1 : data for the section of fig 8

\- \-

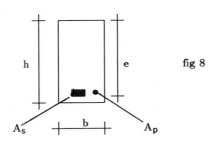

fig 8

$e = 0.95\ h$
$f_{cd} = 13.2\ \text{N mm}^{-2}$ (concrete design strength)
$f_{sd} = 339\ \text{N mm}^{-2}$ (steel design strength)
$f_{ptk} = 1800\ \text{N mm}^{-2}$ (prestressing steel characteristic rupture strength)
$f_{pd} = 1500\ \text{N mm}^{-2}$ (prestressing steel design strength)
$f_{pi} = 0.6\ f_{ptk}$ (initial prestressing tension)
$A_t = A_s\ /\ be$ (percentage of non-prestressed reinf., variable between 0 and 1.5 %)
A_t = Area of non-prestressed reinforcement
A_p = Area of prestressed reinforcement

$$\alpha = 1.10$$

\- \-

The value of α assigned in the example is obtained as a mean value of those obtained in a numerical simulation of the behaviour of a number of rectangular beams with different span dimension and reinforcement [2]

In the following graphs the resistance domain together with the comparison factor Γ is reported. In fig 15 a synthesis of results is given.

In observing the results some general consideration can be offered:

a) $\underline{\Gamma \text{ is less than 1}}$
At first the differences in the ultimate behaviour are due to the fact that the ultimate tension in the bonded tendon is different from the unbonded one. Thus M_{du}^{bo} is always greater than M_{du}^{un} and Γ greater than 1. If α increases than Γ tends to 1.

b) Asympthotic behaviour of Γ
If the prestressing force is very high than both bonded and unbonded section reach the rupture on concrete side, due to the concrete crushing. In such case the increase in the ultimate tension is negligible and the prestressing force for the unbonded and bonded techniques tends to be similar. Therefore the comparison factor Γ is near to 1. In the graphs even very high values of initial prestressing force (higher than those usually adopted) are considered to underline this asymphotic beaviour.

c) Section without non-prestressed reinforcement

For sections without non-prestressed reinforcement the rupture of unbonded beams always happens on the concrete side. The compressed zone in the unbonded beams is less wide than in the bonded ones because the force in the reinforcement to equilibrate is less. Therefore the compression value at the extradox for the unbonded beams is greater or equal to the bonded ones. Then at the ultimate state the lever arm of the internal resultants is greater for the unbonded section. We indicate by z_{du} the lever arm at the ultimate limit state and we have:

$$M_{du}^{bo} = z_{du}^{bo} \cdot N_{pd}^{bo} \quad ; \quad M_{du}^{un} = z_{du}^{un} \cdot N_{pd}^{un}$$

$$\text{as} \quad z_{du}^{bo} < z_{du}^{un}$$

finally, we obtain:

$$\boxed{\Gamma = \frac{M_{du}^{bo}}{M_{du}^{un}} < \frac{N_{pd}^{bo}}{N_{pd}^{un}} = \beta}$$

Hence, as a rule, the comparison factor for beams without non-prestressed reinforcement is always less than the ratio between the ultimate prestressing force.
If N_{pi} approaches to 0 the lever arm approaches to the value e both for bonded and unbonded sections. In such hypothesis the value for Γ tends to the value of β as the lever arm tends to be equal for both technologies.
For sections without non-prestressed reinforcement therefore Γ decreases with N_{pi} and tends to its maximum value for N_{pi} approaching to 0.
For N_{pi} approaching to infinite than Γ tends to 1.
The value β has on the other hand the following upper limitation:

$$N_{pd}{}^{bo} \leq A_p \cdot f_{pd} \quad ; \quad N_{pd}{}^{un} = \alpha \cdot N_{pi} = \alpha \cdot A_p \cdot f_{pi} \quad \Rightarrow$$

$$\boxed{\beta \leq \frac{f_{pd}}{f_{pi}} \cdot \frac{1}{\alpha}}$$

For istance, following Italian Recommendations [4], for a controlled steel we have:

$$f_{pi} = 0.6 \cdot f_{ptk}, \qquad f_{pd} = \frac{f_{ptk}}{1.15} \quad \Rightarrow \quad \frac{f_{pd}}{f_{pi}} = 1.449$$

and hence:

$$\Gamma \leq \beta \leq \frac{1.449}{\alpha}$$

Therefore the $1.449/\alpha$ is actually the upper value for the comparison factor Γ.

d) Beams with non-prestressed reinforcement.

In the case of sections with non-prestressed reinforcement, both unbonded or bonded structures could fail on bonded reinforcement side: this happens with sections with small amounts of prestressing.
With very small initial prestressing (and hence small A_p), both bonded and unbonded sections behave like the normal reinforced concrete section. Therefore, if N_{pi} approaches to 0 the comparison factor approaches to 1. We have already seen that Γ approaches to 1 even for high values of N_{pi}, therefore there must be a maximum of Γ for intermediate values of N_{pi}. In the graphs one can see that effectively a maximum of Γ exists and that it usually is within the N_{pi} zone for which one has the change of rupture pattern. It is in this zone, where ususal prestressed sections lie, that we have the maximum reduction of the ultimate moment of the unbonded beams compared to the same beams with bonded tendons.
The more the non-prestressed reinforcement is increased, the lower becomes the percentage of prestressing action on the whole amount of tensile force. Therefore, as we increase the non-prestressed reinforcement, the behaviour of the bonded or unbonded beams show less and less difference, whilst Γ decreases and approaches to 1. This can be clearly seen from the graph, in which the curve of the comparison factor for different reinforcement percentages is shown.

The non-prestressed reinforcement amounts for which the reduction of the comparison factor is considerable are surely unusual for totally prestressed beams, but are reasonble for newly built partially prestressed beams. As a matter of fact we have a prestressed beam with large amounts of non-prestressed reinforcement when we use external prestressing to repair deteriorated normal reinforced concrete beams. In such a case the percentage of the pre-existing non-prestressed reinforcement will be similar to those we have analysed .

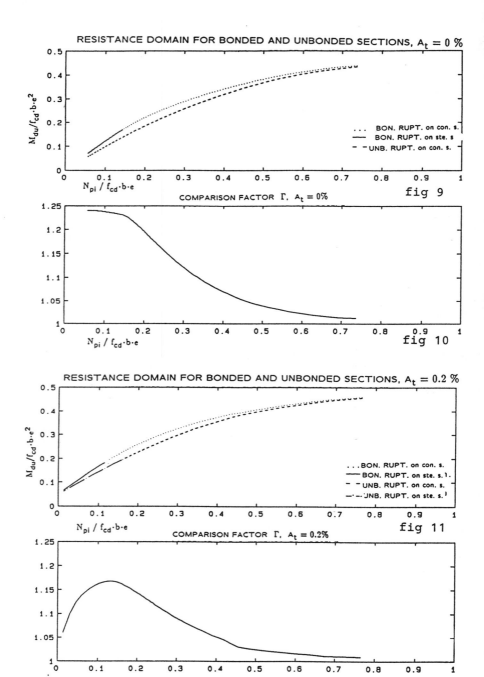

RESISTANCE DOMAIN FOR BONDED AND UNBONDED SECTIONS, $A_t = 0$ %

... BON. RUPT. on con. s.
— BON. RUPT. on ste. s
– – UNB. RUPT. on con. s.

fig 9

COMPARISON FACTOR Γ, $A_t = 0$%

fig 10

RESISTANCE DOMAIN FOR BONDED AND UNBONDED SECTIONS, $A_t = 0.2$ %

...BON. RUPT. on con. s.
— BON. RUPT. on ste. s. .
– – UNB. RUPT. on con. s.
–·– UNB. RUPT. on ste. s.

fig 11

COMPARISON FACTOR Γ, $A_t = 0.2$%

fig 12

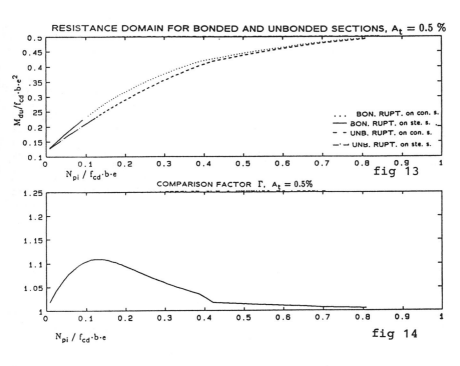

fig 13

fig 14

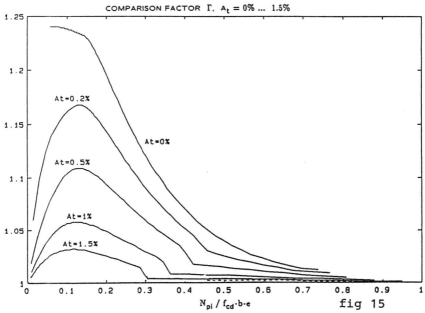

fig 15

APPLICATION EXAMPLE

In the following table and figure the application of external prestressing for the restoration of a deteriorated reinforced concrete deck is shown.
The concrete deck is composed of primary beams and secondary beams with rectangular sections and a slab of 18 cmts. It is located in Rome, near the Old Theater of Massenzio in the Roman Forum, and it was constructed about thirty years ago to protect some historical escavations. During the years, humidity and carbonatation have attacked the concrete and the lower reinforcement has been highly diminished in section. This fact together with the circumstances to guarantee a higher live load on the deck have forced to repair and to reinforce the deck. The technology chosen is to make a new concrete deck on the upper side which substitutes partatially old concrete deck and to use unbonded external tendons for the primary beams to substitute deteriorated steel. At present the application is under construction and has been chosen due to its major efficiency and easier mounting in comparison to other reparation tecniques like beton plaque.

Table: repaired concrete deck data

-overall dimensions: 32 m X 17.70 m
-primary beams span: 8.3 m
-primary beams section dimensions: 740 mm X 350 mm
-secondary beams span :from 3.20 m to 5.60 m
-secondary beams section dimensions: 580 mm X 200 mm , 580 mm X 300 mm
-concrete deck thickness: 180 mm
-dead load: 7 KN/m^2
-live load: 6 KN/m^2

2   external tendons for every primary beams
-diameter: 15.6 mm
-initial tension: 1360 N/mm^2
-design strength: 1080 N/mm^2

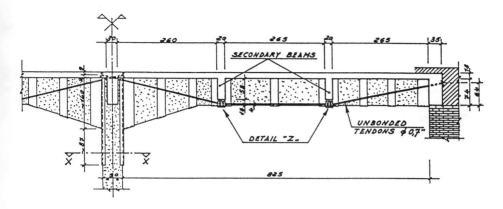

PRIMARY BEAM VIEW   fig 16

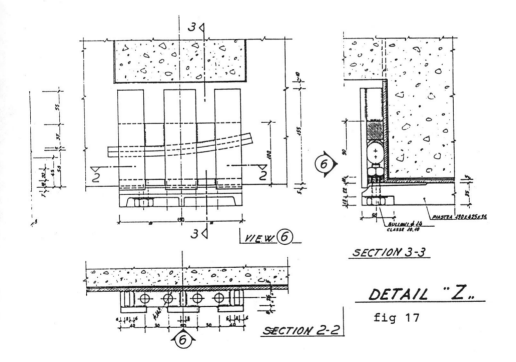

DETAIL "Z"

fig 17

REFERENCE

[1] VIRLOGEUX M. :"La precontrainte exterieure "
Annales de I.T.B.T.P. , N° 420, Decembre 1983

[2] CHIARUGI A., DI STEFANO D., MORANO S., SPINELLI P., "External
prestresssing for beam with non-prestressed reinforcement (in ital.)"
Proceedings of the "Giornate A.I.C.A.P.
Naples May 1989

[3] MORANO S. "Structures prestressed by external tendons (in ital.) "
Civil enginnering thesis, University of Florence, October 1989

[4] ITALIAN RECOMMENDATION D. M. 27 - 7- 1985 :" Norme tecniche per
l'esecuzione delle opere in cemento armato normale e precompresso e per le strutture
metalliche:"

REINFORCED CONCRETE: STRUCTURAL CONSIDERATIONS AND CHOICE OF PRODUCTS FOR REPAIR

J T Williams
A J Parker
Sika Ltd.,

United Kingdom

ABSTRACT. Factors affecting the durability of reinforced concrete are discussed. The repair process is reviewed considering the structure as a whole and the implications of repairs on the safety of the structure. Structural distress and the visible evidence of distress eg excessive cracking, spalling due to steel corrosion and movement are considered. A review of repair products based on Portland cement, polymer modified Portland cement, epoxy resin and polyester resin systems is made, comparing and contrasting their properties. The requirements for protective surface coatings are reviewed followed by a summary of the various repair systems and a table of typical properties. The aims and objectives of FeRFA are stated.

Keywords : Concrete, Durability, Repair, Cement, Polymer, Epoxide, Polyester, Coatings.

J.T. Williams BSc(Eng)C.Eng MICE MIStructE,
            Technical Manager, Sika Limited
After gaining experience with major civil engineering contractors and consulting engineers, John Williams specialised in the use of lightweight concrete and cement replacements prior to joining Sika Limited where he is involved with all aspects of the repair and refurbishment of structures.

A.J. Parker ARICS, Joint Managing Director, Sika Contracts Ltd.
After working in private practice and for Consulting Engineers, gaining experience on large scale civil engineering contracts both at home and abroad, John Parker joined Sika Contracts Limited specialising in waterproofing, concrete repair and refurbishment.

INTRODUCTION

Reinforced Concrete - Factors affecting its durability:

Concrete is weak in tension and it is for this reason
that the tensile properties of the material
were considerably improved by the introduction of
embedded steel at the end of the last century.

Reinforced concrete has in the main proved very
successful in both structural performance and long term
durability. There are, however, many examples of early
damage to concrete units primarily as a result of
premature corrosion of the steel reinforcing bars.

Concrete will normally provide an ideal environment for
the embedded steel reinforcement due to the high level
alkalinity produced by hydration of the cement when first
placed. The main requirements for the concrete are that
it is dense (impermeable to water and atmospheric gas
ingress), well compacted around the embedded steel
reinforcement and so placed as to give an adequate
thickness to cover the steel.

When the concrete is permeable, this allows the ingress
of atmospheric gases, primarily carbon dioxide, together
with water and these react with the alkaline components
in the concrete thus reducing the alkalinity of the
concrete below the level needed to protect the steel from
corrosion - a process known as **carbonation** of the
concrete.

Reinforced concrete can therefore suffer premature damage
by cracking and spalling due to corrosion of the embedded
steel reinforcement as corrosion (rusting) is accompanied
by considerable expansive forces.

Over the years the depth of carbonation, appearing
initially at the surface of the concrete, will gradually
increase until it reaches the position of the embedded
steel reinforcement, destroying the protective alkaline
environment for the steel and causing the onset of
corrosion.

The presence of medium and high levels of **chlorides**
in the concrete can also stimulate corrosion of steel
reinforcement. Chloride may have been introduced as an
additive at the time of initial placing, perhaps as the

set accelerator, calcium chloride. Calcium chloride was extensively used in pre-cast reinforced concrete work during the nineteen fifties and sixties to help speed up the process of manufacture of pre-cast units in the factory for delivery to site at a pace to keep abreast with the shortened timescale allowed for the construction of, typically, high rise blocks of flats and offices.

Chlorides are also introduced subsequent to the initial placing and construction, for example by the use of de-icing salts on concrete bridge structures, car parks and the like, or in marine environments.

Certain aggregates used in the original concrete mix have created problems of **alkali aggregate reaction.** In the UK the most common form of this has been alkali silica reaction where reactive silica has been occasionally present as a part of certain aggregates. In the presence of water a hygroscopic gel is formed which take up water and swells. This may then create internal stresses in the concrete sufficient to cause cracking and thus create a situation leading to even further deterioration by reinforcement corrosion -particularly in exposed environments where ingress of water in to the cracked structure can continue. Reinforced concrete can also deteriorate as a result of **physical damage,** from a number of causes such as -fire damage, explosive damage, impact (handling, construction, vehicular, etc.), frost damage. Physical damage is often very obvious and should not normally lead to reinforcement corrosion unless the damaged structure remains unprotected or not repaired for some time after the damage is caused.

Deteriorated reinforced concrete therefore needs repair, and, as can be appreciated from the brief introduction correct diagnosis of the reasons for deterioration can be a complex exercise, but it is essential if an effective and durable repair is to be carried out.

The Repair Process

The method of repair may depend upon whether there is a need to strengthen the concrete unit in addition to what can be termed general renovation to the unit. The repair can therefore be classified either as structural repair and strengthening or non-structural repair.

Structural repairs may involve the use of sprayed reinforced concrete or complete replacement by re-casting of a structural unit such as a beam or column. Crack injection may also be required using special resins injected under pressure by pumps or by the force of gravity. Other methods include the glueing-on of steel places to the external face of the concrete unit and the insertion and grouting in of steel sections and dowels.

Non-structural repairs are a very common requirement, where there has been general deterioration to the concrete structure through cracking and spalling caused by carbonation of the concrete and corrosion of the embedded steel reinforcement.

All cracked, spalled and loose concrete has to be cut away and corroded steel reinforcement fully exposed, loose rust removed, the steel cleaned and given a protective coating to reintroduce an intense alkaline environment around the steel. The area of concrete cut out is then filled in layers with special cementitious or resin mortar to reform the concrete unit to its original profiles. In many cases a protective/decorative coating is then applied to the surface of the patch repairs and the surrounding surfaces of original concrete.

STRUCTURAL CONSIDERATIONS FOR REPAIRS

Reinforced concrete is a composite material in which the properties of both steel and concrete are utilised to resist the application of external forces. The strength of concrete in compression is allied to the strength of steel in tension or compression so as to balance the forces created by the dead and imposed loadings. The dimensions and strength of the concrete, and the strength and disposition of reinforcement are selected to create members which can safely withstand the expected conditions. Although there is always a margin of tolerance it is possible for structures to become unserviceable due to a variety of causes and repair is then needed to restore the original strength.

The forces to be resisted are shear, bending, direct compression and combinations of these three. All require different arrangements of reinforcement and failure due to each type of force can be recognised by different crack patterns. Normal reinforced concrete elements are rarely found to be severely cracked simply due to excessive load. It is generally a combination of circumstances which leads to a lack of serviceability. Deterioration due to faulty materials, faulty workmanship and corrosion lead to loss of strength and hence the inability to withstand and generate cracks.

Cutting back on the tension side of a member may be safe
if it merely exposes the main reinforcement since it is
only the concrete cover which is being removed. If the
reinforcement is badly rusted and is to be replaced then
propping becomes essential. An exception to this is a
cantilever situation where propping is always advisable.
Cutting in the compression zone is more dangerous since
compression failures can be explosive. It is worth
repeating that this can apply to beams, columns, walls
and slabs.

Remember that elements which span use the concrete to
take compression and the reinforcement to take tension
except for very heavily reinforced sections which may use
compression reinforcement as well. Columns use both the
oncrete and the reinforcement to take compressive loads.

Binders in columns and links in beams are used to carry
bursting forces and shear respectively and are as
important as main reinforcement in structural integrity.
The only reinforcement likely to be found with a
non-structural function is light mesh which is installed
for anti-crack purposes or to enhance fire protection.
This can safely be cut away and restored later provided
an expert opinion is given that there is no loadbearing
function.

Always find the reason for any deterioration and
determine if there are fundamental defects which cannot
be cured by simple repair. If the defect can be cured by
reinstatement then assess whether any cutting out will
cause a further structural reaction. If it will then
arrange to have a propping scheme devised and have it
approved by the owner or his representative.

The design and execution of propping are as important as
building the original structure and should be tackled
with due care. Props must not be removed before the
repair has cured and the member is capable of carrying
its full load. Pay special attention to cantilevers and
columns where failure can be sudden. A well designed
repair which is carried out meticulously will restore a
structure to its original condition and give many years
of life. If possible, the cause of the deterioration
should be removed at the same time to prevent a
recurrence.

Visual examination will indicate, from crack patterns and
deflected shape, how the structure is behaving, but no
firm decisions should be made without the formal report

of a suitably qualified and experienced structural
engineer. This will incorporate comments on the original
drawings (if available), the use of the structure,
cover-meter checks on reinforcement, chemical tests on
the concrete and possibly the examination of cores and
other samples taken from the affected area.

Distinction has to be made between cracks due to
overloading, cracks due to inadequate cover or local
corrosion, and cracks due to inadequate curing, incorrect
water/cement ratios and shrinkage. A valuable guide is
the deflected shape of a structure since most cases of
excessive bending strain lead to bowed beams or slabs
with consequent ponding of water or cracked finishes.
Genuine shear failures are rare but can be without
warning.

The most dangerous failures are in cantilevers, which
might be beams or slabs or retaining walls, because no
re-distribution of stress is possible.

Most repairs entrail cutting back the concrete and
possibly the reinforcement so an adequate propping scheme
is essential. Props must be capable of taking the load
carried by the member being treated and the design should
be carried out by an appropriate expert. The load must
be transmitted by the props to sound members or to the
ground and the props must be kept in place until the
repair is cured and has loadbearing capability.

PRODUCTS

Repair Products based on unmodified Portland Cement

These are generally mixes of Portland cement and a
suitable aggregate which can be placed either by hand or
sprayed by dry or wet mix techniques.

Portland cement based materials have the advantages of
being readily available, easily prepared and relatively
cheap. They also have the major advantage that they
provide a similar eletrochemical environment to the
concrete substrate. Disadvantages may include poor
adherence to the substrate material (particularly when
repairing underhung and vertical surfaces), their
characteristic of high shrinkage after setting (which can
be overcome to a large extent by applying carefully
designed mixes in thin layers, and good curing), and the
long term problems of bond breakdown.

System Products

Repair materials based on Portland cement modified by polymer latex.

The addition of a polymer latex to Portland cement-based repair materials gives some advantages: much reduced permeability; improvement of the bond between the repair material and substrate concrete, particularly when a slurry coat of polymer modified Portland cement containing aggregate is used to prime surfaces; increase in tensile strength of the repair; improved workability through plasticising of the mix while allowing reduction of water/cement ratio; general improvements in chemical resistance of the repair material; and improvements in curing. There are a number of different types of polymer modified latex available including polyvinyl acetate (PVA), styrene butadiene (SBR) polyvinylidene chloride (PVDC), acrylic and styrene acrylic latices. Polyvinyl acetates are not suitable for external use in damp conditions. In the UK, the styrene butadiene latices have been widely used while the acrylic latices have been extensively used on the Continent and now widely used in the UK also. All these materials are more expensive than unmodified Portland cement but this disadvantage is more than balanced by improved workability, lower shrinkage characteristics and better bonding. Now also available are similar materials based on preblended ordinary Portland cement, graded aggregates and special redispersible polymer powders which require the addition only of water on site.

Epoxy resin mortars and grouts

These are generally more difficult to use than materials based on Portland cement. Where fine aggregate is added to the resin to act as a filler and prevent material from slumping, it is important to ensure there is sufficient resin to fill the voids between the aggregate particles or the materials may not form an impermeable barrier, which is essential because the materials do not provide inhibitive protection. The use of primer coats to both concrete substrate and steel is necessary to ensure good bonding and the timing of application of the main repair to the bond coat is critical, being dependent upon temperature. Some epoxy resin systems have poor adherence to damp surfaces and their performance may be jeopardised in continuously wet situations; however, special formulations are available for both wet

situations and those outside normal temperature ranges.
In these conditions it is essential to follow closely the
manufacturer's instruction. Epoxy resin systems develop
high strength and have a shorter cure time than
materials based on OPC (especially useful in adverse
environmental conditions); they also have minimal
shrinkage during and after curing.

They are considerably more expensive than repair systems
based on Portland cement.

Polyester resin mortars and grouts

Polyesters are more difficult to mix and handle than
materials based on OPC; the proportion of resin to
hardener is less critical than those based on epoxy
resins. Most polyesters cannot be bonded to damp
surfaces reliably. A major disadvantage is that they
have a high shrinkage on setting and further long-term
shrinkage but this can be overcome to some extent by
building up the repair in thin layers. Their use is
generally restricted to small repairs. Curing time is
faster than for epoxy resin systems, particularly at low
temperatures, and formulations are available that can
match the strength of good quality concrete in two hours
at an ambient temperature of $4°C$. Their cost is somewhat
lower than epoxy resin systems.

Surface Coatings

The coating should be compatible with both the original
and repaired concrete surfaces in respect of flow,
penetration, film formation, adhesion and flexibility.
With the more specialised polymer repair materials,
appropriate coatings have been developed by the
manufacturers. For compatibility it is generally
preferable to obtain them from the same source. When
cementitious repair materials have been used there is a
wide range of coatings for protective and decorative
purposes.

The choice for greatest protection is limited for
building facades for example, to coatings which inhibit
the ingress of carbon dioxide and water, yet permit the
passage of water vapour. In other situations, high
chemical resistance and water vapour resistance may be of
prime concern may be of prime concern.

All coatings will require eventual maintenance themselves
and this aspect is one that is most often overlooked when

a coating is chosen.   Resistance to abrasion is very
relevant in some structures, for example car parks.

Alkyd resin-based paints are not generally suitable for
use on alkaline surfaces.   Reaction curing coatings - eg
two-pack epoxies, moisture curing polyurethanes - will
usually require mechanical preparation prior to
overcoating thus increasing maintenance costs.
Physically drying coatings can usually be overcoated
easily, due to solvent reactivation of the existing film
by the fresh coating, thus maintenance costs will be a
minimum.

**Advantages and Disadvantages of Different System
Products.**

Portland cement modified by polymer latex

Advantages
1)   Polymers act as an effective water-reducing admixture
     producing workable mortars at low water:cement
     ratios.
2)   Improved bond with the concrete substrate
     especially under typical site conditions.
3)   Reduced permeability of the mortar to water,
     oils, salt solutions etc., and also increased
     resistance to mild chemicals.
4)   Improved abrasion resistance of the mortar,
     screed etc.
5)   Enables modified cement mortars to be laid at
     thicknesses (typical 12mm) significantly thinner than
     are practical with unmodified materials.

6)   Polymers act as an integral curing aid to a degree,
     reducing the need for efficient curing but not
     eliminating it in drying conditions.   The latter is
     essential for unmodified cementitious repair mortars.
7)   Polymer modified mortars are cost effective
     repair mediums.

Disadvantages
1)   Site-batching errors can occur resulting in poorly
     designed and less durable repair material.
2)   Moisture sensitivity in service of some polymers
     e.g.   poly-vinyl acetates (PVA).
3)   Cannot be applied in underwater environments.
4)   Inconvenience factor or constituents required for
     site-batching/storage, although now being rectified
     by wider availability of pre-bagged, ready to use
     system products.

5) Specialist mixing equipment required e.g. forced action mixers, not free fall types.
6) Unsuitable for application at low temperatures e.g. below 4°C.

Epoxy resin mortars and grouts

Advantages
1) Large surface area repairs possible.
2) Moisture tolerant.
3) Low exotherm in filled mortar systems.
4) No shrinkage in short or long term.
5) Tenacious bond.
6) Resilience - ability to accommodate thermal movement.
7) High compressive-tensile strengths.
8) Good water resistance.
9) Impervious allowing thin repairs to be undertaken.
10) Good chemical resistance.

Disadvantages
1) Cost - high.
2) Not tolerable to poor mixing.
3) Some formulations have high odour at time of application.
4) Heat distortion.
5) Health & Safety - operatives need to take additional precautions.
6) Restrictions on application at low temperatures e.g. below 4°C.

Polyester resin mortars and grouts

Advantages
1) Speed of cure.
2) Early return to service following repair works.
3) Lower cost to epoxy resin mortars.
4) More tolerant of poor mixing.
5) Good chemical resistance.
6) When applied, polyester resin mortars are impervious thus enabling thin repair works to be undertaken.

Disadvantages
1) System subject to some shrinkage on short and long term.
2) Sensitivity to damp/wet substrates during application can result in poor adhesion.
3) Not generally suitable for large surface repair works.
4) When cured the system can be brittle with low flexural strengths.
5) Should not be applied at temperatures below 4°C.

TYPICAL PROPERTIES OF DIFFERENT PRODUCT GROUPS

	Unmodified Cementitious Mortars	Polymer Modified Cementitious Systems	Epoxy Resin Mortars & Grouts	Polyester Resin Mortars & Grouts
Compressive strength N/mm^2	20-70	10-60	55-110	55-110
Compressive Modulus E-value, KN/mm^2	20-30	1-30	0.5-20	2-10
Flexural strength N/mm^2	2-5	6-15	25-50	25-30
Tensile strength N/mm^2	1.5-3.5	2-8	9-20	8-17
Elongation at break %	0	0-5	0-15	0-2
Linear coefficient of thermal expansion per °C	$7-12 \times 10^{-6}$	$8-20 \times 10^{-6}$	$25-30 \times 10^{-6}$	$25-35 \times 10^{-6}$
Water absorption, 7 days at 25°C, %	5-15	0.1-0.5	0-1	0.2-0.5
Maximum service temperature under load °C	In excess 300° dependent upon mix design	100-300	40-80	50-80
Rate of development of strength at 20°C	1-4 weeks	1-7 days	6-48 hours	2-6 hours

FeRFA

'FeRFA' was formed in 1969 by Formulators of epoxy resin compositions and contractors. The basic aims and objectives of the Federation are:

(1)   To set and maintain minimum standards of product quality, health and safety, technical competence and application capabilities.

(2)   To promote and develop understanding and knowledge of the benefits of resin, polymer and cement based compositions, their use and applications.

(3)   To liaise with professional bodies and similar associated trade associations, to develop, modify and improve materials and application techniques to meet modern requirements.

The choice of a product manufacturer and specialist repair contractor who are members of recognised trade associations, such as FeRFA and the Concrete Repair Association will assist a client in achieving a high quality, cost effective repair and maintenance contract.

ACKNOWLEDGEMENTS:

1.0   BRE Digests 'The Durability of Steel in Concrete'

Part 1 No 263   July 1982
Part 2 No 264   August 1982
Part 3 No 265   September 1982

2.0   Concrete Society Technical Report No 26

October 1984 'Repair of Concrete damaged by reinforcement corrosion'

3.0   Bingham Cotterell, Structural Engineers Private communication, June 1981.

PROTECTION THROUGH CONSTRUCTION

Session A

Chairman

Mr K A L Johnston,

Engineering Director,

Fairclough Civil Engineering Ltd,

United Kingdom

MEASUREMENT OF QUALITY

J H Bungey

University of Liverpool,

United Kingdom

ABSTRACT. Quality of concrete may be defined in a variety of ways, and these are considered in relation to specifications and perform- ance requirements. Current procedures are outlined, and the range of available quality testing approaches briefly reviewed. Particular attention is concentrated on methods of insitu assessment of hardened concrete including strength, composition, permeability, surface zone parameters, internal uniformity and integrity, causes and extent of deterioration, and structural performance. Consideration is given to test planning, execution and interpretation with emphasis upon limitations including environmental influences. Possibilities for changes in testing practice are identified, with reference to developments in other countries, especially in terms of specification compliance procedures relating to durability performance.

Keywords: Concrete, Quality, Insitu Testing, Specification, Compliance, Durability, Test Planning, Interpreta- tion.

Dr. John H. Bungey is a Senior Lecturer in the Department of Civil Engineering, University of Liverpool, where he teaches constructional materials and structural concrete design. He has been active in research and consultancy in the field of non-destructive testing and insitu assessment of concrete for more than 15 years. He currently chairs the B.S.I. sub-committee and Concrete Society working party on N.D.T. of concrete. Current research includes investigating techniques for assessment of reinforcement corrosion, and the use of radar for testing concrete.

INTRODUCTION

Quality of concrete, and construction with concrete, does not just
equate to strength. Whilst strength may arguably be the simplest
property to measure, and usually forms the basis of design calculations and specifications, quality must inevitably include appearance,
integrity and durability performance. These, together with insitu
strength (as opposed to that of a standard specimen) all depend both
on mix characteristics and workmanship.

Whilst routine tests on fresh and hardened concrete should detect
major mix deficiencies, measurement of other aspects of quality
during construction are usually restricted to checks of line, level
and reinforcement before concreting, and visual surface inspection
after striking of shutters. Although a wide range of test techniques
are available for examination of the finished product, these are not
in widespread routine use in the UK because of the arguments that
testing will cause delays, cost money and that methods are not
sufficiently reliable. Although examples of quality testing during
construction do exist, the general practice is to wait until
something is seen to go wrong before embarking on testing of the
in-place concrete.

This paper considers current practice, outlines the range of testing
techniques which may be of potential value for assessing quality, and
examines ways in which they may be used to better effect.

SPECIFICATION COMPLIANCE

Current practice with regard to testing may be considered in three
distinct phases, of which the first two relate to the concrete
supplied to the point of placing as defined by BS 5328[1] which is
shortly to be published in revised form.

Routine Testing of Fresh Concrete

Workability testing is usually based on slump, or possibly compacting
factor or Vebe, and is aimed at identifying changes in properties
from those known to apply to the specified or agreed mix. Test
procedures, together with that for measurement of air content which
is the only other commonly used test on fresh concrete, are defined
in the appropriate parts of BS 1881[2] whilst acceptance criteria are
established by BS 5328[1]. Other factors of the mix affecting quality
including chloride content, alkali content, admixtures, cement
content and water/cement ratio are commonly judged on the basis of
certified or declared values with detailed composition analysis of
the fresh concrete unusual. Techniques for assessing cement content

and water/cement ratio are described by BS DD83[3] although application is not widespread.

Routine Testing of Hardened Concrete

Predominantly based on cubes or other standard specimens made, cured, and tested in a standard manner, this is again aimed at identifying, in this case retrospectively, concrete which differs in terms of compressive strength from that specified or agreed. Testing and acceptance procedures are again defined by BS 1881[2] and BS 5328[1] respectively, and the premise is that strength variations will reflect mix changes likely to influence other characteristics including overall quality. Testing of samples for other hardened properties including density and tensile strength may be undertaken in specific circumstances. European pre-standard ENV 206[4] dealing with Performance, Production, Placing and Compliance which has recently been accepted recommends water penetration testing of samples for water impermeable concrete using the ISO 7031 test which is based on the established German DIN 1048[5] procedures. Temperature matched curing of specimens[6] is a further approach which may be particularly worthwhile when monitoring of insitu strength development is required. Rebound hammer and ultrasonic pulse velocity measurements are sometimes used (separately or combined), especially in checking quality of precast concrete elements.

Testing of hardened concrete in-place is sometimes, but not commonly, used for structures deemed to be critical in some respect. Examples include:
- a) use of covermeters to check correct positioning of reinforcing steel.
- b) use of cast-in pull-out inserts, or similar, to assess insitu strength at specified ages from casting.
- c) use of initial surface absorbtion or similar tests to assess surface quality, including effectiveness of curing.
- d) microscopic analysis of thin sections cut from small cores to check cement type and content.
- e) tests for abrasion resistance.

Retrospective Measurements of Insitu Quality

These will almost invariably be initiated as a result of adverse routine test results or observed deterioration, and will aim to establish the cause and extent of the problem. Contractural disputes or potential litigation will usually be involved. Insitu strength assessment will often form an early part of such testing based on cores or 'near to surface' partially destructive techniques, but may usefully be preceeded by comparative rebound hammer or ultrasonic pulse velocity surveys according to circumstances. If reinforcement corrosion is suspected, then measurements of cover and analysis of

samples by chemical or microscopic methods will be used to examine
the reinforcement location and concrete composition for comparison
with specified requirements. Surface absorbtion or permeability
measurements may well provide supplementary evidence of a general or
comparative nature relating to quality of the concrete mix as well as
compaction and curing, but unless specified values are available they
are unable to directly prove or disprove compliance.

The principal problem with measurements of insitu quality is that in
many instances the accuracy with which parameters can be assessed
limits conclusive comparisons with specifications to clearly
satisfactory or grossly inadequate cases. Unless they have been
established during construction, correlations are seldom available
for particular test methods for the specific mix involved, and their
establishment represents an additional problem and introduces further
uncertainties. Results of any testing also relate to specific
locations only and there is inevitably a broad 'grey' area within
which conclusive decisions about specification compliance are
impossible.

TESTING FOR OTHER REASONS

This will probably be associated with assessment of structural
adequacy or future durability performance. Whilst it may be linked
to, or follow from, specification testing it may also arise as a
result of proposed change of use, purchase, insurance purposes, from
accidental damage or overload, or from the need to assess or monitor
long term materials characteristics. Testing will also form part of
routine maintenance inspections and will be necessary as a preliminary to design of repair schemes.

Many of the techniques used will be similar to those previously
mentioned but additionally include electrical techniques for
assessment of reinforcement corrosion risk, and integrity assessment
techniques such as dynamic response, thermography and radar. These,
together with other techniques, are outlined below and described in
detail elsewhere[7].

TEST METHODS AVAILABLE FOR HARDENED CONCRETE

The range of principal test methods is summarised in Table 1, and all
these have been described in detail by the Author elsewhere[7] whilst
many are covered by various parts of BS 1881[2]. It should be noted
that the categories of application used to classify the various
methods in Table 1 are not well defined in some instances, and there
is inevitably overlap so that results achieved by all these methods
may be considered to relate to some aspect of quality.

Table 1   Principal test methods

Property under investigation	Test	Equipment type
Concrete Strength	Cores	Mechanical
	Pull-out	Mechanical
	Pull-off	Mechanical
	Break-off	Mechanical
	Internal fracture	Mechanical
	Penetration resistance	Mechanical
	Maturity	Chemical/electrical
	Temperature-matched curing	Electrical/ electronic
Corrosion of embedded steel	Half-cell potential	Electrical
	Resistivity	Electrical
	Cover depth	Electromagnetic
	Carbonation depth	Chemical/microscopic
	Chloride concentration	Chemical/electrical
Concrete quality, durability and deterioration	Surface hardness	Mechanical
	Ultrasonic pulse velocity	Electronic
	Radiography	Radioactive
	Radiometry	Radioactive
	Neutron absorption	Radioactive
	Relative humidity	Chemical/electronic
	Permeability	Hydraulic
	Absorption	Hydraulic
	Petrographic	Microscopic
	Sulphate content	Chemical
	Expansion	Mechanical
	Air content	Microscopic
	Cement type and content	Chemical/microscopic
	Abrasion resistance	Mechanical
Integrity and performance	Tapping	Mechanical
	Pulse-echo	Mechanical/ electronic
	Dynamic response	Mechanical/ electronic
	Acoustic emission	Electronic
	Thermoluminescence	Chemical
	Thermography	Infra-red
	Radar	Electromagnetic
	Reinforcement location	Electromagnetic
	Strain or crack measurement	Optical/mechanical/ electrical
	Load test	Mechanical/ electronic/ electrical

Strength Tests

Cores[8], although causing a larger damage zone than most other techniques, provide the most reliable method of insitu strength assessment as well as enabling visual inspection of the interior of a member and may also provide samples for many other tests including chemical and microscopic analysis and permeability measurement. Near-to-surface strength tests[9], such as pull-out, pull-off and penetration resistance, provide results relating only to the quality of surface zone concrete but this may well be the most important aspect for durability performance. Although their accuracy of strength estimation may only be of the order of ± 20%, they only cause limited damage and provide instant results enabling testing to be more widespread than is possible with cores.

Permeation Characteristics

Surface zone permeability or absorbtion characteristics represent an important aspect of quality assessment in view of the well established links to durability. A wide range of testing techniques are available[5] including Initial Surface Absorbtion, Figg air and water tests, and variations including the CLAM and Covercrete Absorbtion Test[10]. Of these, Initial Surface Absorbtion is the only standardised method. These tests all permit general classification of surface zone concrete quality although parameters measured in each case are different. Insitu tests of permeation characteristics are all influenced considerably by the moisture content of the concrete, and this is extremely difficult to measure using available techniques with sufficient accuracy to permit corrections to measured permeation values. This represents a significant problem which limits more precise classification of insitu concrete on the basis of such results. Water Absorbtion tests may also be undertaken on samples removed from the concrete element, as may water flow tests to determine absolute permeability values.

Electrical Methods

These techniques, including Half-Cell Potential, Resistivity and A.C. Impedance are associated with identification of reinforcement corrosion risks. Of these, only resistivity measurements are related directly to the concrete quality although Half-Cell Potential measurements will be used to identify zones of high corrosion risk that will usually result from poor quality of concrete materials or workmanship. In both cases there have been significant equipment developments in recent years, including automatic data storage facilities. Four probe surface zone resistivity measurements may now be made without the need for drilling of the concrete surface.

Chemical and Microscopic Methods

Chemical analysis of the hardened concrete[11] may permit assessment of features such as carbonation, chloride content, sulphate content, cement type and content, aggregate/cement ratio, alkali content, alkali reactivity and water/cement ratio to varying degrees of accuracy. With the exception of simple site techniques to identify carbonation depths and chloride contents, both of which are important for assessing reinforcement corrosion risk, most chemical tests are complex and expensive. Considerable care over sampling is always required to provide reliable results.

Microscopic examination may be applied to the polished cut surface of a concrete specimen, possibly after impregnation by a fluorescent dye and assisted by illumination by ultra-violet light. Properties that can be assessed include entrained air content, cement and aggregate content, and major internal crack patterns caused by alkali - silica reactions. The other major category of microscopic examination involves the use of thin sections which may be used to identify cement type and content, mix components, carbonation, and causes of deterioration. This is considered in some detail in the recent Concrete Society Technical Report No. 32[11].

Comparative Tests for Other Physical Properties of Concrete

Comparative surveys of concrete quality may be obtained in the form of surface hardness measurements by rebound hammer, and dynamic elastic modulus measurements by ultrasonic pulse velocity tests. Unfortunately, surface skin hardening reduces the value of rebound hammer measurements on concrete more than about 3 months old, whilst the need for low frequency through-transmission ultrasonic techniques with concrete will impose practical limitations. The most significant of these is the need to have access to two opposite or adjacent surfaces, but heavy reinforcement may also cause problems. Penetration resistance tests have been found to provide a convenient approach to comparative surveys of concrete quality where access is difficult and restricted to one surface only. Backscatter radiometry may be used to obtain an indication of surface zone concrete density where only one surface is available, and in such circumstances ultrasonic techniques may also be used to estimate the depth of surface cracks in dry concrete, although reliability is poor.

Where the quality of construction is under examination, voids and poor compaction of the concrete can be identified using ultrasonics and radiography provided that two opposite faces are accessible. Short wave radar is also being developed for this purpose where access is restricted to one surface only. Gamma radiography may indicate the position and quantity of reinforcement and detect voids in grouted post-tensioning ducts but is restricted to concrete thicknesses of less than about 450 mm and is an expensive operation. Covermeters also have an important role in assessing quality of construction, although it must be recognised that insitu cover thickness values cannot be expected to be better than within ± 5 mm and calibrations on meters should always be checked for the circumstances of use by direct physical measurements of cover.

Tests of Internal Condition and Integrity

Experience has shown that the human ear used in conjunction with surface tapping is the most efficient and economical method of determining major delamination. The approach however relies upon a subjective assessment by the operator to differentiate between sound and unsound regions and results cannot readily be quantified. It is not simple to monitor the response electronically since factors such as uniformity of applied blow become significant, but commerically available equipment has now been developed and is especially useful on a comparative basis. A plot of surface temperature contours obtained by an infra-red thermographic survey may also be used to identify sound and delaminated areas and the presence of voids or ducts. Surface temperature differentials exist during heating and cooling as a result of variations in internal heat transfer characteristics. Unfortunately these differences are very small so that great care is needed to avoid extraneous effects, and equipment is very expensive. Within the past two years there has been a significant growth in usage of sub-surface scanning radar[12] to detect hidden features such as delamination, voids and reinforcing bars. Efforts have also been made to assess the quality of insitu joints between precast elements. This technique offers considerable potential and is the subject of a major current research programme at Liverpool University.

Dynamic response testing of entire structures is commercially available and may be used to monitor stiffness changes due to cracking and deterioration. The 'dynamic signature' of a structure may be obtained by monitoring the response at critical points to a known imposed impulse or vibration applied to another part of the structure. This may be particularly useful to monitor the effects of a suspected overload or deterioration with time.

Full Scale Load Tests

These tend to be rather expensive undertakings as a result of the extensive preparatory and safety work that is required, but may be used to demonstrate structural performance when subject to an overload. They are usually at the end of the line of testing, where materials tests have proved inconclusive, or else used for proving the capacity of politically sensitive structures or those of a novel nature where there is doubt about structural actions. Performance is usually judged by deflections and recovery augmented by visual observations, but no information is yielded about the reserve of strength remaining beyond the test load.

TEST SELECTION, PLANNING AND INTERPRETATION

These three activities are closely inter-related and must be
considered in a systematic manner. The basic steps are identified in
simple form in Figure 1, and these will apply whether the testing is
to be undertaken during construction as part of acceptance criteria
or at a later stage for other reasons.

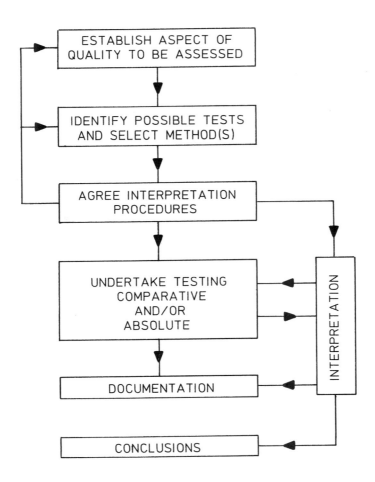

Figure 1 :   Test programme planning outline

Determine Aspect of Quality to be Assessed

This is clearly the fundamental initial stage of any measurement programme and requires a clear identification of the purpose involved. As identified previously, basic classification is likely to be in terms of strength, appearance, durability, or integrity, and it is possible that more than one of these aspects is involved in which case it may be necessary to employ more than one assessment method.

If strength is to be measured it will be necessary to consider whether comparative or absolute values are needed, whether results should relate to the surface zone or interior, and whether a value is required at one specific point in time or strength development is to be monitored. As has been seen, durability embraces the performance of the concrete in terms of chemical resistance, surface density, permeabilty and abrasion resistance, as well as the likelihood of corrosion of embedded steel. Integrity considerations may range from assessment of cracking and delamination to location of hidden honeycombing or voids, as well as checking of thickness and support conditions for ground slabs.

Test Selection

It is likely that several methods are available to measure the particular quality property identified. Selection of that which is most appropriate will be influenced by circumstances of use and the purpose of the testing. Important considerations will include:

a)   The availability and reliability of calibrations; which may be required to relate measured values to the required properties.

b)   The effect of damage; which will relate to both the surface appearance of the test member and the risk of structural damage resulting from the testing of small selection members.

c)   Practical limitations; including the member size and type, surface condition, depth of the test zone, location of reinforcement and access to test points. Other factors may also include ease of transport of equipment, effect of environment on test methods, and safety of test personnel and the general public during testing.

d)   The accuracy required; which will not only influence the choice of test method but also the number of test points.

e)   Economics; the value of the work under examination and the cost of delays must be carefully related to the likely cost of a particular test programme.

In the case of existing structures, a preliminary visual inspection
is important at this stage. In some circumstances it may be helpful
to use more than one method, either to provide corroborative evidence
thereby increasing the confidence levels of conclusions, or to
minimise cost or damage by sequential testing with quick cheap
comparative methods as a preliminary to more precise but expensive or
damaging methods.

Interpretation Procedures

Agreement prior to testing is essential if future disputes are to be
avoided and must take account not only of the capabilities and
accuracies possible from the test methods, but also the effects of
environmental and practical features upon their results.

Examples of environmental factors include differential weathering and
carbonation rates between parts of existing structures dependent upon
exposure, and the influence of insitu moisture conditions upon
results. This latter point is particularly critical in many cases
including permeability, integrity, and strength testing, and it must
be remembered that calibrations prepared in a laboratory may not
relate to site conditions unless careful allowance is made.
Assessment of moisture conditions internally within concrete is not
easy and may pose a major limitation. Practical factors include well
established variations of in-place concrete strength[7] and related
properties, variations in mix proportions within a load as discharged
from a truck-mixer, and the influence of reinforcing steel on test
results.

These factors will affect the locations and numbers of individual
tests to be performed. This will be affected by the purpose to which
results are to be put (i.e. specification compliance or structural
adequacy), and a compromise between accuracy, comprehensiveness, and
cost is inevitable. It should be noted in particular that execution
of testing on site may be more difficult than in a laboratory due to
weather, access, and extraneous influences so that accuracy may be
reduced. It is essential that testing is performed by an experienced
skilled person who is aware of factors which may influence measured
values.

Having established and agreed procedures to be adopted, interpretation should be ongoing throughout testing and be undertaken by an
appropriately experienced Engineer. In this way the programme may be
modified, curtailed or extended as necessary in the light of the
results obtained.

Testing

This should be undertaken in accordance with the agreed procedures by
skilled personnel with ongoing interpretation as identified above. A

distinction should be made between comparative and absolute results,
and in some cases calibration or verification may be required for the
insitu conditions. Examples of this include checks on covermeters
and verification of integrity surveys or corrosion risk surveys, all
of which may involve localised breaking out for visual inspection and
assessment. Core cutting may also be necessary to provide a strength
correlation for other less damaging methods to be used on a more
widespread basis, and in some circumstances destructive testing (e.g.
of precast units) may be used to calibrate non-destructive methods
for insitu use. Visual inspection always forms an important and
informative aspect of testing.

Documentation

This must be thorough, even if litigation is not anticipated, and
should include records of environmental conditions as well as
detailed results, locations and equipment. Numerical results should
always be augmented by sketches and photos wherever possible.

Conclusions

Realism is an essential component of the formulation of conclusions
following testing. The accuracy of prediction for many methods
permits only generalised classification, or if absolute values of
parameters are produced they may be subject to relatively large
ranges of confidence limits. Some claims that have been made for
particular test methods or equipment are unjustifiable for site usage
when factors which have been identified above are taken into account.
Failure to recognise the need for realistic allowance for these
limitations will lead to conclusions which are at best disappointing
or, in the worst case, dangerous.

POSSIBILITIES FOR CHANGE

Use of many of the methods identified is growing steadily in the
United Kingdom as far as existing structures are concerned as a
result of increased interest in durability, maintenance and repair.
Few changes have however taken place in terms of testing during
construction, and it is here that the greatest scope lies.

Insitu strength testing has been recognised as important, to take
account of compaction and curing effects, in several countries. In
Denmark, for example, cast-in pull-out inserts may be used in lieu of
standard strength specimens with results compared with pre-estab-
lished acceptance limits. This approach is now being used on major
structures, especially where long term durability is critical.
Pull-out testing is also used in Canada and the U.S.A., for
monitoring strength development in cooling towers, and the author has
undertaken S.E.R.C. funded research in collaboration with the

C.E.G.B. to confirm the reliability of this approach[13]. Maturity measurements, and temperature-matched curing, are also very useful in monitoring insitu strength development. These methods have all been used successfully in the U.K. but only on a very limited number of occasions.

Assessment of the composition of fresh concrete is an area which has for several years been recognised as being of potential value. Available techniques have however only found limited application. Again it may be worthwhile to look to Denmark where specifications may call for routine microscopic analysis of thin sections of newly cast hardened concrete to assess cement type and content. Insitu measurement of surface absorbtion values have been specified in a small number of structures in the U.K.[5] to ensure satisfactory surface zone durability characteristics taking into account compaction and curing, but usage has not been widespread. It is likely that test complexity or difficulty is a factor here.

The use of covermeters as a matter of routine on construction sites in the U.K. is growing but there is a long way to go. The potential benefits as a quick cheap test to check reinforcement location and cover after formwork removal would seem to be considerable in terms of durability performance and avoidance of need for future repair. There is a wide range of easily portable equipment available and procedures are well documented and standardised[2], nevertheless suspicion still persists and the level of knowledge of this approach amongst many Engineers is poor. In this particular case, the need is for education. The psychological benefits of routine use should not be overlooked.

Pulse echo, radar, thermography and other integrity tests may be useful for confirming the construction quality of major critical structural elements, but scope for increased usage during construction would appear to be limited. They are however enjoying growing popularity in maintenance situations. The routine use of load testing, although used for concrete bridges in some countries[7] seems unlikely to become widely established in the U.K., but the use of large scale dynamic response testing has potential for further development in assessing the effectiveness of large scale repairs. Localised testing of bonding of repairs has been undertaken with pull-off techniques, and may well increase in use in the future.

CONCLUSIONS

Testing of insitu concrete can offer much more than simple strength assessment and may provide useful information about other aspects of quality including durability and integrity. A wide range of tests is available but the majority of current applications are confined to troubleshooting situations.

Many of the tests available will provide comparative results. Where absolute values of insitu properties are required it will usually be necessary to develop correlations for the mix involved. Particular care is needed to allow for environmental factors affecting the insitu concrete, especially the effects of moisture. Whilst accuracies may not be high, a well planned programme of testing which realistically takes account of limitations may still be extremely valuable.

Possibilities would seem to exist for increased routine usage of insitu testing during construction as part of specification compliance procedures, especially in respect of insitu strength development and durability performance. This would help remove uncertainties about compaction and curing and is particularly relevant to surface zones. Increased use of cover measurement would also be most useful in avoiding longer term durability problems.

Measurement of concrete composition and surface permeability are both hampered by problems with current techniques. Scope for increased usage exists but would be greatly improved by test method developments and standardisation.

REFERENCES

1. BRITISH STANDARDS INSTITUTION. BS 5328 : 1981. Methods for Specifying concrete, including ready-mixed concrete. London, pp 15.

2. BRITISH STANDARDS INSTITUTION. BS 1881. Testing Concrete. London.

3. BRITISH STANDARDS INSTITUTION. DD 83 : 1983. Assessment of the composition of fresh concrete. London, pp48.

4. EUROPEAN COMMITTEE FOR STANDARDISATION (CEN). ENV 206, Concrete -Performance, Production Placing and Compliance Criteria. Anticipated Publication September 1990.

5. CONCRETE SOCIETY. Technical Report 31. Permeability of Concrete -a review of testing and experience. London, 1988, pp 95.

6. BUNGEY J H. Temperature Matched curing of Concrete. Reliability in Non-Destructive Testing, Pergamon, Oxford, 1989, pp 95-102.

7. BUNGEY J H. Testing of Concrete in Structures. 2nd Edn. Blackie & Son Ltd. Glasgow, 1989, pp 228.

8. CONCRETE SOCIETY. Technical Report 11. Concrete core testing for strength. London, 1987, pp 59.

9.   BRITISH STANDARDS INSTITUTION.   DC/16059   Recommendations for
      the assessment of concrete strength by near-to-surface tests
      (Draft for Public Comment).   London, 1989, pp 20.

10.   DHIR R K, HEWLETT P C and CHAN Y N.   Near-surface charac-
      teristics of concrete : assessment and development of in-situ
      test methods.   Magazine of Concrete Research, Volume 39, Number
      141, December 1987, pp 183-195.

11.   CONCRETE SOCIETY.   Technical Report 32.   Analysis of Hardened
      Concrete - A guide to tests, performance and interpretation of
      results. Slough, 1990, pp 117.

12.   BAILEY M.   The use of ground penetrating impulse radar in the
      investigation of structural faults.   Proc. Struc. Faults &
      Repairs 89, Eng. Tech. Press, Edinburgh, 1989, pp 11-18.

13.   BUNGEY J H.   Monitoring concrete early age strength development.
      Non-destructive Testing. ed. Boogaard J and Van Dijk G M.
      Elsevier, Amsterdam, 1989, pp 1243-1248.

QUALITY CONTROL AND PROTECTION OF PRODUCTION AND CONSTRUCTION BUILDINGS OF CONCRETE

A.T Semagin

Marijski Politechnical Institute,

USSR

ABSTRACT. Quality control and protection of reinforced concrete constructions include, inspection, study and checking on the stress-strain state with the help of test equipment, flaw detection and flaw systemization of construction, the use of different methods to study the state of construction and the factors promoting the beginning of cracks and corrosion processes and the development of protective measures and carrying out reconditioning of reinforced concrete structural elements.

Reinforced concrete constructions were studied in a process of their production, mounting and use. During the investigation a systematical method of approach was used to efficient correction, production and use control with the purpose of providing favourable stress-strain state of construction, non-admission of critical stress values, strains and cracking. The possible external action and internal state were also taken into account. In excess of critical values of the stress-strained state and in presence of flaws and corrosion, the reconditioning is offered to increase quality and durability of construction.

Keywords: Reinforced Concrete Constructions, Stress-Strained State, Quality Control, Cracks, Flaws, Corrosion, Reconditioning.

Dr A T Semagin is a Dean of the Faculty of Civil Engineering, Assistant Professor, A Candidate of Sciences at Marijski Polytechnical Institute, Yoshkar-Ola, USSR. He is active in the field of quality control in production of reinforced concrete constructions and he is engaged in inspection and estimation of stress-strain state of buildings.

INTRODUCTION

Quality and durability control of reinforced concrete constructions,
buildings and structures gains in importance at present. It is
conditioned by the above mentioned constructions in the buildings
and structures. The problem of raising the quality of reinforced
concrete constructions is always in view in our country. But the
quality of the buildings and structures erected with the use of the
constructions does not meet the modern requirements. Maintenance
expenditures are increasing, a great volume of reconditioning work
is being conducted to keep serviceability of constructed objects
going. Quality increase of reinforced concrete constructions is
closely associated with the improvement of all technical and
economic indices of construction owing to lowering of resources in
flaw correction that takes place in production of buildings and
their maintenance. Low quality of construction and buildings made
of reinforced concrete tells on maintenance expenditures increases
and formation of losses due to the capacity decrease of enterprises
during extraordinary reconditioning.

The most important tasks in solving the problems of quality increase
and durability of reinforced concrete constructions are:

a) creation of an organizational and technological system, the main
   purposes of which are to expose operating conditions of objects
   and ties between designed and built constructions and changes in
   them in time.
b) confirmation of rightful recommendations given in specification
   documentation for the quality indices.

Concrete and reinforced concrete are and will remain, at least for
the nearest decades, the main building materials. The volumes of
materials resources demand rational, economical and effective use,
and the objects build with the use of reinforced concrete should be
reliable and durable owing to the high quality of constructions.

METHODS OF QUALITY CONTROL OF REINFORCED CONCRETE CONSTRUCTIONS

Quality control is the most important line in the system of quality
securing of reinforced concrete constructions and estimation of
their stress-strained state. In the process of control activity we
check the conformity of constructions quality indices stated by the
specification, their fitness to maintenance conditions. For the
conduction of quality control one must choose quality estimation
criteria of constructions and objects.

There are the following methods of quality estimation in practice:

a) percentage of flawless products on the first presentation
b) marks
c) an alternative method
d) a complex method

Methods a) and c) are based on acceptance of work and constructive elements in full accordance of the specification documents with many single quality indices and prevent the appearance of defectives. Differential complex methods in some cases permit a decrease of single quality indices in production of finished products with generalized main indices complying with the demands of the specification documents. These generalized indices depict the main purpose of products. Carrying capacity is the main index for the load carrying structures that determines reliability. For many structural elements the determining indices are rigidity and crack resistance.

Compressive strength for many years has been the main quality index in concrete and reinforced concrete technology. The drawback of this index is a passivity because it permits to registrate the correspondence of strength to the planned one, but it does not provide strength in the process.

The purpose of this work is the increase of the active quality control, ie we use the results of quality control for production and existence of reinforced concrete constructions.

In a control system the choice of a checking plan is important. An active role of quality control in a production process appears with feedback between control results and designing process, production process parameters and organization of work, operation processes.

The demands to quality indices of reinforced concrete constructions develop in two directions:

- indices characterizing physical and mechanical properties and deformation properties of compound materials of a construction
- indices determining geometrical parameters of a construction

Requirements for the quality indices as well as the quality control of reinforced concrete constructions should base on researches of probability parameters of elements using the main principles of mathematical statistics, probability and reliability theories. The state of a construction at any stage of its existence can be characterized by finite number of independent random variables given by:

$$y(x_1, x_2, \ldots, x_n) = 0 \qquad (1)$$

Equation (1) described domain boundaries of allowable states of constructions. Some of the parameters characterize inner properties of constructions and they are, in their turn, the isolated indices of quality, others present external actions. Hence serviceability of a product can by given by:

$$A(x_1, x_2, \ldots, x_i) - B(x_{i+1}, x_{i+2}, \ldots, x_n) \geq 0 \qquad (2)$$

where: $A(x_1, x_2, \ldots x_n)$ – random function of inner properties of a construction

$B(x_{i+1}, x_{i+2}, \ldots x_n)$ – random function of external actions

The main requirements for the initial failure-free performance of constructions can be described by:

$$R = P(A-B \geq 0) \geq R_0 \qquad (3)$$

where: R_0 – the present reliability level.

Equations (2) and (3) are solved by three groups of methods:

- theoretical methods of investigation
- methods based on natural investigation (including inspection)
- methods based on complex investigations

On the ways of making a decision to promote a construction quality there is the use of a correlation – regression analysis of influence of technical factors on a quality level. A precondition for true and objective results of the analysis of factor influence on the quality level of construction is the choice of significant factors.

According to the Building Regulations of the USSR concrete and reinforced concrete elements are calculated in correspondence with the carrying capacity (first group limited states) and serviceability (second group limited states). The demands of the first group are given by:

$$M \leq M_1 \qquad (4)$$
$$Q \leq Q_1 \qquad (5)$$
$$N \leq N_1 \qquad (6)$$

of the second group by:

$$a_{crc} \leq [a_{crc}] \qquad (7)$$

$$f \leq [f] \qquad (8)$$

where: M, Q, N –   calculated (maximum feasible) bending moments, lateral and longitudinal forces of external loads

M_1, Q_1, N_1 –   limited (maximum feasible) carrying capacities of elements exposed to bending, shearing, compression or tension

$a_{crc}, [a_{crc}]$ –   design and maximum feasible parameter of crack opening

f, [f] –   design and maximum parameter of construction deflection

Expressions 4-8 present both quantitative and qualitative characteristics of reinforced concrete constructions because among numerous parameters defining the quality of elements we can single out a group of parameters that are controlled during production, operation and they are included into a design formula.

SYSTEM OF QUALITY CONTROL OF REINFORCED CONCRETE CONSTRUCTIONS

The system of quality control with regard for economy of resources at the construction industry plants producing reinforced concrete constructions is shown in Figure 1.

The following principles lie in the basis of the system:

- The principle of complexness, ie inclusion of all stages of reinforced concrete constructions production process and all factors defining functional properties of products.
- The principle of optimality, ie provision of minimum production expenses
- The principle of dynamism, ie continuous progress connected with production modernization, adaptability to the range of structural elements production.
- The system must define a condition of the production process at any moment, efficiently define a location and a cause of controlled variables deviations.
- The system must realize a feedback of technological parameters with the single and finite quality indices of reinforced concrete constructions.

Beside these main principles the system, in contrast to the existing methods of quality securing, must possess the following properties:

- Flexibility, ie requirements for numerical values of quality indices must be defined depending on stability and accuracy of the concrete production process.
- Economy of resources, ie an ability to reduce material and labour expenditure at each stage of reinforced concrete constructions production.

The diagram in Figure 1 consists of two subsystems A and B. Subsystem A is aimed at the analysis of the production state of a concrete plant and at the development of the necessary production forms and records, paying attention to real feasibility of production lines and characteristic properties of structural elements. Subsystem B is aimed at the efficient quality control, analysis and making a decision concerning provisions securing quality and economy of resources.

The stages are accomplished as follows. The calculated quality level (Section 1) is defined on the basis of statistic modelling. Here we deal with non-maximum states of reinforced concrete constructions. State Standards are taken as the principle of

modelling. The following purpose indices are taken into account -
probability of securing carrying capacity, crack resistance and
rigidity.

When choosing a critical limiting state (Section 2), that will
define the demands to single out quality indices and become an
object to be* controlled, we analyse the results of the previous
Section. We call critical a limiting state - expressions 4-8 -
which has the most probable onset.

The calculated value of the starting failure-free performance is
compared with the normative one (Section 3). The normative level of
quality reliability is defined with the probability of 0.9987.
After comparing we have two possible states: if $R_c < R_s$ (where R_c, R_s -
calculated and normative reliabilities) we must revise design plans
and production forms (Section 4). As a result of this comparison we
can evaluate the quality of a design and find errors which can reach
27% of all deviations from specifications. In case $R_c \geq R_n$ we pass on
to Section 5 in which we construct a regressive model of influence
of single quality indices on the starting failure-free performance
of reinforced concrete constructions according to the criterion
which has been chosen in Section 2.

To estimate a production process state (Section 6) we process
operation control data. If we do not have such data we measure
single quality indices with the following processing. The results
are used to analyse accuracy and stability of production operations.

A statistical regularity based on a mathematical model of
reliability makes it possible to define actual starting failure-
free performance of the reinforced concrete constructions on this
very production line. If we compare an actual failure-free
performance with a calculated and a normative one (Section 7) we
shall have 3 conditions. Thus we can refer all production lines
(processes) to the following three groups

- lines with low stability and accuracy of production operations, ie
  they have $R_f < R_c$, $R_f < R_n$
- lines with satisfactory stability and accuracy, $R_n \leq R_f < R_c$
- lines with high stability and accuracy, $R_f > R_n$ and $R_f \geq R_c$.

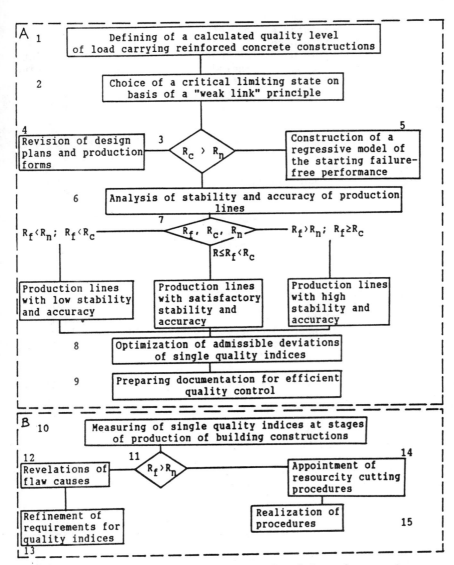

Figure 1: Diagram of quality control of reinforced concrete
constructions

No matter which group the production lines are referred to,
optimization of admissible deviations of single quality indices
(Section 8) considering real production potentialities and
structural features of produced elements. As the main limitation of
the optimization task one should use the starting normative failure-
free performance.

At the following stage (Section 9) we come across provisions
securing quality of reinforced concrete constructions. Operation
control sheets for all production process stages are drawn up. To
raise a human factor, in operation sheets alongside with production
and organization provisions, we included economic grounds concerning
this or that parameter tolerance of constructions and material
incentives for higher accuracy in production processes. After
adoption of proper regulations we pass on to the efficient quality
control, ie to subsystem B.

On the basis of the statistical data obtained as a result of
production processes control (Section 10) we define an actual
failure-free performance. This is a basis to make a decision that
construction products correspond to determined standards (Section
11).

In case $R_f < R_n$ we find out the cause of fall of the starting failure-
free performance outside the limits, define the degree of influence
of quality indices outside the limits on reinforced concrete
constructions serviceability (Section 11) with the following
correction of taking place deviations (Section 12).

From the point of view of quality indices failures effect on
reliability of structural elements, in general, it can be referred
to the class of absolute systems, ie a refusal of single quality
indices can lead a reinforced concrete construction either to its
complete failure or to deterioration of its quality state, or having
some functional redundancy it does not effect the quality.

In case $R_f > R_n$ (Section 13) there appears a serviceability reserve of
a reinforced concrete element that can be used to cut production
costs or it can be left for possible cases of single quality indices
deviations.

Thus we have developed the system of quality control and economy of
resources in production of reinforced concrete constructions. But
for the successful realization of this system we must solve some
scientific-methodical problems to which we refer the development of:

• general methodical approaches of statistical modelling of
  technological factors
• a technique of correlational dependence of single quality indices
  influence on the starting failure-free performance
• optimization of quality indices with regard for real production
  possibilities

CORROSION AND RECONDITIONING OF SOME REINFORCED CONCRETE CONSTRUCTIONS

A cause of losses of the carrying capacity of reinforced concrete
constructions is corrosion. Corrosion is a result of a corrosive
medium action and flaws in concrete. Cracking accelerates corrosion

in reinforced concrete elements. Cracks are a kind of conductor of
corrosive action on reinforcement. Roofing is the most subjected to
corrosion elements of a building.

We refer ribbed slabs, beams and girders to the roofing reinforced
concrete elements. Durability of these elements appears to be low
especially of ribbed slabs. Their life is often 3-4 times shorter
than the normative one, and they need to be strengthened or changed.
The analysis of volumetric-planning solutions of different objects
does not envisage their reconditioning using common hoisting and
transporting mechanisms. Some structures are often beyond the reach
of such mechanisms and demand the use of unique operations on
reconditioning of several elements that need to be changed. As a
result we can sufficiently raise the use of labour, materials and
other resources for reconditioning.

As an example we shall illustrate the reconditioning of a shop
roofing. The shop was build over by other adjoining shops (Figure
2). Inspection of the reinforced concrete constructions showed that
their state was quite satisfactory and they needed only minor
repairs. But the ribbed slabs of the roofing had lost their normal
state and they did not meet the normative requirements.

A progressive method of reconstruction of the shop in straightened
conditions was offered. A helicopter was used as a hoisting
machine. This reconstruction with reconditioning of slabs was
carried out on basis of operative estimation of the stress-strain
condition of a large-panel construction with the help of modern
means and apparatus (Figure 3).

Thy system "helicopter-load" is a complex two-mass system with a
compliant rope tie. Helicopter's movement is conditioned by:

● the pilot's activity on handling the helicopter with the aim of
  placing it above the mounting vertically
● the load's behaviour on an external suspension
● the wind exposure

The use of a helicopter instead of a crane is substantially
different due to the helicopter's specificity as a hoisting machine
and the number of mounted and dismounted elements considerably
exceeds the volume of work that has been done before.

This method provides:

● reconstruction under the conditions of a functioning production
● carrying out the work in straightened conditions
● independence of a mounting machine (a helicopter) of a
  construction site conditions

The efficiency can be expressed as follows - mounting/dismounting
takes 3-4 days instead of 40-45 as usual.

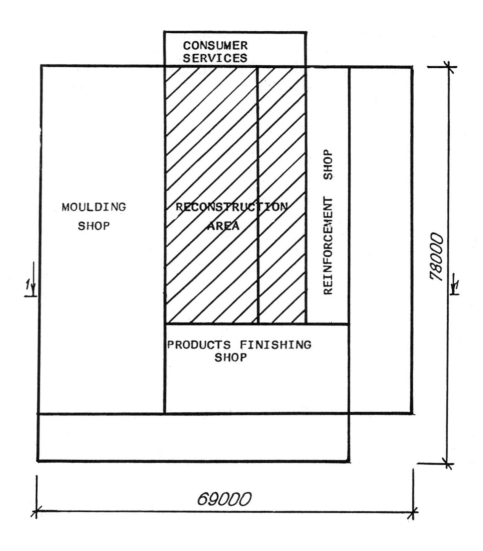

Figure 2: Location of the reconstruction area among other buildings

1 - 1

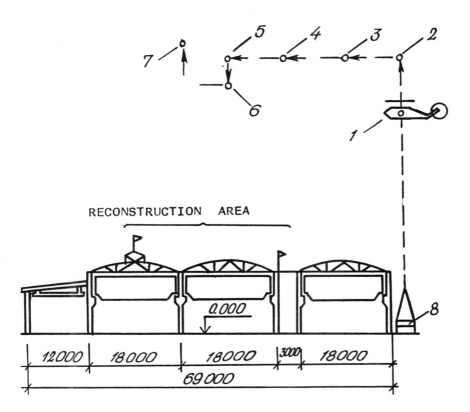

Figure 3: Diagram of the roofing slabs mounting

1 - slinging
2 - the first control hanging
3 - transporting of a slab
4 - the second control hanging
5 - mounting hanging
6 - the third control hanging and slab de-slinging
7 - helicopter's departure from the mounting area
8 - slabs store

CONCLUSIONS

Quality control, protection of objects made of reinforced concrete
is a many-sided problem. It is solved by perfection of the quality
control methods, introduction of the quality securing system.
Reinforced concrete constructions should be protected against
corrosion and their crack resistance must be secured with the
purpose of building equally reliable and equally durable
constructions in the system of the object. If the above mentioned
conditions are not secured, at the designing stage the
reconditioning must be envisaged with minimum economic and material
losses for the object operation.

DEVELOPING QUALITY CONTROL FURTHER IN THE PRODUCTION AND SUPPLY OF CONCRETE

B V Brown

RMC Technical Services Ltd.,

United Kingdom

ABSTRACT. Quality control techniques continue to develop to meet the demands for improved quality of concrete. The paper gives examples of control methods used within the concrete industry, and considers principles for evaluating quality control and design standards.

Keywords: Quality Control, Concrete, Design Criteria, Acceptance Standards.

Mr Bev V Brown is Divisional Technical Executive of RMC Technical Services, which is a subsidiary Company of Ready Mixed Concrete (UK) Ltd, RMC House, High Street, Feltham, Middlesex, England, UK. He has been active in concrete technology for over 25 years and is responsible for the maintenance and development of quality assurance systems, technical operating and advisory services to Ready Mixed Concrete (UK) Ltd. He serves on a number of BSI, CEN and concrete industry technical committees, working parties and task groups related to the use of aggregates and concrete.

INTRODUCTION

ENV 206 The European Prestandard for Concrete [1] establishes the importance of quality control for production and compliance requirements and states:

> "Concrete production, placing and curing shall be subject to quality control procedures".

Quality control (QC) is defined in BS 4778 : Part 1 : 1987 [2] (ISO 8402) as:

> "the operational techniques and activities that are used to fulfil requirements for quality".

This is a vital element in ensuring achievement of specification and some of the tools are identified in this paper. However, it should be recognised this is only part of the full quality assurance necessary for overall quality. It provides the information and prompts action whilst full quality assurance (QA) is defined in BS 4778 : Part 1 : 1987 [2] (ISO 8402) as:

> "All those planned and systematic actions necessary to provide adequate confidence that a product or service will satisfy given requirements for quality".

Thus quality assurance looks at the overall management system and resources for achieving quality. BS 5750 :1987 (ISO 9000/EN 29000) [3] identifies all the necessary elements of QA and the requirements for systematic and documented procedures. QA also demands an ongoing regular audit and review of the standards achieved and available resources for economic fulfilment of quality.

Quality Assurance including its quality control elements does not automatically achieve total quality but provides the management tools and mechanisms for continuing improvement in standards. It is also a means of providing confidence in the supply of concrete to the purchaser. In the short term it is to be hoped it may avoid unnecessary testing for lower grade concretes of less structural significance. These are often treated to a less than professional appraisal by untrained personnel more interested in adequate construction technique than testing proficiency. In the longer term this could also extend to a reduction in testing for higher grades and more structurally significant concrete as confidence can be built in the effectiveness of such quality assurance certification. If QA does improve quality standards through regular review and upgrading this would be a justifiable reward.

TYPES OF QUALITY CONTROL

Assessment tools and acceptable standards need to be defined for quality control which for simplicity can be divided into forward and backward control:

– Forward control consists of identifying the needs for specification compliance, the quality of materials, design, production and delivery procedures.

– Backward control reviews quality achievement and is used to identify actions necessary to improve future quality attainment.

The ability to influence overall quality by forward control is limited to the extent that resources are available. For example a ready mixed concrete plant has an effective life of some 20–25 years and can cost between £0.3 and 1 million pounds dependent on the production capacity required. This together with transport costs is a significant capital investment and unlikely to be rapidly replaced. Material availability is limited to that made available through planning controls. However, advances are possible through upgrading plant and establishing controls on materials, designs and production.

On the other hand backward control, which in many ways is easier to operate, has been the area where much analytical thinking has been used to determine changes to mix design from measurements of properties and responses to customer complaints of appearance or performance.

MATERIALS

The quality control of materials involves:

(i)   Overall appraisal of the material supply source capability.
(ii)   Assessment of the required properties and their average or typical level.
(iii) Identification of acceptable variations and agreement of standards for supply.
(iv)   Monitoring variations and trends from typical average quality.
(v)   Action procedures.

Proper quality control requires these five elements to be defined and monitored. In many cases requirements can be determined from past performance and established technology e.g. the effect of grading on strength performance (see Figure 1). In other cases and for new materials experimentation will be necessary to evaluate acceptable variations. The history of the material should also be used to define frequency of testing or assessment or, in its absence, increased initial testing is necessary to promote rapid appraisal of variability.

The range of properties which require appraisal continue to increase and the effects for both the physical and chemical properties of concrete need careful consideration. For example, chloride and alkali

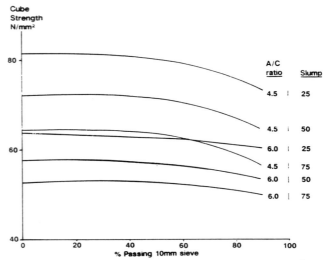

Figure 1 : 28 day cube strength variation with increasing percentage of
10 - 5 mm fraction in 20 - 5 mm coarse aggregate
(same shape material)

contents effects are equally as important to the long term durability
of concrete as effects on strength or water content.

Continuous records and the monitoring of properties are thus necessary
to ensure compliance with accepted or agreed standards for identified
critical properties for all materials.

Two courses of action are possible where material properties fall
outside the accepted or agreed standards:

-   rejection of material
-   acceptance by a waiver instruction which should include a
    consideration and application of any necessary design change to
    maintain specification performance requirement e.g. a change in
    sand grading where a modification to the proportioning of a mix
    could maintain the quality of production.

Both courses of action should be accompanied by corrective action with
the supplier and/or agreed modification to the standards of supply
where applicable. However, where final concrete quality can still be
achieved it may be possible to consider waiver instruction for minor
variations. Responsibility for issue of such waivers should be
defined and restricted to identified quality managers. The waivers
should also be documented to enable recourse for future use and
quality reviews.

Test data provided by suppliers also needs careful appraisal to assess
effects and critical requirements affecting performance. Many tests
are carried out under standardised conditions and these do not always

relate to practical use. For example cement is tested for strength
with standard aggregates at a fixed water/cement ratio. Concrete, on
the other hand, is produced to a fixed workability with varying
aggregate sources. Technologists have come to realise that in terms
of strength performance the way in which materials pack together has a
very significant effect.

The introduction of newer materials such as pulverised fuel ash (pfa),
ground granulated blastfurnace slag (ggbs) and microsilica have
affected performance in a variety of ways including chemical and
pozzolanic reactions, water content adjustments for the required
workability and packing effects. Thus the basis of standard test
assessment may need some correlation with use in practice via
laboratory trials with the actual material combinations. Standard
tests are useful for identification of trends but may mask the
magnitudes of changes in concrete.

MIX PROPORTIONING

Proper initial design and proportioning of mixes is fundamental but
equally important is the ongoing control to ensure such designs
continue to be effective.

Proportioning of mixes can be controlled in an ongoing manner by both
spot checks and routine regular reappraisal. Adjustments are only
necessary if the properties of the materials change. Three systems
are regularly used in controlling these aspects and standards for
redesign established allowing for testing tolerances and natural day
to day variations in materials:

(i)   Monitoring the properties of materials within defined
       tolerances. For example redesign is considered if the
       tolerances on loose bulk density or relative density
       measurements exceed $\pm$ 30 kg/m^3 and $\pm$ 0.04 respectively.

       Alternatively using material properties the designs can be
       checked by calculation or computerised checks using the
       mathematical mix design model produced by J D Dewar [4].

(ii)   Spot check trial mixes. Design tolerances in this case being,
       (say) $\pm$ 1½% for plastic density and/or $\pm$ 0.03 for water/cement
       ratio.

(iii) Appraisal of 28 day cube densities. An equation has been
       formulated which relates 28 day cube densities to plastic
       concrete densities using a correction to allow for hydration
       effects [5]. To avoid individual testing errors the average of
       10 cubes from different batches are examined using the
       following assessment:

28 day cube density - <u>cement content</u> = plastic density $\pm$ 30kg/m^3
                             20

CONTROL METHODS FOR CONCRETE PERFORMANCE DESIGNS

Ongoing control needs to monitor the designs adopted for the range of
specification requirements. This can involve statistical and/or
simple no/no go systems. Dependent on the property being appraised
one or other of these techniques will be the more appropriate.

Concrete producers typically operate systematic testing and monitoring
regimes for the following properties:

- strength
- workability
- air content
- temperature
- density (see previous section on mix proportioning)
- chloride content of concrete
- alkali content of concrete

Where properties are normally distributed (e.g. strength, chloride and
alkali contents), these lend themselves to statistical appraisal
Other properties (e.g. workability and air content) having skew
distributions could be assessed using alternative statistical methods
but more commonly go/no go type controls are applied. In these cases
the variation of a fixed quantity of one of the principal controlling
effects has larger positive effects than negative effects or vice
versa, e.g. ± 8 litres of water could result in an increase of 25mm
slump above a target average of 50mm, but only a 15mm decrease below
50mm.

STRENGTH

Strength is treated as a normally distributed variable and compliance
criteria are often based on a statistical appraisal of groups of
results. For example criteria in ENV 206 [1] are defined in the
following form with an overall philosophy that there should be a 5 per
cent fractile below the characteristic strength f_{ck}.

$$\bar{x}_n \geq f_{ck} + k_1 S_n \qquad \ldots\ldots (1)$$

$$x_{min} \geq f_{ck} - k_2 \qquad \ldots\ldots (2)$$

where

x_{min} is the lowest individual value of the set of n samples from
separate batches

$\bar{x}_n$ is the mean strength of the set of n samples

S_n is the standard deviation of the set of n strength results

k_1 and k_2 are constant

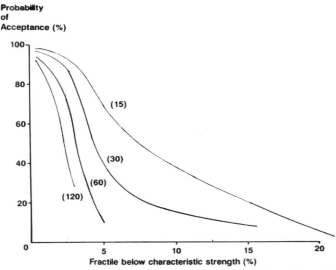

Figure 2 : ENV 206 - k = 1.48 - Non-overlapping sets - Independent results

The probability of acceptance of concrete plotted against the actual
design fractile is termed the operating characteristic curve and
provides an assessment of producer and consumer risks.

Figure 2 [6] shows the probability of acceptance of concrete from such
a curve based on 15 as the number of results assessed for the ENV
criterion. The effect of repeated applications to 30, 60 and 120
results are also shown. It can be seen persistent or regular
application of such statistical criteria increase risk of a potential
costly rejection to a concrete producer. For example a 5% risk for
one application approaches a 10% risk following two applications etc.
Examination of the operating characteristic curves for 120 results
in Figure 2 shows that the producer will commercially need to design
for an order of 1% or less defectives to satisfy the criterion (1)
above without excessive failure. Hence accredited certification of
production of this nature can provide considerable consumer
protection.

Control Methods for Strength

Irrespective of the design level, quality control systems need to
rapidly monitor any variations away from design target values. This
has lead concrete producers to develop systems which monitor
performance relative to target design intention. Two such systems
have gained wide application:

- RMC cusum system and modified versions
- BRMCA counting rules related to results above and below
  target

To speed detection these systems use predictions from early age
results (3 or 7 day) and the correlation with 28 day strengths as well
as trends in the value of the standard deviation are also monitored.
In addition to these systems detecting trends they can also help
identify their magnitude and when the trend commenced.

Cusum

Cusum techniques are described in the Concrete Society Digest No.6 [7]
and BS 5703 [8].

Essentially test results are compared with the designed target mean
values and checks are made to confirm whether they are consistent with
the required levels for compliance. In applying the technique the
target value is subtracted from each of the measured results giving
positive and negative differences. These differences when added
together for a number of results form a cumulative sum (cusum). When
this cumulative sum is graphically plotted against the sequence of
results, a visual presentation of the trend relative to the target
level is produced.

TABLE 1: Example of cusum applied to mean strength

Result No.	28 day strength N/mm^2	Difference from target 40 N/mm^2	Cusum N/mm^2
1	44	+4	+4
2	38	-2	+2
3	45	+5	+7
4	41	+1	+8
5	36	-4	+4
6	44	+4	+8
7	35	-5	+3
8	43	+3	+6
9	47	+7	+13
10	34	-6	+7
11	38	-2	+5
12	33	-7	-2
13	42	+2	0
14	36	-4	-4
15	39	-1	-5
16	35	-5	-10
17	41	+1	-9
18	34	-6	-15

As an example Table 1 gives a series of measured 28 day strengths
where a producer aims for an average 40 N/mm^2 to achieve the required
characteristic strength of 30 N/mm^2. Each result is compared with the
target value of 40 N/mm^2. After two results, one 4 N/mm^2 above and one
2 N/mm^2 below 40, the cusum is shown as +4 -2 = +2. Between the ninth
and eighteenth results it is apparent that a negative trend is
occurring. No result has fallen below the characteristic strength but
there is a trend towards lower strength than design intentions.
Figure 3 is a graphical plot of this cusum against the result number
which visually portrays this trend.

Statistical action limits can be applied to such trends using prepared
action masks. Figure 3 shows such a mask based on a production
variability for a 3.5 N/mm^2 standard deviation. A trend can be seen
to a pending change in mean strength which occurs when the plot
crosses the mask. The design of the mask is complex and depends on
the order of changes expected and the level of production variability.
Calculation of the actual magnitude of the change requires additional
allowances for oscillatory or autocorrelated effects which can
exaggerate apparent trends.

Similar cusums determine trends in standard deviation and/or
effectiveness of the prediction system.

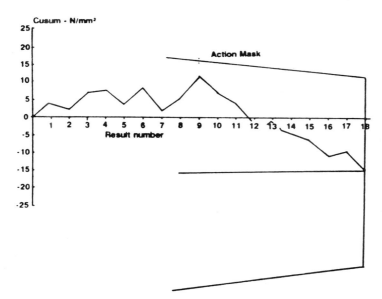

Figure 3 : Cusum plot of mean strength data from
            table 1

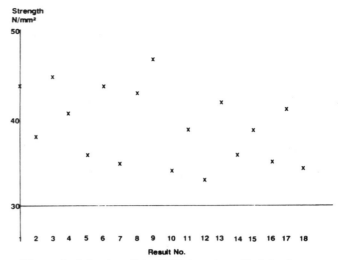

Figure 4: Calendar plot of 28 day strength data from
table 1

Figure 4 shows the corresponding calendar plot of results which may
have resulted in a less severe appraisal of the actual results. Also
calculation of the ENV compliance criteria for 15 results where
k_1 = 1.48 and k_2 = 4 shows criteria (1) and (2) both satisfied i.e.

    (1)   38.5 > 36.4   .... (30 + 1.48 x S.D of 4.3)
    (2)   No result < 26 .... (30 - 4)

Thus the cusum system enables observation of the trend and possible
corrective action in this case before either the calendar chart or ENV
206 criteria are failed.

Counting Rule Method

An alternative more simplistic version based on the cusum system is
the BRMCA counting rule method (9) (10). In this case the number of
results above or below target are counted. Thus in Table 1 only the +
or - signs are identified in the difference column and a change is
signalled when 9 of the last 10 results have the same sign. Similar
appraisals are made for standard deviation and predictive system
checks. This system is slightly less sensitive than a full cusum but
has the benefit of simplicity.

AIR CONTENT/WORKABILITY

Although not normally distributed, plotting of histograms of data
enables a rapid appraisal of the distribution of results. Two forms
of criteria are used to appraise data normally linked to a given
production period and plant:

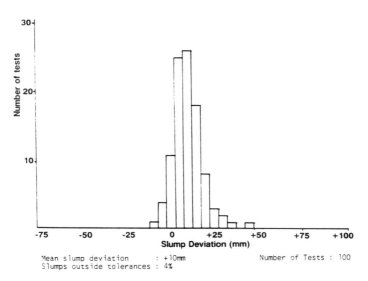

Mean slump deviation   : +10mm   Number of Tests : 100
Slumps outside tolerances : 4%

Figure 5 : Histogram of measured slumps

Mean deviation from the specified value
Number of results outside defined tolerances

Figure 5 shows an example of a computerised analysis for workability
variation.

ALKALI AND CHLORIDE CONTENTS

Control for these parameters is achieved by summing the contributions
from each material based on average data and adding statistical
margins. In some cases producers are prepared to define maximum
levels which will not be exceeded without notice, and these can then
be used in calculations. Acceptance criteria are based on
specification or accepted practice e.g. for alkali contents - Concrete
Society Report No.30 [11].

Concrete producers can then identify the limiting cement contents for
compliance with the range of mixes supplied from a plant. These take
the form of maximum cement content for limiting alkali content and
minimum cement contents for maximum chloride ion contents expressed by
weight of cement.

TESTING

General Test Variability

Quality control data derived from testing is as good as the testing

itself. Many laboratories now accept given test standards set down by
the National Measurement Accreditation Scheme (NAMAS). This requires
regular calibration of all equipment, maintenance and control of the
environment, as well as the proper training of operatives.
Calibration is required to be traceable to National Standards so that
the level of precision can be determined. This form of third party
certification is commended as providing a good base standard on the
following contributors to differences between tests on a batch of
concrete:

>   The operator
>   The equipment
>   The calibration of the equipment
>   The environment (temperature, humidity, air pollution etc.)

Means of assessing the effects of these differences are identified in
BS 5497 : Part 1:1987 [12] (ISO 5725 : 1986). Obviously one means of
reducing the variabilities is to ensure proper and adequate
calibration of test equipment together with standardisation of
equipment and environments.

Published data is becoming available on test reproducibility and
accordingly this should be taken into account when defining acceptance
standards for material and concrete properties. The published values
can be assumed for trained personnel, but in their absence should be
assessed by inter-laboratory comparisons.

Strength Test Variability from Duplicate Cubes

Since concrete cube testing for strength is so important ongoing
regular quality control checks on test variability can be implemented
using duplicate cubes. Testing variations are typified numerically by
the coefficient of variation (C of V) of the strengths of cubes within
sets taken from the same sample e.g.

$$C \text{ of } V = \frac{SD \text{ of set}}{Mean \text{ of set}} \times 100\% = \text{constant for given test conditions.}$$

In practice use is made of the simple estimate of the C of V from
ranges in strength shown by duplicate cubes expressed as fractions of
their mean. The average of 10 such duplicates enables an estimate of
the coefficient of variation to be made as:

$$C \text{ of } V = \text{Average value of } \frac{(Range \text{ of pair})}{1.128} \times \frac{100}{(Mean \text{ of pair})} \text{ per cent}$$

Results are then examined in groups of 10 to ensure that the average
range/mean percentage does not exceed 5% and individual range/mean
percentages in excess of 14% are not being obtained more frequently
than once in forty results.

If these criteria are exceeded an investigation is made into possible causes of excessive variation. This can occur in any part of the testing regime e.g sampling, manufacture, equipment, curing and crushing.

Table 2 shows an example calculation sheet with results from 7 day cube results.

TABLE 2: Test variability data

Cube Ref	Date Cast	Strength at 7 days (N/mm²)	Range (N/mm²)	Mean Strength (N/mm²)	Range - Mean Strength (R/M)%	Average R/M (R/M)%
101	2.7.90	21.0				
102	2	22.0	1.0	21.5	4.7	
103	4	36.0				
104	4	34.0	2.0	35.0	5.7	
105	8	19.0				
106	8	22.5	3.5	20.75	16.8 *	
107	15	40.5				
108	15	44.5	4.0	42.5	9.4	
109	18	33.0				
110	18	34.0	1.0	33.5	3.0	
111	24	31.5				
112	24	35.0	3.5	33.25	10.5	
113	29	38.0				
114	29	38.0	Nil	38.0	Nil	
115	2.8.90	25.5				
116	2	28.0	2.5	26.75	9.3	
117	12	29.0				
118	12	29.5	0.5	29.25	1.7	
119	14	35.0				
120	14	34.0	1.0	34.5	2.9	6.4 +

* Result above 14% - (Allowance 1 in 40 checks)
+ This result fails satisfactory testing requirement and should be investigated

ACCEPTANCE STANDARDS FOR QUALITY CONTROL

General

In setting acceptance standards for quality control purposes or indeed specification requirements the inherent production variability (p) and the reproducibility of the test procedure (t) need accommodation. Overall variability can be assessed from the sums of squares of these individual variabilities i.e. $\sqrt{p^2 + t^2}$ for normally distributed variables.

Similarly for batching variations if the overall operator plus equipment tolerance is ± 3% and equipment capability of ± 2% is identified, this leaves the operator with a tolerance of ± $\sqrt{3^2 - 2^2}$ = ± $\sqrt{5}$ = ± 2.2%

This type of appraisal should be used in setting acceptance standards for both materials and product control parameters. Two examples of its use are given below and many others can be conjectured.

Constituents of Concrete Variability

Identification of individual causes of poor quality control may require multiple appraisal to decide on necessary action. For example records of quality control for individual variables can also be used to identify the contributing factors in the overall production variation.

Table 3 gives suggested constituents of the standard deviation of concrete cube strengths resulting from materials, production and testing.

Workability Control Acceptance Standards

Tolerances for workability control can be defined after converting testing and production capability into terms of a normally distributed variable e.g. water adjustment. This is necessary since slump itself is not a normally distributed variable. The following variables for slump achievement at 95% confidence levels at an average 50mm slump are assumed for this example:

- Visual appraisal achievable is -10mm to +15mm equivalent to say 8 litres
- Batching tolerance of ± 3%. Based on 133 litres/m³ added this is equivalent to a range of say 8 litres
- Test reproducibility of 20mm, say 6 litres

Overall variability = $\sqrt{8^2 + 8^2 + 6^2}$ = $\sqrt{164}$ = 13 litres i.e ± 6½ litres. This equates to a slump range of 30mm to 75mm and suggests that technically and statistically tolerances of ± 25mm are reasonable. In practice users may be prepared to accept 75mm for a

TABLE 3:   Suggested constituents of measured standard deviation of[13] concrete cube strengths

Source of Variation	Concrete Variability		
	Very Consistent	Average Variability	Very Variable
	N/mm^2	N/mm^2	N/mm^2
Cement Strength	1.5	2.5	3.5
Sand Grading	0.5	1.0	1.5
Coarse Aggregate (shape and grading)	1.0	1.5	2.5
Batching Accuracy (plant and personnel)	1.5	2.5	3.5
Within Batch Variations (batching and mixing) procedure	1.0	2.0	3.0
Workability	1.5	2.0	2.5
Testing (Note 1)	1.0	1.5	2.0
Overall Standard Deviation	3.0	5.1	7.2

(Note 1)   Testing is actually typified by coefficient of variation and the Standard deviation in this case is based on a mean strength of 40 N/mm^2.

50mm specification, but would find 30mm inadequate for compaction. Where this is the case setting acceptance standards becomes complex linked to process variability and user expectations. Combining these criteria the following practical control levels could be realistic:

   Mean deviation from nominal 50mm slump no greater than -10mm to +15mm
   No more than 5% of results outside tolerances of -10mm to +40mm

Such tolerances would result in slump ranges typically of 40mm to 90mm for 95% of results with a mean slump of the order of 60mm.

EXTENDED QUALITY CONTROL

Total quality control involves using operational activities and techniques to satisfy quality requirements. This paper has sought to extend the vision by identifying techniques for both monitoring quality and also defining acceptability standards for assessment and design.

Quality assurance will extend the need for effective quality control data and analysis. The requirement in quality assurance to review such standards regularly will enhance this requirement and the role of quality control.

REFERENCES

1. COMITÉ EUROPÉEN DE NORMALISATION. ENV 206. Concrete – Performance, production, placing and compliance criteria.

2. BRITISH STANDARDS INSTITUTION. BS 4778 : Part 1 : 1987. Quality vocabulary. (Corresponding International Standard ISO 8402).

3. BRITISH STANDARDS INSTITUTION. BS 5750 : 1987. Quality Systems (Corresponding International Standard ISO 9000 and European Standard EN 29000).

4. DEWAR J D . The Structure of Fresh Concrete – Frederick Lea Memorial Lecture. Institute of Concrete Technology 1986. BRMCA Reprint.

5. CONCRETE SOCIETY. Concrete core testing for strength. Technical Report No.11 : 1987. Page 15.

6. ROBERTS M. Private Technical Note produced on behalf of C & G Concrete Ltd and the BRMCA. A comparison of the strength compliance rules in BS 5328 and ENV 206. February 1988.

7. BROWN B V. Monitoring concrete by the cusum system. Concrete Society Digest No. 6. 1984.

8. BRITISH STANDARDS INSTITUTION. BS 5703 : Parts 1 and 2 : 1980 and Part 3 : 1981 – Guide to data analysis and quality control using cusum techniques.

9. BARBER P M and SYM R. An assessment of a "Target Value" method of quality control. Paper W13A(1). 7th ERMCO Congress. London 22-26 May , 1983. 1-7.

10. DEWAR J D and ANDERSON R. Manual of ready mixed concrete. December, 1987. Blackie. pp 135-139.

11.   CONCRETE SOCIETY. Alkali Silica Reaction : Minimising the risk
      of damage to concrete. Technical Report No.30. 1987.

12.   BRITISH STANDARDS ORGANISATION. BS 5497 : Part 1 : 1987. Guide
      for the determination of repeatability and reproducibility for a
      standard test method by inter-laboratory test. (Corresponding
      International Standard ISO 5725 : 1986).

13.   BROWN B V. Quality control inspection, measuring and testing.
      Paper presented to course on "Quality Assurance for the
      Construction Industry" University of Dundee. February 1989.

FACTORS OF CONSTRUCTION ERROR AT PROTECTION

B Kucuk
H Kaplan
Dokuz Eylul University
Turkey

ABSTRACT. In Civil Engineering, the protection of concrete, is very important. The purpose of this protection is to keep the compressive strength of the concrete above the characteristic strength determined by worldwide standards. The desired concrete characteristics should be achieved during production and protected while it is being used. Since this is an extensive subject, the study has been diversified and divided into three parts. Firstly the factors which have an effective role in the quality of production have been examined. This stage was achieved by research undertaken at the regional quarries of Denizli, where the samples were taken. Secondly, constructional defects, were researched at the reinforced concrete buildings in Konya. Finally, the influence of negative environmental effects on reinforced concrete buildings was investigated and protection methods are discussed.

Keywords: Protection of Concrete, Production Quality, Construction defect, Environmental effects.

Bahattin Kucuk is a lecturer at the Department of Civil Engineering, Dokuz Eylul University, Faculty of Engineering of Denizli, Turkey He is very active in constructional materials and site techniques.

Hasan Kaplan is a research assistant at the Department of Civil Engineering, Dokuz University, Faculty of Engineering, Denizli, Turkey. He has been studying concrete technology and structural analysis.

INTRODUCTION

Building techniques have improved due to better production of
construction materials and the development of sectional
construction. Building codes, standards and regulations have been
prepared as a result of recently applied tests and experiments. The
relationship between human beings and their environment has also
been established. The successful survival of this relationship is
only possible if the buildings in which people live and continue
their social activities are adequately protected. Such buildings
must perform their functions for a long time.

There may be constructional errors in many of the official and
private buildings. Failures may occur in some of them because of
immediate uses, and in some others as a result of an earthquake.
These errors are called, constructional element problems and are
caused either by poor quality constructional materials or by the
primitiveness of the construction techniques.

Protection of all reinforced concrete buildings is possible by
protecting the individual elements made from reinforced concrete.
Protection can be conferred at the design stage by elimination of
false applications. Post construction protection is also important
because of the use of the structure for alternative purposes, the
changing of service loads on earthquake or some un-estimated impact
force.

SAFETY IN CONSTRUCTION AND FAILURE REASONS

Good quality of construction can be achieved by the application of
legitimate principles and techniques. By this means the protection
problem can be reduced to a minimum and an easier assembly and
greater economy can be achieved. It is not merely enough to ensure
safety against the collapse of the structure, it is also necessary
to avoid repairing and renewing protection methods after
construction as this will incur heavy expenses. Protection methods
must be considered at the design stage of the concrete. Whether a
reinforced concrete building is artistic or not, the constructional
sections must be adequately protected against strength reducing
factors in order to maintain a minimum quality. It has been
established that the reasons for failure of existing protection of
reinforced concrete buildings are as follows:

a) Application Errors in the Production of Construction Elements

Generally these occur during the construction stage due to the
unsuitability of local construction techniques and non application
of the appropriate technical construction systems and apparatus.
Apart from this insufficient competence and knowledge of employees,
and weakness or carelessness in obeying the standards, by the staff
of the technical committee, may contribute to these errors.

b) The Errors Existing In The Stucture Immediately Prior To Use

Loads imposed inconsistent with those taken account of in the design, can cause deformation and cracks. At the start of the project, when the base carrying capacity, service loads and material properties are being chosen, the section weaknesses which occur in construction due to dynamic effects must be considered. It is also true that, not working within the safety limits, failure to use an approved method, or construction material, during the construction of the project, or failure to adhere to the values of standards can also cause these problems.

c) Failure In Reinforced Concrete Due to the Effect of Negative Environment Conditions

Rain water, frost damage, attack by chlorides, alkali-aggregate reactions, carbonation (which causes steel reinforcement failure and hence reduces strength) are accepted as chemical effects. Because chemical substances added during chemical mixing (calcium chloride) increase the corrosion of reinforcement, they are generally accepted to reduce the strength and increasing the possibility of failure.

THE INVESTIGATION OF CONSTRUCTION ERRORS

The protection of reinforced concrete is obtained by protecting the system elements. This is possible when materials are appropriate, and technical standards are adhered to during the production of these elements. So we must know why we have to follow the building codes, standards and regulations.

Application Errors and Protection Precautions

During the course of the project design the sections and reinforcements are calculated according to the effective horizontal and vertical loads. We use the standards prepared by the Turkish standardisation Institute, such as TS 500, the standards for calculation and codes for construction of reinforced buildings. In the preparation of concrete and its protection we must follow certain building codes. It was observed that major cracks may occur in construction sections which are below the standard limit during the assembly of the structure.

The errors in some buildings in Konya, one of the largest cities in Turkey, resulting from the non compliance with building codes, were investigated. The results of these errors were characterised photographically. The interpretation and effect of each of these errors is discussed below.

Figure 1: Base structure construction errors.

During the construction of the base structure one is generally faced
with problems of loose ground conditions, and the pouring of
concrete of at least 5 cm depth direct onto such a ground state.
Under these conditions, because of ground water loss, holes appear
in the concrete, moreover, because of loose soil covering the steel
surfaces, adherence of the reinforcement with the concrete is
impeded. In this case protection is obtained by applying 250 dose
concrete on the ground. If there is underground water in the base
region, the concrete is protected from adverse water effects by
drainage and insulation. If the base region includes jibs or is of
low load bearing capacity, a calculated thickness must be removed
and replaced with a suitable material.

It is known that the concrete hardens quickly and if it is
maintained in a humid environment its strength increases, reaching a
specified level after 28 days. To reproduce this situation it is
necessary to eliminate external effects. This is possible by using a
suitable aggregate, compaction into the formwork, and so on. Voids
causing a decrease in cross-section may occur due to the use of
oversize aggregate and poor compaction. Examples of this were seen
on building columns (Figure 2) and on the beams (Figure 3). In
order to restore corrosion protection and reinforce these members it
is necessary that the holes are repaired with a high quality
concrete. It must be emphasized that during this process a good
bond must be achieved between the repair material and the existing
concrete.

The aggregate used should have a convenient size distribution, and
their mineralogic - petrographic structure should be sound. All of
these factors have a positive effect on the strength.

Figure 2: Holes in the column

Figure 3: Hole in the beam

For this reason, the aggregate quarries in Denizli and its vicinity
were investigated (Table 1). The properties of specimens taken from
these quarries were determined using standard tests in the Materials
Laboratory of the Engineering Faculty at Denizli. The size
distribution curves of the concrete aggregate, used either in these
tests of in the buildings in Denizli, were determined in accordance
with the TS 706-707 standards, and are given in Figure 4.

Table 1   The properties of aggregates taken from the quarries at
Denizil and its vicinity

No	QUARRY NAME (Tuvenen materials)	Fine Modulus	Unit Weight (kg/m³)	Specif. gravity (kg/m³)	Absorb. tion %	Clay frac- tion %	Wear Lose %	Aggreg- ate dm³	28th day strength N/mm²
1	Tavas Vakif Ocagi	4.39	1730	2630	8	6	48.9	1128	12.6
2	Cameli-Duzencam	3.61	1780	2850	6	14	44.4	1188	22.4
3	Acipayam Bedirbey-Da	4.22	1830	2560	6	3	22.2	1040	22.6
4	Yenice-Dogankoy	3.92	1620	2560	8	7	35.0	1085	15.6
5	Buldan Kadikoy Deresi	3.30	1760	2700	1	1	48.7	1138	16.0
6	Korucuk	4.96	1900	2700	8	1	20.8	1054	27.5
7	Civril Karayahsiler	3.54	1480	2500	7	6	34.3	1253	18.3
8	Civril Kocak Deresi	3.07	1750	2440	6	1	33.6	1034	19.0
9	Cardak Sulatma Deresi	3.94	1840	2700	7	7	33.8	1089	26.3

Note 1: These materials stabilised as each other and some of the mixture results are as
follows:

Cameli Duzencam	547	dm³	Yenice	398	dm³	Tavas Vakif	455	dm³
Korucuk	568	dm³	Korucuk	692	dm³	Korucuk	629	dm³
28th day strength	24.7	N/mm²	28th day strength	28.6	N/mm²	28th day strength	27.4	N/mm²

Note 2: The reason of being low of the examination results for crushed stone produced at
Camlik Crushed Stone Foundation is that consisting of capilary cracks on itself.

Example is: 0-7mm 534 dm³, 15mm 146 dm³, 25mm 481 dm³
28th day strength 24.5 N/mm²

Water = 158 lt   Cement = 350kg   Water/Cement = 0.45

Some test results fall below the available strength. This may be
attributed to non-optimal size distribution, clay-silt value is
lower than the standard value, low aggregate strength,
mineralogical-petrographic structure, and non-optimal particle
shape.

Tests made with an N-type test hammer to measure the strength at 28
days of building elements constructed using suitable materials
showed lowered levels. Although the measured strengths compared to
BS16, they were considerable below the design strength. The causes
of this are as follows:

a)   Primitive preparation transport and placement of concrete.

b)   Bad working technique combined with careless and insensitive
      production attributable to untrained workmen.

c)   The water/cement ratio being above the 0.6 in the concrete.
      This may be called liquid sensitivity.

d)   Badly controlled concrete mix preparation and poor quality
      control during concrete placement.

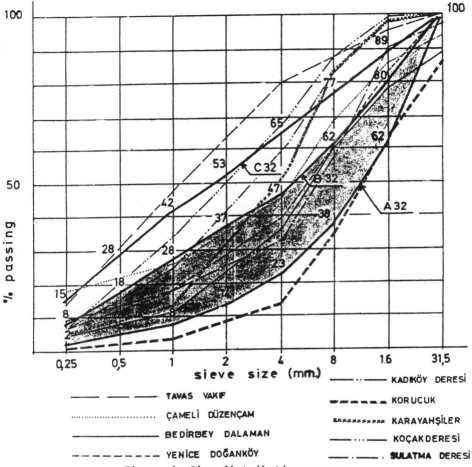

Figure 4: Size distribution curves

e)   Poor application of water or stream during repair work, poor
     protection of fresh concrete from external factors, and the
     occurrence of shrinkage cracks.

f)   Occurrence of capillary cracks caused by the application of
     a high load above the design load due to a change in the
     function of the structure.

g)   Crack occurring as a result of torsion and deflection,
     disregarded design dimensions and quantities and a deficient
     assembly arrangement.

Since it is not possible to increase the strength of existing
concrete, to realise the potential available strength it is
necessary for us to conform to building codes in transportation,
placement, preparation and protection of concrete.

Figure 5: The incorrect preparation and transportation of concrete.

Also the aggregate has to be of optimum size distribution, suitable
mineral structure and angular shape, free from foreign substances.
The proper concrete mixture ratios must be maintained during
batching, and adequate quality control must be maintained during
placement. These factors should lead to the provision of adequate
protection for the concrete construction.

Precautions Against Mechanical Factors

Structural deformation can occur as a result of a change in the
function resulting in an overload compared with the design load more
over, thermal expansion, impact and wear cracks and a reduction in
the concrete cover of reinforcement due to the effects of concrete
structures. The resulting cracks can cause the diffusion of
external chemical effects in the concrete. The reinforcement must
be cleaned, coated with a flexible layer and then a thin plaster
coat applied in order to prevent corrosion.

Precautions Against Chemical Failures

Chemical failures can be attributed to rainwater effects, industrial
chemical pollution, chloride exacerbated freezing effects, alkali
silica reactions and carbonation. Sulphate attack from ground or
aggregate sources is particularly problematical as it eventually
leads to the breakdown and total failure of the concrete. In order
to prevent such attack, the sulphate source should be eliminated and
a sulphate resistant cement used. Similarly, concrete elements

should be insulated with bituseal emulsion, geotextile insulation materials etc, in order to protect them from the effects of acidic and underground waters.

Chloride ions in the form of calcium chloride have long been added to concrete as an accelerator at quite high local levels. These chloride ions increase the electrical conductivity of the pure water, and breakdown the passive film protecting the steel reinforcement, thus causing it to corrode. The resulting stress on the concrete will cause steel to crack, thus allowing greater access of the corrosive environment. Ingression of the chloride ions may be prevented by using dense, compact concrete, ensuring adequate coverage over the reinforcement, and by coating an elastomeric membrane with high resistance to chloride ions permeability.

As concrete is porous, it will absorb water at the surface. If this water freezes, the resulting stress increases the volume and overcomes the bonding strength of the concrete, and causes it to spall. This cyclic process will repeat itself to destroy the concrete. The only solution to this process is to inhibit the absorption of water. Another chemical failure is alkali-aggregate reaction. Silicaceous aggregates may react with the alkali fluids within the pores to form a gel. This gel takes up water and swells with the resultant stress being sufficient to spall the concrete. This can be prevented by reducing the water ingress to the concrete so that the gel will not expand.

An additional reason for failure is reinforcement corrosion initiated by carbonation. Hydrated concrete consists mainly of calcium hydroxide and silicates with a pH of about 12.5. A reduction of this pH value causes the corrosion of reinforcement by acidic species. There is 0.003% CO_2 in the earth's atmosphere, which despite being a small quantity, is sufficient to react with alkaline calcium hydroxide to form the neutral calcium carbonate. This calcium carbonate accumulates on the concrete surface and forms a hard layer. However the reaction continues towards the interior of the concrete, until eventually the alkaline environment surrounding the reinforcement is destroyed, allowing corrosive reaction to take place. As the steel corrodes its volume expands and the resulting pressure in the concrete spalls. Concrete structures can be protected against carbonation by facing with marble, synthetic coating materials or acrylic based protective painting materials. The occurrence of surface cracks may be prevented by coating with epoxy resin systems or other proprietary protective materials.

CONCLUSIONS

We are now facing problems with concrete, which originate from the use of incorrect materials in the mixture. The purpose of the protection of concrete is to maintain its compressive strength above the accepted world standard levels. This can only be achieved by adhering to the rules during production.

The primary causes of strength reduction and cracking beyond acceptable limits is the production of concrete with a high water cement ratio. The use of non-standard aggregates, a lower amount of cement than design values and assembly errors all compound the effect. The cavities resulting from the use of a primitive production system, and the use of oversized aggregates can be prevented by using modern techniques and vibration compaction during placing in the formwork. The bearing capacity of the base region must be well established and if the level of underground water is high, drainage must be prepared and soil amelioration methods must be applied.

During the preparation of the concrete, materials must never be used without previous experimental investigation under structural conditions. The first requisite for good protection is the choice of a good aggregate. To protect reinforcement from corrosion, reinforcement must be embedded in high quality concrete.

By means of these protective measures the buildings to which people entrust their life will be safe from risks. The workers must be educated about such techniques and effective control mechanisms must be established.

REFERENCES

1   ERIC, M. Materials Science and the Problems of Construction. Maket Corp. Istanbul, 1982.

2   KOCATASKIN, F. Construction Materials Science. Arpaz Press. Istanbul, 1975.

3   POSTACIOGLU, B. Materials Science Problems. Caglayan Corp. Istanbul, 1975.

4   IMAR ve ISKAN BAKANLIGI DEPREM ARASTIRMA ENSTITUSU BASKANLIGI. Regulations for the Buildings to be Constructed at the Disastrous Regions. Ministry of Improvements and Settlements. Earthquake Research Institute. Ankara 1975.

5   BAYINDIRLIK ve ISKAN BAKANLIGI YUKSEK FEN KURULU BASKANLIGI.. General Technical Regulations. Ministry of Public Works. Higher Scientific Executive, Volume 23, Ankara, 1985.

6   WATSON, D A. Construction Materials and Processes. Mc Graw-Hill Book Company, 1972.

PROTECTION OF CONCRETE IN DEVELOPING COUNTRIES THROUGH IMPROVED CONSTRUCTION PRACTICES

K Mahmood

King Abdulaziz University,

Saudi Arabia

Z Mian

University of Engineering and Technology, Lahore,

Pakistan

ABSTRACT. Limited availability of suitable materials, equipment and skilled labour, absence of relevant specifications and codes of practice and severe climatic conditions may influence the performance of concrete structures in a developing country. Economical and feasible methods for the protection of concrete in developing countries must be developed by considering the judicious use of local materials and techniques. The paper briefly describes the results of some studies on construction practices and the performance of concrete in developing countries. Based on these studies a number of suggestions have been put up in the paper for the protection of concrete through improvements in the prevalent construction practices.

Keywords: Concrete, Reinforced Concrete, Developing Countries, Materials, Construction Practices, Hot Weather, Performance of Concrete.

Prof. Khalid Mahmood has been Professor at the Civil Engineering Department, King Abdulaziz University, Jeddah, Saudi Arabia, since 1985. He obtained PhD degree from the University of New South Wales, Sydney, on an Australian award under Commonwealth Scholarship and Fellowship Plan. Prior to his present position, he was Professor and Chairman, Civil Engineering Department, University of Engineering & Technology, Lahore, Pakistan, where he was also engaged in consultancy services for design, construction, condition survey and repair of concrete structures. His current research interests are in design and construction in developing countries and behaviour of reinforced concrete structures.

Prof. Zia-Ud-Din Mian is Professor at the Civil Engineering Department, University of Engineering & Technology, Lahore, Pakistan. He holds an MSc degree from the University of Surrey, UK. Besides teaching and research into concrete behaviour, he is actively engaged in consultancy services through the University of Engineering. His current research interests are in the use of locally available materials for making concrete.

INTRODUCTION

Construction activity in a developing country may be affected by a number of environmental and infrastructural restrictions including shortage of suitable materials, equipment and labour and absence of relevant specifications and codes of practice (1). Further influence on concrete construction may be exerted by extreme climatic conditions of the region. Construction in a developing country with disregard to the environmental and infrastructural restrictions and climatic conditions will therefore affect the quality, performance and durability of concrete structures.

The measures for the protection of concrete in developing countries need to be considered separately from those in the developed countries for various reasons. Firstly, the protection may be required against some causes of disorder/ deterioration which may be peculiar to a region. Secondly, in most cases a method for the protection of concrete may have to be developed by taking into account the local materials and techniques as the use of imported materials and equipment may not be economically and practically possible.

This paper briefly describes the results of laboratory and field studies carried out in some developing countries on construction practices and the performance of concrete. Keeping in view the results of these studies measures have been suggested in the paper for concrete protection in developing countries through improvements in construction practices.

STUDIES ON CONSTRUCTION PRACTICES

Building construction in most developing countries is generally carried out by using indigenous materials, labour-intensive methods and cast-in-place reinforced concrete. In many cities of the Gulf countries, ready-mixed concrete and the related equipment are now extensively used. Nevertheless, a large volume of construction in these countries is still undertaken by using site-mixed concrete and labour-intensive methods.

Cast-in-place concrete construction is hardly ever carried out under ideal conditions and therefore the resulting structure may develop defects for a number of reasons (2). Cracks in horizontal surfaces due to plastic shrinkage, cracks over reinforcing bars due to plastic settlement, surface texture defects, colour variation and lack of cover to reinforcement are the common type of defects which may occur during in-situ construction (2). However, in the case of concrete structures located in a developing region different other defects, including structural defects and deterioration, may occur due to inadequacies in design and construction.

In the absence of national code of practice of a developing country, the design of concrete structures is generally carried out by using the code of practice of some developed country. Experience has shown

that in many cases where the design had been correctly carried out by
using the relevant provisions of the code of practice of developed
country, the resulting structure showed numerous defects because of
inadequacies/ errors in construction. Some of the sources of these
inadequacies in terms of materials, site practices and effect of hot
weather have been discussed in the following paragraphs.

Materials

Workability, strength and durability are three important properties of
concrete which need to be evaluated when selecting the constituent
materials (3). Type of aggregate, its maximum size, shape and
grading, water-cement ratio and degree of compaction are some of the
factors which affect the above-mentioned properties of concrete.

Rivers in the alluvial plains of the Punjab, Pakistan, deposit a large
amount of fine sand after the monsoon period every year. This fine
sand is commonly used for making concrete in the region. However,
this sand does not fall within the specified grading limits and can
drastically affect the workability and strength of the concrete if mix
proportions and water-cement ratio are not carefully controlled (4).
Dune and coastal sands are also too fine and have poor grading. The
use of these sands increases the water requirement in a concrete mix
with decrease in strength and increase in drying shrinkage (5).

Field studies (5) on deterioration of concrete structures in Eastern
Saudi Arabia show that corrosion of reinforcing steel, sulphate
attack, salt weathering and environmental cracking occur due to a
number of factors including the use of low quality aggregates
contaminated with sulphate and chloride salts.

As mentioned earlier, most concrete used in developing countries is
produced near the construction site. Data on site-mixed concrete from
different developing regions show that, due to poor quality control
procedures and supervision, large variations occur in workability and
strength of these concretes. On-site production also results in
wastage of materials. In 1983 the wastage of cement in India because
of poor construction practices, for example, was estimated to be more
than one million tonnes of cement a year (3).

Variations may also occur in the properties of reinforcing steel. Data
on steel reinforcing bars from various building sites around Lahore
showed (6) that although the bars of different sizes fulfilled the
specified requirements with respect to the mechanical properties, they
did not meet the requirements with respect to weight per unit length,
cross-sectional area and surface deformations. It is obvious that the
use of under-sized bars in a member designed with a low steel ratio
can appreciably affect the actual capacity and stiffness of the
member leading to cracking and visible deformation under service
loads.

Site Practices

Construction of large projects in developing countries is generally
carried out by following the universally accepted practices of
concrete design, quality control and inspection. However, proper
practices and procedures are sometimes not adopted in the construction
of small residential and industrial buildings, particularly in the
private sector.

Required structural shape, surface finish and dimensions within
acceptable tolerances can be achieved by the use of proper formwork.
Good quality timber is not available in many developing countries and
therefore, reuse of wooden formwork is a common practice. In many
cases this results in poor finishes with difference in actual
dimensions and designed values exceeding the tolerances specified in
the standards of most developed countries.

Adequately detailed drawings are often not available for the
construction of small projects. Acceptable procedures are also not
followed for the placement of reinforcement in such cases. Slab
reinforcement, for example, is quite often placed directly on the
forms thus providing inadequate concrete cover. In the case of beams a
reduction in effective depth may result due to improper placement of
bars and a cover thicker than the indicated value.

Volume batching of concrete mixes, without any control of water-cement
ratio is usually carried out at the site for small construction
projects. In such cases, the quality control requirements given in
the code of practice used for designing the structure are either
incompletely specified or missing, and if specified, they are not
fully implemented at the work site because of inadequate testing
facilities and lack of understanding of the quality control
procedures (6). This results in poor quality and low strength
concrete.

Ready-mixed concrete offers many advantages in comparison with the
site-mixed concrete. However, data are not presently available to
indicate the degree of uniformity of properties of ready-mixed
concrete being used in developing countries.

Hot Weather

Most countries of Asia and the Arabian Peninsula have hot climate for
a major part of the year. Concrete construction in hot weather
without appropriate measures may impair the properties of fresh as
well as hardened concrete.

A laboratory study (7) on the performance of concrete mixes in hot
weather shows that higher initial temperature of dry ingredients and
higher air temperature result in greater water loss due to evaporation
in the initial period after mixing. The results of a field study (8)
on the production of ready-mixed concrete in the hot weather of Jeddah

indicated slump loss of zero to 25 mm in morning and night shifts and
10 mm to 35 mm in noon and afternoon shifts during the time (25 to 35
minutes) of transportation and delivery to construction site.

The study (8) further indicated that besides weather components, the
time of casting, difference of concrete and air temperatures and
moisture condition of the concrete surface influence the rate of
evaporation from freshly placed concrete surfaces. For the concrete
specimens, with surface area of 0.1 sq. m, cast in the morning, noon
or early afternoon time in hot and humid weather, the maximum rate of
evaporation exceeded (8) the limiting value of 1.0 kg/sq. m/hr, from
view point of susceptibility to plastic shrinkage cracking. For the
specimens cast in the noon and afternoon time, the maximum rate of
evaporation generally occured in the first hour after casting when a
layer of bleed water was available on the surface. However, for the
specimens cast in the morning time, the maximum rate of evaporation
occured at noon time, 3 to 5 hours after casting, when the concrete
had partly set and had no free moisture on the surface. Figure 1
shows typical variations in the rate of evaporation from the concrete
surfaces cast at different time of the day (8).

Moist curing plays an important part in the development of concrete
strength in hot weather. The results of laboratory-cast specimens
exposed to hot weather, without curing, show that their strength gain
was arrested after about 90 days with slight reduction in strength
under continued hot weather exposure (7).

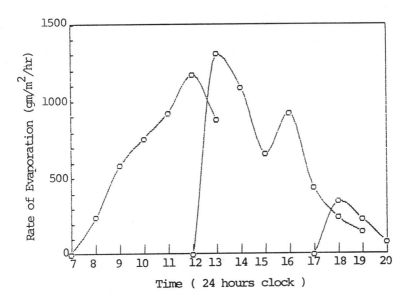

Figure 1 : Variations in rate of evaporation with time (8)

A study (9) on different practical methods of curing reports that for
hot and dry weather, curing for 14 days by sprinkling water twice a
day and keeping the concrete covered by plastic sheets, is the most
effective method from the view point of strength gain.

Experimental results on corrosion of reinforcement (5) show that the
curing period has a distinct effect on the durability of the concrete.
The concrete specimens subjected to moist curing for 28 days performed
4.4 and 12.2 times better than those cured for 7 days and 3 days
respectively. The increase in the curing period also improved the
resistance against sulphate attack (5).

IMPROVEMENTS IN CONSTRUCTION PRACTICES

In order to ensure adequate performance of concrete structures, built
under various infrastructural and environmental restraints and severe
climatic conditions, appropriate measures of concrete protection can
be developed by considering improvements in construction practices.
These measures may range from selection of suitable constituent
materials to moist curing for a definite period. It is not possible
to cover all aspects of concrete construction in one paper. The
suggestions for improvements in construction practices given in this
section have been drawn from the results of various studies reported
in the paper.

The suitability of locally available aggregate, which do not conform
to the relevant specifications, should be determined in terms of
strength and durability of the concretes from trial mixes (4).
Separate provisions for concretes made from these materials may need
to be included in the specifications (4). The use of such materials
should be decided by considering anticipated savings in cost of
construction vis-a-vis expected useful life of structure.

Aggregate contaminated with salts, silt and clay should be washed to
remove these materials before using the aggregate for making
concrete (3). Experimental results (5) show that by using washed
aggregate in concrete the initiation of corrosion of reinforcement,
under aggressive environment, is delayed when compared with concrete
having same mix proportions, water-cement ratio and unwashed
aggregate. Because of low solubility, calcium sulphate is difficult
to remove by washing and therefore, it is suggested (3) that the
amount of residual sulphate should be determined by testing after
washing the aggregate.

The proportioning of a mix, by taking into account the required level
of workability, strength and durability and the job control
conditions, provides the best means of concrete protection. The use
of statistical methods for mix design may not be successful in some
developing countries, for the time being, because of inadequate data
on concretes made from locally available materials, unsatisfactory
batching and sampling procedures and the nature of assumptions for
statistical deductions (6). However, the use of statistical approach

of mix design should be encouraged in developing regions, wherever possible.

A number of steps can be suggested for the production, quality control, and placement of concrete, fabrication of formwork and positioning of reinforcement. However, all these steps will be successful only if properly trained supervisory staff and adequate facilities for sampling and testing are available for construction projects. It is therefore suggested that facilities for the training of supervisory staff and adequately equipped testing laboratories should be set up in different regions of developing countries.

The initial temperature of concrete is an important consideration in hot weather concreting. Depending on the climatic conditions, the initial temperature of concrete should be limited to a value between 24C and 38C by precooling the ingredients or by adding ice to the mixing water (10). The temperature of aggregate can be kept low during hot weather by shading the stockpiles. Shading can also provide an effective, cheap and easily manageable method to protect freshly placed concrete surfaces during the day time. The results of field study (8) show that shading can reduce the rate of evaporation by 50 percent or more when compared with the rate of evaporation from freshly placed concrete surfaces without any protective measures.

The normal curing time for concrete construction ranges from 3 to 10 days and not many structures are cured beyond 14 days (3). The method and duration of curing for a particular area should, therefore, be decided by considering the weather conditions, type of structure and water supplies at the site.

CONCLUDING REMARKS

The results of studies reported in the paper show that concrete construction in a developing country may be influenced by numerous environmental and infrastructural restraints and climatic conditions. The effect of various factors on the performance of concrete in a particular region should be evaluated and accordingly, improvements in the prevalent construction practices should be made.

In order to provide adequate protection to structural concrete in developing countries, on a long-term basis, there is a need to develop codes of practice, standards and specifications for those countries. These codes and standards should be prepared taking into account the socio-economic and climatic conditions, the technical know-how, construction materials and the methods available in a developing country (6).

REFERENCES

1 MAHMOOD, K. Building under Extreme Environmental and Infrastructural Restrictions. Introductory Report (Vienna 1980),

International Association for Bridge and Structural Engineering, Zurich, 1979, pp 105-112.

2 ALLEN, R T L and EDWARDS S C. The Repair of Concrete Structures. Blackie & Son Limited, Glasgow, 1987, pp 197.

3 SPENCER, R J S and COOK, D J. Building Materials in Developing Countries. John Wiley & Sons, Chichester, 1983, pp 327.

4 MAHMOOD, K, KHALIFA, S and CHAUDHRY, M Y. Use of Locally Available Aggregates in Concrete. Proceedings, 2nd Australian Conference on Engineering Materials, Sydney, 1981, pp 545-556.

5 RASHEEDUZZAFAR, AL-GAHTANI, A S and AL-SAADOUN, S S. Influence of Construction Practices on Concrete Durability. ACI Materials Journal. November-December 1989, pp 566-575

6 MAHMOOD, K, MIAN, Z and MIRZA, M S. Towards a Code of Practice for Design & Construction of Concrete Structures in Developing Countries. Proceedings, International Colloquium on Concrete in Developing Countries, Lahore, 1985, pp 891-911.

7 MIRZA W H and AL-NOURY S I. Performance of Construction Materials in Hot-Weather Conditions. Bentonwerk + Fertigteil-Technik, 1/1988 pp 48-52.

8 HASANAIN, G S, KHALLAF, T A and MAHMOOD, K. Water Evaporation from Freshly Placed Concrete Surfaces in Hot Weather. Cement and Concrete Research, Volume 19, No. 3, 1989, pp 465-475.

9 TARYAL, M S, CHOWDHURY M K and MATALA, S. The Effect of Practical Methods Used in Saudi Arabia on Compressive Strength of Plain Concrete. Cement and Concrete Research, Volume 16, No. 5, 1986, pp 633-645.

10 AMERICAN CONCRETE INSTITUTE COMMITTEE 305. Hot Weather Concreting. ACI 305R-77 Revised, ACI Manual of Concrete Practice.

PROTECTION THROUGH CONSTRUCTION

Session B

Chairman

Dr M C Alonso

Department of Civil Engineering,

Instituto Eduardo Torroja,

Spain

GETTING THE CONCRETE RIGHT IN THE FIRST PLACE

M Levitt

Laing Technology Group Limited,

United Kingdom

ABSTRACT. Emphasis if placed on achieving the planned
performance properties of concrete by getting it right in the
first place. Particular importance is placed upon design and
workmanship aspects with the material characteristic receiving
the minimal comment that it deserves. Discussion is centred upon
the routes by which one can target the right concrete and proving
that the aim has been achieved. It needs to be borne in mind
that additional costs would accrue if one had to do something to
the concrete after casting to achieve a specified performance or
to return at a later date to repair and/or demolish prematurely.
It is argued that for a large range of concrete applications
these necessities or risks can be virtually eliminated and that
the extra cost involved on achieving this performance is
cost-effective in both money terms as well as enhancing the
reputation of the construction industry.

Keywords: Supervision, Quality Assurance, Design, Workmanship,
         Materials, Compliance Testing.

Dr Maurice Levitt is an Associate in the Laing Technology Group
and has specialised in concrete and cast stone, especially
durability properties. He has written many papers on the subject
and is the sole author of the book 'Precast Concrete'. His
current work includes construction development projects,
litigation, QA and CEN-associated activities. He serves on
several BS Committees and is a member of the Concrete Society's
Materials Group and is also a Fellow of the Institute of Quality
Assurance.

INTRODUCTION

It is rather a sad state of affairs to have to comment that at
the time of preparation of this paper the remedial and repair
market in the UK constituted about 40% of the annual turnover.
Contractual work involving up-grading or remedial works of
concrete buildings and civil engineering works is considered to
account for a considerable proportion of this sum. Most of the
problems arise out of design and/or workmanship aspects and these
aspects are considered to have arisen because of the lack of
inter-disciplinary appreciation amongst the various professions
involved coupled with a large degree of ignorance concerning the
properties of materials.

Having said all this one could well respond with a comment
referring to organisations, exemplified by the one to which the
author of this paper belongs, benefiting by obtaining contracts
involving site survey, testing, preparing remedial specifications
etc. The other side of the coin could equally and possibly more
forcibly be argued that such organisations should spend more time
on development work, basic research, quality-related matters etc.
if the UK is to be competitive in the EEC. This paper takes a
multi-disciplinary philosophical view of the whole process of
concrete and aims to show that for the vast majority of concrete
applications there is no need to undertake protective treatment
prior to the handover nor to expect any more than a normal
maintenance programme would require. Concrete structures
undergoing extensive remedial works tend to get a bad press
coupled with a prominent position in the news media. All this
reflects very badly on the construction industry and it rarely
matters who the contractor was as the bad name permeates through
to all in the construction field.

A not too dissimilar situation applies in the concrete remedial
field because repair work badly undertaken and subject to early
failure serves as a double indictment. In effect these will be
the problem that the structure needs repair and the added problem
that there is a likelihood that any attempt at a repair would
possibly fail. If anything one should try and establish that the
term "irreparable" means that the concrete never needs repairing
in the planned lifetime of the construction and not that it means
that it cannot be repaired when something goes wrong.

DESIGN

It is either impossible or difficult to get 'it' right first time
when the bases on which contracts are placed leave too many avenues
open to interpretation, ambiguity and/or even irrelevance. Where
technical clauses are performance-based it is only reasonable to rely
on the producer to meet these requirements without telling them how

to achieve the levels so specified. However, there needs to be a
method or methods that can be used contractually to prove or
disprove compliance.

The concept stage.

It does not matter if the construction is a building or a civil
engineering venture nor does it matter if the work is a plain,
managed, fee or sub-contract oriented; in all cases the product
to be used requires a specific reference or references and its
development in association with other products in the building
needs careful identification. In order to analyse the desirable
chronological steps one immediately becomes involved in the
multi-discipline scene and, possibly, the best way to exemplify
this is to take an office block which has been sketched by an
architect and approved by the client. This of course, is subject
to the client's acceptance of the full scheme in detail and
budget price or target.

The application.

1.   Appointment of consulting engineers (civil, mechanical and
     electrical) in order to form a liaison body with the
     architect in both preliminary works and the contract per se.

2.   Submission of planned construction drawings, tender
     documents and Bills to selected contractors who may submit
     both a tender to the Bill as well as an alternative tender
     where they are free to vary both document and Bill items on
     the understanding that each of the variations be identified.

3.   Clients representative discuss with each contractor whom
     they feel has submitted acceptable tenders and selects one
     as the nominated contractor. This acceptance could be on
     the originally submitted documents, the contractor's
     proposed variance or other agreed changes with accompanying
     prices.

The problems

Since this paper concentrates on the subject of concrete a number
of typical examples based upon the author's experience may be
selected illustrating some of the pitfalls in 1 & 2 above.

1a. The ground survey did not cover sulphate contents down to
     20m, the maximum pile depth, and only dealt with the top 10m
     where the sulphate was relatively low. From 10-20m depth
     the concrete needed to be based upon sulphate-resisting
     cement or needed to be tanked. The consulting civil
     engineer was at risk in accepting survey data from another
     source not within his control.

1b. In selecting the exterior cladding system and the window
     frames no account was taken of the method of fixing the
     window frames which were inserted from the interior face and
     levered back into position. The drip detail on the exterior
     concrete cladding fouled this rotational intention.

2a. A diversion of detail occurred in a part of the technical
     specification not tallying with the part of the drawing to
     which it referred namely that the words "minimum cover"
     occurred in the text and the word "cover" alone on the
     drawings. This was picked out by one of the tenderers who
     stated that his submission was based upon "nominal cover" as
     defined in the Code and applicable to both text and
     drawings.

2b. Although one of the tenderers know that there was a
     detriment in one of the contract details he had no
     alternative to offer but submitted an estimated Bill part
     for the work as required in the contract. In order to
     comply with a contractor's duty to warn he submitted a
     separate letter with his tender stating that although the
     item in question was priced he would have liked to have an
     alternative.

     One could wax at length on one's experiences but, in the
     following Sections, the concentration on concrete becomes
     more emphatic as there is abundant milage for 'getting it
     right' with this one material. However, there are two
     aspects of design with concrete that, in the author's
     opinion, merit discussion.

Concrete design

BS5328[1] covers two basic methods of specifying concrete, one by
prescription and the other by design. Before discussing the pros
and cons of either approach one could well ask the philosophical
question - what other construction product does the client tell
the contractor how to make it? The enigma is further compounded
on designed mixes by asking the question what product used on
site is accepted on the basis of testing something (eg. a cube)
that is not used in construction? Perhaps the philosophy that
lies behind all this is that the materials used in concrete are
variable and one needs to play either the design or the
prescription games by over-designing or over-prescribing.
Furthermore, it is not uncommon to find a specification for a
contract where both approaches simultaneously appear.

Concerning prescription mix design it is probably easier to check
the weights of all the materials constituting the concrete rather
than the potential hardened properties of a fresh plastic
material. The designer might still feel that there is a

necessity for having a strength figure but a degree of trust in
the Standards for the raw materials and the way they are put
together would not be a bad thing. After all, in the design of
foundation work the designer places a lot of faith in the
properties of so-called undisturbed soil samples when,
practically speaking, it is virtually impossible to leave soil
undisturbed and sample it at the same time. Yet, a whole lot of
essential calculations concerning things like ground expansion on
top soil removal, compaction in construction load, friction on
piles, are based upon empirical data. In addition, a large range
of components are accepted on the construction site without the
specification saying anything other than invoking, where
possible, a Standard reference. Probably, the reason why
concrete gets singled out for the lion's share of specification
words is that it is the only construction material, other than
bituminous products, that has a different 'as built' to
'as-supplied' state except for factory-made concrete products.

WORKMANSHIP

It is difficult to separate workmanship from design in a sharp
demarcation as the border area is rather blurred. Suffice it to
say that it does not matter how soundly concrete is designed
because, on site, it is relatively easy to produce poor quality
sections with the best of materials. Possibly the best way of
dealing with workmanship is to list a number of typical problems
and follow with a discussion on how one may deal with the
problems.

(1)   Poor compaction - voidage, honeycombing.
(2)   Poor formwork assembly - grout leakage, fins, sand runs.
(3)   Poor curing - surface dusting, increased permeability,
      cracking.
(4)   Inadequate cover to steel - corrosion, spalling.
(5)   Poor storage of materials - set retardation, reduced
      strength, reduced bond strength to steel.
(6)   Over indulgence in the use of mould release agent - surface
      retardation, staining.

It would be an ideal world if every aspect of site work covering
workmanship could be encased within a specification clause that
was supported by some method of certification or attestation.
Unfortunately, this is not possible, and the fallback position is
one of adequate supervision reinforced by motivation of all
involved in the construction. Motivation is probably the stage
on which all processes are played out, whether they are product
and/or service based and, in the author's view, is or should be
the main framework in construction. This applies equally to
specifiers, contractors, supervisors and manual workers. Perhaps
a series of teach-in points, so popular on management courses, is

the best way of promoting the goal:-

A. Show the operative how to do it yourself. No respect nor
    motivation will evolve if someone tells someone else what
    to do without demonstration.
B. Tell the operative why you are carrying out the operation
    in a certain way.
C. Show the operative what happens if it is not done that way.
D. Encourage questions, allow oneself to be criticised and
    reward suggestions that result in economies and/or improved
    efficiency.

Having said all this one may now adopt a deeply pessimistic view
as motivation in and by all parties is only possible if everyone
is sure that they are in an industry where long-term views may be
taken, training can be applied and there is a reasonable
assurance of job security. Unfortunately, this is not the case
as the easy-to-adjust construction industry has suffered huge
transfusions of cash, both in and out over decades. Thus, people
who have acquired the "know-how" leave the industry and potential
newcomers feel that there is little attraction. This distills
down to workmanship being in the province of the self-motivating
personnel only. The writing of workmanship clauses into Codes is
only part of the solution; such clauses have to be applied
practically and proven so whenever possible.

The latest Code[2] has gone quite a way to dealing with
workmanship. Ten parts of this fifteen part Code had been
published at the time of preparation of this paper. One of the
five unpublished parts is "Concrete" which, understandably, has
been causing a lot of discussion in the BS Code Committee. It is
the author's view, taking Part 9 (Screeds & Toppings) as an
example, that the Code is of help to specifiers and supervisors
but is not amenable to operatives.

PRODUCT TESTING

There are two types of concrete one can test :-

(1) A sample made from the concrete such as a cube or cylinder.
     This is a 'type' test or, as is commonly known today
     'labcrete'.

(2) The hardened or fresh concrete on site and/or the precast
     concrete product. This is a 'proof' test or, similar to the
     above, 'realcrete'.

     and two methods of testing :-

(b)   Surface permeability tests.
(c)   Cement content and cement gradient.
(d)   Chloride content and chloride gradient.
(e)   Examination for segregation, voidage etc.
(f)   Air entrainment bubble spacing and size.

It may be seen that the above 6 applications of the core can
relate to a number of design and workmanship factors.

Covermeter

These devices, usually based upon electromagnetic principles,
are very useful tools for determining the geometry of the
reinforcement, the cover depth and estimating bar diameters and,
as mentioned earlier, for indicating where not to drill for core
extraction.

However, they can be dangerous in the hands of inexperienced
personnel. The following are some of the drawbacks :-

(1)   Their accuracy in cover depth determination can be as high
      as ±5mm.
(2)   The cover indication gives one no idea of the quality of the
      cover.
(3)   In congested reinforced sections there is interference from
      neighbouring bars that results in a lower cover indication
      than the actual cover.
(4)   At large depths of cover (viz +100mm) they have difficulty
      in locating bars, especially bars of diameter 10mm or less.
      This might appear unimportant from the point of view of
      specification but cover to the other side of the section,
      which might be inaccessible to test, could be relevant.
(5)   Bar diameter estimation for one type of covermeter is
      inaccurate for mild steel bars but fairly accurate for high
      tensile bars.
(6)   Where the type of steel (mild or high tensile) in the
      concrete is unknown higher errors than ±5mm in cover depth
      measurement can occur.

Pull-out tests

There are a number of proprietary systems available and they are
generally based upon placing a fixing into the concrete then
pulling it out under tension. Their use in providing a
relationship with strength can be limited by the type and size of
coarse aggregate and the position of the fixing which could butt
up against a piece of large coarse aggregate. Thus, a realistic
number of fixings need to be tested so that 'outrider' results
can be dealt with separately if necessary. Like the rebound
hammer, the accuracy of the test is vastly improved by setting up
a specific pull-out strength/compressive strength/age

relationship for the concrete in question.

Surface and near-surface permeability tests

The ISAT and the Figg-air tests have been shown to relate to
curing efficacy and durability risks such as weathering,
resistance to sulphates, acids and freeze/thaw cycling without
the de-icing or anti-frost chemicals. The ISAT has a longer
track record with its own Standard[5] and is involved with limits
in the cast stone Standard[6]. It has also been used in several
contracts to great effect[7]. Curing is of great importance in
concrete work and it is a pity that the tests are not involved
more in specifications as they relate to workmanship - especially
where membrane curing compounds are used (or mis-used).

Half-cell

This is the 'in' test for condition survey and is simply a
measure of the potential set up between the reinforcement and the
surface of the concrete. A common way of expressing the data is
in the form of potential maps. Although potentials less than
about -120mV are considered to indicate corrosion one needs to be
careful in interpretation. Potential gradients and background
potential are more important factors as one can have large
negative potentials without corrosion if the background potential
is large and gradients are small.

Rebound hammer

Like the pull-off test this is a useful tool for indicating
strength and its value is reinforced by having a correlation from
cubes tested at specific ages. The value of the test decreases
for concretes above 3 months' old as the surface hardening effect
gives optimistic results.

Ultrasonic pulse velocity

This is a method of measuring the velocity of high frequency
sound either across the section, semi-direct or along a surface.
It is very useful in determining consistency of the concrete,
locating defects such as voids and cracks and the crack depth in
concrete. It can also be used for determining the E-value but
can only be related to strength when details of the materials
used in the concrete are known. In general its main value is in
telling one when and where there might be a problem but this
suspicion needs to be backed up by additional assessments (viz
core test).

Radar

This method measures the reflected signal of pulses fired into the concrete and is useful in detecting voids, rebar positions at large depths and other defects which are associated with changes in the homogeneity of the concrete. The 'maps' that are produced are difficult to interpret and such an exercise should be left to the expert. Like the ultrasonic test the data could well result in the need for additional assessments.

Gamma radiography

This is a high energy radiation photographic technique commonly used either for detecting voids in post-tensioned cable ducts or for an overall view of the reinforcement 'shadow'. The method requires statutory notification of movement of a radio-active source from the origin to site as well as partial site evacuation and area roping.

Carbonation depth

This is a simple site or laboratory test for determining, by virtually non-destructive means, the depth where the pH-value of the matrix is reduced to 8 or less. It is only in high alkaline environments that the reinforcement will remain passive to corrosion. The exception to the rule is in the case of high chloride levels, especially where chloride gradients along the steel are high enough to set up corrosion-inducing galvanic reactions. A useful example of the tenuous relationship between strength and corrosion durability is a building being monitored by the author. This consists of a lightweight coarse aggregate, a cement content of 590 kg/m^3 and a 28 day average cube strength of about 30 N/mm^2. At 15 years' old the carbonation depth averages at 1mm.

QUALITY

This is the 'in' thing now in the construction industry and many organisations have applied or are currently applying for BS 5750 'Quality Assurance' registrations. One could write a tome on the subject and many papers and books on the subject have been published. However, to be brief, the whole subject is tied up in the Manual that has to be prepared to cover the particular activity which covers, inter alia, the following main points :-

(a) The scope of the activity and the company set-up.
(b) Nominated senior personnel and their responsibilities.
(c) Staff training and induction.
(d) Process control and testing.
(e) Calibration requirements.

(f)   Reporting, drawings control and archiving.
(g)   Health & Safety.
(h)   Auditing procedures.

Quality, quality assurance, improvement, motivation etc. are all good things as they cover (or should) all aspects of the contract or supply works.   Probably the most difficult part of the exercise in implementing the system is in motivating and creating enthusiasm in the operatives.   The bottom line is really to tell each person that they are their own quality manager and if the whole team does not get it right first time it will do the business no good at all.

PLANT AND EQUIPMENT

All that needs to be said about this subject is to recommend that everything is working properly and accurately.   For ready-mixed concrete, as an example, a visit to the plant to inspect the weigh-batch equipment and water (and, possibly, admixture) dispensers is not a bad exercise.   One can always demand to see when the equipment was last calibrated and to what accuracy it works.   The same applies to site-batched concrete in that weigh-batch equipment should be assessed with a number of cement bags (not one) or with known and calibrated weights.

The most important thing is maintenance and it is felt that not enough time is spent in the construction industry because this is misappropriately considered to be non-productive.   It is a good thing and causes less hold-ups if one starts a shift using clean equipment knowing, also, that it will start at the first touch of the controls.

CONCLUSIONS

In both summarising the foregoing sections and taking a 'crystal ball' view of the future it may be seen that getting the concrete right in the first place is relatively easy exercise and the following are the main points of this treatise :-

(1)   Inter-disciplinary liaison should occur at the concrete concept stage so that what is specified is right for the job, can be supplied and can be assessed for properties.

(2)   It is preferable to specify concrete on a performance basis coupled with acceptable and practical means of assessing this performance.

(3) Workmanship will only be at an optimum level when the construction industry becomes more static, motivation and enthusiasm is general and there is an overall effort to obtain high quality in all matters.

(4) Destructive and non-destructive concrete tests have their uses and limitations and it is essential that test data relates to the job performance.

(5) As far as the future is concerned it is possible with the arrival of the end of 1992 that contractual working conditions will have to be upgraded to match the requirements of longer-term maintenance contracts. It is also probable that the application of BS 5750 in all aspects (including concrete) will become the norm.

REFERENCES

1. BRITISH STANDARD INSTITUTION. BS 5328:1981. Methods for specifying concrete, including ready-mixed concrete. London. pp.16.

2. BRITISH STANDARDS INSTITUTION. BS 8000:1989 Parts 1-15. Workmanship on building sites. London.

3. LEVITT.M. The philosophy of testing concrete. Concrete. December 1985. pp.4-5.

4. BRITISH STANDARDS INSTITUTION. BS 1881:1983 Part 120. Testing concrete. Method for determination of the compressive strength of concrete cores. London. pp.5.

5. BRITISH STANDARDS INSTITUTION. BS 1881:1970. Part 5. Methods of testing hardened concrete for other than strength. Clause 6. Test for determining the initial absorption of concrete. London. pp.27-35.

6. BRITISH STANDARDS INSTITUTION. BS.1217:1986. Cast stone. London. pp.4.

7. CONCRETE SOCIETY TECHNICAL REPORT NO.31. Permeability testing of site concrete. London. pp.95.

ASSESSMENT OF THE EFFICIENCY OF CHEMICAL MEMBRANES TO CURE CONCRETE

P J Wainwright

J G Cabrera

N Gowripalan

University of Leeds,

United Kingdom

ABSTRACT The assessment of curing efficiency of chemical curing membranes is normally carried out using a method based on monitoring the water evaporation of a mortar specimen kept under controlled conditions. There is no relation between the measurements carried out and the properties of cured mortar. Because the object of curing is to produce a "good quality concrete" a method of assessing efficiency should ideally also give a measure of quality of the concrete.

The efficiency of four different types of curing membranes are assessed by measuring oxygen permeability, total porosity, percentage of hydration and also rate of moisture loss of mortar and concrete specimens kept at 35°C and 45% R.H. The results show that the most sensitive method to assess curing efficiency and the quality of mortars and concretes is the oxygen permeability test. The most effective curing membrane was a wax emulsion. They also show that only the top 50mm of the slabs were influenced by curing conditions.

Keywords: Concrete, Curing, Curing Membranes, Curing Efficiency, Permeability, Porosity, Moisture Loss, Compressive Strength.

Dr Peter J Wainwright is a Senior Lecturer in Concrete Materials at the Department of Civil Engineering at the University of Leeds. His main research interests include the use of waste and "by-product" materials in concrete, thermal cracking, and durability aspects of concrete.

Mr Joe G Cabrera is a Senior Lecturer in Engineering Materials in the Department of Civil Engineering at the University of Leeds, where he has lectured since 1967. His main research field is the performance and durability of concrete and other engineering materials. He is also interested in the development of techniques to use local materials in developing countries.

Dr N Gowripalan is a Research Fellow in the Department of Civil Engineering at the University of Leeds, where he also obtained his PhD in 1987. His current research interests are curing of concrete, durability of concrete, and polymer reinforcement for concrete.

INTRODUCTION

The object of curing is to provide an environment in which the concrete can be kept as close to saturation as possible until the volume of water-filled spaces in the fresh cement paste has become substantially reduced.

Both concrete strength and more importantly durability can be significantly impaired if correct attention is not paid to curing. The durability of concrete is largely affected by its ability to resist ingress of a variety of deleterious substances from the environment. It has been shown (1) that it is the surface layer (i.e. the top 30 - 50mm) of the concrete that is most sensitive to lack of curing. The resulting high rate of evaporation of water from the surface leads to insufficient hydration resulting in a porous, permeable and weak surface layer.

The curing process (when implemented) is usually continued for three to seven days after casting and involves either the supply of additional water to the concrete surface (e.g. by ponding) or the prevention of evaporation of the pore water by curing with polythene or wet hessian or by spraying with chemical membranes (2-6). The technique of ponding is the most effective method but is only suitable for flat slabs and the more traditional methods using sheets of polythene or wet hessian are often difficult for contractors to implement. As a result chemical membrane curing compounds are becoming more widely used because they are economical, easy to apply and are maintenance free.

Although the importance of curing is well recognised it is frequently neglected or carried out inadequately by contractors on site. The reasons for this are many and include:
(i)   lack of supervision on site
(ii)   not costed as a separate item in the bill of quantities
(iii)   there is no simple field test available that can determine
        whether or not curing has been carried out correctly.
(iv)   the effects of poor curing are not apparent at early ages.

Even in the laboratory there are no standard test methods available for determining the curing efficiency of such traditional methods as wet hessian and burlap. Curing membranes, on the other hand, are assessed in the laboratory using a test based on the evaporation of water from the surface of cement mortar slabs stored under controlled environmental conditions (7,8). This method is however often criticised as it does not measure any property of the mortar that is being cured.

The work reported here forms part of an extensive research programme currently being undertaken at Leeds into curing of concrete and its influence on concrete durability. Tests have been carried out to assess the efficiency of four generic types of chemical curing membranes. This was achieved by measuring oxygen permeability, total porosity, percentage hydration and rate of moisture loss on concrete and mortar slabs. In addition a comparison is made between the oxygen permeability test and that proposed in the Draft British Standard (7) for measuring the efficiency of curing membranes.

EVALUATION OF THE OXYGEN PERMEABILITY TEST

Before beginning the main programme of work it was necessary to assess the inherent variability of the oxygen permeability test and to compare its performance with that test specified in the Draft British Standard (7). This latter test essentially measures the amount of water evaporated from the surface of cement mortar specimens stored under controlled environmental conditions for a given period of time.

A full description of the oxygen permeability cell developed at Leeds has been given before (9) as have the details of tests designed to measure its variability. The Leeds permeability cell was used to measure the oxygen permeability of 50mm diameter cylindrical specimens 35mm in height subjected to an absolute pressure of 2 bars. The specimens were made and cured in accordance with the draft British Standard (7) on testing of curing compounds.

In parallel with this work an independent testing laboratory was asked to measure the curing efficiency of the same curing compounds based on moisture loss measurements following exactly the procedure laid down in the draft standard (7). Throughout all these tests four generic types of curing membranes were used the properties of which are given in Table 1.

Table 1   Physical properties of curing membranes

CURING COMPOUND	SPECIFIC GRAVITY	SURFACE TENSION (N/m)
Solvent-borne resin	0.873	0.029
Wax emulsion	0.981	0.043
Solvent-borne acrylic	0.865	0.032
Acrylic emulsion	1.007	0.039

Table 2   Oxygen permeability results for the assessment of curing
efficiency of solvent-borne resin membrane

Test number	$k_1(\times 10^{-16} m^2)$	$k_2(\times 10^{-16} m^2)$	$k_3(\times 10^{-16} m^2)$	$E = \dfrac{k_1 - k_2}{k_1 - k_3}$ (%)
1	3.70	1.01	0.50	84.1
2	3.99	1.12	0.48	81.8
3	3.63	0.98	0.45	83.3
4	3.34	0.95	0.46	83.0
5	3.71	1.09	0.48	81.1
6	3.60	0.92	0.45	85.1
7	3.95	1.11	0.46	81.4
8	3.40	0.96	0.47	83.3
Mean	3.67	1.02	0.47	82.8
SD	0.230	0.079	0.017	1.45
CV: %	6.28	7.72	3.68	1.75

k_1 = intrinsic permeability of non-cured

k_2 =   "   "   " membrane-cured

k_3 =   "   "   " water-cured

Table 2 shows the results from eight nominally identical tests
carried out using the solvent-borne resin membranes. Each of the
oxygen permeabilities k_1, k_2 and k_3 reported are the average of
three specimens.

The coefficient of variation (CV) of individual permeability values
of non-cured, membrane-cured and water-cured specimens varied
between 3.7% to 7.7%. The curing efficiency calculated from the
permeability values was reported with the remarkably low coefficient
of variation of 1.8%. The oxygen permeability results of all four
curing membranes showed a high degree of repeatability (Table 3).
The curing efficiency of the chemical membranes obtained from both
evaporation measurements and from oxygen permeability measurements
are compared in Table 4. Both methods are equally good for higher
efficiency rated membranes such as solvent-borne resin and wax
emulsion. The curing efficiencies of acrylic and solvent-borne
acrylic emulsions obtained from moisture loss, however, give larger
variations than those obtained from oxygen permeability
measurements. This may be due to the difficulties in achieving
consistent results for moisture loss from mortar slabs cured using
membranes, even under controlled laboratory environmental
conditions.

Table 3   Statistical analysis of oxygen permeability results for different types of membrane

Curing membrane	$k_1 (x10^{-16}m^2)$		$k_2 (x10^{-16}m^2)$		$k_3 (x10^{-16}m^2)$	
	Mean	CV:%	Mean	CV:%	Mean	CV:%
Solvent-borne resin	3.67	6.28	1.02	7.72	0.47	3.68
Wax emulsion	3.83	6.92	0.72	6.07	0.48	3.90
Solvent-borne acrylic	3.78	5.96	0.98	5.30	0.47	4.62
Acrylic emulsion	3.91	7.1	2.11	6.32	0.49	5.91

Table 4   Comparison of curing efficiency results

Curing membrane	Curing efficiency from moisture loss:%		Curing efficiency from oxygen permeability:%	
	Mean	CV:%	Mean	CV:%
Solvent-borne resin	87.0	3.8	82.8	1.8
Wax emulsion	90.1	3.0	92.8	2.2
Solvent-borne acrylic	81.3	7.2	84.6	2.9
Acrylic emulsion	56.2	11.1	52.6	6.8

To determine whether or not the residue of the membrane itself had any effect on the permeability measurements the top 2mm of some specimens were removed by grinding. A comparison of permeability measurements on similar specimens with and without the residue showed no significant differences. The results obtained were therefore a measure of the permeability of the mortar itself and any variations were indicative of the curing regime uses.

In order to verify the results from the oxygen permeability test some measurements of degree of hydration were made on samples of mortar taken from the top 10mm of the specimens. Hydration measurements were made using differential thermogravimentry (DTG). Typical results from both tests are given in Figures 1 and 2 and show clearly that non-cured surface specimens tend to hydrate less and carbonate more than specimens cured either by ponding or by protecting them with curing membranes.

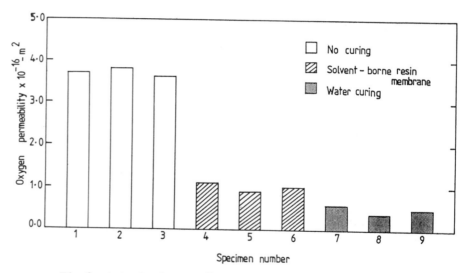

Fig 1   A typical set of oxygen permeability results
for the assessment of curing efficiency

From the results discussed above it is concluded that curing efficiency assessment based on oxygen permeability measurements gives results which are statistically better than the results obtained from moisture loss measurements. Furthermore the method has the advantage that it measures directly a property related to performance, and that it can be used for the assessment of traditional methods of curing.

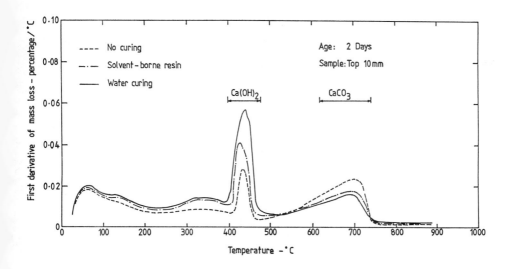

Fig 2   Differential thermogravimetry (DTG) of cement mortar
cured using different methods

MAIN PROGRAMME OF EVALUATION OF CURING MANUFACTURES

MATERIALS

Ordinary Portland cement and quartzitic gravel (maximum size 10mm) and
fines were used throughout the main test programme.   Three concrete
mixes with nominal cement contents of 270, 360 and 450 kg/m^3, constant
fine to total aggregate ratio (0.39) and constant w/c ratio (0.48)
were used for this investigation.   Corresponding mortar mixes were
made by eliminating the coarse aggregate fraction while keeping the
same w/c ratio from each of the concrete mixes.   The proportions of
the concrete mixes are given in Table 5.   The curing membranes were
those listed in Table 1.

Table 5   Concrete mix proportions (by weight)

MIX	CEMENT	SAND	COARSE AGG.	W/C RATIO
1	1	2.9	4.50	0.48
2	1	2.0	3.16	0.48
3	1	1.5	2.35	0.48

EXPERIMENTAL DETAILS

Concrete slabs of 600 x 300 x 100mm and mortar slabs of 300 x 150 x 50mm were cast in steel moulds. Immediately after casting the specimens were transferred to an environmental chamber kept at 35 ± 1°C and 45 ± 5% R.H. The surfaces of the specimens were finished with a soft brush to give an even texture and then to some of the specimens the curing membrane was applied evenly at a rate of 0.2 litre/m² with an automatic spraying equipment specifically developed for this project. Of those remaining specimens some were left with their top surface untreated and exposed and the others were cured by ponding of the top surface for three days. All the specimens were exposed to a simulated wind with volecity of 3m/s.

Moisture loss due to evaporation from the mortar slabs was measured periodically up to an age of 7 days and compressive strength was measured at 3, 7 and 28 days using 100 mm cubes stored in the same environment. The cubes were treated in the same way as the slabs with the top as cast face being cured with the membranes or by ponding and the specimens remaining in the moulds until tested.

After 3, 7 and 28 days the specimens were demoulded, 50mm and 25mm diameter cores were cut from the concrete and mortar slabs respectively and tested for porosity, pore size distribution and permeability. Initially porosity profile measurements on 50 mm diameter concrete specimens were found not to be satisfactory and as a result 150 x 150 x 20 mm concrete specimens were cast for this test. Porosity of concrete was determined using a vacuum saturation method (10) in which the oven-dried specimens were evacuated dry for 1 hour and a further evacuation was carried out for 1 hour after introducing water in order to saturate the specimens. The porosity is calculated from the oven dry weight, saturated weight and submerged weight of the specimens.

Pore size distribution measurements were carried out using a 'Micromeritics' mercury intrusion porosimeter on 25 mm diameter x 12.5 mm thick mortar discs cut from the cores. The oxygen permeability test was used on 50 mm diameter cores 25 mm thick cut from the concrete slabs.

All the specimens for the measurement of porosity, pore size distribution and oxygen permeability were oven-dried at 105 °C for 24 hours and stored in a desiccator until testing.

RESULTS AND DISCUSSIONS

Compressive Strength

Because of its simplicity the compressive strength test is frequently used to measure the quality of concrete and has been used by some researchers to assess the effectiveness of curing (11). The results obtained in this study of compressive strength development of concrete under different curing conditions are shown in Figure 3.

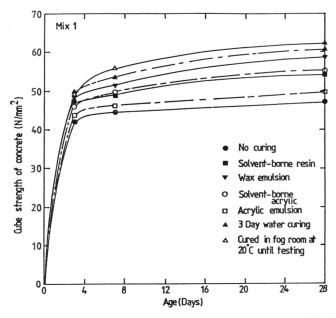

Fig 3   Effect of method of curing on cube strength of concrete

It can be seen that the 28 day strength of the non-cured specimens is about 22% lower than those water cured for 3 days. This reduction is similar to that of between 20-30% reported by other workers (11) who compared the 28 day cylinder strengths of non-cured and water cured specimens. Such reductions are, however, unlikely to be as great in a practical situation where the rate of drying of the bulk of the concrete will be far less than that of the 100 mm cubes. Nevertheless, under laboratory conditions the test is capable of distinguishing between the different curing regimes and between the different curing membranes.

Moisture Loss

Moisture loss due to evaporation of water from the top surface was measured by monitoring the change of weight of the specimens with time. For practical reasons it was not possible to perform this test on the large concrete slabs and the results presented in Fig. 4 are from measurements made on the smaller mortar slabs. Since all the curing membranes have a volatile component the values have been corrected to take account of the loss in weight resulting from the evaporation of the curing compounds. The results show that any of the curing methods used produced significant reductions in moisture loss. Of the curing membranes, the wax emulsion performed the best and was shown to be almost as efficient as 3 days ponding. The trends are similar to those of compressive strength but the magnitude of the differences is far greater. For example, the moisture loss of the non-cured specimens was almost three times that of those cured by ponding. It must be borne in mind, however, that one is not comparing like with like since the strength tests were conducted on concrete specimens whilst mortar specimens were used for the evaporation tests.

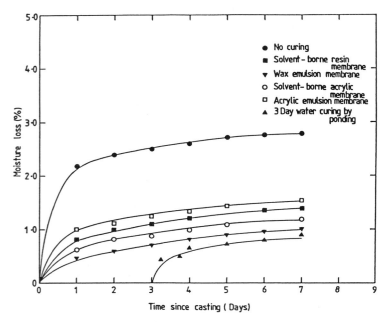

Fig 4 Moisture loss due to evaporation from mortar slabs cured using different methods

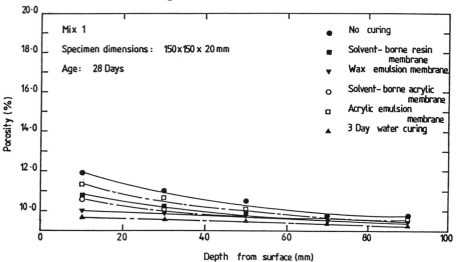

Figure 5: Variation of porosity with Depth of concrete slabs cured using different methods

Porosity

Porosity measurements were made at various depths beneath the surface of the concrete slabs made with all three mixes investigated. The porosity profiles for these slabs made from mix 1 and cured under all the curing regimes are shown in Fig. 5.

It is apparent that the ranking order of the different curing methods is the same as that observed with the previous two tests. However, it can also be seen that the differences decrease with increasing depth and that at depths greater than about 50 mm the differences become insignificant. This infers that the concrete more than 50 mm from an exposed surface will be little affected by the curing conditions at the surface. The results in Fig. 5 show that at 10 mm depth the concrete which had not been cured had a porosity approximately 30% more than that cured by ponding for 3 days.

Similar trends were observed with slabs made from the other two mixes.

Pore Size Distribution

Pore size distribution measurements were carried out on the mortar slabs in preference to the concrete slabs. Samples were taken near the surface (ie. within 12 mm from the top) of the slabs cured under all regimes and the results for one of the mixes are shown in Fig. 6.

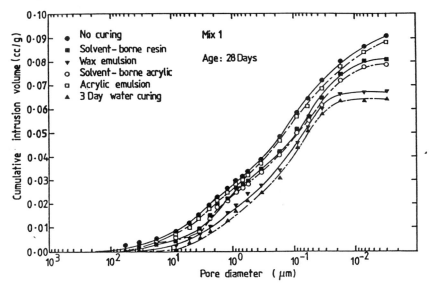

Fig 6   Pore size distribution near the surface (top 12 mm) of mortar slabs cured using different methods

This shows that the volume of the larger capillary pores (ie. greater than 0.01 μm) is significantly reduced when the mortar has been adequately cured. The medium pore diameter (based on volume) showed a reduction from 0.2 μm for a non-cured specimen to 0.1 μm for a specimen cured for 3 days in water. Other researchers (12, 13) have also reported a significant reduction in total porosity of well-cured cement paste specimens at relative humidities above 80% and shown that storing the specimens below this relative humidity produced a coarsened pore structure with a large-diameter porosity (pore size greater than 37 ηm) three times greater than that obtained with saturated curing.

Oxygen Permeability

As with porosity oxygen permeability measurements were made on samples taken from various depths beneath the surface of the concrete slabs. The results for mix 1 are presented in Fig.7 and show a similar trend to the porosity results (Fig. 5) although the differences observed in the permeability tests are far greater.

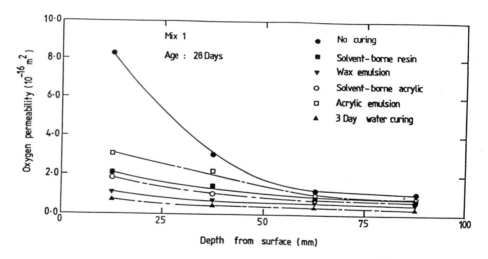

Fig 7   Variation of oxygen permeability with depth of
concrete slabs cured using different methods

For example, the concrete near the surface of the slab which had not been cured was as much as 8-10 times more permeable than that cured by ponding for 3 days. The corresponding difference for porosity was only 30%. Again, as with porosity, it is interesting to note that the curing conditions only affected the concrete in about the top 50 mm of. the slabs. Similar trends to those shown in Fig. 7 were observed with all other specimens made from the other two mixes.

CONCLUSIONS

1. A test method based on oxygen permeability has been developed for assessing the curing efficiency of chemical curing compounds. This test has been shown to be more repeatable than the standard test based on moisture loss measurements.

2. The influence of curing conditions on concrete and mortar slabs stored in a controlled environment (ie. 35 ^{0}C, 45% RH, wind speed 3 m/s) was assessed using the following tests: compressive strength, moisture loss, porosity/pore size distribution and oxygen permeability. All test showed similar trends but oxygen permeability was shown to be the most sensitive.

3. Of the curing membranes used wax emulsion was found to be the most efficient and was almost as effective as 3 days ponding. The acrylic emulsion membrane was the least efficient of all those tested.

4. The concrete more than about 50 mm below the exposed surface of a slab is little affected by the curing conditions.

5. The durability of any one concrete is thought to be reflected to a large extent by its permeability. Based on this premise it is now evident that many researchers and engineers are implementing measurements of permeability as a parameter to assess the potential performance of concrete. In this work it has been shown that permeability is highly affected by the method of curing and therefore it is suggested that permeability measurements are very useful not only to assess potential durability but also to evaluate the efficiency of a curing compound.

REFERENCES

1. FOSROC Technology, UK, CONCURE Curing Membrane, pp 15.

2. BIRT J C, Curing Concrete - An appraisal of attitudes, practices and knowledge, CIRIA Report No 43, February 1973, pp 35.

3. SHAW J D N, Curing membranes, Precast Concrete, Vol 10, No 2, February 1979, pp 73-77.

4. HEIMAN J L, Curing and curing compounds, Seminar on Transporting, Placing and Curing - How they affect the properties of concrete. Sydney, Australia, 3-4 August 1982, pp 159-182.

5. WAINWRIGHT P J, CABRERA J G and ALAMRI A, Durability aspects of cement mortars related to mix proportions and curing conditions. First International Conference in deterioration and repair of reinforced concrete in the Arabian Gulf, Proceedings, Vol 1, Bahrain, October 1985, pp 453-465.

6. GOWRIPALAN N, CABRERA J G, CUSENS A R and WAINWRIGHT P J. Effect of curing on durability. Concrete International, Vol 12, No 2, February 1990, pp 47-54.

7. DD 147, DRAFT FOR DEVELOPMENT. Method of testing for curing compounds for concrete, British Standards Institution, 1987.

8. AMERICAN SOCIETY FOR TESTING AND MATERIALS. Standard test method for water retention by concrete curing materials, C156-80a, November 1980.

9. CABRERA J G and LYNSDALE C J. A new gas permeameter for measuring the permeability of mortar and concrete, Magazine of Concrete Research, Vol 40, No 144, September 1988, pp 177-182.

10. CABRERA J G and LYNSDALE C J. Measurement of chloride permeability in superplasticised ordinary Portland cement and pozzolanic cement mortars. Proceedings of the International Conference on Measurements and Testing in Civil Engineering, Lyon, France, Vol 1, 13-16 September 1988, pp 279-291.

11. COMMONWEALTH EXPERIMENTAL BUILDING STATION, AUSTRALIA. Curing compounds and bonding agents for concrete and their effects on strength development and adhesion. Technical Record TR52/75/409, February 1973, pp 22.

12. PATEL R G, KILLOH D C, PARROTT L J and GUTTERIDGE W G. Influence of curing at different relative humidities upon compound reactions and porosity in Portland cement paste. Materials and Structures: Research and testing, Vol 21, No 123, May 1988, pp 192-197.

13. PARROTT L J. Measurements and modelling of moisture, microstructure and properties in drying concrete. From materials science to construction materials engineering. Proceedings, First International RILEM Congress, Paris, 1987, London, Chapman and Hall, 1987, Vol 1, pp 135-142.

ON THE MERITS OF FLY ASH ADDITIONS TO CONCRETE FOR REDUCING MICROSTRUCTURAL DETERIORATION DUE TO FATIGUE

P C Taylor

R B Tait

University of Cape Town,

South Africa

ABSTRACT. Although it is acknowledged that concrete exhibits fatigue in the form of microcracking degradation, the understanding of the long term response of concrete to fatigue is still not fully developed. One of the techniques used to improve concrete performance, including durability, is the addition of fly ash as a partial cement replacement. However the effect of fly ash on the fatigue behaviour of concrete as well as the mechanisms of any such effects are not well understood. This paper reports on the results of a comparative research program aimed at evaluating the relative fatigue response of mortars containing varying amounts of fly ash. The approach has been to use the double torsion specimen to provide data for the plotting of fatigue crack velocity - stress intensity (V-K) curves. The results suggest that fly ash mortar is slightly worse than OPC mortars in resisting fatigue microcrack degradation, but there are some hydration age effects. SEM micrographs of a 25% fly ash mortar also suggest that the extent of hydration of fly ash particles after seven days is limited, and this may elucidate some of the observed fatigue effects of fly ash in concrete with time.

Keywords: Fly Ash, Fatigue, Mortar, Concrete, Double Torsion Test, V-K Curve, Crack Growth Rate

Mr P.C. Taylor is a Senior Lecturer in the Department of Civil Engineering at the University of Cape Town, South Africa where he is pursuing PhD research into the fatigue characteristics of fly ash mortars.

Prof R.B. Tait is an Associate Professor in the Department of Mechanical Engineering at the University of Cape Town, South Africa. His present research work concerns various aspects of fracture mechanics and slow crack growth in brittle materials including cement mortars, tungsten carbide, ceramics and rock.

INTRODUCTION

Fly ash is formed from the incombustible (rock) portion of the
pulverized fuel burned in coal fired power stations. This material is
usually predominantly silicon and aluminium oxides that have fused in
the furnace to form glassy spherical particles typically from $1\mu m$ to
$100\mu m$ in diameter. Fly ash also exhibits pozzolanic behaviour and is
thus beneficial to many concrete properties when used as a cement
replacement or extender [1-3], however the performance is a function of
the chemical and physical make up of the fly ash. Due to its
dependence on the geophysical properties of the coal source, the fly
ash produced at any one power station is unique, thus making it
necessary to investigate and characterise the behavior of the ash from
each station. In South Africa at present the only ash that is marketed
for use in concrete is from Matla Power station where a high quality,
consistent material is produced.

The quality of a fly ash (from the concrete technologists point of
view) is dependent on the fineness, the Loss-on-Ignition (LOI) and the
amount of calcium oxide (CaO) present in its chemical composition.
Generally, quality improves with increasing particle fineness and with
decreasing amounts of LOI and CaO.

Some of the reported effects on the properties of concrete, both in the
plastic and hardened states, arising from the addition of fly ash
include the improvement in workability [3]; sightly increased density;
and reduced bleeding rates due to the smaller particles settling more
slowly, (but because the setting times are delayed the total bleed
volume may not be significantly less). The heat of hydration of
concrete containing fly ash is considerably reduced [3], whilst the
strength gain in mixes which contain fly ash is initially slower than
for similar OPC mixes but continues for a longer period. This makes
discussion about the comparative advantages or disadvantages of using
ash delicate because the relative strengths of the mixes at the time of
testing have to be taken into account. Amtsbuchler has reported that
the ash particles only begin to take part in the hydration process some
two weeks after the concrete has been mixed [4], which would also
account for the observed initial slower strength gain. The
permeability of mixes containing fly ash have also been reported [3] to
be somewhat less than OPC and thus there is a small increase in
corrosion resistance due to the reduced penetration by corrosive
agents. It has also been reported [5] that the addition of fly ash to
a mix prevents or retards alkali aggregate reaction (AAR), apparently
due to the dilution of the alkalis.

From the foregoing, it can be seen that in general, the use of fly ash
appears to be advantageous from the civil engineer's point of view.
However an aspect that has not been subjected to extensive study or
comparison, is the effect of cyclic (fatigue) loading on concrete
performance and this is partially addressed here. The purpose of this
paper is to report on some results obtained in an ongoing experimental
study comparing the fatigue behavior of cement mortars containing
varying amounts of local fly ash as a cement extender.

BACKGROUND

Fatigue in Fly Ash Concretes

From a reasonably extensive literature review [6] it appears that the comparative fatigue behaviour of fly ash concretes or mortars is far from being completely understood. Tse et al [7] conducted research using compression tests to produce comparative S-N curves of concrete containing high and low calcium ashes. Mixes with a range of cement replacements from 0 to 75%, all with the same slump and raw materials were made and tested at twenty eight days with a variety of compressive strengths from 4 to 40 MPa. Their results indicated that for high calcium ashes with 25% and 75% replacements there is a reduction in fatigue life, but in all other cases there was no great variation from the OPC control mix.

Tait [8] conducted a series of double torsion tests using American (Indiana) ashes in small cement paste specimens. Using a replacement value of 20% but without changing the water/cement ratio, he reported a small reduction in the fatigue resistance of the fly ash compared with the OPC control at early ages. However the mixes being compared were not comparable in terms of compressive strength and hydration ages were typically seven to fourteen days.

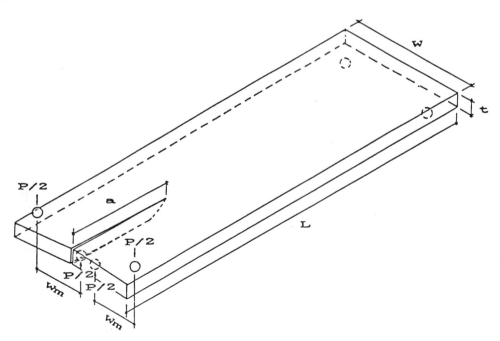

Figure 1: The Double Torsion test configuration showing the specimen shape and the loading points. Note the absence of grooves in the DT specimen.

The Double Torsion Test Configuration

For the so called "double torsion" (DT) test configuration the geometry of the specimen is such that for about the middle half of the specimen length the stress intensity (K) is constant with changing crack length [9]. Thus the measurement of crack velocity (V) at a known K is facilitated, because K is dependent only on the applied load and the specimen dimensions.

In the DT test, a thin plate (without grooves) is subjected to flexural point loads as shown in Figure 1, which results in a crack that, ideally, grows down the center-line of the specimen. Thus it is possible to measure V by means of measuring the crack length at known intervals. When crack lengths (a) are plotted against the number of cycles (n) a straight line is frequently obtained (at constant K), the slope of which is the rate of crack growth as illustrated in Figures 2(a) and 2(b).

The resulting linear relationship between a log-log plot of fatigue crack velocity (V) versus applied maximum stress intensity (K) provides one means of comparing crack velocity under given loading conditions for different materials or experimental conditions. For many materials there is a straight line region of the V-K curve for brittle materials (including cementitious systems) that is of interest, where the relationship between the velocity and stress intensity may be described by equation 1 [10]. In this equation A and n may be regarded as material constants and the comparison of these values provides a means of comparing slow crack growth or the "fatigue" behaviour of similar materials.

$$V = AK^n \qquad\qquad (1)$$

One of the particular advantages of the DT test is that it is possible to change the parameters during a test and immediately observe the resultant change in crack velocity as a change of slope of a crack length vs number of cycles plot as shown in Figure 2(a). It is possible to change the test parameters a number of times and thus obtain a number of points on a V-K curve from a single specimen.

In the present research project the double torsion test has been used with effect for testing a number of mortar specimens containing various amounts of fly ash. These were fatigue tested under a variety of different conditions in an attempt to compare the fatigue behaviour of mortars containing fly ash with that of mortars with no fly ash. Explanatory details of the tests are given in the next section.

EXPERIMENTAL DETAILS

Mix Design and Specimen Preparation

In the design of the mortar mixes a number of criteria were considered, these being that the mixes had to contain as wide a range of ash

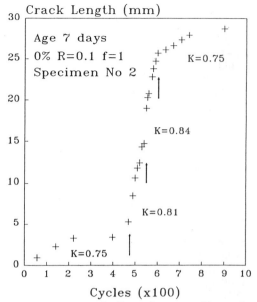

Figure 2(a): A typical crack length – number of cycles (a-n) plot showing the change in velocity corresponding to changes in applied stress intensity.

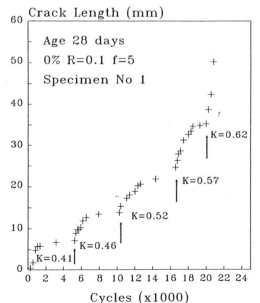

Figure 2(b): A crack length versus number of cycles (a-n) plot showing the crack velocity slowing and stopping several times under successively increased loads.

percentages as possible, without becoming unrealistic from an engineering viewpoint. They were to be of the same approximate strength at a given age, namely the age at which the predominance of tests were conducted; and that the sand contents of the mixes was to be similar as was workability. The testing age was selected as seven days for convenience. The research was undertaken on mortars as opposed to concretes in order to minimise the effects of having large particles in the relatively thin (8mm) specimens when studying microstructural behaviour.

The maximum ash percentage was fixed at 25% due to the fact that a higher percentage would cause difficulty in attaining the target strength of approximately 30 MPa at seven days even with the rapid hardening (Type III) cement used. The other mixes contained 0% and 15% replacement fly ash so that a range of comparative results could be obtained. The sand used was a plaster sand (<600μm), once again to minimise the scale effects.

The mix design method used was one recommended by the Portland Cement Institute [11] which involves reading off the cementitious/water ratio required to obtain a given strength for a given ash replacement from a graph. The final design mixes used c/w ratios for the 0, 15% and 25% mixes of 1.8, 2.1, 2.3, respectively and with ratios of cementitious materials/aggregate of 0.54, 0.61 and 0.69. In this test series the specimen dimensions were 225 x 75 x 8mm which are within the normally accepted proportions for the DT test [12]. Loading was achieved by means of an electro servo hydraulic testing machine under load control, in a purpose made rig.

Specimens were demoulded twenty four hours after casting, followed by trimming to size and the introduction of a starter notch cut into the specimen before curing in a lime saturated tank. It was critical that the notch was cut precisely in the centre of the specimen and that the specimen was very carefully aligned in the testing rig, or else the crack would not grow straight. Three of the four test cubes cast with the fifteen DT specimens were tested at seven days to monitor the quality of the batching, and the fourth was crushed on the same day as the DT testing to provide an indication of the relationship between toughness and cube strength.

Ramp and Fatigue Tests and The Test Matrix.

From a batch of 15 DT specimens, 5 were tested to failure under monotonically increasing displacement (i.e. ramped), three at the beginning and two at the end of a batch, in order that the nominal toughness of the batch could be evaluated and monitored over the time taken for testing. Six specimens were cyclically tested at a frequency of five hertz and the remaining four at either one or ten hertz, with the maximum load being changed periodically for each specimen to give as much comparative data as possible. Results were not obtained from every specimen because some were destroyed during preparation or testing, and on some others the crack grew skew invalidating the data.

TABLE 1   Matrix of Test Parameters

Frequency	(Hz)	1,	5,	10	
Load Ratio (P_{max}/P_{min})		0.1,	0.4		
Specimen Age	(days)	7,	28,	90,	180
Percentage Ash	(%)	0,	15,	25	

The crack velocity was measured by recording the position of the crack
tip at known intervals using a travelling microscope. However
observation of the crack tip is difficult when working with
cementitious materials because the true crack "tip" is not easily
defined in the zone of microcracking [13]. This difficulty was also
exacerbated by the surface conditions of the specimen, where if the
surface was coarse or wet the crack became almost indistinguishable.
The method, however, involved measuring the difference in crack length
with time and as long as the observer was consistent in his choice of
the so called "crack tip", the results were self consistent and
acceptable.

Four parameters have been included in the test matrix to date; these
being the cyclic frequency, load ratio R (P_{max}/P_{min} in compression),
specimen age and percentage ash. The range of values used for each of
these parameters is given in Table 1. One complete batch was used for
each different parameter, except that of frequency as described above,
and the results for the successful tests are given below. One set of
tests was also conducted with the aim of investigating if the material
properties were constant through the depth of the specimen or if the
material got "tougher" with depth (with respect to the mould), thus
causing the cracks to slow down.

RESULTS

Crack Length (a) versus Number of Cycles (n) Plots

From plots of the strength against the measured "toughness" (or
ultimate K) measured in the ramp tests for the same batches, it is
interesting to note that there seems to be an approximately linear
proportional relationship between the toughness (K_c) and strength; in
contrast to the trend for metals which is normally of decreasing
toughness with increasing strength.

In the double torsion tests it was noted that the crack velocity
appeared to slow down and stop in a number of specimens under a variety
of test conditions (Figure 2(b)). As a result of this effect a range
of velocities could be read off the a-n plot for any given value of K
resulting in a wide scatter in the V-K plots.

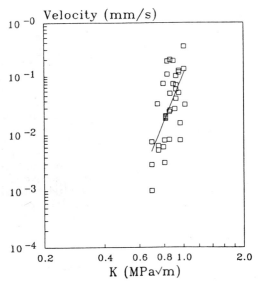

Figure 3: A typical set of V-K data points showing the extent of the scatter and the calculated best fit line.

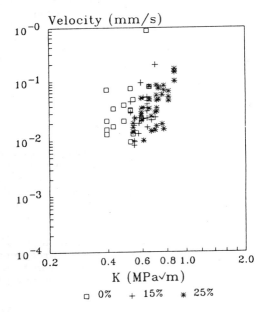

Figure 4(a). A V-K plot for tests at 7 days and 5 Hz as a function of the percentage ash contents.

V-K Plots

The velocity (V) obtained from the a-n plots was then plotted on a log-log scale against the applied stress intensity (K), which was calculated from the specimen dimensions and the applied load (P). A typical set of data points on a V-K plot for a given set of test conditions is shown in Figure 3. The straight line through the points is the best fit line found by regression for Equation 1 given above.

For each set of data points the values of A and n can be calculated using a power regression; however the scatter is such that the correlation coefficients are low, which means that statistically there was only a questionable relationship between the variables. Therefore at this stage it has only been attempted to compare the location of each "cloud" of data and to draw a conclusion as to whether a given data set is "better", "worse" or "similar to" another. In this respect, data which leads to faster crack growth rates is regarded as "worse" and slower crack growth rates as "better", for a given stress intensity. This is the approach that is used below, with the effect of changing one parameter at a time being discussed.

Effect of Various Parameters

For the 7 day old 5 Hz 0.1R tests (Figure 4(a)) it would appear that the no ash (0%) mix exhibited a slightly worse performance than the ash mixes in that there is a predominance of 0% data points to the left of the "cloud". The plots for 28 and 90 days show less scatter and no significant difference in crack growth rates between the three mixes. The 180 day plot (Figure 4(b)) shows a slight reversal of the 7 day trend behaviour which is consistent with the results found by Tait reported earlier [8].

For all three mixes it would seem that there is an improvement in fatigue crack resistance up to approximately 7 days but thereafter there is no continued observable change (as indicated in Figure 4(c)). This is apparent from the clustering of 7 day data points to the left of the rest of the data for other ages which is in a single cloud.

For all the sets of test parameters there is no significant observable trend or difference in crack growth behaviour due to changes in the cyclic frequency. In Figure 5(a), a plot of crack velocity in terms of da/dt, the data for all three frequencies tended to occur in the same general region of the V-K graph, whereas the frequencies were readily distinguished when plotted on a da/dn basis (Fig 5(b)). There is also possibly a marginally slower crack growth in the specimens loaded with an R ratio (P_{max}/P_{min}) of 0.4 than those loaded at R=0.1 as shown in Figure 6. These aspects are discussed more fully in the Discussion section to follow.

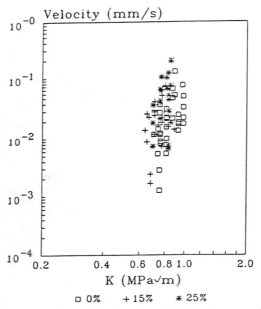

Figure 4(b): A V-K plot for tests at 180 days and 5 Hz as a function of the percentage ash contents.

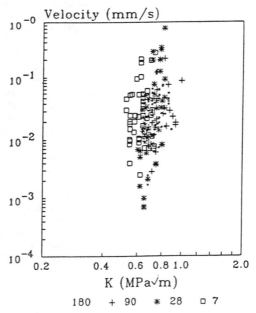

Figure 4(c): A V-K plot for the 15% mix tested at 5Hz and as a function of hydration age.

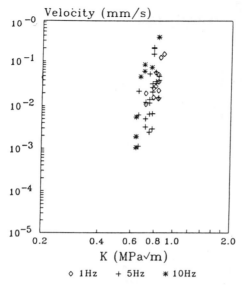

Figure 5(a) V (da/dt) versus K.

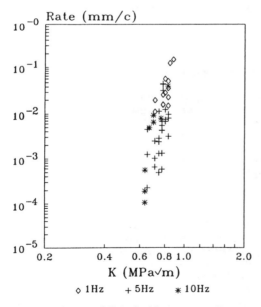

Figure 5(b) da/dn versus K.

Figures 5(a) and (b): Two plots of the same set of data as a function of frequency showing that the rate of crack growth is predominantly a function of time at these stress levels.

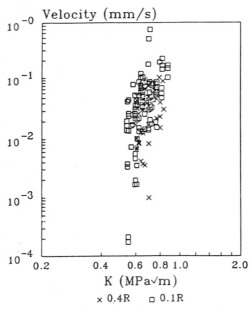

Figure 6: A V-K plot showing the small difference in crack
growth rate under the influence of different R Ratios for the
25% mix.

Scanning Electron Microscope (SEM) Observations

In order to evaluate the supposition [4] that fly ash only plays a
significant part in the hydration reaction after a period of
approximately two weeks, a series of small (12 mm diameter) cylindrical
specimens containing 25% fly ash were cast over a number of days.
After various hydration ages ranging from twelve hours to seven days
they were broken in two to reveal a fracture surface. These specimens
were then freeze-dried to arrest hydration and in order to draw off all
the evaporable water with the least amount of microstructural damage
[14].

The specimens were mounted and coated with gold-palladium and examined
under a Cambridge S180 scanning electron microscope. The micrographs
in Figure 7 show the observed changes occurring in a 25% fly ash mortar
as a function of hydration age, from 1 day and progressing through 2,
4, and 7 days. It is interesting to note that whilst there is a
progressive growth in the length of the calcium silicate hydrate (CSH)
whiskers, there are still a significant number of seemingly "untouched"
fly ash spheres at an age of 7 days. It should however also be noted
that in all the micrographs there are what appear to be concave
mouldings of gel that have been moulded around a sphere that has
subsequently pulled away. From this it would seem that the ash is

providing a source of "weak links" in the matrix at these ages, in that
the bond between a sphere and the gel is not as strong as the gel
itself.   It would also appear that not all the particles are taking
part in the reaction before seven days as there are a still a number of
untouched spheres visible at this age.   Although this work is very
preliminary, it may be surmised that the fly ash does not contribute
significantly to the strength development before 7 days.   This point
needs clarification and is under further investigation.

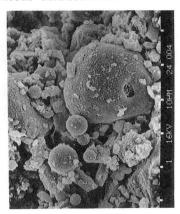

Figure 7(a): 1 day

Figure 7(b): 2 days

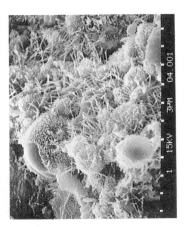

Figure 7(c): 4 days

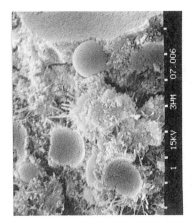

Figure 7(d): 7 days

Figure 7: A series of micrographs showing the morphological
changes in a 25% mortar as hydration proceeds over periods from
1 day to 7 days.   Note the increasing amounts of calcium
silicate hydrate with increasing age and that there appear to
be debonded unhydrated fly ash spheres even at seven days.

DISCUSSION

From the preliminary results presented above, several interesting
observations may be made. The first is to note the effect of the ash
percentage and the age of the mixes at testing. At early ages the
presence of fly ash seems to have a marginal beneficial effect on
fatigue crack growth behaviour in that the cloud of data for the
control (OPC) mix is to the left of the ash data points (Figure 4(a)),
meaning that for a given applied stress intensity OPC mixes exhibit a
higher crack growth rate. However at six months the trend is reversed
(Figure 4(b)), although the difference between the two data clouds is
smaller. Thus a longer life for an OPC concrete structure than a fly
ash concrete structure is implied. The reason for the change with time
is consistent with the fact that fly ash is known to affect the
hydration reaction with time as indicated by the changes in the rate of
strength gain. However the reason for the difference between the two
sets of data is more difficult to explain. The present results
partially contradict earlier work undertaken by Tait [8] who, although
reporting a decreased fatigue crack growth performance for fly ash
mixes, found this phenomenon at early rather than later hydration ages.
His mixes however did not have comparable strengths at the ages at
which he tested them, but merely a consistent cementitious/water ratio.

What is also noticeable is the reduction in the amount of scatter
between the 7 day and the 180 day tests. This is thought to be due to
the fact that the mix is hydrating relatively rapidly at an age of
seven days and is sensitive to small differences in the environment and
in the exact testing age.

Discussion about the mechanisms of crack growth, centre about the
question of whether the crack is driven by delta K (plasticity)
mechanisms (and described by a Paris formulation) or by K_{max} and
environmental mechanisms (described in terms of V-K curves). The two
plots shown in Figure 5 regarding frequency would seem to indicate
that the da/dt rather than the da/dn approach is more appropriate for
this material. This is indicated by the fact that all the data for the
da/dt plot (Figure 5(a)) is in a single cluster, but the same data
plotted as da/dn (Figure 5(b) tends to separate out into three
different clouds each pertaining to a different test frequency. The
results shown in Figure 6 for the R ratio tests, although exhibiting
substantial scatter and overlap, also tend to indicate that the trend
for the change in velocity with a change in stress ratio is consistent
with that which would be predicted by the Modified Goodman diagram, but
is quantitatively inexact. This would imply that the mechanism is
neither a pure K_{max} controlled nor a pure delta-K controlled one, but
some combination of the two. This point is presently under further
investigation.

The other question raised from the results is that of why the crack
slows down in an apparently constant K field?. It is not due to
changes in the material from the casting orientation or relative crack
direction, as the same effect was observed in specimens cast in other
orientations. Since it is accepted that the double torsion test yields

a constant stress intensity with changing crack length [12], it is thus postulated that this slowing down is because the growing process zone is absorbing an increasing amount of energy as elastic deformation at the multitude of crack tips (crack shielding) thus preventing the same rate of crack extension. As the load is increased the zone also increases in size up to a limit as there is a load above which the crack does not slow down. It is reasonable that the shape and extent of the process zone is a function of the stress field as has previously been reported [13] and an attempt to verify this using numerical modelling will be undertaken at a later stage.

CONCLUSIONS

1. Comparative double torsion fatigue tests were conducted on sets of fly ash mortars in order to ascertain the effect of fly ash on their fatigue performance.

2. It would seem that fly ash seems to have a small but noticeable detrimental effect on the fatigue performance of concrete at practical ages. Fatigue performance of fly ash concretes at very young ages (less than seven days) may however, be slightly better than OPC concretes.

3. It would also appear that fly ash does not play a significant part in the strength or toughness development of a mortar for at least the first seven days as evidenced by the SEM studies. Further studies of this point and indeed the fatigue design aspects of fly ash concrete need to be undertaken, particularly from a microstructural viewpoint.

ACKNOWLEDGMENTS

The assistance of Prof M O de Kock and Prof M G Alexander is gratefully acknowledged in their co-supervision of this project. Thanks are also due to Mr V Appleton for his assistance with the photographs. The financial support of the Foundation for Research Development (CSIR), Waste Management Group, is also gratefully acknowledged.

REFERENCES

1   RAVINA D. Fly ash performance in plastic concrete. ASH - A Valuable Resource, Proceedings, Volume 2, 1987.

2   CABRERA J G and WOOLLEY. A study of twenty-five year old pulverised fuel ash concrete used in concrete structures. Proc Instn Civ Engrs, 2, 1985, 79, Mar, pp 149-165.

3   BERRY E E and MALHOTRA V M. Fly ash for use in concrete - a critical review. ACI Journal, March-April 1980, pp 59-73.

4   AMTSBUCHLER R. Effect of different curing regimes on fly ash
        concrete. ASH - A Valuable Resource, Proceedings, Volume 2,
        1987.

5   OBERHOLSTER R E. Inhibiting alkali-silica reaction: the role of
        portland cement, milled granulated blastfurnace slag, fly ash and
        silica fume. Symposium: Practical Guidelines On The Selection
        And Use Of Portland Cement, MGBS, Fly Ash And Silica Fume In
        Concrete, Portland Cement Institute, 1987.

6   TAYLOR P C and TAIT R B. Fatigue and fracture of cement mortars
        containing fly ash - literature survey, University of the
        Witwatersrand, Department of Metallurgy and Materials
        Engineering, FRG/88/8, June 1988.

7   TSE E W, LEE D Y and KLAIBER F W. Fatigue behaviours of concrete
        containing fly ash. 2nd International Conference on Fly Ash,
        Silica Fume and Slag, Proceedings, 1986, SP91-12.

8   TAIT R B. Comparative slow crack behaviour in hardened cement
        paste containing PFA. ASH - A Valuable Resource, Proceedings,
        Volume 2, 1987.

9   FRASSINE R, RICCO T, RINK M and PAVAN A. An evaluation of double
        torsion testing of polymers by visualisation and recording of
        curved crack growth. Journal of Materials Science, 23, 1988, pp
        4027-4036.

10   BINSHENG Z, ZHAOHONG Z and KERU W. Fatigue rupture of plain
        concrete analysed by fracture mechanics. SEM/RILEM International
        Conference on Fracture of Concrete and Rock, Houston Texas, June
        1987, Edited by S.P. Shah and S.E. Swartz.

11   GRIEVE G R H. The proportioning of concrete mixes made with
        portland cement and MGBS, fly ash or silica fume. Symposium:
        Practical Guidelines On The Selection And Use Of Portland Cement,
        MGBS, Fly Ash And Silica Fume In Concrete, Portland Cement
        Institute, 1987.

12   TAIT R B, FRY P R and GARRETT G G. Review and evaluation of the
        double torsion technique for fracture toughness and fatigue
        testing of brittle materials. Experimental Mechanics, pp 14-22,
        March 1987.

13   TAIT R B and ALEXANDER M G. Characterisation of Microcracked
        zone development in cementitious materials as a function of
        specimen type and stress state. In Soc for Experimental
        Mechanics Spring Conference. Boston, June 1989.

14   TAIT R B and GARRETT G G. In situ double torsion fracture
        studies of cement mortar and cement paste inside a scanning
        electron microscope. Cement and Concrete Research. 1986, 16, pp
        143-155.

DESIGN OF PATCH REPAIRS: MEASUREMENTS OF PHYSICAL AND MECHANICAL PROPERTIES OF REPAIR SYSTEMS FOR SATISFACTORY STRUCTURAL PERFORMANCE

N K Emberson
G C Mays

Royal Military College of Science, Cranfield Institute of Technology,
United Kingdom

ABSTRACT. Damaged reinforced concrete members are often reinstated by the application of patch repair materials. This is only successful if the cause of the original damage has been eliminated, appropriate materials are selected and these are applied in a suitable manner.

As a preliminary to a wider study to investigate the structural consequences of property mismatch between patch repair materials and the substrate concrete, the relevant physical and mechanical properties of repair systems are identified and discussed. These properties include strength, modulus, coefficient of thermal expansion, tensile adhesion, Poisson's ratio, early curing shrinkage, long term creep and shrinkage.

Available test methods to determine these properties were examined and several found to be deficient or to provide unrealistic results. Those found questionable have been revised or new test methods developed. Tests for the measurement of flexural strength, flexural modulus and creep in compression have been submitted to the British Standards Institution for public comment.

The combination of existing standards, non-standard test methods and newly devised test procedures are discussed in the context of the measurement of the above properties for nine patch repair systems. These represent the range of generically different systems on the market and include resin mortars, polymer modified cementitious mortars and cementitious mortars.

Keywords: Concrete, Repair materials, Test methods, Patch repairs, Structural repairs, Design requirements, Damaged concrete.

Dr N K Emberson is a research officer with RMCS (Cranfield). During the last seven years he has been investigating the significance of property mismatch in the repair of structural concrete.

Dr G C Mays is Head of Civil Engineering at RMCS (Cranfield) and has research interests in the repair and strengthening of concrete structures and their design to resist blast loading.

INTRODUCTION.

Patch Repairs to Structural Members.

Current annual expenditure in the United Kingdom on repairs maintenance and improvements generally within the construction industry is of the order of £15 billion. Of this approximately £500 million is attributable to the repair of concrete members. Despite this large expanding market little fundamental research has been carried out on the structural implications of property mismatch between patch repair materials and the substrate reinforced concrete.

This paper reports results from an investigation on the design of patch repairs for structural members and is primarily concerned with measurement techniques of both physical and mechanical properties of repair systems for assessment of their probable structural performance. A review of previous work related to the subject indicated knowledge in this area was virtually non-existent.

The Structural Consequences of Property Mismatch.

Some of the problems associated with patch repairs, for example, curing contraction (caused by cooling in resinous formulations or drying of cementitious formulations), differing elastic modulus values and strengths have previously been discussed (1).

In addition to these there are other aspects which may be relevant, for example if members are structurally repaired at high or low ambient temperatures with materials having dissimilar coefficients of thermal expansion compared to the substrate, significant stress concentrations can theoretically develop.

Another important consideration is the load state of the structure both during and after repair. Indeed, failure to temporarily support some of a member's load prior to and during the repair procedure will result in stress being transmitted by undamaged parts of the member so that they may become permanently over stressed.

REPAIR MATERIALS.

Desirable Characteristics of Patch Repair Materials for Structural Efficiency.

A theoretical investigation into the desirable characteristics of repair systems for satisfactory structural performance was undertaken and the results of this are summarized in table 1. A selection of representative repair systems were then chosen from a number of different generic categories (table 2) to establish typical mechanical characteristics. It should be noted that within some generic categories there are many hundreds of formulations from which to choose and it should not therefore be assumed that the results presented represent limiting values.

Table 1: Requirements of repair materials for
structural efficiency.

Property	Repair system (R)		Substrate concrete (C)
Compression strength	R	$\geq$	C
Tension strength	R	$\geq$	C
Flexural strength	R	$\geq$	C
Adhesion strength	R	$\geq$	C§
Stiffness	R	$\approx$	C
Coefficient of thermal expan.	R	$\approx$	C
Effect of temperature	R	$\leq$	C
Creep and relaxation	R	$\ll$	C
Fatigue characteristics	R	$\geq$	C
Moisture (adhesion)	R	‡	C
Chemical resistance	R	$\geq$	C
Curing shrinkage	R	$\ll$	C¶
Strain capacity	R	$\geq$	C
Poisson's ratio	R	†	C
Degradation	R	$\leq$	C

§Cohesive strength. ¶Drying shrinkage.
‡Needs to be good on damp surfaces.
†Dependant on stiffness and type of repair.

Table 2: Categories of systems for concrete
patch repair.

Resinous Materials		Polymer modified cementitious materials		Cementitious materials	
A	Epoxy Mortar	D	S.B.R. Modified	G	O.P.C./Sand Mortar
B	Polyester Mortar	E	Vinyl acetate Modified	H	H.A.C. Mortar
C	Acrylic Mortar	F	Magnesium Phosphate Modified	I	Flowing Concrete

Known characteristics of selected products were recorded but it was found that only brief property details are listed on data sheets which are otherwise packed with descriptions of methods for preparation and placement. More detailed information on ingredients and relevant technical data regarding tensile strengths, shear strengths, shrinkage, adhesion strength to both concrete and steel, strain capacity and elastic modulus are invariably unavailable, unresearched or secret, although compressive strengths are often quoted.

MEASUREMENT OF PHYSICAL AND MECHANICAL PROPERTIES.

Review of Current Standards and Test Methods.

Available test methods to determine unknown characteristics were examined and several were found deficient or provided unrealistic results. A combination of existing standards, non standard test methods, and newly devised test procedures allowed many unknown characteristics of repair systems to be established. Several of the procedures detailed in existing standards that were found questionable, have on the basis of this research been revised. In addition, new test methods have been drafted. Both revised and new test methods have been submitted to the British Standards Institution and issued for public comment.

PROPOSED TEST METHODS:

Compression, Tension and Flexural Strengths.

Of these it has been found that existing test methods for 1. resinous materials, i.e. BS 6319:Pt. 2 (40mm cubes) and BS 6319:Pt. 7 (briquettes), and 2. for cementitious materials i.e. BS 4550:1978:Pt. 3, section 3.2 in conjunction with BS 1881:Pt. 116 (70mm cubes) and BS 12:1971:Pt. 2, appendix H (briquettes) can be successfully used to determine the former two strength characteristics. However, existing methods used for measurement and comparison of equivalent flexural strengths have been found suspect, i.e. BS 6319:Pt.3 for resinous materials and EN 196 for cementitious materials.

Two sets of results for the flexural strength values of a number of repair materials are given in table 3. Higher flexural failure strengths were recorded in three point bending specimens because failure in these is constrained to occur at mid-span. Failure at this location though, is not necessarily at the prism's weakest cross section.

Table 3: Flexural strength results for
selected systems. (N/mm^2)

Repair material†	A	B	C	D	E	F	G	H	I
Flexural (4 point)	21.9	27.5	14.3	12.5	7.9	8.6	5.6	6.2	6.5
Flexural (3 point)	27.8	36.0	18.2	14.4	7.7	10.0	10.1	8.5	7.8
†See table 2.									

In four point bending specimens there is a greater volume of material under the maximum stress. Failures were therefore much more likely to occur at the weakest cross section. For testing resinous and cementitious mortars, DIN 1164, EN 196, BS 6319:Pt. 3, ISO 4013, and ISO R679 1968:Pt. 1, all prescribe the less accurate 3 point test. Test results incorporating these methods are approximately 25% higher than flexural results obtained from 4 point methods.

As a result of this work a revised draft of BS 6319:Pt. 3 incorporating a four point bending test has been circulated for public comment (2). The new method makes use of samples to enable polymer and polymer/cement based mortars, i.e. materials containing aggregates up to 5mm, to be tested. These samples are 25mm square in section x 320mm long.

Compression, Tension and Flexural Modulus.

Existing test methods for measurement of compression modulus on resinous and cementitious materials i.e. BS 6319:Pt. 6 and BS 1881:Pt. 121 respectively were found satisfactory and values were found to range from about 18,000N/mm² for material C (a polymethyl methacrylate mortar) to about 50,000N/mm² for material F (a magnesium phosphate mortar). However, a satisfactory method for measurement of tensile modulus on similar materials was not found. In the absence of a suitable procedure and the desire to be able to draw direct comparisons between both categories it was necessary to develop an appropriate test method. It was also required that direct comparison of the results could be made with each material's compression modulus determined in accordance with the standards above and for this reason the most suitable sample size was a 40mm square section prism 160mm long.

The main difficulty of using prismatic samples is the application of tensile load but this was overcome by manufacturing a mould, in which prismatic samples were cast with trapezoidal end sections, figure 1.

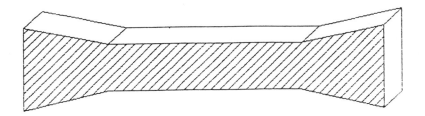

Figure 1: Prismatic samples 40mm square x 160mm long
with trapezoidal ends.

Strains were monitored in accordance with BS 1881:Pt. 121, (based on ISO 6784) using a clip-on extensometer. Values of tensile modulus were found to cover the same range as compression modulus values but the limits for these are slightly greater, i.e. 14,000N/mm^2 for material C to 60,000N/mm^2 for material F.

The procedure for four point and three point flexural modulus tests were identical to those for measurement of flexural strengths but with the addition of deflection being recorded at mid-span. Values of secant modulus were determined from load/deflection graphs at 0.8 times the flexural strength of each material. Table 4 lists the flexural modulus test results obtained from both tests for each material.

Table 4: Flexural modulus test results for
selected systems. (kN/mm^2)

Repair material	A	B	C	D	E	F	G	H	I
Flexural (4 point)	13.9	20.7	10.5	24.8	25.1	37.3	22.6	27.8	31.3
Flexural (3 point)	15.6	18.6	7.5	21.4	13.2	13.5	14.2	16.3	12.4

With one exception all flexural modulus test results were found to be smaller than tension and compression modulus test results which indicate that the materials are apparently less stiff in flexural loading than uniaxial loading. The results also indicate that the materials are apparently more flexible when tested in mid-point loading than third point loading. However, the total deflection of a beam is made up of a bending deflection plus a shear deflection. The shear deflection is negligible at high span to depth ratios but is very significant at low ratios. e.g. in the mid-point bending test the contribution to total deflection by shear deflection is 45% of the bending deflection when $\nu = 0.17$ and this is represented in figure 2 by point 1. For the third point bending test the contribution is 1%, (point 2, figure 2).

Comparison of the results in table 4 shows that mid-point flexural modulus values are about two thirds of third point flexural modulus values. It is clear from this that there is a discrepancy between the two test procedures. In rectangular beams that have span to depth ratios less than 8, shear stresses become relatively high and corresponding deflections become greater percentages of total deflection. For beams with span to depth ratios between 8 and 4 it has been shown that the assumption of linear stress distributions, on which the simple theory of flexure is based, becomes increasingly inaccurate and for span to depth ratios below about 4 the theory is invalid. It is believed to be mainly for this reason that the results shown in table 4 for the mid-point loading test (incorporating prisms with span to depth ratios of 2.5) are significantly lower than those from third point loading tests, (incorporating beams with span to depth ratios of 14.4).

As a result of this work BS 6319:Pt. 3 is being extended to include measurement of flexural modulus using the same specimen and loading configuration as that recommended for flexural strength tests (2).

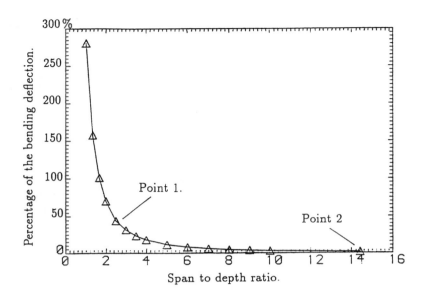

Figure 2: Shear deflection, expressed as a percentage of the bending deflection, versus span to depth ratios for a beam in mid-point loading. (Sample depth = 40mm, $\nu = 0.17$).

Coefficient of Thermal Expansion.

Standard test procedures for measurement of this property are documented and an example is ASTM C531–81. In this procedure specimens 25mm square x 254mm long are required to be heated in an oven to 22^0C, and then to 100^0C. At both temperatures the specimens are removed from the oven and measured with a demec gauge. The difference between recorded values allows the coefficient of thermal expansion to be calculated.

Materials used for structural patch repairs are normally subjected to in service temperatures within the range -30^0C to $+30^0$C. This paper is concerned with the structural performance of patch repairs which are not post cured prior to service and it is therefore appropriate to consider test methods which allow the performance of materials to be investigated over a range of temperatures following an initial period of curing at a fixed temperature.

Figure 3 provides details of suitable test equipment and specimen configuration within. The specimen cross section is small to allow uniform temperatures to be quickly obtained. The length of the specimen is as long as possible to enable reasonable length changes with modest temperature changes to be monitored by a single linear variable displacement transformer, (L.V.D.T.). The L.V.D.T. is positioned on the outside of the chamber and connected to the cross head of a rigid universal testing machine. Thus it and the testing machine framework are not subjected to temperature changes and continuous strain measurements can be monitored with time. During tests the upper and lower prisms of steel have temperature gradients along their length and these are monitored by thermocouples at each of their ends which enables the equipment to be calibrated with a standard steel bar specimen of the same material.

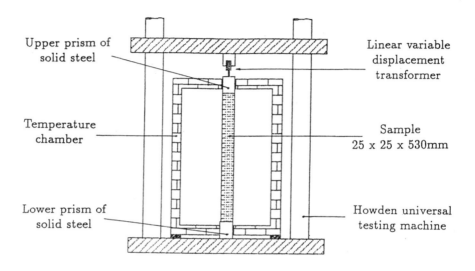

Figure 3: Arrangement of sample, solid prisms of steel, and L.V.D.T
to measure coefficient of thermal expansion.

Tests undertaken on materials A to I in the temperature range –60°C to +60°C at 20°C intervals revealed that some materials have different coefficient values above and below their cure temperature. It also revealed that some resinous mortars and some cementitious mortars continue to shrink long after their initial cure period at temperatures above their cure temperature due to further curing or drying shrinkage respectively. Finally, and of significant importance, it showed that the coefficient of some materials is not compatible for use in structural repairs subject to ambient temperature variations.

Overall the results indicate resinous materials have significantly higher coefficients of thermal expansion compared to non-resinous materials. Non-resinous materials tend to have values about equal to that of many unmodified concretes and the addition of polymers to unmodified materials has little effect on their coefficients of thermal expansion.

Tensile Adhesion Test.

Over the last 25 years several methods have been proposed for the measurement of bond strength between materials by compression - slant shear techniques. In recent times revision of these methods (3, 4) helped to produce the standard compression slant - shear test methods outlined in BS 6319:Pt. 4 and more briefly in AASHTO-T-237-73. Bond strength is a term for adhesion which is defined as the intermolecular forces which hold matter together. These forces are normally measured in terms of shear strength and tensile strength. The disadvantage of BS 6319:Pt. 4 is that it measures neither of these characteristics and for this reason a non standard direct tension test method was used in this research.

It was decided to assess adhesion strengths between a cured concrete and the materials selected from the categories shown in table 2, using the *Limpet* device (5). Prismatic samples of concrete 90mm(deep) x 100mm(wide) x 500mm(long) were manufactured by under filling an all steel mould 100mm square in section x 500mm long. After curing, prisms were air dried for several days and then dry grit blasted on their proposed adhesion face before being replaced in their moulds. A repair material was then applied to their grit blasted surface in accordance with the manufacturers instructions. Where appropriate this included application of primers and curing membranes. Following full cure of each repair material, 4 or 5 annular voids were cored into the layer of repair material which penetrated at least 10mm into the substrate. The exposed adhesion face of the repair materials and one side of the *Limpet's* aluminium disks were then dry grit blasted and stuck together with an epoxy resin. Following full cure of the glue, tensile adhesion tests were undertaken.

Table 5 lists averaged *Limpet* pull-off test results for control samples and all other systems. Data provided by control samples indicated the tensile strength of the substrate concrete to be 1.68N/mm², 20% less than measured in splitting tension tests to BS 1881:Pt. 117 on concrete cylinders from the same mix. The actual cause of this difference is unknown but splitting tensile strengths are generally known to be slightly higher than direct tensile strength measurements on mortar briquettes and it may be for the same reason that slightly lower values were recorded with the *Limpet* equipment.

Resinous systems have tensile adhesion strengths sufficient for most concrete repair applications but little of their tensile strength will be mobilized due to the low tensile strength of most concretes. In contrast, some non-resinous materials (i.e. materials F and H) indicate significantly poorer performance. There is only a little advantage to be gained from the use of modified O.P.C. mortars, i.e. materials D and E with their associated primers, compared to an unmodified O.P.C. mortar without a primer, i.e. material G, and this has also been found by others.

However, it is apparent that by appropriate modification of mortars better adhesion strengths can be obtained. This is demonstrated by material I which had the least tensile strength but greater adhesion strength than all other materials with the exception of material A.

Table 5: Adhesion strength of repair materials to concrete
measured by the Limpet test.

Material	Average Pull-off Load KN	Location of failure plane‡			Average Pull-off Stress N/mm^2
		C	I	R	
Control.† Concrete.	3.72	100%			1.68
A	3.87	100%			1.75
B	2.97	100%			1.35
C	2.85	85%	15%		1.29
D	3.09	45%	55%		1.40
E	3.43		95%	5%	1.56
F	1.57		100%		0.71
G	2.66	65%	35%		1.21
H	1.82	7%	93%		0.83
I	3.87	100%			1.75

‡C = in the concrete, I = at the interface,
R = in the repair material. †Cube strength = 38N/mm²
Splitting tension strength = 2.1N/mm²

Poisson's Ratio.

Although there are standard methods for determining Poisson's ratio in static compression and dynamically, e.g. ASTM C215 and C469, the values obtained are not thought to be equal to that measured in tension. This is mainly due to the fact that concrete and many patch repair materials are not isotropic. It was appropriate for this research to establish typical results from tension tests and furthermore, Poisson's ratio is defined in tension only (6). Experimental tests for Poisson's ratio were therefore undertaken in tension and on identical samples to those that had been cast for tension modulus tests, (figure 1). Longitudinal and lateral strains were recorded by chart recorders connected to electrical resistance strain gauges bonded to the surface of each specimen.

Values of Poisson's ratio were found to be equal to about 0.2 for most materials with the exceptions of materials H and I which have values of about half this, table 6. Values for stiffer non-filled epoxy and polyester formulations are generally in the region 0.4 to 0.5 and it is apparent from this that addition of fillers to resinous formulations significantly reduces their values.

Table 6: Poisson's ratio of selected systems.

Repair material	Poisson's ratio	Repair material	Poisson's ratio
A	0.228	F	0.176
B	0.198	G	0.238
C	0.219	H	0.114
D	0.195	I	0.080
E	0.165	–	–

Linear Shrinkage (or Expansion) 0 to 24 Hours.

The majority of repairs to concrete structures are undertaken on members aged between a few months and many years. Most of the plastic shrinkage and early thermal contraction associated with curing in these members starts immediately after casting and continues for several days. It is followed by slower long term shrinkage as drying out continues. This shrinkage is small by comparison with the plastic shrinkage and early thermal contraction and in older structures can be insignificant.

At the time of repairing many members a large percentage of the ultimate shrinkage is likely to have taken place. Products used for structural repairs that shrink following placement may mobilize some if not all of their adhesion strength, and this following re-application of load can subsequently produce interfacial failures. For these reasons it is necessary to ensure that products used for structural repair do not shrink appreciably following placement.

One method for measuring this property is outlined in A.S.T.M. C531–81 but in this standard specimens are not completely unrestrained during their initial curing period. As a result of this absolute values of early shrinkage can not be determined. Since early shrinkage (or expansion) is an important point to consider an alternative method developed by Staynes (7) was investigated. Details of his method have been presented in draft form to the British Standards Institution. The method monitors, (via twin L.V.D.T's) early linear shrinkage of freshly mixed resinous and cementitious mortars placed in a single thin layer on the base of a shallow mould, figure 4.

Values of linear shrinkage strain (negative) and expansion strain (positive) recorded for each material at the end of a 24 hour curing period are shown in table 7 and adjacent to these are peak values of temperature recorded for the materials during the same period. The binders of resinous materials are known to have varying degrees of exothermic reaction during cure periods and it is usual for large volumes of ordinary Portland cement based products to become warm following hydration. However, only resinous material B and non-resinous materials F and H were found to change temperature during tests.

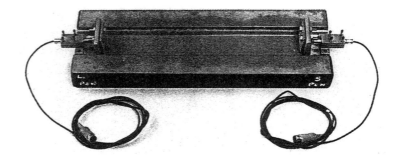

Figure 4: Test equipment developed by Staynes to monitor early linear curing shrinkage of freshly mixed mortars, (0 - 24 hours).

Table 7: 0 - 24 hours expansion (contraction) strain results.

Repair material	Microstrain ($\times 10^{-6}$)	Temperature (°C)	
A	− 170	21	constant
B	− 4680	44	peak
C	− 80	21	constant
D	− 920	21	constant
E	− 2400	21	constant
F	+ 830	32	peak
G	− 710	21	constant
H	− 350	24	peak
I	− 1120	21	constant

From table 7 it is apparent that, with the exception of material F, all materials shrank to a greater or lesser extent. Variation of shrinkage strain with time for materials B and C, and that of temperature with time for material B, are shown in figures 5 and 6.

It can be seen from figure 5 that material B, the polyester mortar, rapidly increased in temperature shortly after placement and that nearly all this products very high early shrinkage takes place during the first 15 to 20 minutes. The reduction in temperature is nearly as rapid as the original increase and this is because the sample is thin. For a 300mm long patch repair incorporating material B the shrinkage (based on these results) 30 minutes after placement would be 1.4mm and this would require the product to have very high interfacial adhesion to maintain bond with stiff substrates.

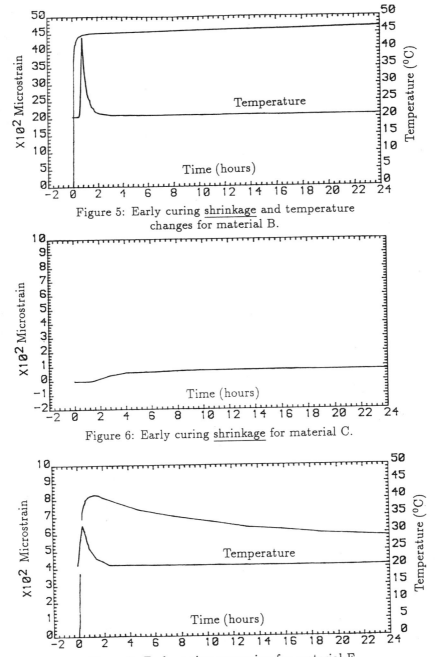

Figure 5: Early curing shrinkage and temperature
changes for material B.

Figure 6: Early curing shrinkage for material C.

Figure 7: Early curing expansion for material F.
N.B. the ordinate scales on these graphs are not the same.

In contrast to material B the shrinkage associated with material C, the vinyl acetate mortar, is small and most of this takes place between 1 and 3 hours after placement. Shrinkage versus time for resinous material A, the epoxy mortar, closely resembles the trend shown in figure 6 for material C but its 24 hour shrinkage strain is twice that of material C at $170\mu\epsilon$.

In contrast to all other results material F expanded $830\mu\epsilon$ within 1.3 hours of placement and within 0.46 hours of placement, a significant increase in temperature was monitored, figure 7. The reduction in temperature, similar to material B, is relatively rapid and this is again because the sample is thin. After 1.3 hours the material starts to shrink, some of this shrinkage is associated with the specimen cooling but after 2.5 hours shrinkage continues at room temperature at a reducing rate.

More suitable products for patch repairs are those that develop lesser values of shrinkage during cure periods. It appears that some modified products, e.g. material E, have greater early shrinkage than some unmodified mortars, e.g. material G.

Linear Shrinkage (or Expansion) from 24 Hours.

Tests for this property are not standardized but a suitable method under ambient conditions was devised and undertaken on prismatic samples 40mm square x 160mm long cast in an identical manner to compression modulus samples. At the age of approximately 17 hours each sample was carefully removed from its mould and on opposite cast faces 140mm long acoustic vibrating wire strain gauges were attached. Samples were then transported to an out-building to cure under polyethylene covers for the first month. The temperature and humidity of air in the out-building varied between 12^{0}C and 22^{0}C, and 50% and 90%, respectively. Results indicated that shrinkage of the materials is not significantly affected by these temperature or humidity variations. The results can therefore be considered to represent the behaviour of the materials at a mean temperature of 16^{0}C and a mean relative humidity of 68%.

Typical values of longitudinal shrinkage strain (negative) and expansion strain (positive) for all materials at various ages are given in table 8. Results for an age of 487 days (16 months) are presented in the right column. These have been corrected to a mean temperature of 16^{0}C and a mean relative humidity of 68%. Strains ranged between $-1140\mu\epsilon$ for material G, the sand cement mortar, to $+700\mu\epsilon$ for material F and material A has a low and stable value of shrinkage strain, i.e. $-50\mu\epsilon$ after 20 days. It is apparent some materials continue to shrink appreciably e.g. materials E, G, H, and I while others are stabilising e.g. materials B, D and F.

Total shrinkage (expansion) strains obtained by summing the early shrinkage (expansion) strains given in table 7 with the values given in table 8 are shown in table 9 for the age of 16 months. Points to note from this table are: 1. Materials B, E and I undergo substantial shrinkage strain at a very early age. 2. Values of total shrinkage (expansion) strain at 16 months indicate there is no advantage to be gained with modified cementitious materials E and I compared to unmodified cementitious material G. Material H has the lowest total shrinkage value of the cementitious materials and material F is expansive. 3. Resinous materials A and C have particularly low values of shrinkage. 4. In view of these results, statements by manufacturers which claim their products are non-shrink should be treated with caution.

Table 8: Average shrinkage (expansion)
microstrains at the ages given.

Material	20 days $\mu\epsilon$	30 days $\mu\epsilon$	40 days $\mu\epsilon$	120 days $\mu\epsilon$	487 days $\mu\epsilon$
A	−50	−50	−50	−50	−50
B	−250	−270	−280	−360	−280
C	−240	−260	−270	−360	−360
D	−500	−540	−560	−700	−740
E	−580	−640	−680	−600	−1060
F	+1040	+1020	+960	+860	+700
G	−600	−700	−760	−760	−1140
H	−460	−480	−480	−480	−760
I	−420	−440	−460	−460	−650

Table 9: Total shrinkage (expansion) test
results for the materials at an age of 16 months.

Material	0 - 24 hours. $\mu\epsilon$	From 24 hours to approximately 16 months. $\mu\epsilon$	Total shrinkage (expansion). ($\mu\epsilon$)
A	−170	−50	−220
B	−4680	−280	−4960
C	−80	−360	−440
D	−920	−740	−1660
E	−2400	−1060	−3460
F	+830	+700	+1530
G	−710	−1140	−1850
H	−350	−760	−1110
I	−1120	−650	−1770

Compression Creep Tests.

No standard test procedures for this property are known to exist but a suitable method under ambient conditions was devised and this has been submitted to the British Standards Institution as a draft for public comment (8). The tests were undertaken on prismatic specimens 40mm square x 160mm long cast and cured in an identical manner to that of compression modulus specimens. Following full cure, each specimen was fitted with two 140mm long acoustic vibrating wire strain gauges and individually placed in axial alignment above or below the flat jacks of a loading apparatus. The specimens ranged in age between 28 and 33 days at the start of the test which was undertaken in the same out-building as that previously referred to. All specimens were loaded to 10N/mm² at 1N/mm²/minute. Records of creep were regularly taken over a period of 15 months, during which time minor pressure fluctuations were corrected.

Initial strains induced at the time of loading and values of total strain at 7 months and 15 months are given in table 10. The strains have been corrected for the mean temperature of 16°C and mean relative humidity of 68%. Table 11 gives values of net creep strain over the 15 month test period, calculated by subtracting the shrinkage strain over the age range 1 to 16 months (table 8) from the gross creep strain over 15 months in table 10.

Referring to table 10, it is apparent some materials creep more than others. For example, the specimen manufactured from resinous material C had an elastic deformation of 540$\mu\epsilon$ after it was loaded. Seven months later this value increased 12 fold to 2750$\mu\epsilon$ and at fifteen months it was nearly twice this at 4750$\mu\epsilon$. Material E, (the vinyl acetate modified cementitious mortar) is the second most prominent product to be significantly affected by constant load. Material F, which expanded during cure periods, has the lowest value of gross creep strain.

Table 10: Compression creep strain - Time characteristics for
materials A to I at a constant stress of 10N/mm².

Material	Initial loading strain. Sample age 1 month. $\mu\epsilon$	7 Month strain $\mu\epsilon$	15 Month strain $\mu\epsilon$	Gross creep strain $\mu\epsilon$
A	−580	−1050	−1300	−720
B	−420	−1200	−1800	−1380
C	−540	−2750	−4750	−4210
D	−270	−550	−700	−430
E	−580	−2100	−2600	−2020
F	−220	−450	−550	−330
G	−360	−900	−1200	−840
H	−310	−1100	−1550	−1240
I	−260	−700	−1050	−790

It is apparent that some modified cementitious materials have significantly lower values of gross creep strain than unmodified cementitious products, i.e. materials D and F compared to material G, and that some resinous materials, i.e. materials B and C have values which are greater. Clearly all these products demonstrate that gross creep strain is an important characteristic to consider in relation to the ability of a structural patch repair to transmit load for a long period of time.

With regard to the individual components of gross creep, i.e. net shrinkage and net creep, it is apparent, from table 11 that creep is the dominant characteristic for most products when loaded at an age of one month. However, for material F, it is the long term shrinkage that follows its large early expansion which is dominant. Probably the most interesting result revealed is that material D, the SBR modified mortar, has a low net creep strain which may make this product suitable for repairs to older concrete members, providing its early shrinkage strain can be accommodated.

Table 11: Gross compression creep strain less shrinkage
(expansion) strain. (at 16°C)

Material	Shrinkage after 1st month †	Shrinkage after 16th month †	Net shrinkage between 1st & 16th mth	Gross creep between 1st & 16th mth	Net creep between 1st & 16th mth
A	−50	−50	0	−720	−720
B	−270	−280	−10	−1380	−1370
C	−260	−360	−100	−4210	−4110
D	−540	−740	−200	−430	−230
E	−640	−1060	−420	−2020	−1600
F	+1020	+700	−320	−330	−10
G	−700	−1140	−440	−840	−400
H	−480	−760	−280	−1240	−960
I	−440	−650	−210	−790	−580
†Values exclude 0 - 24 hour shrinkage (expansion) strain.					

CONCLUSIONS.

Those mechanical and physical properties of mortars for patch repair which may be of significance to the subsequent structural behaviour of the repaired member have been identified. Where possible existing standard methods of test have been used to measure these properties. Elsewhere new or revised methods have been developed. In the case of flexural strength, flexural modulus and creep in compression the methods have been submitted to the relevant British Standard Institution committee for public comment.

The results of the tests applied to a range of nine generically different patch repair systems reveals a wide range of values for each property. Each of the property values and their interaction will be an important consideration in the selection of a repair system and in the design of a successful patch repair.

REFERENCES.

1. Emberson, N. K. and Mays, G. C. Polymer mortars for the repair of structural concrete: the significance of property mismatch. In: The Production Performance and Potential of Polymers in Concrete. 5th International Congress on Polymers in Concrete. 22nd - 24th September 1987, pp. 335 - 341. (Ed. B. Staynes.)

2. British Standards Institution. Draft BS 6319. Testing of resin compositions for use in construction, Part 3, Method for measurement of modulus of elasticity in flexure and flexural strength. Technical committee CSB/20, October 1989.

3. Kriegh, J. D. Arizona slant shear test: A method to determine epoxy bond strength. A.C.I. Journal Vol. 73, No. 7, July 1976, pp. 372 - 373.

4. Tabor, L. J. The evaluation of repair systems for concrete repair. Magazine of Concrete Research. Vol. 30, No. 105, December 1978. pp. 221 - 225.

5. Long, A. E. and Murray, A. McC. Application of the pull-off test for the reduction of structural faults. Proceedings 2nd. International conference on structural faults and repair. (Ed. M. C. Forde and B. H. V. Topping.) pp. 323 - 331, 30th April - 2nd May 1985.

6. Chambers dictionary of science and technology. W and R Chambers Ltd. Edinburgh. Reprinted 1978.

7. Staynes, B. Draft method of measurement for shrinkage (0 - 24 hours) of polymer mortars and resins. BS.6319, February 1989, Brighton Polytechnic.

8. British Standards Institution. Draft BS 6319. Testing of resin compositions for use in construction, Part 11, Method of determination of creep in compression and in tension. Technical committee CSB/20, November 1989.

IMPACT ON CURRENT PRACTICE OF THE INTEGRATED EUROPEAN MARKET

Chairman

Dr J Wood
Director, Special Services Division,
Mott MacDonald Group,
United Kingdom

CONCRETE AND CONCRETE RELATED EUROPEAN DOCUMENTS

T A Harrison

British Cement Association,

United Kingdom

ABSTRACT. The replacement of national standards with European standards is an essential link in the creation of a single market. This will not be achieved by 1992 and is likely to take at least a decade. When it is completed, all major design, construction and materials standards will be European and conflicting national standards withdrawn.

European standardization work is in progress on all the major codes and standards and some have reached the stage of publication as full (EN) or voluntary (ENV) standards.

Concrete durability is mainly covered in ENV 206: Concrete – Performance, production, placing and compliance criteria and EC2: Design of concrete structures, both of which should be available as ENV's in 1990. If durable concrete structures are to be designed and built without significant contractual difficulties, ENV 206 will need to be supplemented by a National Annex plus additional specification clauses.

Keywords: Standardization, Specification, Durability

Dr Tom Harrison is the Standards Manager at the British Cement Association. He is a member of CAB/4, CAB/4/3, CAB/4/-/1, CAB/4/3/1 to 7, CSB/39/1 and the UK representative to CEN/TC104:WG1:TG7. Prior to his present appointment, he spent 17 years on construction research and its application to practice and five years on site construction.

INTRODUCTION

The Single European Act (SEA) of 1987 was a commitment to establish a single market by the end of 1992. A single market is an area without internal frontiers in which the free movement of goods, persons, services and capital is ensured in accordance with the provisions of the Treaty of Rome. To achieve this, the Act has various provisions including a change to weighted majority voting for the adoption of European Standards.

There are many barriers to free trade including those classed as institutional, legal and technical and the Commission of the European Communities (CEC) is working on a broad front trying to eliminate these barriers. National standards were clearly identified by the CEC as one of the key technical barriers to trade and therefore they set about removing them. Their concept was to replace national standards with European standards and as they had neither the ability nor resources to draft the hundreds of new European standards, they set up the Comité Européan de Normalisation (CEN) comprising the national standards bodies of the EEC (and EFTA) countries. In theory the CEC mandates CEN to produce standards in a specified area and then CEN sets up a technical committee and starts working on these standards. In practice CEN frequently works with provisional mandates or it has anticipated areas in which mandates may be forthcoming and commenced work whilst hoping that when the mandates arrive they will not be counter to the progress already achieved.

The Community also issues Regulations, Directives, Decisions, Recommendations and Opinions. A directive is binding as to the result to be achieved in a stated period but it is left to the individual governments to implement them. Directives address a particular subject and are written in general terms. The 'New Approach' directives, i.e. any written since May 1985, comprise the following elements: scope, essential requirements, methods of satisfying the 'essential requirements', attestation, transitional arrangements, free circulation and safeguard procedure. In the future, all products available in the EEC will have to comply with the requirements of one and frequently more directives.

As compliance with a European standard will be deemed to be compliance with the relevant directives, there is an obligation on drafting panels to produce a standard that complies with the appropriate directives. This is easier said than done as a directive is written in general terms and how and to what products each essential requirement is applied is defined in "Interpretative Documents". Many of the key interpretative documents are still not available. When they do become available, industry will still have the task of translating them into practice.

A number of directives are of particular significance to construction. These include the Construction Products Directive, the Works Directive 71/305 and the Supplies Directive 77/62.

The Construction Products Directive (CPD) has been enacted to prevent
barriers to trade within the Community for construction products that
carry an EC mark and the EEC Member States have to implement their
provisions before June 1991. A 'construction product' is any product
which is produced for incorporation in the permanent works. In time,
all construction products will have to satisfy the essential
requirements of the CPD. Compliance with a European standard will be
sufficient, as all European construction products standards have to
satisfy the essential requirements of the CPD. The essential
requirements relate to the performance of the <u>finished works</u> and
cover:

1. Mechanical resistance and stability
2. Safety in fire
3. Hygiene, health and environment
4. Safety in use
5. Protection against noise
6. Energy, economy and heat retention

Each of these requirements has to be satisfied throughout an
economically reasonable working life.

The 'Works Directive' and the 'Supplies Directive' apply to government
departments and local authorities for all contracts over defined cost
thresholds. In essence they require 'open' tendering for works and
supplies in all but exceptional circumstances. This requires
specifications to be written in a form that does not exclude products
that conform with techncial specifications approved by the Commission.
The approved technical specifications may include CEN or ISO
standards, European Technical approvals, acknowledged rules of
technology or some national standards. Without a single common EN,
this leaves the manufacturer with the opportunity to produce their
products to the least onerous of the approved technical specifications
and have them accepted throughout Europe on public works contracts.

In August 1989, the European Commission submitted a modified proposal
to apply a similar, but rather more flexible, regime to the transport,
energy, water and telecommunications sectors, which are excluded from
the 'Supplies' Directive and, except for telecommunications, from the
'Works' Directive.

In July 1989 the Council adopted a 'common position' (that is,
agreement subject to the European Parliament proposing amendments) on
a draft Directive which aims to ensure that suppliers and contractors
can pursue complaints about discrimination, and that action can be
taken against offending purchasers, in the context of alleged breaches
of the 'Works' and 'Supplies' Directives. The European Commission has
promised proposals on the enforcement of the proposed regime for the
excluded sectors by the end of 1989.

EUROPEAN COMMITTEE FOR STANDARDIZATION (CEN)

CEN comprises the national standards bodies of 18 EEC and EFTA Member
States. BSI is the UK member of CEN. Anyone may put forward a
proposal for a standard but the CEC has become an increasingly
important source of proposals.

Three types of standards are being produced:

European Standards (EN)

This is a Standard which must be implemented, normally within six
months, as a full national standard by each of the Member States
regardless of whether they voted for or against it. Any conflicting
national standard must be withdrawn.

Harmonization Document (HD)

A Harmonization Document (HD) is drawn up and adopted in the same way
as an EN but, because it relates to an area where there are national
differences of a technical or legal nature, its application is more
flexible so that the differences can be taken into account. It need
not be reproduced nationally but entails the same commitment to
withdraw any conflicting national standards within the agreed time.
Once the national situations have been harmonized, HDs can be changed
to ENs.

Voluntary European Standard (ENV)

This has equivalent status to a BSI DD (draft for development) and it
operates in parallel with existing national standards. It has a
limited life of 3 to 5 years and within this time scale it must either
be upgraded to an EN or withdrawn.

In conjunction with an ENV or EN there can also be a National Annex
(sometimes called a National Implementation Document) which gives
national rules where required or permitted. A National Annex cannot
override or change the requirements of an ENV or EN.

As a general rule, most Codes and specifications are likely to be
produced as ENV's before conversion to an EN and most test methods are
likely to be produced directly as EN's. This is because they are
often taken without significant change from an existing national test
method.

When a draft is completed, it is sent to the CEN members for a vote of
acceptance or rejection. The weighted voting rules are complex and
the vote is assessed using the following rules:

1. A simple majority in favour
2. Not more than 3 members vote negatively
3. At least 25 affirmative weighted votes are recorded
4. At most 22 negative weighted votes are given

The votes are weighted as follows.

EEC			EFTA	
Belgium	5		Austria	3
Denmark	3		Finland	3
France	10		Iceland	1
Germany	10		Norway	3
Greece	5		Sweden	5
Ireland	3		Switzerland	5
Italy	10			
Luxemburg	2			
Netherlands	5			
Portugal	5			
Spain	8			
UK	10			

If a negative vote is given and the standard was mandated by the CEC, a re-count is taken of the EEC votes. This is to prevent a standard required by the EEC being blocked by EFTA.

PROGRESS WITH KEY EUROPEAN STANDARDS

Cements

Pre ENV 197: Cement: Composition, specifications and conformity criteria received a negative vote at the end of 1989. The reasons given by the countries voting negatively were that the draft did not contain specifications for all cements available in the EEC and the limits on composition for composite cements needed to be extended.

How these issues will be resolved has not been settled at the time of writing. Preparing standards for all cements is likely to take several years as, for example, there are no agreed test methods for sulphate resisting cements. A probable outcome will be an agreement to standardize the first list of cements; Table 1, and to progress the other cement standards as rapidly as possible.

Cement is produced in a continuous process and it is not practical to produce cements to two standards, ENV 197 and, say, BS 12. It is therefore the intention that the British standards on cement will be revised to comply with the corresponding sections of ENV 197.

The main differences between current British standards on cement and the draft European standards are:

a)   the introduction of two strength classes with lower and upper limits on strength;
b)   narrower ranges on the proportions of ggbs, flyash and filler in specific composite cements;
c)   the introduction of conformity criteria.

TABLE 1. Cement types and composition covered by ENV 197 (1989)

Cement type	Proportion by mass[1] %					
	Main constituents					
	Portland cement clinker	Granulated blastfurnace slag	Natural pozzolana	Fly ash	Filler	Minor additional constituents[2]
CE 1	95 to 100	-	-	-	-	0 to 5
CE 11	65 to 94	0 to 27	0 to 23	0 to 23	0 to 16	-
CE 11-S	65 to 94	6 to 35	-	-	-	0 to 5
CE 11-Z	72 to 94	-	6 to 28	-	-	0 to 5
CE 11-C	72 to 94	-	-	6 to 28	-	0 to 5
CE 11-F	80 to 94	-	-	-	6 to 20	0 to 5
CE 111	20 to 64	36 to 80	-	-	-	0 to 5
CE IV	≥60	-	-	≤40	-	0 to 5

1) The values in the table refer to cement nucleus, excluding calcium sulphate and additives.

2) Minor additional constituents may be one or more of granulated blastfurnace slag, natural pozzolana, fly ash or filler unless included as main constituents in the cement.

Note: Cements are also classified by class. The two permitted classes are 42.5 and 32.5 with sub divisions of 42.5R and 32.5R which indicate high early strength.

d) the inclusion of all types (compositions) of cement in a single standard

The main differences in performance between the cements will be between Class 32.5 and Class 42.5 cements. For equal cement content and equal w/c ratio (i.e. the requirements of ENV 206), a Class 42.5 cement will give higher concrete strength than a Class 32.5 cement. Differences in durability performance due to the type of cement may be less significant than class, as the cements in one class all have similar 28 day mortar prism strengths. Type of cement has more significance for constructional reasons, such as workability, finishing times, rates of strength development (formwork striking times) and curing periods. Shortcomings in workmanship may produce differences in durability performance between cement types.

When European cement standards replace British Standards, most users of concrete will not notice a difference in the properties or performance of their concrete if Class 42.5 cements are specified.

Additions

This is the term used to describe products such as Fly ash, ggbs, silica fume and limestone filler. Additions are sub-divided into Type 1 which are nearly inert additions, and Type 2 which have pozzolanic or latent hydraulic properties. According to ENV 206, national rules may modify the specified minimum cement content and maximum water/cement ratio if Type 2 additions are used.

A draft EN: Flyash for concrete is being prepared. The quality level being defined in this standard is basically that for a product that does no harm to concrete and at its lowest level, does not reduce at 90 days the standard mortar prism strength by more than 15% (tested at 25% flyash and same W/C ratio). The UK has been pressing, so far without success, for a two tier standard with the upper quality level being equivalent to BS 3892: Part 1 pfa. The minimum quality level defined in the addition standards will affect how additions may be used in concrete.

A CEN/TC 104: WG1 Task Group is studying the way in which additions may be used in EN 206. A reasonable starting point would be that any addition must have its own product standard and, therefore, it is likely that in the near future, task groups will be formed to draft European standards on ggbs and silica fume.

Admixtures

European standards on admixtures for concrete are being prepared by CEN/TC104: WG3. Approval for the draft test methods will be voted upon in the spring of 1990 but the work on 'requirements' has not been started. Work on admixtures for mortar and grout has also not been started but it is likely that these specifications will call up many of the tests being developed for concrete admixtures.

Aggregates

Pr ENV: Aggregates for concrete and pr ENV: Lightweight aggregates, are being drafted in a style that calls up national test methods. At the same time, work is in progress on about 50 EN test methods and when agreed, they will replace the national test methods. As the UK holds the convenorship and secretariat of CEN/TC 154: Aggregates and Sub-committee 6: Test methods, they are insisting that EN test methods contain a precision statement, i.e. as required by BSI. The need for precision trials to obtain these data are likely to delay the introduction of EN test methods for aggregates.

Reinforcing and prestressing steel

These standards are being produced by the European Committee for Iron and Steel Standardization (ECISS) and not by CEN, although it is linked to CEN for administrative and legal purposes.

Euronorm 80: Reinforcing steel, was produced directly by the CEC and it has not been widely accepted. Work is in progress on drafting EN 10080: Reinforcing steel (not for prestressing), and the first draft is expected in May 1990. The main problem with completing this standard is the selection of grade(s) of steel. This has important commercial implications for the production of reinforcing steel, but from the standard viewpoint has less importance as most structures could be designed to any of the steel grades being considered. In a similar way, Euronorm 138: Prestressing steel, is being re-drafted as a European standard, EN 10138. When the committee has agreed the re-draft, it will be issued for public comment.

Both of these standards have been mandated by the CEC and are expected by the end of 1992.

Concrete

ENV 206: Concrete - Performance, production, placing and compliance criteria, received a positive vote in the Autumn of 1989. The document is being edited and it will be available in 1990. As the document is only suitable for use with EC2 and with a national annex, it is likely that these documents will be issued as a 'package' in the Autumn of 1990.

The UK have strong reservations about many of the details of ENV 206 and require these to be changed before it is up-graded to an EN. In the meanwhile, ENV 206 is likely to have only a very minor impact on UK practice. A few public bodies may use it in an experimental way, but even with a national annex, its use would still present formidable contractual problems. At present, seven task groups are reviewing aspects of ENV 206, with the aim of finding agreement on clauses to be included in EN 206. Other task groups are likely to be formed in the next year to resolve other problem areas.

ENV 206 covers both concrete as a product and concrete workmanship. Its scope therefore covers the same ground as BS 5328, part of BS 8000, and section six of BS 8110. The real importance of ENV 206 lies in its scope and principles, as these are unlikely to be changed during its upgrading into an EN. The detail is important in the sense that objectors now have to provide a case for changing the ENV 206. As agreement could not be reached on many key issues, ENV 206 permits the use of 'National Rules'. The extent to which this approach can be continued in the EN is open to debate. It is the author's view that pressure will be put on technical committees to eliminate 'National Rules'. This may be achieved by increasing the options but great care is needed before rules that are specific to one area, or one set of materials, are generalised.

An introduction to the approach to durability taken in ENV 206 is given later in this paper.

Precast concrete

As precast concrete is used for such a wide range of products, it is inevitable that the way in which it is handled in European standards is a bit messy. In some cases, precast concrete is treated as one of the materials used to produce a product and the Technical Committee deals with all the materials used to manufacture these products. CEN/TC 125: Masonry, is a good example of this approach. The advantages of this approach are that any product will have to satisfy the same interpretation of the Construction Products Directive for the same economically reasonable working life and that, where possible, common test methods will be used.

The other approach is to deal solely with precast concrete products. A Technical Committee is being formed to cover a wide range of precast products. Its brief is to work, where applicable, to the requirements of EC2 and ENV206. How it co-ordinates its activities with those of similar products made with other materials has still to be resolved.

Within the next decade, almost all British precast concrete product standards will be replaced by European standards.

Design standards

The CEC had commissioned experts to produce a range of design codes. Good progress was made and a draft of Eurocode No. 2: Design of concrete structures: Part 1: General Rules and Rules for Buildings, was agreed. In 1989, it was agreed to transfer the work on Eurocodes from the CEC to CEN. A Technical Committee is being formed but at the time of writing, its structure has not been agreed.

In addition to Part 1, the following Parts are in preparation or planned:

    Part 1A - Plain or lightly reinforced concrete structures
    Part 1B - Precast concrete structures

Part 1C - The use of lightweight aggregate concrete
Part 1D - The use of unbonded and external prestressing tendons
Part 1E -
....

Part X - Fire resistance of concrete structures

Part 2 - Reinforced and prestressed concrete bridges
Part 3 - Concrete foundations and piling
Part 4 - Liquid-retaining structures
Part 5 - Temporary structures, structures having a short design
    life
Part 6 - Massive civil engineering structures.

All design Eurocodes will use the same "Actions on structures
(loading)" , which is being drafted in 20 Parts.

EC2 is strongly linked with ENV 206 and it would be very difficult to
use it with the existing British Standards on concrete. It would be
sensible to use the 'package' concept of BS 8110 with BS 5328 or EC2
with ENV 206 and not attempt to mix the packages. With respect to
durability, it should be noted that EC2 specifies <u>minimum</u> covers to
which a margin is added to reflect the level of workmanship. Whilst
the tabulated values may be modified for some circumstances, it does
not have the same degree of trade-off between concrete quality and
cover given in BS 8110.

Comment

The political drive to remove barriers to trade will result in the
replacement of the bulk of the existing portfolio of British
Standards with European standards. This will not be achieved by 1992
but is likely to take at least the next decade.

THE TREATMENT OF DURABILITY IN ENV 206

Introduction

The way in which ENV 206 covers concrete durability is best considered
at two levels: how it compares with existing British Standards and,
more importantly, how does it reflect the state of technical
knowledge. Comparisons are difficult as the starting point (exposure
conditions) are not exactly the same and one should consider the whole
package and not just specific aspects. Another problem is that
Standards tend to treat durability in a global way rather than
considering each aspect on its own[1], Table 2. The development of
Table 2 could lead to a more rational approach to durability design
but this is outside the scope of this paper.

TABLE 2   Summary of aspects of durability

Aspects	Environmental conditions for a significant durability risk	Key factors in providing durability
Corrosion of re-inforcement due to carbonation	Varying RH (the critical range for carbonation is 30 to 80 and for corrosion 70+) [1]. Chimneys and similar where steam and CO_2 are carried	Actual cover Concrete quality
Corrosion of re-inforcement due to external chlorides	Offshore and coastal regions; areas treated with de-icing salts or areas where saline water, spray or vehicle carried de-icing salts may reach; (rate of chloride penetration is greatest when saline water comes into contact with relatively dry concrete)	Physical barriers Actual cover Concrete quality
Freeze/thaw damage	Freeze/thaw cycles whilst the concrete is at a high level of saturation. (Note: Saturation can be obtained by capillary action); worse when freezing with de-icing salts.	Entrained air at the correct spacing
Abrasion/ cavitation	Volume and nature of trafficking; speed of water flow and content of solids	Curing Finishing operations Concrete quality
Sea water attack	Exposed to sea water	Concrete quality Materials
External chemical attack	The concentration of aggressive chemical(s) and its rate of replenishment; temperature	Physical barriers Concrete quality Materials

Exposure class

The exposure classes given in ENV 206 and EC2 are more detailed than those given in BS 8110. For example, classes 3, 4a and 4b all lie within the BS 8110 classification of very severe. However, ENV 206 does not contain an equivalent to the 'extreme' exposure. It is interesting to compare the approach in Table 2 with that in ENV 206. For significant volumes of concrete, e.g. that used in a dry environment and buried concrete in non-aggressive soils, no durability requirements are identified in Table 2. In these cases, the quality of concrete should be determined by structural and constructional considerations.

ENV 206 classifies non-aggressive soil as 2a or 2b and requires a relatively high quality of concrete but 'National rules' can be used for simple structures and, by implication, less onerous requirements. It is technically illogical that the quality of concrete in the same environment should depend on the nature of the structure, but it is a pragmatic engineering approach to reduce the consequences of an incorrect assessment of the soil conditions in important structures.

W/C ratio and cement type

Table 3 is a summary of the key table on durability in ENV 206.

For reinforced concrete, the requirements for W/C ratio are equal to or 0.05 lower than those specified in BS 8110. However, the BS 8110 values were determined for an OPC to BS 12 which is generally equivalent to a CE1 42.5 cement. Equivalent strength requirements were then specified in BS 8110 and all cements or mixer-blends have to satisfy these strengths. The effect of this approach is that for concrete supplied to BS 8110 strength generally controls the mix design rather than W/C ratio.

ENV 206 does not have a strength requirement, although it permits the National Annex to include one and the UK National Annex is almost certain to contain such a requirement. The classes and types of cement to be permitted in each exposure class are also subject to National rules. Without the protection of grade, the implication is that all cements give an adequate performance regardless of class or type. Figure 1 illustrates the difference in performance between Class 42.5 and Class 32.5 cements. Cement type will influence performance to a lesser extent, provided the concrete is handled correctly on site. A priority issue for the revision of the durability requirements in ENV 206 must be a recognition of the differences in performance between Class 42.5 and 32.5 cements.

ENV 206 also requires that 'National rules' are used to define 'sulphate resisting cements' and this causes problems as the approach used in the UK has been to produce sulphate resisting concretes from a range of cements with differing sulphate resisting properties. To compensate for a moderately sulphate resisting cement, a lower W/C ratio and higher cement content are required by British Standards.

TABLE 3 Summary of durability requirements (from ENV 206)

Requirements	Exposure class								
	1	2a	2b	3	4a	4b	5a	5b	5c[1]
max w/c ratio[2] for									
- plain concrete	-	0.70							
- reinforced	0.65	0.60	0.55	0.50	0.55	0.50	0.55	0.50	0.45
- prestressed	0.60	0.60							
min cement content[2] kg/m^3									
- plain concrete	150	200	200				200		
- reinforced	260	280	280	300	300	300	280	300	300
- prestressed	300	300	300				300		
min air content % for nominal max aggregate size[3]									
- 32 mm	-	-	4	4	-	4	-	-	-
- 16 mm	-	-	5	5	-	5	-	-	-
- 8 mm	-	-	6	6	-	6	-	-	-
frost resistant aggregates	-	-	yes	yes	-	yes	-	-	-
impermeable concrete according to ISO 7031	-	-	yes	yes	yes	yes	yes	yes	yes
types of cement for plain and reinforced concrete	National rules				sulphate resisting cement for sulphate contents > 500 mg/kg in water > 3000 mg/kg in soil				

1) Plus protection

2) National rules may modify these values when Type II additions are used.

3) Plus a spacing factor of < 0.20 mm

INDOOR EXPOSURE 3 DAY CURE

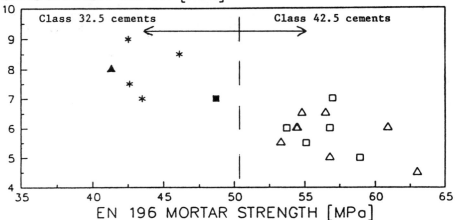

OUTDOOR EXPOSURE 3 DAY CURE

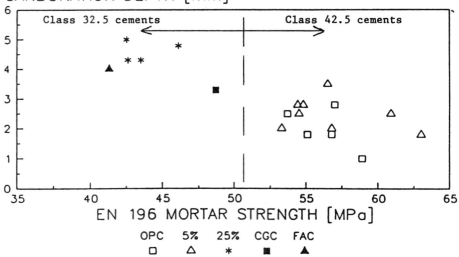

Figure 1: Carbonation depth for mix C v 28 d mortar strength (Figure taken from MOIR, G K. and KELHAM, S. : BRE Seminar Paper on the performance of limestone-filled cements : Durability 1)

Care is needed in translating Continental experience with sodium sulphate conditions to the more aggressive magnesium sulphate conditions found in the UK.

Work is in progress on tests for sulphate resisting cement and until these are agreed, it would be best if the UK National Annex specified the continuation of existing UK practice which can be used with confidence to produce sulphate resisting concrete.

Minimum cement content

The reasons for specifying a minimum cement content are:

a) to provide sufficient paste to fill the voids between the aggregate;
b) to provide a reserve of alkalinity such that the rate of carbonation is sufficiently slow;
c) to provide some binding of chloride ion;
d) to ensure a high probability of achieving the specified W/C ratio.

The minimum cement contents in ENV 206 are 15 to 45 kg/m^3 lower than those given in BS 8110. The BS 8110 cement contents are based on a water demand of 180 litres/m^3 and the specified W/C ratios, whilst the ENV 206 cement contents are reasonably compatible with the specified W/C ratios if a water reducer is used.

Some important issues still have to be resolved:

a) to what cement class do these minimum cement contents apply;
b) what adjustment is needed for the other class of cement;
c) how can the cement contents and W/C ratios be modified when Type II additions are used?

Freeze/thaw durability

In a similar manner to British Standards, ENV 206 requires entrained air in conditions where the concrete will be frozen whilst wet. The detail differs in that minimum air contents are specified and there is an additional requirement for a spacing factor of less than 0.20 mm in the hardened concrete. The spacing factor is very important for freeze/thaw durability and research[2] has shown that where air entraining agents are used in conjunction with superplasticisers, the air bubbles may coalesce, giving the correct total air content but a high spacing factor and a poor freeze/thaw resistance.

There is also a requirement for the use of aggregates with freeze/thaw resistance. Pop-outs do occur due to poor freeze/thaw durability of some aggregates and therefore there is a need to avoid using these aggregates. A number of approaches to conformity could be taken, ranging from a good track record, freeze/thaw testing of concrete or freeze/thaw testing of aggregates using, for example, the DIN 4226[3] requirements.

Impermeable concrete

For the more severe exposure classes, ENV 206 has a requirement for a resistance to water penetration based on the (draft) ISO 7031 test. This draft is based on the DIN 1048[4] test, which is in the process of being revised. The reason for this test was that German experience indicated that even with the specified cement content and W/C ratio, some concretes had a high permeability and, by implication, a low durability. In practice, most suppliers of concrete in Germany now have little problem in satisfying the requirements of the DIN 1048 test.

Resistance to abrasion

The minimum strength of concrete specified in ENV 206 for abrasion resistance is C37 compared with the C40 in BS 8204: Part 2. It recognises the importance of curing but not the importance of the finishing operations.[5] Yet again, it is the National regulations that are called up for determining the resistance to abrasion.

Cover

The actual quality and quantity of cover are of paramount importance for the protection of reinforcement. Regretfully we are not at a stage where we have accepted specification limits and performance tests for cover concrete quality, but we do have them for the depth of cover. The inclusion of such a requirement for concrete exposed to aggressive environments would be the single most significant step that could be taken immediately to achieve durable concrete structures.

Closing comment

Before ENV 206 can be used with confidence to specify cost effective, durable concrete structures, changes are needed. In addition, a significant amount of work is needed on the conformity aspects and test methods.

REFERENCES

1   PARROTT, L. Characteristics of surface layers that affect the durability of concrete. Proceedings of "Advances in cement manufacturer and use, July 31 - Aug 5, 1988". Published by Engineering Foundation. pp 137-142.

2   SIEBEL, E. Air-void characteristics and freezing and thawing resistance of superplasticized air-entrained concrete with high workability. Proc. 3rd Int. Conf. on 'Superplasticizers and other chemical admixtures in concrete; Ohawa, 1989. ACI SP-119, pp 297-319.

3   DEUTSCHE NORM. Aggregates for concrete DIN 4226, Part 1, 1983.

4   DEUTSCHE NORMEN. Prüfuerfahren für Beton (Test methods for
     concrete), Teil 1. DIN 1048, 1978.

5   CHAPLIN, R G. The influence of ggbs and pfa additions and other
     factors on the abrasion resistance of industrial concrete floors.
     (To be published in 1990 by the British Cement Association).

TECHNICAL HARMONISATION OF CONCRETE PRODUCTS

J B Menzies

Building Research Establishment,

United Kingdom

ABSTRACT. Technical harmonisation is an important requirement for the successful implementation of the Integrated European Market. For concrete products, as for other construction materials, the achievement of technical harmonisation is presenting challenges across the whole field from the technology of the constituent materials of concrete through to the structural design of concrete construction.

The paper reviews, against the background of the developing European legal framework progress in the preparation of European standards relating to concrete construction. Limited comparative studies of practice in concrete production and use in Germany, France, and the United Kingdom are described to illustrate the wide ranging nature of differences in practice. Finally the impact of European technical harmonisation on current practice locally is discussed.

Keywords: Concrete, Concrete Products, Construction Products Directive, European Technical Standards

Dr John B Menzies, BSc(Eng), PhD, FEng, FIStructE is Director, Geotechnics and Structures of the Building Research Establishment (BRE) Garston, Watford WD2 7JR, UK. He is concerned primarily with research and development in the geotechnical and structural engineering of buildings and other constructions.
He is currently taking part in the preparation for the European Commission of Interpretive Documents of the Construction Products Directive and Codes of Practice for the design of structures and has been concerned particularly with preparation of the Eurocode for Concrete Structures (EC2)

INTRODUCTION

The Commission of European Communities (CEC) started the task of
technical harmonisation of the construction industry in 1976 with the
objective of removing barriers to trade. The availability of European
technical base documents, such as prepared by the Comité
Euro-International du Beton (CEB) together with the existence of well
established procedures for preparation of codes of practice in the
Member States led to work commencing in the area of structural
engineering. In 1981, with advice from the Member States, the
Commission advanced a programme for the development of eight
'Structural' Eurocodes with a view to them being adopted nationally
with an equivalent legal status as national codes of practice. One of
these Eurocodes was Eurocode EC2: Common unified rules for concrete
structures.

The Government Heads of Member States called upon the CEC in 1985 to
present proposals for the completion of the Internal Market by 1992.
The result was the White Paper: 'Completing the Internal Market'
adopted at the Milan summit in 1985. The White Paper contained a
programme of about 300 measures and a timetable to meet a range of
objectives essential for completing the market including removal of
technical barriers and liberalisation of public procurement.

The White Paper provided the basis for the Single European Act which
was adopted in June 1987. This Single Act defines the 'Internal
Market' as 'space without internal frontiers, in which free movement
of goods, persons, services and capital..... is ensured'. Thus the
scope of the Internal Market was extended to include services as well
as goods. Also the procedure for the adoption of Community Directives
was changed to qualified majority voting in the Council of Ministers
instead of unanimity.

Another important change introduced by Council Resolution in 1985 was
the 'new approach' to technical harmonisation and standards. This
approach allows only the essential requirements of products to be
specified in Community Directives leaving the details to be covered in
standards.

The result of these changes is that:

(1)   Community Directives now contain less technical detail
(2)   Voting them through is easier
(3)   European standards become much more important

THE CONSTRUCTION PRODUCTS DIRECTIVE

Amongst the measures introduced for the achievement of the Internal
Market, the Construction Products Directive is one of the most
important for the construction industry. It is designed to remove
technical barriers to trade due to nationally differing technical

specifications, standards and approvals, as well as inspection and control procedures, so as to achieve a free movement of construction products. It is one of the new approach Directives and therefore includes the definition of Essential Requirements for safety and other aspects important for general well-being.

The Essential Requirements relate to works, not construction products. The Directive applies to products insofar as the Essential Requirements for works relate to them.

Six Essential Requirements are defined:

- Mechanical resistance and stability
- Safety in case of fire
- Hygiene, health and the environment
- Safety in use
- Protection against noise
- Energy economy and heat retention

Products must be suitable for permanent incorporation into construction works which as a whole and in their separate parts are fit for their intended use, account being taken of economy. Essential Requirements must, subject to normal maintenance, be satisfied for an economically reasonable working life.

The Directive defines Interpretative Documents which are to provide the links between the Essential Requirements for works and the resulting requirements for products. These documents, which are being drafted at the time of writing, will be the basis for mandates by the CEC to the Comité Européen de Normalisation (CEN) for harmonised technical standards and also for guidelines for European technical approvals. The documents will take account of the different levels of requirements for works in Member States by establishing classifications of requirements and performance levels.

Construction products shall be presumed to be fit for their intended use if they bear the EC mark. The EC mark indicates that the product complies with the 'technical specifications' which are primarily -

European harmonised technical standards, eg CEN standards
European technical approvals.

There are some products which have less influence on safety with respect to the Essential Requirements. For such products a simplified method of verification of fitness for use based on a test certificate of an approved body is available for this purpose. Other products which have little or no influence on safety may be placed on the market with only a declaration of compliance with the 'acknowledge rule of technology' made by the manufacturer.

European technical approvals are favourable technical assessments of the fitness for use of a product incorporated into a works meeting the

essential requirements. Approval bodies will be designated by the
Member States, eg the British Board of Agrément in the United Kingdom.
Such approvals relate only to products for which there is no European
harmonised standard, national standard, or for which it is considered
no standard can be prepared.

TECHNICAL CODES AND STANDARDS

The 'new approach' to European technical harmonisation and standards
and the Construction Products Directive have provided a major impetus
to the development of the codes and standards for concrete
construction and products. CEN is gearing itself up to adapt and
expand its work in response to the mandates arising from the
Construction Products Directive. Already the transfer of
responsibility for the preparation of the 'Structural' Eurocodes has
been negotiated and a new CEN Technical Committee (TC) is being
established to undertake the work.

EUROCODE EC2: CONCRETE STRUCTURES

Eurocode EC2 applies to the design of buildings and civil engineering
works in plain, reinforced and prestressed concrete. It is concerned
only with stability, strength, serviceability and durability of
structures. Its use will be dependent, amongst other things, on
reference to technical standards for construction products such as
concrete, cement, aggregates, reinforcing steel and prestressing
steel.

The preparation of Part 1 of the Eurocode EC2 was completed in the
work programme of the European Commission prior to the transfer to
CEN. Part 1 gives the general basis for the design of buildings and
civil engineering works in reinforced and prestressed concrete made
with normal weight aggregates. It gives, in addition, detailed rules
for the design of ordinary buildings.

Part 1 does not cover resistance to fire, special types of buildings
or civil engineering works such as bridges, dams and pressure vessels
nor does it cover no-fines concrete and aerated concrete components,
or components containing heavy aggregate or structural steel. Parts
dealing with these and other topics are foreseen for preparation in
the future work on Eurocode EC2 to be undertaken by CEN.

In the meantime the Commission has requested that CEN publish Part 1
of EC2, without technical modification, in the form of a European
Prestandard ENV:EC2 with its appropriate National Implementation
Document. This document will provide:

- a list of interim supporting standards on, for example, concrete
  products to be used with ENV:EC2

- partial safety factors and other data needed to provide levels of safety compatible with current UK national practice, and

- Advice on use.

Where CEN standards for concrete products are available, even if only in draft, Part 1 of EC2 refers to them, eg ENV 206 'Concrete', or it includes drafts as complements, eg for reinforcing and prestressing steel.

TECHNICAL STANDARDS FOR CONCRETE PRODUCTS

CEN already has several Technical Committees (TC) preparing standards on concrete products following CEC mandate to CEN during 1987-89. The more important are:

TC51   Cement and building limes
TC94   Ready-mixed concrete production and delivery
TC104 Concrete (performance, production, placing and compliance criteria)
TC125 Masonry
TC154 Aggregates
TC229 Precast concrete products

Standards for reinforcing and prestressing steels are being prepared by the European Committee for Iron and Steel Standardization (ECISS). Its Committee TC19 has two sub committees:

SC1   Reinforcement   (EN 10080)
SC2   Prestressing steel (EN 10138)

These Technical Committees have each established a range of Working Groups and Task Groups to undertake the detailed technical work. Examples are given below:

Technical Committee TC51 has Working Groups:

WG1   Mechanical strength
WG2   Physical tests
WG3   Chemical tests
WG4   Constituent contents
WG6   Definitions and terminology
WG7   Sampling
WG8   Specifications
WG9   Conformity procedures
WG10 Masonry cement
WG11 Building line

Technical Committees TC94 and TC104 meet jointly and have joint Working Groups and Task Groups:

Working Groups:

WG1   Revision of ENV 206 - revise and transform into European Standard

WG3   Admixtures - definitions, test methods, requirements and quality control

WG4   Fly ash for concrete - definitions, test methods, requirements and quality control

WG5   Mixing water - specifications and test methods

WG6   Grouts for prestressing tendons - specifications, test methods and quality control

WG7   Sheaths for prestressing tendons - definitions, test methods, requirements and quality control

WG8   Protection and repairs of concrete structures

These Working Groups may then be subdivided where the workload is large. For example, WG1 has the following Task Groups:

Task Groups:

TG1   Durability - amendments to durability tables and clauses in ENV206

TG2   Chemically aggressive environments - new document to define chemical aggressiveness, annex to ENV206

TG3   Conformity criteria - statistical aspects

TG4   Strength classes - higher strength classes, classes for lightweight concrete, cube/cylinder conversion factor

TG5   Additions - general principles for use, account taken in water/cement ratio and cement content

TG6   Precast elements - needs of special production methods

TG7   Curing - reconsider requirements in ENV206

TG8   ISO standards (not yet set up) - examine ISO standards for transformation into European Standards

Technical Committee TC125 has Working Groups and Task Groups:

WG1   Masonry units
      WG1/1 Clay units
      WG1/2 Calcium silicate
      WG1/3 Aggregate concrete
      WG1/4 Autoclaved aerated concrete
      WG1/5 Manufactured stone
      WG1/6 Gypsum units
      WG1/7 Natural stone

WG2   Mortar
      WG2/1 Masonry mortar
      WG2/2 Rendering and plastering

WG3   Ancillary components
     WG3/1   Wall ties, straps and hangers
     WG3/2   Lintels
     WG3/3   Bed joint reinforcement

WG4   Testing
     WG4/1   Masonry
     WG4/2   Units
     WG4/3   Mortar
     WG4/4   Ancillary components

EUROPEAN CONCRETE PRACTICE

The extensive organisation servicing the CEN and ECISS Technical Committees serves to emphasise the substantial nature of the work of European technical harmonisation of concrete products. Whilst a superficial view of construction may suggest that practice is broadly similar across Member States, a detailed examination reveals that there are considerable differences in virtually all areas.

These differences have to be identified in order to determine routes to acceptable harmonised standards. For this purpose the Building Research Establishment undertook comparative studies during 1989 of concrete practice in Germany and France. Some differences from United Kingdom practice which appear to exist are:

Cements

The most commonly used German cement is ordinary Portland containing 3% of filler to control bleeding. Cements are produced in different strength grades and blastfurnace slag cements are popular.

French ordinary Portland cements also contain 3% filler but the most common cement contains about 20% limestone filler. Cements are produced in different strength grades. Pulverised fuel ash (pfa) is not such a common ingredient of cements as in Germany and the United Kingdom. Pfa is often added at the mixer in Germany but only as a filler; it cannot be counted as part of the cement content. The addition of blastfurnace slag at the mixer is not permitted. However, additions of slag, pfa or limestone at the mixer are permitted in France but may then only be considered to be part of the fine aggregate. Such additions must be part of the cement as manufactured to be treated as cement.

Water/cement ratio

Water/cement ratio is regarded as a laboratory concept in France and is not included in specifications for concrete.

Suitability testing

Data from a suitability test must be available in Germany, where the concrete grade exceeds C25, to show the concrete is fit for its purpose. For concrete for external use, where water impermeability is often a requirement, a water penetration test is used on standard laboratory specimens to measure susceptibility to penetration of water under pressure. This test may also be used for compliance testing of site cast specimens.

Ready-mixed concrete

All ready-mixed concrete in Germany must be mixed in pan-mixers. Workability is measured during mixing by the current required to drive the mixer paddles. High workability concrete is common. In France slumps are generally high also - in the range 100-150mm compared to 75mm usually in the United Kingdom. Admixtures appear not to be generally used in France. In Germany they are only allowed for their primary effect, ie a combined water-reducing and air-training admixture cannot be used as is sometimes the case in the United Kingdom. About 60% of the ready-mixed concrete industry in France is owned by the cement industry and some 90% of plants used pan-mixing. Placing is generally, but not universally, done by pumping in Germany. Concrete pumping is not common in France.

IMPACT OF TECHNICAL HARMONISATION ON PRACTICE

Development of harmonised standards for concrete products is clearly a difficult task in view of the differences in practice across the European Market. Each Member State has its own traditions and expectations of concrete construction. A wide range of concrete construction practice has evolved in response to local circumstances and experience of what works and what does not work. Practice in each Member State still reflects to a considerable degree the types and qualities of materials available locally for concrete construction and the local conditions. The substantial differences in climate across the Market result in a spread of aggressivity to concrete products which is met by a corresponding range of concrete quality.

These considerations suggest that the task of harmonisation should be tackled primarily by resort to test data, measurement and fundamental science. The weight of local experience cannot generally be a sufficient basis for determining harmonised practice in concrete construction. There is therefore a need for Europe-wide research into concrete, the technology of its constituent materials and of its products, its testing and specification and the design and construction of products built with it. Such research offers the best and perhaps only, prospect of achieving technical harmonisation in the near future. But whatever the route to harmonisation the impact on local concrete construction practice is likely to be significant.

ACKNOWLEDGEMENTS

The author would like to thank Dr B K Marsh for assistance in the preparation of this paper which is published with the permission of the Director of the Building Research Establishment.

THE CONSTRUCTION PRODUCTS DIRECTIVE: IMPLICATIONS FOR SPECIFIERS, SUPPLIERS AND CONTRACTORS

J C Kalra

Department of the Environment,

United Kingdom

ABSTRACT. The European Community has set itself until 1 January 1993 to dismantle the remaining barriers to the free movement of goods, services and capital within the Community. Two initiatives taken by the Community have provided the stimulus to the achievement of this goal - the Single European Act and the 'New Approach' to the structure and operation of Directives.

One of the 'New Approach' Directives - called the Construction Products Directive - was adopted on 21 December 1988 and has to be implemented no later than 27 June 1991.

This paper reviews the implications of this Directive for the construction industry and in particular for specifiers, suppliers, and contractors.

Keywords: Single European Act, New Approach Directives, Construction Products Directive, Public Procurement Directives.

Mr J C Kalra, BSc, MA, CEng, MIStructE is Superintending Civil Engineer and head of Structural Engineering Branch, Building Regulations Division, DOE. He represents UK in the CEC Technical Committee responsible for producing the Interpretative Document for Mechanical Resistance and Stability. He participates in the development of BSI and CEN Codes and Standards (e.g. Structural Eurocodes) relevant to the structural aspects of Building Regulations. He is responsible for providing civil and structural engineering advice to Ministers and directorates in DOE.

INTRODUCTION

The European Community Directives are binding on Member States as to the results to be achieved within a stated period but leave the method of implementation to national governments. In itself a Directive does not have legal force in the Member States but particular provisions may take direct effect if the Directive is not duly implemented.

For many years, the European Community has sought to remove technical and regulatory barriers to trade by adopting Directives setting out the detailed technical and other requirements that products must satisfy in order to be traded freely throughout the Community. Not surprisingly, this has been a slow process because of the complexity of the technical and other issues involved and the difficulty of obtaining the unanimity required by Article 100 of the Treaty of Rome.

NEW APPROACH DIRECTIVES

In order to get away from this traditional approach to Directives, the Community Council of Ministers adopted a Resolution on 7 May 1985 setting out a 'New Approach' to technical harmonisation and standards. The following are the four fundamental principles on which the 'New Approach' is based:

-       legislative harmonisation is limited to the adoption of the essential safety requirements (or other requirements in the general interest);

-       the task of drawing up the technical specifications needed for the production and placing on the market of products conforming to the essential requirements is entrusted to organisations competent in the standardisation area (CEN/CENELEC);

-       these technical specifications maintain their status of voluntary standards;

-       but at the same time national authorities are obliged to recognise that products manufactured in conformity with harmonised standards are presumed to conform to the essential requirements established by the Directive. (This signifies that the producer has the choice of not manufacturing in conformity with the standards but that in this event he has an obligation to prove that his products conform to the essential requirements of the Directive.)

SINGLE EUROPEAN ACT

The political commitment to complete the internal market by
1 January 1993 is now enshrined in the Treaty of Rome through
amendments made by the Single European Act in 1985, which came
into effect on 1 July 1987. The Act introduced a number of
fundamental changes to the way Community law is developed. Of
particular importance is the introduction of Article 100A which
extends qualified majority voting to virtually all the major
areas of the single market programme. In addition, this Article
instructs the European Commission, when developing proposals for
legislation, to take as a base a high level of health, safety,
environmental and consumer protection. The idea of enforcing
mutual recognition of testing and certification appears in
Article 100B.

The Single Market and the Construction Industry

The construction industry is of key economic significance to the
UK and to the rest of the Community. The value of this industry
in the UK is about £35bn a year whereas the Community market is
valued at around £260bn a year. Measured by value, the output per
head of population in the UK industry is below that of both West
Germany and France. On the assumption that construction materials
account for about 40% of construction costs, the Community
construction materials industry is worth more then £100bn a
year.

The scale and significance of the industry is such that it is
hardly surprising that it has been identified by the Commission
for priority attention under the single market programme. A
number of crucial measures have already been agreed. Of most
significance are the Construction Products Directive and, because
of the common operational currency, the ongoing measures to
promote increased competition for 'public sector' contracts.

THE CONSTRUCTION PRODUCTS DIRECTIVE

Formal Title

'Council Directive of 21 December 1988 on the approximation of
laws, regulations and administrative provisions of the Member
States relating to construction products (89/106/EEC).'

General

The Construction Products Directive was formally adopted on
21 December 1988 and its provisions will come into effect

throughout the Community by no later than 27 June 1991. This means that individual Governments must now take whatever legislative action is necessary to bring the Directive into effect, ensure that there are no conflicts with existing national legislation and to notify the European Commission what they have done. There is absolutely nothing to stop individual Governments implementing the Directive in advance of the implementation deadline but in doing so they cannot require reciprocal treatment by others. Aside from the legal obligation to implement the Directive by the due date there are likely to be major commercial advantages in ensuring that legislative implementation and the accompanying attestation infrastructure are in place as quickly as possible.

Primary Purpose

The primary purpose of the Construction Products Directive is (a) to establish mechanisms for free trade in qualifying construction products in order to promote competition and increase choice within the market place and (b) to safeguard essential health and safety requirements. The Directive is substantially undeveloped in a number of key areas and a great deal of additional work needs to be done both at Community and also domestic level to facilitate the implementation of the Directive.

Scope

The field of application is construction products, defined as products which are produced for incorporation in a permanent manner in construction works, including both buildings and civil engineering works, in so far as the essential requirements relate to them and when and where the regulations contain such requirements.

The Standing Committee

The Directive will be managed by a Standing Committee of representatives from Member States. The UK is represented by Mr Ian Macpherson and Mr Oliver Palmer from the DOE and Mr Anthony Davies, an industrial advisor who is working with the DOE. The Committee will examine any questions posed by the implementation and the practical application of the Directive.

Main Provisions

Article 2.1 of the Directive uses certain phrases the meaning of which is not so obvious. For ease of reference, the text of Article 2.1 is reproduced here.

'Member States shall take all necessary measures to ensure that the products referred to in Article 1, which are intended for use in works, may be placed on the market only if they are fit for their intended use, that is to say they have such characteristics that the works in which they are to be incorporated, assembled, applied or installed, can, if properly designed and built, satisfy the essential requirements referred to in Article 3 when and where such works are subject to regulations containing such requirements.'

The implication of Article 2.1 is that the essential requirements relate to the works and only indirectly to construction products, but they will influence the characteristics of those products. The Directive lays down no requirement that the works have to be properly designed and built nor that the works have to meet any or all of the essential requirements. The fitness of products for their intended use in the works has to be judged as if the works needed to meet the relevant essential requirements, and were to be properly designed and built.

The Essential Requirements

The essential requirements are given in Annex 1 of the Directive and concern - Mechanical resistance and stability; Safety in case of fire; Hygiene, health and the environment; Safety in use; Protection against noise; Energy economy and heat retention.

Part of the preamble to Annex 1 states:

'Such requirements must, subject to normal maintenance, be satisfied for an economically reasonable working life.'

Interpretative Documents

Much of the confusion over the application of the essential requirements arises out of the indirect relationship between the characteristics of products and the essential requirements applicable to works. In Articles 3 and 12, the Directive provides for the preparation of Interpretative Documents (one for each essential requirement) which would amplify the essential requirements and identify characteristics of products so as to enable mandates for harmonised standards to be issued to CEN/CENELEC and guidelines for European technical approval to the European Organisation for Testing and Certification (EOTC).

Considerable difficulties are being encountered in the
preparation of Interpretative Documents because models for such
documents do not exist in any Member State. The Interpretative
Documents are due to be finalised by the end of 1990.

Presumption to Conformity

For products to be presumed fit for their intended use in the
works in any Member State, they must comply, in accordance with
Article 4 of the Directive, with the relevant national standards
transposing the harmonised standards produced by CEN/CENELEC or a
European technical approval issued by one of the designated
bodies in accordance with the Directive or national technical
specifications recognised at Community level. Such products bear
the CE mark.

Products which play a minor part with respect to health and
safety and in respect of which a declaration of compliance with
acknowledged rule of technology is issued by the manufacturer
may be placed on the market but these products will not bear the
CE mark. A list of 'minor part' products is being prepared by the
Standing Committee.

For products which do not conform to the requirements of
Article 4 of the Directive, onus is on the manufacturer to prove
that such products are 'fit for their intended use'.

The Standing Committee has also identified the need to permit a
Member State to admit onto the market in its territory products
which have only a local or regional importance in the Member
State concerned, without being subject to harmonised technical
specifications.

Content of Technical Specifications

In order to satisfy the requirements of the Directive, the
European technical specifications will satisfy relevant essential
requirements - but they can do more. Specifications which only
satisfy the requirements of national regulators may not address
issues of usability etc and may not be acceptable in the market
place with the result that manufacturers and specifiers would
probably want to add to the specification and thereby recreate
the obstacles to free trade that the Directive is intended to
remove. A Community/EFTA (CEN membership includes EFTA standard
making bodies) solution to this problem has yet to be determined
but it is possible that CEN/CENELEC might produce graded
standards which would include a minimum grade limited to the
essential requirements (mandated by the Commission) with the
development of additional grades, addressing issues of usability

and other factors of vital importance to manufacturers and specifiers, left to the discretion of CEN/CENELEC.

Design and Execution

Since 1976, the European Commission has been sponsoring work on design codes. The Structural Eurocodes are indended to harmonise design rules for structures in the various materials like concrete, steel, timber, masonry etc. Work on eight Structural Eurocodes made up of individual parts dealing with buildings, bridges etc is in progress. In keeping with the spirit of the 'New Approach' Resolution, the Commission transferred the work on further development and issue of these Eurocodes to CEN in March 1990.

The status of the Eurocodes within the terms of the Construction Products Directive needs examination. If the Directive does not require that the works have to meet any or all of the essential requirements, why develop rules for design and execution?

The simple answer is that certain products may cover design or execution aspects when they are placed on the market e.g. precast flooring beams or piles, proprietary trusses or pre-designed components, assemblies or systems. The attestation of conformity procedures laid down by the Construction Products Directive may involve the certification of design or execution aspects of these products. The Structural Eurocodes will be the harmonised standards against which the design aspects of these products will be judged as to their fitness for intended use. Because the products may relate to the superstructure or the substructure of the works, the Structural Eurocodes should cover the complete structure of the works.

It is possible that certain authorities in Member States may require, through regulation or administrative provisions, certain works to meet any or all of the essential requirements. In such cases, compliance with the relevant Structural Eurocodes will offer a way of meeting the essential requirement for Mechanical Resistance and Stability.

Transitional Arrangements

In ideal circumstances, the Interpretative Documents should have preceded the drawing up of harmonised standards and guidelines for European technical approval. In reality, the Interpretative Documents have still not been finalised. When the Directive comes into force on 27 June 1991, hardly any harmonised product standards would achieve an EN status. It may take three to five years before a respectable number of harmonised standards achieve

this status. The situation regarding European technical approvals
is not likely to be much better. In Article 6, the Directive
allows transitional arrangements, whereby, Member States will
allow products to be placed on the market in their territory if
they satisfy national provisions consistent with the
Treaty unless the technical specifications referred to in
Article 4 of the Directive provide otherwise. The use of a mixed
bag of harmonised standards, European Technical approvals and
existing national standards presents risks to specifiers,
suppliers and contractors who will have to ensure their
compatibility in order to safeguard their respective liabilities.
A complete package of harmonised standards and European Technical
approvals necessary for a project may not become available for
eight to ten years.

Testing and Certification

Under present domestic arrangements, independent testing of
product performance and independent assessment of production
systems is an entirely voluntary matter between the supplier and
the client. However, the Construction Products Directive provides
for mandatory 'third party' product assessment in certain, as yet
undefined situations. In addition the existence of a production
management system - described in the Directive as "factory
production control" is going to be a requirement for all products
bearing the CE mark although this will not always be the subject
of third party assessment. The Directive includes a menu of
attestation regimes but decisions on the procedures to apply in
individual cases have yet to be determined.

In order to establish the domestic infrastructure to provide for
mandatory attestation where required, Member States are required
to designate test houses and certification bodies to provide
"third-party" services. The DOE is currently developing
proposals, which are likely to require conformity with the
relevant European Standards, using the NAMAS and NACCB
arrangements wherever possible. The British Board of Agrément
has been designated already as an approved body to issue European
technical approvals under the terms of the Construction Products
Directive and to act as the UK spokesbody in the European
Organisation for Technical Approvals.

It is intended that the UK's voluntary system of third-party
certification should continue to be available to manufacturers of
construction products. Subject to the demands of the market
place, it is likely that certification of management systems will
continue to be offered on the basis of EN 29000. This will by
definition be deemed to satisfy the requirements of the
Construction Products Directive.

Work is currently being done to develop a horizontal approach which will apply to testing and certification requirements of all the Directives, including the Construction Products Directive.

PUBLIC PROCUREMENT DIRECTIVES

Background

Public procurement is a key element of the single market, both on account of the scale of expenditure and also because of its perceived susceptibility to political or other pressures to favour the domestic supplier or contractor. The total spent on public procurement across the Community is in excess of £300bn per annum (15% of Community GDP). Spending on construction accounts for about 30% of this.

The Directives

The Public Works Directive of 1971 and the Public Supplies Directive of 1977 have had some success in ensuring that the contract letting process is more visible but in terms of contracts let to non-national firms the effect of the Directives has been minimal.

Revised Directives have been developed to improve the flow of information available to potential bidders and at the same time to further limit the opportunities for authorities to favour domestic suppliers and contractors. Revisions include a requirement to specify European specifications where appropriate. There is very limited scope for derogation. A revised Public Supplies Directive, incorporating this provision, came into effect on 1 January 1989. A similarly structured revision to the Public Works Directive was adopted on 18 July 1989 and will come into effect in July 1990. These Directives require public sector specifiers to refer to European Standards where available or alternatively European Technical Approvals and only refer to national specifications in default. It is clear that the product specifications developed to support the Construction Products Directive will in effect be mandatory under the provisions of these revised Public Procurement procedures.

The Commission has produced a draft Utilities Directive which, if adopted, would apply some of the provisions of the Works and Supplies Directives including those applying to technical specifications to the mix of public and private bodies operating in the so called Excluded Sectors - energy, transport, water and telecommunications.

Concluding Remarks

Even though the Construction Products Directive has completed its legislative stages through the Community Institutions it is far from being a workable piece of legislation. Much work needs to be done in a number of critical areas including:

-   development of Interpretative Documents

-   development of standardisation mandates

-   development of mandates for European technical approvals

-   determination of attestation procedures

-   operation of safeguard and appeals machinery.

EUROCODE NO. 2 - CONCRETE STRUCTURES

R S Narayanan

S.B. Tietz & Partners,

United Kingdom

ABSTRACT. This paper summarises the design requirements of the proposed EC2. It is concerned with the needs of the designer involved in routine structures. The details are introduced by way of comparison between the Eurocode and the current British Code of Practice BS8110. Some comparative tables are also included.

Keywords: Partial safety factors, redistribution of moments, durability, bending, shear, torsion, punching shear, buckling, deformation, detailing, minimum reinforcement and prestressed concrete.

Nary Narayanan is a Partner of SB Tietz & Partners, Consulting Civil Structural and Traffic Engineers in London. He has been a practising designer for nearly 30 years and his experience includes heavy civil engineering structures in India and a variety of projects in the UK. He is the UK Liaison Engineer on the EC2 Committee and has close knowledge of the development of the Code. He serves on a number of technical committees in the UK including British Standards Committees on Concrete, Masonry and Wind Loading. He is also a member of the British Reinforcement Commission.

INTRODUCTION

By the time this paper is published, EC2 Part 1, is likely to have been published as an ENV (Pre Standard). It is likely to remain an ENV for about three years, when the member nations are expected to familiarize themselves with the Code. Thereafter, it is likely to be published as a full EN (Euronorm).

While EC2 deals with design requirements, it relies on ENV 206 for matters concerned with production and placing of concrete.

This paper highlights the design requirements of EC2. For ease of appreciation comparisons are made with BS8110.

SCOPE

EC2 will be published in a number of parts, dealing with different types of structures and construction as noted below:

Part 1	–	RC and prestressed concrete in buildings
Part 1A	–	Plain concrete
Part 1B	–	Precast concrete
Part 1C	–	Lightweight concrete
Part 1D	–	Prestressed concrete using unbonded or external tendons
Part 2	–	RC & prestressed concrete in bridge structures
Part 3	–	Concrete foundations
Part 4	–	Liquid retaining structures
Part 5	–	Temporary structures
Part 6	–	Massive civil engineering structures

Part 1 will contain 7 chapters. Four appendices will deal with matters only required occasionally in a design office.

GENERAL

Layout

EC2 will comprise principles and rules of application. Principles are general statements, definitions, other requirements and analytical models for which no alternative is permitted. The rules of application are, generally recognised rules which follow the principles and satisfy their requirements. It is permissible to use alternative design rules.

The Clause which are principles are prefixed with 'P'.

A number of numerical values will appear within boxes and these are meant to be for guidance only. Each member state is required to fix the values which will apply within its jurisdiction. The long term aim is to harmonise the values across Europe as much as possible.

Engineers in the UK will need to become familiar with some new expressions. 'Loads' are referred to as 'actions' 'bending moments' and 'shear forces' are called 'internal forces and moments', and 'superimposed loads' are 'variable loads', and 'self-weight' and 'dead loads' are referred to as permanent loads'.

EC2 uses limit state principles.

Concrete strength in EC2 refers to the cylinder strength (f_{ck}). Relationship between cylinder and cube strengths is given at the beginning of the Code.

Frame Analysis

There are slight differences in the partial safety factors for loads. EC2 uses factors of 1.35 for dead Loads and 1.5 for imposed loads. Corresponding values in BS8110 are 1.4 and 1.6 (see Table 1).

In EC2 partial safety factors for materials, are 1.5 for concrete and 1.15 for reinforcement and prestressing steel. BS8110 adopts 1.15 for reinforcement and for concrete it uses 1.5 in bending, 1.25 in shear and 1.4 for bond.

For continuous beams and slabs without cantilevers subjected dominantly to uniformly distributed loads, EC2 recommends consideration of two load cases (viz):

a)   alternate spans carrying the full factored dead and imposed load with other spans carrying the factored dead load;

b)   any two adjacent spans carrying the full factored dead and imposed loads with other spans carrying factored dead load.

BS8110 while adopting the first load case uses 'all spans loaded' as the second load case. Also in BS8110 the load factor on the dead load is reduced to 1 for the spans not loaded with superimposed loading.

BS8110 will give about 8-15% more span moments, while the moment at the penultimate support will be almost identical. Moments over internal supports according to BS8110 will be about 85% of the EC2 moments.

Minimum Design Horizontal Force

There are striking differences here. EC2 wants the designer to assume an artificial inclination of the structure and assess the effect of this imperfection on the overall stability. The deviation from the vertical may be replaced at each floor level by equivalent horizontal forces. The effect of these horizontal loads is then compared with that of other horizontal loads (such as wind loads) to identify the critical load.

BS8110 requires the structure to be designed for 1.5% of the characteristic dead load above any one level or factored horizontal design loads, whichever is greater.

The EC2 approach will generally give about a third of the notional force recommended in BS8110.

Redistribution of Moments in Continuous Structures

EC2 permits redistribution of moments in non-sway structures. The maximum redistribution permitted is related to the ductility of reinforcement ie whether it was high or low ductility. The ratio of the redistributed moment to the moment before the distribution is limited to 70% for the former and 85% for the latter.

Explicit check on rotation capacity of critical sections is not required, provided certain conditions are met. These essentially relate the amount of redistribution to the neutral axis depth. EC2 requirements are more severe in this respect, compared to BS8110.

EC2 also limits the depth of the neutral axis to 0.4d compared to 0.5d in BS8110 (d is the effective depth).

While no redistribution is permitted in sway frames in EC2, up to 10% redistribution is permitted in BS8110.

In a member, designed using plastic analysis, EC2 sets an upper limit to the amount of reinforcement. It restricts the reinforcement (by limiting the neutral axis depth) to ensure that the member has sufficient ductility to realise the assumptions of plastic analysis.

There is no corresponding requirement in BS8110.

Durability

Like BS8110, EC2 has a number of durability requirements. Designer should also refer to ENV 206. Nine exposure classes in ENV 206 correspond to the first four exposure conditions of BS8110. There is no class in EC2 corresponding to the 'Extreme' class of BS8110. ENV 206 stipulates the minimum cement content and maximum water/cement ratio for each exposure class but does not give the concrete grade deemed to satisfy these two separate requirements. Minimum cover to reinforcement and prestressing steel is given and these are boxed. Tolerance to allow for workmanship deficiencies should be added to these values. Unlike BS8110, EC2 does not permit the reduction in cover if concrete of better quality than the specified minimum is used. The general requirements of the two codes are summarised in Table 3.

For the various exposure classes water/cement ratios are the same in both the codes. Minimum cement content and nominal cover to reinforcement are slightly less onerous in EC2. In practice concrete grades are unlikely to be much different when designing to EC2.

Fire Resistance

The current draft does not have any fire requirements. An appendix is to be added in due course.

ULTIMATE LIMIT STATES

Bending and Longitudinal Forces

The basic design assumptions in EC2 are almost identical to those in BS8110. However, the simplified stress blocks in the two codes are different see Figure 1. In EC2 the concrete compression is taken as $(0.85f_{ck}/\gamma_c)$. The depth of the compression block is limited to $(0.8\ X)$, where X is the depth of the neutral axis. These assumptions together with the limitation of the neutral axis to $(0.4d)$ gives rise to discrepancies between the two codes when the moment of resistance of the section is calculated. BS8110 will give about 25% greater capacity compared to EC2.

Shear

Applied shear force (V_{sd}) is compared with three values for the resistance (V_{Rd}).

V_{Rd1} represents the shear capacity of the concrete alone; V_{Rd2} is the shear resistance determined by the capacity of the notional concrete struts; and V_{Rd3} is the capacity of a section with shear reinforcement.

In the calculation of V_{Rd1} account is taken of depth of the member percentage of longitudinal steel, presence of any axial force and concrete strength. There is a requirement in EC2 which is not in BS8110 (viz) that the depth correction factor should be taken as 1 in members which use curtailed reinforcement. Apparently this is meant to cover cases where more than 50% of the bottom bars have meant to cover cases where more than 50% of the bottom bars have been curtailed.

If V_{sd} is less than V_{Rd1} nominal shear reinforcement is required except in slabs and members of minor importance where shear reinforcement may be omitted.

EC2 tends to give lower values for resistance for all but low percentages of reinforcement. This is partly explained by the adoption of a lower load factor in shear in BS8110. Thus a few members which would not have been reinforced for shear according to BS8110 may need reinforcement if designed in accordance with EC2.

Once shear reinforcement has been found to be necessary, EC2 provides two alternative methods for the calculation of the reinforcement. One is called the 'standard method' which is similar to BS8110 where notional concrete struts are assumed to be at 45°. The contribution of the shear reinforcement is calculated in much the same way in both the codes. V_{Rd3} is the sum of the contributions of concrete and steel.

In the alternative, the 'variable strut inclination method', the designer is permitted to assume an inclination between about 22° and 68° for the concrete struts. Adopting this method the shear reinforcement can be reduced depending on the angle assumed. However, in this method all the shear is resisted by reinforcement, ignoring the concrete contribution.

For sections close to the support enhancement of the shear resistance is recognised; but unlike BS8100, rather than enhancing the resistance, the applied shear is reduced.

Even when shear reinforcement is provided, the applied shear should be less than V_{Rd2}. This is similar to limiting the maximum shear stress to $0.8 f_{cu}$ in BS8110. There is good correspondence between the codes in this respect.

Torsion

Like BS8110, EC2 requires torsion to be considered only when static equilibrium of the structure depends on the torsional resistance of the element.

There are differences between the two codes in the calculation of the torsional resistance of members. EC2 requires the cross sections to be transformed into idealised thin walled closed sections and proceeds to give formulae for the calculation of the resistance of thin walled sections. BS8110 uses plastic distribution of shear stress across the whole cross section. Both the codes set limits on the maximum torsion permitted even with reinforcement.

Punching Shear

The method of calculating the punching shear resistance is largely
similar in both the codes. The critical perimeter is taken at 1.5d
from the face of the load in both the codes (d is the effective
depth). EC2 uses rounded corners whereas BS8110 does not. The
method of dealing with moment transfer is presented differently in
the two codes but the effective shear force to be considered is
about the same in both codes. The threshold at which the slab needs
to be reinforced for punching shear in EC2 is about 70% of the
BS8110 values. Again the use of a higher partial safety factor in
EC2 explains part of the discrepancy.

Buckling

This chapter generally corresponds to the chapter on columns in
BS8110 but EC2 also covers slender beams in this chapter. Although
EC2 apparently uses a slightly more complex procedure, the end
result is remarkably similar in both the codes and in fact EC2 is
marginally more economic. The step by step procedure is shown on
Table 3.

SERVICEABILITY LIMIT STATES

Limitation of Stresses and Their Serviceability Conditions

This check is to prevent formation of longitudinal cracks and micro
cracking in members with excessive compressive stress under service
load. There is no corresponding requirement for R.C. structure in
BS8110. Limits on the stresses have been given. There is also a
dispensation for not carrying out these checks if certain conditions
are met.

In general there would be no need to carry out this separate check.

Cracking

EC2 gives a procedure for calculating the minimum reinforcement to
prevent uncontrolled cracking. Distinction is made between those
cracks caused dominantly by restraint and those caused dominantly by
loading. Tables are provided for limiting bar spacings and bar
diameters to limit crack widths without performing the crack width
calculations.

The requirements are slightly more onerous in EC2 compared with
BS8110.

Deformation

As in BS8110, span/depth table is provided. There is also an option

for calculating the deflection and guidance on this is given in an
appendix.

The EC2 span/depth ratio table is simpler than in BS8110. For each
type of member it provides two values, one for highly stressed
members and another for lightly stressed members. In this context
members with less than 0.5% of reinforcement are considered lightly
stresses and members with 1.5% of reinforcement are considered
highly stressed. Interpolation and extrapolation are permitted.

Generally the EC2 values are marginally higher than those given in
the British Code and slightly shallower construction would be
possible.

Detailing

Formulae for calculating basic lap lengths are similar to that in
BS8110 but EC2 uses marginally lower bond stress. It should also be
remembered that the material safety factor in BS8110 is 1.4 whereas
it is 1.5 in EC2. Basic lap lengths in 'good' bond conditions are
virtually identical for type 2 deformed bars in both codes but EC2
demands slightly longer laps for plain bars as it adopts lower bond
stresses. For fabrics EC2 only gives guidance for welded mesh made
of ribbed wires.

EC2 states that good bond conditions will apply: a) to the whole
member if it is 250mm or less in thickness; and b) for members
thicker than 250mm to the lower half (as cast) of the member or to
the zone below 300mm from the top surface if this is greater. Zones
where good bond conditions do not apply are considered to have poor
bond conditions and the calculated lap lengths will need to be
increased by about 40%.

Also EC2 requires additional lap lengths depending on various
conditions related to cover to the bars, spacing of bars and
percentage of bars lapped at any one section. These are somewhat
similar to the BS8110 requirements which are based on different
parameters see Figure 2.

When bars are curtailed, EC2 requires the theoretical section at
which the anchorage begins, to be displaced by a length which is
approximately equal to half the depth of the member. Thus the total
length of bars in EC2 is likely to be marginally greater compared to
BS8110.

Minimum Percentages of Reinforcement

There is hardly any difference between the two codes in the minimum
longitudinal reinforcement in beams and slabs. For shear
reinforcement EC2 gives percentages depending on the concrete
strength. The stronger the concrete the higher the percentage. In
EC2 the maximum spacing of shear reinforcement is related to the
ratio of V_{sd}/V_{Rd2} and it ranges from 0.8d for lower ratios and 0.3d
for higher ratios. In BS8110 there is a blanket maximum spacing of
0.75d.

For columns EC2 requires the reinforcement to be not less than 0.3%
and not greater than 8% (even at laps). BS8110 gives the minimum
value as 0.4% and the maximum value of up to 10% at laps. There is
an overriding requirement in EC2 that the reinforcement alone
should be able to carry at least 15% of the applied load.

Limitation of Damage Due to Accidental Forces

The requirements here are similar to BS8110 although detailed
information on the tie forces is not given in EC2.

PRESTRESSED CONCRETE

EC2 only gives general principles for the design of prestressed
concrete members. BS8110 gives more detailed guidance on a number
of items, eg design of beam.

BS8110 classifies prestressed members with respect to service
conditions. EC2 does not have a similar classification. It
requires the crack width to be limited to 0.2mm for exposure class 2
(moderate and severe classification in BS8110) for post-tensioned
work. For more severe conditions it requires the tendons to lie at
least 25mm within concrete under compression or wants the designer
to consider coating of the tendons and limiting the crack width to
0.2mm.

In the calculation of shear resistance EC2 does not distinguish
between cracked and uncracked sections as does BS8110. It increases
the contribution of concrete to shear resistance by 15% of the
stress caused by the axial force.

Thus in EC2 much is left to the designer and there is more freedom
in design compared to BS8110.

CONCLUSIONS

It is difficult to generalise the likely overall impact of using
EC2. Marginally shallower beams and slabs may be possible with the
higher span/depth ratios for members. Members subject to flexure

may require slightly more reinforcement because of the reduced
moment of resistance of the concrete associated with the simplified
stress block.   Where it is required, shear reinforcement is unlikely
to be significantly different between the codes although some
members not reinforced for shear now may require shear
reinforcement.   Slabs subject to punching shear are likely to need
reinforcing at a lower load compared to BS8110.   Columns are
marginally more economical with EC2.   Lap lengths are essentially
the same for deformed bars and are marginally longer in EC2 for
plain bars.   The shift rule for detailing will increase the total
length of bars used.   Durability requirements are slightly less
onerous in EC2.   Concrete grades to be used are unlikely to be
different.

TABLE 1   Ultimate limit state partial factors to EC2

Load combination		Load Type					
		Permanent		Variable		Earth and Water	Wind	Accidental
		Beneficial	Adverse	Beneficial	Adverse			
1)	Permanent + variable	1	1.35	0	1.5*	1.35	-	-
2)	Permanent + wind	1	1.35	-	-	1.35	1.5	-
3)	Permanent + variable + wind	1	1.35	0	1.35	1.35	1.35	-
4)	Permanent + variable + accidental + wind	1	1.05	0	0.3**	1.05	0.3	1

*   This may be reduced to 1.35, when considering the simultaneous effects of more than one type of variable load on the structure.

**   This should be increased to 1.05, in buildings used predominantly for storage.

Partial factors for materials for the ultimate limit state - EC2

Load Combination	Concrete c	Reinforcement or prestressing steel s
Combinations (1)-(3) in Table above	|1.5|	|1.15|
Combination (4) in Table above	|1.3|	|1.0|

TABLE 2   Durability

| | EC2 | | | | | BS 8110 | | | |
Exposure class	Minimum cement content	Max W/C Content	Nominal cover[1]	Likely Concrete Grade in UK[2]	Environment	Minimum Cement Content	Max W/C ratio	Nominal Cover[1]	Concrete grade
1.	260	0.65	20	C30	Mild	275	0.65	25	C30
2a	280	0.60	25	C35	Moderate	300	0.60	35	C35
2b	280	0.55	30	C40	Severe	325	0.55	40	C40
3.	300	0.50	45	C40	Very				
4a	300	0.55	45	C40	Severe	325	0.55	50	C40
4b	300	0.50	45	C40					

1) 5mm tolerance has been assumed.

2) Based on the June 1988 draft of ENV 206 and on the assumption that in the UK OP cements fall into CE 42.5 class.

TABLE 3   Buckling: Flow chart for design

1) Classify the structure as braced or non-braced.

2) Classify the structure as sway or non-sway.

3) Calculate the effective height of columns.

4) Test for slenderness - column slender if $\lambda > \lambda crit$

   Note: In EC2 slenderness is based on radius of gyration.

5) If column is non-slender, design for axial load and first order eccentricity but not less than (h/20).

6) If column is slender, use the following procedure.

7) Calculate the various eccentricities:

   1 -Additional eccentricity to allow for imperfections;

   2 -First order eccentricities;

   -Effective first order eccentricity for calculation at mid-height;

   3 -Second order eccentricity.

8) 1 -Design the column for axial load and BM at top or bottom.

   2 -Design the column for axial load and BM at the middle of the column.

   3 -For rectangular column bent about one axis only, bending about the other axis should also be checked separately.

   4 -For columns subjected to biaxial bending, separate checks about each axis are permissible under certain conditions.

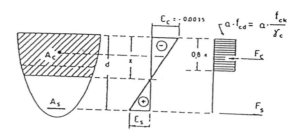

Rectangular diagram - EC 2

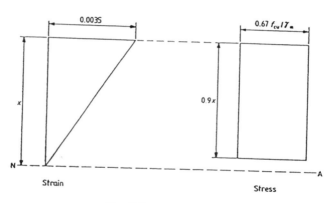

Strain   Stress

Simplified stress block - BS 8110

Stress block and strain

EC2		BS 8110	
fck	$\dfrac{\alpha fck}{rc}$	fcu	$\dfrac{.67fcu}{rm}$
20	11.33	25	11.17
25	14.17	30	13.40
30	17.00	37	16.53

Concrete resisting moment

EC2		BS8110	
fck	$\dfrac{M}{bd^2}$	fcu	$\dfrac{M}{bd^2}$
20	3.05	25	3.90
25	3.81	30	4.67
30	4.57	37	5.76

Figure 1: Simplified stress blocks

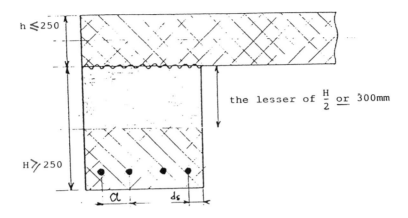

 good bond conditions   poor bond conditions

Tension bars

Condition 1 : ds > 5·D
Condition 2 : a > 10·D
Condition 3 : less than 30% of the bars in the section are lapped in one location.

Condition 1	Yes	Yes	Yes	No	No
Condition 2	Yes	Yes	No	Yes	No
Condition 3	Yes	No	Yes	Yes	No
	1·lb	1.4·lb	1.4·lb	1.4·lb	2·lb
	1.4·lb	2·lb	2·lb	2·lb	2.8·lb

Bars in compression

Lap length = normal lap length

Figure 2: Modification of basic lap lengths - EC2

CARBONATION, CORROSION AND STANDARDIZATION

L J Parrott

British Cement Association

United Kingdom

ABSTRACT. Carbonation and corrosion were examined in a wide range of concretes cured and exposed in a variety of ways. Water/cement ratio was a critical variable but cement type, curing and exposure had substantial effects. Carbonation correlated with various measures of strength and with initial weight changes during exposure. It was controlled by diffusion of carbon dioxide through the pores in the carbonated concrete surface that had been emptied by drying. The rate of reinforcement corrosion was related to the unneutralized remainder i.e. the distance between the steel and the plane of phenolphthalein neutralization. As the unneutralized remainder approached zero and became negative the rate of corrosion greatly increased. Control in design codes of carbonation and related corrosion would be more effective if based upon a measure of concrete performance rather than on specification of mix proportions, particularly if cements contain pulverised fuel ash or ground granulated blastfurnace slag.

Keywords: Carbonation, corrosion, mix proportions, cement type, curing, exposure, strength, standards.

Leslie Parrott received his bachelors and doctors degrees in civil engineering from London University. Over the last 20 years he has produced about 90 publications on various aspects of concrete performance including creep, shrinkage, elasticity, strength, internal moisture, permeability and carbonation. Many of the property measurements have been paralleled by examinations of cement hydration, porosity and other aspects of microstructure. Dr Parrott is currently active on several international committees concerned with permeability, durability, moisture and modelling of microstructure.

INTRODUCTION

Standards

The European draft code for concrete structures, EC2(1) deals with durability by indicating minimum cover requirements for each of nine exposure classes. The choice of cement and concrete for a particular exposure class is delegated to another document ENV206, Concrete : Performance, production, placing and compliance criteria(2). Currently ENV206 specifies the concrete requirements for durability mainly in terms of a maximum water/cement ratio and a minimum cement content. Due to lack of information the choice of cement for a given exposure class may be based upon local regulations. The current range of cements is detailed in prENV197, Cement : Composition, specifications and conformity criteria(3). The position for durability is summarized in Table 1; some of the broader issues of European codification for concrete are discussed in references 4 to 6.

Table 1 European durability requirements

Exposure class	EC2		EN206			
	Min cover (mm)		Max w/c ratio		Min cement kg/m^3	
	Reinf.	Prestr.	Reinf.	Prestr.	Reinf.	Prestr.
1 Dry	15	25	0.65	0.60	260	300
2 Humid, a) no frost	20	30	0.60	0.60	280	300
b) frost	25	35	0.55	0.55	280	300
3 2b + Deicing salts	40	50	0.50	0.50	300	300
4[+] Seawater, a) no frost	40	50	0.55	0.55	300	300
b) frost	40	50	0.50	0.50	300	300
5[+] Aggressive a) slightly chemical	25	35	0.55	0.55	280	300
b) moderately	30	40	0.50	0.50	300	300
c) highly	40	50	0.45	0.45	300	300

[+] Sulphate resistant cement if sulphate > 500 mg/kg in water.

The wide range of cements that can be used for each exposure class might lead the user of EC2 to believe that for given mix proportions concretes made with each cement yield comparable performance with regard to durability. The results in the present report plus other data from the literature suggest that this may not be true for carbonation and that some type of performance criteria would be desirable. Furthermore the results suggest that classification of exposure may require refinement if carbonation and corrosion are to be realistically accounted for.

The treatment of Durability in EC2 and ENV206 will not be dictated by data on carbonation alone and, during the three year period in which ENV206 has provisional status, additional test data relating to other aspects of durability will emerge and contribute to a balanced treatment of durability in the final documents.

Carbonation(7)

A considerable amount of data on carbonation has been published but a recent review (7) leads to a fairly simple overview. Carbonation involves diffusion of atmospheric carbon dioxide through the pores in carbonated concrete and a reaction with the underlying uncarbonated or partially carbonated concrete. The rate of carbon dioxide diffusion is greater in more porous concrete but is retarded if the concrete is moist and the pores are partially blocked. However carbonation requires a certain amount of moisture to dissolve the carbon dioxide and thus a maximum rate can be observed for exposure relative humidities around 60%. The rate of carbonation is also a function of the quantity of cement hydrates that are able to react with the incoming carbon dioxide and act as a chemical buffer. The engineering consequences of these processes are that carbonation increases with an increase in water/cement ratio, a reduction in curing time and with drying, at least where relative humidities do not drop to very low levels. The position with regard to cements depends upon how test programmes were formulated. If concretes are compared on the basis of equal strength then concretes made with Portland cements plus mineral admixtures of the type permitted in ENV197 carbonate at a rate similar to or slightly greater than those made with a pure Portland cement. The equal strength basis for comparison is representative of current practice in the UK Construction Industry but the proposed European code implies comparison on the basis similar mix proportions. Concretes made with cements that contain mineral admixtures then carbonate at a greater rate (7).

The carbonation of concrete causes a reduction of alkalinity in the pore liquid and this leaves embedded steel susceptible to corrosion. The process of steel corrosion in carbonated concrete is a function of concrete resistivity or moisture content and when the relative humidity around the steel is below 70% corrosion rates are small. The rate of corrosion rises with an increase of internal relative humidity until pores become sufficiently blocked with condensed liquid to prevent diffusion to the steel of the oxygen necessary for the corrosion reaction. The relative humidity may then be in excess of 95%.

EXPERIMENTAL

Scope

The experimental work reported here is one part of a BCA programme on the carbonation of concrete. The reported work examines the effects of four cements, five water/cement ratios, three periods of moist curing and five exposure conditions upon carbonation and corrosion using thirty combinations of the main variables, as summarized in Table 2. Apart from carbonation and corrosion measurements, parallel data relating to compressive strength development, porosity and moisture changes during drying were obtained. Carbonation and corrosion measurements were undertaken at ages of 180 and 545 days. Some of the 545 day tests have not yet been completed so the main focus of attention is on the 180 day results.

The cements used were blended from the materials detailed in Table 3a. Table 3b indicates the blend proportions and the cement types according to prENV197; the cements were each in the 42.5 strength class due to the high reactivity of the Portland cement that was used.

Measurements

Twenty one, 100 mm concrete cubes were cast for each series. Five of the cubes were sealed on five faces at the end of the curing period so that moisture exchange was uniaxial. One of these partially sealed cubes was made with 7mm diameter cavities parallel to the exposed face. The cavities housed small cement paste prisms made with the same water/cement ratio as the concrete so that the moisture content of the binder phase could be estimated gravimetrically. A small probe was used to measure relative humidity in the cavities at appropriate times. These techniques have been described in more detail in reference 8.

Four of the partially sealed cubes were cast with four, 6.4mm diameter mild steel reinforcing rods parallel to and at 4, 8, 12 and 20mm from the exposed face. In most series the steel rods were treated with a steel cleaning fluid prior to embedment. The cubes were weighed periodically during the exposure following curing and then, at the appropriate test age, they were wetted to allow measurement of the rate of absorption and to stimulate corrosion of the embedded steel. The wetting involved an initial period of four days when the exposed face was immersed in water to a depth of about 1mm and absorption data could be obtained. The second stage involved 24 days in a saturated water vapour environment. After this second stage the cubes were dried and split so that six estimates of the depth of carbonation could be made after spraying a phenolphthalein indicator solution on the fractured concrete surfaces. The phenolphthalein method of measurement will not detect regions of partial carbonation that can sometimes be observed using thermal analysis (9,10). The steel reinforcing rods were then removed from the concrete and cleaned with a fluid containing hydrochloric acid and inhibitors. Weight losses of each rod during cleaning were extrapolated back to zero time in order to calculate the corrosion loss.

Unsealed cubes were tested after curing in water for 3, 28 and 545 days and after exposure for 545 days. Three cubes were crushed for each condition. The remaining four unsealed cubes were tested for weight changes and carbonation.

Table 2   Data for each test series

Series	Water/Binder Ratio*	Cement	Length of cure (days)	Exposure **	Cube strength N/mm² at	
					End of cure	28 days
A & L	0.59	opc	3	Lab	22.6	42.1
C	0.59	pfa	3	Lab	13.3	29.5
E	0.59	ggbfs	3	Lab	10.0	33.9
B	0.59	opc	3	OS	22.3	42.8
D	0.59	pfa	3	OS	13.3	30.0
F	0.59	ggbfs	3	OS	9.8	33.1
G	0.59	opc	3	OV	20.4	40.5
H	0.59	opc	3	OH	21.0	39.7
I	0.59	opc	3	Office	21.0	41.6
M	0.59	opc	1	Lab	11.0	41.7
N	0.59	pfa	1	Lab	5.8	29.5
O	0.59	ggbfs	1	Lab	3.0	30.9
P	0.59	opc	28	Lab	44.0	44.0
Q	0.59	pfa	28	Lab	28.2	28.2
R	0.59	ggbfs	28	Lab	30.7	30.7
S	0.71	opc	3	Lab	13.6	28.5
T	0.71	pfa	3	Lab	7.0	18.8
U	0.71	ggbfs	3	Lab	6.6	21.1
V	0.47	opc	3	Lab	29.7	55.0
W	0.47	pfa	3	Lab	20.8	40.5
X	0.47	ggbfs	3	Lab	17.2	46.6
Y	0.83	opc	3	Lab	10.9	19.6
Z	0.83	pfa	3	Lab	5.4	12.3
1	0.83	ggbfs	3	Lab	4.9	16.0
2	0.35	opc	3	Lab	53.8	70.8
3	0.35	pfa	3	Lab	34.0	56.9
4	0.35	ggbfs	3	Lab	31.2	60.5
5	0.59	filler	3	Lab	21.0	40.7
6	0.59	filler	3	OS	20.6	39.9

*   Free water content of concretes = 188 kg/m³
**   Lab - 60% relative humidity, 20°C
        OS - Outside, concrete face sheltered
        OV - Outside, concrete face vertical
        OH - Outside, concrete face horizontal

Table 3a Analyses of materials used in cements

	F	P	A	S
	Limestone filler	Portland cement	Pulverized fly ash	Ground granul. blastfurnace slag
% by weight of				
SiO_2	2.5	20.3	47.7	34.5
Al_2O_3	0.3	5.02	26.6	12.6
Fe_2O_3	0.1	3.23	9.1	0.6
CaO	54.1	64.8	1.7	41.6
MgO	-	1.30	1.3	6.9
SO_3	0.04	2.96	0.8	0.81 sulphide
K_2O (Total)	0.05	0.86	3.4	0.68
Insoluble residue	-	1.19	-	0.43
LOI	42.5	1.59	5.8	0.55
Free lime	-	2.06	-	-
Apparent particle density (kg/m^3)	2700	3110	2400	2920
SSA (m^2/kg)	1180	395	375	455
45 μm % residue	8.3	-	7.0	11.6
150 μm	-	0.14	-	-
90 μm	-	-	-	1.9

Table 3b Cement types

Cement name	Components by weight %	ENV 197 Cement	
		Type	Class
OPC	100% P	CEI	42.5R
pfa	70% P + 30% A	CEIV*	32.5R
ggbfs	50% P + 50% S	CEIII	42.5
filler	95% P + 5% F	CEI	42.5R

* Fly ash near 28% limit for CEIIC

RESULTS AND DISCUSSION

General

The depth of carbonation after 545 days correlated closely with that after 180 days but their ratio, 1.45 on average, was slightly smaller than the value of 1.74 expected from the usual square root time function. There was no obvious effect of cement, water/cement ratio, curing or exposure condition on this ratio. The maximum depth of carbonation for each test was found to be 0 to 6mm greater than the average value. The depths of carbonation of the unsealed cubes were consistently greater than those of the partially sealed cubes.

Curing and cement

It can be observed from Table 4 that a longer period of moist curing consistently leads to a reduced depth of carbonation. The effect of curing seemed more pronounced for testing at 180 days than for testing at 545 days. The replacement of Portland cement with pulverised fuel ash or ground granulated blastfurnace slag generally led to an increased depth of carbonation. The increase averaged 46 and 69% for pfa and ggbfs respectively.

Table 4   Effect of curing and cement on carbonation depth
             (lab exposure, 0.59 w/c)

Cure	Test Age	Carbonation depth in mm.		
(d)	(d)	OPC	Pfa	ggbfs
1	180	5.2 (100)	8.1 (156)	9.0 (173)
3	180	3.9 (100)	6.4 (164)	6.3 (162)
28	180	1.1 (100)	4.2 (382)	5.8 (527)
1	545	9.6 (100)	8.9 (93)	11.0 (115)
3	545	5.7 (100)	8.0 (140)	10.5 (179)
28	545	4.3 (100)	7.9 (184)	7.8 (181)
	mean	4.97 (100)	7.25 (146)	8.40 (169)

Water/cement ratio and cement

It can be seen from Figure 1 that the depth of carbonation measured at an age of 180 days generally increases as the water/cement ratio increases. A similar effect is observed with each cement although the data for the OPC concrete with a 0.83 water/cement ratio seems anomalous. The replacement of Portland cement with pfa or ggbfs generally led to similar increases in depth of carbonation. The relative advantage of the OPC was greater at the lower water/cement ratios. The average depths of carbonation for the five water/cement ratios were ranked 100, 155 and 185 for the OPC, pfa and ggbfs concretes, respectively. This ranking was similar to that obtained from Table 4.

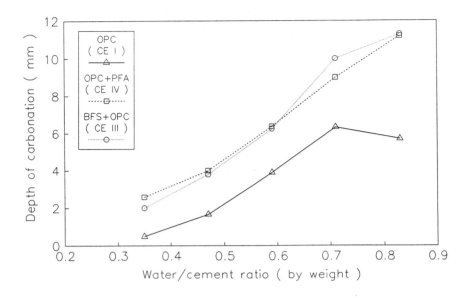

Figure 1   180 day carbonation versus water/cement ratio

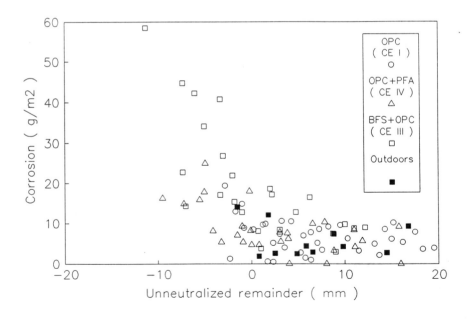

Figure 2   Corrosion of steel in moist, carbonated concrete

Exposure

Test specimens were exposed in the laboratory at 65% relative humidity, in an office, outside sheltered from rain and outside not sheltered from rain. The general trends of relative humidity within the 15mm surface layer of exposed concrete were as follows:

Laboratory	–	steady drop from 100 to 65%
Office	–	steady drop from 100 to 50%
Outside sheltered	–	fluctuates mainly between 80 and 60%
Outside horizontal	–	fluctuates between 100 and 80%
Outside vertical	–	fluctuates between 100 and 80%

The results in Table 5 indicate that outside exposure is associated with lower depths of carbonation than those for indoor exposure. In general the wetter the concrete the smaller is the depth of carbonation but carbonation of outdoor concrete sheltered from rain averages about 80% of the value for laboratory exposure. The OPC concrete benefited more from outdoor exposure than did the pfa and ggbfs concretes, as indicated by the ratios of outdoor sheltered to laboratory carbonation depths.

Table 5   Effect of exposure upon depth of carbonation (mm)
          (0.59 water/cement ratio)

Exposure	Cement	180d	545d
Laboratory	OPC	3.9 (100)	6.3 (100)
Outside sheltered	OPC	2.2 (56)	3.7 (59)
Laboratory	pfa	6.4 (100)	8.0 (100)
Outside sheltered	pfa	5.6 (88)	8.5 (106)
Laboratory	ggbfs	6.3 (100)	10.5 (100)
Outside sheltered	ggbfs	5.5 (87)	8.2 (78)
Office	OPC	4.8	
Outside vertical	OPC	3.2	
Outside horizontal	OPC	2.2	

Corrosion

A simple way of assessing carbonation-induced corrosion is to plot it against the unneutralized remainder(11). The unneutralized remainder is the distance between the outermost surface of the reinforcing rod and the plane of phenolphthalein neutralization; it is positive before carbonation reaches the surface of the reinforcing rod and negative thereafter. Figure 2 illustrates corrosion data for tests where the reinforcing rods were cleaned prior to embedment. In spite of the preliminary cleaning a small weight loss was observed even when the unneutralized remainder exceeded 10mm. It can also be seen that as the plane of phenolphthalein neutralization approaches and passes the reinforcing rods the degree of corrosion generally rises. The results are in broad agreement with the in situ data of Kashino(11) and Fukushi(12) although their results suggest that more significant corrosion can occur before the unneutralized remainder reaches zero. The observed rates of corrosion, equivalent to 15 to 70μm/year, are also comparable to in situ corrosion values reported by Muller (13).

The results in Figure 2 suggest that the rate of corrosion is largely controlled by the unneutralized remainder. However, for a given value of unneutralized remainder there was a tendency for the corrosion rates in concretes containing ggbfs to be higher than those in OPC or pfa concretes. There were no obvious effects of curing, water/cement ratio or exposure condition. The results for tests where the rods had not been cleaned before embeddment were comparable to those shown in Figure 2 but were more scattered. The weight loss component associated with the corrosion of the steel in the "as-received" condition was the main source of scatter.

Carbonation correlations and durability control

The carbonation data presented earlier in this report were broadly consistent with other data reported in the published literature from many countries(7). It is thus of interest to examine various strategies for unifying the data as a basis for understanding carbonation and for possible control of durability by codes.

The present version of ENV 206 indicates that for reinforced concrete under dry or humid exposure (Classes 1 and 2a) there is no need to restrict or consider cement type : it is only necessary to control concrete mix proportions. Figure 3 illustrates data where results for concretes made with OPC (type CEI) are paralleled by results for concretes made with other cements and where mix proportions, curing and exposure are the same. It is evident that the approach advocated by ENV 206 will not lead to consistent levels of resistance to carbonation. The possibility of higher corrosion rates for a given depth of carbonation in concretes containing ground granulated blastfurnace slag (type CE III) further widens the range of possible durability performance for concretes subjected to a particular ENV 206 exposure class.

An alternative approach to control of concrete durability is to specify, for a given exposure condition, a minimum 28 day strength of water cured concrete (i.e. the concrete grade). Figure 4 indicates that this approach goes some what towards accounting for the effects of cement type and mix proportions although it obviously cannot account for wide variations of curing and exposure. A simple and workable extension of this approach is to use the strength at the end of the curing period instead of the 28 day strength. Figure 5 shows that this extension yields an improved carbonation/strength correlation. A less workable but apparently promising alternative approach is demonstrated in Figure 6 where carbonation depths are seen to correlate well with the long-term strength of exposed cubes. The results in Figure 6 are incomplete so judgement on the use of long-term strength data should be reserved until the remaining results are reported. A similarly promising correlation was observed with the absorption data that have been collected so far.

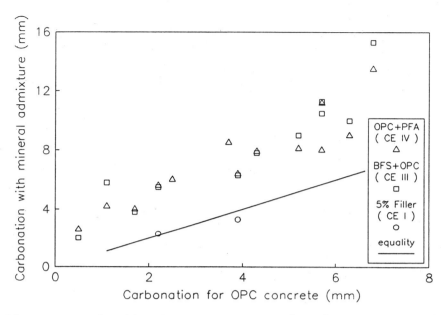

Figure 3   Carbonation for concretes made with different cements

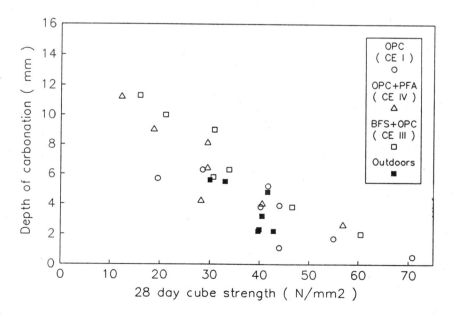

Figure 4   180 day carbonation versus 28 day cube strength

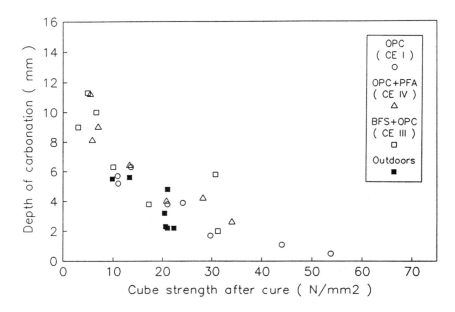

Figure 5   180 day carbonation versus strength at end of curing

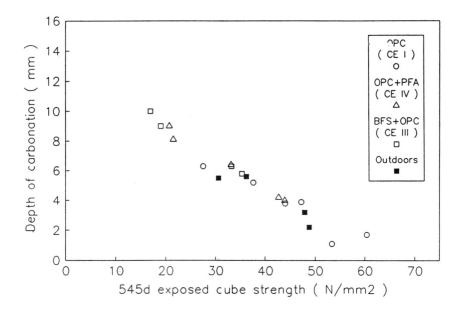

Figure 6   180 day carbonation versus 545 day strength of exposed cubes

Internal relative humidity measurements of the type reported in
reference 8 suggest that during the initial stages of drying, water is
lost primarily from the cover concrete and that relative humidity
profiles are not greatly affected by the different experimental
conditions detailed in Table 2. Thus it was possible to view the
weight loss after 18 days of exposure as an approximate measure of the
empty porosity in the carbonating region of the concrete. The results
in Figure 7 suggest that this measure of weight loss correlates
reasonably well with carbonation depth. Weight loss results from the
small paste prism stored in cavities' cast into the concretes also
correlated with carbonation depths and supported the idea that
carbonation rates depend upon the empty porosity of the matrix phase
in cover concrete.

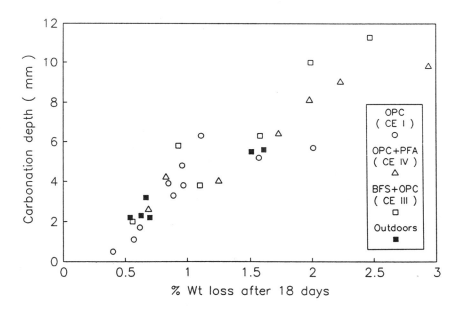

Fig 7   180 day carbonation versus 18 day weight loss of exposed cubes

The present work and projects elsewhere will produce more results that
will contribute to the selection of a suitable control method for
durability. Also control methods based upon absorption rate and air
permeability are currently under investigation at BCA and in other
laboratories. However it is already clear that specifications based
on simple measurements of concrete performance offer a better way of
controlling carbonation than the mix proportion method advocated by
ENV 206. Ideally a method of measurement should reflect a wide range
of durability characteristics and be useable on site.

The experimental results suggest that carbonation-induced corrosion is
likely to be greatest where exposure conditions are dry enough to

permit deep carbonation but wet enough to stimulate corrosion of reinforcement. Such conditions could arise where there is drying and wetting due to fluctuating climatic conditions or where an initially dry structure leaks or is later subjected to moist conditions. Drying and wetting is not specifically considered in EC2 and ENV 206 and it could be argued that selection from any of the existing exposure classes would not lead to the expectation of carbonation-induced corrosion : dry conditions would prevent corrosion and humid conditions would limit carbonation. Since the occurence of carbonation-induced corrosion is in no doubt (7), the addition of an appropriate exposure class should be considered.

CONCLUSIONS

Carbonation of concrete at ages of 180 and 545 days was investigated using water/cement ratios in the range 0.35 to 0.83, type CE I, CE III and CE IV cements, initial moist curing periods of 1, 3 or 28 days and various internal or external exposure conditions. Ignoring minor interactions of variables the following broad conclusions were drawn :

1. Water/cement ratio was a critical variable but cement type, curing and exposure condition each had a substantial effect upon the depth of carbonation.

2. Carbonation depths correlated well with various measures of strength and with initial weight losses during exposure.

3. The results supported the hypothesis that carbonation was controlled by diffusion of carbon dioxide through the pores in the carbonated, concrete surface that had been emptied by drying.

4. Corrosion of reinforcement in moist, carbonated concrete was mainly related to the unneutralized remainder, i.e. the distance between the steel surface and the plane of phenolphthalein neutralization. As the unneutralized remainder approached zero and became more negative the rate of corrosion greatly increased. There was some evidence that corrosion rates in concrete made with type CE III (Class 42.5) cement were greater than those in concretes made with type CE I (Class 42.5R) or CE IV (Class 32.5R) cements.

5. Control in design codes of carbonation and related corrosion would be more effective if it was based upon an appropriate measure of concrete performance rather than upon mix proportions, particularly if cement type is not considered. Several measures of concrete performance are currently viable (e.g. 28 day strength, strength after curing, long-term exposed strength and initial weight loss on exposure) and it seems possible that other measures such as absorption and permeability will also prove usable. Selection of a concrete test for control of durability in codes and standards would require consideration of properties other than carbonation and carbonation-induced corrosion.

ACKNOWLEGEMENTS

The assistance of Susan Poole, David Olerenshaw, Peter Pearson and Jim Martin with the experimental work is gratefully acknowledged.

REFERENCES

1.   EUROCODE No. 2, Design of Concrete Structures, Part 1 : General Rules and Rules for Buildings.   Final Draft, October 1989.

2.   ENV 206, Concrete - Performance, Production, Placing and Compliance Criteria, Final Draft, February 1989.

3.   pr ENV 197, Cement - Composition, Specifications and conformity criteria, Final Draft, June 1989.

4.   LITZNER, H. "Harmonized European Concrete Construction Standards - Present Status and Tendencies", Betonwerk & Fertigteil - Technik, 1988, No. 12, 21-26.

5.   KIRKBRIDE, T.   "Concrete codes move into a European context", Concrete 1989, February, 36-38.

6.   SOMERVILLE, G.   "Eurocode No. 2 - design of concrete structures", British Cement Association Bulletin, 1989, No. 5, 1-2.

7.   PARROTT, L.   "A review of carbonation in reinforced concrete", C&CA/BRE Report C/1-0987, July 1987, 126 pages.

8.   PARROTT, L.   "Moisture profiles in drying concrete", Advances in Cement Research, 1988, Vol.1, No. 3 164-170.

9.   PARROTT, L. and KILLOH, D.   "Carbonation in a 36 year old, in situ concrete".   Cement and Concrete Research, 1989, Vol.19, No. 4, 649-656.

10.   RAHMAN, A. and GLASSER, F.   "Comparative studies of the carbonation of hydrated cements", Advances in Cement Research, 1989, Vol.2, No.6, 49-54.

11.   KASHINO, N.   "Investigation into the limit of initial corrosion occurrence in existing reinforced concrete structures". Proceedings Conference Durability of building materials and components, Espoo, Finland 1984, Vol.3, 176-186.

12.   FUKUSHI, I. et al.   "Carbonation and Durability of Reinforced Concrete Buildings", Cement and Concrete, 1985, No. 461, 8-16.

13.   MULLER, K.   "Possibility of predicting the service life of reinforced concrete structures", Proceedings RILEM International Symposium on Long-Term Observation of Concrete Structures, Budapest 1984, Vol.1, 9-20.

ADDITIONAL PAPERS

THE USE OF EPOXY SYSTEMS IN CIVIL ENGINEERING : DAM REPAIRS

S Paz Abuin

M Pazos Pellin

Gairesa Laboratories

J A Lopez Portillo

Shell Spain,

Spain

ABSTRACT. The San Esteban dam is situated on the River Sil, in Galicia NW region of Spain. Constructed in 1955, the gravity arch concrete dam is 115 m high and comprises of 16 independent blocks 15-22 meters wide. The development of cracks in the dam, probably caused by alkali - agregate reactions, many of them running through from the upstream to the downstream face, prompted a thorough investigation of the structure. As a result of the investigation the decision to undertake a major programme of repair was taken. This was planned with two main objectives, to regenerate the damaged areas by injecting suitable materials into the cracks and voids and to prevent the penetration of water into the concrete structure. With the help of several analitical techniques such as DSC, FT-IR and TGA, we carried out our studies to investigate properties and the aging conditions for the materials to be used.

Keywords: Epoxy Resin, Differential Scanning Calorimetry, FT-IR Spectrometry, Kinetic Equations, Activation Energy, Preexponential Factor, Order of Reaction, Thermal Degradation, Thermogravimetric Analysis.

S. Paz Abuin is a Chemistry plastics specialist, with almost twenty years of experience in investigation on epoxy resins. He has published several laboratory papers within Spain and abroad in different fields, such as aging, kinetics, thermodinamic studies on epoxy resins. He is the technical manager of his own company, Gairesa.

M. Pazos Pellin is a Biologist, specialising in analitical techniques, such as FT-IR, DSC and TGA. She has colaborated in several laboratory jobs either for publication or as development of new products. She is the laboratory manager of Gairesa.

J.A. Lopez Portillo marketing department head for resins and polymers, intermediate in Shell Espana S.A. Madrid head office. He was assigned for 2 years in London in Shell International.

INTRODUCTION

With the progressive apparition of cracks along the San Esteban dam, a serious study using different methods was carried out from 1976 to 1986. The deformations measured showed a movement of the dam towards the right margin of almost 1mm/year at the crest of the dam.

After several studies to determine the possible factors causing the movements, the conclusion reached was that the concrete itself was the first risk factor, so a plan was laid down to analyze the aggregates and the concrete using specimen tests extracted from the wall of the dam and the quarry. Although there was not a contrasted theory, in the majority of the cases it coincided that these deformations were caused by the reaction between the aggregate and the cement in presence of water (ALKALI-AGGREGATE REACTION), which could produce expansion phenomena.

SOLUTION TO THE PROBLEM

The owners of the dam with the help of the laboratory results, reached the decision that a waterproofing system was needed to prevent free water circulation through the dam.

The conditions of the contract specified the technical characteristics of the job and the products to be used as well as the following surfaces to be treated:
- 5200m² of cracks.
- 6400m² of concrete joints, average thickness 1-2 mm.
- 11800m² of waterproofing of the water-side of the dam.

Waterproofing of the walls of the dam.

Different systems of waterproofing and other solutions were offered. The first possibility was to use no products at all but to rely on the waterproof properties of the concrete itself, that is to say to repair the damage (cracks and joints) and to let the concrete act freely as a water barrier. This solution had a great economical advantage, but due to the permeability of the concrete and the practically total impossibility of stopping the water by simply sealing the cracks, this solution was immediately discarded. Due to the fact that different types of waterproofing systems could be offered, a pre-selection was needed. Adhering and non-adhering solutions offered both advantages and disadvantages. The adhering solutions assured a major water-tightness but followed the strains of the wall, possibly originating cracks with simple thermal movements.

Furthermore, losses of adhesion might have been generated by sub-pression phenomena (e. g. rapid emptying of reservoir). On the other hand non-adhering solutions presented a major risk of water loss through tearing of the membrane (impacts of floating materials in the river, etc.).

One of the highest risks are the adhesion points but they do not
follow the structural movements nor the subpression phenomena.

After considering the above mentioned factors, an adhering waterproof
membrane made "in situ" with a very good crack bridging capacity, high
tear resistance, a very low elastic modulus and a very good water
resistance with easy repair, was chosen.

The organic coatings were the most satisfactory materials to be used
for the adhering membranes with the above mentioned requirements,
because when adequately formulated, tough soft products which are
mechanically and chemically resistent, could be produced. Furthermore
these can give good adhesion on hydraulic supports. Epoxy, polyester,
acrylic and polyurethane resins were studied, not only under the
previously mentioned requirements, but also taken into account were
other factors such as the ecological impact, the possibility to
reformulate on site in accordance with the climatical conditions and
the whole repair capacity. The formulations based on epoxy resins
were chosen for their best overall balance of properties.

Crack sealing material.

In a similar way, the material to seal the cracks had to fulfil the
following requirements: low shrinkage during the drying/curing
process, good dimension stability, good dry/wet adhesion and a very
good water resistance. Here again epoxy resins were selected. The
epoxy resin compounds were injected via drilling holes alongside the
galleries and downstream.

STUDY OF THE CANDIDATE EPOXY RESIN FORMULATIONS WITH THE HELP OF MODERN TECHNIQUES.

Even though different formulations related to crack and joint
characteristics were to be used in the approved project, we have
chosen two formulations to present here, due to both the quantity used
and practical importance; one associated with low elastic modulus
waterproofing and the other with high elastic modulus injection.

Although in most cases the development and adjusting of a formulation
is carried out via experimental data we considered that it was very
convenient and quicker to start from theoretical considerations.

Study of Formulation.

According to the rubber kinetic theory[1,2,3], the elastic modulus E is
defined as:

$$E = \frac{3dRT}{Mc} \qquad \text{(Eq. 1)}$$

Where:
d= Polymer density
R= Gases constant
T= Kelvin temperature
Mc= Molecular weight between cross-link units

The above equation is applicable for values above the Tg (glass
transition temperature) and at a 100% elongation; for example under
these conditions a system based on a DGEBA (Mw = 370), cured with a
diamine of short chain length such as EDA (Mw = 60), will produce a Mc
= 246 and E ~ 3550 N/mm² this value can be considered as the highest
possible one for polyamine-epoxy cured systems. Using other equations
based on this theory it is possible to estimate tensile strength
values[4], with which we would have a theoretical estimate for two very
important parameters such as: ELASTIC MODULUS AND TENSILE RESISTANCE
of the epoxy systems to be used. However there are other parameters
that need to be studied.

When using two component epoxy systems, the knowledge of values for
polymerisation time at different temperatures is of vital importance
especially with the existence of several injection steps, where the
control of the gel-time at different temperatures assures arrival of
the compound through the drilled holes to the forecast points.

Differential Scanning Calorimetry, DSC, and FTIR spectrometry are
extensively used to determine the kinetic parameters to know the time
-temperature - conversion relationships.

By using the well-known kinetic model based on the study of a dynamic
curve[5], which uses the capacity of the DSC calorimeter to obtain
simultaniously both the rate and the enthalpy of reaction (fig 1), the
reaction rate, $d\alpha/dt$, is mathematically expressed by:

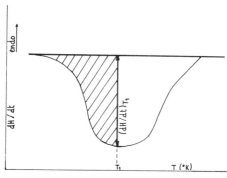

$$d\alpha/dt = 1/\Delta H_o * dH/dt \qquad (Eq. 2)$$

Where:
ΔH_o = Max. value of enthalpy of
             reaction
dH/dt = Heat flow

By combination of the Eq. 2 with
the general equation rate $d\alpha/dt =
K(1-\alpha)^n$ (K constant rate, $\alpha = \Delta H_t/
\Delta H_o$ degree of conversion, n order
of reaction), we obtained the
conversion values against
temperature.

Figure 1: Typical DSC dynamic curve.

With a DSC 7 Perkin Elmer, we carried out our tests for both
formulations (F-520 waterproofing resin and IM-100 injection resin)
the values of which are reflected in table 1 obtained with the help of
a kinetic software also of Perkin Elmer. The values of Ea and A are
calculated from Arrhenius equation, similar to that as later seen in
our degradation studies.

TABLE 1: Kinetic parameters (DSC study)

RESIN	Ea (Kj/mol)	A(1/sec)	n
IM-100	67.12	$4.07*10^7$	1.25
F-520	48.71	$4.75*10^5$	1.25

Ea= Activation energy
A = Preexponencial factor
n = Order of reaction

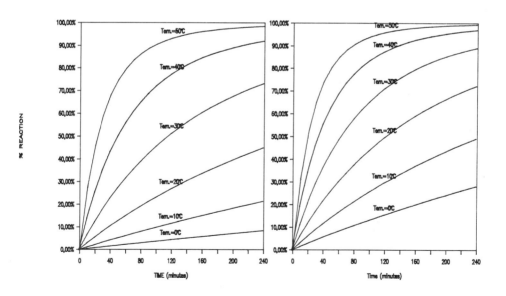

Figures 2 and 3: Degree of Cure vs. Time, IM-100 and F-520

Figures 2 and 3 show the relation conversion-time at different
temperatures of reaction obtained with the data from table 1.
Pot-life values, at different temperatures, can be easily calculated.
According to Flory theory[6] the gel point value is close to conversion
values $\alpha=60\%$. If we take into account that the running temperatures
during the application were 30°C outside and 20°C inside the concrete,
gel-time values could be obtained.

We carried out other measurements such as kinematic viscosity, surface
tensions and adhesion, both dry and wet; flexure, tensile and tear
strain; obtaining values matching and even higher than those that were
required.

AGING: THEORETICAL CONSIDERATIONS

Evidently the behaviour of an organic matrix will be determined by its
resistence to the property changes in function with time. These
irreversible changes are referred to as aging. Figure 4 shows a

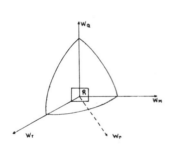

closed thermodynamic system formed by the
epoxy formulation, R, and its
surroundings. Assuming an ideal state,
R will follow the thermodynamic laws for
the reversible processes, the mechanical
work will follow the Hook law, $Wm = T\epsilon/2$
(T stress and ϵ deformation) the thermal
answer as given by $Wr = KT$ (K is the
Bolztman constant, T Kelvin temperature)
and the chemical work as $Wq = KT(\ln A_2/A_1)$
where A_1 and A_2 are the chemical
activities for the epoxy system and medium
the radiation energy can be calculated
from $Wr = h\mu$ (h the Plank constant and μ
frequency of the radiation.

Figure 4: Thermodinamic scheme material/surroundings.

These equations show energetical values that can act together or
separately, and as a consequence they produce physical-chemical
changes which cause the aging. In accordance to this, and supposing
that factors such as bacteriological attacks were negligible, we
commenced our study. The data on the stresses to which both
formulations could be submitted varied between 0.5-1 N/mm^2, which we
considered of little importance for aging[7]. Wr was taken into account
for the formulation F-520, and was studied only for its surface
effect. The chemical activity and temperature were the main factors
to be considered, this hypothesis being very well documented in the
bibliography. In fact, the water absorption in the polymeric
materials is judged to be the main cause for aging[8,9,10].

EXPERIMENTAL

Following ASTM D-570-81 we took a series of specimens for both
formulations, one part was immersed in desionized water at 23°C,
another at 60°C, and the last conserved under dry conditions. We
measured the weight variation daily till the equilibrium was reached
(table 2), the values of which were not significatively changed under
thermical effects, even though the equilibrium times were sensibly
shorter at t=60°C, in good agreement with the diffusion laws of Fick.
As a consequence of this, and taking into account that the presence of
some soluble products (at 60°C) present in the formulations could
significately affect the final results, all the posterior tests were
carried out with the formulations at 23°C.

TABLE 2: Water absortion

	% Water gained 23°C	60°C	% Soluble 23°C	60°C	Equilibrium time/days 23°C	60°C
F-520	1.5	1.61	---	6.4	395	56
IM-100	2.2	2.35	---	1.7	415	60

Even though there are several methods to determine the aging degree,
it is our belief that most of them (either via mechanical or/and
chemical measurement of properties) do not predict the deterioration
of a product at different temperatures, so we have preferred to obtain
estimative measurements for the aging at different degrees of
degradation at several temperatures.

MEASUREMENT OF AGING

Tg versus water absorption.

One of the most important features of the thermosetting polymers is
the glass transition temperature, Tg. Its close relationship with
some mechanical properties has been widely documented, therefore its
variations will originate variations of the mechanical properties.

By the means of a DSC-7 Perkin Elmer we carried out the Tg
measurements (scanning rate: 10°C/min) for both systems, before and
after reaching the equilibrium in water at 23°C (Table 3).

TABLE 3: Tg measurement (°C)

	Before water immersion	After water equilibrium
F-520	2	1.5
IM-100	48	38.5

The decrease of Tg after immersion observed in Table 3 is attributed
to internal plasticizing phenomena, the loss in Tg values is
equivalent to a loss in elastic modulus. The Ueberreiter equation
establishes the relationship between the Tg (Tg$_0$ before the
crosslinking) and the Mc that is related to the elastic modulus
through Eq.1

$$Tg - Tg_0 = 3.8 * 10^9/Mc \qquad \text{(Eq. 3)}$$

Thermal degradation before and after water immersion.

As previously stated, one of the most important objectives in the
studies on aging is the prediction of the degradation time-
temperature. Nowadays with the help of the new analitical techniques
some theoretical studies can be carried out with an acceptable limit
of theoretical-experimental agreement.

The thermogravimetric analysis, TGA, is a useful technique for
degradation studies of polymeric materials. We carried out our
experiments with a TG7 Perkin Elmer calibrated according to manual
procedure by measuring of Curie points between the range of
temperatures used (25-700°C). Six tests for each resin, dry and wet,
at different scanning rates (2, 5, 10, 15, 20 and 25°C/min) were
necesary to obtain, in agreement with Flynn and Wall[11], Doyle[12] and
Zsako[13], the values of activation energy, Ea, and preexponencial
factor, A. This treatment combines the general equation rate
$d\alpha/dt = K(1-\alpha)^n$, with the well-known Arrhenius relation:

$$K=A*e^{-Ea/RT} \qquad \text{(Eq. 4)}$$

where:
$d\alpha/dt$= rate of reaction
K= constant rate
α= percent of degradation
n= order of reaction
A= preexponencial factor
R= gas constant
Ea= activation energy
T= Kelvin temperature

Which leads to:

$$\ln(d\alpha/dt) = \ln A - Ea/RT + n \ln(1-\alpha) \qquad \text{(Eq. 5)}$$

Table 4 shows the different values of Ea and A for formulation, both before and after water equilibrium.

TABLE 4: Kinetic parameters (TGA study)

	Ea Kj/mol	A(1/min)	n(*)
IM-100 Dry	77.7	9.244	1
IM-100 Wet	71	6.459	1
F-520 Dry	61.5	74.7	1
F-520 Wet	58.7	62.6	1

(*) estimated value not determinated

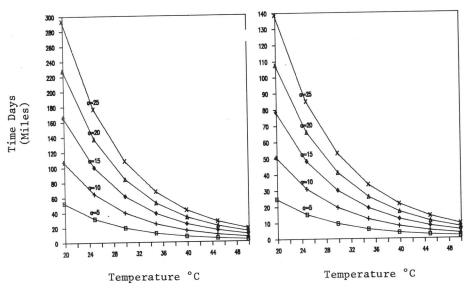

Figures 5 and 6: Time against temperature, IM-100 dry and IM-100 wet, at different conversions.

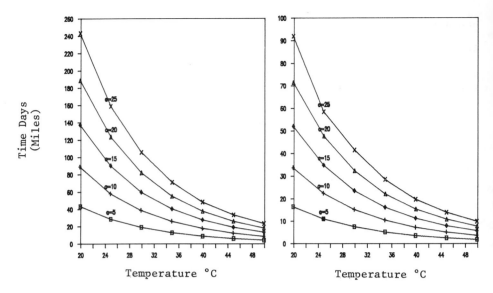

Figures 7 and 8: Time against temperature, F-520 dry and F-520 wet, at different conversions.

Although the determination of the Ea is straight-forward, one cannot carry out further kinetic calculations without further data or assuming the form of the kinetic equations. It has often been observed that the initial portion of a TGA curve can usually taken as a first order reaction equation.

Figures 5, 6, 7 and 8 obtained by integration of Eq.5 taking the data from table 4 show the estimated lifetimes at different selected temperatures and for different values of α for both formulations dry and wet.

The fall in lifetime in both formulations IM-100 and F-520 after water equilibrium is indeed very strong. In fact the water reduces significately the epoxy lifetime in the formulations studied. However the lifetime after water equilibrium is still long enough, but it shows the need to carry out this kind of test. A value of $\alpha=5$ originates a lost of impact strength between 20-40% (Izod measurement) which was the maximum value allowed in our case.

PRACTICAL WORK

The preparation work for the job started at the beginning of June 1987, to be finished before the end September of the same year. The dam wall and the cracks treated were the top 50 m, being the affected area.

The previously studied formulations, IM-100 and F-520, were used as a reference, but in some cases they were adapted to the irregularities of the job, so the viscosity of the IM-100 was modified to be used for the different thicknesses of joints and cracks. Almost 300 Tn of different epoxy compounds were used in less than four months.

The repair work included:
- Injection of epoxy compounds into the cracks and joints.
- Surface treatment of these repaired cracks and joints with glass fibre reinforced epoxy compounds.
- Coating of the upstream face of the dam with a glass fibre reinforced epoxy composition.

Different steps.

The injection into the cracks and joints was carried out inside the galleries, from the dam top and from scaffolding erected on both faces of the dam. The treatment was done in two phases; verifying in the second the quality of the previous phase with the extraction of test specimens. The injection pressures varied in relation to the width of the crack and viscosity of the compound.

The waterproofing of the wall on the upstream face of the dam which obtained a thickness of 7mm, was done as follows:
- Surface preparation by water blasting.
- Application of an epoxy primer (of a very low viscosity).
- Levelling of the surface by using a thixotropic surfacer.
- Double laminate using F-520 as the binder (glass fibre mat of 250 g/m²).
- Finishing coat by applying a urethane aliphatic isocyanate based-hydroxil acrylic resin composition.

A small laboratory was installed on the site to exhaustively control all the material used and its fitness by means of destructive mechanical tests in more than six hundred controls.

CONCLUSION

We have tried to give a methodology for the study and prediction of long term behaviour of formulations, to which the difficult task of prevention of the seeping of water through the walls of a dam has been given. Even though we have at our disposition the data for adhesion, compression, flexural, tensile, fatigue and impact tests, these have been deliberately omitted because although of importance, they are aging dependent.

Now, two years later, even though it may be too early to pass judgement, it can be said that the results are very satisfactory. The filtrations on the inner side of the dam and joints have been reduced by 98%, and the movements in cracks and joints are non-existent.

The specimens recently extracted show that the theorical previsions were correct and with this we hope that our studies have been of some use.

ACKNOWLEDGMENTS

First of all to the San Esteban Dam owners, Iberduero, they have given permission to publish reserved data. We also give thanks to those companies that applied our products, Rodio and Teimper, and Evelyne for help with the translation.

REFERENCES

1 P.J. Flory, Principles of Polymer Chemistry, Cornell Univ. Press, Ithaca, N.Y., (1953), p.432
2 A.V. Tobolsky, Properties and Structure of Polymers, Wiley, N.Y., (1960)
3 A.V. Tobolsky, D.W. Carlson and N. Indictor, J. Polymer Sci., 54, 175, (1960)
4 J.L. Gardon, Treatise on Adhesion and Adhesives, R.L. Patrick, ed. vol.1, Dekker, N.Y., (1967), p. 314
5 R.B. Prime, Polym. Eng. Sci. 13, 365 (1973)
6 P.J. Flory, Principles of Polymer Chemistry, Cornell Univ. Press, Ithaca, N.Y., (1953), p. 348
7 S. Paz Abuin, Quimica Hoy 6, 52 (1989)
8 C. Browning, Polymer Engineering and Science, January, (1978) vol. 18 n·1
9 C.D. Shirrell, Sampe 23, 174 (1978)
10 G.S. Springer. Developments in Reinforced Plastics 2, G. Pritchard ed., vol. 2, Applied Sci., Publishers, London, (1982), p. 43
11 J.H. Flynn, and L.A. Wall, Polym. Letter 4, 323 (1966)
12 C.D. Doyle, J. Appl. Polym. Sci. 5, 285 (1961)
13 J. Zsako and J.Jr. Zsako, J. Therm. Anal. 19, 333 (1980)

HIGH STRENGTH CONCRETE USING DOLOMITE AS COARSE AGGREGATE

H H Bahnasawy
F E El-Refai

General Organisation for Housing Building & Planning Research,

Egypt

ABSTRACT. High strength concrete has become of special significane to be used in special structures, precast concrete and prestressed concrete. In this work about 12 concrete mixes have been investigated using crushed dolomite as coarse aggregate, control mixes for comparsion. Various cement contents, water/cement ratios, coarse/fine agrregate ratio and different curing methods have been studied to evaluate the potentiality of using this type of coarse aggregate to produce high strength concrete to be used for some special structures erected in Suez Canal area. Plasticizer and early strength admixture with different dosage are used in order to reduce the amount of mixing water to compensate the relatively high water absorption of the crushed dolomite. Fresh and hardened concrete properties are studied and compared with contral mixes having gravel as coarse aggregates. The increase in concrete strength when using crushed dolomite as coarse aggregate is tangible. Analysis, discussions and conclusions for test results of the basic properties and their control in both fresh and hardened stages were presented.

Keywords: Dolomite, Gravel, Coarse Aggregate, Cement Contents, Plasticizer, Early Strength Admixture, Curing Methods, Compressive Strength, Tensile Strength, Bond Strength.

Dr. Heba H Bahnasawy is a Researcher at Strength of Materials Research and Quality Control Division, General Organization for Housing, Building and Planning Research Cairo, Egypt. She has participated in evaluation and optimum utilization of building materials, load bearing masonry and quality control.

Dr. Fatma E El-Refai is a Professor, Strength of Materials Research and Quality Control Division, General Organization for Housing, Building and Planning Research Cairo, Egypt. She is actively engaged in quality control and quality assurance, inspection of existing building and formulation of Codes of practice and Specifications.

INTRODUCTION

Concrete, being the most commonly used engineering material in construction industry, is being produced under different environmental conditions.
Here in Egypt especially in Suez Canal area there are many dolomite
quarries. Recently, and because of the many projects for reconstruction
that area trails began to study the possibility of using crushed
dolomite as coarse aggregate in concrete, and also in precast concrete
units.

In this research work, fresh and hardened properties for 12 different
concrete mixes have been studied. A comparison between gravel concrete
and dolomite concrete mixes as a local product in that area was carried
out. Test results and evaluation of this type of concrete are presented. Studying the effect of adding superplasticizer with different
percentages to concrete mixes was adopted.

Different five curing methods were applied for all dolomite concrete
mixes with and without saperplasticizer. Conclusions having the
remarkable results are also presented in this work.

MATERIALS USED

Cement

Ordinary Portland Cement is used. The physical, mechanical and
chemical properties of this cement are in complete compliance with the
limits of the Egyptian standard specifications(1).

Aggregates

Natural dry gravel from Elyahmoum quarries at Cairo and crushed dolomite
from Ataka near Suze Canal zone, are used as coarse aggregate.   The
properties of both types of coarse aggregates used are shown in (Table
1). Natural sand is used as fine aggregate and it complies with
Egyptian standard specificans (2).

Admixture

Early strength super-plasticiser for concrete is used to reduce water
content and increase early strength. It is used with two dosage percent
1, 2 % by weight of cement content.

Table 1 Properties of Coarse Aggregates

PROPETY	DOLOMITE	GRAVEL
Maximum Nominal Size(mm)	20	20
Percentage of fines	2	3
Unit Weight	2.5	2.4
Mass Density (Kg/m³)	1540	1720
Percentage of Voids	38.4	28.3
Percentage of water absorption	5.5	2.5
Aggregate Crushing Value	17.4	15.1
Impact Resistance Value	16.22	14.8
Abrasion Resistance Value	30	22

EXPERIMENTAL INVESTIGATION

The different investigated concrete mixes with different mix proportions are shown in (Table 2). The dolomite and gravel were used as
coarse aggregate with different cement contents 300, 400 and 500 Kg/m³.
The ratios between coarse and fine aggregate were 1 : 1.5, 1 : 2 and
1 : 2.5. Superplasticizer was added to dolomite mixes with the percentage 1, 2 by weight of cement. The concrete specimens were cast at
room temperature (18 - 20°C). The concrete specimens used were compressive strength at 3, 7, 28 days age, 15 x 30 cm cylinders for indirect
tensile strength (Brazilian test) at 28 days and cylinders with steel
bars 16 mm diameter for bond strength. They were kept at room temperature for 24 hours. The moulds were taken off and the specimens were
stored in water (17 - 19°C) till the day of testing. The dolomite
concrete mixes with 0, 1, 2 % super plasticizer were cured in different
wasy as shown below:

1. Room temperature (18 - 20°C) for 3, 7 and 28 days

2. Submerged in water (≈ 17°C) for 3, 7 and 28 days

3. Submerged in hot water (55°C) for 3, 7 and 28 days

4. Dry oven (55°C)   for 3, 7 and 28 days

5. Moisture Cabinet at (55?C and 50% Relative Humidity for 3,7&28 days
    age.

Table 2   Dolomite and Gravel Concrete Mixes

Mix No.	Type of coarse Aggregat	Cement content kg/m³	Fine to coarse Aggregate	Water cement ratio	Super plasti- cizer %	Slump (mm)	Curing method
D₁		300	1 : 2	0.5	–	43	2
D₂			1 : 1.5	0.41	–	40	2
D₃			1 : 2	0.44	–	30	1,2,3,4,5
D₄	Dolomite	400	1 : 2	0.37	1	40	1,2,3,4,5
D₅			1 : 2	0.34	2	35	1,2,3,4,5
D₆			1 : 2.5	0.51	–	42	2
D₇		500	1 : 2	0.35	–	38	2
G₁		300	1 : 2	0.5	–	45	2
G₂			1 : 1.5	0.44	–	40	2
G₃	Gravel	400	1 : 2	0.46	–	42	2
G₄			1 : 2.5	0.48	–	42	2
G₅		500	1 : 2	0.38	–	35	2

TEST RESULTS AND DISCUSSION

Fresh Concrete Properties

The workability for the different investigated mixes were measured by slump tests as shown in (Table 2). It was found that the slump range for dolomite concrete mixes was from 30 - 43 mm while the slump range for gravel concrete mixes was from 35 - 45 mm. for different dolomite concrete mixes water-cement ratios were between 0.35 - 0.51, while for gravel concrete mixes it was between 0.38 - 0.5. By adding superplasticizer with 1 and 2 % of cement weight to dolomite concrete mixes the reduction in water-cement ratio was about 16 and 23 % respectively.

Mass Density

The test results given in (Table 3) show that the mass density for different dolomite concrete mixes ranges between 2375 to 2700 kg/m³. Also for different gravel concrete mixes the man density ranges between 2400 to 2715 kg/m³. It can be seen that the range for both gravel and dolomite concrete mixes are nearly the same.

Compressive Strength

The test results for the relation between compressive strength and
cement content for both dolomite and gravel concrete mixes are shown in
(Table 3) and (Figure 1). It can be noticed that compressive strength
of dolomite concrete mixes are higher than those of gravel for all
cement contents and at all ages of test. This increase ranged beteen
15-48 %, 22-37% and 18-52 % for cement contents 300, 400 and 500kg/m³
respectively at the different ages adopted in this work. The test
results for the relation between compressive strength and the ratio of
fine-coarse aggregate for both dolomite and gravel concrete mixes are
shown in (Table 3) and (Figure 2). It can be seen that both dolomite
and gravel concrete mixes with the ratio of fine/coarse aggregate 1 : 2
and M.NS 20 mm gives the highest value for the compressive strength at
the different ages. The ranges for this increase are 7 - 24 % for
dolomite concrete mixes and 7 - 22 % for gravel concrete mixes at
different ages of testing. It can be noticed also that for the ratios
of 1 : 1.5, 1 : 2 and 1 : 2.5 fine/coarse aggregate the compressive
strength of dolomite concrete mixes are higher than those of gravel
concrete mixes by about 40 % , 27 % and 21 % at 28 days age.

Splitting Tensile Strength

(Figure 3) and (Table 3) show the testing results of splitting tensile
strength for the dolomite and gravel concrete mixes. All specimens
have been tested at 28 days. It can be seen that the dolomite concrete
mixes give higher values for the splitting tensile strength for all
fine/coarse aggregate ratios adopted in this work. For the same fine/
coarse aggregate ratio 1 : 2 the increase in tensile strength of
dolomite concrete than gravel concrete was 54 %, 77 % and 50 % for
cement contents 300, 400 and 500 kg/m³ respectively.

For cement content 400 Kg/m³ it was found that the splitting tensile
strength for dolomite concrete increase than the gravel concrete by
68 %, 77 % and 43 % for fine/coarse aggregate ratios 1 : 1.5 , 1 : 2
and 1 : 2.5 respectively.

The dolomite concrete mix with fine/course aggregate 1 : 2 and cement
content 400 kg/m³ has been chosen to add superplasticizer by 1 % and 2%
of cement weight. It has been found that the increase in splitting
tensile strength due to admixture adding 3 % and 4.8 % respectively.

Bond Strength

(Figure 4) and (Table 3) show the testing results of bond strength for
the dolomite and gravel concrete mixes. All specimens have been tested
at 28 days. It can be seen that the dolomite concrete mixes give
higher values for the bond strength for all fine/coarse aggregate
ratios adopted in this work. For the same fine/coarse aggregate ratio
1 : 2 the increase in bond strength of dolomite concrete than gravel
concrete was 10 %, 45 % and 50 % for cement contents 300, 400 and 500
kg/m³ respectively.

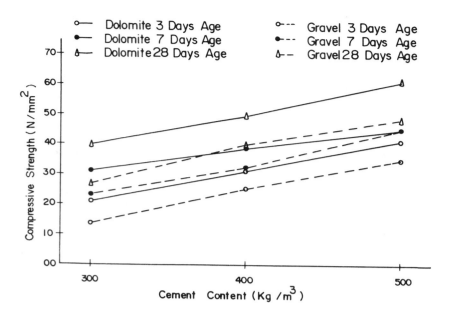

Fig. 1 : COMPRESSIVE STRENGTH OF DOLOMITE AND GRAVEL
CONCRITE MIXES FOR DIFFERENT CEMENT CONTENTS.

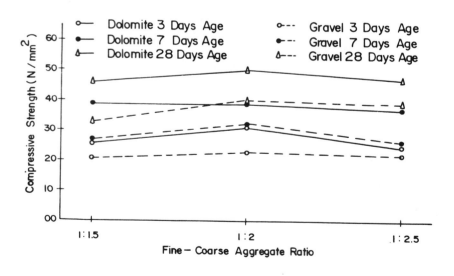

Fig. 2 : COMPRESSIVE STRENGTH OF DOLOMITE AND GRAVEL
CONCRETE OF DIFFERENT COARSE-FINE AGGREGATE
RATIO WITH CEMENT CONTENT 400 Kg/m³

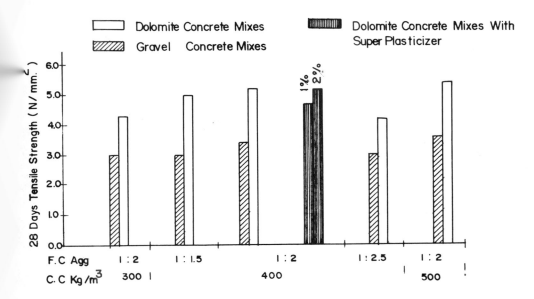

Fig. 3: SPLITTING TENSILE STRENGTH FOR DIFFERENT DOLOMITE AND GRAVEL CONCRETE MIXES.

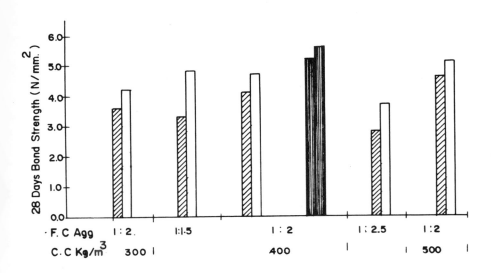

Fig. 4: BOND STRENGTH FOR DIFFERENT DOLOMITENT AND GRAVEL CONCRETE MIXES.

For cement content 400 kg/m³ it was found that bond strength for dolomite concrete increase than the gravel concrete by 51 %, 45 % and 43 % for fine/coarse aggregate ratios 1 : 1.5 , 1 : 2 and 1 : 2.5 respectively.

The dolomite concrete mix with fine/coarse aggregate 1 : 2 and cement content 400 kg/m³ has been chosen to add superplasticizer by 1 % and 2 % of cement weight. It has been found that the increase in bond strength due to admixture adding 11 % and 19 % respectively.

Table 3 Mass Density, Compressive Strength at 3,7,28 days age, Splilting Tensile and Bond Strength

Nix	Mass density(kg/m³) at ages (days)			Copressive Strength (N/mm²)at ages(days)			Splitting Tensile Strength (N/mm²)	Bond Strength (N/mm²)
No.	3	7	28	3	7	28		
D_1	2543	2700	2625	21.2	30.8	40.3	4.3	4.2
D_2	2560	2500	2490	26.3	38.6	45.6	5.0	4.79
D_3	2540	2550	2455	31.0	39.0	50.4	5.0	4.7
D_4	2585	2420	2465	35.8	50.3	64.1	5.15	5.2
D_5	2660	2635	2550	38.9	54.5	63.3	5.24	5.6
D_6	2440	2405	2475	24.9	36.5	47.0	4.25	3.7
D_7	2488	2500	2375	40.9	44.9	60.9	5.4	5.1
G_1	2665	2565	2715	14.3	25.1	34.9	2.97	3.58
G_2	2440	2410	2400	21.1	26.6	32.5	2.98	3.31
G_3	2405	2450	2575	22.6	31.9	39.6	3.44	4.11
G_4	2500	2485	2420	22.0	26.2	38.9	2.98	2.82
G_5	2415	2480	2500	26.9	38.0	47.6	3.61	4.64

Effect of Superplasticizer and Curing Methods.

In this work, the effect of adding superplasticizer and also the effect of the curing method adopted on the compressive strength of dolomite concrete mixes have been studied.

Figure 5 and Table 4 show the test results of this phase in this work. Superplasticizer with percentages 0 % , 1 % and 2 % of cement weight has been added to dolomite concrete mixes to study its effect on the compressive strength at different ages 3, 7 and 28 days. At the same time five curing methods have been adopted to study the effect of them on the compressive strength of dolomite concrete with superplasticizer. It can be noticed that at the age of 3 days, the compressive strength of specimens cured in hot water (55°C) has the higher values 39.3, 46.6 and 57.5 N/mm² with the mixes having 0 % , 1 % and 2 % superplasticizer. At 28 days age, the compressive strength of specimens cured in cabinet (55°C and R.H 50 %) give the higher values 53.6, 75 and 75.1 N/mm² for mixes having 0 % , 1 % and 2 % superplasticizer. For comparison we consider the dolomite concrete mix without superplasticizer as a control mix to study the effect of the different curing methods adopted in this work. It has been found that higher results abtained with the curing by cabinet (55°C + R.H. 50 %), then curing by hot water (55°C), then curing by submengh in water (17°C) and at last the specimens left in air (18 – 20°C).

Table 4 Compressive Strength of Dolomite Concrete Using Superplasticizer and Different Curing Mehtods.

Super-plasticizer %	Age of test (days)	Compressive Strength (N/mm²) (Different Curing Methods)				
		Water (17°C)	Water (55°C)	Air (18–20°C)	Oven (55°C)	Oven (55°C) R.H(50 %)
0	3	31.0	39.3	28.2	32	34.3
	7	39.0	41.6	36.1	35.4	41.6
	28	50.4	49.0	41.1	47.1	53.6
1	3	35.8	46.6	37.9	34.1	45.4
	7	50.3	53.5	48.7	49.7	58.7
	28	63.3	61.1	58.4	52.4	75
2	3	38.9	57.5	51.1	38.7	48.2
	7	54.5	66.1	64.1	52.1	62.2
	28	64.1	71.2	70.7	66.3	76.1

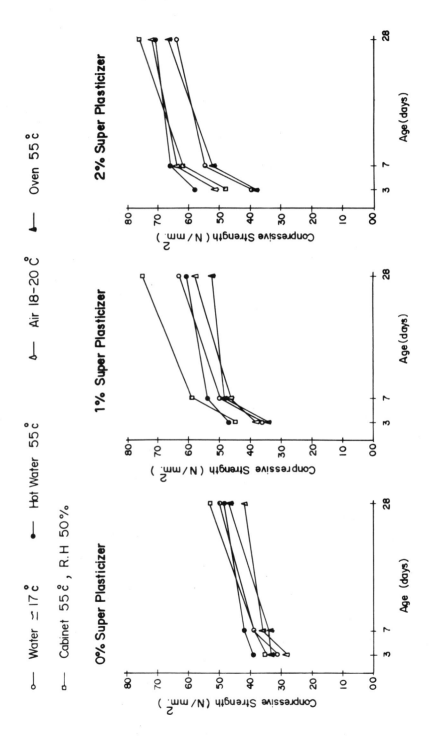

Fig.5 : COMPRESSIVE STRENGTH OF DOLOMITE CONCRETE MIXES FOR DIFFERENT CONCRETE MIXES AND DIFFERENT CURING METHODS.

CONCLUSIONS

From this study the following conclusions can be considered:

1. Water.cement ratio for both dolomite and gravel concrete mixes are nearly the same for different cement contents and different fine/coarse aggregate ratios.

2. Adding superplasticier as a percentage of cement weight has great effect on reducing water-cement ratio for all studied dolomite concrete mixes.

3. Maso density for both dolomite and gravel concrete mixes has found to be nearly the same for different concrete mixes.

4. Compressive strength of dolomite concrete mixes are higher than those of gravel concrete mixes for all cement contents and for all fine/coarse aggregate ratios at different ages of test (3, 7 and 28 days).

5. Splitting tensile strength results for dolomite concrete specimens is significantly higher than the gravel concrete specimens for all fine/coarse aggregate ratios (1 : 1.5, 1 : 2 and 1 : 2.5) cement content 400 kg/m³ at 28 days age.

6. Bond strength for dolomite concrete as measured by pulling out test is higher than that of gravel concrete for all fine/coarse aggregate ratios (1 : 1.5 , 1 : 2 and 1 : 2.5) and cement content 400 kg/m³ at 28 days age

7. Adding superplasticizer to dolomite concrete mixes greatly improves its consistency, compressive strength, splitting tensile strength and bond strength.

8. Using different curing methods with dolomite concrete specimens, it has been concluded that the most effective curing method was the moisture cabinet method (55°C + 50 % R.H).

REFERENCES

1. ESS 373, 1984, Ordinary Portland Cement and Rapid Hardening Portland Cement.

2. ESS 1109, 1973, "Aggregate from Natural Sources"

3. OMER Z. CEBCI, Strength of Concrete in warm and dry environment. Materials and Structures.July 1987,Volume 20, pp. 270-272.

4. R.G.PATEL, D.C.KILLOH, L.J.PARROTT, W.A.GUTTER IDGE, Influence of curing at different relative humidities upon compound reactions and porosity in Portland cement paste. Materials and Structures, May 1988, Volume 21, pp. 192 - 197.

5. Omaima A.Salah El-Din, Samir H.Okba.,Superplasticizers effect on behaviour of concrete with different cement types. Proceedings and Arab Conference on Structural Engineering,Jordan, April 1987.

6. F.D.LYDON, Concrete Mix Design. Galliard Limited, England, 1979.

EFFECT OF HEAVY LATERAL REINFORCEMENT ON THE STRENGTH OF CONCENTRICALLY LOADED R C COLUMNS

H H Shaheen
S N El-Ibiari
M A Saleh

Building Research Center,

Egypt

ABSTRACT. Lateral reinforcement is used in columns to prevent longitudinal reinforcement from buckling and to confine concrete. There are different types of lateral reinforcement such as circular, continuous spirals, rectangular ties, continuous rectangular spirals and mesh reinforcement. In this research, the effect of heavy lateral reinforcement. i.e. mesh reinforcement, on the behavior of reinforced concrete centrically loaded columns is studied. The behavior of twenty five reinforced concrete columns having mesh reinforcement was studied considering the effect of changing the dimensions of the mesh reinforcement and the percentage of the lateral steel. A certain factor named the confinement factor has been considered to evaluate the change in both the ultimate strength and the ultimate strain. An equation has been deduced a certain formula was found calculating the gain in the concrete strength due to the existence of lateral reinforcement in the form of a mesh.

keywords: Concrete columns, mesh reinforcement, ductility, lateral reinforcement, concrete confinement,

Dr Hamdy H.Shaheen is a professor and head of the Department of Reinforced Concrete Structures, General Organization for Housing, Building and Planning Research, Cairo, Egypt. He has been most active in concrete technology by means of research. His current research work includes: use of high tensile steel and repair and strengthening of reinforced concrete structures.

Dr Shadia N.Elibiari is an associate professor at the Department of Reinforced Concrete Structures, General Organization for Housing, Building and Planning Research, Cairo, Egypt. She is actively engaged in research into various aspects of concrete behaviour.

Eng. Mohsen A.Saleh is an assistant lecturer at the Department of Reinforced Concrete Structures, General Organization for Housing, Building and Planning Research, Cairo, Egypt. He is actively engaged in research into various aspects of concrete behaviour.

INTRODUCTION

Most national codes[1] disregard the effect of lateral reinforcement on bound concrete behavior, even though they all contain clauses limiting the size and spacing of ties. It appears that most code recommendations and some of the previous investigations are based on the assumptions that the main function of lateral reinforcement is to prevent longitudinal reinforcement from buckling rather than to confine the concrete core.

The lateral reinforcement creates lateral confinement for reinforced concrete columns. Therefore, lateral confinement creates a triaxial state of compression which allows the column to have more strain. The ductility is the property which allows the material to undergo a large plastic change under applied load. Lateral reinforcement increases the ductility of the reinforced concrete columns[2].

Many investigators[3,4] have studied the effect of lateral reinforcement on the behavior of the reinforced concrete columns. There are few investigations[5] on mesh reinforcement as lateral reinforcement.

EXPERIMENTAL PROGRAM

In order to study the effect of the change of the dimensions of mesh reinforcement and the percentage of mesh reinforcement on the behavior of the reinforced concrete columns, twenty-five reinforced concrete columns of 120 cms height and of 25x25 cms cross-section are tested. The longitudinal reinforcement of each of these columns is 4 O 4 for fixing the stirrups. The details of the steel reinforcement and the concrete compressive strength for the tested columns are given in table (1). The steel mesh used for lateral reinforcement are formed by welding plain steel bars of diameters 4,6,8 and 10 mms. The spacing of the ties are reduced in the last 10 cms at both ends of the columns in order to reduce the effect of the stress concentration at this area.

Figure (1) shows the shape of the reinforcement for some of the tested specimens. Concrete mix used was made of ordinary Portland cement, sand and gravel. Three different concrete mix proportions were designed for concrete of average compressive strength equal to 25, 30 and 40 N/mm2 at 28 days age. The ties used as lateral reinforcement were welded at their four corners to the four vertical steel plain bars at the corners of the columns. Figure (2) shows the measurements taken during the test.

TEST RESULTS AND DISCUSSION

Figures (3 to 6) give the test results for of the columns showing the effect of both the change of the dimensions of mesh reinforcement and the percentage of mesh reinforcement on the behavior of the reinforced concrete columns.

Table (1) The detils of the tested columns

Group No.	Mesh (axa)	Spacing cm	Fc kg/cm	μ	Column No.
A	2.5x2.5	20	250	0.539	A-3-1
			300	0.375	A-3-2
			400	1.212	A-3-5
				0.875	A-3-4
				1.212	A-3-3
		15	300	1.167	A-2-4
				1.616	A-2-3
		10	400	1.750	A-1-4
				2.693	A-1-5
B	5.0x5.0	20	300	0.875	B-3-4
			400	0.875	B-3-3
				1.212	B-3-2
		15	400	1.167	B-2-2
				1.616	B-2-5
				1.616	B-2-3
		10	300	.750	B-1-4
			400	2.693	B-1-3
C	10 x10	20	250	0.539	C-3-1
			300	.875	C-3-3
			400	1.875	C-3-4
				1.212	C-3-5
		15	300	1.167	C-2-2
				1.616	C-2-5
		10	400	1.750	C-1-2
				2.693	C-1-3

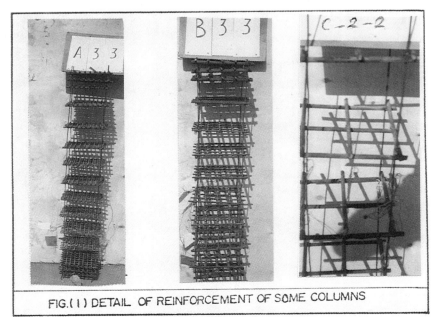

FIG. (1) DETAIL OF REINFORCEMENT OF SOME COLUMNS

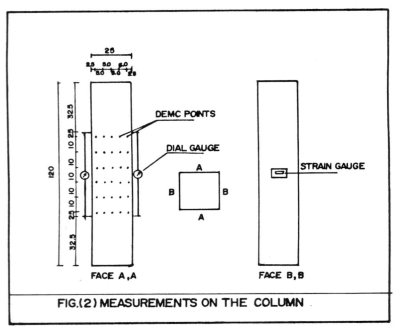

FIG.(2) MEASUREMENTS ON THE COLUMN

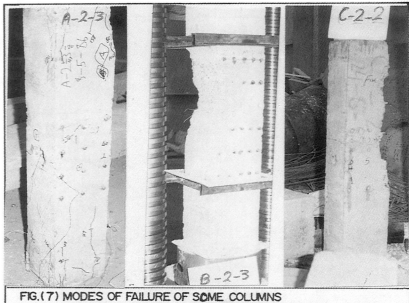

FIG.(7) MODES OF FAILURE OF SOME COLUMNS

Effect of dimensions of mesh reinforcement on the behavior of reinforced concrete columns
Figure (3) shows the stress-strain relations by measuring the over all strain on the column. These strains are measured by vertical dial gauges. It may be seen that for the same load, the strain recorded for a column in group (C) is more than that recorded for a similar column in group (B) which is in accordance more than that recorded for similar column in group (A). Figure (4) shows the stress-strain relations by measuring the strain on the concrete surface. These strains were measured by using a deformeter of gauge length 10cm. The above figure, it shows higher strain for the column in group (C) under the same applied load. In accordance, the strain of column in group (B) is more than that for the similar column in group (A). It may be also summarized that small dimensions of mesh reinforcement caused more confinement for the column and in accordance give more ultimate strain.

Effect of the percentage of mesh reinforcement on the behavior of reinforced concrete columns
Figures (5 & 6) show the stress-strain relations by measuring the over all strain on the column and measuring the strain on the concrete surface respectively. The comparison was made between two similar columns which differ only in the percentage of mesh reinforcement. It can be seen that the columns having higher percentage of mesh reinforcement indicated lower strain values for the same load.

Mode of Failure

Figure (7) shows the modes of failure of some of the tested columns. It may be noticed that the first sign of failure in all the specimens was the appearance of vertical cracks in the concrete cover. As the applied load increased, the cracks spread rapidly in the concrete cover. However, by the loss of the concrete cover the concrete core has been confined by the arch action of the lateral reinforcement. In general, failure happened within the mid height of the columns, in most of the tested columns.

Effect of lateral reinforcement on the ultimate strain

In evaluating the effect of the percentage of lateral reinforcement (μ on the ultimate strain, ($\bar{\mu}$) is calculated according to the following equation:

$$\mu = \frac{\text{volume of lateral reinforcement}}{\text{volume of concrete calculated over area of cross-section}}$$

In order to measure the degree of confinement given to concrete by lateral reinforcement, a confinement factor (α_c) has been taken as:

$$\alpha_c = \mu \ (F_y/F'_c)$$

where; μ : the percentage of lateral reinforcement, F_y : the yield strength of lateral reinforcement, F'_c : 0.7 * concrete cube strength.

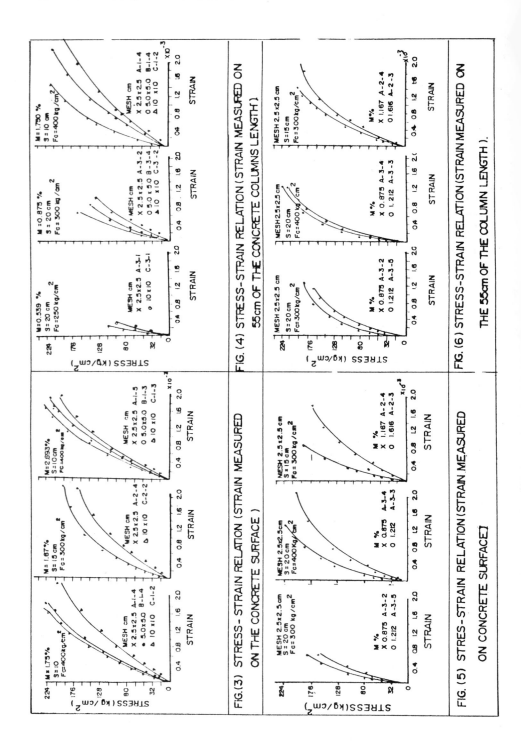

FIG. (3) STRESS - STRAIN RELATION (STRAIN MEASURED ON THE CONCRETE SURFACE)

FIG. (4) STRESS-STRAIN RELATION (STRAIN MEASURED ON 55cm OF THE CONCRETE COLUMNS LENGTH)

FIG. (5) STRES-STRAIN RELATION (STRAIN MEASURED ON CONCRETE SURFACE)

FIG. (6) STRESS-STRAIN RELATION (STRAIN MEASURED ON THE 55cm OF THE COLUMN LENGTH).

Test results were analyzed to evaluate the increase in the ultimate concrete strain resulting from the confinement. It may be seen that when the ultimate strain is plotted against the confinement factor (α_c) the straight line obtained did not pass through the origin as shown in Figure (8). The equation of this straight line is:

$\epsilon u = \epsilon o + 0.0007 * \alpha C$

where; ϵu : the ultimate strain for confined concrete, ϵo :the ultimate strain for unconfined concrete taken as constant equal to 0.002 (i.e. according to several investigation[4,7,8]).

Effect of the lateral reinforcement on the ultimate strength

Table (2) shows the values of the increase in the ultimate strength due to confinement Δp) for the 25 tested specimens. The increase in the ultimate load due to confinement may be calculated due to the following equation:

$\Delta p = K * Fy * \alpha C$, where; K : is constant.

Table (3) shows the values of αc) for each specimen and the values of (K) due to the previous equation. By using the fitting curves, the relation between (K) and the confinement αc can be expressed as the following equation: $K = (3.5 + \alpha c) / (1.0 + 5.50 * \alpha_c)$.The USSR code[9] gives the following relation between (K) and the confinement (α_c): $K = (5.0 + \alpha c) / (1.0 + 4.50 * \alpha_c)$

From Figure (9), it can be seen that the relation between (K) and the confinement (αc) given by the USSR code[9] and that derived in this work have the same character.

CONCLUSIONS

From the above results, it may be concluded that:

(1) The presence of the cross bars welded to the stirrups increased the flexural rigidity of these stirrups which in turn increased the degree of confinement. Thus, the ductility of the column reinforced with mesh reinforcement as lateral reinforcement, has been increased in comparison with the columns having ordinary rectangular stirrups.

(2) The dimensions of mesh reinforcement affects the behavior of the reinforced concrete columns. As the spacing of the bars in the mesh reinforcement decreases, the flexural rigidity of the stirrups increases. Decreasing the spacing of the bars in the mesh from 5.0 to 2.5 cm increased both the ultimate column stress and the maximum value of the measured concrete strain by about 30%, while decreasing the spacing of the bars from 10 to 2.5 cm increased the ultimate column stress by about 40% and the maximum value of the measured concrete strain by about 45%.

(3) The confining effect of the lateral reinforcement increased considerably by increasing the percentage of lateral reinforcement

Table(2) Calculation of the increasing in the ultimate strengh (ΔP) due to confinement.

Column No.	P ultimate (ton)	P unconfined (ton)	ΔP (ton)
A-1-4	167.20	113.39	53.84
A-1-5	167.80	113.28	63.47
A-2-3	126.40	86.94	39.43
A-2-4	120.80	86.50	34.32
A-3-1	75.40	64.54	10.86
A-3-2	103.50	81.18	22.14
A-3-3	145.60	110.65	35.01
A-3-4	137.50	107.80	29.99
A-3-5	110.00	84.30	25.74
B-1-3	166.40	110.43	55.99
B-1-4	155.00	109.51	45.48
B-2-2	137.40	106.90	30.53
B-2-3	107.10	78.70	28.38
B-2-5	142.70	107.08	35.65
B-3-2	122.70	99.08	23.65
B-3-3	127.00	101.14	25.79
B-3-4	96.00	75.61	20.43
C-1-2	134.60	94.48	40.15
C-1-3	150.30	100.04	50.26
C-2-2	92.20	71.50	20.64
C-2-5	100.40	74.55	25.85
C-3-1	69.80	59.97	9.76
C-3-2	88.40	68.95	19.45
C-3-4	115.70	94.48	21.24
C-3-5	119.40	96.03	23.41

Table(3) Calculation of the (k) factor and the confinement factor.

Column No.	f of lateral steel	Area of core	K	α_c
A-1-4	1.750	380.24	3.41	0.139
A-1-5	2.693	376.36	2.69	0.208
A-2-3	1.616	376.36	2.79	0.163
A-2-4	1.167	380.24	3.26	0.122
A-3-1	0.539	384.16	2.17	0.078
A-3-2	0.875	380.24	2.81	0.097
A-3-3	1.212	376.36	3.30	0.096
A-3-4	0.875	380.24	3.80	0.073
A-3-5	1.212	376.36	2.43	0.126
B-1-3	2.693	368.60	2.27	0.224
B-1-4	1.750	372.48	2.74	0.152
B-2-2	1.167	372.48	2.76	0.104
B-2-3	1.616	368.60	1.92	0.188
B-2-5	1.616	368.60	2.40	0.138
B-3-2	0.875	372.48	2.85	0.084
B-3-3	1.212	368.60	2.32	0.110
B-3-4	0.875	372.48	2.46	0.120
C-1-2	1.750	364.80	2.32	0.183
C-1-3	2.693	361.00	1.95	0.247
C-2-2	1.167	364.80	1.79	0.161
C-2-5	1.616	361.00	1.67	0.207
C-3-1	0.539	372.48	1.91	0.085
C-3-2	0.785	364.80	2.25	0.125
C-3-4	0.875	364.80	2.46	0.092
C-3-5	1.212	361.00	2.02	0.121

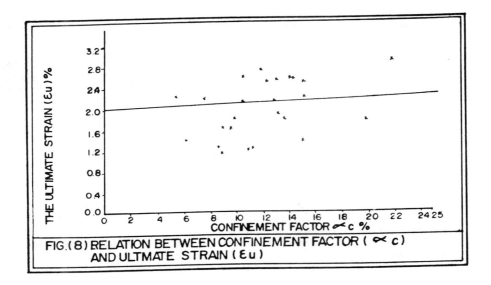

FIG.(8) RELATION BETWEEN CONFINEMENT FACTOR (∝ c)
AND ULTMATE STRAIN (εu)

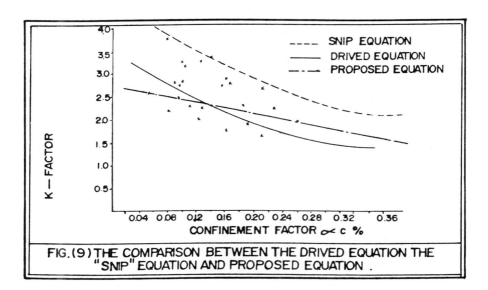

FIG.(9) THE COMPARISON BETWEEN THE DRIVED EQUATION THE
"SNIP" EQUATION AND PROPOSED EQUATION .

The confinement increases the maximum value of measured concrete strain and consequently, increases the ductility of the column. Increasing the percentage ratio (μ) by 30% increased the ultimate concrete stress by about 15% and the maximum value of concrete strength by about 25%.

(4) The maximum value of measured concrete strain is found to be increased with increasing the confinement factor. The following equation was found to be the best fitting with the available experimental results:

$$\epsilon_u = \epsilon_D + 0.0007 * \alpha C$$

(5) The gain in concrete strength due to the existence of lateral reinforcement in the form of mesh can be expressed as:

$$F'c = Fc + K . Fy$$

where; k = 3.14 - 5.23 $* \alpha C$

(6) The deduced equation for calculating the gain in the concrete strength due to the confinement may be applied within the following limits:

- Spacing between lateral reinforcement is not less than 6 cm and not more than (b/3) or 15 cms.
- The area of steel in one direction of the mesh is not more than 1.5 the area of the steel in its other direction.
- The minimum diameter of the mesh reinforcement is 5 mms and not more than 14 mms.

REFERENCES

1. American Concrete Institute. The Building Requirements for Reinforced Concrete. ACI 318-83, Detroit, PP46-47.
2. Sargin M. Stress-strain Relationships for Concrete and the Analysis of Structural Concrete Sections. Soil Mechanics Division, University of Waterloo, Waterloo, Ontario, Canada, Study No.4, December 1970.
3. King J.W.H. The Effect of Lateral Reinforcement in Reinforced Concrete Columns. Structural Engineer, London, Vol.24, July 1946, pp.335-388.
4. Soliman M.T.M. Ultimate Strength and Plastic Rotation Capacity of Reinforced Concrete Members. Ph.D. thesis, Faculty of Engineering, University of London, July 1966.
5. Burdette E.G. The Effect of Lateral Reinforcement on the Behavior of Axially Compression Members. Ph.D. thesis, Department of Civil Engineering, University of Illinois, January 1969.
6. Ramaley D. and Mc Henry D. Stress-strain Curves for Concrete strained Beyond the Ultimate Load. Laboratory Report No. SP-12, U.S. Breau of Reclamation, Denver, March 1947, 22 pp.
7. Hognestad E., Hanson N.W. and Mc Henry D. Concrete Stress Distribution in Ultimate Strength Design. ACI Journal, Proceeding Vol.52, No.4, December 1955, pp455-479.
8. Rusch H. Researches Toward a General Flexural Theory for Structural Concrete. ACI Journal, Proceeding Vol.57, No.1, July 1960, pp 1-28.
9. The USSR code of Practice (SNIP No. II-21-75).

SILANE TREATMENT OF CONCRETE BRIDGES

B J Brown

STATS Scotland Ltd.,

United Kingdom

ABSTRACT. The application of silane to concrete bridges is required by the Department of Transport and Scottish Development Department for repair, new construction and bridges less than 6 years old. An investigation into two bridges in the marine environment has indicated that silane penetration may be minimal and that if applied after a significant period of marine exposure and chloride ingress, no significant differences in chloride content between treated and untreated areas can be perceived; however half cell potentials show a slight reduction in activity in treated areas.

Keywords: Silane Treatment, Concrete, Marine Conditions, Bridges, Investigation, Corrosion Protection.

Dr. Brian J Brown is Managing Director of STATS Scotland Ltd, Consulting Engineers and Materials Scientists of East Kilbride, Scotland, UK and Honorary Lecturer in the Department of Civil Engineering of Dundee University. He is actively engaged in the assessment of construction materials and structures especially the inspection and investigation of concrete bridges. Dr. Brown is currently Chairman of the Glasgow and West of Scotland Section of the Society of Chemical Industry.

INTRODUCTION

The treatment of structures with alkyl alkoxysilane (silane) to combat
chloride ingress is advocated by the Department of Transport for
repair (1), new construction and for bridges less than 6 years old (2,
3).   The Scottish Development Department (which is the Scottish
equivalent of the Department of Transport) has recently commissioned a
series of silane treatment trials to establish the effectiveness of
the silane in combating reinforcement corrosion and in enhancing
concrete durability.

Three recently constructed bridges, two in the marine environment and
one inland, were chosen for the investigation.   Only the two marine
structures will be dealt with here.   The bridges in question are both
in the Highland Region of Scotland and are Cromarty Bridge and Kessock
Bridge (see Figure 1).

BRIDGE DETAILS

Cromarty Bridge

Cromarty Bridge is a low-level, multi-span reinforced concrete
structure (see Figures 2 and 3).   The r/c deck slab with permanent
inverted glass reinforced plastic formwork is on a superstructure of
prestressed r/c beams on r/c capping beams supported on columns with
piled foundations (see Figures 3,4 and 5).   The bridge is 0.9 miles
long, carries the A9 Trunk Road over the Cromarty Firth and was
completed in 1979.

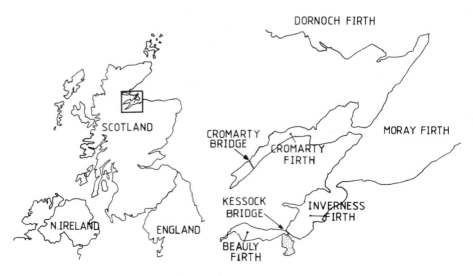

Figure 1: Location Map

Figure 2:   Cromarty Bridge

The columns of the bridge are within the submerged, tidal and splash
zones with the remaining support and superstructures generally marine
atmospheric although within the splash zone in periods of heavy
storms.

The specification for the in-situ concrete was for class 37.5/20 with
60mm minimum cover to all exposed surfaces. The prestressed precast
concrete of the deck beams was specified as class 52.5/20 with the
same minimum cover. As the inverted GRP formwork to the deck soffit
was permanent the depth of cover was only 30mm.

Kessock Bridge

Kessock Bridge is a high level cable stayed structure with a steel
deck supported on twelve reinforced concrete piers (see Figures 6 and
7). All piers have twin columns constructed by slipforming but only
the two main centre span piers have an r/c crosshead spanning between
the columns. The bridge is 0.6 miles long, also carries the A9 Trunk
Road on the Beauly Firth at Inverness and is 5 miles south of Cromarty
Bridge. It was completed in 1982. The pier cutwaters are within the
submerged, tidal and splash zones with the remaining parts of the
columns marine atmospheric.
The specification for the in-situ concrete at Kessock was the same as
at Cromarty with class 37.5/20 concrete and 60mm minimum cover.

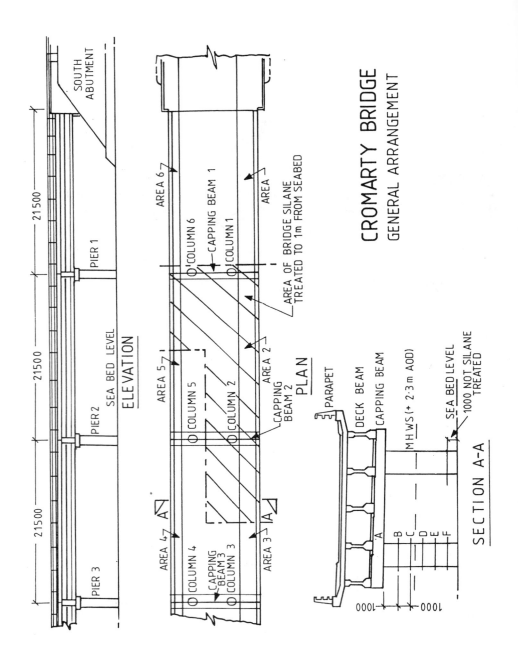

Figure 3

Figure 4:   Deck beams, Cromarty Bridge

Figure 5:   Columns and capping beams, Cromarty Bridge

Figure 6: Kessock Bridge

PRELIMINARY INVESTIGATION PROGRAMME

Methods of Investigation

Prior to any silane treatment trials, selected areas of each bridge were investigated to assess their material properties. This included the following site and laboratory testing:

SITE TESTING

Visual survey
Half cell potential survey
Depth of cover survey
Depth of carbonation survey
Core and drill sampling
Resistivity survey

LABORATORY TESTING

Visual assessment
Compressive strength
Density
Cement content and w/c ratio
Chloride content
Petrographic examination
Water absorption
Permeability

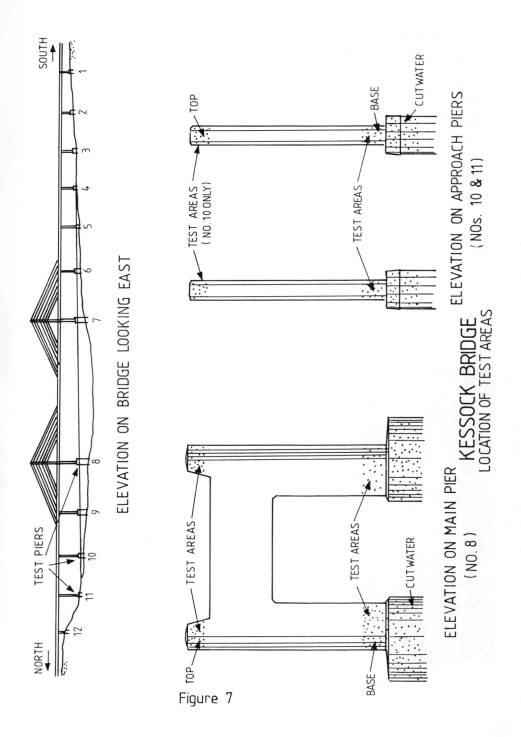

ELEVATION ON BRIDGE LOOKING EAST

KESSOCK BRIDGE
LOCATION OF TEST AREAS

ELEVATION ON MAIN PIER
(NO. 8)

ELEVATION ON APPROACH PIERS
(NOs. 10 & 11)

Figure 7

Extent of Investigation

At Cromarty Bridge all structural elements excepting the internal
support beams were investigated in the first three spans at the
southern end (see Figure 3); access was by use of an underbridge
machine. At Kessock, 1 no. main span pier, 1 no. approach span pier
and 1 no. land based approach span pier were investigated (see Figure
7) with only the cutwaters and column bases tested utilising boat
access or by working at low tide (Figure 8). The pier tops at deck
level were also investigated for the two marine based piers using
abseiling techniques for access (see Figure 9).

Timescale

The Cromarty Bridge investigation was carried out in 1986 which is
some 7 years after construction. The Kessock Bridge investigation was
conducted in 1987 some 5 years after completion.

RESULTS OF THE PRELIMINARY INVESTIGATION

Detailed results of the two investigations are contained in Tables 1
and 2 and are summarised below.

Figure 8: Investigation of cutwater at low tide,
Kessock Bridge

Figure 9: Abseiling investigation at column top,
Kessock Bridge

Cromarty Bridge

Parapet upstands and deck cantilever soffits

The moderately good quality concrete has occasional transverse cracks
which may have been due to shrinkage or early age thermal effects.
The concrete fails to comply with the current British recommendations
(4) to ensure durability due to a lack of air-entrainment.

Low chloride contents and low half cell activity indicate that
corrosion of steel reinforcement is unlikely except in areas of joint
between deck spans where it is uncertain.

TABLE 1   Summary of results

CROMARTY BRIDGE

		SOUTH ABUTMENT	PARAPET & CANTILEVER	DECK BEAMS	CAPPING BEAMS	COLUMNS
HALF CELL POTENTIAL (%)	< -200mV		95	96	16	4
	-200 to -350 mV	NT	5	4	79	38
	> -350mV		0	0	5	58
DEPTH OF COVER (%)	< 20mm		0	0	0	0
	20-40mm	NT	0	0	24	0
	> 40mm		100	100	76	100
CHLORIDE CONTENT at 0-30mm (Cl⁻ % CEMENT)	max	1.17	0.09	0.34	1.81	3.64
	min	0.64	0.09	0.02	0.02	0.16
	mean	0.98	0.09	0.22	0.82	1.40
DEPTH OF CARBONATION (mm)	mean					
	max	6.0	1.3	1.5	4.1	4.2
	mean					
	min	2.3	0	0.1	1.4	1.7
CEMENT CONTENT (kg/m³)	max	430	445	410	480	570
	min			410	320	400
W/C RATIO (C as unity)	max	0.46	0.43	0.42	0.57	0.43
	min			0.40	0.40	0.38
ESTIMATED IN-SITU CUBE STRENGTH (N/mm²)	max	31.5	57.0	83.0	52.0	60.0
	min			71.0	34.0	44.0
	mean			77.5	43.0	52.0
SATURATED DENSITY (kg/m³)	max	2340	2370	2410	2380	2410
	min			2390	2340	2360
	mean			2410	2360	2380

NT = NOT TESTED

Deck Beams

The concrete is generally of good quality and complies with current BS 5400 durability recommendations (4). However, microcracking was present in all of the cores sampled which may have been due to drying shrinkage.

The concrete complies with current cover recommendations (4) to ensure durability due to its high strength even though it lacks air-entrainment.

Chloride contents and half cell activity are generally low indicating that corrosion of steel reinforcement is unlikely. The exception is in localised areas of de-icing salt ingress from gully pots and sea-spray ingress at the cracks on the outer face of the external beams (some of which were measured at a depth greater than the concrete cover) where corrosion is more probable.

Capping Beams

The concrete sampled varies from low to moderate quality. Some of the concrete thus fails to comply with current BS 5400 durability recommendations.

The depth of cover varies from 20 to 90mm and thus many areas fail to comply with current cover recommendations to ensure durability (4).

Although in general, the depth of carbonation is not a problem, in the worst situation of low cover and relatively high carbonation penetration, it can be predicted that the carbonation front will reach the steel reinforcement relatively soon.

There are substantial amounts of chloride ingress in areas of poor quality concrete. This, together with relatively high half cell activity associated with areas of low concrete cover, indicate that corrosion of steel reinforcement may be a problem.

Columns

The concrete is visually sound and is of moderately good quality and as such complies with current BS 5400 durability recommendations (4).

The concrete fails to comply with current BS 5400 cover recommendations (4) to ensure durability due to its moderate strength and lack of air-entrainment.

The chloride ingress is considerable, particularly in the tidal zone where chloride content results, at the level of steel reinforcement, are substantial.

The high half cell activity in the tidal zone together with relatively low concrete resistivity and considerable chloride ingress indicate that corrosion of steel reinforcement is probable.

South Abutment

The concrete sampled is low to moderate in quality and does not seem to have reached the design strength. As such, it fails to comply with current durability recommendations.

The chloride content results are substantial and indicate that corrosion of the steel reinforcement is likely.

Kessock Bridge

Columns

The concrete is of moderate quality and generally well compacted. Compressive strengths and saturated densities are variable and many cores, especially in the cutwater, fail to meet the class 37.5 specification even after 5 years.

Notable visual defects are horizontal cracks and bands of map cracking in the columns, map cracked patch repairs at the top of columns, vertical cracks at the base of columns and occasional isolated rust spots and exposed reinforcement.

Horizontal colour banding caused by the slipforming is much in evidence.

Depth of cover is very variable and while values in excess of 80mm are not uncommon, many areas fail to meet the 60mm specified cover, some exceedingly so.

Depth of carbonation is minimal and generally less than 3mm; it should not be a problem within the design life of the bridge.

Chloride contents are high within the tidal and splash zones decreasing internally into the structure; values at the level of reinforcement in the lowermost areas of cutwater exceed the 0.20% Cl⁻ by mass of cement threshold value for steel corrosion.

Chloride contents of the land based pier (Pier 11) are nil thus indicating that it is not affected by sea spray.

Areas above the splash zone in the two marine based piers examined are variable but low with values greater than the 0.20% Cl⁻ by mass of cement at the surface only.

TABLE 2   Summary of results

KESSOCK BRIDGE

| | | PIER 8 | | | PIER 10 | | | PIER 11 | |
		C/WATER	BASE	TOP	C/WATER	BASE	TOP	C/WATER	BASE
HALF CELL	< -200mV	100	100	100	100	98	100	91	100
POTENTIAL	-200 to	0	0	0	0	2	0	9	0
(%)	-350 mV								
	> -350mV	0	0	0	0	0	0	0	0
DEPTH OF	< 20mm	0	1	11	0	0	2	0	2
COVER	20-40mm	2	1	11	0	0	17	0	38
(%)	> 40mm	98	98	78	100	100	81	100	60
CHLORIDE	max	1.76	1.65	0.32	3.04	0.72	0.21	<0.03	<0.03
CONTENT	min	0.03	<0.03	0.14	0.16	<0.03	0.05	<0.03	<0.03
at 0-30mm	mean	0.55	0.42	0.21	1.27	0.33	0.11	<0.03	<0.03
(Cl⁻ % CEMENT)									
DEPTH OF	mean								
CARBONATION	max	2.1	2.8	1.8	1.4	1.4	1.0	1.1	1.7
(mm)	mean								
	min	0.6	1.2	0.3	0.1	0.3	0	0.1	0.2
CEMENT CONTENT	max	350	NT	NT	335	365	NT	NT	475
(kg/m³)	min	345			300				
W/C RATIO	max	0.43	NT	NT	0.49	0.49	NT	NT	0.36
(C as unity)	min	0.42			0.46				
ESTIMATED	max	37.0	NT	NT	51.5	NT	NT	42.0	59.5
IN-SITU CUBE	min	37.0			28.0			33.0	
STRENGTH	mean	37.0			42.0			37.5	
(N/mm²)									
SATURATED	max	2360	2400	NT	2420	NT	NT	2410	2420
DENSITY	min	2330	2300		2380			2400	2410
(kg/m³)	mean	2350	2350		2410			2410	2420

NT = NOT TESTED

The variability in chloride content of the concrete above the tidal
zone will partly be related to exposure to sea spray; however, the
dominant factor is probably variability in the quality of the concrete.

Half cell potential values are numerically low in all areas tested
indicating a low corrosion probability at present.

Resistivity values are generally low to moderate, however, indicating
that a significant rate of corrosion may be possible.

SILANE TREATMENT TRIAL

Cromarty Bridge

Following the preliminary investigation of the first three spans of the
structure in July/August 1986, an area of the bridge within the
investigation area was treated with silane in October 1986 (see Figure
3). All exposed surfaces except pier faces within one metre of the sea
bed were treated with two coats of commercially available 100% silane
applied by low pressure spray.

TABLE 3   Silane Penetration

BRIDGE	LOCATION	DEPTH OF PENETRATION (mm)	REMARKS	APPLICATION
CROMARTY	Deck Beams - Vertical Faces	< 0.5-1.5	3-6mm down cracks	
	Deck Beams - Soffits	< 0.5-1.5		All site applied
	Parapet Upstands - Horizontal Face	0.5-1.0	10 to 100mm down cracks	
	Capping Beams - Vertical Faces	< 0.5-2.0		
	Columns - Vertical Faces	< 0.5-2.0		
KESSOCK	Columns - Cut- waters and Bases Vertical and Horizontal Faces	< 0.5-3 1.0-3.5 < 0.5-3.0	Surface Concretes	Lab applied - brush - immersion - vacuum
		No test < 0.5-11.5 1-13	Internal Concretes	- brush - immersion - vacuum

A few weeks after the application, 6 months and $2^3/_4$ years later, cores were extracted from the various treated structural elements to determine depth of silane penetration. The results which are given in Table 3 indicate low penetration depths of between < 0.5 and 2.0mm; penetration down cracks was dependant on surface crack width and was between 3 and 110mm.

The outer 20mm of cores from the preliminary investigation were tested for water absorption and capillary porosity by BS 1881 methods (5). The core slices were subsequently immersed in silane for half an hour and the tests re-done to assess the effect of the silane impregnation. The results are given in Table 4. The average water absorption figure prior to treatment was 3.22% which reduced to an average value of 0.33% after treatment, a reduction of 90%. Capillary porosity reduced by only 24% from an average of 4.47% prior to an average of 3.42% after treatment. The silane treatment was thus confirmed as a water repellant rather than a pore blocker.

TABLE 4   Comparison of concrete before and after silane treatment

BRIDGE	LOCATION	WATER ABSORPTION (%)		CAPILLARY POROSITY (%)	
		BEFORE	AFTER	BEFORE	AFTER
CROMARTY	South Abutment	3.60	0.36	5.23	4.30
	Deck Beams	3.44	0.32	3.78	2.83
		3.20	0.33	4.27	3.12
		2.99	0.00	3.48	2.96
		3.38	0.73	4.28	2.96
		4.61	0.00	4.80	4.47
	Capping Beams	2.58	0.50	4.56	3.09
		2.96	0.29	4.10	3.12
		3.03	0.33	4.61	3.60
		3.55	0.28	6.00	4.71
	Columns	2.99	0.29	4.67	3.59
		2.31	0.63	3.83	2.83
			a b		
KESSOCK	Columns	2.96	0.31/0.42	No results	
		2.94	0.45/0.41		
		3.20	0.39/0.72		

a. ½ hour immersion
b. ½ hour vacuum immersion

Half cell potential, chloride content and depth of carbonation tests
were carried out on selected treated and untreated areas of the bridge
at six months, 1 year, $2^3/_4$ and 3 years after treatment and the results
compared with the results of the preliminary investigation. The
results after $2^3/_4$ or 3 years are given in Tables 5 to 8 and in Figure
10.

A comparison of results $2^3/_4$ or 3 years after silane treatment with the
results from the untreated concrete indicates the following:

'Half Cell' potential values in the untreated areas are exhibiting
similar values to the original values; however, in the treated areas
values have become more positive indicating a drying out of the
concrete and a lessening of corrosion risk. Equipotential contour
patterns before and after treatment are similar in both treated and
untreated areas.

Chloride contents in both untreated and treated areas are generally not
rising but are remaining reasonably constant or are showing movement
within the concrete without additional external ingress of chloride.

Depth of carbonation results are similar in both treated and untreated
areas and show no significant increase from the original values.

All of the above results are not conclusive and a lengthier time period
is required to assess the effectiveness of the silane treatment; a
further investigation 5 years after treatment has been recommended.

Kessock Bridge

As yet no site silane treatment trial has been carried out on Kessock
Bridge. However, testing of laboratory treated cores from the
preliminary investigation has been carried out in the same manner as
that undertaken with the Cromarty Bridge cores. Core tops or internal
slices from Kessock Bridge were either treated with two brush applied
coats of silane, immersed in silane for half an hour or immersed in
silane under vacuum for half an hour. Water absorption and silane
impregnation tests were then carried out. The results are given in
Tables 3 and 4.

As with the Cromarty Bridge concrete, silane penetration to the outer
surface was minimal. Larger penetrations were achieved with cut
internal surfaces. The water absorption of silane treated material was
reduced by up to 87%. Permeabilities of treated surface concretes were
within the range of permeabilities for untreated concretes with no
marked reductions.

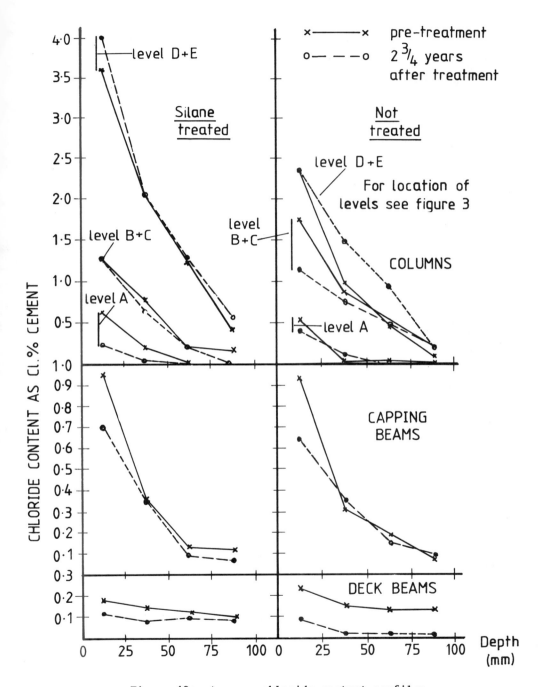

Figure 10:   Average chloride content profiles

TABLE 5   Summary of chloride content results - untreated areas

Pre-application Depth (mm)	0-25	25-50	50-75	75-100	2 $^3/_4$ Years After Application 0-25	25-50	50-75	75-100

Parapet Upstands & Cantilever Soffit

	0-25	25-50	50-75	75-100	0-25	25-50	50-75	75-100
number	0	0	0	0	2	2	2	2
mean					0.09	<0.02	<0.02	<0.02
s.d								

Deck Beams

number	12	12	12	12	11	11	11	11
mean	0.23	0.25	0.13	0.13	0.08	0.02	0.02	0.02
s.d.	0.09	0.10	0.08	0.09	0.05	0.04	0.04	0.04

Capping Beams

number	8	8	8	8	8	8	8	8
mean	0.93	0.31	0.18	0.07	0.64	0.34	0.15	0.09
s.d.	0.68	0.52	0.44	0.12	0.45	0.34	0.22	0.11

Columns

Level A

number	2	2	2	2	2	2	2	2
mean	0.52	0.02	0.03	<0.02	0.36	0.12	<0.02	<0.02
s.d								

Level B + C

number	2	2	2	2	2	2	2	2
mean	1.74	0.89	0.56	0.23	1.16	0.77	0.54	0.22
s.d.								

Level D + E

number	3	3	3	3	3	3	3	3
mean	2.36	0.98	0.48	0.07	2.34	1.47	0.92	0.22
s.d.	0.86	0.75	0.10	0.05	0.11	0.35	0.60	0.20

TABLE 6   Summary of chloride content results - treated areas

Depth (mm)	Pre-application				2 $^3/_4$ Years After Application			
	0-25	25-50	50-75	75-100	0-25	25-50	50-75	75-100
Parapet Upstands & Cantilever Soffit								
number	1	1	1	1	2	2	2	2
mean	0.09	<0.02	<0.02	<0.02	<0.02	<0.02	<0.02	<0.02
s.d								
Deck Beams								
number	7	7	7	7	7	7	7	7
mean	0.18	0.14	0.12	0.10	0.11	0.08	0.09	0.08
s.d.	0.12	0.11	0.10	0.10	0.06	0.09	0.08	0.06
Capping Beams								
number	8	8	8	8	8	8	8	8
mean	0.94	0.37	0.13	0.12	0.70	0.36	0.09	0.07
s.d.	0.58	0.38	0.21	0.15	0.45	0.26	0.17	0.20
Columns								
Level A								
number	3	3	3	3	3	3	3	3
mean	0.61	0.19	0.03	<0.02	0.23	0.04	<0.02	<0.02
s.d	0.39	0.31	0.06	0.01	0.27	0.08		
Level B + C								
number	3	3	3	3	3	3	3	3
mean	1.27	0.79	0.21	0.16	1.26	0.67	0.20	<0.02
s.d.	0.50	0.16	0.12	0.16	0.91	0.26	0.20	
Level D + E								
number	2	2	2	2	2	2	2	2
mean	3.60	2.05	1.26	0.44	3.98	2.04	1.29	0.59
s.d.								

TABLE 7 Summary of depth of carbonation results

| | Pre-application | | 2 $^3/_4$ Years After Application | | | |
| | | | Untreated Area | | Treated Area | |
	max	min	max	min	max	min
Parapet Upstands & Cantilever Soffit						
number	3	3	2	2	2	2
mean	1.3	0	1.5	0.5	1.0	0
s.d.	–	–	–	–	–	–
Deck Beams						
number	33	33	14	14	10	10
mean	1.5	0.1	2.1	0.6	2.2	0.5
s.d.	0.6	0.3	1.3	0.8	1.1	0.5
Capping Beams						
number	39	39	8	8	8	8
mean	4.1	1.4	2.6	0.8	2.6	1.4
s.d.	2.4	0.9	1.7	0.7	0.9	0.7
Columns						
Level A						
number	12	12	2	2	3	3
mean	6.3	2.8	3.5	2.5	3.7	1.3
s.d	3.1	1.2	–	–	–	–
Level B + C						
number	15	15	2	2	3	3
mean	3.9	1.5	4.5	2.0	3.0	1.3
s.d.	2.0	1.0	–	–	–	–
Level D + E						
number	6	6	3	3	2	2
mean	0.7	0	1.0	0	0.5	0
s.d.	0.5	–	–	–	–	–

TABLE 8 Summary of half cell potential survey results
(all results in mV, negative sign omitted)

ELEMENT	Deck Beams				Capping Beams		Columns	
LOCATION	Area 4	Area 5S	Area 5N	Area 6	2 W	2 E	2 A-E	5 A-E
SILANE TREATMENT	No	Yes	No	No	Yes	No	Yes	No
Pre-Treatment								
mean	67.6	77.5	97.9	78.6	304.9	271.7	382.4	395.2
s.d.	59.0	71.4	75.9	71.7	62.1	42.8	122.1	115.6
6 Months After								
mean	NT	77.7	97.2	NT	241.5	243.2	361.7	NT
s.d		77.1	74.6		68.1	61.5	114.6	
1 Year After								
mean	NT	43.1	87.9	NT	NT	240.7	NT	425.5
s.d.		46.4	62.7			49.1		159.3
2 $^3/_4$ Years After								
mean	70.5	50.7	89.0	65.2	NT	244.8	NT	486.1
s.d.	54.8	50.3	67.2	64.7		46.8		190.7
3 Years After								
mean	NT	NT	NT	NT	182.3	NT	346.7	NT
s.d.					56.4		154.0	

NT = NOT TESTED

CONCLUSIONS

A site trial of silane treatment at Cromarty Bridge and laboratory trials on cores extracted from Cromarty and Kessock Bridges have shown that penetration of silane into the concrete is minimal and may be confined to the surface laitance only. However, water absorption tests before and after laboratory silane treatment indicate that very significant reductions of circa 90% can be achieved.

The long term durability of such minimal silane penetrations is of concern especially in the tidal zone.

A comparison of treated and untreated concretes at Cromarty Bridge before silane treatment and $2^3/_4$ to 3 years after treatment indicates that a trend may be developing with reduction in half cell potential in the treated areas whereas values in the untreated areas are generally remaining constant. Chloride contents in both treated and untreated areas are remaining reasonably constant, not rising from original pretreatment values.   However, a much greater time period is required before any positive conclusions on the effect of the silane can be made.

ACKNOWLEDGEMENTS

The work was undertaken by STATS for the Scottish Development Department who gave permission for the publication of this data.   The help and assistance of J. Steel, M. Sharma and B. Malcolm - Scottish Development Department, D. McKenzie - Highland Regional Council and various colleagues in STATS is gratefully acknowledged.

REFERENCES

1. DEPARTMENT OF TRANSPORT. BD 27/86.   Material for the repair of concrete highway structures, 1986.

2. DEPARTMENT OF TRANSPORT. BD 43/90.   Criteria and material for the impregnation of concrete highway structures, 1990.

3. DEPARTMENT OF TRANSPORT. BA 33/90.   Impregnation of concrete highway structures, 1990.

4. BRITISH STANDARDS INSTITUTION. BS 5400.   Steel, concrete and composite bridges, Part 4: 1984; Code of practice for the design of concrete bridges.   Part 7: 1978; Specification for materials and workmanship, concrete reinforcement and prestressing tendons.   Part 8: 1978; Recommendations for materials and workmanship, concrete, reinforcement and prestressing tendons.

5. BRITISH STANDARDS INSTITUTION. BS 1881.   Testing concrete.   Part 122: 1983; Method for determination of water absorption.   Part 124: 1988; Methods for analysis of hardened concrete.

CLOSING ADDRESS

Peter C Hewlett

Director, British Board of Agrément,

United Kingdom

INDEX OF AUTHORS

INDEX

This index has been compiled from the keywords provided by the authors of the papers. The numbers are the page numbers of the first pages of the relevant paper.